FOUNDATIONS OF EARTH SCIENCE

FOUNDATIONS OF

EARTH SCIENCE

FIFTH EDITION

Frederick K. Lutgens

Edward J. Tarbuck

Illustrated by

Dennis Tasa

Upper Saddle River, NJ 07458

Library of Congress Cataloging-in-Publication Data

Lutgens, Frederick K.
Foundations of earth science / Frederick K. Lutgens, Edward J. Tarbuck ; illustrated by Dennis Tasa.
—5th ed.
p. cm.
ISBN 0-13-240135-5
1. Earth sciences—Textbooks. I. Tarbuck, Edward J. II. Tasa, Dennis. III. Title.
QE28.L96 2008
550—dc22

2007013800

Publisher, Geosciences: *Daniel Kaveney*
Editor in Chief, Science: *Nicole Folchetti*
Project Manager: *Crissy Dudonis*
Executive Managing Editor: *Kathleen Schiaparelli*
Assistant Managing Editor: *Beth Sweeten*
Production Editor: *Patty Donovan, Pine Tree Composition*
Media Production Editor: *Richard Barnes*
Media Editor: *Andrew Sobel*
Assistant Editor: *Sean Hale*
Editorial Assistant: *John DeSantis*
Marketing Manager: *Amy Porubsky*
Marketing Assistant: *Ilona Samouha*
Director of Logistics, Operations, and Vendor Relations: *Barbara Kittle*
Senior Operations Supervisor: *Alan Fischer*
Art Director: *Kenny Beck*
Interior and Cover Designer: *Michael J. Fruhbeis*
Senior Managing Editor, Art Production and Management: *Patricia Burns*
Manager, Production Technologies: *Matthew Haas*
Managing Editor, Art Management: *Abigail Bass*
AV Project Manager: *Rhonda Aversa*
Director, Image Resource Center: *Melinda Patelli*
Manager, Rights and Permissions: *Zina Arabia*
Interior Image Specialist: *Beth Brenzel*
Cover Image Specialist: *Karen Sanatar*
Image Permission Coordinator: *Debbie Hewitson*
Color Scanning Supervisor: *Joseph Conti*
Composition: *Pine Tree Composition*
Cover Photo: *Cedar Wright and Keving Thaw climbing the South Buttress of Kaga Tondo. Hand of Fatima, Mali, Africa. Photographer: Jimmy Chin.*
Title Page Photo: *La Sal Mountains beneath Mesa Arch, Canyonlands National Park, Utah. Photographer: Jack Dykinga Photography.*

Pearson Prentice Hall
Pearson Education, Inc.
Upper Saddle River, NJ 07458

Printed in the United States of America

10 9 8 7 6 5 4 3 2 1

ISBN-10: 0-13-240135-5
ISBN-13: 978-0-13-240135-7

Pearson Education Ltd., *London*
Pearson Education Australia Pty., Limited, *Sydney*
Pearson Education *Singapore*, Pte. Ltd.
Pearson Education North Asia Ltd., *Hong Kong*
Pearson Education Canada, Ltd., *Toronto*
Pearson Educación de Mexico, S.A. de C.V.
Pearson Education—Japan, *Tokyo*
Pearson Education Malaysia, Pte. Ltd.

To Nancy and Joanne

BRIEF CONTENTS

Preface xv

Introduction 1

UNIT I
EARTH MATERIALS 16

1 Minerals: Building Blocks of Rocks 17

2 Rocks: Materials of the Solid Earth 37

UNIT II
SCULPTURING EARTH'S SURFACE 64

3 Landscapes Fashioned by Water 65

4 Glacial and Arid Landscapes 101

UNIT III
FORCES WITHIN 128

5 Plate Tectonics: A Scientific Theory Unfolds 129

6 Restless Earth: Earthquakes, Geologic Structures, and Mountain Building 159

7 Fires Within: Igneous Activity 193

UNIT IV
DECIPHERING EARTH'S HISTORY 222

8 Geologic Time 223

UNIT V
THE GLOBAL OCEAN 244

9 Oceans: The Last Frontier 245

10 The Restless Ocean 267

UNIT VI
EARTH'S DYNAMIC ATMOSPHERE 292

11 Heating the Atmosphere 293

12 Moisture, Clouds, and Precipitation 321

13 The Atmosphere in Motion 349

14 Weather Patterns and Severe Weather 368

UNIT VII
EARTH'S PLACE IN THE UNIVERSE 394

15 The Nature of the Solar System 395

16 Beyond the Solar System 431

APPENDIX A
Metric and English Units Compared 452

APPENDIX B
Mineral Identification Key 453

APPENDIX C
Relative Humidity and Dew-Point Tables 455

APPENDIX D
Earth's Grid System 457

Glossary 466

Index 475

GEODe: Earth Science v.2

A copy of *GEODe: Earth Science* is packaged with each copy of *FOUNDATIONS OF EARTH SCIENCE,* Fifth Edition— This dynamic learning aid reinforces key concepts by using tutorials, animations, and interactive exercises.

UNIT 1: Earth Materials

A. Minerals
   1. Introduction
   2. Major Mineral Groups
   3. Properties Used to Identify Minerals
   4. Mineral Identification
   5. Minerals Quiz

B. Rock Cycle

C. Igneous Rocks
   1. Introduction
   2. Igneous Textures
   3. Naming Igneous Rocks
   4. Igneous Rocks Quiz

D. Sedimentary Rocks
   1. Introduction
   2. Types of Sedimentary Rocks
   3. Sedimentary Rocks Quiz

E. Metamorphic Rocks
   1. Introduction
   2. Agents of Metamorphism
   3. Textural and Mineralogical Changes
   4. Common Metamorphic Rocks
   5. Metamorphic Rocks Quiz

UNIT 2: Sculpturing Earth's Surface

A. Hydrologic Cycle

B. Running Water
   1. Stream Characteristics
   2. Hydrological Cycle and Running Water Quiz

C. Groundwater
   1. Groundwater and its Importance
   2. Springs and Wells
   3. Groundwater Quiz

D. Glaciers
   1. Introduction
   2. Budget of a Glacier
   3. Glaciers Quiz

E. Deserts
   1. Distribution and Causes of Dry Lands
   2. Common Misconceptions About Deserts
   3. Deserts Quiz

UNIT 3: Forces Within

A. Earthquakes
   1. What is an Earthquake?
   2. Seismology: Earthquake Waves
   3. Locating an Earthquake
   4. Earth's Layered Structure
   5. Earthquakes and Earth's Interior Quiz

B. Plate Tectonics
   1. Introduction
   2. Plate Boundaries
   3. Plate Tectonics Quiz

C. Igneous Activity
   1. The Nature of Volcanic Eruptions
   2. Materials Extruded During an Eruption
   3. Volcanoes
   4. Intrusive Igneous Activity
   5. Igneous Activity Quiz

UNIT 4: Deciphering Earth's History

A. Geologic Time Scale

B. Relative Dating

C. Radiometric Dating

D. Geologic Time Quiz

UNIT 5: The Global Ocean

A. Floor of the Ocean
   1. Mapping the Ocean Floor
   2. Features of the Ocean Floor
   3. Floor of the Ocean Quiz

B. Coastal Processes
   1. Waves and Beaches
   2. Wave Erosion
   3. Coastal Processes Quiz

UNIT 6: Earth's Dynamic Atmosphere

A. Heating the Atmosphere
   1. Solar Radiation
   2. What Happens to Incoming Solar Radiation?
   3. The Greenhouse Effect
   4. Temperature Structure of the Atmosphere
   5. Heating the Atmosphere Quiz

B. Moisture and Cloud Formation
   1. Water's Changes of State
   2. Humidity: Water Vapor in the Air
   3. Cloud Formation
   4. Moisture and Cloud Formation Quiz

C. Air Pressure and Wind
   1. Measuring Air Pressure
   2. Factors Affecting Wind
   3. Highs and Lows
   4. Air Pressure and Wind Quiz

D. Basic Weather Patterns
   1. Air Masses
   2. Fronts
   3. Introducing Middle-Latitude Cyclones
   4. Examining a Mature Middle-Latitude Cyclone
   5. Basic Weather Patterns Quiz

UNIT 7: Earth's Place in the Universe

A. The Planets: An Overview

B. Calculating Your Age and Weight on Other Planets

C. Earth's Moon

D. A Brief Tour of the Planets

E. Solar System Quiz

This *GEODe: Earth Science* icon appears throughout the book wherever a text discussion has a corresponding module on the CD-ROM.

CONTENTS

Preface xv

Introduction 1

FOCUS ON LEARNING 1

What Is Earth Science? 2
Earth's Spheres 3
- Hydrosphere 4
- Atmosphere 5
- Geosphere 5
- Biosphere 6

Earth as a System 6
Scales of Space and Time in Earth Science 8
Resources and Environmental Issues 8
- Resources 9
- Environmental Problems 11

The Nature of Scientific Inquiry 11
- Hypothesis 12
- Theory 12
- Scientific Methods 12

Studying Earth Science 13
The Chapter in Review 13
Key Terms 14
Questions for Review 15
Online Study Guide 15

UNIT I
EARTH MATERIALS 16

1 Minerals: Building Blocks of Rocks 17

FOCUS ON LEARNING 17

Minerals: The Building Blocks of Rocks 18
Elements: Building Blocks of Minerals 19
- Atoms 20

Why Atoms Bond 21
- Ionic Bonds: Electrons Transferred 22
- Covalent Bonds: Electrons Shared 23

Isotopes and Radioactive Decay 23
Properties of Minerals 23
- Optical Properties 24
- Crystal Shape or Habit 25
- Mineral Strength 25
- Density and Specific Gravity 26
- Other Properties of Minerals 27

Mineral Groups 28
- Silicate Minerals 29
- Important Nonsilicate Minerals 31

Mineral Resources 32
The Chapter in Review 33
Key Terms 34
Questions for Review 34
Online Study Guide 35

2 Rocks: Materials of the Solid Earth 37

FOCUS ON LEARNING 37

Earth as a System: The Rock Cycle 38
- The Basic Cycle 38
- Alternative Path 40

Igneous Rocks: "Formed by Fire" 40
- Magma Crystallizes to Form Igneous Rocks 41
- Igneous Textures 42
- Igneous Compositions 43
- Classifying Igneous Rocks 43
- How Different Igneous Rocks Form 45

Weathering of Rocks to Form Sediment 46
- Mechanical Weathering of Rocks 46
- Chemical Weathering of Rocks 47

Sedimentary Rocks: Compacted and Cemented Sediment 50
- Classifying Sedimentary Rocks 50
- Lithification of Sediment 54
- Features of Sedimentary Rocks 54

Metamorphic Rocks: New Rock from Old 55
  Agents of Metamorphism 57
  Metamorphic Textures 59
  Common Metamorphic Rocks 60
The Chapter in Review 62
Key Terms 62
Questions for Review 63
Online Study Guide 63

UNIT II
SCULPTURING EARTH'S SURFACE 64

3 Landscapes Fashioned by Water 65

FOCUS ON LEARNING 65

Earth's External Processes 66
Mass Wasting: The Work of Gravity 66
  Mass Wasting and Landform Development 67
  Controls and Triggers of Mass Wasting 67
The Water Cycle 69
Running Water 70
  Drainage Basins 71
  River Systems 71
Streamflow 72
  Gradient and Channel Characteristics 72
  Discharge 72
  Changes from Upstream to Downstream 73
The Work of Running Water 74
  Erosion 74
  Transportation 74
  Deposition 75
Stream Channels 75
  Bedrock Channels 75
  Alluvial Channels 75
Base Level and Stream Erosion 77
Shaping Stream Valleys 77
  Valley Deepening 78
  Valley Widening 79
  Changing Base Level and Incised Meanders 80
Depositional Landforms 80
  Deltas 80
  Natural Levees 80
Drainage Patterns 82
Floods and Flood Control 82
  Causes of Floods 82
  Flood Control 83
Groundwater: Water Beneath the Surface 85
  The Importance of G[illegible]
  Groundwater's Geologic Roles 86
Distribution and Movement of Groundwater 86
  Distribution 86
  Factors Influencing the Storage and Movement of Groundwater 87
  Groundwater Movement 88
Springs 88
  Hot Springs 89
  Geysers 89
Wells 89
Artesian Wells 90

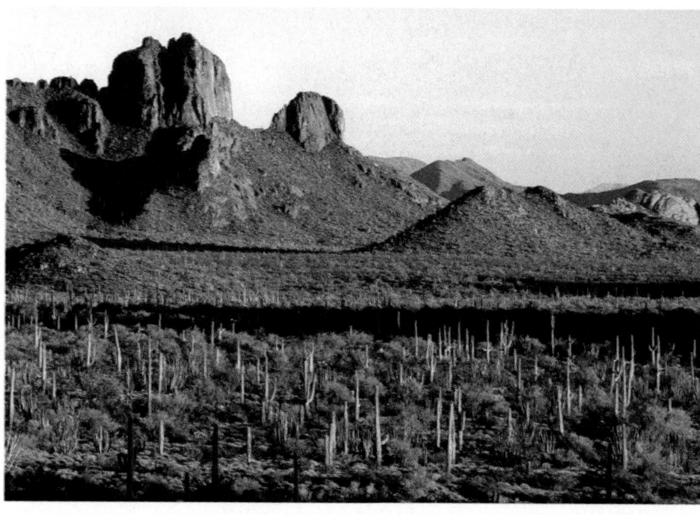

Environmental Problems of Groundwater 91
  Treating Groundwater as a Nonrenewable Resource 91
  Land Subsidence Caused by Groundwater Withdrawal 91
  Groundwater Contamination 92
The Geologic Work of Groundwater 93
  Caverns 94
  Karst Topogrqaphy 96
The Chapter in Review 96
Key Terms 97
Questions for Review 98
Online Study Guide 99

4 Glacial and Arid Landscapes 101

FOCUS ON LEARNING 101

Glaciers: A Part of Two Basic Cycles 102
  Valley (Alpine) Glaciers 102
  Ice Sheets 103
  Other Types of Glaciers 103
How Glaciers Move 104
  Observing and Measuring Movement 104
  Budget of a Glacier: Accumulation versus Wastage 104
Glacial Erosion 106
  How Glaciers Erode 107
  Landforms Created by Glacial Erosion 107
Glacial Deposits [illegible]
  Types of Glacial Drift 110
  Moraines, Outwash Plains, and Kettles 111
  Drunlines, Eskers, and Kames 112
Glaciers of the Ice Age 113
Other Effects of Ice Age Glaciers 114
Deserts 116
  Distribution and Causes of Dry Lands 116
  The Role of Water in Arid Climates 117
Basin and Range: The Evolution of a Desert Landscape 118
Wind Erosion 119
  Deflation, Blowouts, and Desert Pavement 121
  Wind Abrasion 123

Wind Deposits 123
Loess 123
Sand Dunes 124
The Chapter in Review 125
Key Terms 126
Questions in Review 126
Online Study Guide 127

UNIT III
FORCES WITHIN 128

5 Plate Tectonics: A Scientific Theory Unfolds 129
FOCUS ON LEARNING 129
Continental Drift: An Idea Before Its Time 130
Evidence: The Continental Jigsaw Puzzle 131
Evidence: Fossils Match Across the Seas 131
Evidence: Rock Types and Structures Match 133
Evidence: Ancient Climates 133
The Great Debate 134
Plate Tectonics: The New Paradigm 134
Earth's Major Plates 135
Plate Boundaries 138
Divergent Boundaries 138
Oceanic Ridges and Seafloor Spreading 138
Continental Rifting 139
Convergent Boundaries 141
Oceanic-Continental Convergence 141
Oceanic-Oceanic Convergence 143
Continental-Continental Convergence 143
Transform Fault Boundaries 143
Testing the Plate Tectonics Model 145
Evidence: Ocean Drilling 146
Evidence: Hot Spots 147
Evidence: Paleomagnetism 148
The Breakup of Pangaea 151
What Drives Plate Motions? 152
Forces That Drive Plate Motion 152
Models of Plate-Mantle Convection 153
The Chapter in Review 155
Key Terms 156
Questions in Review 156
Online Study Guide 157

6 Restless Earth: Earthquakes, Geologic Structures, and Mountain Building 159
FOCUS ON LEARNING 159
What Is an Earthquake? 161
Earthquakes and Faults 161
Discovering the Cause of Earthquakes 162
Foreshocks and Aftershocks 163
Seismology: The Study of Earthquake Waves 164
Locating an Earthquake 166
Measuring the Size of Earthquakes 167
Intensity Scales 167
Magnitude Scales 168
Destruction from Earthquakes 169
Structural Damage from Seismic Vibrations 170
What is a Tsunami? 172
Fire 173
Landslides and Ground Subsidence 173
Probing Earth's Interior: "Seeing" Seismic Waves 175
Formation of Earth's Layered Structure 176
Earth's Internal Structure 176
Geologic Structures 177
Rock Deformation 177
Folds 178
Faults 179
Mountain Building 182
Mountain Building at Subduction Zones 183
Collisional Mountain Ranges 185
The Chapter in Review 188
Key Terms 189
Questions in Review 190
Online Study Guide 191

7 Fires Within: Igneous Activity 193
FOCUS ON LEARNING 193
The Nature of Volcanic Eruptions 194
Factors Affecting Viscosity 194
Importance of Dissolved Gases in Magma 196
What Is Extruded During Eruptions? 197
Lava Flows 197
Gases 198
Pyroclastic Materials 198
Volcanic Structures and Eruptive Styles 199
Anatomy of a Volcano 200
Shield Volcanoes 200
Cinder Cones 202
Composite Cones 204
Living in the Shadow of a Composite Cone 205
Nuée Ardente: A Deadly Pyroclastic Flow 205
Lahars: Mudflows on Active and Inactive Cones 206

Other Volcanic Landforms 207
- Calderas 207
- Fissure Eruptions and Lava Plateaus 209
- Volcanic Pipes and Necks 209

Intrusive Igneous Activity 210
- Dikes 210
- Sills and Laccoliths 211
- Batholiths 212

Plate Tectonics and Igneous Activity 213
- Igneous Activity at Convergent Plate Boundaries 216
- Igneous Activity at Divergent Plate Boundaries 216
- Intraplate Igneous Activity 217

Living with Volcanoes 218
- Volcanic Hazards 218
- Monitoring Volcanic Activity 219

The Chapter in Review 219

Key Terms 220

Questions in Review 220

Online Study Guide 221

UNIT IV
DECIPHERING EARTH'S HISTORY 222

8 Geologic Time 223

FOCUS ON LEARNING 223

Geology Needs a Time Scale 224

Some Historical Notes About Geology 224
- Catastrophism 225
- The Birth of Modern Geology 225
- Geology Today 225

Relative Dating—Key Principles 226
- Law of Superposition 226
- Principle of Original Horizontality 227
- Principle of Cross-Cutting Relationships 227
- Inclusions 228
- Unconformities 228
- Using Relative Dating Principles 229

Correlation of Rock Layers 229

Fossils: Evidence of Past Life 230
- Types of Fossils 231
- Conditions Favoring Preservation 234
- Fossils and Correlation 234

Dating with Radioactivity 235
- Reviewing Basic Atomic Structure 236
- Radioactivity 236
- Half-Life 237
- Radiometric Dating 237
- Dating with Carbon-14 238
- Importance of Radiometric Dating 239

The Geologic Time Scale 239
- Structure of the Time Scale 239
- Precambrian Time 241

Difficulties in Dating the Geologic Time Scale 241

The Chapter in Review 242

Key Terms 242

Questions in Review 242

Online Study Guide 243

UNIT V
THE GLOBAL OCEAN 244

9 Oceans: The Last Frontier 245

FOCUS ON LEARNING 245

The Vast World Ocean 246
- Geography of the Ocean 246
- Comparing the Oceans to the Continents 247

Composition of Seawater 247
- Salinity 247
- Sources of Sea Salts 247
- Processes Affecting Seawater Salinity 248

The Ocean's Layered Structure 249

An Emerging Picture of the Ocean Floor 250
- Mapping the Seafloor 250
- Viewing the Ocean Floor from Space 251
- Provinces of the Ocean Floor 252

Continental Margins 252
- Passive Continental Margins 252
- Active Continental Margins 253

Deep-Ocean Basin 257
- Deep-Ocean Trenches 257
- Abyssal Plains 258
- Seamounts, Guyots, and Oceanic Plateaus 258

The Oceanic Ridge 258

Seafloor Sediments 259
- Types of Seafloor Sediments 260
- Distribution of Seafloor Sediments 261
- Seafloor Sediments and Climate Change 262

The Chapter in Review 262

Key Terms 263

Questions in Review 264

Online Study Guide 265

10 The Restless Ocean 267

FOCUS ON LEARNING 267

Surface Circulation 268
- Ocean Circulation Patterns 268

Ocean Currents and Climate 269
Upwelling 270
Deep-Ocean Circulation 271
The Shoreline: A Dynamic Interface 271
Waves 272
Wave Characteristics 273
Circular Orbital Motion 273
Waves in the Surf Zone 274
Beaches and Shoreline Processes 274
Wave Erosion 275
Sand Movement on the Beach 276
Shoreline Features 278
Erosional Features 278
Depositional Features 280
The Evolving Shore 281
Stabilizing the Shore 282
Hard Stabilization 282
Alternatives to Hard Stabilization 284
Erosion Problems Along U.S. Coasts 284
Coastal Classification 285
Emergent Coasts 285
Submergent Coasts 286
Tides 286
Causes of Tides 286
Monthly Tidal Cycle 287
Tidal Patterns 287
Tidal Currents 288
The Chapter in Review 289
Key Terms 290
Questions in Review 290
Online Study Guide 291

UNIT VI
EARTH'S DYNAMIC ATMOSPHERE 292

11 Heating the Atmosphere 293

FOCUS ON LEARNING 293
Weather and Climate 294
Composition of the Atmosphere 295
Major Components 295
Variable Components 295
Ozone Depletions—A Global Issue 296
Height and Structure of the Atmosphere 298
Pressure Changes 298
Temperature Changes 298
Earth—Sun Relationships 300
Earth's Motions 300
Seasons 300
Earth's Orientation 301
Solstices and Equinoxes 302
Energy, Heat, and Temperature 304
Mechanisms of Heat Transfer 305
Conduction 305
Convection 305
Radiation 305
The Fate of Incoming Solar Radiation 307
Reflection and Scattering 307
Absorption 308
Heating the Atmosphere: The Greenhouse Effect 308
Global Warming 309
CO_2 Levels Are Rising 309
The Atmosphere's Response 310
Some Possible Consequences 311
For the Record: Air Temperature Data 312
Why Temperatures Vary: The Controls of Temperature 313
Land and Water 313
Altitude 314
Geographic Position 314
Cloud Cover and Albedo 315
World Distribution of Temperature 316
The Chapter in Review 317
Key Terms 318
Questions for Review 319
Online Study Guide 319

12 Moisture, Clouds, and Precipitation 321

FOCUS ON LEARNING 321
Water's Changes of State 322
Ice, Liquid Water, and Water Vapor 322
Latent Heat 322
Humidity: Water Vapor in the Air 324
Saturation 324
Mixing Ratio 324
Relative Humidity 325
Dew-Point Temperature 326
Measuring Humidity 327
The Basis of Cloud Formation: Adiabatic Cooling 328
Fog and Dew versus Cloud Formation 328
Adiabatic Temperature Changes 328
Adiabatic Cooling and Condensation 328
Pressures That Lift Air 329
Orographic Lifting 329
Frontal Wedging 330
Convergence 330
Localized Convective Lifting 331
The Weathermaker: Atmospheric Stability 331
Types of Stability 331
Stability and Daily Weather 334

Condensation and Cloud Formation 334
Types of Clouds 335
Fog 336
Fogs Caused by Cooling 336
Evaporation Fogs 339
Precipitation 340
How Precipitation Forms 340
Forms of Precipitation 341
Measuring Precipitation 343
The Chapter in Review 345
Key Terms 346
Questions in Review 346
Online Study Guide 347

13 The Atmosphere in Motion 349

FOCUS ON LEARNING 349

Understanding Air Pressure 350
Measuring Air Pressure 351
Factors Affecting Wind 352
Pressure-Gradient Force 352
Coriolis Effect 353
Friction with Earth's Surface 355
Highs and Lows 356
Cyclonic and Anticyclonic Winds 357
Weather Generalizations About Highs and Lows 357
General Circulation of the Atmosphere 359
Circulation on a Nonrotating Earth 359
Idealized Global Circulation 359
Influence of Continents 360
The Westerlies 360
Local Winds 362
Land and Sea Breezes 362
Mountain and Valley Breezes 362
Chinook and Santa Ana Winds 362
How Wind Is Measured 363
The Chapter in Review 365
Key Terms 365
Questions in Review 366
Online Study Guide 366

14 Weather Patterns and Severe Weather 369

FOCUS ON LEARNING 369

Air Masses 370
What Is an Air Mass? 370
Source Regions 371
Weather Associated with Air Masses 372
Fronts 373
Warm Fronts 373
Cold Fronts 374
Stationary Fronts and Occluded Fronts 375
The Middle-Latitude Cyclone 375
Idealized Weather of a Middle-Latitude Cyclone 376
The Role of Airflow Aloft 377
What's in a Name? 378
Thunderstorms 379
Thunderstorm Occurrence 379
Stages of Thunderstorm Development 380
Tornadoes 381
Tornado Occurrence and Development 382
Tornado Destruction 383
Tornado Forecasting 384
Hurricanes 386
Profile of a Hurricane 386
Hurricane Formation and Decay 388
Hurricane Destruction 389
The Chapter in Review 392
Key Terms 392
Questions in Review 393
Online Study Guide 393

UNIT VII
EARTH'S PLACE IN THE UNIVERSE 394

15 The Nature of the Solar System 395

FOCUS ON LEARNING 395

Ancient Astronomy 396
Early Greeks 396
Ptolemy's Model 397
The Birth of Modern Astronomy 398
Nicolaus Copernicus 398
Tycho Brahe 399
Johannes Kepler 400
Galileo Galilei 401
Sir Isaac Newton 403
The Planets: An Overveiw 404
Formation of the Planets 405
Terrestrial and Jovian Planets 407
The Compositions of the Planets 408
The Atmospheres of the Planets 408
Earth's Moon 409
The Lunar Surface 409
Impact Hypothesis for Moon's Origin 411

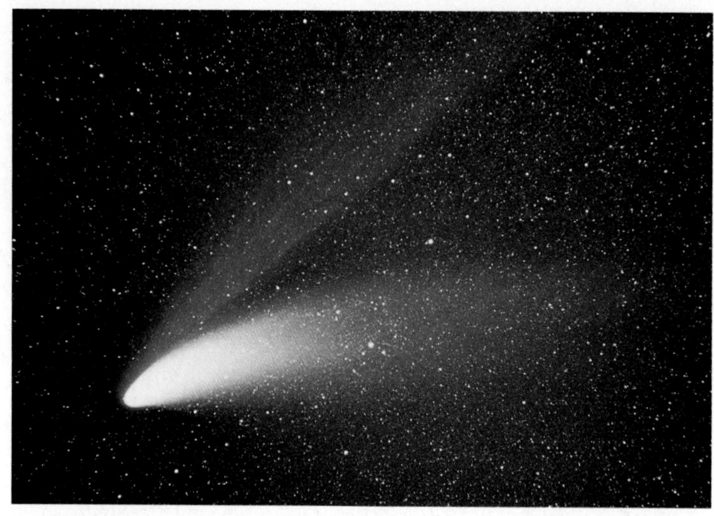

The Planets: A Brief Tour 412
Mercury: The Innermost Planet 412
Venus: The Veiled Planet 413
Mars: The Red Planet 414
Jupiter: Lord of the Heavens 416
Saturn: The Elegant Planet 418
Uranus and Neptune: The Twins 420
Minor Members of the Solar System: Asteroids, Comets, Meteoroids, and Dwarf Planets 421
Asteroids: Planetesimals 421
Comets: Dirty Snowballs 422
Meteoroids: Visitors to Earth 424
Dwarf Planets 425
The Chapter in Review 427
Key Terms 428
Questions in Review 428
Online Study Guide 429

16 Beyond the Solar System 431

FOCUS ON LEARNING 431

Properties of Stars 432
Measuring Distances to the Stars 432
Stellar Brightness 433
Stellar Color and Temperature 434
Binary Stars and Stellar Mass 434
Hertzsprung-Russell Diagram 435
Variable Stars 435
Interstellar Matter 437
Stellar Evolution 437
Star Birth 438
Protostar Stage 439
Main-Sequence Stage 439
Red Giant Stage 440
Burnout and Death 440
H-R Diagrams and Stellar Evolution 443
Stellar Remnants 443
White Dwarfs 443
Neutron Stars 443
Black Holes 444
The Milky Way Galaxy 444
Structure of the Milky Way Galaxy 445
Galaxies 445
Types of Galaxies 445
Galactic Clusters 447
Red Shifts 448
Expanding Universe 448
The Big Bang 448
The Chapter in Review 450
Key Terms 451
Questions in Review 451
Online Study Guide 451

APPENDIX A
Metric and English Units Compared 452

APPENDIX B
Mineral Identification Key 453

APPENDIX C
Relative Humidity and Dew-Point Tables 455

APPENDIX D
Earth's Grid System 457

Glossary 460

Index 469

PREFACE

Foundations of Earth Science, Fifth Edition, like its predecessors, is a college-level text designed for an introductory course in Earth science. It consists of seven units that emphasize broad and up-to-date coverage of basic topics and principles in geology, oceanography, meteorology, and astronomy. The book is intended to be a meaningful, nontechnical survey for undergraduate students with little background in science. Usually these students are taking an Earth science class to meet a portion of their college's or university's general requirements.

In addition to being informative and up-to-date, a major goal of *Foundations of Earth Science* is to meet the need of beginning students for a readable and user-friendly text, a book that is a highly usable "tool" for learning basic Earth science principles and concepts.

Distinguishing Features

Readability. The language of this book is straightforward and *written to be understood.* Clear, readable discussions with a minimum of technical language are the rule. The frequent headings and subheadings also help students follow discussions and identify the important ideas presented in each chapter. In the fifth edition, improved readability was achieved by examining chapter organization and flow and by writing in a more personal style. Many sections were substantially rewritten in an effort to make the material more understandable.

Focus on Learning. To assist student learning, every chapter opens with a series of questions. Each question alerts the reader to an important idea or concept in the chapter. Upon completion of a chapter, four useful devices help students review. First, a helpful summary—*The Chapter in Review*—recaps all of the major points. Next is a checklist of *Key Terms* with page references. Learning the language of Earth science helps students retain the material. This is followed by *Questions for Review,* that helps students examine their knowledge of significant facts and ideas. Each chapter closes with a reminder to visit the Web site for *Foundations of Earth Science 5e* (http://www.prenhall.com/lutgens). This valuable tool works as an on-line study guide that gives students one more opportunity to test their understanding of key concepts, ideas, and terms.

Illustrations and Photographs. The Earth sciences are highly visual. Therefore, photographs and artwork are a very important part of an introductory book. *Foundations of Earth Science,* Fifth Edition, contains dozens of new high-quality photographs that were carefully selected to aid understanding, add realism, and heighten the reader's interest.

The illustrations in each new edition of *Foundations of Earth Science* keep getting better and better. In the fifth edition more than 100 pieces of line art are new or redesigned. The new art illustrates ideas and concepts more clearly and realistically than ever before. The art program was carried out by Dennis Tasa, a gifted artist and respected Earth science illustrator.

Focus on Basic Principles and Instructor Flexibility. The main focus of the Fifth Edition remains the same as in the first four—to foster student understanding of basic Earth science principles. With this in mind, we have attempted to provide the reader with a sense of the observational techniques and reasoning processes that characterize the Earth sciences.

Whereas student use of the text is a primary concern, the book's adaptability to the needs and desires of the instructor is equally important. Realizing the broad diversity of Earth science courses in both content and approach, we have continued to use a relatively nonintegrated format to allow maximum flexibility for the instructor. Each of the major units stands alone; hence, they can be taught in any order. A unit can be omitted entirely without appreciable loss of continuity, and portions of some chapters may be interchanged or excluded at the instructor's discretion.

Interactive CD-Rom. Each copy of *Foundations of Earth Science,* Fifth Edition, comes with an extremely interactive CD-ROM: *GEODe: Earth Science.* In addition to excellent treatment of topics in geology, the CD-ROM includes broad coverage of the oceans, basic meteorology, and the solar system. *GEODe: Earth Science* includes many tutorials, interactive exercises, animations, and video clips. Every unit in the text has a corresponding unit in *GEODe: Earth Science.* A special *GEODe: Earth Science* icon appears throughout the book wherever a text discussion has a corresponding *GEODe: Earth Science* activity.

More About the Fifth Edition

The Fifth Edition of *Foundations of Earth Science* represents a thorough revision. *Every* part of the book was examined carefully with the dual goals of keeping topics current and improving the clarity of text discussions.

People familiar with previous editions of *Foundations of Earth Science* will notice that the trim size of the fifth edition is larger. The result is a less cluttered appearance and room to expand the size of many of the diagrams and photos.

The list of specifics is long. Examples include the following:

- The Introduction, an unnumbered "minichapter" includes a revised, expanded, easier-to-follow discussion of Earth as a system and an all new section on "Scales of Space and Time in Earth Science."
- Much is new in the chapters on Earth materials. Chapter 1 "Minerals: Building Blocks of Rocks" has a completely rewritten section on the geologic definition of minerals and an all new discussion of elements that more clearly explains atomic structure and bonding. Chapter 2 "Rocks: Materials of the Solid Earth" includes an expanded treatment of the rock cycle including new line art and also a significant revision of the section on igneous rocks.
- The portion of Chapter 3 "Landscapes Fashioned by Water" that focuses on running water has been reorganized and almost entirely rewritten so that the discussion of streams progresses in a manner that is clearer and more logical for the beginning student. The portion of this chapter that pertains to groundwater includes new sections on the storage and movement of groundwater.
- Chapter 6 "Restless Earth: Earthquakes, Geologic Structures, and Mountain Building" has an expanded treatment of earthquake destruction that includes a close look at the devastating tsunami that struck the Indian Ocean region in December 2004. The chapter also includes an all new section on "Earth's Interior." Coverage is not only broader and more up-to-date, but explanations are written in an easier to understand style. The chapter concludes with the new heading on "Collisional Mountain Ranges" (formerly "Continental Collisions"). The new discussion and accompanying art present a clearer, more up-to-date view of this complex topic.
- Chapter 7 "Fires Within: Igneous Activity" concludes with a new section called "Living With Volcanoes" that examines volcanic hazards and the monitoring of volcanic activity.
- Chapter 8 "Geologic Time" has new discussions on "Types of Fossils" and "Dating With Carbon-14" that provide clear descriptions of these sometimes misunderstood topics.
- In Unit V The Global Ocean, Chapter 9 includes new information on mapping the ocean floor and the oceanic ridge system. In Chapter 10, there is a new section entitled "The Shoreline: A Dynamic Interface" that focuses on the coastal environment and a revised treatment of "Sand Movement on the Beach."
- Unit VI Earth's Dynamic Atmosphere, has experienced many changes that have strengthened and improved discussions and updated coverage. Chapter 11 has an expanded section contrasting "Weather and Climate," and a new discussion that explains the differences between "Energy, Heat, and Temperature." This chapter also has updated coverage of "Global Warming" and a new explanation of isotherms (with art) in the section titled "For the Record: Air Temperature Data" (formerly "Temperature Measurement and Data").
- Other changes to the unit on the atmosphere include a revised discussion of "Water's Changes of State" in Chapter 12, a clearer discussion on "Understanding Air Pressure" in Chapter 13, and an expanded and updated examination of hurricanes (including a discussion and images of Hurricane Katrina) in Chapter 14.
- Chapter 15 "The Nature of the Solar System" has received a major overhaul. There are new discussions on solar system evolution, lunar history, and the dynamic geologic evolution of Mars. Also included is a new section on dwarf planets (Pluto now has this status) and the latest on the Kuiper Belt and Oort cloud.

Additional Highlights

- The text's interactive CD-ROM GEODe *Earth Science*, 2/E v.2 is an even better learning tool than before. Each section now concludes with a Review Quiz consisting of randomly generated questions that helps students test their understanding of basic facts and concepts.
- NEW FEATURE: *Did You Know?* This involves the placement of interesting facts and ideas at various places throughout each chapter that are intended to add interest and relevance to text discussions.

Acknowledgments

Writing a college textbook requires the talents and cooperation of many people. Working with Dennis Tasa, who is responsible for all of the outstanding illustrations and much of the development work on *GEODe: Earth Science*, is always special for us. We not only value his outstanding artistic talents and imagination but his friendship.

Our sincere appreciation goes to those colleagues who prepared in-depth reviews of the manuscript. Their critical comments and thoughtful input helped guide our work and clearly strengthened the text. Special thanks to:

Walter Childs, III, Livingstone College
Paul A. Horton, Indian River Community College
Kristine Larsen, Central Connecticut State University
Donald W. Lovejoy, Palm Beach Atlantic University
Randal Mandock, Clark Atlanta University
Gustavo Morales, Valencia Community College
Roger Podewell, Olive Harvey College
Frank A. Revetta, SUNY College at Potsdam
Erwin Selleck, SUNY Canton College of Technology
John Tacinelli, Rochester Community and Technical College
Joseph L. Tinsley, Clemson University
R. Aileen Yingst, University of Wisconsin, Green Bay

We also want to acknowledge the team of professionals at Prentice Hall. We sincerely appreciate the company's continuing strong support for excellence and innovation. Special thanks to our geosciences publisher, Dan Kaveney. We value his leadership and appreciate his attention to detail, excellent communication skills, and easygoing style. The production team, led by Patty Donovan at Pine Tree Composition, Inc., has done an outstanding job. All are true professionals with whom we are very fortunate to be associated.

Frederick K. Lutgens
Edward J. Tarbuck

The Teaching and Learning Package

For the Instructor

Instructor's Resource Center (IRC) on DVD

The IRC puts all of your lecture resources in one easy-to-reach place.

- *All* of the line art, tables, and photos from the text in .jpg files (Are illustrations central to your lecture? Check out the *Student Lecture Notebook* under "For the Student".)
- Animations of over one hundred key geological processes
- *Images of Earth* photo gallery
- PowerPoint™ presentations
- *Instructor's Manual* in Microsoft Word
- *Test Item File* in Microsoft Word
- TestGen EQ test generation and management software

Animations

The *Prentice Hall Geoscience Animation Library 4e* includes over 100 animations illustrating many difficult-to-visualize Earth science concepts. Created through a unique collaboration among seven of Prentice Hall's leading geoscience authors, these animations represent a significant leap forward in lecture presentation aids. They are provided both as Flash files and, for your convenience, pre-loaded into PowerPoint™ slides.

PowerPoint™ Presentations

Found on the IRC are *three* PowerPoint™ files for each chapter. Cut down on your preparation time, no matter what your lecture needs:

1. Exclusive Art—All of the photos, art, and tables from the text, in order, loaded into PowerPoint™ slides.
2. Lecture Outline—Authored by Stanley Hatfield of Southwestern Illinois College, this set averages 35 slides per chapter and includes customizable lecture outlines with supporting art.
3. Animations—Each animation is preloaded into PowerPoint™ for easy integration into your presentation.

All art is modified for projection—labels are enlarged, colors brightened, contrast sharpened. The slides are designed for *clear viewing*, even in the largest lecture halls, and *brightened* so you need not dim the lights as much.

"Images of Earth" Photo Gallery

Supplement your personal and text-specific slides with this amazing collection of over 300 geologic photos contributed by Marli Miller (University of Oregon) and other professionals in the field. The photos are grouped by geologic concept and available on the Instructor's Resource Center.

Instructor's Manual with Tests

Authored by Stanley Hatfield (Southwestern Illinois College), the *Instructor's Manual* contains: learning objectives, chapter outlines, answers to end-of-chapter questions and suggested, short demonstrations to spice up your lecture. The *Test Item File* incorporates art and averages 75 multiple-choice, true/false, short answer and critical thinking questions per chapter.

Transparencies

Every table and most of Dennis Tasa's illustrations in *Foundations of Earth Science, Fifth Edition* are available on full-color projection-enhanced transparencies. (Are illustrations central to your lecture? Check out the *Student Lecture Notebook* under "For the Student").

For the Laboratory

Applications and Investigations in Earth Science, Fifth Edition. Written by Ed Tarbuck, Fred Lutgens, and Ken Pinzke, this full-color laboratory manual contains 22 exercises that provide students with hands-on experiences in geology, oceanography, meteorology, astronomy, and Earth science skills.

For the Student

GEODe: Earth Science v.2

GEODe EARTH SCIENCE

Somewhere between a text and a tutor, *GEODe: Earth Science* reinforces key concepts using animations, video, narration, interactive exercises, and practice quizzes. A copy of *GEODe: Earth Science* is automatically included in every copy of the text purchased from Prentice Hall.

MyGeologyPlace

www.prenhall.com/lutgens This Web site has been designed to provide students with all the tools needed for online study and review. A self-quiz is available for each key concept within a chapter, and hints and feedback are provided for guidance. Links to other resources are also included for further study. An access code for MyGeologyPlace is bound into the front of every new copy of this textbook.

Student Lecture Notebook

All of the line art from the transparency set is reproduced in this full-color notebook, with space for notes. Students can now fully focus on the lecture and not be distracted by attempting to replicate figures. Each page is three-hole punched for easy integration with other course materials.

Introduction to Earth Science

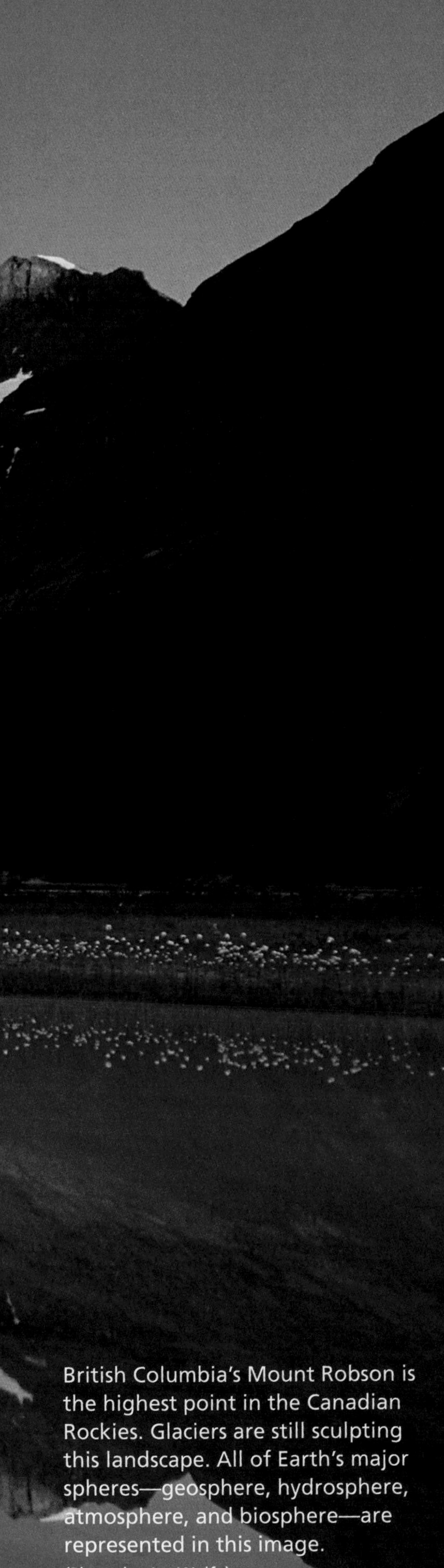

British Columbia's Mount Robson is the highest point in the Canadian Rockies. Glaciers are still sculpting this landscape. All of Earth's major spheres—geosphere, hydrosphere, atmosphere, and biosphere—are represented in this image.
(Photo by Art Wolfe)

FOCUS ON LEARNING

To assist you in learning the concepts in the Introduction, you will find it helpful to focus on the following questions:

1. What are the sciences that collectively make up Earth science?
2. What are the four "spheres" that comprise Earth's natural environment?
3. What are the principal divisions of the solid Earth?
4. Why should Earth be thought of as a system?
5. What are the sources of energy that power the Earth system?
6. What are some of Earth's important environmental issues?
7. How is a scientific hypothesis different from a scientific theory?

The spectacular eruption of a volcano, the magnificent scenery of a rocky coast, and the destruction created by a hurricane are all subjects for the Earth scientist. The study of Earth science deals with many fascinating and practical questions about our environment. What forces produce mountains? Why is our daily weather so variable? Is climate really changing? How old is Earth and how is it related to the other planets in the solar system? What causes ocean tides? What was the Ice Age like? Will there be another? Can a successful well be located at this site?

The subject of this text is *Earth science.* To understand Earth is not an easy task, because our planet is not a static and unchanging mass. Rather, it is a dynamic body with many interacting parts and a long and complex history.

What Is Earth Science?

Earth science is the name for all the sciences that collectively seek to understand Earth and its neighbors in space. It includes geology, oceanography, meteorology, and astronomy.

In this book, Units 1 through 4 focus on the science of **geology,** a word that literally means "study of Earth." Geology is traditionally divided into two broad areas—physical and historical.

Physical geology examines the materials comprising Earth and seeks to understand the many processes that operate beneath and upon its surface. Earth is a dynamic, ever-changing planet. Internal forces create earthquakes, build mountains, and produce volcanic structures. At the surface, external processes break rock apart and sculpture a broad array of landforms. The erosional effects of water, wind, and ice result in a great diversity of landscapes. Because rocks and minerals form in response to Earth's internal and external processes, the study of Earth materials is basic to an understanding of our planet.

In contrast to physical geology, the aim of *historical geology* is to understand the origin of Earth and the development of the planet through its 4.5-billion-year history. It strives to establish an orderly chronological arrangement of the multitude of physical and biological changes that have occurred in the geologic past (Figure I.1). The study of physical geology logically precedes the study of Earth's history because we must first understand how Earth works before we attempt to unravel its past.

Unit 5, *The Global Ocean,* is devoted to **oceanography.** Oceanography is actually not a separate and distinct science. Rather, it involves the application of all sciences in a comprehensive and interrelated study of the oceans in all their aspects and relationships. Oceanography integrates chemistry, physics, geology, and biology. It includes the study of the composition and movements of seawater, as well as coastal processes, seafloor topography, and marine life.

Unit 6, *Earth's Dynamic Atmosphere,* examines the mixture of gases that is held to the planet by gravity and thins rapidly with altitude. Acted on by the combined effects of Earth's motions and energy from the Sun, the formless and invisible atmosphere reacts by producing an infinite variety of weather, which in turn creates the basic pattern of global climates. **Meteorology** is the study of the atmosphere and the

Figure I.1 Paleontologists study ancient life. Here scientists are excavating the remains of *Albertasaurus,* a large carnivore similar to *Tyrannosaurus rex,* which lived during the late Cretaceous Period. The excavation site is near Red Deer River, Alberta, Canada. The aim of historical geology is to understand the development of Earth and its life through time. Fossils are essential tools in that quest. (Photo by Richard T. Nowitz/Science Source/Photo Researchers, Inc.)

processes that produce weather and climate. Like oceanography, meteorology involves the application of other sciences in an integrated study of the thin layer of air that surrounds Earth.

Unit 7, *Earth's Place in the Universe,* demonstrates that an understanding of Earth requires that we relate our planet to the larger universe. Because Earth is related to all of the other objects in space, the science of **astronomy**—the study of the universe—is very useful in probing the origins of our own environment. As we are so closely acquainted with the planet on which we live, it is easy to forget that Earth is just a tiny object in a vast universe. Indeed, Earth is subject to the same physical laws that govern the many other objects populating the great expanses of space. Thus, to understand explanations of our planet's origin, it is useful to learn something about the other members of our solar system. Moreover, it is helpful to view the solar system as a part of the great assemblage of stars that comprise our galaxy, which in turn is but one of many galaxies.

To understand Earth science is challenging because our planet is a dynamic body with many interacting parts and a complex history. Throughout its long existence, Earth has been changing. In fact, it is changing as you read this page and will continue to do so into the foreseeable future. Sometimes the changes are rapid and violent, as when severe storms, landslides, or volcanic eruptions occur. Just as often, change takes place so gradually that it goes unnoticed during a lifetime. Scales of size and space also vary greatly among the phenomena studied in Earth science.

Earth science is often perceived as science that is done in the out-of-doors, and rightly so. A great deal of what Earth scientists study is based on observations and experiments performed in the field. But Earth science is also conducted in the laboratory, where, for example, the study of various Earth materials provides insights into many basic processes, and the creation of complex computer models allows for the simulation of our planet's complicated climate system. Frequently, Earth scientists require an understanding and application of knowledge and principles from physics, chemistry, and biology. Geology, oceanography, meteorology, and astronomy are sciences that seek to expand our knowledge of the natural world and our place in it.

Earth's Spheres

A view such as the one in Figure I.2A provided the *Apollo 8* astronauts as well as the rest of humanity with a unique perspective of our home. Seen from space, Earth is breathtaking in its beauty and startling in its solitude. Such an image reminds us

A.

B.

Figure I.2 **A.** View that greeted the *Apollo 8* astronauts as their spacecraft emerged from behind the Moon. **B.** Africa and Arabia are prominent in this classic image of Earth taken from *Apollo 17.* The tan cloud-free zones over the land coincide with major desert regions. The band of clouds across central Africa is associated with a much wetter climate that sustains tropical rain forests. The dark blue of the oceans and the swirling cloud patterns remind us of the importance of the oceans and the atmosphere. Antarctica, a continent covered by glacial ice, is visible at the South Pole. (Both photos courtesy of NASA/Science Source/Photo Researchers, Inc.)

Did You Know?

In 1492, when Columbus set sail, many Europeans thought that Earth was flat and that Columbus would sail off the edge. However, more than 2000 years earlier, ancient Greeks realized that Earth was spherical because it always cast a curved shadow on the Moon during a lunar eclipse. In fact, Erastosthenes (276–194 B.C.) calculated Earth's circumference and obtained a value close to the modern measurement of 40,075 km (24,902 mi).

Did You Know?

The volume of ocean water is so large that if Earth's solid mass were perfectly smooth (level) and spherical, the oceans would cover Earth's entire surface to a uniform depth of more than 2000 m (1.2 mi)!

that our home is, after all, a planet—small, self-contained, and in some ways even fragile.

As we look closely at our planet from space, it becomes clear that Earth is much more than rock and soil. In fact, the most conspicuous features in Figure I.2A are not continents, but swirling clouds suspended above the surface and the vast global ocean. These features emphasize the importance of air and water to our planet.

The closer view of Earth from space, shown in Figure I.2B, helps us appreciate why the physical environment is traditionally divided into three major parts: the water portion of our planet, the hydrosphere; Earth's gaseous envelope, the atmosphere; and, of course, the solid Earth, or geosphere.

It should be emphasized that our environment is highly integrated and not dominated by water, air, or rock alone. Rather, it is characterized by continuous interactions as air comes in contact with rock, rock with water, and water with air. Moreover, the biosphere, the totality of life forms on our planet, extends into each of the three physical realms and is an equally integral part of the planet. Thus, Earth can be thought of as consisting of four major spheres: the hydrosphere, atmosphere, geosphere, and biosphere.

The interactions among Earth's four spheres are incalculable. Figure I.3 provides an easy-to-visualize example. The shoreline is an obvious meeting place for rock, water, and air. In this scene, ocean waves that were created by the drag of air moving across the water are breaking against the rocky shore. The force of the water can be powerful, and the erosional work that is accomplished can be great.

Hydrosphere

Earth is sometimes called the *blue planet.* Water, more than anything else, makes Earth unique. The **hydrosphere** is a dynamic mass of water that is continually on the move, evapo-

Figure I.3 The shoreline is one obvious example of an *interface*—a common boundary where different parts of a system interact. In this scene, ocean waves *(hydrosphere)* that were created by the force of moving air *(atmosphere)* break against California's rocky Big Sur shore *(geosphere).* The force of the water can be powerful, and the erosional work that is accomplished can be great. (Photo by Carr Clifton)

rating from the oceans to the atmosphere, precipitating to the land, and flowing back to the ocean again. The global ocean is certainly the most prominent feature of the hydrosphere, blanketing nearly 71 percent of Earth's surface to an average depth of about 3800 meters (12,500 feet). It accounts for about 97 percent of Earth's water. However, the hydrosphere also includes the freshwater found underground and in streams, lakes, and glaciers. Moreover, water is an important component of all living things.

Although these latter sources constitute just a tiny fraction of the total, they are much more important than their meager percentage indicates. In addition to providing the freshwater that is so vital to life on the continents, streams, glaciers, and groundwater are responsible for sculpturing and creating many of our planet's varied landforms.

Atmosphere

Earth is surrounded by a life-giving gaseous envelope called the **atmosphere.** When we watch a high-flying jet plane cross the sky, it seems that the atmosphere extends upward for a great distance (Figure I.4). However, when compared to the thickness (radius) of the solid Earth, (about 6400 kilometers or 4000 miles), the atmosphere is a very shallow layer. One-half lies below an altitude of 5.6 kilometers (3.5 miles), and 90 percent occurs within just 16 kilometers (10 miles) of Earth's surface. This thin blanket of air is nevertheless an integral part of the planet. It not only provides the air that we breathe, but it also acts to protect us from the dangerous radiation emitted by the Sun. The energy exchanges that continually occur between the atmosphere and Earth's surface and between the atmosphere and space produce the effects we call *weather* and *climate.*

If, like the Moon, Earth had no atmosphere, our planet would be lifeless because many of the processes and interactions that make Earth's surface such a dynamic place could not operate. Without weathering and erosion, the face of our planet might more closely resemble the surface of the Moon, which has not changed appreciably in nearly 3 billion years.

Figure I.4 This jet is flying high in the atmosphere at an altitude of more than 9000 meters (30,000 feet). More than two-thirds of the atmosphere is below this height. To someone on the ground, the atmosphere seems to extend for a great distance. However, when compared to the thickness (radius) of the solid Earth, the atmosphere is a very shallow layer. (Photo by Warren Faidley/Weatherstock)

Did You Know?

During the twentieth century, the global average surface temperature increased by about 0.6°C (1°F). By the end of the twenty-first century, globally averaged surface temperature is projected to increase by an additional 1.4 to 5.8°C (2.5 to 14.5°F).

Geosphere

Lying beneath the atmosphere and the ocean is the solid Earth or **geosphere.** The geosphere extends from the surface to the center of the planet, a depth of nearly 6400 kilometers (4000 miles), making it by far the largest of Earth's spheres. Much of our study of the solid Earth focuses on the more accessible surface and near-surface features, but it is worth noting that many of these features are linked to the dynamic behavior of Earth's interior. Figure I.5 provides a diagrammatic view of Earth's interior. As you can see, it is not uniform but is divided into various layers. Based on compositional differences, three principal layers are identified: a dense inner sphere called the **core;** the less dense **mantle;** and the **crust,** which is the light and very thin outer skin of Earth. The crust is not a layer of uniform thickness. It is thinnest beneath the oceans and thickest where continents exist. Although the crust may seem insignificant when compared with the other layers of the geosphere, which are much thicker, it was created by the same general processes that formed Earth's present structure. Thus, the crust is important in understanding the history and nature of our planet.

The solid Earth is also divided based on how materials behave when various forces and stresses are applied. The term **lithosphere** refers to the rigid outer layer that includes the crust and uppermost mantle. Beneath the rigid rocks that compose the lithosphere, the rocks of the **asthenosphere** are weak and are able to slowly flow in response to the uneven distribution of heat deep within Earth.

The two principal divisions of Earth's surface are the continents and the ocean basins. The most obvious difference between these two diverse provinces is their relative levels. The average elevation of the continents above sea level is about 840 meters (2750 feet), whereas the average depth of the oceans is about 3800 meters (12,500 feet). Thus, the continents stand on the average 4640 meters (about 4.6 kilometers or nearly 3 miles) above the level of the ocean floor.

Did You Know?

We have never sampled the mantle or core directly. The structure of Earth's interior is determined by analyzing seismic waves from earthquakes. As these waves of energy penetrate Earth's interior, they change speed and are bent and reflected as they move through zones having different properties. Monitoring stations around the world detect and record this energy.

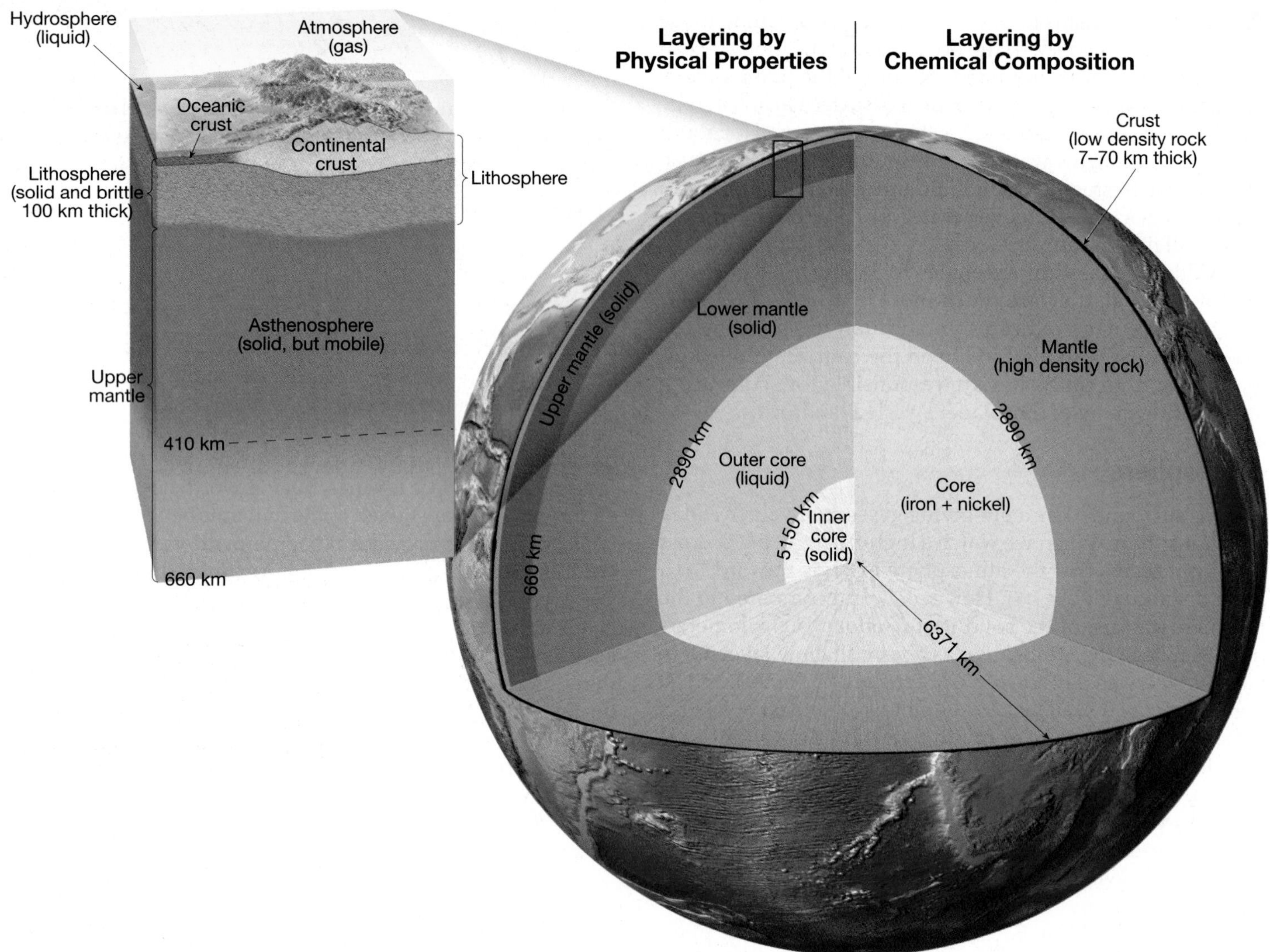

Figure I.5 The right side of the globe shows that Earth's interior is divided into three different layers based on compositional differences—the crust, mantle, and core. The left side of the globe shows the five main layers of Earth's interior based on physical properties and mechanical strength—the lithosphere, asthenosphere, lower mantle, outer core, and inner core. The block diagram shows an enlarged view of the upper portion of Earth's interior.

Biosphere

The fourth "sphere," the **biosphere,** includes all life on Earth. Ocean life is concentrated in the sunlit waters of the sea. Most life on land is also concentrated near the surface, with tree roots and burrowing animals reaching a few meters underground and flying insects and birds reaching a kilometer or so above the surface. A surprising variety of life forms are also adapted to extreme environments. For example, on the deep ocean floor where pressures are extreme and no light penetrates, there are places where vents spew hot, mineral-rich fluids that support communities of exotic life forms. On land, some bacteria thrive in rocks as deep as 4 kilometers (2.5 miles) and in boiling hot springs. Moreover, air currents can carry microorganisms many kilometers into the atmosphere. But even when we consider these extremes, life still must be thought of as being confined to a narrow band very near Earth's surface.

Plants and animals depend on the physical environment for the basics of life. However, organisms do more than just respond to this environment. Through countless interactions, life forms help maintain and alter their physical environment. Without life, the makeup and nature of the geosphere, hydrosphere, and atmosphere would be very different.

Earth as a System

Anyone who studies Earth soon learns that our planet is a dynamic body with many separate but interacting parts or spheres. The hydrosphere, atmosphere, biosphere, and geosphere and all of their components can be studied separately. However, the parts are not isolated. Each is related in some way to the others to produce a complex and continuously interacting whole that we call the **Earth system.**

A simple example of the interactions among different parts of the Earth system occurs every winter as moisture evaporates from the Pacific Ocean and subsequently falls as rain in the hills of southern California, triggering destructive landslides. The processes that move water from the hydrosphere to the atmosphere and then to the geosphere have a profound impact on the plants and animals (including humans) that inhabit the affected regions.

Scientists have recognized that in order to more fully understand our planet, they must learn how its individual components (land, water, air, and life forms) are interconnected. This endeavor, called *Earth system science,* aims to study Earth as a *system* composed of numerous interacting parts, or *subsystems.* Using an interdisciplinary approach, those who practice Earth system science attempt to achieve the level of understanding necessary to comprehend and solve many of our global environmental problems.

A **system** is a group of interacting, or interdependent, parts that form a complex whole. Most of us hear and use the term frequently. We may service our car's cooling *system,* make use of the city's transportation *system,* and be a participant in the political *system.* A news report might inform us of an approaching weather *system.* Further, we know that Earth is just a small part of a large system known as the *solar system,* which, in turn, is a subsystem of an even larger system called the Milky Way Galaxy.

The parts of the Earth system are linked so that a change in one part can produce changes in any or all of the other parts. For example, when a volcano erupts, lava from Earth's interior may flow out at the surface and block a nearby valley. This new obstruction influences the region's drainage system by creating a lake or causing streams to change course. The large quantities of volcanic ash and gases that can be emitted during an eruption might be blown high into the atmosphere and influence the amount of solar energy that can reach Earth's surface. The result could be a drop in air temperatures over the entire hemisphere.

Where the surface is covered by lava flows or a thick layer of volcanic ash, existing soils are buried. This causes the soil-forming processes to begin anew to transform the new surface material into soil (Figure I.6). The soil that eventually forms will reflect the interactions among many parts of the Earth system—the volcanic parent material, the type and rate of weathering, and the impact of biological activity. Of course, there would also be significant changes in the biosphere. Some organisms and their habitats would be eliminated by the lava and ash, whereas new settings for life, such as the lake, would be created. The potential climate change could also impact sensitive life forms.

The Earth system is characterized by processes that vary on spatial scales from fractions of millimeters to thousands of kilometers. Time scales for Earth's processes range from milliseconds to billions of years. As we learn about Earth, it becomes increasingly clear that despite significant separations in distance or time, many processes are connected and that a change in one component can influence the entire system.

The Earth system is powered by energy from two sources. The Sun drives external processes that occur in the atmosphere, hydrosphere, and at Earth's surface. Weather and climate, ocean circulation, and erosional processes are driven by energy from the Sun. Earth's interior is the second source of energy. Heat remaining from when our planet formed and heat that is continuously generated by radioactive decay power the internal processes that produce volcanoes, earthquakes, and mountains.

Humans are *part of* the Earth system, a system in which the living and nonliving components are entwined and interconnected. Therefore, our actions produce changes in all of the other parts. When we burn gasoline and coal, build seawalls along the shoreline, dispose of our wastes, and clear the land, we cause other parts of the system to respond, often in unforeseen ways. Throughout this book, you will learn about many of Earth's subsystems: the hydrologic system, the tectonic (mountain-building) system, and the climate

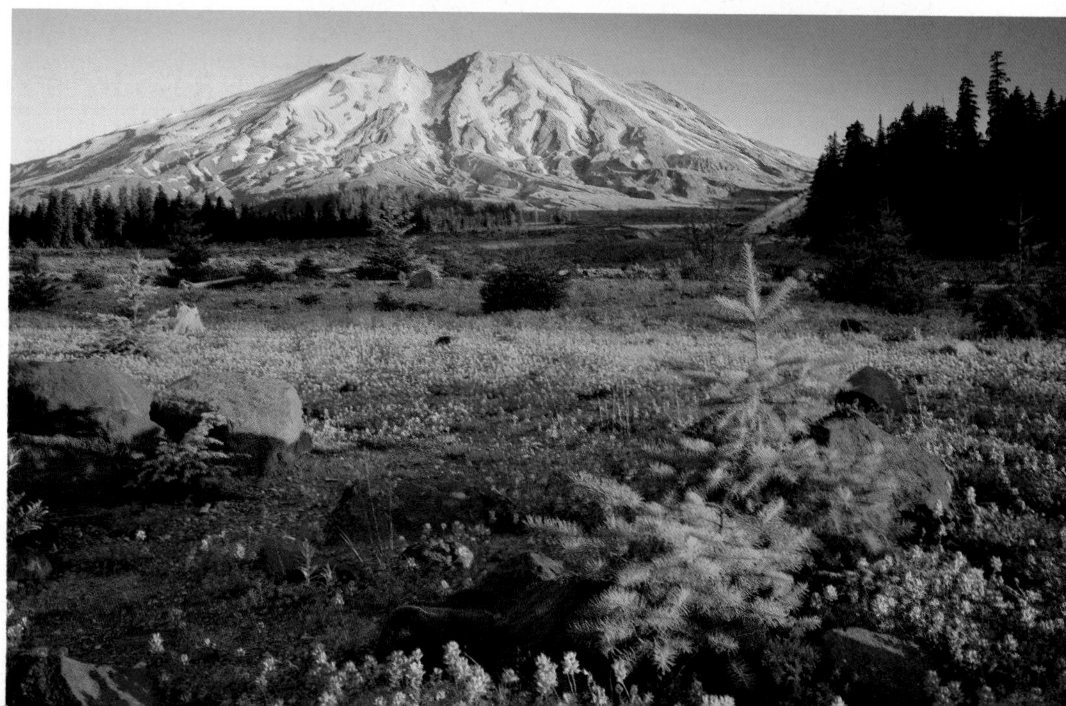

Figure I.6 When Mount St. Helens erupted in May 1980, the area shown here was buried by a volcanic mudflow. After just a few years, plants were reestablished and new soil was forming. (Photo by Jack W. Dykinga Associates)

system, to name a few. Remember that these components *and we humans* are all part of the complex interacting whole we call the Earth system.

Scales of Space and Time in Earth Science

When we study Earth, we must contend with a broad array of space and time scales (Figure I.7). Some phenomena are relatively easy for us to imagine, such as the size and duration of an afternoon thunderstorm or the dimensions of a sand dune. Other phenomena are so vast or so small that they are difficult to imagine. The number of stars and distances in our galaxy (and beyond!) or the internal arrangement of atoms in a mineral crystal are examples of such phenomena.

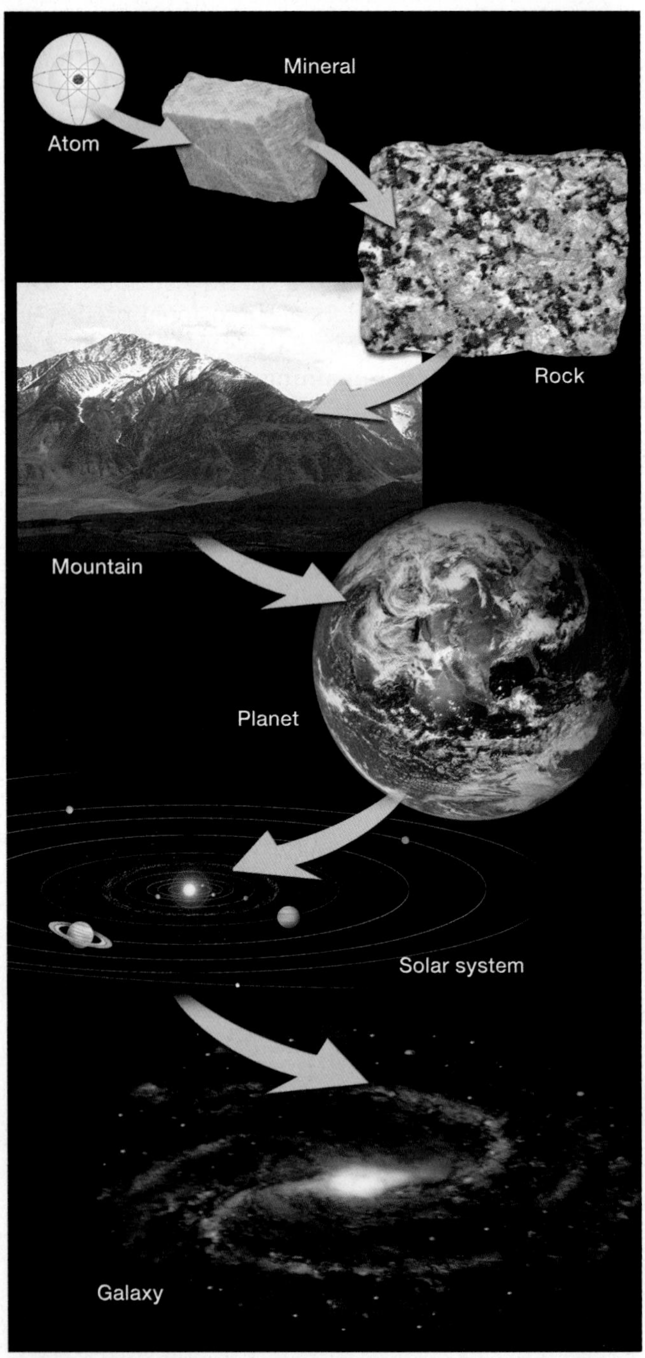

Figure I.7 Earth science involves investigations of phenomena that range in size from atoms to galaxies and beyond.

Some of the events we study occur in fractions of a second. Lightning is an example. Other processes extend over spans of tens or hundreds of millions of years. The lofty Himalaya Mountains began forming about 45 million years ago, and they continue to develop today.

The concept of geologic time is new to many nonscientists. People are accustomed to dealing with increments of time that are measured in hours, days, weeks, and years. Our history books often examine events over spans of centuries, but even a century is difficult to appreciate fully. For most of us, someone or something that is 90 years old is *very old,* and a 1000-year-old artifact is *ancient.*

By contrast, those who study Earth science must routinely deal with vast time periods—millions or billions (thousands of millions) of years. When viewed in the context of Earth's 4.5-billion-year history, an event that occurred 100 million years ago may be characterized as "recent" by a geologist, and a rock sample that has been dated at 10 million years may be called "young."

An appreciation for the magnitude of geologic time is important in the study of our planet because many processes are so gradual that vast spans of time are needed before significant changes occur.

How long is 4.5 billion years? If you were to begin counting at the rate of one number per second and continued 24 hours a day, seven days a week and never stopped, it would take about two lifetimes (150 years) to reach 4.5 billion!

Over the past 200 years or so, Earth scientists have developed the **geologic time scale** of Earth history. It subdivides the 4.5-billion-year history of Earth into many different units and provides a meaningful time frame within which the events of the geologic past are arranged (Figure I.8). The principles used to develop the geologic time scale are examined at some length in Chapter 8.

Resources and Environmental Issues

Environment refers to everything that surrounds and influences an organism. Some of these things are biological and social, but others are nonliving. The factors in this latter category are collectively called our physical environment. The **physical environment** encompasses water, air, soil, and rock, as well as conditions such as temperature, humidity, and sunlight. The phenomena and processes studied by the Earth sciences are basic to an understanding of the physical environment. In this sense, most of Earth science may be characterized as environmental science.

However, when the term *environmental* is applied to Earth science today, it usually means relationships between people and the physical environment. Application of the Earth sciences is necessary to understand and solve problems that arise from these interactions.

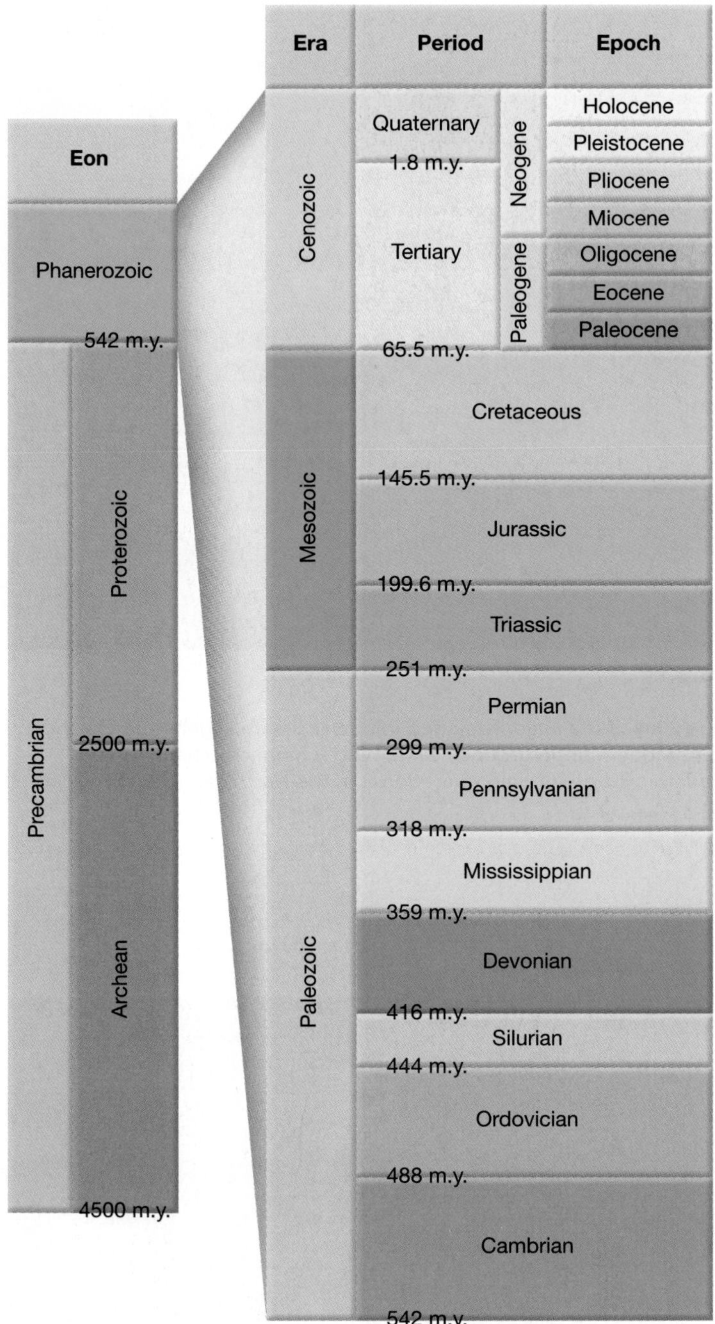

Figure I.8 The geologic time scale divides the vast 4.5-billion-year history of Earth into eons, eras, periods, and epochs. We presently live in the Holocene epoch of the Quaternary period. This period is part of the Cenozoic era, which is the latest era of the Phanerozoic eon. There is much more about geologic time in Chapter 8.

We can dramatically influence natural processes. For example, river flooding is natural, but the magnitude and frequency of flooding can be changed significantly by human activities such as clearing forests, building cities, and constructing dams. Unfortunately, natural systems do not always adjust to artificial changes in ways we can anticipate. Thus, an alteration to the environment that was intended to benefit society may have the opposite effect.

Resources

Resources are an important focus of the Earth sciences that is of great value to people. Resources include water and soil, a great variety of metallic and nonmetallic minerals, and energy. Together, they form the very foundation of modern civilization (Figure I.9). The Earth sciences deal not only with the formation and occurrence of these vital resources but also with maintaining supplies and the environmental impact of their extraction and use.

Few people who live in highly industrialized nations realize the quantity of resources needed to maintain their present standard of living. For example, the annual per capita consumption of metallic and nonmetallic mineral resources for the United States is nearly 10,000 kilograms (11 tons). This is each person's prorated share of the materials required by industry to provide the vast array of products modern society demands. Figures for other highly industrialized countries are comparable.

Resources are commonly divided into two broad categories. Some are classified as **renewable,** which means that they can be replenished over relatively short time spans. Common examples are plants and animals for food, natural fibers for clothing, and forest products for lumber and paper. Energy from flowing water, wind, and the Sun is also considered renewable.

By contrast, many other basic resources are classified as **nonrenewable.** Important metals such as iron, aluminum, and copper fall into this category, as do our most important fuels: oil, natural gas, and coal. Although these and other resources continue to form, the processes that create them are so slow that significant deposits take millions of years to accumulate. In essence, Earth contains *fixed* quantities of these substances. When the present supplies are mined or pumped from the ground, there will be no more. Although some nonrenewable resources, such as aluminum, can be used over and over again, others, such as oil, cannot be recycled.

How long will the remaining supplies of basic resources last? How long can we sustain the rising standard of living in today's industrial countries and still provide for the growing needs of developing regions? How much environmental deterioration are we willing to accept in pursuit of basic resources? Can alternatives be found? If we are to cope with an increasing demand and a growing world population, it is important that we have some understanding of our present and potential resources.

Did You Know?

Each year, an average American requires huge quantities of Earth materials. Imagine receiving your annual share in a single delivery. A large truck would pull up to your home and unload 12,965 lbs. of stone, 8945 lbs. of sand and gravel, 895 lbs. of cement, 395 lbs. of salt, 361 lbs. of phosphate, and 974 lbs. of other nonmetals. In addition, there would be 709 lbs. of metals, including iron, aluminum, and copper.

A.

B.

Figure I.9 Coal represented about 22 percent of U.S. energy consumption in 2004. It is a major fuel used in power plants to generate electricity. By comparison, petroleum accounts for about 40 percent of U.S. energy use. **A.** Modern underground coal mining is highly mechanized and relatively safe. (Photo by Melvin Grubb/Grubb Photo Services, Inc.) **B.** This modern offshore drilling operation is in the North Sea. (Photo by Peter Bowater/Photo Researchers, Inc.)

A.

B.

Figure I.10 Urban air pollution and soil erosion represent two of the environmental problems Earth scientists deal with. **A.** A severe air pollution episode at Kuala Lumpur, Malaysia, on August 10, 2005. Fuel combustion by motor vehicles and power plants provides a high proportion of the pollutants. Meteorological factors determine whether pollutants remain "trapped" in the city or are dispersed. (Photo by Viviane Moos/CORBIS) **B.** Gully erosion in poorly protected soil, southern Colombia. Just 1 millimeter of soil lost from a single acre of land amounts to about 5 tons. (Photo by Carl Purcell/Photo Researchers, Inc.)

Environmental Problems

In addition to the quest for adequate mineral and energy resources, the Earth sciences must also deal with a broad array of other environmental problems. Some are local, some are regional, and still others are global in extent. Serious difficulties face developed and developing nations alike. Urban air pollution, acid rain, ozone depletion, and global warming are just a few that pose significant threats (Figure I.10). Other problems involve the loss of fertile soils to erosion, the disposal of toxic wastes, and the contamination and depletion of water resources. The list continues to grow.

In addition to human-induced and human-accentuated problems, people must also cope with the many *natural hazards* posed by the physical environment (Figure I.11). Earthquakes, landslides, volcanic eruptions, floods, and hurricanes are just five of the many risks. Others, such as drought, although not as spectacular, are nevertheless equally important environmental concerns. In many cases, the threat of natural hazards is aggravated by increases in population as more people crowd into places where an impending danger exists or attempt to cultivate marginal lands that should not be farmed.

It is clear that as world population continues its rapid growth, pressures on the environment will increase as well. Therefore, an understanding of Earth is not only essential for the location and recovery of basic resources but also for dealing with the human impact on the environment and minimizing the effects of natural hazards. Knowledge about our planet and how it works is necessary to our survival and well-being. Earth is the only suitable habitat we have, and its resources are limited.

It took until about the year 1800 for the world population to reach 1 billion. Since then, the planet has added nearly 6 billion more people.

The Nature of Scientific Inquiry

As members of a modern society, we are constantly reminded of the benefits derived from science. But what exactly is the nature of scientific inquiry? Developing an understanding of how science is done and how scientists work is an important theme in this book. You will explore the difficulties in gathering data and some of the ingenious methods that have been developed to overcome these difficulties. You will also see examples of how hypotheses are formulated and tested, as well as learn about the evolution and development of some major scientific theories.

All science is based on the assumption that the *natural world behaves in a consistent and predictable manner.* The overall goal of science is to discover the underlying patterns in the natural world and then to use this knowledge to predict what will or will not happen, given certain facts or circumstances.

The development of new scientific knowledge involves some basic logical processes that are universally accepted. To determine what is occurring in the natural world, scientists collect *facts* through observation and measurement (Figure I.12). Because some error is inevitable, the accuracy of a particular measurement or observation is always open

A.

B.

Figure I.11 Natural hazards are a part of living on Earth. **A.** Aerial view of earthquake destruction at Muzaffarabad, Pakistan, January 31, 2006. (Photo by Danny Kemp/AFP/Getty Images, Inc.) **B.** Satellite image of Hurricane Katrina in late August 2005 shortly before it devastated the Gulf Coast. (NASA)

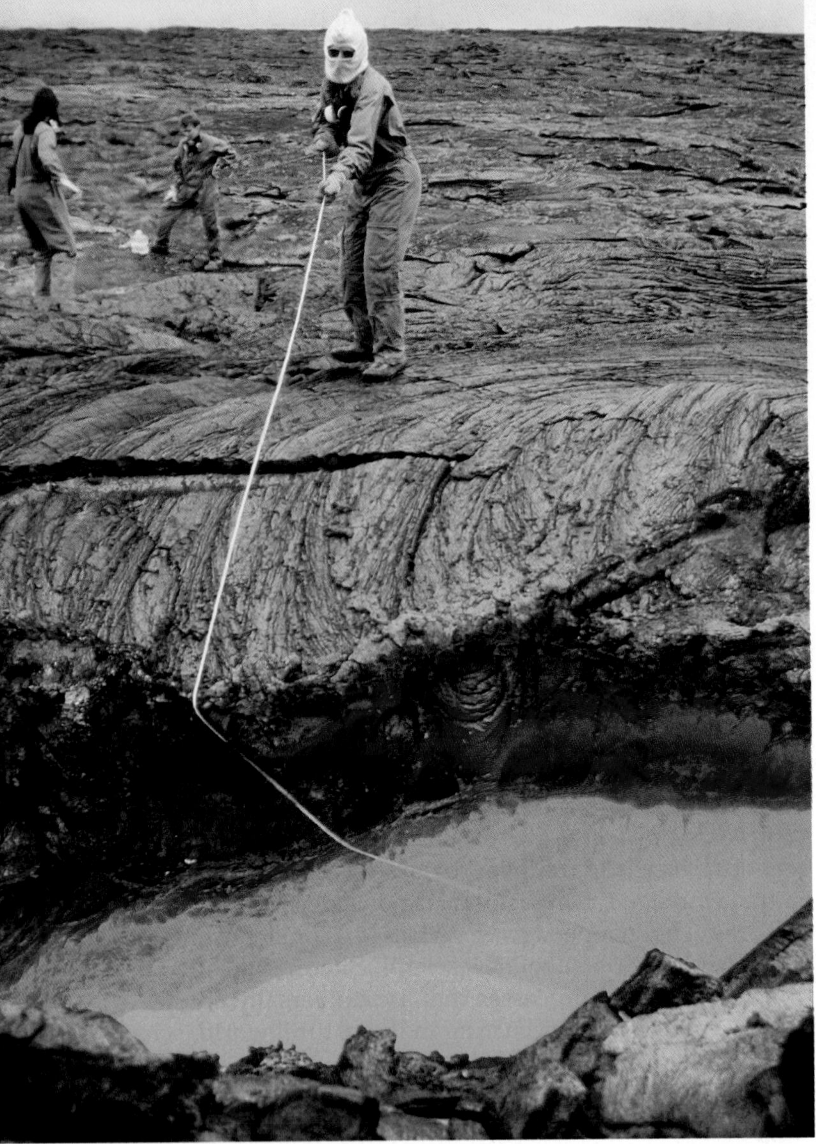

Figure I.12 This geologist is taking a lava sample from a "skylight" in a lava tube near Kilauea Volcano, Hawaii. (Photo by G. Brad Lewis/Getty Images Inc.—Stone Allstock)

to question. Nevertheless, these data are essential to science and serve as the springboard for the development of scientific theories.

Hypothesis

Once facts have been gathered and principles have been formulated to describe a natural phenomenon, investigators try to explain how or why things happen in the manner observed. They can do this by constructing a preliminary, untested explanation, which we call a scientific **hypothesis** or **model.** (The term *model,* although often used synonymously with hypothesis, is a less precise term because it is sometimes used to describe a scientific theory as well.) It is best if an investigator can formulate more than one hypothesis to explain a given set of observations. If an individual scientist is unable to devise multiple hypotheses, others in the scientific community will almost always develop alternative explanations. A spirited debate frequently ensues. As a result, extensive research is conducted by proponents of opposing models, and the results are made available to the wider scientific community in scientific journals.

Before a hypothesis can become an accepted part of scientific knowledge, it must pass objective testing and analysis. (If a hypothesis cannot be tested, it is not scientifically useful, no matter how interesting it might seem.) The verification process requires that *predictions* be made based on the hypothesis being considered and that the predictions be tested by comparing them against objective observations of nature. Put another way, hypotheses must fit observations other than those used to formulate them in the first place. Those hypotheses that fail rigorous testing are ultimately discarded. The history of science is littered with discarded hypotheses. One of the best known is the Earth-centered model of the universe—a proposal that was supported by the apparent daily motion of the Sun, Moon, and stars around Earth. As the mathematician Jacob Bronowski so ably stated, "Science is a great many things, but in the end they all return to this: Science is the acceptance of what works and the rejection of what does not."

Theory

When a hypothesis has survived extensive scrutiny, and when competing hypotheses have been eliminated, a hypothesis may be elevated to the status of a scientific **theory.** In everyday language, we may say, "That's only a theory." But a scientific theory is a well-tested and widely accepted view that scientists agree best explains certain observable facts.

Theories that are extensively documented are held with a very high degree of confidence. Theories of this stature that are comprehensive in scope have a special status. They are called **paradigms** because they explain a large number of interrelated aspects of the natural world. One well-known example is the theory of evolution, which explains the development of life on Earth. Another example is the theory of plate tectonics, which is a paradigm of the geological sciences that provides the framework for understanding the origin of mountains, earthquakes, and volcanic activity. In addition, plate tectonics explains the evolution of the continents and the ocean basins through time—a topic we will consider later in this book.

Scientific Methods

The process just described, in which scientists gather facts through observations and formulate scientific hypotheses that may eventually become theories, is termed the *scientific method.* Contrary to popular belief, the scientific method is not a standard recipe that scientists apply in a routine manner to unravel the secrets of our natural world. Rather, it is an endeavor that involves creativity and insight. Rutherford and Ahlgren put it this way: "Inventing hypotheses or theories to imagine how the world works and then figuring out how

Did You Know?

A scientific *law* is a basic principle that describes a particular behavior of nature that is generally narrow in scope and can be stated briefly—often as a simple mathematical equation.

they can be put to the test of reality is as creative as writing poetry, composing music, or designing skyscrapers."*

There is no fixed path that scientists always follow that leads unerringly to scientific knowledge. Nevertheless, many scientific investigations involve the following steps: (1) the collection of scientific facts through observation and measurement (Figure I.13); (2) the development of one or more working hypotheses to explain these facts; (3) development of observations and experiments to test the hypothesis; and (4) the acceptance, modification, or rejection of the hypothesis based on extensive testing.

Other scientific discoveries represent purely theoretical ideas, which stand up to extensive examination. Some researchers use high-speed computers to simulate what is happening in the "real" world. These models are useful when dealing with natural processes that occur on very long time scales or take place in extreme or inaccessible locations. Still other scientific advancements have been made when a totally unexpected happening occurred during an experiment. These serendipitous discoveries are more than pure luck, for as Louis Pasteur said, "In the field of observation, chance favors only the prepared mind."

Scientific knowledge is acquired through several avenues, so it might be best to describe the nature of scientific inquiry as the *methods* of science rather than the scientific method. In addition, it should always be remembered that even the most compelling scientific theories are still simplified explanations of the natural world.

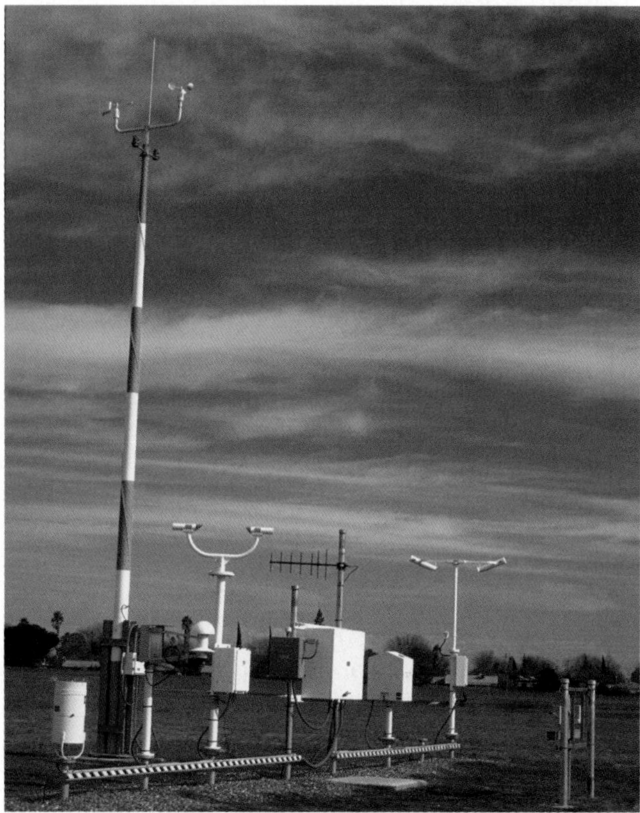

Figure I.13 Gathering data is a basic part of the scientific method. This Automated Surface Observing System (ASOS) installation is one of nearly 900 in use for data gathering as part of the U.S. primary surface observing network.

Studying Earth Science

In this book, you will discover the results of centuries of scientific work. You will see the end product of millions of observations, thousands of hypotheses, and hundreds of theories. We have distilled all of this to give you a "briefing" in Earth science.

But realize that our knowledge of Earth is changing daily, as thousands of scientists worldwide make satellite observations, analyze drill cores from the seafloor, study earthquake waves, develop computer models to predict climate, examine the genetic codes of organisms, and discover new facts about our planet's long history. This new knowledge often updates hypotheses and theories. Expect to see many new discoveries and changes in scientific thinking in your lifetime.

*F. James Rutherford and Andrew Ahlgren, *Science for All Americans* (New York: Oxford University Press, 1990), p. 7.

The Chapter in Review

1. Earth science is the name for all the sciences that collectively seek to understand Earth and its neighbors in space. It includes geology, oceanography, meteorology, and astronomy. Geology is traditionally divided into two broad areas—physical and historical.

2. Earth's physical environment is traditionally divided into three major parts: the solid Earth, the geosphere; the water portion of our planet, the hydrosphere; and Earth's gaseous envelope, the atmosphere. In addition, the biosphere, the totality of life on Earth, interacts with each of the three physical realms and is an equally integral part of Earth.

3. Although each of Earth's four spheres can be studied separately, they are all related in a complex and continuously interacting whole that we call the Earth system. Earth system science uses an interdisciplinary approach to integrate the knowledge of several academic fields in the study of our planet and its global environmental problems.

4. The two sources of energy that power the Earth system are (1) the Sun, which drives the external processes that occur in the atmosphere, hydrosphere, and at Earth's surface, and (2) heat from the Earth's interior that powers the internal processes that produce volcanoes, earthquakes, and mountains.

5. Environment refers to everything that surrounds and influences an organism. These influences can be biological, social, or physical. When applied to Earth science today, the term *environmental* is usually reserved for those aspects that focus on the relationships between people and the natural environment.

6. Resources are an important environmental concern. The two broad categories of resources are (1) renewable, which means that they can be replenished over relatively short time spans, and (2) nonrenewable. As population grows, the demand for resources expands as well.

7. Environmental problems can be local, regional, or global. Human-induced problems include urban air pollution, acid rain, ozone depletion, and global warming. Natural hazards include earthquakes, landslides, floods, and hurricanes. As world population grows, pressures on the environment also increase.

8. All science is based on the assumption that the natural world behaves in a consistent and predictable manner. The process by which scientists gather facts through observation and careful measurement and formulate scientific hypotheses and theories is called the scientific method. To determine what is occurring in the natural world, scientists often (1) collect facts, (2) develop a scientific hypothesis, (3) construct experiments to validate the hypothesis, and (4) accept, modify, or reject the hypothesis on the basis of extensive testing. Other discoveries represent purely theoretical ideas that have stood up to extensive examination. Still other scientific advancements have been made when a totally unexpected happening occurred during an experiment.

Key Terms

asthenosphere (p. 5)
atmosphere (p. 5)
astronomy (p. 3)
biosphere (p. 6)
core (p. 5)
crust (p. 5)
Earth science (p. 2)
Earth system (p. 6)
geologic time scale (p. 8)
geology (p. 2)
geosphere (p. 5)
hydrosphere (p. 3)
hypothesis (p. 12)
lithosphere (p. 5)
mantle (p. 5)
meteorology (p. 3)
model (p. 12)
nonrenewable resource (p. 9)
oceanography (p. 2)
paradigm (p. 12)
physical environment (p. 8)
renewable resource (p. 9)
system (p. 7)
theory (p. 12)

Questions for Review

1. Name the specific Earth science described by each of the following statements:
   a. This science deals with the dynamics of the oceans.
   b. This word literally means "the study of Earth."
   c. An understanding of the atmosphere is the primary focus of this science.
   d. This science helps us understand Earth's place in the universe.
2. List and briefly define the four "spheres" that constitute our environment.
3. Name the three principal divisions of the solid Earth based on compositional differences. Which one is thinnest?
4. The oceans cover nearly _____ percent of Earth's surface and contain about _____ percent of the planet's total water supply.
5. What are the two sources of energy for the Earth system?
6. Contrast renewable and nonrenewable resources. Give one or more examples of each.

Online Study Guide

The *Foundations of Earth Science* Web site uses the resources and flexibility of the Internet to aid in your study of the topics in this chapter. Written and developed by Earth science instructors, this site will help improve your understanding of Earth science. Visit **http://www.prenhall.com/lutgens** and click on the cover of *Foundations of Earth Science 5e* to find:

- Online review quizzes.
- Critical thinking exercises.
- Links to chapter-specific Web resources.
- Internet-wide key-term searches.

http://www.prenhall.com/lutgens

UNIT I EARTH MATERIALS

CHAPTER

1 Minerals: Building Blocks of Rocks

Topaz crystal with albite. (Photo by Jeff Scovil)

FOCUS ON LEARNING

To assist you in learning the important concepts in this chapter, you will find it helpful to focus on the following questions:

1. What are minerals, and how are they different from rocks?
2. What are the smallest particles of matter? How do atoms bond together?
3. How do isotopes of the same element vary from one another, and why are some isotopes radioactive?
4. What are some of the physical and chemical properties of minerals? How can these properties be used to distinguish one mineral from another?
5. What are the eight elements that make up most of Earth's continental crust?
6. What is the most abundant mineral group? What do all minerals within this group have in common?
7. When is the term *ore* used with reference to a mineral? What are the common ores of iron and lead?

Earth's crust and oceans are the source of a wide variety of useful and essential minerals. Most people are familiar with the common uses of many basic metals, including aluminum in beverage cans, copper in electrical wiring, and gold and silver in jewelry. But some people are not aware that the lead in pencils contains the greasy-feeling mineral graphite and that bath powders and many cosmetics contain the mineral talc. Moreover, many do not know that drill bits impregnated with diamonds are used by dentists to drill through tooth enamel, or that the common mineral quartz is the source of silicon for computer chips. In fact, practically every manufactured product contains materials obtained from minerals. As the mineral requirements of modern society grow, the need to locate additional supplies of useful minerals also grows, becoming more challenging as well.

In addition to the economic uses of rocks and minerals, all of the processes studied by geologists are in some way dependent on the properties of these basic Earth materials. Events such as volcanic eruptions, mountain building, weathering and erosion, and even earthquakes involve rocks and minerals. Consequently, a basic knowledge of Earth materials is essential to the understanding of all geologic phenomena.

Minerals: The Building Blocks of Rocks

Earth Materials
▼ Minerals

We begin our discussion of Earth materials with an overview of **mineralogy** (*mineral* = mineral, *ology* = the study of), because minerals are the building blocks of rocks (Figure 1.1). In addition, minerals have been employed by humans for both useful and decorative purposes for thousands of years.

Figure 1.1 Collection of well-developed quartz crystals found near Hot Springs, Arkansas. (Photo by Jeff Scovil)

The first minerals mined were flint and chert, which people fashioned into weapons and cutting tools. As early as 3700 B.C., Egyptians began mining gold, silver, and copper; and by 2200 B.C., humans discovered how to combine copper with tin to make bronze, a tough, hard alloy. The Bronze Age began its decline when the ability to extract iron from minerals such as hematite was discovered. By about 800 B.C., iron-working technology had advanced to the point that weapons and many everyday objects were made of iron rather than copper, bronze, or wood. During the Middle Ages, mining of a variety of minerals was common throughout Europe, and the impetus for the formal study of minerals was in place.

The term *mineral* is used in several different ways. For example, those concerned with health and fitness extol the benefits of vitamins and minerals. The mining industry typically uses the word when referring to anything taken out of the ground, such as coal, iron ore, or sand and gravel. The guessing game known as *Twenty Questions* usually begins with the question, *Is it animal, vegetable, or mineral?* What criteria do geologists use to determine whether something is a mineral?

Geologists define a **mineral** as *any naturally occurring inorganic solid that possesses an orderly crystalline structure and a well-defined chemical composition.* Thus, those Earth materials that are classified as minerals exhibit the following characteristics:

1. **Naturally occurring**—This tells us that a mineral forms by natural, geologic processes. Consequently, synthetic diamonds and rubies, as well as a variety of other useful materials produced by chemists are not considered minerals.
2. **Solid**—In order to be considered a mineral, substances must be solids at temperatures normally experienced at Earth's surface. Thus, ice (frozen water) is considered a mineral, whereas liquid water is not.
3. **Orderly crystalline structure**—Minerals are crystalline substances, which means their atoms are arranged in an orderly, repetitive manner as shown in Figure 1.2. This orderly packing of atoms is reflected in the regularly shaped objects we call crystals (Figure 1.2D). Some naturally occurring solids, such as volcanic glass (obsidian), lack a repetitive atomic structure and are referred to as *amorphous* (without form) and are not considered minerals.
4. **Well-defined chemical composition**—Most minerals are chemical compounds made up of two or more

Did You Know?

Archaeological finds show that more than 2000 years ago, the Romans used lead for pipes to transport water within their buildings. In fact, Roman smelting of lead and copper ores between 500 B.C. and A.D. 300 caused a small but significant rise in atmospheric pollution, as recorded in Greenland ice cores.

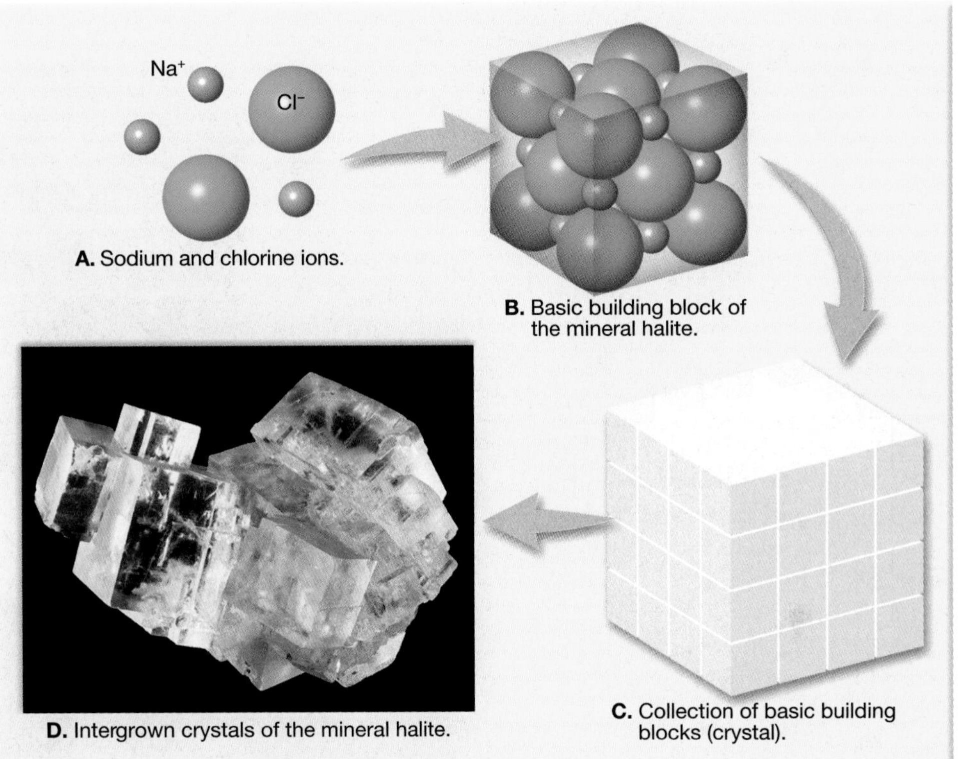

Figure 1.2 Illustration of the orderly arrangement of sodium and chloride atoms (ions) in the mineral halite. The arrangement of atoms into basic building blocks having a cubic shape results in regularly shaped cubic crystals. (Photo by M. Claye/Jacana Scientific Control/Photo Researchers, Inc.)

elements. A few, such as gold and silver, consist of only a single element. The common mineral quartz consists of two oxygen (O) atoms for every silicon (Si) atom, giving it a chemical composition expressed by the formula SiO_2. Thus, no matter what the environment is, whenever atoms of oxygen and silicon join together in the ratio of 2 to 1, the product is always quartz. However, it is not uncommon for elements of similar size and electrical charge to substitute freely for one another. As a result the chemical composition of a mineral may vary somewhat from sample to sample.

5. **Generally inorganic**—Inorganic crystalline solids, as exemplified by ordinary table salt (halite), that are found naturally in the ground are considered minerals. Organic compounds, on the other hand, are generally not. Sugar, a crystalline solid like salt, comes from sugar cane or sugar beets and is a common example of such an organic compound. However, many marine animals secrete inorganic compounds, such as calcium carbonate (calcite), in the form of shells and coral reefs. These materials are considered minerals by most geologists.

Rocks, on the other hand, are more loosely defined. Simply, a **rock** is any solid mass of mineral or mineral-like matter that occurs naturally as part of our planet. A few rocks are composed almost entirely of one mineral. A common example is the sedimentary rock *limestone,* which consists of impure masses of the mineral calcite. However, most rocks, like the common rock granite (shown in Figure 1.3), occur as aggregates of several kinds of minerals. The term *aggregate* implies that the minerals are joined in such a way that the properties of each mineral are retained. Note that you can easily identify the mineral constituents of the sample of granite shown in Figure 1.3.

A few rocks are composed of nonmineral matter. These include the volcanic rocks *obsidian* and *pumice,* which are noncrystalline glassy substances, and *coal,* which consists of solid organic debris.

Although this chapter deals primarily with the nature of minerals, keep in mind that most rocks are simply aggregates of minerals. Because the properties of rocks are determined largely by the chemical composition and crystalline structure of the minerals contained within them, we will first consider these Earth materials. Then, in Chapter 2, we will take a closer look at Earth's major rock groups.

Elements: Building Blocks of Minerals

You are probably familiar with the names of many elements, including copper, gold, oxygen, and carbon. An **element** is a substance that cannot be broken down into simpler substances by chemical or physical means. There are about 90 different elements found in nature, and, consequently, about 90 different kinds of atoms. In addition, scientists have succeeded in creating about 23 synthetic elements.

The elements can be organized into rows so that those with similar properties are in the same column. This arrangement, called the **periodic table,** is shown in Figure 1.4. Notice that symbols are used to provide a shorthand way of representing an element. Each element is also known by its atomic number, which is shown above each symbol on the periodic table.

Some elements, such as copper (number 29) and gold (number 79), exist in nature with single atoms as the basic

Figure 1.3 Rocks are aggregates of one or more minerals. (Top and left photos courtesy E. J. Tarbuck; center and right photos by GeoScience Resources/ American Geological Institute)

unit. Thus, native copper and gold are minerals made entirely from one element. However, many elements are quite reactive and join together with atoms of one or more other elements to form **chemical compounds.** As a result, most minerals are chemical compounds consisting of two or more different elements.

Atoms

To understand how elements combine to form compounds, we must first consider the atom. The **atom** (*a* = not, *tomos* = cut) is the smallest particle of matter that retains the essential characteristics of an element. It is this extremely small particle that does the combining.

Protons, Neutrons, and Electrons The central region of an atom is called the **nucleus.** The nucleus contains protons and neutrons. **Protons** are very dense particles with positive electrical charges. **Neutrons** have the same mass as a proton but lack an electrical charge.

Orbiting the nucleus are **electrons,** which have negative electrical charges. For convenience, we sometimes diagram atoms to show the electrons orbiting the nucleus, like the orderly orbiting of the planets around the Sun (Figure 1.5A). However, electrons move in a less predictable way than do planets. As a result, they create a sphere-shaped negative zone around the nucleus. A more realistic picture of the positions of electrons can be obtained by envisioning a cloud of negatively charged electrons surrounding the nucleus (Figure 1.5B).

Studies of electron configurations predict that individual electrons move within regions around the nucleus called **principal shells,** or **energy levels.** Furthermore, each of these shells can hold a specific number of electrons, with the outermost principal shell containing the **valence electrons.** Valence electrons are important because, as you will see later, they are involved in chemical bonding.

Atomic Number Some atoms have only a single proton in their nuclei, while others contain more than 100 protons. The number of protons in the nucleus of an atom is called the **atomic number.** For example, all atoms with six protons are carbon atoms; hence, the atomic number of carbon is 6. Likewise, every atom with eight protons is an oxygen atom.

Atoms in their natural state also have the same number of electrons as protons, so the atomic number also equals

Did You Know?

Although wood pulp is the main ingredient for making newsprint, many higher grades of paper contain large quantities of clay minerals. In fact, each page of this textbook consists of about 25 percent clay (the mineral kaolinite). If rolled into a ball, this much clay would be roughly the size of a tennis ball.

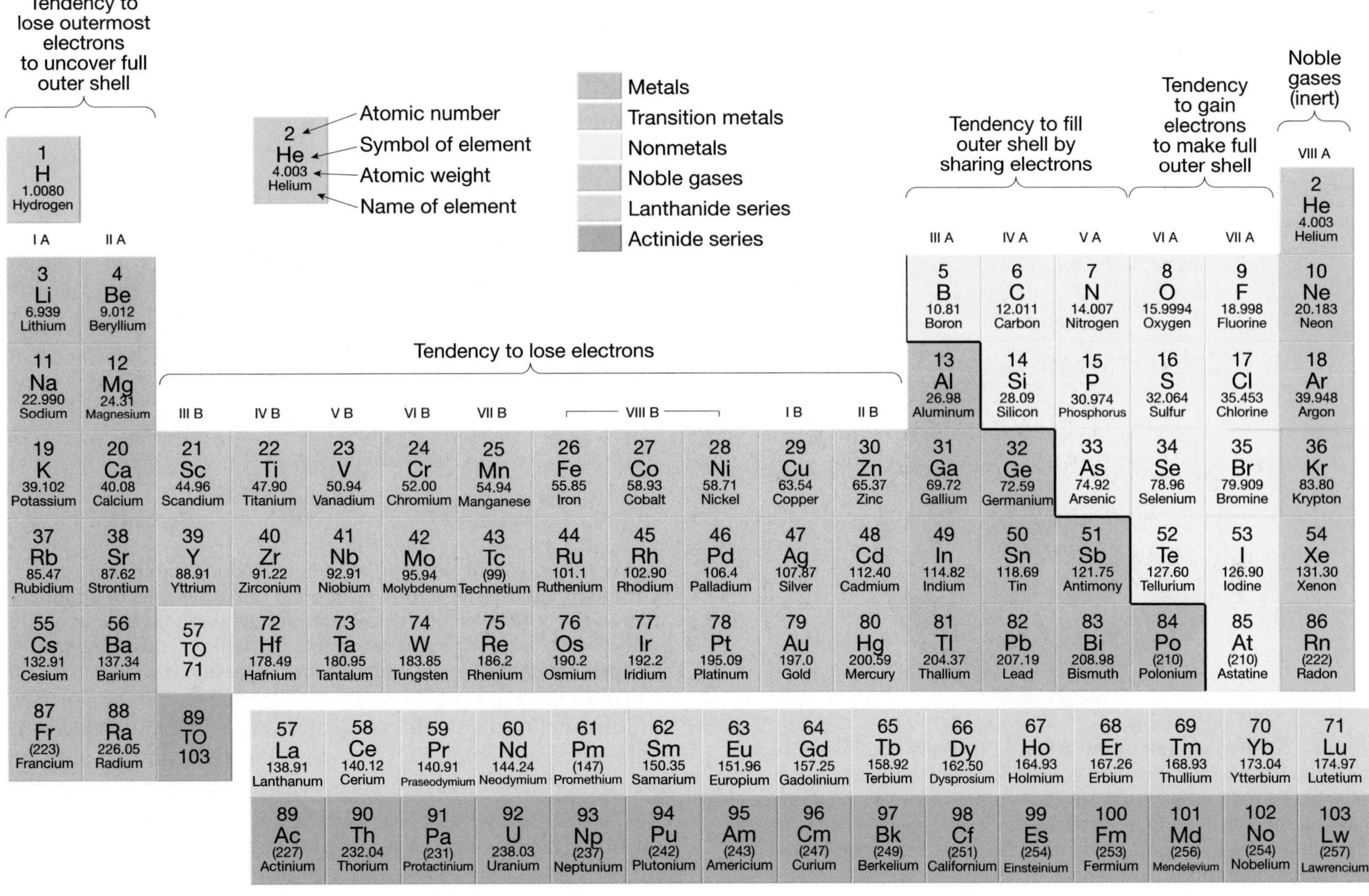

Tendency to lose outermost electrons to uncover full outer shell
2 He 4.003 Helium
Atomic number
Symbol of element
Atomic weight
Name of element
Metals
Transition metals
Nonmetals
Noble gases
Lanthanide series
Actinide series
Tendency to fill outer shell by sharing electrons
Tendency to gain electrons to make full outer shell
Noble gases (inert)
VIII A
I A
II A
III A
IV A
V A
VI A
VII A
Tendency to lose electrons
III B
IV B
V B
VI B
VII B
VIII B
I B
II B
1 H 1.0080 Hydrogen
2 He 4.003 Helium
3 Li 6.939 Lithium
4 Be 9.012 Beryllium
5 B 10.81 Boron
6 C 12.011 Carbon
7 N 14.007 Nitrogen
8 O 15.9994 Oxygen
9 F 18.998 Fluorine
10 Ne 20.183 Neon
11 Na 22.990 Sodium
12 Mg 24.31 Magnesium
13 Al 26.98 Aluminum
14 Si 28.09 Silicon
15 P 30.974 Phosphorus
16 S 32.064 Sulfur
17 Cl 35.453 Chlorine
18 Ar 39.948 Argon
19 K 39.102 Potassium
20 Ca 40.08 Calcium
21 Sc 44.96 Scandium
22 Ti 47.90 Titanium
23 V 50.94 Vanadium
24 Cr 52.00 Chromium
25 Mn 54.94 Manganese
26 Fe 55.85 Iron
27 Co 58.93 Cobalt
28 Ni 58.71 Nickel
29 Cu 63.54 Copper
30 Zn 65.37 Zinc
31 Ga 69.72 Gallium
32 Ge 72.59 Germanium
33 As 74.92 Arsenic
34 Se 78.96 Selenium
35 Br 79.909 Bromine
36 Kr 83.80 Krypton
37 Rb 85.47 Rubidium
38 Sr 87.62 Strontium
39 Y 88.91 Yttrium
40 Zr 91.22 Zirconium
41 Nb 92.91 Niobium
42 Mo 95.94 Molybdenum
43 Tc (99) Technetium
44 Ru 101.1 Ruthenium
45 Rh 102.90 Rhodium
46 Pd 106.4 Palladium
47 Ag 107.87 Silver
48 Cd 112.40 Cadmium
49 In 114.82 Indium
50 Sn 118.69 Tin
51 Sb 121.75 Antimony
52 Te 127.60 Tellurium
53 I 126.90 Iodine
54 Xe 131.30 Xenon
55 Cs 132.91 Cesium
56 Ba 137.34 Barium
57 TO 71
72 Hf 178.49 Hafnium
73 Ta 180.95 Tantalum
74 W 183.85 Tungsten
75 Re 186.2 Rhenium
76 Os 190.2 Osmium
77 Ir 192.2 Iridium
78 Pt 195.09 Platinum
79 Au 197.0 Gold
80 Hg 200.59 Mercury
81 Tl 204.37 Thallium
82 Pb 207.19 Lead
83 Bi 208.98 Bismuth
84 Po (210) Polonium
85 At (210) Astatine
86 Rn (222) Radon
87 Fr (223) Francium
88 Ra 226.05 Radium
89 TO 103
57 La 138.91 Lanthanum
58 Ce 140.12 Cerium
59 Pr 140.91 Praseodymium
60 Nd 144.24 Neodymium
61 Pm (147) Promethium
62 Sm 150.35 Samarium
63 Eu 151.96 Europium
64 Gd 157.25 Gadolinium
65 Tb 158.92 Terbium
66 Dy 162.50 Dysprosium
67 Ho 164.93 Holmium
68 Er 167.26 Erbium
69 Tm 168.93 Thullium
70 Yb 173.04 Ytterbium
71 Lu 174.97 Lutetium
89 Ac (227) Actinium
90 Th 232.04 Thorium
91 Pa (231) Protactinium
92 U 238.03 Uranium
93 Np (237) Neptunium
94 Pu (242) Plutonium
95 Am (243) Americium
96 Cm (247) Curium
97 Bk (249) Berkelium
98 Cf (251) Californium
99 Es (254) Einsteinium
100 Fm (253) Fermium
101 Md (256) Mendelevium
102 No (254) Nobelium
103 Lw (257) Lawrencium

Figure 1.4 Periodic table of the elements.

the number of electrons surrounding the nucleus. Therefore, carbon has six electrons to match its six protons, and oxygen has eight electrons to match its eight protons. Neutrons have no charge, so the positive charge of the protons is exactly balanced by the negative charge of the electrons. Consequently, atoms in the natural state are neutral electrically and have no overall electrical charge.

Why Atoms Bond

Why do elements join together to form chemical compounds? It has been learned from experimentation that the forces holding the atoms together are electrical. Further, it is known that chemical bonding results in a change in the electron configuration of the bonded atoms. As we noted earlier, it is the valence electrons (outer-shell electrons) that are generally involved in chemical bonding. Figure 1.6 shows a shorthand way of representing the number of electrons in the outer principal shell (valence electrons). Notice that the elements in group I have one valence electron, those in group II have two, and so forth up to group VIII, which has eight valence electrons.

Other than the first shell, which can hold a maximum of two electrons, *a stable configuration occurs when the valence shell contains eight electrons.* Only the noble gases, such as neon and argon, have a complete outermost principal shell. Hence, the noble gases are the least chemically reactive and are designated as "inert."

When an atom's outermost shell does not contain the maximum number of electrons (8), the atom is likely to chemically bond with one or more other atoms. A *chemical bond* is the sharing or transfer of electrons to attain a stable electron configuration among the bonding atoms. If the electrons are transferred, the bond is an *ionic bond.* If the electrons are shared, the bond is called a *covalent bond.* In either case, the bonding atoms get stable electron configurations, which usually consist of eight electrons in the outer shell.

Did You Know?

The purity of gold is expressed by the number of *karats.* Twenty-four karats is pure gold. Gold less than 24 karats is an alloy (mixture) of gold and another metal, usually copper or silver. For example, 14-karat gold contains 14 parts gold (by weight) mixed with 10 parts of other metals.

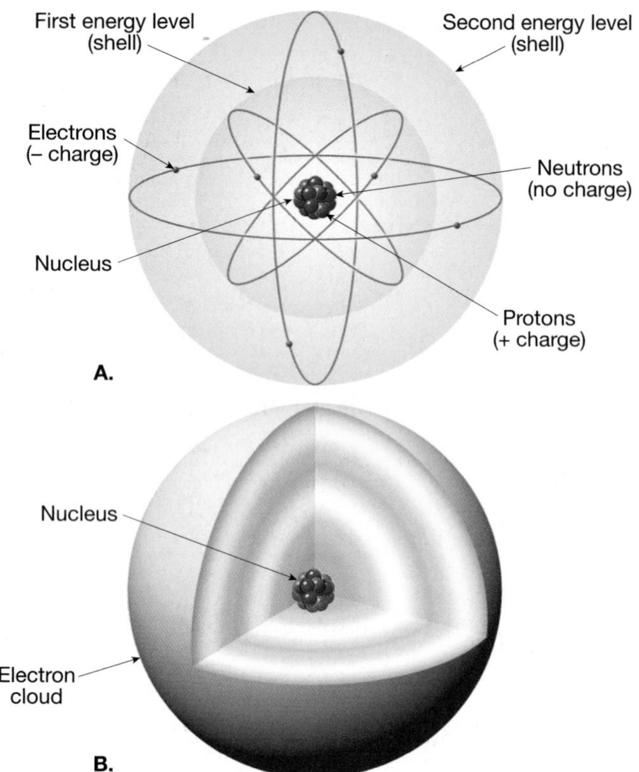

Figure 1.5 Two models of the atom. **A.** This very simplified view of the atom has a central nucleus, consisting of protons and neutrons, encircled by high-speed electrons. **B.** Another model of the atom showing spherically shaped electron clouds (energy-level shells). Note that these models are not drawn to scale. Electrons are minuscule in size compared to protons and neutrons, and the relative space between the nucleus and electron shells is much greater than illustrated.

Ionic Bonds: Electrons Transferred

Perhaps the easiest type of bond to visualize is an **ionic bond.** In ionic bonding, one or more valence electrons are transferred from one atom to another. Simply, one atom gives up its valence electrons, and the other uses them to complete its outer shell. A common example of ionic bonding is sodium (Na) and chlorine (Cl) joining to produce sodium chloride (common table salt). This is shown in Figure 1.7A. Notice that sodium gives up its single valence electron to chlorine. As a result,

Electron Dot Diagrams for Some Representative Elements

I	II	III	IV	V	VI	VII	VIII
H							He
Li	Be	B	C	N	O	F	Ne
Na	Mg	Al	Si	P	S	Cl	Ar
K	Ca	Ga	Ge	As	Se	Br	Kr

Figure 1.6 Dot diagrams for some representative elements. Each dot represents a valence electron found in the outermost principal shell.

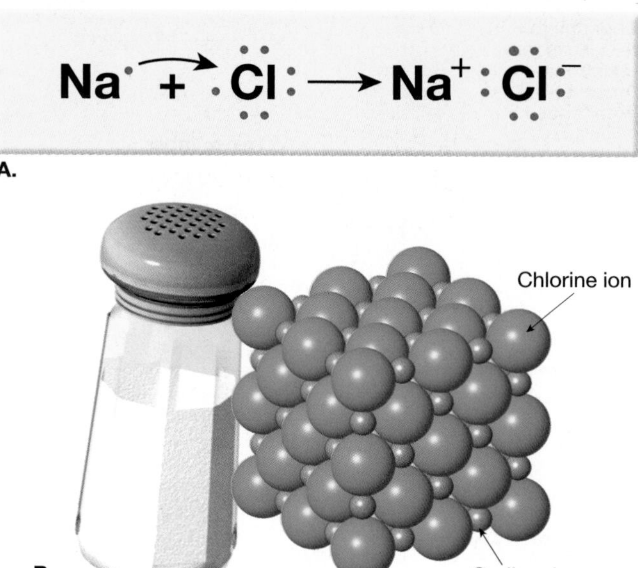

Figure 1.7 Chemical bonding of sodium chloride (table salt). **A.** Through the transfer of one electron in the outer shell of a sodium atom to the outer shell of a chlorine atom, the sodium becomes a positive ion and chlorine a negative ion. **B.** Diagram illustrating the arrangement of sodium and chlorine ions in table salt.

sodium achieves a stable configuration having eight electrons in its outermost shell. By acquiring the electron that sodium loses, chlorine—which has seven valence electrons—gains the eighth electron needed to complete its outermost shell. Thus, through the transfer of a single electron, both the sodium and chlorine atoms have acquired a stable electron configuration.

Once electron transfer takes place, atoms are no longer electrically neutral. By giving up one electron, a neutral sodium atom becomes *positively charged* (11 protons/10 electrons). Similarly, by acquiring one electron, the neutral chlorine atom becomes *negatively charged* (17 protons/18 electrons). Atoms such as these, which have an electrical charge because of the unequal numbers of electrons and protons, are called **ions.**

We know that ions with like charges repel, and those with unlike charges attract. Thus, an *ionic bond* is the attraction of oppositely charged ions to one another, producing an electrically neutral compound. Figure 1.7B illustrates the arrangement of sodium and chlorine ions in ordinary table salt. Notice that salt consists of alternating sodium and chlorine ions, positioned in such a manner that each positive ion is attracted to and surrounded on all sides by negative ions, and vice versa. This arrangement maximizes the attraction between ions with unlike charges while minimizing the repulsion between ions with like charges. Thus, *ionic compounds consist of an orderly arrangement of oppositely charged ions assembled in a definite ratio that provides overall electrical neutrality.*

The properties of a chemical compound are *dramatically different* from the properties of the elements comprising it. For example, sodium, a soft, silvery metal, is extremely reactive and poisonous. If you were to consume even a small amount of elemental sodium, you would need immediate medical attention. Chlorine, a green poisonous gas, is so toxic it was used as a chemical weapon during World War I. Together, however, these elements produce sodium chloride, a harmless flavor enhancer that we call table salt. When elements combine to form compounds, their properties change dramatically.

Covalent Bonds: Electrons Shared

Not all atoms combine by transferring electrons to form ions. Some atoms *share* electrons. For example, the gaseous elements oxygen (O_2), hydrogen (H_2), and chlorine (Cl_2) exist as stable molecules consisting of two atoms bonded together, without a complete transfer of electrons.

Figure 1.8 illustrates the sharing of a pair of electrons between two chlorine atoms to form a molecule of chlorine gas (Cl_2). By overlapping their outer shells, these chlorine atoms share a pair of electrons. Thus, each chlorine atom has acquired, through cooperative action, the needed eight electrons to complete its outer shell. The bond produced by the sharing of electrons is called a **covalent bond.**

A common analogy may help you visualize a covalent bond. Imagine two people at opposite ends of a dimly lit room, each reading under a separate lamp. By moving the lamps to the center of the room, they are able to combine their light sources so each can see better. Just as the overlapping light beams meld, the shared electrons that provide the "electrical glue" in covalent bonds are indistinguishable from each other. The most common mineral group, the silicates, contains the element silicon, which readily forms covalent bonds with oxygen.

Isotopes and Radioactive Decay

Subatomic particles are so incredibly small that a special unit, called an *atomic mass unit,* was devised to express their mass. A proton or a neutron has a mass just slightly more than one atomic mass unit, whereas an electron is only about one two-thousandth of an atomic mass unit. Thus, although electrons play an active role in chemical reactions, they do not contribute significantly to the mass of an atom.

The **mass number** of an atom is simply the total of its neutrons and protons. Atoms of the same element always have the same number of protons. But the number of neutrons for atoms of the same element can vary. Atoms with the same number of protons but different numbers of neutrons are **isotopes** of that element. Isotopes of the same element are labeled by placing the mass number after the element's name or symbol.

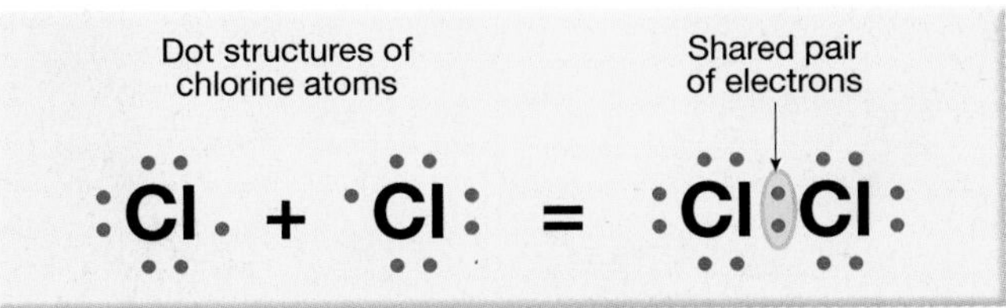

Figure 1.8 Dot diagrams used to illustrate the sharing of a pair of electrons between two chlorine atoms to form a chlorine molecule. Notice that by sharing a pair of electrons, both chlorine atoms achieve a full outer shell (8 electrons).

Did You Know?

One of the world's heaviest cut and polished gemstones is the 22,892.5-carat golden-yellow topaz. Currently housed in the Smithsonian Institution, this roughly 10-pound gem is about the size of an automobile headlight and could hardly be used as a piece of jewelry, except perhaps by an elephant.

For example, carbon has three well-known isotopes. One has a mass number of 12 (carbon-12), another has a mass number of 13 (carbon-13), and the third, carbon-14, has a mass number of 14. Since all atoms of the same element have the same number of protons, and carbon has six, carbon-12 also has *six neutrons* to give it a mass number of 12. Likewise, carbon-14 has six protons plus *eight neutrons* to give it a mass number of 14.

In chemical behavior, all isotopes of the same element are nearly identical. To distinguish among them is like trying to differentiate identical twins, with one being slightly heavier. Because isotopes of an element react the same chemically, different isotopes can become parts of the same mineral. For example, when the mineral calcite forms from calcium, carbon, and oxygen, some of its carbon atoms are carbon-12 and some are carbon-14.

The nuclei of most atoms are stable. However, many elements do have isotopes in which the nuclei are unstable. "Unstable" means that the isotopes disintegrate through a process called **radioactive decay.** Radioactive decay occurs when the forces that bind the nucleus are not strong enough.

During radioactive decay, unstable atoms radiate energy and emit particles. Some of this energy powers the movements of Earth's crust and upper mantle. The rates at which unstable atoms decay are measurable. Therefore, certain radioactive atoms can be used to determine the ages of fossils, rocks, and minerals. A discussion of radioactive decay and its applications in dating past geologic events is found in Chapter 8.

Properties of Minerals

Earth Materials
▼ Minerals

Each mineral has a definite crystalline structure and chemical composition that give it a unique set of physical and chemical properties shared by all samples of that mineral. For example, all specimens of halite have the same hardness and the same density and break in a similar manner. Because the internal structure and chemical composition of a mineral are difficult to determine without the aid of sophisticated tests and equipment, the more easily recognized physical properties are frequently used in identification.

The diagnostic physical properties of minerals are those that can be determined by observation or by performing a simple test. The primary physical properties that are commonly used to identify hand samples are luster, color, streak, crystal shape (habit), tenacity, hardness, cleavage, fracture, and density or specific gravity. Secondary (or "special") properties that are exhibited by a limited number of minerals include magnetism, taste, feel, smell, double refraction, and chemical reaction to hydrochloric acid.

Optical Properties

Of the many optical properties of minerals, four—luster, the ability to transmit light, color, and streak—are most frequently used for mineral identification.

Luster The appearance or quality of light reflected from the surface of a mineral is known as **luster.** Minerals that have the appearance of metals, regardless of color, are said to have a *metallic luster* (Figure 1.9). Some metallic minerals, such as native copper and galena, develop a dull coating or tarnish when exposed to the atmosphere. Because they are not as shiny as samples with freshly broken surfaces, these samples are often said to exhibit a *submetallic luster.*

Most minerals have a *nonmetallic luster* and are described using various adjectives such as *vitreous* or *glassy.* Other nonmetallic minerals are described as having a *dull* or *earthy luster* (a dull appearance like soil), or a *pearly luster* (such as a pearl, or the inside of a clamshell). Still others exhibit a *silky luster* (like satin cloth), or a *greasy luster* (as though coated in oil).

The Ability to Transmit Light Another optical property used in the identification of minerals is the ability to transmit light. When no light is transmitted, the mineral is described as *opaque;* when light but not an image is transmitted through a mineral it is said to be *translucent.* When light and an image are visible through the sample, the mineral is described as *transparent.*

Color Although **color** is generally the most conspicuous characteristic of any mineral, it is considered a diagnostic property of only a few minerals. Slight impurities in the common mineral quartz, for example, give it a variety of tints including pink, purple, yellow, white, gray, and even black. Other minerals, such as tourmaline, also exhibit a variety of hues, with multiple colors sometimes occurring in the same sample. Thus, the use of color as a means of identification is often ambiguous or even misleading.

Streak Although the color of a sample is not always helpful in identification, **streak**—the color of the powdered mineral—is often diagnostic. The streak is obtained by rubbing the mineral across a piece of unglazed porcelain, termed a *streak plate,* and observing the color of the mark it leaves (Figure 1.10). Although the color of a mineral may vary from sample to sample, the streak usually does not.

Streak can also help distinguish minerals with metallic luster from those having nonmetallic luster. Metallic minerals generally have a dense, dark streak, whereas minerals with nonmetallic luster have a streak that is typically light colored.

Did You Know?

The name *crystal* is derived from the Greek (*krystallos,* meaning "ice") and was applied to quartz crystals. The ancient Greeks thought quartz was water that had crystallized at high pressures deep inside Earth.

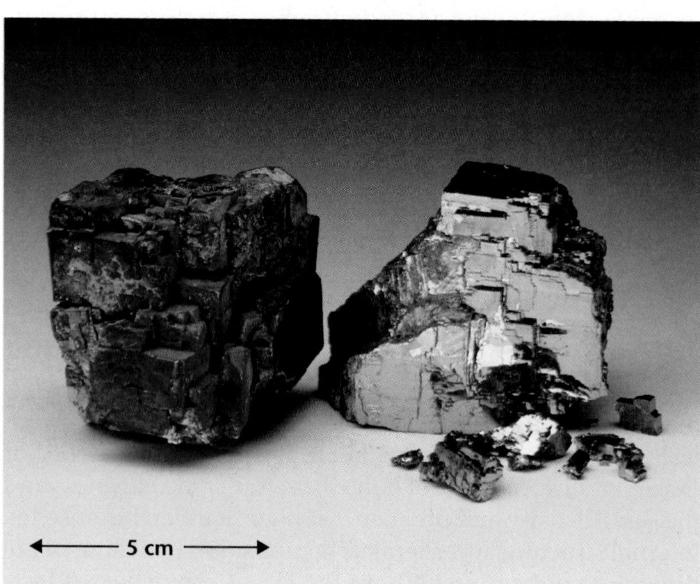

Figure 1.9 The freshly broken sample of galena (right) displays a metallic luster, while the sample on the left is tarnished and has a submetallic luster. (Photo courtesy of E. J. Tarbuck)

Figure 1.10 Although the color of a mineral is not always helpful in identification, the streak, which is the color of the powdered mineral, can be very useful. (Photo by Dennis Tasa)

It should be noted that not all minerals produce a streak when using a streak plate. For example, if the mineral is harder than the streak plate, no streak is observed.

Crystal Shape or Habit

Mineralogists use the term **habit** to refer to the common or characteristic shape of a crystal or aggregate of crystals. A few minerals exhibit somewhat regular polygons that are helpful in their identification. While most minerals have only one common habit, a few such as pyrite have two or more characteristic crystal shapes (Figure 1.11).

By contrast, some minerals rarely develop perfect geometric forms. Many of these do, however, develop a characteristic shape that is useful for identification. Some minerals tend to grow equally in all three dimensions, whereas others tend to be elongated in one direction, or flattened if growth in one dimension is suppressed. Commonly used terms to describe these and other crystal habits include *equant* (equidimensional), *bladed, fibrous, tabular, prismatic, platy, blocky,* and *botryoidal* (Figure 1.12).

Figure 1.12 Some common crystal habits. **A.** *Bladed.* Elongated crystals that are flattened in one direction. **B.** *Prismatic.* Elongated crystals with faces that are parallel to a common direction. **C.** *Banded.* Minerals that have stripes or bands of color or texture. **D.** *Botryoidal.* Groups of intergrown crystals resembling a bunch of grapes.

Mineral Strength

How easily minerals break or deform under stress relates to the type and strength of the chemical bonds that hold the crystals together. Mineralogists use terms including *tenacity, hardness, cleavage,* and *fracture* to describe mineral strength and how minerals break when stress is applied.

Tenacity The term **tenacity** describes a mineral's toughness, or its resistance to breaking or deforming. Minerals that are ionically bonded, such as fluorite and halite, tend to be *brittle* and shatter into small pieces when struck. By contrast, minerals with metallic bonds, such as native copper, are *malleable,* or easily hammered into different shapes. Minerals, including gypsum and talc, that can be cut into thin shavings are described as *sectile.* Still others, notably the micas, are *elastic* and will bend and snap back to their original shape after the stress is released.

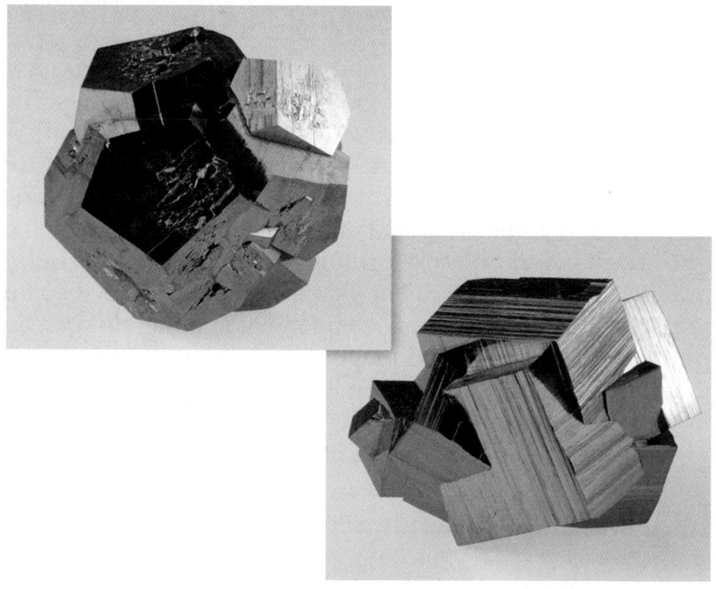

Figure 1.11 Although most minerals exhibit only one common crystal shape, some such as pyrite have two or more characteristic habits. (Photo by Dennis Tasa)

Hardness One of the most useful diagnostic properties is **hardness,** a measure of the resistance of a mineral to abrasion or scratching. This property is determined by rubbing a mineral of unknown hardness against one of known hardness, or vice versa. A numerical value of hardness can by obtained by using the **Mohs scale** of hardness, which consists of 10 minerals arranged in order from 1 (softest) to 10 (hardest), as shown in Figure 1.13A. It should be noted that the Mohs scale is a relative ranking, and it does not imply that mineral number 2, gypsum, is twice as hard as mineral 1, talc. In fact, gypsum is only slightly harder than talc as shown in Figure 1.13B.

In the laboratory, other common objects can be used to determine the hardness of a mineral. These include a human fingernail, which has a hardness of about 2.5, a copper penny 3.5, and a piece of glass 5.5. The mineral gypsum, which has a hardness of 2, can be easily scratched with a fingernail. On the other hand, the mineral calcite, which has a hardness of 3, will scratch a fingernail but will not scratch glass. Quartz, one of the hardest common minerals, will easily scratch glass. Diamonds, hardest of all, scratch anything.

Cleavage Some minerals have atomic structures that are not the same in every direction and chemical bonds that vary in strength. As a result, these minerals tend to break, so that

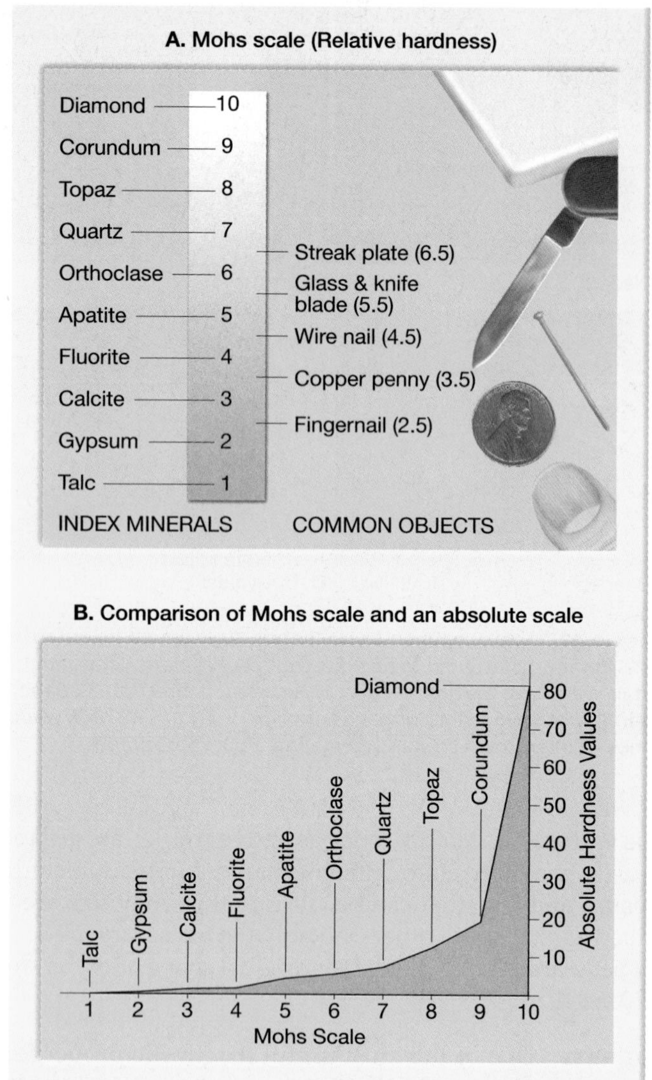

Figure 1.13 Hardness scales. **A.** Mohs scale of hardness, with the hardness of some common objects. **B.** Relationship between Mohs relative hardness scale and an absolute hardness scale.

the broken fragments are bounded by more or less flat, planer surfaces, a property called **cleavage** (*kleiben* = carve). Cleavage can be recognized by rotating a sample and looking for smooth, even surfaces that reflect light like a mirror. Note, however, that cleavage surfaces can occur in small, flat segments arranged in stair-step fashion.

The simplest type of cleavage is exhibited by the micas (Figure 1.14). Because the micas have much weaker bonds in one direction than in the others, they cleave to form thin, flat sheets. Some minerals have excellent cleavage in several directions, whereas others exhibit fair or poor cleavage, and still others have no cleavage at all. When minerals break evenly in more than one direction, cleavage is described by the *number of cleavage planes and the angle(s) at which they meet* (Figure 1.15).

Do not confuse cleavage with crystal shape. When a mineral exhibits cleavage, it will break into pieces that have the same geometry as one another. By contrast, the smooth-

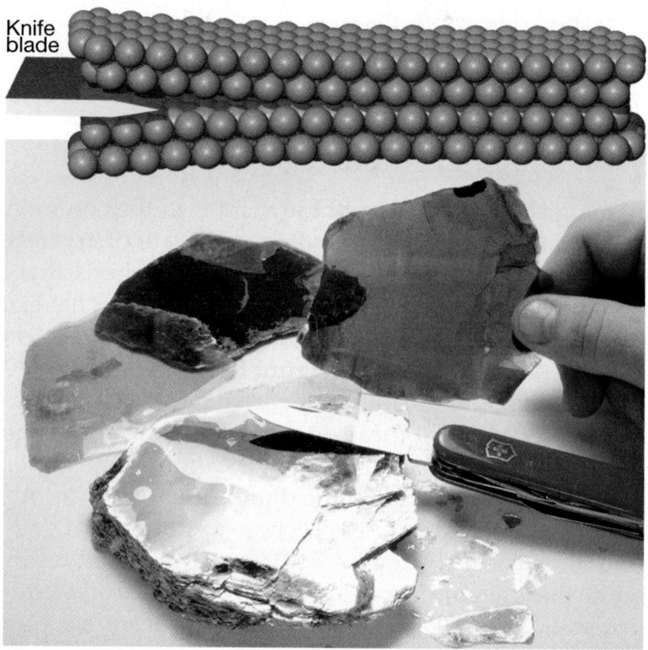

Figure 1.14 The thin sheets shown here were produced by splitting a mica (muscovite) crystal to its perfect cleavage. (Photo by Breck P. Kent)

sided quartz crystals shown in Figure 1.1 (p. 18) do not have cleavage. If broken, they fracture into shapes that do not resemble one another or the original crystals.

Fracture Minerals that have structures that are equally, or nearly equally, strong in all directions **fracture** to form irregular surfaces. Those, such as quartz, that break into smooth curved surfaces resembling broken glass exhibit a *conchoidal fracture* (Figure 1.16). Others break into splinters or fibers, but most minerals display an irregular fracture.

Density and Specific Gravity

Density is an important property of matter defined as mass per unit volume usually expressed as grams per cubic centimeter. Mineralogists often use a related measure called *specific gravity* to describe the density of minerals. **Specific gravity** is a unitless number representing the ratio of a mineral's weight to the weight of an equal volume of water.

Most common rock-forming minerals have a specific gravity between 2 and 3. For example, quartz has a specific gravity of 2.65. By contrast, some metallic minerals such as pyrite, native copper, and magnetite are more than twice as dense as quartz. Galena, which is an ore of lead, has a specific gravity of roughly 7.5, whereas the specific gravity of 24-karat gold is approximately 20.

Did You Know?

Clay minerals have been used as an additive to thicken milkshakes in fast-food restaurants.

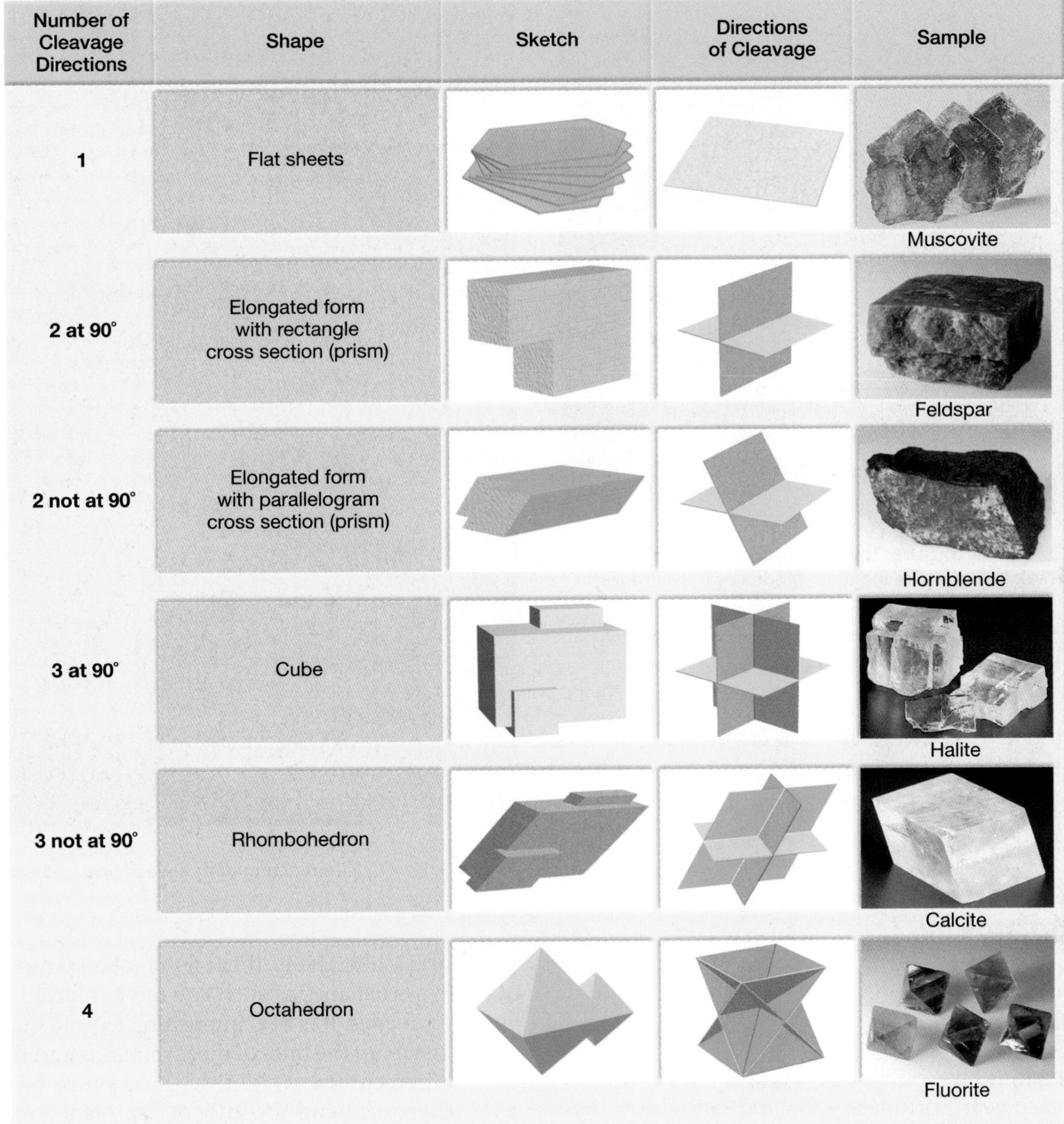

Number of Cleavage Directions	Shape	Sketch	Directions of Cleavage	Sample
1	Flat sheets			Muscovite
2 at 90°	Elongated form with rectangle cross section (prism)			Feldspar
2 not at 90°	Elongated form with parallelogram cross section (prism)			Hornblende
3 at 90°	Cube			Halite
3 not at 90°	Rhombohedron			Calcite
4	Octahedron			Fluorite

Figure 1.15 Common cleavage directions exhibited by minerals. (Photos by E. J. Tarbuck and Dennis Tasa)

With a little practice, you can estimate the specific gravity of a mineral by hefting it in your hand. Ask yourself, does this mineral feel about as "heavy" as similar sized rocks you have handled? If the answer is "yes," the specific gravity of the sample will likely be between 2.5 and 3.

Other Properties of Minerals

In addition to the properties already discussed, some minerals can be recognized by other distinctive properties. For example, halite is ordinary salt, so it can be quickly identified through taste. Talc and graphite both have distinctive feels: Talc feels soapy, and graphite feels greasy. Further, the streak of many sulfur-bearing minerals smells like rotten eggs. A few minerals, such as magnetite, have a high iron content and can be picked up with a magnet, while some varieties (lodestone) are natural magnets and will pick up small iron-based objects such as pins and paper clips (see Figure 1.23A, p. 32).

Moreover, some minerals exhibit special optical properties. For example, when a transparent piece of calcite is placed over printed material, the letters appear twice. This optical property is known as *double refraction* (Figure 1.17).

One very simple chemical test involves placing a drop of dilute hydrochloric acid from a dropper bottle onto a freshly broken mineral surface. Certain minerals, called carbonates, will effervesce (fizz) as carbon dioxide gas is released (Figure 1.18). This test is especially useful in identifying the common carbonate mineral calcite.

Figure 1.16 Conchoidal fracture. The smooth, curved surfaces result when minerals break in a glasslike manner. (Photo by E. J. Tarbuck)

Did You Know?

The mineral pyrite is commonly known as "fool's gold," because its golden-yellow color can lead to its being mistaken for gold. The name *pyrite* is derived from the Greek *pyros* ("fire"), because it gives off sparks when struck sharply.

Mineral Groups

Earth Materials
▼ Minerals

Nearly 4000 minerals have been named, and several new ones are identified each year. Fortunately, for students who are beginning to study minerals, no more than a few dozen are abundant! Collectively, these few make up most of the rocks of Earth's crust and, as such, are often referred to as the *rock-forming minerals.* It is also interesting to note that *only eight elements* make up the bulk of these minerals and represent more than 98 percent (by weight) of the continental crust (Figure 1.19). These elements, in order of abundance, are oxygen (O), silicon (Si), aluminum (Al), iron (Fe), calcium (Ca), sodium (Na), potassium (K), and magnesium (Mg).

As shown in Figure 1.19, silicon and oxygen are by far the most common elements in Earth's crust. Furthermore, these two elements readily combine to form the framework for the most common mineral group, the **silicates,** which account for more than 90 percent of Earth's crust.

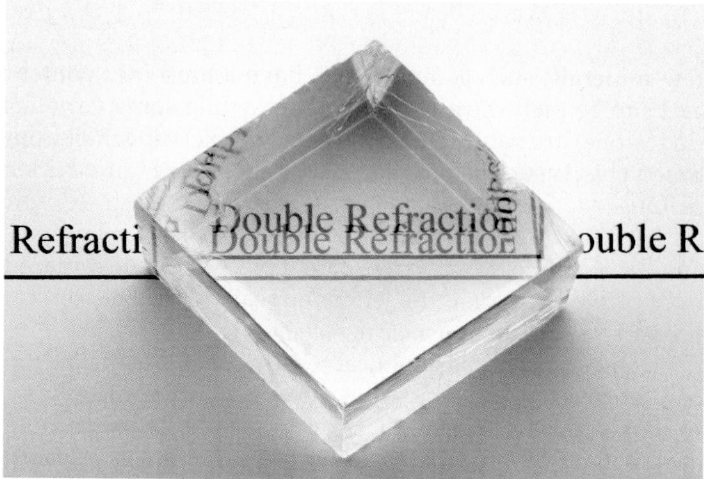

Figure 1.17 Double refraction illustrated by the mineral calcite. (Photo by Chip Clark)

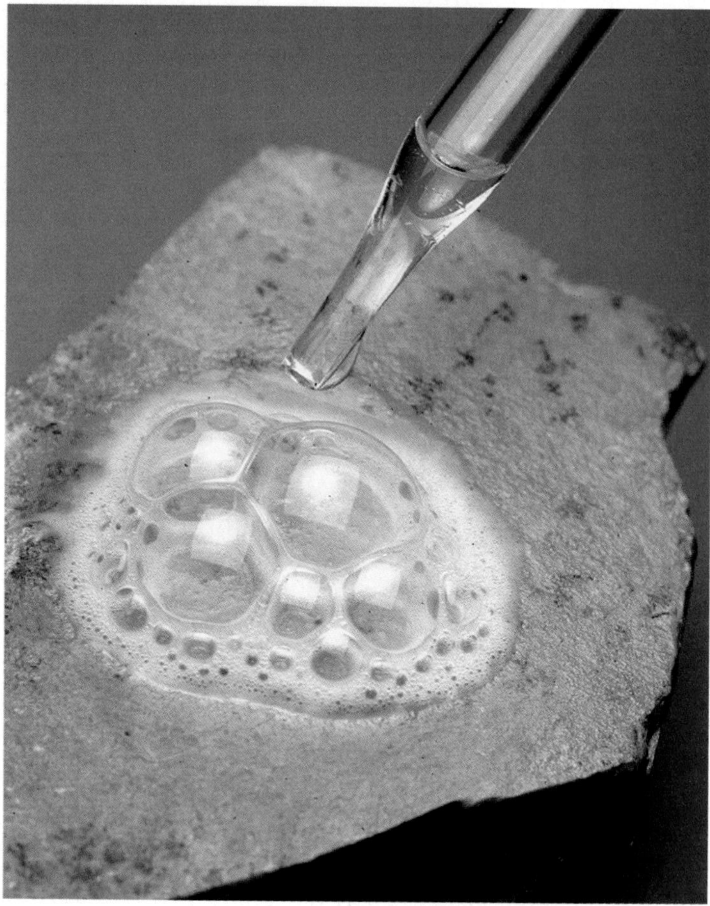

Figure 1.18 Calcite reacting with a weak acid. (Photo by Chip Clark)

Did You Know?

The names of precious gems often differ from the names of parent minerals. For example, *sapphire* is one of two gems that are varieties of the same mineral, *corundum.* Tiny amounts of the elements titanium and iron in corundum produce the most prized blue sapphires. When the mineral corundum contains chromium, it exhibits a brilliant red color and the gem is called *ruby.*

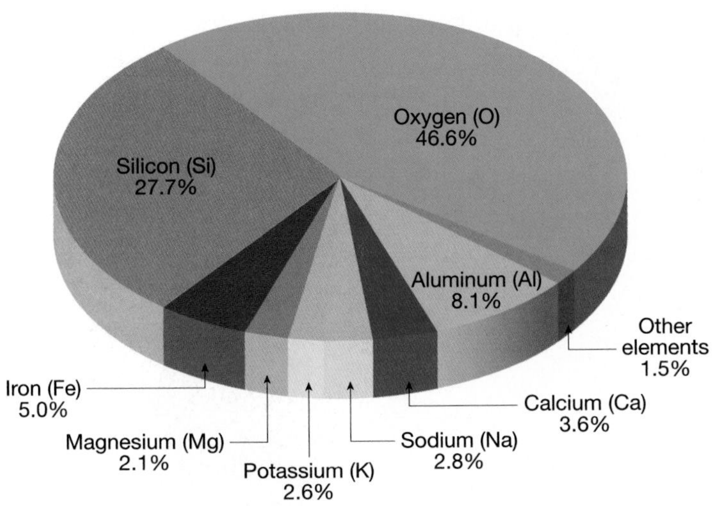

Figure 1.19 Relative abundance of the eight most common elements in the continental crust.

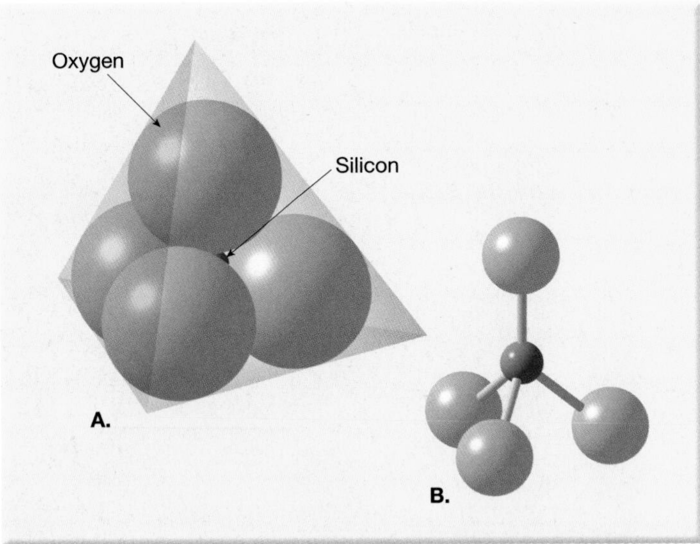

Figure 1.20 Two representations of the silicon-oxygen tetrahedron. **A.** The four large spheres represent oxygen atoms, and the blue sphere represents a silicon atom. The spheres are drawn in proportion to the radii of the atoms. **B.** A model of the tetrahedron using rods to depict the bonds that connect the atoms.

Because other mineral groups are far less abundant in Earth's crust than the silicates, they are often grouped together under the heading **nonsilicates.** Although not as common as the silicates, some nonsilicate minerals are very important economically. They provide us with the iron and aluminum to build our automobiles, gypsum for plaster and drywall to construct our homes, and copper for wire to carry electricity and to connect us to the Internet. Some common nonsilicate mineral groups include the carbonates, sulfates, and halides. In addition to their economic importance, these mineral groups include members that are major constituents in sediments and sedimentary rocks.

We will first discuss the most common mineral group, the silicates, and then consider some of the prominent nonsilicate mineral groups.

Silicate Minerals

Each of the silicate minerals contains oxygen and silicon atoms. Except for a few silicate minerals, such as quartz, most silicate minerals also contain one or more additional elements that join together to produce an electrically neutral compound. These elements give rise to the great variety of silicate minerals and their varied properties.

All silicates have the same fundamental building blocks, the **silicon-oxygen tetrahedron.** This structure consists of four oxygen atoms surrounding a much smaller silicon atom, as shown in Figure 1.20. Thus, a typical hand-size silicate mineral specimen contains millions of these silicon-oxygen tetrahedra, joined together in a variety of ways.

In some minerals, the tetrahedra are joined into chains, sheets, or three-dimensional networks by sharing oxygen atoms (Figure 1.21). These larger silicate structures are then connected to one another by other elements. The primary elements that join silicate structures are iron (Fe), magnesium (Mg), potassium (K), sodium (Na), and calcium (Ca).

Major groups of silicate minerals and common examples are given in Figure 1.21. The *feldspars* are by far the most plentiful group, comprising more than 50 percent of Earth's crust. *Quartz,* the second most abundant mineral in the continental crust, is the only common mineral made completely of silicon and oxygen.

Notice in Figure 1.21 that each mineral *group* has a particular silicate *structure.* A relationship exists between this internal structure of a mineral and the *cleavage* it exhibits. Because the silicon-oxygen bonds are strong, silicate minerals tend to cleave between the silicon-oxygen structures rather than across them. For example, the micas have a sheet structure and thus tend to cleave into flat plates (see Figure 1.14, p. 26). Quartz, which has equally strong silicon-oxygen bonds in all directions, has no cleavage but fractures instead.

How do silicate minerals form? Most crystallize from molten rock as it cools. This cooling can occur at or near Earth's surface (low temperature and pressure) or at great depths (high temperature and pressure). The *environment* during crystallization and the *chemical composition of the molten rock* mainly determine which minerals are produced. For example, the silicate mineral olivine crystallizes at high temperatures (about 1200°C [2200°F]), whereas quartz crystallizes at much lower temperatures (about 700°C [1300°F]).

In addition, some silicate minerals form at Earth's surface from the weathered products of other silicate minerals. Still other silicate minerals are formed under the extreme

Did You Know?

Most "scoopable" kitty litters sold on the market today contain a naturally occurring material called *bentonite.* Bentonite is largely composed of highly absorbent clay minerals that swell and clump up in the presence of moisture. This allows kitty waste to be isolated and scooped out, leaving behind clean litter.

	Mineral/Formula	Cleavage	Silicate Structure	Example
	Olivine group $(Mg, Fe)_2SiO_4$	None	Independent tetrahedron 	 Olivine
	Pyroxene group (Augite) $(Mg,Fe)SiO_3$	Two planes at right angles	Single chains 	 Augite
	Amphibole group (Hornblende) $Ca_2(Fe,Mg)_5Si_8O_{22}(OH)_2$	Two planes at 60° and 120°	Double chains 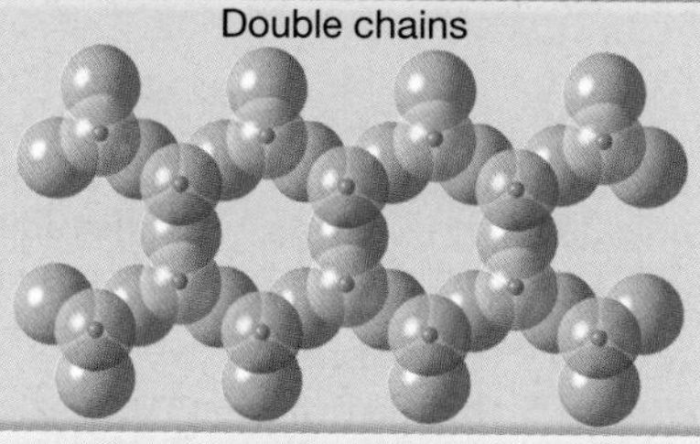	 Hornblend
Micas	Biotite $K(Mg,Fe)_3AlSi_3O_{10}(OH)_2$	One plane	Sheets 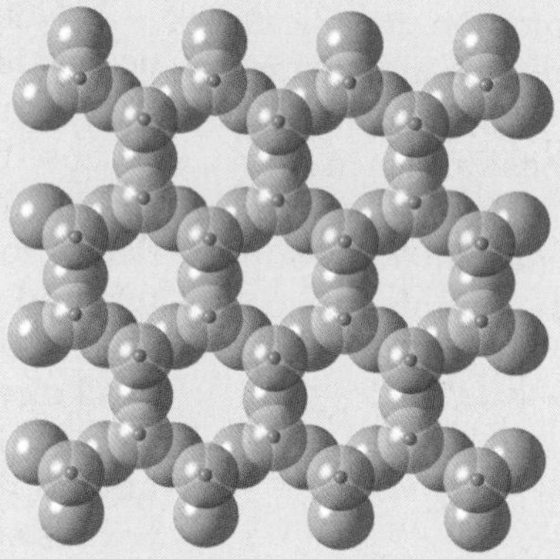	 Biotite
Micas	Muscovite $KAl_2(AlSi_3O_{10})(OH)_2$	One plane	Sheets	 Muscovite
Feldspars	Potassium feldspar (Orthoclase) $KAlSi_3O_8$	Two planes at 90°	Three-dimensional networks 	 Potassium feldspar
Feldspars	Plagioclase feldspar $(Ca,Na)AlSi_3O_8$	Two planes at 90°	Three-dimensional networks	
	Quartz SiO_2	None	Three-dimensional networks	 Quartz

Figure 1.21 Common silicate minerals. Note that the complexity of the silicate structure increases down the chart.

pressures associated with mountain building. Each silicate mineral, therefore, has a structure and a chemical composition that *indicate the conditions under which it formed.* Thus, by carefully examining the mineral makeup of rocks, geologists can often determine the circumstances under which the rocks formed.

Important Nonsilicate Minerals

Although nonsilicates make up only about 8 percent of Earth's crust, some minerals, such as gypsum, calcite, and halite, are major constituents in sedimentary rocks. Furthermore, many others are important economically. Table 1.1 lists some of the nonsilicate mineral classes and a few examples of each. Some of the most common nonsilicate minerals belong to one of three classes of minerals—the carbonates (CO_3^{2-}), the sulfates (SO_4^{2-}), and the halides (Cl^{1-}, F^{1-}, B^{1-}).

The carbonate minerals are much simpler structurally than the silicates. This mineral group is composed of the carbonate ion (CO_3^{2-}) and one or more kinds of positive ions. The most common carbonate mineral is *calcite,* $CaCO_3$ (calcium carbonate). This mineral is the major constituent in two well-known rocks: limestone and marble. Limestone has many uses, including as road aggregate, as building stone, and as the main ingredient in portland cement. Marble is used decoratively.

Two other nonsilicate minerals frequently found in sedimentary rocks are *halite* and *gypsum.* Both minerals are commonly found in thick layers that are the last vestiges of ancient seas that have long since evaporated (Figure 1.22). Like limestone, both are important nonmetallic resources. Halite is the mineral name for common table salt (NaCl). Gypsum ($CaSO_4 \cdot 2H_2O$), which is calcium sulfate with water bound into the structure, is the mineral of which plaster and other similar building materials are composed.

Most nonsilicate mineral classes contain members that are prized for their economic value. This includes the oxides, whose members hematite and magnetite are important ores of iron (Figure 1.23). Also significant are the sulfides, which are basically compounds of sulfur (S) and one or more metals.

Did You Know?

Gypsum, a white-to-transparent mineral, was first used as a building material in Anatolia (present-day Turkey) around 6000 B.C. It is also found on the interiors of the great pyramids in Egypt, which were erected in about 3700 B.C. Today, an average new American home contains more than 7 metric tons of gypsum in the form of 6000 square feet of wallboard.

Table 1.1 Common nonsilicate mineral groups

Mineral Group	Name	Chemical Formula	Economic Use
Oxides	Hematite	Fe_2O_3	Ore of iron, pigment
	Magnetite	Fe_3O_4	Ore of iron
	Corundum	Al_2O_3	Gemstone, abrasive
	Ice	H_2O	Solid form of water
Sulfides	Galena	PbS	Ore of lead
	Sphalerite	ZnS	Ore of zinc
	Pyrite	FeS_2	Sulfuric acid production
	Chalcopyrite	$CuFeS_2$	Ore of copper
	Cinnabar	HgS	Ore of mercury
Sulfates	Gypsum	$CaSO_4 \cdot 2H_2O$	Plaster
	Anhydrite	$CaSO_4$	Plaster
	Barite	$BaSO_4$	Drilling mud
Native elements	Gold	Au	Trade, jewelry
	Copper	Cu	Electrical conductor
	Diamond	C	Gemstone, abrasive
	Sulfur	S	Sulfa drugs, chemicals
	Graphite	C	Pencil lead, dry lubricant
	Silver	Ag	Jewelry, photography
	Platinum	Pt	Catalyst
Halides	Halite	NaCl	Common salt
	Fluorite	CaF_2	Used in steelmaking
	Sylvite	KCl	Fertilizer
Carbonates	Calcite	$CaCO_3$	Portland cement, lime
	Dolomite	$CaMg(CO_3)_2$	Portland cement, lime

Figure 1.22 Thick bed of halite (salt) at an underground mine in Grand Saline, Texas. (Photo by Tom Bochsler)

Examples of important sulfide minerals include galena (lead), sphalerite (zinc), and chalcopyrite (copper). In addition, native elements, including gold, silver, and carbon (diamonds), plus a host of other nonsilicate minerals—fluorite (flux in making steel), corundum (gemstone, abrasive), and uraninite (a uranium source)—are important economically.

Mineral Resources

Mineral resources are Earth's storehouse of useful minerals that can be recovered for use. Resources include already identified deposits from which minerals can be extracted profitably, called **reserves,** as well as known deposits that are not yet recoverable under present economic conditions or technology. Deposits inferred to exist, but not yet discovered, are also considered mineral resources.

The term **ore** denotes useful metallic minerals that can be mined at a profit. In common usage, the term is also applied to some nonmetallic minerals, such as fluorite and sulfur. However, materials used for such purposes as building stone, road paving, abrasives, ceramics, and fertilizers are not usually called ores; rather, they are classified as *industrial rocks and minerals.*

Recall that more than 98 percent of Earth's crust is composed of only eight elements. Except for oxygen and silicon, all other elements make up a relatively small fraction of common crustal rocks (see Figure 1.19, p. 29). Indeed, the natural concentrations of most elements in crustal rocks are exceedingly small. A rock containing the average crustal percentage of a valuable element such as gold has no economic value, because the cost of extracting it greatly exceeds the value of the gold that could be recovered.

To have economic value, an element must be concentrated above the level of its average crustal abundance. For example, copper makes up about 0.0135 percent of the crust. For a deposit to be considered an ore of copper, it must contain a concentration that is about 100 times this amount. Aluminum, on the other hand, represents 8.13 percent of the crust and can be extracted profitably when it is found in concentrations only about four times its average crustal percentage.

It is important to realize that economic changes affect whether or not a deposit is profitable to extract. If demand for a metal increases and prices rise sufficiently, the status of a previously unprofitable deposit changes and it becomes an ore. The status of unprofitable deposits may also change if a technological advance allows the material to be extracted at a lower cost than before.

Conversely, changing economic factors can turn a once profitable ore deposit into an unprofitable deposit that can

A.

B.

Figure 1.23 Two important ores of iron. **A.** Magnetite and **B.** Hematite. (Photos by E. J. Tarbuck)

Figure 1.24 Aerial view of Bingham Canyon copper mine near Salt Lake City, Utah. This huge open-pit mine is about 4 kilometers (2.5 miles) across and 900 meters (nearly 3000 feet) deep. Although the amount of copper in the rock is less than 1 percent, the huge volumes of material removed and processed each day (about 200,000 tons) yield, significant quantities of metal. (Photo by Michael Collier)

no longer be called an ore. This situation was illustrated recently at the copper-mining operation located at Bingham Canyon, Utah, one of the largest open-pit mines on Earth (Figure 1.24). Mining was halted there in 1985 because outmoded equipment had driven the cost of extracting the copper beyond the current selling price. The owners responded by replacing an antiquated 1000-car railroad with conveyor belts and pipelines for transporting the ore and waste. These devices achieved a cost reduction of nearly 30 percent and returned this mining operation to profitability.

The Chapter in Review

1. A *mineral* is a naturally occurring inorganic solid that possesses an orderly crystalline structure and a definite chemical composition. Most *rocks* are aggregates composed of two or more minerals.

2. The building blocks of minerals are *elements.* An *atom* is the smallest particle of matter that still retains the characteristics of an element. Each atom has a *nucleus,* which contains *protons* and *neutrons.* Orbiting the nucleus of an atom are *electrons.* The number of protons in an atom's nucleus determines its *atomic number* and the name of the element. Atoms bond together to form a *compound* by either gaining, losing, or sharing electrons with another atom.

3. *Isotopes* are variants of the same element but with a different *mass number* (the total number of neutrons plus protons found in an atom's nucleus). Some isotopes are unstable and disintegrate naturally through a process called *radioactive decay.*

4. The properties of minerals include *crystal shape (habit), luster, color, streak, hardness, cleavage, fracture,* and *density* or *specific gravity.* In addition, a number of special physical and chemical properties (*taste, smell, elasticity, feel, magnetism, double refraction,* and *chemical reaction to hydrochloric acid*) are useful in identifying certain minerals. Each mineral has a unique set of properties that can be used for identification.

5. Of the nearly 4000 minerals, no more than a few dozen make up most of the rocks of Earth's crust and, as such, are classified as rock-forming minerals. Eight elements (oxygen, silicon, aluminum, iron, calcium, sodium, potassium, and magnesium) make up the bulk of these minerals and represent more than 98 percent (by weight) of Earth's continental crust.

6. The most common mineral group is the *silicates.* All silicate minerals have the *silicon-oxygen tetrahedron* as their fundamental building block. In some silicate minerals, the tetrahedra are joined in chains; in others, the tetrahedra are arranged into sheets or three-dimensional networks. Each silicate mineral has a structure and a chemical composition that indicates the conditions under which it formed. The *nonsilicate* mineral groups include the *oxides* (e.g., magnetite, mined for iron), *sulfides* (e.g., sphalerite, mined for zinc), *sulfates* (e.g., gypsum, used in plaster and frequently found in sedimentary rocks) *native elements* (e.g., graphite, a dry lubricant), *halides* (e.g., halite, common salt, and frequently found in sedimentary rocks), and *carbonates* (e.g., calcite, used in portland cement and a major constituent in two well-known rocks: limestone and marble).

7. The term *ore* is used to denote useful metallic minerals, such as hematite (mined for iron) and galena (mined for lead), that can be mined at a profit, as well as some nonmetallic minerals, such as fluorite and sulfur, that contain useful substances.

Key Terms

atom (p. 20)
atomic number (p. 20)
chemical compound (p. 20)
cleavage (p. 26)
color (p. 24)
covalent bond (p. 23)
density (p. 26)
electron (p. 20)
element (p. 19)
energy levels (p. 20)
fracture (p. 26)
habit (p. 25)
hardness (p. 25)
ionic bond (p. 22)
ions (p. 22)
isotope (p. 23)
luster (p. 24)
mass number (p. 23)
mineral (p. 18)
mineralogy (p. 18)
mineral resource (p. 32)
Mohs scale (p. 25)
neutron (p. 20)
nonsilicate (p. 29)
nucleus (p. 20)
ore (p. 32)
periodic table (p. 19)
principal shell (p. 20)
proton (p. 20)
radioactive decay (p. 23)
reserve (p. 32)
rock (p. 19)
silicates (p. 28)
silicon-oxygen tetrahedron (p. 29)
specific gravity (p. 26)
streak (p. 24)
tenacity (p. 25)
valence electrons (p. 20)

Questions for Review

1. Briefly describe the five characteristics an Earth material should have in order to be considered a mineral.
2. Define the term *rock.*
3. List the three main particles of an atom and explain how they differ from one another.
4. If the number of electrons in an atom is 35 and its mass number is 80, calculate the following:
   a. the number of protons
   b. the atomic number
   c. the number of neutrons
5. What occurs in an atom to produce an ion?
6. What is an isotope?
7. Although all minerals have an orderly internal arrangement of atoms (crystalline structure), most mineral samples do not visibly demonstrate their crystal form. Why?
8. Why might it be difficult to identify a mineral by its color?
9. If you found a glassy-appearing mineral while rock hunting and had hopes that it was a diamond, what simple test might help you make a determination?
10. Table 1.1 (p. 31) lists a use for corundum as an abrasive. Explain why it makes a good abrasive in terms of the Mohs hardness scale.
11. Gold has a specific gravity of almost 20. If a five-gallon pail of water weighs about 40 pounds, how much would a five-gallon pail of gold weigh?
12. What are the two most common elements in Earth's crust?
13. What is the term used to describe the basic building block of all silicate minerals?
14. What are the two most common silicate minerals in Earth's crust?
15. List three nonsilicate minerals that are commonly found in rocks.
16. Contrast a mineral *resource* and a mineral *reserve.*
17. What might cause a mineral deposit that had not been considered an ore to become reclassified as an ore?

Online Study Guide

The *Foundations of Earth Science* Web site uses the resources and flexibility of the Internet to aid in your study of the topics in this chapter. Written and developed by Earth science instructors, this site will help improve your understanding of Earth science. Visit **http://www.prenhall.com/lutgens** and click on the cover of *Foundations of Earth Science 5e* to find:

- Online review quizzes.
- Critical thinking exercises.
- Links to chapter-specific Web resources.
- Internet-wide key-term searches.

http://www.prenhall.com/lutgens

GEODe: Earth Science

GEODe: Earth Science makes studying more effective by reinforcing key concepts using animation, video, narration, interactive exercises, and practice quizzes. A copy is included with every copy of *Foundations of Earth Science 5e.*

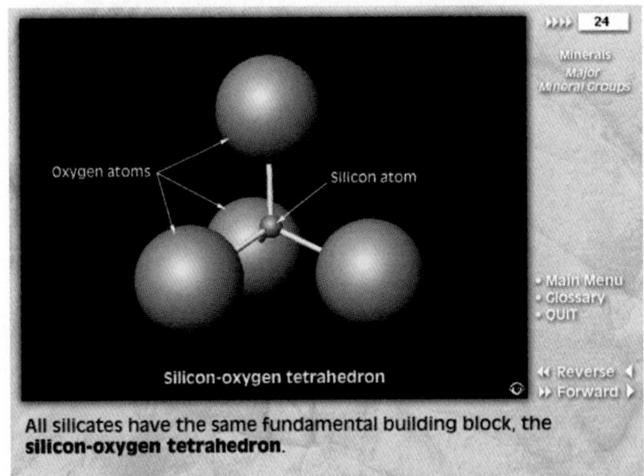

Climbers scaling the vertical face of El Capitan in Yosemite National Park, California. (Photo by Ron Niebruggel/Mira)

Rocks: Materials of the Solid Earth

FOCUS ON LEARNING

To assist you in learning the important concepts in this chapter, you will find it helpful to focus on the following questions:

1. What are the three groups of rocks and the geologic processes involved in the formation of each?
2. What two criteria are used to classify igneous rocks?
3. What are the two major types of weathering and the processes associated with each?
4. What are the names and environments of formation for some common detrital and chemical sedimentary rocks?
5. What are the names, textures, and environments of formation for some common metamorphic rocks?

Why study rocks? You have already learned that rocks and minerals have great economic value. Furthermore, all Earth processes in some way depend on the properties of these basic materials. Events such as volcanic eruptions, mountain building, weathering, erosion, and even earthquakes involve rocks and minerals. Consequently, a basic knowledge of Earth materials is essential to understanding Earth phenomena.

Every rock contains clues about the environment in which it formed. For example, some rocks are composed entirely of small shell fragments. This tells Earth scientists that the particles making up the rock originated in a shallow marine environment. Other rocks contain clues that indicate they formed from a volcanic eruption or deep in the Earth during mountain building (Figure 2.1). Thus, rocks contain a wealth of information about events that have occurred over Earth's long history.

We divide rocks into three groups, based on their mode of origin. The groups are igneous, sedimentary, and metamorphic. Before examining each group, we will view the *rock cycle,* which depicts the interrelationships among these rock groups.

Earth as a System: The Rock Cycle

Earth Materials
▼ The Rock Cycle

Earth is a system. This means that our planet consists of many interacting parts that form a complex whole. Nowhere is this idea better illustrated than when we examine the rock cycle (Figure 2.2). The **rock cycle** allows us to view many of the interrelationships among different parts of the Earth system. It helps us understand the origin of igneous, sedimentary, and metamorphic rocks and to see that each type is linked to the others by the processes that act upon and within the planet. Learn the rock cycle well; you will be examining its interrelationships in greater detail throughout this chapter and many other chapters as well.

The Basic Cycle

We begin at the top of Figure 2.2. **Magma** is molten material that forms inside Earth. Eventually, magma cools and solidifies. This process, called **crystallization,** may occur either beneath

Figure 2.1 Rocks contain information about the processes that produce them. This large exposure of igneous rocks located in the Sierra Nevada, California, was once a molten mass found deep within Earth. (Photo by Brian Bailey/Getty Images)

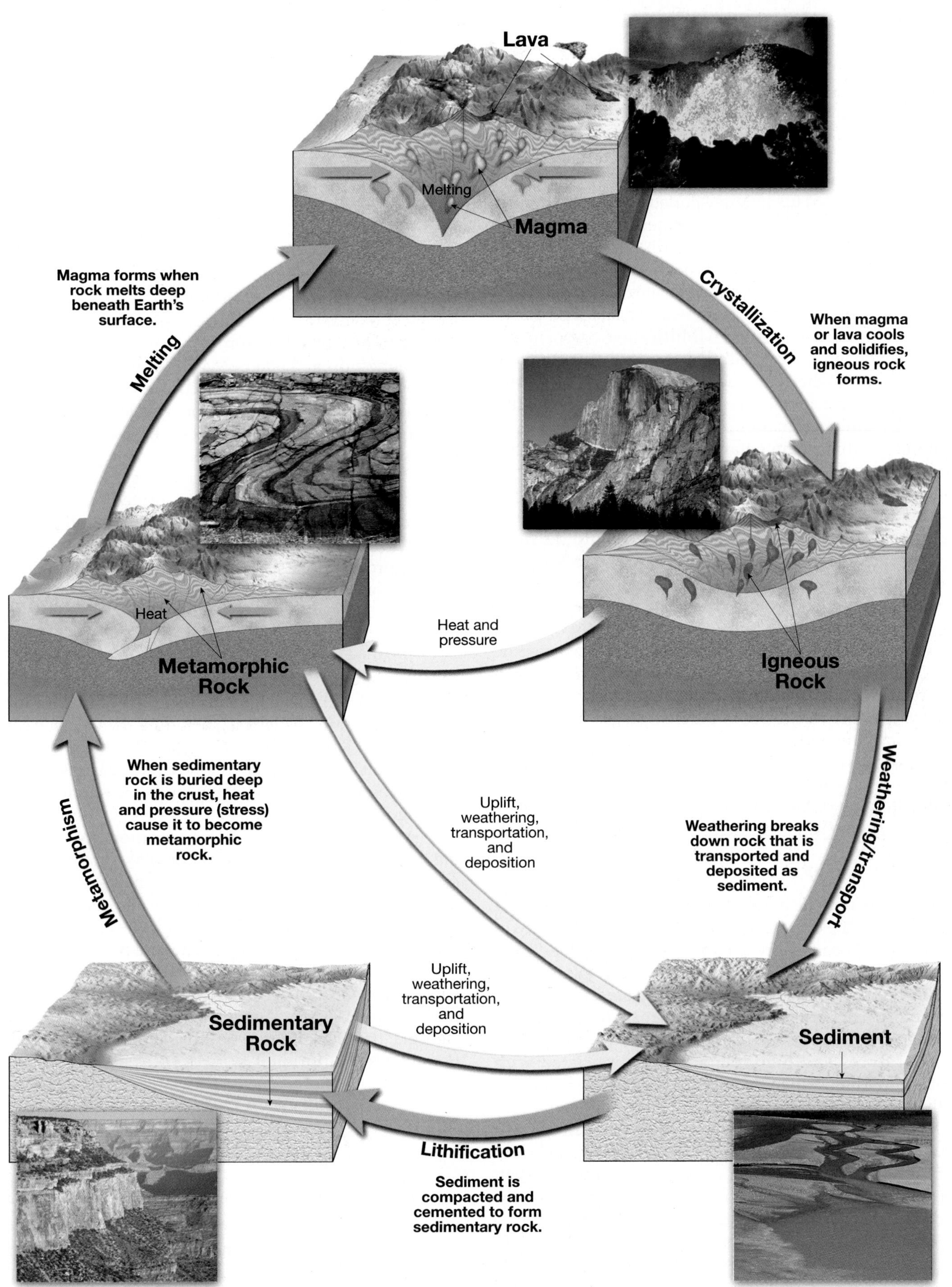
Lava
Melting
Magma
Magma forms when rock melts deep beneath Earth's surface.
Melting
Crystallization
When magma or lava cools and solidifies, igneous rock forms.
Heat
Metamorphic Rock
Heat and pressure
Igneous Rock
When sedimentary rock is buried deep in the crust, heat and pressure (stress) cause it to become metamorphic rock.
Metamorphism
Uplift, weathering, transportation, and deposition
Weathering breaks down rock that is transported and deposited as sediment.
Weathering/transport
Uplift, weathering, transportation, and deposition
Sedimentary Rock
Sediment
Lithification
Sediment is compacted and cemented to form sedimentary rock.

Figure 2.2 Viewed over long spans, rocks are constantly forming, changing, and reforming. The rock cycle helps us understand the origin of the three basic rock groups. Arrows represent processes that link each group to the others.

the surface or, following a volcanic eruption, at the surface. In either situation, the resulting rocks are called **igneous rocks.**

If igneous rocks are exposed at the surface, they will undergo **weathering,** in which the day-in and day-out influences of the atmosphere slowly disintegrate and decompose rocks. The materials that result are often moved downslope by gravity before being picked up and transported by any of a number of erosional agents, such as running water, glaciers, wind, or waves. Eventually, these particles and dissolved substances, called **sediment,** are deposited. Although most sediment ultimately comes to rest in the ocean, other sites of deposition include river floodplains, desert basins, swamps, and sand dunes.

Next, the sediments undergo **lithification,** a term meaning "conversion into rock." Sediment is usually lithified into **sedimentary rock** when compacted by the weight of overlying layers or when cemented as percolating groundwater fills the pores with mineral matter.

If the resulting sedimentary rock is buried deep within Earth and involved in the dynamics of mountain building or intruded by a mass of magma, it will be subjected to great pressures and/or intense heat. The sedimentary rock will react to the changing environment and turn into the third rock type, **metamorphic rock.** If metamorphic rock is subjected to still higher temperatures, it will melt, creating magma, which will eventually crystallize into igneous rock, starting the cycle all over again.

Although rocks may seem to be unchanging masses, the rock cycle shows that they are not. The changes, however, take time—great amounts of time. In addition, the rock cycle is operating all over the world, but in different stages. Today, new magma is forming under the island of Hawaii, while the Colorado Rockies are slowly being worn down by weathering and erosion. Some of this weathered debris will eventually be carried to the Gulf of Mexico, where it will add to the already substantial mass of sediment that has accumulated there.

Alternative Paths

The paths shown in the basic cycle are not the only ones that are possible. To the contrary, other paths are just as likely to be followed as those described in the preceding section. These alternatives are indicated by the blue arrows in Figure 2.2.

Igneous rocks, rather than being exposed to weathering and erosion at Earth's surface, may remain deeply buried. Eventually, these masses may be subjected to the strong compressional forces and high temperatures associated with mountain building. When this occurs, they are transformed directly into metamorphic rocks.

Metamorphic and sedimentary rocks, as well as sediment, do not always remain buried. Rather, overlying layers may be eroded away, exposing the once buried rock. When this happens, the material is attacked by weathering processes and turned into new raw materials for sedimentary rocks.

Where does the energy that drives Earth's rock cycle come from? Processes driven by heat from Earth's interior are responsible for forming igneous and metamorphic rocks. Weathering and the movement of weathered material are external processes powered by energy from the Sun. External processes produce sedimentary rocks.

Igneous Rocks: "Formed by Fire"

Earth Materials
▼ Igneous Rocks

In our discussion of the rock cycle, we pointed out that igneous rocks form as *magma* cools and crystallizes. But what is magma and what is its source? Magma is molten rock generated by partial melting of rocks in Earth's mantle and in much smaller amounts, in the lower crust. This molten material consists mainly of the elements found in the silicate minerals. Silicon and oxygen are the main constituents in magma, with lesser amounts of aluminum, iron, calcium, sodium, potassium, magnesium, and others. Magma also contains some gases, particularly water vapor, which are confined within the magma body by the weight of the overlying rocks.

Once formed, a magma body buoyantly rises toward the surface because it is less dense than the surrounding rocks. Occasionally molten rock reaches the surface, where it is called **lava.** Sometimes, lava is emitted as fountains that are produced when escaping gases propel molten rock skyward. On other occasions, magma is explosively ejected from a vent, producing a spectacular eruption such as the 1980 eruption of Mount St. Helens. However, most eruptions are not violent; rather, volcanoes more often emit quiet outpourings of lava (Figure 2.3).

Igneous rocks that form when molten rock solidifies *at the surface* are classified as **extrusive** or **volcanic** (after the fire god Vulcan). Extrusive igneous rocks are abundant in western portions of the Americas, including the volcanic cones of the Cascade Range and the extensive lava flows of the Columbia Plateau. In addition, many oceanic islands, typified by the Hawaiian Islands, are composed almost entirely of volcanic igneous rocks.

Most magma, however, loses its mobility before reaching the surface and eventually crystallizes at depth. Igneous rocks that *form at depth* are termed **intrusive** or **plutonic** (after Pluto, the god of the lower world in classical mythology). Intrusive igneous rocks would never be exposed at the surface if portions of the crust were not uplifted and the overlying rocks stripped away by erosion. Exposures of intrusive igneous rocks occur in many places, including Mount Washington, New Hampshire; Stone Mountain, Georgia; the Black Hills of South Dakota; and Yosemite National Park, California.

Did You Know?

During the catastrophic eruption of Vesuvius in A.D. 79, the entire city of Pompeii (near Naples, Italy) was completely buried by several meters of pumice and volcanic ash. Centuries passed, and new towns sprang up around Vesuvius. It was not until 1595, during a construction project, that the remains of Pompeii came to light. Today, thousands of tourists stroll amongst the excavated remains of Pompeii's shops, taverns, and villas.

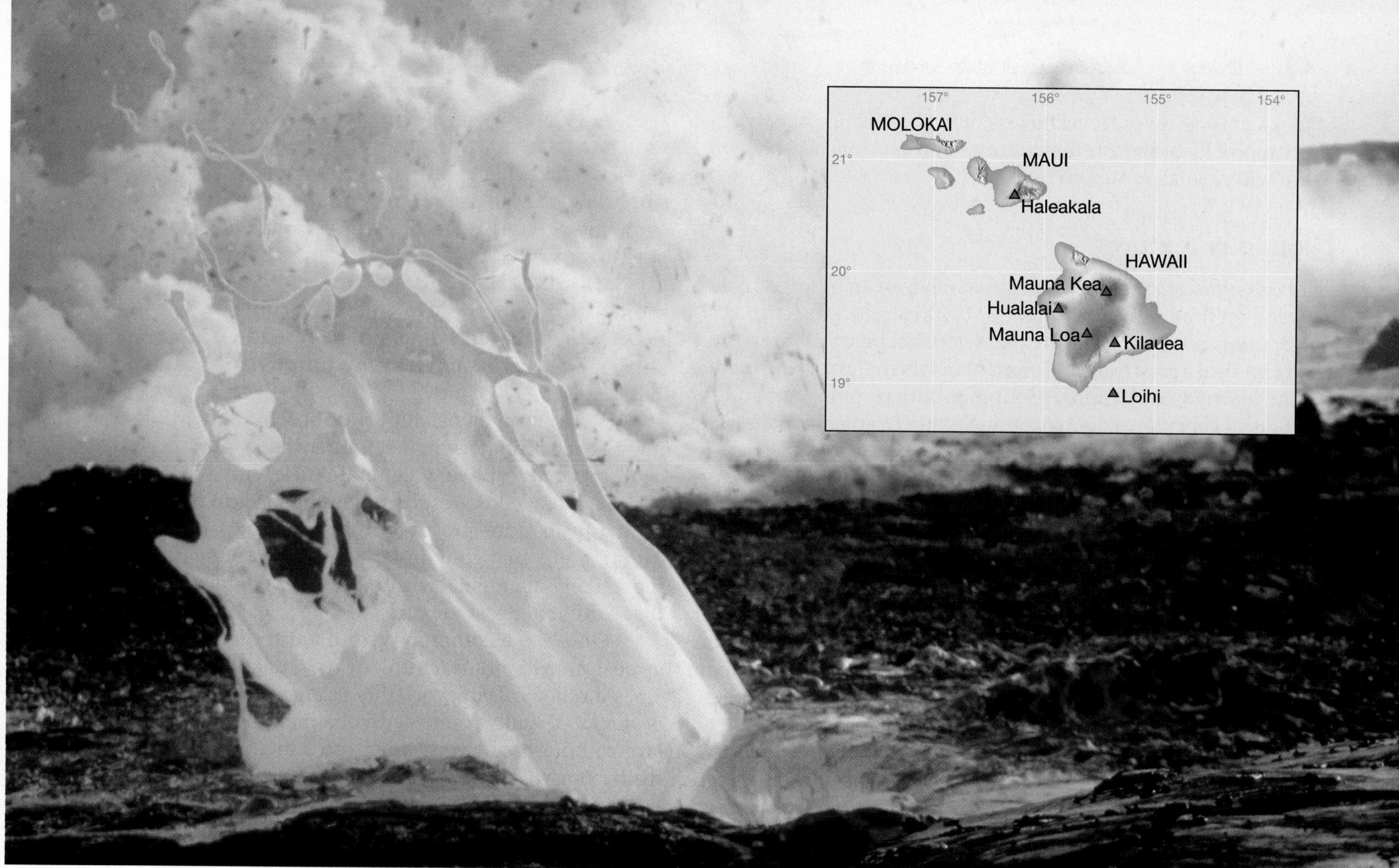

Figure 2.3 Fluid basaltic lava moves down the slopes of Hawaii's Kilauea Volcano. (Photo by G. Brad Lewis/Getty Images—Liaison)

Magma Crystallizes to Form Igneous Rocks

Magma is basically a very hot, thick fluid, but it also contains solids and gases. The solids are mineral crystals. The liquid portion of a magma body is composed of ions that move about freely. However, as magma cools, the random movements of the ions slow, and the ions begin to arrange themselves into orderly patterns. This process is called *crystallization.* Usually, the molten material does not all solidify at the same time. Rather, as it cools, numerous small crystals develop. In a systematic fashion, ions are added to these centers of crystal growth. When the crystals grow large enough for their edges to meet, their growth ceases for lack of space, and crystallization continues elsewhere. Eventually, all of the liquid is transformed into a solid mass of interlocking crystals.

The rate of cooling strongly influences crystal size. If a magma cools very slowly, relatively few centers of crystal growth develop. Slow cooling also allows ions to migrate over relatively great distances. Consequently, *slow cooling results in the formation of large crystals.* On the other hand, if cooling occurs quite rapidly, the ions lose their motion and quickly combine. This results in a large number of tiny crystals that all compete for the available ions. Therefore, *rapid cooling results in the formation of a solid mass of small intergrown crystals.*

Thus, if a geologist encounters igneous rock containing crystals large enough to be seen with the unaided eye, it means the molten rock from which it formed cooled quite slowly. But if the crystals can be seen only with a microscope, the geologist knows that the magma cooled very quickly.

If the molten material is quenched almost instantly, there is not sufficient time for the ions to arrange themselves into a crystalline network at all. Therefore, solids produced in this manner consist of randomly distributed ions. Such rocks are called *glass* and are quite similar to ordinary manufactured glass. "Instant" quenching occurs during violent volcanic eruptions that produce tiny shards of glass called *volcanic ash.*

In addition to the rate of cooling, the composition of a magma and the amount of dissolved gases influence crystallization. Because magmas differ in each of these aspects, the physical appearance and mineral composition of igneous

Did You Know?

During the Stone Age, volcanic glass (obsidian) was used for making cutting tools. Today, scalpels made from obsidian are being employed for delicate plastic surgery because they leave less scarring. "The steel scalpel has a rough edge, where the obsidian scalpel is smoother and sharper," explains Lee Green, MD, an associate professor at the University of Michigan Medical School.

rocks vary widely. Nevertheless, it is possible to classify igneous rocks based on their *texture* and *mineral composition.* We will now look at both features.

Igneous Textures

Texture describes the overall appearance of an igneous rock, based on the *size* and *arrangement* of its interlocking crystals. Texture is a very important characteristic, because it reveals a great deal about the environment in which the rock formed. You learned that rapid cooling produces small crystals, whereas very slow cooling produces much larger crystals. As you might expect, the rate of cooling is slow in magma chambers lying deep within the crust, whereas a thin layer of lava extruded upon Earth's surface may chill to form solid rock in a matter of hours. Small molten blobs ejected into the air during a violent eruption can solidify almost instantly.

Igneous rocks that form rapidly at the surface or as small masses within the upper crust have a **fine-grained texture,** with the individual crystals too small to be seen with the unaided eye (Figure 2.4A). Common in many fine-grained igneous rocks are voids, called *vesicles,* left by gas bubbles that formed as the lava solidified (Figure 2.5).

When large masses of magma solidify far below the surface, they form igneous rocks that exhibit a **coarse-grained texture.** These coarse-grained rocks have the appearance of a mass of intergrown crystals, which are roughly equal in size and large enough that the individual minerals can be identified with the unaided eye. Granite is a classic example (Figure 2.4B).

A large mass of magma located at depth may require tens of thousands, even millions, of years to solidify. Because all materials within a magma do not crystallize at the same rate or at the same time during cooling, it is possible for some crystals to become quite large before others even start to form. If magma that already contains some large crystals suddenly erupts at the surface, the remaining molten portion of the lava would cool quickly. The resulting rock, which has large crystals embedded in a matrix of smaller crystals, is said to have a **porphyritic texture** (Figure 2.4D).

During some volcanic eruptions, molten rock is ejected into the atmosphere, where it is quenched very quickly. Rapid cooling of this type may generate rock with a **glassy texture** (Figure 2.4C). Glass results when the ions do not have sufficient time to unite into an orderly crystalline structure. In addition, melts that contain large amounts of silica (SiO_2) are more likely than melts with a low silica content to form rocks that exhibit a glassy texture.

Obsidian, a common type of natural glass, is similar in appearance to a dark chunk of manufactured glass (Figure 2.6). Another volcanic rock that often exhibits a glassy texture is *pumice.* Usually found with obsidian, pumice forms when large amounts of gas escape from a melt to generate a gray, frothy mass (Figure 2.7). In some samples, the vesicles are quite noticeable, whereas in others, the pumice resembles fine shards of intertwined glass. Because of the large volume of air-filled voids, many samples of pumice will float in water.

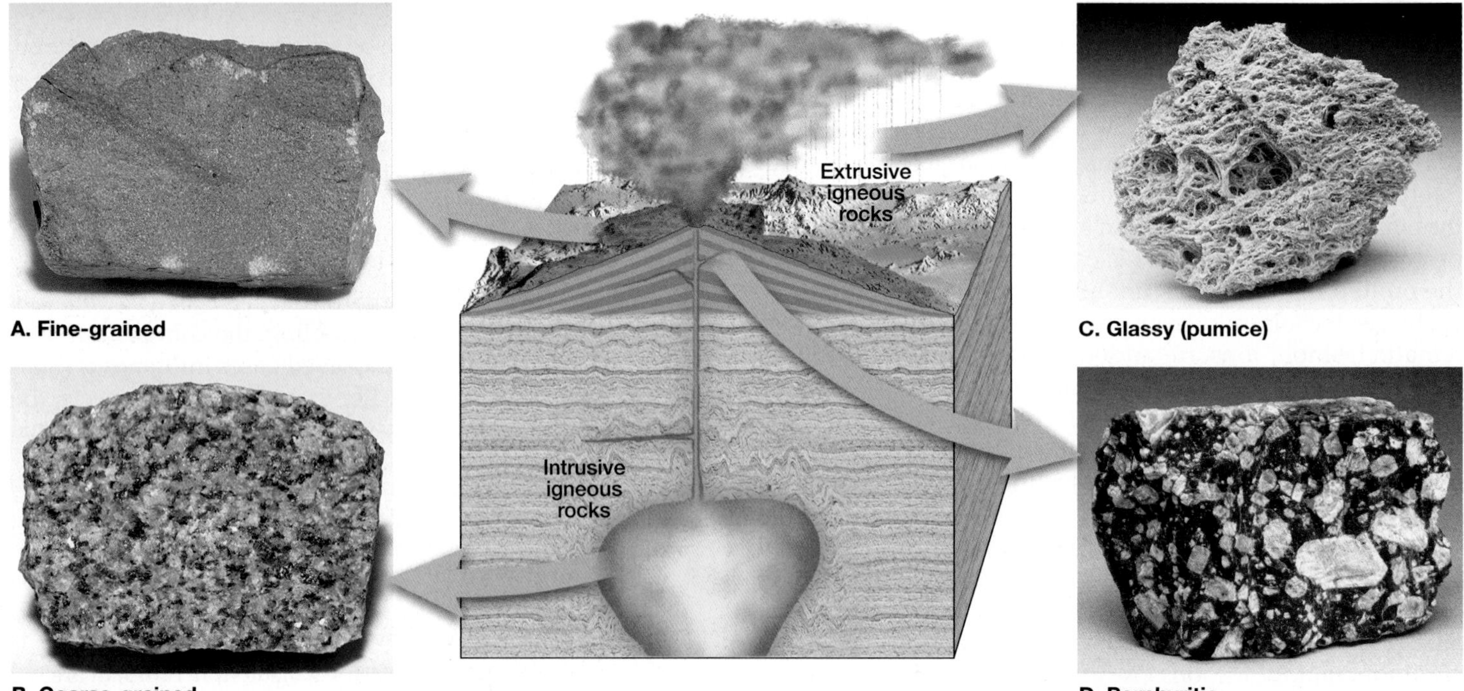

Figure 2.4 Igneous rock textures. **A.** Igneous rocks that form at or near Earth's surface cool quickly and often exhibit a fine-grained texture. **B.** Coarse-grained igneous rocks form when magma slowly crystallizes at depth. **C.** During a volcanic eruption in which silica-rich lava is ejected into the atmosphere, a frothy glass called pumice may form. **D.** A porphyritic texture results when magma that already contains some large crystals migrates to a new location where the rate of cooling increases. The resulting rock consists of large crystals embedded within a matrix of smaller crystals. (Photos courtesy of E. J. Tarbuck)

Figure 2.5 Scoria is a volcanic rock that is vesicular. Vesicles form as gas bubbles escape near the top of a lava flow. (Photo from GeoScience Resources/American Geological Institute)

Figure 2.7 Pumice, a glassy rock, is very lightweight because it contains numerous vesicles. (Inset photo by Chip Clark)

Igneous Compositions

Igneous rocks are mainly composed of silicate minerals. Furthermore, the mineral makeup of a particular igneous rock is ultimately determined by the chemical composition of the magma from which it crystallizes. Recall that magma is composed largely of the eight elements that are the major constituents of the silicate minerals. Chemical analysis shows that silicon and oxygen (usually expressed as the silica [SiO_2] content of a magma) are by far the most abundant constituents of igneous rocks. These two elements, plus ions of aluminum (Al), calcium (Ca), sodium (Na), potassium (K), magnesium (Mg), and iron (Fe), make up roughly 98 percent by weight of most magmas.

Figure 2.6 Obsidian, a natural glass, was used by Native Americans for making arrowheads and cutting tools. (Photo by E. J. Tarbuck; inset photo by Jeffrey Scovil)

As magma cools and solidifies, these elements combine to form two major groups of silicate minerals. The *dark silicates* are rich in iron and/or magnesium and are relatively low in silica. *Olivine, pyroxene, amphibole,* and *biotite mica* are the common dark silicate minerals of Earth's crust. By contrast, the *light silicates* contain greater amounts of potassium, sodium, and calcium rather than iron and magnesium. As a group, these minerals are richer in silica than the dark silicates. The light silicates include *quartz, muscovite mica,* and the most abundant mineral group, the *feldspars.* The feldspars make up at least 40 percent of most igneous rocks. Thus, in addition to feldspar, igneous rocks contain some combination of the other light and/or dark silicates listed earlier.

Classifying Igneous Rocks

Igneous rocks are classified by their texture and mineral composition. Various igneous textures result from different cooling histories, while the mineral compositions are a consequence of the chemical makeup of the parent magma and the environment of crystallization.

Did You Know?

Quartz watches actually contain a quartz crystal to keep time. Before quartz watches, timepieces used some sort of oscillating mass or tuning fork. Cogs and wheels converted this mechanical movement to the movement of the hand. It turns out that if voltage is applied to a quartz crystal, it will oscillate with a consistency that is hundreds of times better for timing than a tuning fork. Because of this property, and modern integrated-circuit technology, quartz watches are now built so cheaply they are sometimes given away in cereal boxes. Modern watches that employ mechanical movements are very expensive indeed.

Despite their great compositional diversity, igneous rocks can be divided into broad groups according to their proportions of light and dark minerals. A general classification scheme based on texture and mineral composition is provided in Figure 2.8.

Granitic (Felsic) Rocks Near one end of the continuum are rocks composed almost entirely of light-colored silicates—quartz and potassium feldspar. Igneous rocks in which these are the dominant minerals have a **granitic composition.** Geologists also refer to granitic rocks as being **felsic,** a term derived from *fe*ldspar and *si*lica (quartz). In addition to quartz and feldspar, most granitic rocks contain about 10 percent dark silicate minerals, usually biotite mica and amphibole. Granitic rocks are rich in silica (about 70 percent) and are major constituents of the continental crust.

Granite is a coarse-grained igneous rock that forms where large masses of magma slowly solidify at depth. During episodes of mountain building, granite and related crystalline rocks may be uplifted, whereupon the processes of weathering and erosion strip away the overlying crust. Pikes Peak in the Rockies, Mount Rushmore in the Black Hills, Stone Mountain in Georgia, and Yosemite National Park in the Sierra Nevada are all areas where large quantities of granite are exposed at the surface.

Granite is perhaps the best-known igneous rock (Figure 2.9). This is partly because of its natural beauty, which is enhanced when polished, and partly because of its abundance. Slabs of polished granite are commonly used for tombstones and monuments and as building stones.

Rhyolite is the extrusive equivalent of granite and, like granite, is composed essentially of the light-colored silicates (Figure 2.9). This fact accounts for its color, which is usually buff to pink or light gray. Rhyolite is fine-grained and frequently contains glass fragments and voids, indicating rapid cooling in a surface environment. In contrast to granite, which is widely distributed as large plutonic masses, rhyolite deposits are less common and generally less voluminous. Yellowstone Park is one well-known exception. Here rhyolite lava flows and thick ash deposits of similar composition are extensive.

Basaltic (Mafic) Rocks Rocks that contain substantial amounts of dark-colored silicate minerals (mainly pyroxene), and calcium-rich plagioclase feldspar are said to have a **basaltic composition** (Figure 2.9). Because basaltic rocks contain a high percentage of dark silicate minerals, geologists also refer to them as **mafic** (from *ma*gnesium and *fe*rrum, the Latin name for iron). Because of their iron content, basaltic rocks are typically darker and denser than granitic rocks.

Basalt is a very dark green to black fine-grained volcanic rock composed primarily of pyroxene, olivine, and plagioclase feldspar. Basalt is the most common extrusive igneous rock. Many volcanic islands, such as the Hawaiian Islands and Iceland, are composed mainly of basalt. Further, the upper layers of the oceanic crust consist of basalt. In the United States, large portions of central Oregon and Washington were the sites of extensive basaltic outpourings.

The coarse-grained, intrusive equivalent of basalt is called *gabbro* (Figure 2.9). Although gabbro is not commonly exposed on the surface, it makes up a significant percentage of the oceanic crust.

Andesitic (Intermediate) Rocks As you can see in Figure 2.9, rocks with a composition between granitic and basaltic rocks are

Chemical Composition		Granitic (Felsic)	Andesitic (Intermediate)	Basaltic (Mafic)	Ultramafic
Dominant Minerals		Quartz Potassium feldspar Sodium-rich plagioclase feldspar	Amphibole Sodium- and calcium-rich plagioclase feldspar	Pyroxene Calcium-rich plagioclase feldspar	Olivine Pyroxene
TEXTURE	Coarse-grained	Granite	Diorite	Gabbro	Peridotite
	Fine-grained	Rhyolite	Andesite	Basalt	Komatiite (rare)
	Porphyritic	"Porphyritic" precedes any of the above names whenever there are appreciable phenocrysts			Uncommon
	Glassy	**Obsidian** (compact glass) **Pumice** (frothy glass)			
Rock Color (based on % of dark minerals)		0% to 25%	25% to 45%	45% to 85%	85% to 100%

Figure 2.8 Classification of the major groups of igneous rocks based on their mineral composition and texture. Coarse-grained rocks are plutonic, solidifying deep underground. Fine-grained rocks are volcanic, or solidify as shallow, thin plutons. Ultramafic rocks are dark, dense rocks, composed almost entirely of minerals containing iron and magnesium. Although relatively rare on Earth's surface, these rocks are believed to be major constituents of the upper mantle.

Figure 2.9 Common igneous rocks. (Photos by E. J. Tarbuck)

said to have an **andesitic** or **intermediate composition** after the common volcanic rock *andesite*. Andesitic rocks contain a mixiture of both light- and dark-colored minerals, mainly amphibole and plagioclase feldspar. This important category of igneous rocks is associated with volcanic activity that is typically confined to the margins of continents. When magma of intermediate composition crystallizes at depth, it forms the coarse-grained rock called *diorite* (Figure 2.9).

Ultramafic Rocks Another important igneous rock, *peridotite,* contains mostly the dark-colored minerals olivine and pyroxene and thus falls on the opposite side of the compositional spectrum from granitic rocks (see Figure 2.8). Because peridotite is composed almost entirely of dark silicate minerals, its chemical composition is referred to as **ultramafic.** Although ultramafic rocks are rare at Earth's surface, peridotite is believed to be the main constituent of the upper mantle.

How Different Igneous Rocks Form

Because a large variety of igneous rocks exist, it is logical to assume that an equally large variety of magmas must also exist. However, geologists have observed that a single volcano may extrude lavas exhibiting quite different compositions. Data of this type led them to examine the possibility that magma might change (evolve) and thus become the parent to a variety of igneous rocks. To explore this idea, a pioneering investigation into the crystallization of magma was carried out by N. L. Bowen in the first quarter of the twentieth century.

Bowen's Reaction Series In a laboratory setting, Bowen demonstrated that unlike a pure compound, such as water, which solidifies at a specific temperature, magma with its diverse chemistry crystallizes over a temperature range of at least 200 degrees. Thus, as magma cools, certain minerals crystallize first, at relatively high temperatures (top of Figure 2.10). At successively lower temperatures, other minerals crystallize. This arrangement of minerals, shown in Figure 2.10 became known as **Bowen's reaction series.**

Bowen discovered that the first mineral to crystallize from a mass of magma is olivine. Further cooling results in the formation of pyroxene, as well as plagioclase feldspar. At intermediate temperatures the minerals amphibole and biotite begin to crystallize.

During the last stage of crystallization, after most of the magma has solidified, the minerals muscovite and potassium feldspar may form (Figure 2.10). Finally, quartz crystallizes from any remaining liquid. As a result, olivine is not usually found with quartz in the same igneous rock, because quartz crystallizes at much lower temperatures than olivine.

Evidence that this highly idealized crystallization model approximates what can happen in nature comes from the analysis of igneous rocks. In particular, we find that minerals that form in the same general temperature range on Bowen's reaction series are found together in the same igneous rocks. For example, notice in Figure 2.10 that the minerals quartz, potassium feldspar, and muscovite, which are located in the same region of Bowen's diagram, are typically found together as major constituents of the igneous rock *granite.*

Magmatic Differentiation Bowen demonstrated that different minerals crystallize at different temperatures. But how do Bowen's findings account for the great diversity of igneous rocks? During the crystallization process, the composition of the melt (the liquid portion of magma excluding the solid crystals) continually changes because it gradually becomes depleted in those elements used to make the earlier formed minerals. This process, coupled with the fact that at one or more stages during crystallization, a separation of the solid and liquid components of magma can occur creates different mineral assemblages. One way this happens is called **crystal settling.** This process occurs

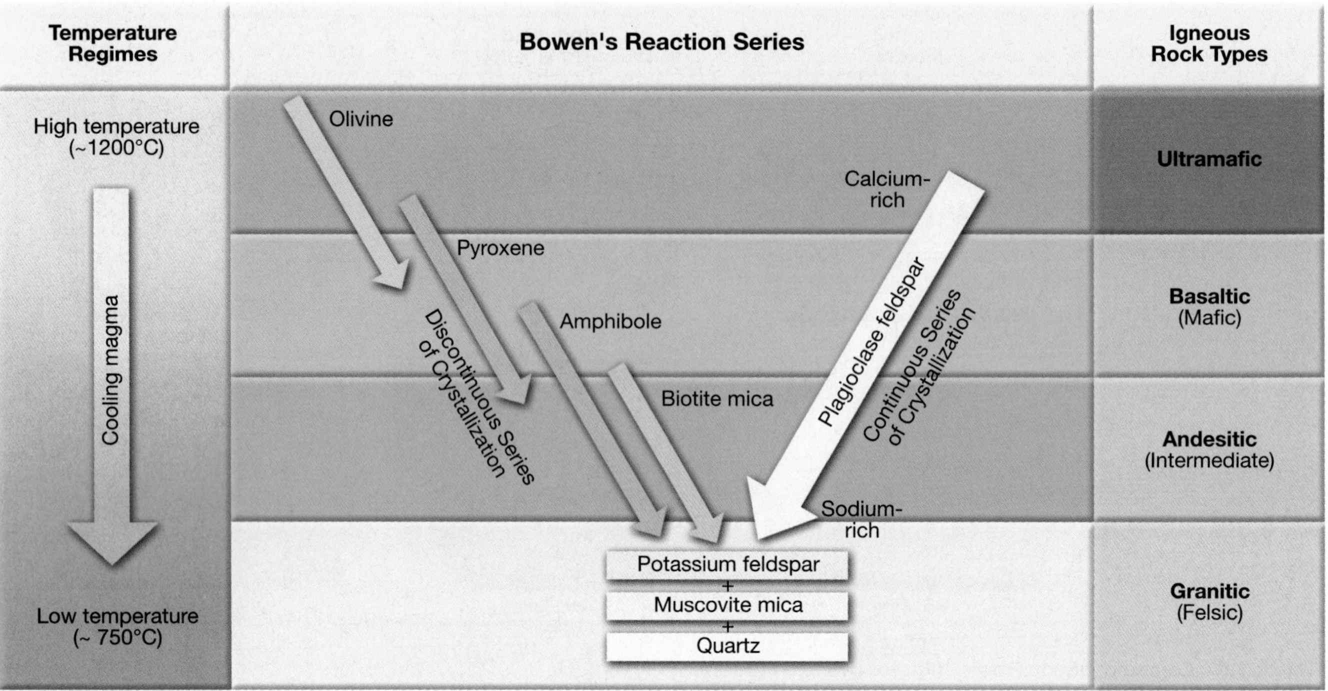

Figure 2.10 Bowen's reaction series shows the sequence in which minerals crystallize from a magma. Compare this figure to the mineral composition of the rock groups in Figure 2.8. Note that each rock group consists of minerals that crystallize at the same time.

when the earlier formed minerals are denser (heavier) than the liquid portion and sink toward the bottom of the magma chamber, as shown in Figure 2.11. When the remaining melt solidifies—either in place or in another location if it migrates into fractures in the surrounding rocks—it will form a rock with a chemical composition much different from the parent magma (Figure 2.11). The formation of one or more secondary magmas from a single parent magma is called **magmatic differentiation.**

At any stage in the evolution of a magma, the solid and liquid components can separate into two chemically distinct units. Further, magmatic differentiation within the secondary melt can generate additional chemically distinct fractions. Consequently, magmatic differentiation and separation of the solid and liquid components at various stages of crystallization can produce several chemically diverse magmas and ultimately a variety of igneous rocks.

Weathering of Rocks to Form Sediment

Earth Materials
▼ Sedimentary Rocks

All materials are susceptible to weathering. Consider, for example, the synthetic rock we call concrete. A newly poured concrete sidewalk is smooth, but many years later, the same sidewalk will appear chipped, cracked, and rough, with pebbles exposed at the surface. If a tree is nearby, its roots may grow under the sidewalk, heaving and buckling the concrete. The same natural processes that eventually break apart a concrete sidewalk also act to disintegrate natural rocks, regardless of their type or strength.

Why does rock weather? Simply, weathering is the natural response of Earth materials to a *new environment.* For instance, after millions of years of erosion, the rocks overlying a large body of intrusive igneous rock may be removed. This exposes the igneous rock to a whole new environment at the surface. This mass of crystalline rock, which formed deep below ground, where temperatures and pressures are high, is now subjected to very different and comparatively hostile surface conditions. In response, this rock mass will gradually change until it is once again in equilibrium, or balance, with its new environment. Such transformation of rock is what we call *weathering.*

In the following sections, we will discuss the two kinds of weathering—mechanical and chemical. Mechanical weathering is the physical breaking up of rocks. Chemical weathering actually alters what a rock is, changing it into a different substance. Although we will consider these two processes separately, keep in mind that they usually work simultaneously in nature. Furthermore, the activities of erosional agents—wind, water, and glaciers—that transport weathered rock particles are important. As these mobile agents move rock debris, they relentlessly disintegrate it further.

Mechanical Weathering of Rocks

When a rock undergoes **mechanical weathering,** it is broken into smaller and smaller pieces. Each piece retains the characteristics of the original material. The end result is many small pieces from a single large one. Figure 2.12 shows that breaking a rock into smaller pieces increases the surface area available for chemical attack. An example is adding sugar to water. A chunk of rock candy will dissolve much more slow-

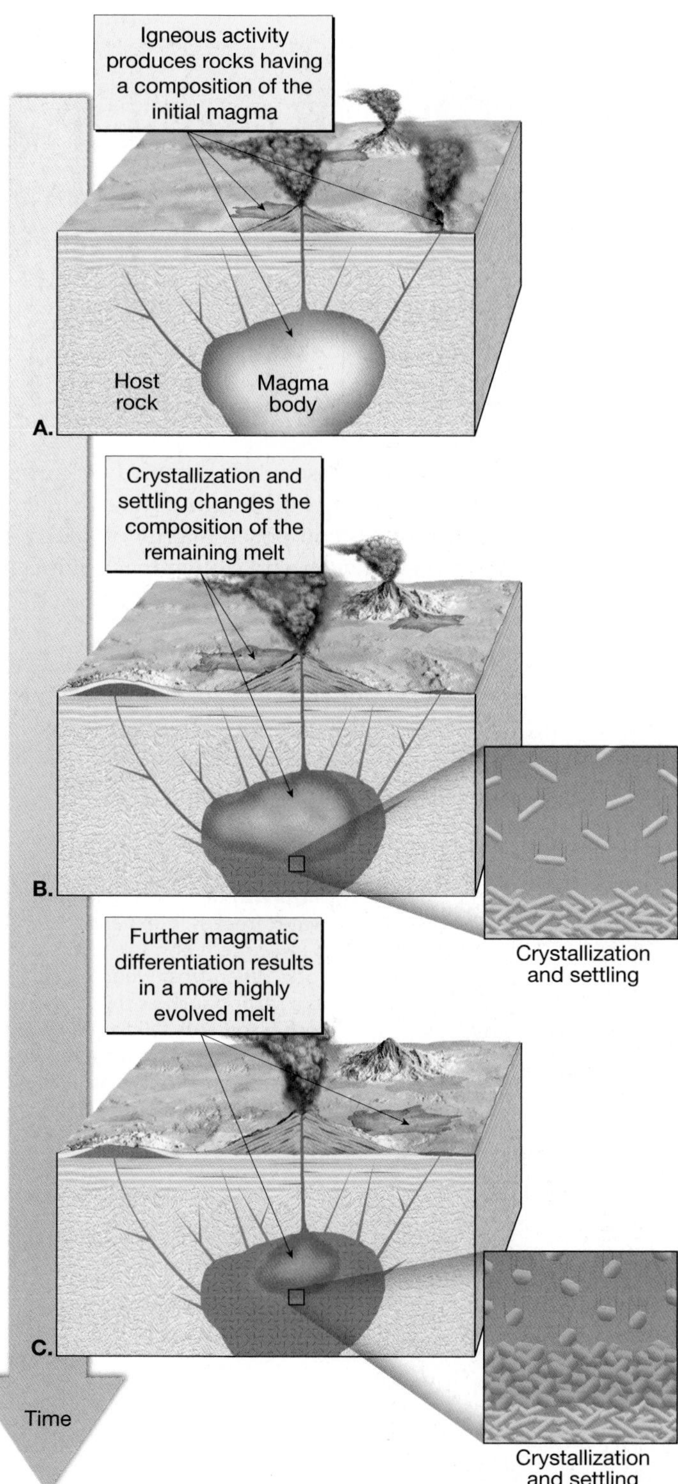

Figure 2.11 Illustration of how a magma evolves as the earlier formed minerals (those richer in iron, magnesium, and calcium) crystallize and settle to the bottom of the magma chamber, leaving the remaining melt richer in sodium, potassium, and silica (SiO_2). **A.** Emplacement of a magma body and associated igneous activity generates rocks having a composition similar to that of the initial magma. **B.** After a period of time, crystallization and settling change the composition of the melt, while generating rocks having a composition quite different from the original magma. **C.** Further magmatic differentiation results in another more highly evolved melt with its associated rock types.

ly than will an equal volume of sugar granules because of the vast difference in surface area. Hence, by breaking rocks into smaller pieces, mechanical weathering increases the amount of surface area available for chemical weathering.

In nature, three important physical processes break rocks into smaller fragments: frost wedging, expansion resulting from unloading, and biological activity.

Frost Wedging Alternate freezing and thawing of water is one of the most important processes of mechanical weathering. Water has the unique property of expanding about 9 percent when it freezes. This increase in volume occurs because, as ice forms, the water molecules arrange themselves into a very open crystalline structure. As a result, when water freezes, it expands and exerts a tremendous outward force. Here is everyday proof: Water in a car's cooling system will freeze in winter, expanding and cracking the engine block. This is why antifreeze is added; it lowers the temperature at which the solution freezes.

In nature, water works its way into every crack or void in rock and, upon freezing, expands and enlarges the opening. After many freeze-thaw cycles, the rock is broken into pieces. This process is appropriately called *frost wedging* (Figure 2.13). Frost wedging is most pronounced in mountainous regions in the middle latitudes where a daily freeze-thaw cycle often exists. Here, sections of rock are wedged loose and may tumble into large piles called *talus* or *talus slopes* that often form at the base of steep rock outcrops (Figure 2.13).

Unloading When large masses of igneous rock are exposed by erosion, entire slabs begin to break loose, like the layers of an onion. This *sheeting* is thought to occur because of the great reduction in pressure when the overlying rock is eroded away. Accompanying the unloading, the outer layers expand more than the rock below and thus separate from the rock body. Granite is particularly prone to sheeting.

Continued weathering eventually causes the slabs to separate and spall, causing *exfoliation domes.* Excellent examples of exfoliation domes include Stone Mountain, Georgia, and Liberty Cap Half Dome in Yosemite National Park (Figure 2.14).

Biological Activity Weathering is also accomplished by the activities of organisms, including plants, burrowing animals, and humans. Plant roots in search of water grow into fractures, and as the roots grow, they wedge the rock apart (Figure 2.15). Burrowing animals further break down the rock by moving fresh material to the surface, where physical and chemical processes can more effectively attack it.

Chemical Weathering of Rocks

Chemical weathering alters the internal structure of minerals by removing and/or adding elements. During this transformation, the original rock is altered into substances that are stable in the surface environment.

Water is the most important agent of chemical weathering. Oxygen dissolved in water will *oxidize* some materials. For example, when an iron nail is found in the soil, it will have a coating of rust (iron oxide), and if the time of exposure

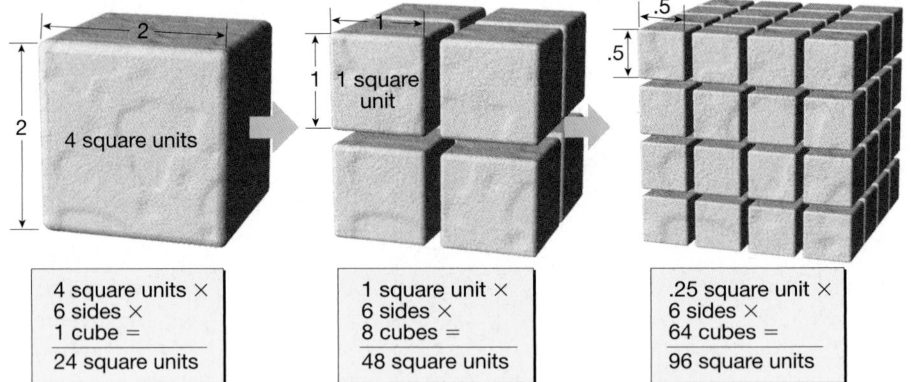

Figure 2.12 Chemical weathering can occur only to those portions of a rock that are exposed to the elements. Mechanical weathering breaks rock into smaller and smaller pieces, thereby increasing the surface area available for chemical attack.

has been long, the nail will be so weak that it can be broken as easily as a toothpick. When rocks containing iron-rich minerals (such as hornblende) oxidize, a yellow to reddish-brown rust will appear on the surface.

Carbon dioxide (CO_2) dissolved in water (H_2O) forms carbonic acid (H_2CO_3). This is the same weak acid produced when soft drinks are carbonated. Rain dissolves some carbon dioxide as it falls through the atmosphere, so normal rainwater is mildly acidic. Water in the soil also dissolves carbon dioxide released by decaying organic matter. The result is that acidic water is everywhere on Earth's surface.

How does rock decompose when attacked by carbonic acid? Consider the weathering of the common igneous rock, granite. Recall that granite is composed mainly of quartz and potassium feldspar. As the weak acid slowly reacts with crystals of potassium feldspar, potassium ions are displaced. *This destroys the mineral's crystalline structure.*

The most abundant products of the chemical breakdown of feldspar are clay minerals. Because clay minerals are the end product of chemical weathering, they are very stable under surface conditions. Consequently, clay minerals make up a high percentage of the inorganic material in soils.

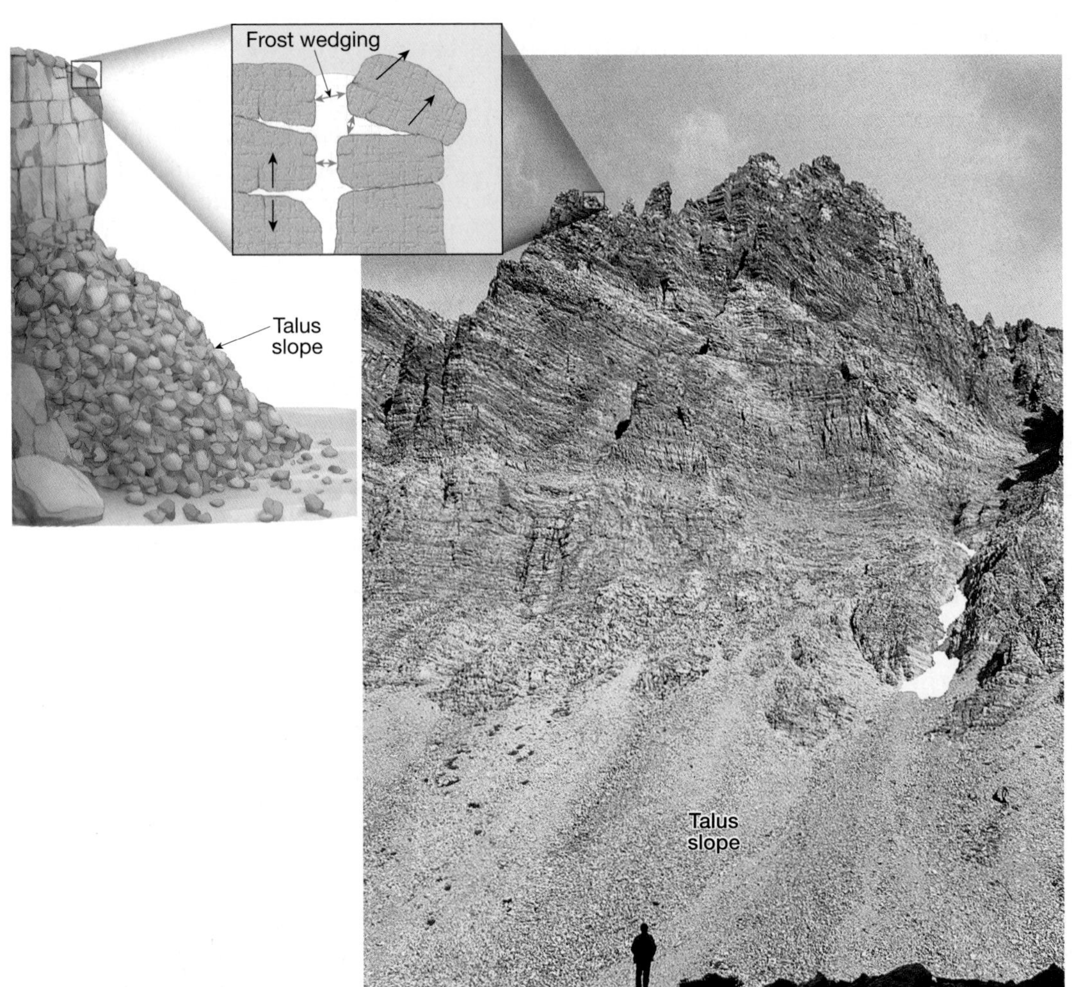

Figure 2.13 Frost wedging. As water freezes, it expands, exerting a force great enough to break rock. When frost wedging occurs in a setting such as this, the broken rock fragments fall to the base of the cliff and create a cone-shaped accumulation known as talus. (Photo by Tom & Susan Bean, Inc.)

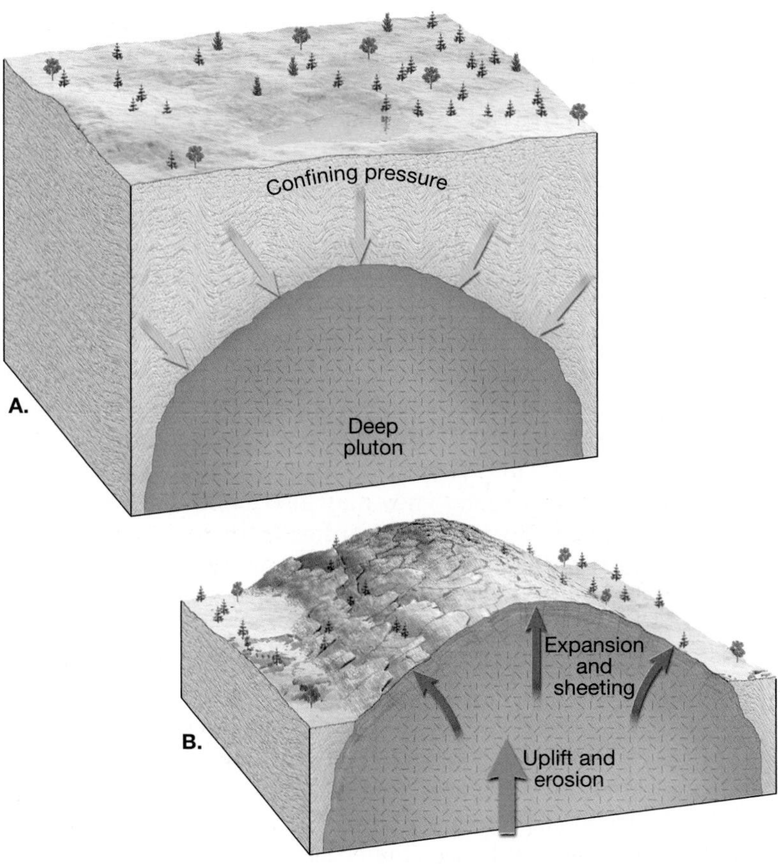

Figure 2.14 Sheeting is caused by the expansion of crystalline rock as erosion removes the overlying material. When the deeply buried pluton (**A**) is exposed at the surface following uplift and erosion (**B**), the igneous mass fractures into thin slabs. The photo (**C**) is of the summit of Half Dome in Yosemite National Park, California. It is an exfoliation dome and illustrates the onionlike layers created by sheeting. (Photo by Breck P. Kent)

Figure 2.15 Root wedging widens fractures in rocks and aids the process of mechanical weathering. (Photo by Tom Bean/DRK Photo)

In addition to the formation of clay minerals, some silica (SiO_2) is dissolved from the feldspar structure and is carried away by groundwater. The dissolved silica will eventually precipitate to produce a hard, dense sedimentary rock (chert), fill pore spaces between mineral grains, or be carried to the ocean, where microscopic animals will build silica shells from it.

Quartz, the other main component of granite, is very resistant to chemical weathering. Because it is durable, quartz remains substantially unaltered when attacked by weak acid. As granite weathers, the feldspar crystals become dull and slowly turn to clay, releasing the once interlocked quartz grains, which still retain their fresh, glassy appearance. Although some quartz remains in the soil, much is transported to the sea and other sites, where it becomes sandy beaches and sand dunes.

To summarize, *the chemical weathering of granite produces clay minerals along with potassium ions and silica, which enters into solution.* In addition, durable quartz grains are freed.

Table 2.1 lists the weathered products of some of the most common silicate minerals. Remember that silicate minerals make up most of Earth's crust and are composed primarily of just eight elements (see Figure 1.14, p. 26). When chemically weathered, the silicate minerals yield sodium, calcium, potassium, and magnesium ions. These may be used by plants or removed by groundwater. The element iron combines with oxygen to produce iron-oxide compounds that give soil a red-

Table 2.1 Products of weathering

Original Mineral	Weathers to Produce	Released into Solution
Quartz	Quartz grains	Silica (SiO_2)
Feldspar	Clay minerals	Silica (SiO_2)
		Ions of potassium, sodium, and calcium
Hornblende	Clay minerals	Silica (SiO_2)
	Iron minerals (limonite and hematite)	Ions of calcium and magnesium
Olivine	Iron minerals (limonite and hematite)	Silica (SiO_2)
		Ions of magnesium

dish-brown or yellowish color. The three remaining elements—aluminum, silicon, and oxygen—join with water to produce clay minerals that become an important part of the soil. Ultimately, the products of weathering form the raw materials for building sedimentary rocks, which we consider next.

Sedimentary Rocks: Compacted and Cemented Sediment

Earth Materials
▼ Sedimentary Rocks

Recall the rock cycle, which shows the origin of sedimentary rocks. Weathering begins the process. Next, gravity and erosional agents (running water, wind, waves, and glacial ice) remove the products of weathering and carry them to a new location where they are deposited. Usually, the particles are broken down further during this transport phase. Following deposition, this *sediment* may become lithified, or "turned to rock." Commonly, *compaction* and *cementation* transform the sediment into solid sedimentary rock.

The word *sedimentary* indicates the nature of these rocks, for it is derived from the Latin *sedimentum,* which means "settling," a reference to a solid material settling out of a fluid. Most sediment is deposited in this fashion. Weathered debris is constantly being swept from bedrock and carried away by water, ice, or wind. Eventually, the material is deposited in lakes, river valleys, seas, and countless other places. The particles in a desert sand dune, the mud on the floor of a swamp, the gravels in a streambed, and even household dust are examples of sediment produced by this never-ending process.

The weathering of bedrock and the transport and deposition of the weathering products are continuous. Therefore, sediment is found almost everywhere. As piles of sediment accumulate, the materials near the bottom are compacted by the weight of the overlying layers. Over long periods, these sediments are cemented together by mineral matter deposited from water in the spaces between particles. This forms solid sedimentary rock.

Geologists estimate that sedimentary rocks account for only about 5 percent (by volume) of Earth's outer 16 kilometers (10 miles). However, the importance of this group of rocks is far greater than this percentage implies. If you sampled the rocks exposed at Earth's surface, you would find that the great majority are sedimentary (Figure 2.16). Indeed, about 75 percent of all rock outcrops on the continents are sedimentary. Therefore, we can think of sedimentary rocks as comprising a relatively thin and somewhat discontinuous layer in the uppermost portion of the crust. This makes sense because sediment accumulates at the surface.

It is from sedimentary rocks that geologists reconstruct many details of Earth's history. Because sediments are deposited in a variety of different settings at the surface, the rock layers that they eventually form hold many clues to past surface environments. They may also exhibit characteristics that allow geologists to decipher information about the method and distance of sediment transport. Furthermore, it is sedimentary rocks that contain fossils, which are vital evidence in the study of the geologic past.

Finally, many sedimentary rocks are important economically. Coal, which is burned to provide a significant portion of U.S. electrical energy, is classified as a sedimentary rock. Other major energy resources (petroleum and natural gas) occur in pores within sedimentary rocks. Other sedimentary rocks are major sources of iron, aluminum, manganese, and fertilizer, plus numerous materials essential to the construction industry.

Classifying Sedimentary Rocks

Materials accumulating as sediment have two principal sources. First, sediments may originate as solid particles from weathered rocks, such as the igneous rocks earlier described. These particles are called *detritus,* and the sedimentary rocks that they form are called **detrital sedimentary rocks** (Figure 2.17).

The second major source of sediment is soluble material produced largely by chemical weathering. When these dissolved substances are precipitated back as solids, they are called *chemical sediment,* and they form **chemical sedimentary rocks.** We will now look at detrital and chemical sedimentary rocks. (Figure 2.17)

Detrital Sedimentary Rocks Though a wide variety of minerals and rock fragments may be found in detrital rocks, clay minerals and quartz dominate. As you learned earlier, clay minerals are the most abundant product of the chemical weathering of silicate minerals, especially the feldspars. Quartz, on the other hand, is abundant because it is extremely durable and very resistant to chemical weathering. Thus, when igneous rocks such as granite are weathered, individual quartz grains are set free.

Geologists use particle size to distinguish among detrital sedimentary rocks. Figure 2.17 presents the four size categories for particles making up detrital rocks. When gravel-size particles predominate, the rock is called *conglomerate* if the sediment is rounded (Figure 2.18A) and *breccia* if the pieces

Figure 2.16 Sedimentary rocks exposed in Canyonlands National Park, Utah. Sedimentary rocks occur in layers called strata. About 75 percent of all rock outcrops on the continents are sedimentary rocks. (Photo by Jeff Gnass)

are angular (Figure 2.18B). Angular fragments indicate that the particles were not transported very far from their source prior to deposition and so have not had corners and rough edges abraded. *Sandstone* is the name given rocks when sand-size grains prevail (Figure 2.18C). *Shale,* the most common sedimentary rock, is made of very fine-grained sediment (Figure 2.18D). *Siltstone,* another rather fine-grained rock, is sometimes difficult to differentiate from rocks such as shale, which are composed of even smaller clay-size sediment.

Particle size is not only a convenient method of dividing detrital rocks; the sizes of the component grains also provide useful information about the environment in which the sediment was deposited. Currents of water or air sort the particles by size. The stronger the current, the larger the particle size carried. Gravels, for example, are moved by swiftly flowing rivers, rockslides, and glaciers. Less energy is required to transport sand; thus, it is common in windblown dunes, river deposits, and beaches. Because silts and clays settle very slowly, accumulations of these materials are generally associated with the quiet waters of a lake, lagoon, swamp, or marine environment.

Although detrital sedimentary rocks are classified by particle size, in certain cases the mineral composition is also part of naming a rock. For example, most sandstones are predominantly quartz-rich, and they are often referred to as quartz sandstone. In addition, rocks consisting of detrital sediments are rarely composed of grains of just one size. Consequently, a rock containing quantities of both sand and silt can be correctly classified as sandy siltstone or silty sandstone, depending on which particle size dominates.

Did You Know?

The most important and common material used for making glass is silica, which is usually obtained from the quartz in "clean," well-sorted sandstones.

Chemical Sedimentary Rocks In contrast to detrital rocks, which form from the solid products of weathering, chemical sediments are derived from material that is carried in solution to lakes and seas. This material does not remain dissolved in the water indefinitely. When conditions are right, it precipitates to form chemical sediments. This precipitation may occur directly as the result of physical processes, or indirectly through life processes of water-dwelling organisms. Sediment formed in this second way has a *biochemical* origin.

An example of a deposit resulting from physical processes is the salt left behind as a body of saltwater evaporates. In contrast, many water-dwelling animals and plants extract dissolved mineral matter to form shells and other hard

Detrital Sedimentary Rocks

Clastic Texture Particle Size		Sediment Name	Rock Name
Coarse (over 2 mm)		Gravel (Rounded particles)	**Conglomerate**
		Gravel (Angular particles)	**Breccia**
Medium (1/16 to 2 mm)		Sand (If abundant feldspar is present the rock is called **Arkose**)	**Sandstone**
Fine (1/16 to 1/256 mm)		Mud	**Siltstone**
Very fine (less than 1/256 mm)		Mud	**Shale**

Chemical Sedimentary Rocks

Composition	Texture	Rock Name	
Calcite, $CaCO_3$	Nonclastic: Fine to coarse crystalline	**Crystalline Limestone**	
		Travertine	
	Clastic: Visible shells and shell fragments loosely cemented	**Coquina**	**Biochemical Limestone**
	Clastic: Various size shells and shell fragments cemented with calcite cement	**Fossiliferous Limestone**	
	Clastic: Microscopic shells and clay	**Chalk**	
Quartz, SiO_2	Nonclastic: Very fine crystalline	**Chert (light colored) Flint (dark colored)**	
Gypsum $CaSO_4 \cdot 2H_2O$	Nonclastic: Fine to coarse crystalline	**Rock Gypsum**	
Halite, NaCl	Nonclastic: Fine to coarse crystalline	**Rock Salt**	
Altered plant fragments	Nonclastic: Fine-grained organic matter	**Bituminous Coal**	

Figure 2.17 Identification of sedimentary rocks. Sedimentary rocks are divided into two major groups, detrital and chemical, based on their source of sediment. The main criterion for naming detrital rocks is particle size, whereas the primary basis for distinguishing among chemical rocks is their mineral composition.

parts. After the organisms die, their skeletons may accumulate on the floor of a lake or ocean.

Limestone is the most abundant chemical sedimentary rock. It is composed chiefly of the mineral calcite ($CaCO_3$). Ninety percent of limestone is biochemical sediment. The rest precipitates directly from seawater.

One easily identified biochemical limestone is *coquina*, a coarse rock composed of loosely cemented shells and shell fragments (Figure 2.19). Another less obvious but familiar example is *chalk*, a soft, porous rock made up almost entirely of the hard parts of microscopic organisms that are no larger than the head of a pin (Figure 2.20).

Inorganic limestones form when chemical changes or high water temperatures increase the concentration of calcium carbonate to the point that it precipitates. *Travertine*, the type of limestone that decorates caverns, is one example. Groundwater is the source of travertine that is deposited in caves. As water drops reach the air in a cavern, some of the carbon dioxide dissolved in the water escapes, causing calcium carbonate to precipitate.

Dissolved silica (SiO_2) precipitates to form varieties of microcrystalline quartz (Figure 2.21). Sedimentary rocks composed of microcrystalline quartz include chert (light color), flint (dark), jasper (red), and agate (banded). These chemical sedimentary rocks may have either an inorganic or biochemical origin, but the mode of origin is usually difficult to determine.

Very often, evaporation causes minerals to precipitate from water. Such minerals include halite, the chief component of *rock salt,* and gypsum, the main ingredient of *rock gypsum.* Both materials have significant commercial importance. Halite is familiar to everyone as the common salt used in cooking and seasoning foods. Of course, it has many other uses and has been considered important enough that people have sought, traded, and fought over it for much of human history. Gypsum is the basic ingredient of plaster of Paris. This material is used most extensively in the construction industry for "drywall" and plaster.

Did You Know?

Each year, about 30 percent of the world's supply of salt is extracted from seawater. The seawater is pumped into ponds and allowed to evaporate, leaving behind "artificial evaporites," which are harvested.

A.

B.

C.

D.

Figure 2.18 Common detrital sedimentary rocks. **A.** Conglomerate (rounded particles). **B.** Breccia (angular particles). **C.** Sandstone. **D.** Shale with plant fossil. (Photos by E. J. Tarbuck)

Figure 2.19 This rock, called *coquina*, consists of shell fragments; therefore, it has a biochemical origin. (Photos by E. J. Tarbuck)

In the geologic past, many areas that are now dry land were covered by shallow arms of the sea that had only narrow connections to the open ocean. Under these conditions, water continually moved into the bay to replace water lost by evaporation. Eventually, the waters of the bay became saturated and salt deposition began. Today, these arms of the sea are gone, and the remaining deposits are called **evaporite deposits.**

On a smaller scale, evaporite deposits can be seen in such places as Death Valley, California. Here, following rains or periods of snowmelt in the mountains, streams flow from surrounding mountains into an enclosed basin. As the water evaporates, *salt flats* form from dissolved materials left behind as a white crust on the ground (Figure 2.22).

Coal is quite different from other chemical sedimentary rocks. Unlike other rocks in this category, which are calcite- or silica-rich, coal is made mostly of organic matter. Close examination of a piece of coal under a microscope or magnifying glass often reveals plant structures such as leaves, bark, and wood that have been chemically altered but are still identifiable. This supports the conclusion that coal is the end product of the burial of large amounts of plant material over extended periods (Figure 2.23).

The initial stage in coal formation is the accumulation of large quantities of plant remains. However, special conditions

Figure 2.20 White Chalk Cliffs, East Sussex, England. (Photo by Art Wolfe)

are required for such accumulations, because dead plants normally decompose when exposed to the atmosphere. An ideal environment that allows for the buildup of plant material is a swamp. Because stagnant swamp water is oxygen-deficient, complete decay (oxidation) of the plant material is not possible. At various times during Earth history, such environments have been common. Coal undergoes successive stages of formation. With each successive stage, higher temperatures and pressures drive off impurities and volatiles, as shown in Figure 2.23.

Lignite and bituminous coals are sedimentary rocks, but anthracite is a metamorphic rock. Anthracite forms when sedimentary layers are subjected to the folding and deformation associated with mountain building.

In summary, we divide sedimentary rocks into two major groups: detrital and chemical. The main criterion for classifying detrital rocks is particle size, whereas chemical rocks are distinguished by their mineral composition. The categories presented here are more rigid than is the actual state of nature. Many detrital sedimentary rocks are a mixture of more than one particle size. Furthermore, many sedimentary rocks classified as chemical also contain at least small quantities of detrital sediment, and practically all detrital rocks are cemented with material that was originally dissolved in water.

Lithification of Sediment

Lithification refers to the processes by which sediments are transformed into solid sedimentary rocks. One of the most common processes is *compaction.* As sediments accumulate through time, the weight of overlying material compresses the deeper sediments. As the grains are pressed closer and closer, pore space is greatly reduced. For example, when clays are buried beneath several thousand meters of material, the volume of the clay may be reduced as much as 40 percent. Compaction is most significant in fine-grained sedimentary rocks such as shale, because sand and other coarse sediments compress little.

Cementation is another important means by which sediments are converted to sedimentary rock. The cementing materials are carried in solution by water percolating through the pore spaces between particles. Over time, the cement precipitates onto the sediment grains, fills the open spaces, and joins the particles. Calcite, silica, and iron oxide are the most common cements. Identification of the cementing material is simple. Calcite cement will effervesce (fizz) with dilute hydrochloric acid. Silica is the hardest cement and thus produces the hardest sedimentary rocks. When a sedimentary rock has an orange or red color, this usually means iron oxide is present.

Features of Sedimentary Rocks

Sedimentary rocks are particularly important evidence of Earth's long history. These rocks form at Earth's surface, and as layer upon layer of sediment accumulates, each records the nature of the environment at the time the sediment was deposited. These layers, called **strata,** or **beds,** are the *single most characteristic feature of sedimentary rocks* (see Figure 2.16).

The thickness of beds ranges from microscopically thin to tens of meters thick. Separating the strata are *bedding planes,* flat surfaces along which rocks tend to separate or break. Generally, each bedding plane marks the end of one episode of sedimentation and the beginning of another.

Sedimentary rocks provide geologists with evidence for deciphering past environments. A conglomerate, for example, indicates a high-energy environment, such as a rushing stream, where only the coarse materials can settle out. By contrast, black shale and coal are associated with a low-energy,

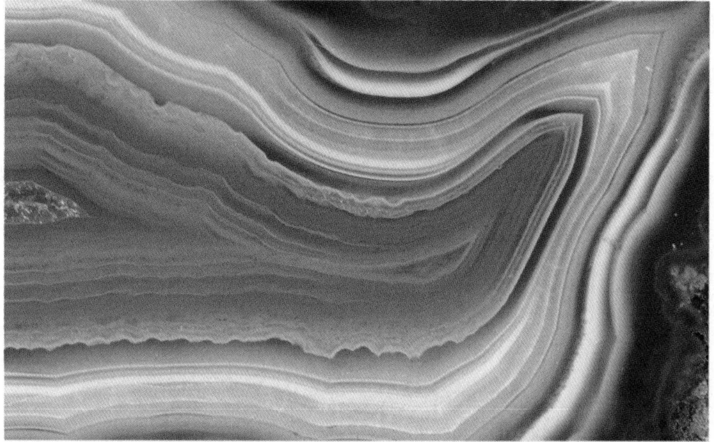

A. Agate

B. Flint

C. Jasper

D. Chert arrowhead

Figure 2.21 *Chert* is a name used for a number of dense, hard rocks made of microcrystalline quartz. Three examples are shown here. **A.** *Agate* is the banded variety. (Photo by Jeffrey A. Scovil) **B.** The dark color of *flint* results from organic matter. (Photo by E. J. Tarbuck) **C.** The red variety, called *jasper,* gets its color from iron oxide. (Photo by E. J. Tarbuck) **D.** Native Americans frequently made arrowheads and sharp tools from chert. (Photo by LA VENTA/CORBIS/SYGMA)

organic-rich environment, such as a swamp or lagoon. Other features found in some sedimentary rocks also give clues to past environments (Figure 2.24).

Fossils, the traces or remains of prehistoric life, are perhaps the most important inclusions found in some sedimentary rock. Knowing the nature of the life forms that existed at a particular time may help answer many questions about the environment. Was it land or ocean, lake or swamp? Was the climate hot or cold, rainy or dry? Was the ocean water shallow or deep, turbid or clear? Furthermore, fossils are important time indicators and play a key role in matching up rocks from different places that are the same age. Fossils are important tools used in interpreting the geologic past and will be examined in some detail in Chapter 8.

Metamorphic Rocks: New Rock from Old

Earth Materials

▼ Metamorphhic Rockss

Recall from the discussion of the rock cycle that metamorphism is the transformation of one rock type into another. Metamorphic rocks are produced from preexisting igneous, sedimentary, or even other metamorphic rocks. Thus, every metamorphic rock has a *parent rock*—the rock from which it was formed.

Metamorphism, which means to "change form," is a process that leads to changes in the mineralogy, texture (for example, grain size), and often the chemical composition of rocks. Metamorphism takes place when preexisting rock is subjected to a physical or chemical environment that is significantly different from that in which it initially formed. In response to changes in temperature and pressure (stress) and to the introduction of chemically active fluids, the rock gradually changes until a state of equilibrium with the new environment is reached. Most metamorphic changes occur at the elevated temperatures and pressures that exist in the zone beginning a few kilometers below Earth's surface and extending into the upper mantle.

Metamorphism often progresses incrementally, from slight changes *(low-grade metamorphism)* to substantial changes *(high-grade metamorphism).* For example, under low-grade metamorphism, the common sedimentary rock *shale* becomes the more compact metamorphic rock called *slate.* Hand samples of these rocks are sometimes difficult to distinguish, illustrating that the transition from sedimentary to metamorphic rock is often gradual and the changes subtle.

In more extreme environments, metamorphism causes a transformation so complete that the identity of the parent rock cannot be determined. In high-grade metamorphism, such features as bedding planes, fossils, and vesicles that may have existed in the parent rock are obliterated. Further, when rocks at depth (where temperatures are high) are subjected to directed pressure, they slowly deform to produce a variety of textures as well as large-scale structures such as folds (Figure 2.25). In the most extreme metamorphic environments, the temperatures approach those at which rocks melt. However, *during*

Figure 2.22 Bonneville Salt Flats, Utah. (Photo by Tom & Susan Bean, Inc.)

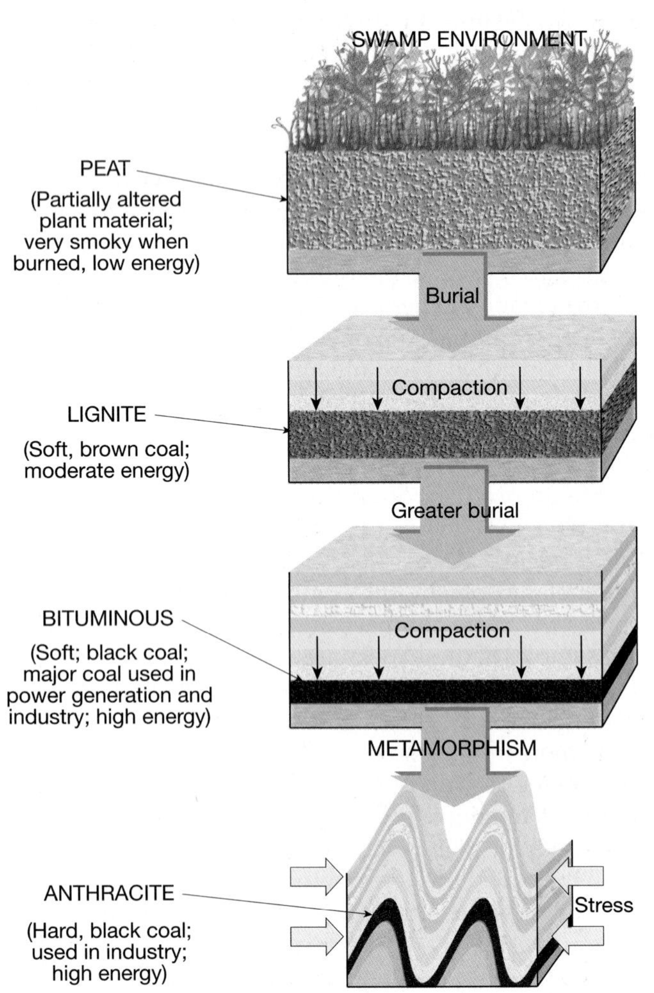

Figure 2.23 Successive stages in the formation of coal.

metamorphism the rock must remain essentially solid, for if complete melting occurs, we have entered the realm of igneous activity.

Most metamorphism occurs in one of two settings:

1. When rock is intruded by a magma body, **contact** or **thermal metamorphism** may take place. Here, change is driven by a rise in temperature within the host rock surrounding a molten igneous body.
2. During mountain building, great quantities of rock are subjected to directed pressures and high temperatures associated with large-scale deformation called **regional metamorphism.**

Extensive areas of metamorphic rocks are exposed on every continent. Metamorphic rocks are an important component of many mountain belts, where they make up a large portion of a mountain's crystalline core. Even the stable continental interiors, which are generally covered by sedimentary rocks, are underlain by metamorphic basement rocks. In all of these settings, the metamorphic rocks are usually highly deformed and intruded by igneous masses. Indeed, significant parts of Earth's continental crust are composed of metamorphic and associated igneous rocks.

Did You Know?

Some low-grade metamorphic rocks actually contain fossils. When fossils are present in metamorphic rocks, they provide useful clues for determining the original rock type and its depositional environment. In addition, fossils whose shapes have been distorted during metamorphism provide insight into the extent to which the rock has been deformed.

A B

Figure 2.24 **A.** Ripple marks preserved in sedimentary rocks may indicate a beach or stream channel environment. (Photo by Stephen Trimble) **B.** Mud cracks form when wet mud or clay dries and shrinks, perhaps signifying a tidal flat or desert basin. (Photo by Gary Yeowell/Getty Images Inc.—Stone Allstock)

Agents of Metamorphism

The agents of metamorphism include *heat, pressure (stress),* and *chemically active fluids.* During metamorphism, rocks are usually subjected to all three metamorphic agents simultaneously. However, the degree of metamorphism and the contribution of each agent vary greatly from one environment to another.

Heat as a Metamorphic Agent The most important agent of metamorphism is *heat* because it provides the energy to drive chemical reactions that result in the recrystallization of existing minerals and/or the formation of new minerals. The heat to metamorphose rocks comes mainly from two sources. First, rocks experience a rise in temperature when they are intruded by magma rising from below. This is called *contact or thermal metamorphism.* Here, the adjacent host rock is "baked" by the emplaced magma.

Second, rocks that formed at Earth's surface will experience a gradual increase in temperature as they are transported to greater depths. In the upper crust, this increase in temperature averages between 20°C and 30°C per kilometer. When buried to a depth of about 8 kilometers (5 miles), where temperatures are between 150°C and 200°C, clay minerals tend to become unstable and begin to recrystallize into other minerals, such as chlorite and muscovite, that are stable in this environment. (Chlorite is a micalike mineral formed by the metamorphism of iron- and magnesian-rich silicates.)

Figure 2.25 Folded and metamorphosed rocks in Anza Borrego Desert State Park, California. (Photo by A. P. Trujillo/APT Photos)

Did You Know?

The temperature of Earth's crust increases with depth, an idea that can be expressed as *the deeper one goes, the hotter it gets.* This causes considerable problems for underground mining efforts. In the Western Deep Levels mine in South Africa, which is 4 kilomenters (2.5 miles) deep, the temperature of the rock is hot enough to scorch human skin. Here, the miners work in groups of two: one to mine the rock, and the other to operate the fan that keeps them cool.

However, many silicate minerals, particularly those found in crystalline igneous rocks—quartz, for example—remain stable at these temperatures. Thus, metamorphic changes in these minerals occur at much higher temperatures.

Pressure (Stress) as a Metamorphic Agent Pressure, like temperature, also increases with depth as the thickness of the overlying rock increases. Buried rocks are subjected to *confining pressure,* which is analogous to water pressure, where the forces are applied equally in all directions (Figure 2.26A). The deeper you go in the ocean, the greater the confining pressure. The same is true for rock that is buried. Confining pressure causes the spaces between mineral grains to close, producing a more compact rock having a greater density. Further, at great depths, confining pressure may cause minerals to recrystallize into new minerals that display a more compact crystalline form.

During episodes of mountain building, large rock bodies become highly crumpled and metamorphosed (Figure 2.26B). The forces that generate mountains are unequal in different directions and are called *differential stress.* Unlike confining pressure, which "squeezes" the rock equally from all directions, differential stresses are greater in one direction than in others. As shown in Figure 2.26B, rocks subjected to differential stress are shortened in the direction of greatest stress, and elongated, or lengthened, in the direction perpendicular to that stress. The deformation caused by differential stresses plays a major role in developing metamorphic textures.

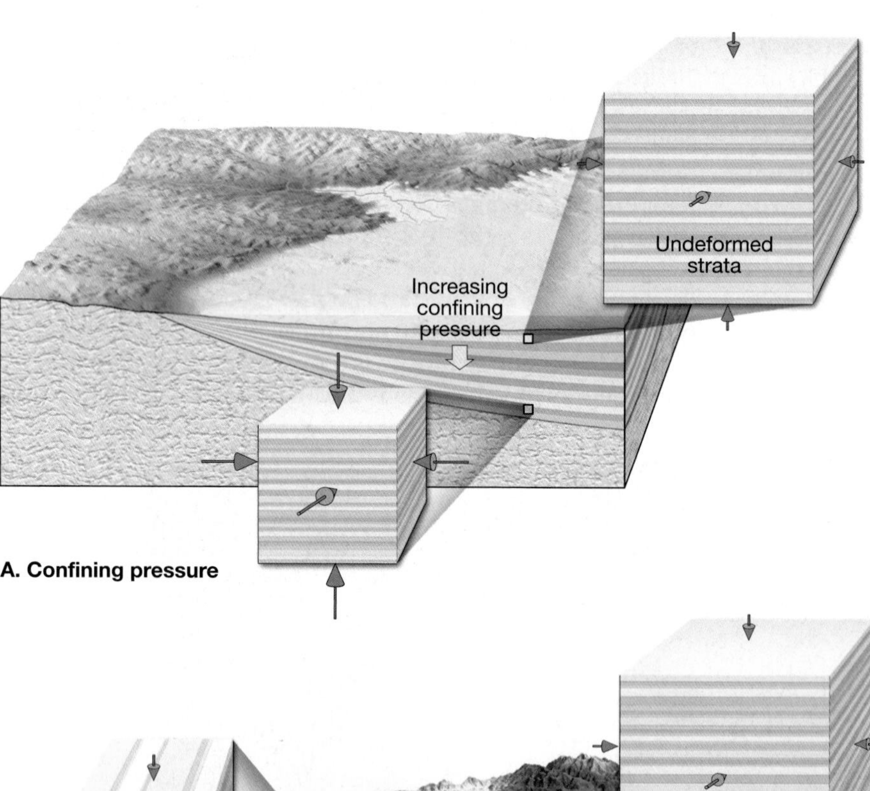

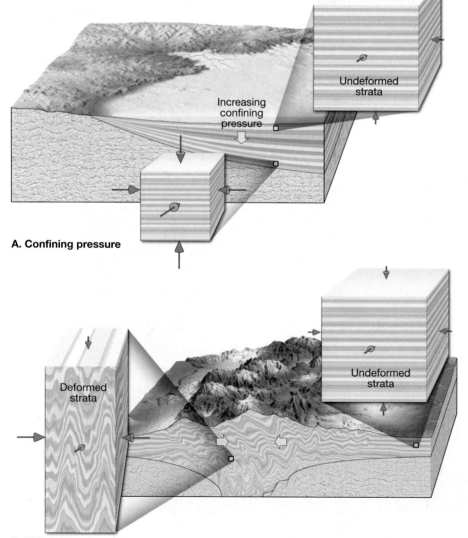

Figure 2.26 Pressure (stress) as a metamorphic agent. **A.** In a depositional environment, as confining pressure increases, rocks deform by decreasing in volume. **B.** During mountain building, rocks subjected to differential stress are shortened in the direction that pressure is applied, and lengthened in the direction perpendicular to that force.

In surface environments where temperatures are comparatively low, rocks are *brittle* and tend to fracture when subjected to differential stress. Continued deformation grinds and pulverizes the mineral grains into small fragments. By contrast, in high-temperature environments, rocks are *ductile*. When rocks exhibit ductile behavior, their mineral grains tend to flatten and elongate when subjected to differential stress. This accounts for their ability to deform by flowing (rather than fracturing) to generate intricate folds (Figure 2.27).

Chemically Active Fluids Fluids composed mainly of water and other volatiles (materials that readily change to a gas at surface conditions), including carbon dioxide, are believed to play an important role in some types of metamorphism. Fluids that surround mineral grains act as catalysts to promote recrystallization by enhancing ion migration. In progressively hotter environments, these ion-rich fluids become correspondingly more reactive.

When two mineral grains are squeezed together, the parts of their crystalline structures that touch are the most highly stressed. Ions located at these sites are readily dissolved by the hot fluids and migrate along the surface of the grain to the spaces located between individual grains. Thus, hot fluids aid in the recrystallization of mineral grains by dissolving material from regions of high stress and then precipitating (depositing) this material in areas of low stress. As a result, *minerals tend to recrystallize and grow longer in a direction perpendicular to compressional stresses.*

When hot fluids circulate freely through rocks, ionic exchange may occur between two adjacent rock layers, or ions may migrate great distances before they are finally deposited. The latter situation is particularly common when we consider hot fluids that escape during the crystallization of an intrusive igneous mass. If the rocks that surround the mass differ markedly in composition from the invading fluids, there may be a substantial exchange of ions between the fluids and host rocks. When this occurs, a change in the overall composition of the surrounding rock results.

Metamorphic Textures

The degree of metamorphism is reflected in the rock's texture and mineralogy. (Recall that the term *texture* is used to describe the size, shape, and arrangement of grains within a rock.) When rocks are subjected to low-grade metamorphism, they become more compact and thus more dense. A common example is the metamorphic rock slate, which forms when shale is subjected to temperatures and pressures only slightly greater than those associated with the compaction that lithifies sediment. In this case, differential stress causes the microscopic clay minerals in shale to align into the more compact arrangement found in slate.

Under more extreme conditions, stress causes certain minerals to recrystallize. In general, recrystallization encourages the growth of larger crystals. Consequently, many metamorphic rocks consist of visible crystals, much like coarse-grained igneous rocks.

The crystals of some minerals will recrystallize with a preferred orientation, essentially perpendicular to the direction of the compressional force. The resulting mineral alignment usually gives the rock a layered or banded appearance termed **foliated texture** (Figure 2.28). Simply, foliation results whenever the minerals of a rock are brought into parallel alignment.

Not all metamorphic rocks have a foliated texture. Such rocks are said to exhibit a **nonfoliated texture.** Metamorphic rocks composed of only one mineral that forms equidimensional crystals are, as a rule, not visibly foliated. For example, pure limestone, is composed of only a single mineral, calcite. When a fine-grained limestone is metamorphosed, the small calcite crystals combine to form larger interlocking crystals. The resulting rock resembles a coarse-grained igneous rock. This nonfoliated metamorphic equivalent of limestone is called *marble*.

Figure 2.27 Deformed metamorphic rocks exposed in a road cut in the Eastern Highland of Connecticut. Imagine the tremendous force required to fold rock in this manner. (Photo by Phil Dombrowski)

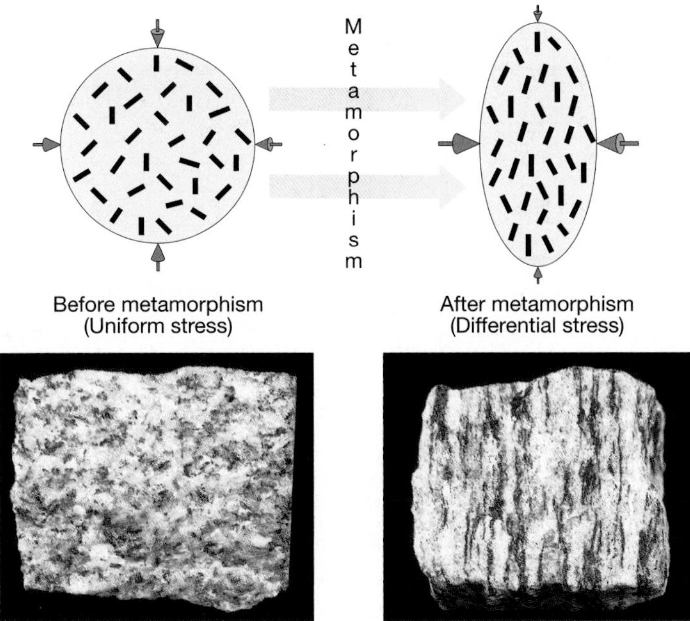

Figure 2.28 Under the pressures of metamorphism, some mineral grains become reoriented and aligned at right angles to the stress. The resulting orientation of mineral grains gives the rock a foliated (layered) texture. If the coarse-grained igneous rock (granite) on the left underwent intense metamorphism, it could end up closely resembling the metamorphic rock on the right (gneiss). (Photos by E. J. Tarbuck)

Common Metamorphic Rocks

To review, metamorphic processes cause many changes in existing rocks, including increased density, growth of larger crystals, foliation (reorientation of the mineral grains into a layered or banded appearance), and the transformation of low-temperature minerals into high-temperature minerals (Figure 2.29). Further, the introduction of ions generates new minerals, some of which are economically important.

Here is a brief look at common rocks produced by metamorphic processes.

Foliated Rocks *Slate* is a very fine-grained foliated rock composed of minute mica flakes (Figure 2.30). The most noteworthy characteristic of slate is its excellent rock cleavage, meaning that it splits easily into flat slabs. This property has made slate a most useful rock for roof and floor tile, chalkboards, and billiard tables (Figure 2.31). Slate is most often generated by the low-grade metamorphism of shale, although less frequently it forms from the metamorphism of volcanic ash. Slate can be almost any color, depending on its mineral constituents. Black slate contains organic material; red slate gets its color from iron oxide; and green slate is usually composed of chlorite, a micalike mineral.

Schists are strongly foliated rocks formed by regional metamorphism (Figure 2.30). They are platy and can be readily split into thin flakes or slabs. Like slate, the parent material from which many schists originate is shale, but in the case of schist, the metamorphism is more intense.

The term *schist* describes the *texture* of a rock regardless of composition. For example, schists composed primarily of muscovite and biotite are called *mica schists.*

Did You Know?

The reason high-quality billiard tables are so heavy is that they have surfaces made of a thick slab of the metamorphic rock, slate. Because slate splits easily into slabs, it is highly prized for use as billiard-table surfaces as well as building materials such as floor and roof tiles.

Gneiss (pronounced "nice") is the term applied to banded metamorphic rocks that contain mostly elongated and granular, as opposed to platy, minerals (Figure 2.30). The most common minerals in gneisses are quartz and feldspar, with lesser amounts of muscovite, biotite, and hornblende. Gneisses exhibit strong segregation of light and dark silicates, giving them a characteristic banded texture. While in a plastic state, these banded gneisses can be deformed into intricate folds.

Nonfoliated Rocks *Marble* is a coarse, crystalline rock whose parent rock is limestone. Marble is composed of large interlocking calcite crystals, which form from the recrystallization of smaller grains in the parent rock.

Because of its color and relative softness (hardness of only 3 on the Mohs scale), marble is a popular building stone. White marble is particularly prized as a stone from which to carve monuments and statues, such as the famous statue of *David* by Michelangelo (Figure 2.32). Often the limestone from which marble forms contains impurities that color the marble. Thus, marble can be pink, gray, green, or even black.

Rock Name		Texture		Grain Size	Comments	Parent Rock
Slate	Increasing Metamorphism	Foliated		Very fine	Excellent rock cleavage, smooth dull surfaces	Shale, mudstone, or siltstone
Phyllite				Fine	Breaks along wavy surfaces, glossy sheen	Slate
Schist				Medium to coarse	Micaceous minerals dominate, scaly foliation	Phyllite
Gneiss				Medium to coarse	Compositional banding due to segregation of minerals	Schist, granite, or volcanic rocks
Marble		Nonfoliated		Medium to coarse	Interlocking calcite or dolomite grains	Limestone, dolostone
Quartzite				Medium to coarse	Fused quartz grains, massive, very hard	Quartz sandstone
Anthracite				Fine	Shiny black organic rock that may exhibit conchoidal fracture	Bituminous coal

Figure 2.29 Classification of common metamorphic rocks.

Figure 2.30 Common metamorphic rocks. (Photos by E. J. Tarbuck)

Quartzite is a very hard metamorphic rock most often formed from quartz sandstone. Under moderate- to high-grade metamorphism, the quartz grains in sandstone fuse. Pure quartzite is white, but iron oxide may produce reddish or pinkish stains, and dark minerals may impart a gray color.

Figure 2.31 Excellent rock cleavage is exhibited by the slate in this quarry near Alta, Norway. (Photo by Fred Bruermmer/DRK Photo) Because slate breaks into flat slabs, it has many uses. In the inset photo, it is used to roof this house in Switzerland. (Photo by E. J. Tarbuck)

Did You Know?

Because marble can be carved readily, it has been used for centuries for buildings and memorials. Examples of important structures whose exteriors are clad in marble include the Parthenon in Greece, the Taj Mahal in India, and the Washington Monument in the United States.

Figure 2.32 Replica of the statue of *David* by Michelangelo. Like the original, this sculpture was created from a large block of white marble. (Photo by Andrew Ward/Getty Images, Inc./Photo Disk)

The Chapter in Review

1. The three rock groups are igneous, sedimentary, and metamorphic. *Igneous rock* forms from *magma* that cools and solidifies in a process called *crystallization. Sedimentary rock* forms from the *lithification* of sediment. *Metamorphic rock* forms from rock that has been subjected to great pressure and heat in a process called *metamorphism.*

2. Igneous rocks are classified by their *texture* and *mineral composition.*

3. The rate of cooling of magma greatly influences the size of mineral crystals in igneous rock and thus its texture. The four basic igneous rock textures are (1) *fine-grained,* (2) *coarse-grained,* (3) *porphyritic,* and (4) *glassy.*

4. Igneous rocks are divided into broad compositional groups based on the percentage of dark and light silicate minerals they contain. *Felsic rocks* (e.g., granite and rhyolite) are composed mostly of the light-colored silicate minerals potassium feldspar and quartz. Rocks of *intermediate* composition (e.g., andesite) contain plagioclase feldspar and amphibole. *Mafic rocks* (e.g., basalt) contain abundant pyroxene, and calcium-rich plagioclase feldspar.

5. The mineral makeup of an igneous rock is ultimately determined by the chemical composition of the magma from which it crystallized. N. L. Bowen showed that as magma cools, minerals crystallize in an orderly fashion at different temperatures. *Magmatic differentiation* changes the composition of magma and causes more than one rock type to form from a common parent magma.

6. *Weathering* is the response of surface materials to a changing environment. *Mechanical weathering,* the physical disintegration of material into smaller fragments, is accomplished by *frost wedging,* expansion resulting from *unloading,* and *biological activity. Chemical weathering* involves processes by which the internal structures of minerals are altered by the removal and/or addition of elements. It occurs when materials are *oxidized* or *react with acid,* such as carbonic acid.

7. *Detrital sediments* originate as solid particles derived from weathering and are transported. *Chemical sediments* are soluble materials produced largely by chemical weathering that are precipitated by either inorganic or organic processes. *Detrital sedimentary rocks,* which are classified by particle size, contain a variety of mineral and rock fragments, with clay minerals and quartz the chief constituents. *Chemical sedimentary rocks* often contain the products of biological processes or mineral crystals that form as water evaporates and minerals precipitate. *Lithification* refers to the processes by which sediments are transformed into solid sedimentary rocks.

8. Common detrital sedimentary rocks include *shale* (the most common sedimentary rock), *sandstone,* and *conglomerate.* The most abundant chemical sedimentary rock is *limestone,* consisting chiefly of the mineral calcite. *Rock gypsum* and *rock salt* are chemical rocks that form as water evaporates.

9. Some features of sedimentary rocks that are often used in the interpretation of Earth history and past environments include *strata* or *beds* (the single most characteristic feature), *bedding planes,* and *fossils.*

10. Two types of metamorphism are (1) *regional metamorphism* and (2) *contact or thermal metamorphism.* The agents of metamorphism include *heat, pressure* (stress), and *chemically active fluids.* Heat is perhaps the most important because it provides the energy to drive the reactions that result in the *recrystallization* of minerals. Metamorphic processes cause many changes in rocks, including *increased density,* growth of *larger mineral crystals, reorientation of the mineral grains* into a layered or banded appearance known as *foliation,* and the formation of *new minerals.*

11. Some common metamorphic rocks with a *foliated texture* include *slate, schist,* and *gneiss.* Metamorphic rocks with a *nonfoliated texture* include *marble* and *quartzite.*

Key Terms

andesitic (intermediate) composition (p. 45)
basaltic composition (p. 44)
Bowen's reaction series (p. 45)
chemical sedimentary rock (p. 50)
chemical weathering (p. 47)
coarse-grained texture (p. 42)
contact (thermal) metamorphism (p. 56)
crystallization (p. 38)
crystal settling (p. 45)
detrital sedimentary rock (p. 50)
evaporite deposit (p. 53)
extrusive (volcanic) (p. 40)
felsic (p. 44)
fine-grained texture (p. 42)
foliated texture (p. 59)
fossil (p. 55)
glassy texture (p. 42)
granitic composition (p. 44)
igneous rock (p. 40)
intrusive (plutonic) (p. 40)
lava (p. 40)
lithification (p. 40)
mafic (p. 44)
magma (p. 38)
magmatic differentiation (p. 46)
mechanical weathering (p. 46)
metamorphic rock (p. 40)
metamorphism (p. 55)
nonfoliated texture (p. 59)
porphyritic texture (p. 42)
regional metamorphism (p. 56)
rock cycle (p. 38)
sediment (p. 40)
sedimentary rock (p. 40)
strata (beds) (p. 54)
texture (p. 42)
ultramafic composition (p. 45)
weathering (p. 40)

Questions for Review

1. Explain the statement "One rock is the raw material for another" using the rock cycle.
2. If a lava flow at Earth's surface had a basaltic composition, what rock type would the flow likely be (see Figure 2.8)? What igneous rock would form from the same magma if it did not reach the surface but instead crystallized at great depth?
3. What does a porphyritic texture indicate about the history of an igneous rock?
4. How are granite and rhyolite different? The same? (See Figure 2.8.)
5. Relate the classification of igneous rocks to Bowen's reaction series.
6. If two identical rocks were weathered, one mechanically and the other chemically, how would the products of weathering for the two rocks differ?
7. How does mechanical weathering add to the effectiveness of chemical weathering?
8. How is carbonic acid formed in nature? What are the products when this acid reacts with potassium feldspar?
9. Which minerals are most common in detrital sedimentary rocks? Why are these minerals so abundant?
10. What is the primary basis for distinguishing among various detrital sedimentary rocks?
11. Distinguish between the two categories of chemical sedimentary rocks.
12. What are evaporite deposits? Name a rock that is an evaporite.
13. Compaction is an important lithification process with which sediment size?
14. What is probably the single most characteristic feature of sedimentary rocks?
15. What is metamorphism?
16. List the three agents of metamorphism and describe the role of each.
17. Distinguish between regional and contact metamorphism.
18. Which feature would easily distinguish schist and gneiss from quartzite and marble?
19. In what ways do metamorphic rocks differ from the igneous and sedimentary rocks from which they formed?

Online Study Guide

The *Foundations of Earth Science* Web site uses the resources and flexibility of the Internet to aid in your study of the topics in this chapter. Written and developed by Earth science instructors, this site will help improve your understanding of Earth science. Visit http://www.prenhall.com/lutgens and click on the cover of *Foundations of Earth Science 5e* to find:

- Online review quizzes.
- Critical thinking exercises.
- Links to chapter-specific Web resources.
- Internet-wide key-term searches.

http://www.prenhall.com/lutgens

GEODe: Earth Science

GEODe: Earth Science makes studying more effective by reinforcing key concepts using animation, video, narration, interactive exercises, and practice quizzes. A copy is included with every copy of *Foundations of Earth Science 5e.*

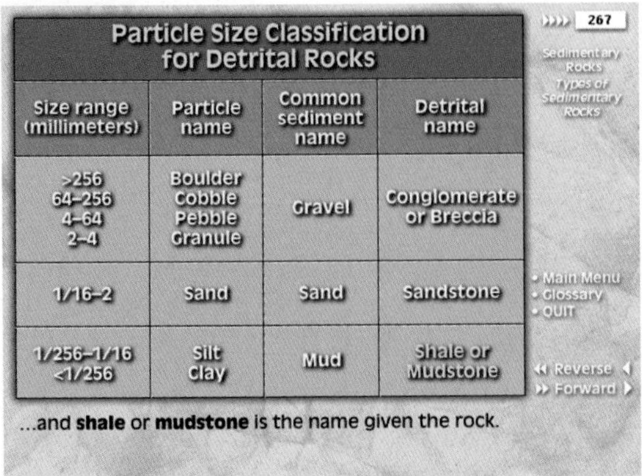

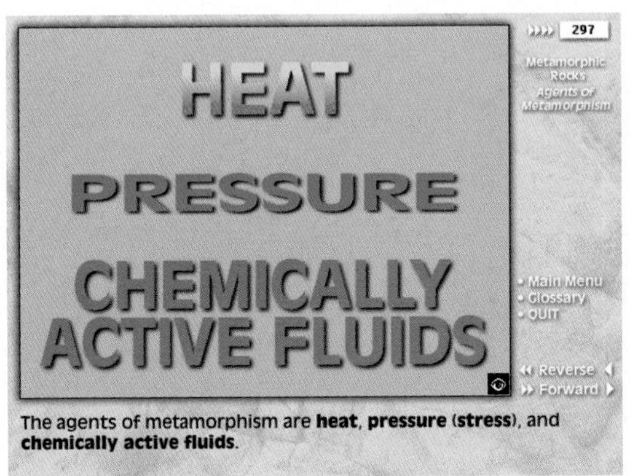

UNIT II SCULPTURING EARTH'S SURFACE

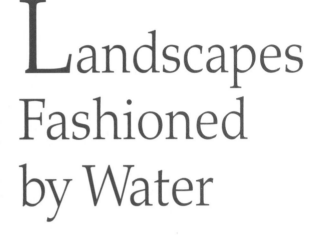

Hudson River in New York State's Adirondack State Park. (Photo by Carr Clifton Photography)

Landscapes Fashioned by Water

FOCUS ON LEARNING

To assist you in learning the important concepts in this chapter, you will find it helpful to focus on the following questions:

1. What is the role of *external processes* in the rock cycle?
2. What role does mass wasting play in the development of landforms?
3. What is the controlling force of mass wasting, and what factors might trigger mass-wasting processes?
4. What is the water cycle? What is the source of energy that powers the cycle?
5. What are the factors that determine the velocity of water in a stream?
6. The work of a stream involves what three processes?
7. How does base level influence a stream's ability to erode?
8. What common drainage patterns do streams produce?
9. What is the importance of groundwater as a resource and as a geologic agent?
10. What is groundwater, and how does it move?
11. What are some common natural phenomena associated with groundwater?
12. What are some environmental problems associated with groundwater?

Earth is a dynamic planet. Volcanic and other internal forces elevate the land, while opposing external processes continually wear it down. The Sun and gravity drive external processes occurring at Earth's surface. Rock is disintegrated and decomposed, moved to lower elevations by gravity, and carried away by water, wind, or ice. In this manner, the physical landscape is sculptured.

Earth's External Processes

Weathering, mass wasting, and erosion are called **external processes** because they occur at or near Earth's surface and are powered by energy from the Sun. External processes are a basic part of the rock cycle because they are responsible for transforming solid rock into sediment.

To the casual observer, the face of Earth may appear to be without change, unaffected by time. In fact, 200 years ago most people believed that mountains, lakes, and deserts were permanent features of an Earth that was thought to be no more than a few thousand years old. Today we know Earth is 4.5 billion years old and that mountains eventually succumb to weathering and erosion, lakes fill with sediment or are drained by streams, and deserts come and go with changes in climate.

Earth is a dynamic body. Some parts of Earth's surface are gradually elevated by mountain building and volcanic activity. These **internal processes** derive their energy from Earth's interior. Meanwhile, opposing external processes are continually breaking rock apart and moving the debris to lower elevations (Figure 3.1). The latter processes include:

1. **Weathering**—the physical breakdown (disintegration) and chemical alteration (decomposition) of rock at or near Earth's surface.
2. **Mass wasting**—the transfer of rock and soil downslope under the influence of gravity.
3. **Erosion**—the physical removal of material by a mobile agent such as flowing water, waves, wind, or ice.

Weathering processes were treated in Chapter 2. In this chapter, we will focus on mass wasting and on two important erosional processes—running water and groundwater. Chapter 4 will examine the geologic work of two other significant erosional processes—glacial ice and wind.

Mass Wasting: The Work of Gravity

Earth's surface is never perfectly flat but instead consists of slopes. Some are steep and precipitous; others are moderate or gentle. Some are long and gradual; others are short and abrupt. Some slopes are mantled with soil and covered by vegetation; others consist of barren rock and rubble. Their form and variety are great.

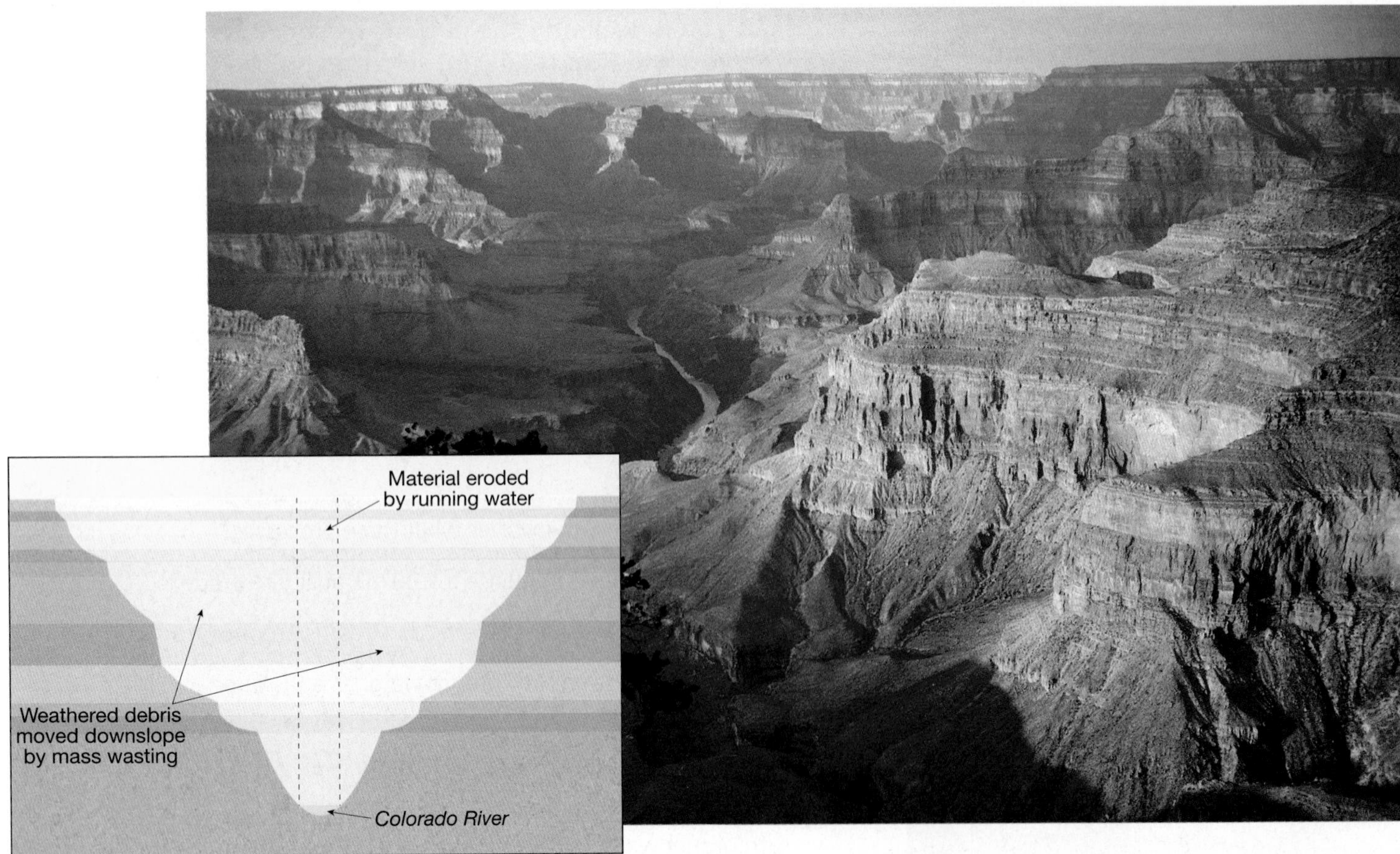

Figure 3.1 The walls of the Grand Canyon extend far from the channel of the Colorado River. This results primarily from the transfer of weathered debris downslope to the river and its tributaries by mass wasting processes. (Photo by Tom & Susan Bean, Inc.)

Did You Know?

Although many people, including geologists, frequently use the word *landslide,* the term has no specific definition in geology. Rather, it is a popular nontechnical term used to describe all relatively rapid forms of mass wasting.

Although most slopes appear to be stable and unchanging, they are not static features, because the force of gravity causes material to move downslope. At one extreme, the movement may be gradual and practically imperceptible. At the other extreme, it may consist of a roaring debris flow or a thundering rock avalanche. Landslides are a worldwide, natural hazard. When these natural processes lead to loss of life and property, they become natural disasters.

Mass Wasting and Landform Development

Events such as those pictured in Figure 3.2 are spectacular examples of a common geologic process called mass wasting. *Mass wasting* is the downslope movement of rock and soil under the direct influence of gravity. It is distinct from the erosional processes because mass wasting does not require a transporting medium such as water, wind, or glacial ice. There is a broad array of mass-wasting processes. Four are illustrated in Figure 3.3.

Mass wasting is the step that follows weathering in the evolution of most landforms. Once weathering weakens and breaks rock apart, mass wasting transfers the debris downslope, where a stream, acting as a conveyor belt, usually carries it away (see Figure 3.1). Although there may be many intermediate stops along the way, the sediment is eventually transported to its ultimate destination: the sea. It is the combined effects of mass wasting and running water that produce stream valleys, which are the most common and conspicuous landforms at Earth's surface and the focus of a later part of this chapter.

If streams alone were responsible for creating the valleys in which they flow, valleys would be very narrow features. However, the fact that most river valleys are much wider than they are deep is a strong indication of the significance of mass-wasting processes in supplying material to streams. The walls of a canyon extend far from the river because of the transfer of weathered debris downslope to the river and its tributaries by mass-wasting processes. In this manner, streams and mass wasting combine to modify and sculpture the surface. Of course, glaciers, groundwater, waves, and wind are also important agents in shaping landforms and developing landscapes.

Controls and Triggers of Mass Wasting

Gravity is the controlling force of mass wasting, but several factors play an important role in overcoming inertia and creating downslope movements. Long before a landslide occurs, various processes work to weaken slope material, gradually making it more and more susceptible to the pull of gravity. During this span, the slope remains stable but gets closer and closer to being unstable. Eventually, the strength of the slope

Figure 3.2 **A.** This rockslide occurred in January 1997 on Highway 140 near the Arch Rock entrance to Yosemite National Park, California. (Photo by Roger J. Wyan/AP/Wide Word Photos) **B.** This debris flow at La Conchita, California, in January 2005, was triggered by extraordinary rains. (Photo by David NcNew/Getty Images, Inc.)

A.

B.

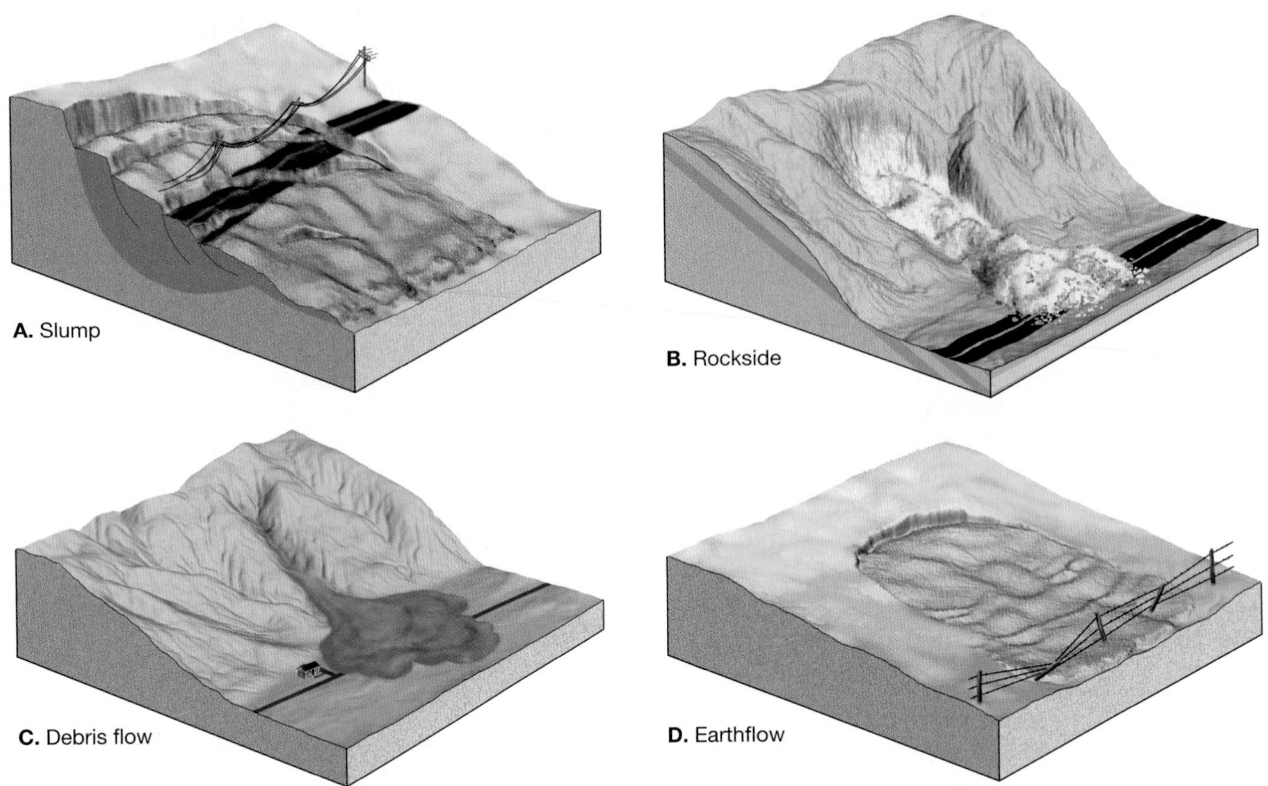

Figure 3.3 The four processes illustrated here are all considered to be relatively rapid forms of mass wasting. Because material in slumps **(A)** and rockslides **(B)** move along well-defined surfaces, they are said to move by sliding. By contrast, when material moves downslope as a viscous fluid, the movement is described as a flow. Debris flow **(C)** and earthflow **(D)** advance downslope in this manner.

is weakened to the point that something causes it to cross the threshold from stability to instability. Such an event that initiates downslope movement is called a *trigger.* Remember that the trigger is not the sole cause of the mass-wasting event, but just the last of many causes. Among the common factors that trigger mass-wasting processes are saturation of material with water, oversteepening of slopes, removal of anchoring vegetation, and ground vibrations from earthquakes.

The Role of Water Mass wasting is sometimes triggered when heavy rains or periods of snowmelt saturate surface materials. Such a situation is shown in Figure 3.2B.

When the pores in sediment become filled with water, the cohesion among particles is destroyed, allowing them to slide past one another with relative ease. For example, when sand is slightly moist, it sticks together quite well. However, if enough water is added to fill the openings between the grains, the sand will ooze out in all directions. Thus, saturation reduces the internal resistance of materials, which are then easily set in motion by the force of gravity. When clay is wetted, it becomes very slick—another example of the "lubricating" effect of water. Water also adds considerable weight to a mass of material. The added weight in itself may be enough to cause the material to slide or flow downslope.

Did You Know?

The cost of damages by all types of mass wasting in the United States is conservatively estimated to exceed $2 billion in an average year.

Oversteepened Slopes Oversteepening of slopes is another trigger of many mass movements. This takes place in many situations in nature. A stream undercutting a valley wall and waves pounding against the base of a cliff are two familiar examples. Furthermore, through their activities, people often create oversteepened and unstable slopes that become prime sites for mass wasting.

Unconsolidated, granular (sand-size or coarser) particles assume a stable slope called the **angle of repose,** the steepest angle at which material remains stable. Depending on the size and shape of the particles, the angle varies from 25 to 40 degrees. The larger, more angular particles maintain the steepest slopes. If the angle is increased, the rock debris will adjust by moving downslope.

Oversteepening is important not only because it triggers movements of unconsolidated granular materials, but it also produces unstable slopes and mass movements in cohesive soils, regolith (weathered rock debris), and bedrock. The response will not be immediate, as with loose, granular material, but sooner or later one or more mass-wasting processes will eliminate the oversteepening and restore stability to the slope.

Removal of Vegetation Plants protect against erosion and contribute to the stability of slopes because their root systems

Did You Know?

The U.S. Geological Survey estimates that between 25 and 50 people are killed by landslides annually in the United States. The worldwide death toll is much higher, more than 600 per year.

bind soil and regolith together. Where plants are lacking, mass wasting is enhanced, especially if slopes are steep and water is plentiful. When anchoring vegetation is removed by forest fires or by people (for timber, farming, or development), surface materials frequently move downslope.

Earthquakes as Triggers Among the more important and dramatic mass-wasting triggers are earthquakes. An earthquake and its aftershocks can dislodge enormous volumes of rock and unconsolidated material. In many areas that are jolted by earthquakes, it is not ground vibrations directly but landslides and ground subsidence triggered by the vibrations that cause the greatest damage.

The Water Cycle

Sculpturing Earth's Surface
▼ Hydrologic Cycle

> All the rivers run into the sea; yet the sea is not full; unto the place from whence the rivers come, thither they return again.
>
> *Ecclesiastes 1:7*

As the perceptive writer of Ecclesiastes indicated, water is continually on the move, from the ocean to the land and back again in an endless cycle. The remainder of this chapter deals with that part of the water cycle that returns water to the sea. Some water travels quickly via a rushing stream, and some moves more slowly beneath the surface. We will examine the factors that influence the distribution and movement of water, as well as look at how water sculptures the landscape. The Grand Canyon, Niagara Falls, Old Faithful geyser, and Mammoth Cave all owe their existence to the action of water on its way to the sea.

Water is everywhere on Earth—in the oceans, glaciers, rivers, lakes, air, soil, and living tissue. All of these "reservoirs" constitute Earth's hydrosphere. In all, the water content of the hydrosphere is an estimated 1.36 billion cubic kilometers (326 million cubic miles). The vast bulk of it, about 97.2 percent, is stored in the global ocean. Ice sheets and glaciers account for another 2.15 percent, leaving only 0.65 percent to be divided among lakes, streams, groundwater, and the atmosphere (Figure 3.4). Although the percentage of Earth's total water found in each of the latter sources is just a small fraction of the total inventory, the absolute quantities are great.

The water found in each of the reservoirs depicted in Figure 3.4 does not remain in these places indefinitely. Water

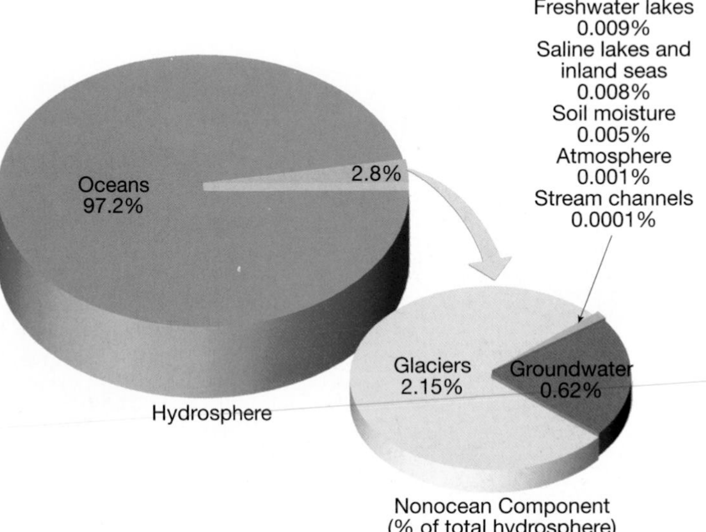

Figure 3.4 Distribution of Earth's water.

can readily change from one state of matter (solid, liquid, or gas) to another at the temperatures and pressures that occur at Earth's surface. Therefore, water is constantly moving among the hydrosphere, the atmosphere, the geosphere, and the biosphere. This unending circulation of Earth's water supply is called the **water cycle.** The cycle shows us many critical interrelationships among different parts of the Earth system.

The water cycle is a gigantic, worldwide system powered by energy from the Sun in which the atmosphere provides the vital link between the oceans and continents (Figure 3.5). Water evaporates into the atmosphere from the ocean and to a much lesser extent from the continents. Winds transport this moisture-laden air, often great distances. Complex processes of cloud formation eventually result in precipitation. The precipitation that falls into the ocean has completed its cycle and is ready to begin another. The water that falls on the continents, however, must make its way back to the ocean.

What happens to precipitation once it has fallen on land? A portion of the water soaks into the ground (called **infiltration**), slowly moving downward, then laterally, finally seeping into lakes, streams, or directly into the ocean. When the rate of rainfall exceeds Earth's ability to absorb it, the surplus water flows over the surface into lakes and streams, a process called **runoff.** Much of the water that infiltrates or runs off eventually returns to the atmosphere because of evaporation from the soil, lakes, and streams. Also, some of the water that soaks into the ground is absorbed by plants, which then release it into the atmosphere. This process is called **transpiration.**

When precipitation falls in very cold places—at high elevations or high latitudes—the water may not immediately soak in, run off, or evaporate. Instead, it may become part of a snowfield or a glacier. In this way, glaciers store large quantities of water on land. If present-day glaciers were to melt and release all their water, sea level would rise by several

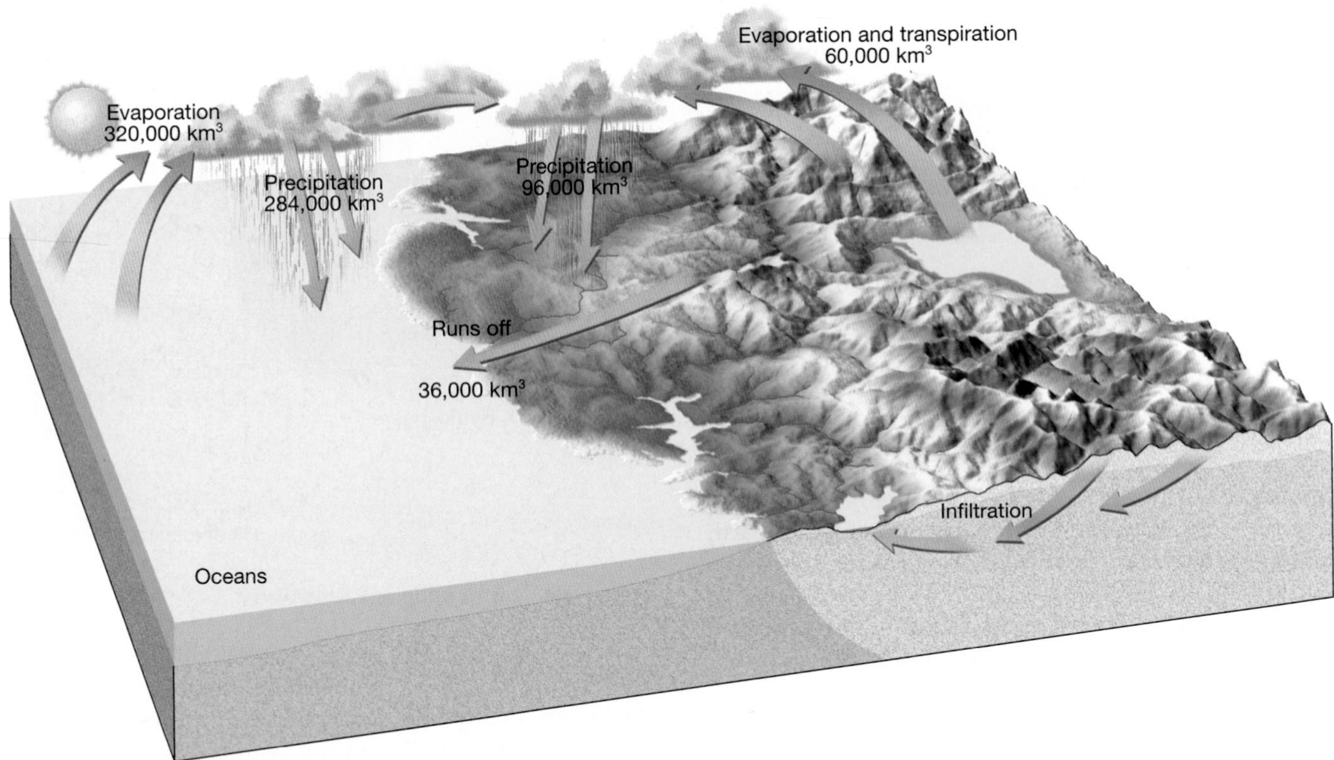

Figure 3.5 Earth's water balance. Each year, solar energy evaporates about 320,000 cubic kilometers of water from the oceans, while evaporation and transpiration from the land (including lakes and streams) contributes 60,000 cubic kilometers of water. Of this total of 380,000 cubic kilometers of water, about 284,000 cubic kilometers fall back to the ocean, and the remaining 96,000 cubic kilometers fall on the land surface. Of those 96,000 cubic kilometers, only 60,000 cubic kilometers of water return to the atmosphere by evaporation and transpiration leaving 36,000 cubic kilometers of water to erode the land during the journey back to the oceans.

dozen meters. This would submerge many heavily populated coastal areas. As you will see in Chapter 4, over the past 2 million years, huge ice sheets have formed and melted on several occasions, each time affecting the balance of the water cycle.

Figure 3.5 also shows Earth's overall *water balance,* or the volume that passes through each part of the cycle annually. The amount of water vapor in the air at any one time is just a tiny fraction of Earth's total water supply. But the *absolute* quantities that are cycled through the atmosphere over a one-year period are immense—some 380,000 cubic kilometers (91,000 cubic miles)— enough to cover Earth's entire surface to a depth of about 1 meter (39 inches).

It is important to know that the water cycle is *balanced.* Because the total amount of water vapor in the atmosphere remains about the same, the average annual precipitation worldwide must be equal to the quantity of water evaporated. However, for all of the continents taken together, precipitation exceeds evaporation. Conversely, over the oceans, evaporation exceeds precipitation. Because the level of the world ocean is not dropping, the system must be in balance. In Figure 3.5, the 36,000 cubic kilometers (8600 cubic miles) of water that annually makes its way from the land to the ocean causes enormous erosion. In fact, this immense volume of moving water is *the single most important agent sculpturing Earth's land surface.*

To summarize, the water cycle is the continuous movement of water from the oceans to the atmosphere, from the atmosphere to the land, and from the land back to the sea. The land-back-to-the-sea step is the primary action that wears down Earth's land surface. In the rest of this chapter, we will observe the work of water running over the surface, including floods, erosion, and the formation of valleys. Then we will look underground at the slow labors of groundwater as it forms springs and caverns and provides drinking water on its long migration to the sea.

Did You Know?

Each year, a field of crops may transpire the equivalent of a water layer 60 centimeters (2 feet) deep over the entire field. The same area of trees may pump twice this amount into the atmosphere.

Running Water

Sculpturing Earth's Surface
▼ Running Water

Running water is of great importance to people. We depend on rivers for energy, transportation, and irrigation. Their fertile floodplains have been favored sites for agriculture and industry ever since the dawn of civilization. As the dominant agent of erosion, running water has shaped much of our physical environment.

Drainage Basins

All water in rivers ultimately had its source as rain and snow, although there can be considerable lag time before this water actually enters a river system. Upon reaching the surface, some precipitation soaks into the ground, while the remaining water flows over the surface. This *overland flow* is routed across the landscape by hillslopes into rills, gullies, streams, and finally large rivers that usually flow to the ocean.

The land area that contributes water to a river system is called a **drainage basin** (Figure 3.6). The drainage basin of one stream is separated from the drainage basin of another by an imaginary line called a **divide** (Figure 3.6). Divides range in scale from a ridge separating two small gullies on a hillside to a *continental divide,* which splits whole continents into enormous drainage basins. The Mississippi River has the largest drainage basin in North America. Extending between the Rocky Mountains in the West and the Appalachian Mountains in the East, the Mississippi River and its tributaries collect water from more than 3.2 million square kilometers (1.2 million square miles) of the continent.

Did You Know?

North America's largest river is the Mississippi. South of Cairo, Illinois, where the Ohio River joins it, the Mississippi is more than 1.6 km (1 mi) wide. Each year the "Mighty Mississippi" carries about a half billion tons of sediment to the Gulf of Mexico.

River Systems

Rivers and streams can be simply defined as water flowing in a channel. They have three important roles in the formation of a landscape: They erode the channels in which they flow, they transport sediments provided by weathering and mass wasting, and they produce a wide variety of erosional and depositional landforms. In fact, in most areas, including many arid regions, river systems have shaped the varied landscape that we humans inhabit.

A river system consists of three main parts in which different processes dominate: a zone of erosion, a zone of sediment transport, and a zone of sediment deposition. It is important to realize that some erosion, transport, and deposition occur in all three zones; however, within each zone, one of these processes is usually dominant. In addition, the parts of a river system are interdependent, so that the processes occurring in one part influence the others.

In large river systems, erosion is the dominant process in the upstream area, which generally consists of mountainous or hilly topography. Here, small tributary streams erode the channels in which they flow and carry material provided by weathering and mass wasting.

The region within a river system that is dominated by deposition is usually located where the stream enters a large body of water. Here, sediments accumulate to form a delta, or

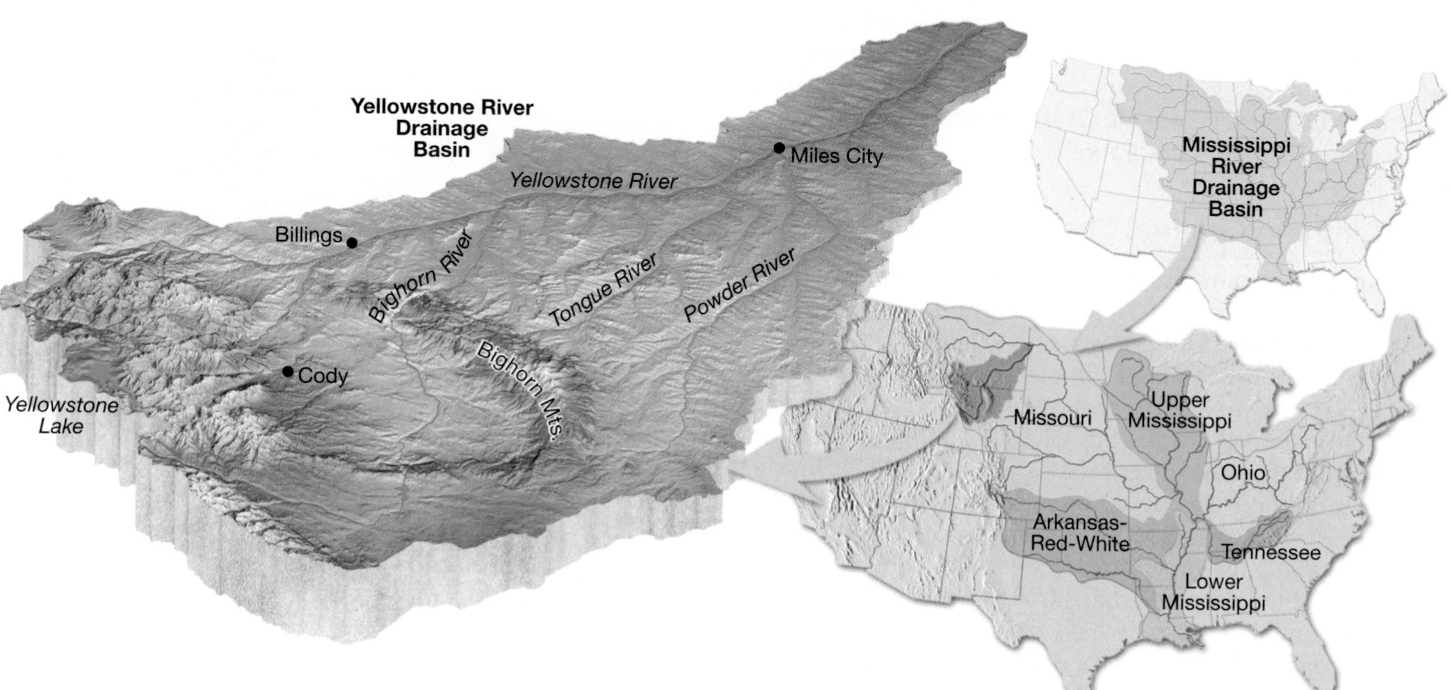

Figure 3.6 A *drainage basin* is the land area drained by a stream and its tributaries. *Divides* are the boundaries separating drainage basins. Drainage basins and divides exist for all streams, regardless of size. The drainage basin of the Yellowstone River is one of many that contribute water to the Missouri River, which, in turn, is one of many that make up the drainage basin of the Mississippi River. The drainage basin of the Mississippi River, North America's largest, covers about 3 million square kilometers.

are reworked by wave action to form a variety of coastal features. Between the zones of erosion and deposition is the *trunk stream* that serves to transport sediments. Taken together, erosion, transportation, and deposition are the processes by which rivers move Earth's surface materials and sculpt landscapes.

Streamflow

Water may flow in one of two ways, either as **laminar flow** or **turbulent flow.** In slow-moving streams, the flow is often laminar, and the water particles move in roughly straight-line paths that parallel the stream channel. However, streamflow is usually turbulent, with the water moving in an erratic fashion that can be characterized as a swirling motion. Strong, turbulent flow may be seen in whirlpools and eddies, as well as roiling whitewater rapids. Even streams that appear smooth on the surface often exhibit turbulent flow near the bottom and sides of the channel. Turbulence contributes to the stream's ability to erode its channel because it acts to lift sediment from the streambed.

Water makes its way to the sea under the influence of gravity. The time required for the journey depends on the velocity of the stream. Velocity is the distance that water travels in a unit of time. Some sluggish streams flow at less than 1 kilometer (0.6 mile) per hour, whereas a few rapid ones may exceed 30 kilometers (19 miles) per hour. Velocities are measured at gaging stations (Figure 3.7A). Along straight stretches, the highest velocities are near the center of the channel just below the surface, where friction is lowest (Figure 3.7B). But when a stream curves, its zone of maximum speed shifts (Figure 3.7C).

The ability of a stream to erode and transport materials depends on its velocity. Even slight changes in velocity can lead to significant changes in the load of sediment that water can transport. Several factors determine the velocity of a stream, including (1) gradient; (2) shape, size, and roughness of the channel; and (3) discharge.

Gradient and Channel Characteristics

The slope of a stream channel expressed as the vertical drop of a stream over a specified distance is **gradient.** Portions of the lower Mississippi River, for example, have very low gradients of 10 centimeters per kilometer or less. By contrast, some mountain stream channels decrease in elevation at a rate of more than 40 meters per kilometer, or a gradient 400 times steeper than the lower Mississippi (Figure 3.8). Gradient varies not only among different streams but also over a particular stream's length. The steeper the gradient, the more energy available for streamflow. If two streams were identical in every respect except gradient, the stream with the higher gradient would obviously have the greater velocity.

A stream's channel is a conduit that guides the flow of water, but the water encounters friction as it flows. The shape, size, and roughness of the channel affect the amount of friction. Larger channels have more efficient flow because a smaller proportion of water is in contact with the channel. A smooth channel promotes a more uniform flow, whereas an irregular channel filled with boulders creates enough turbulence to slow the stream significantly.

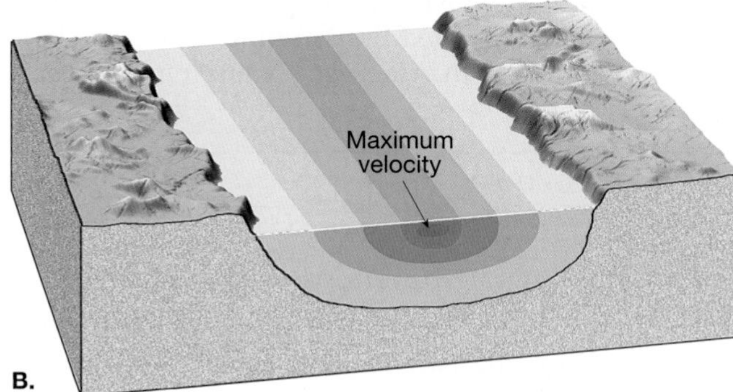

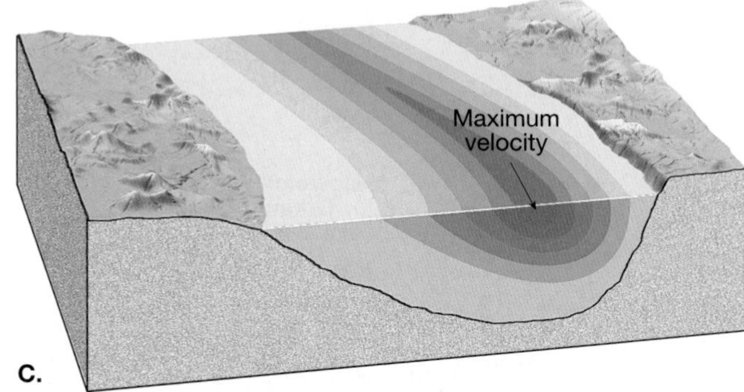

Figure 3.7 **A.** Continuous records of stage and discharge are collected by the U.S. Geological Survey at more than 7000 gaging stations in the United States. Average velocities are determined by using measurements from several spots across the stream. This station is on the Rio Grande south of Taos, New Mexico. (Photo by E. J. Tarbuck) **B.** Along straight stretches, stream velocity is highest at the center of the channel. **C.** When a stream curves, its zone of maximum speed shifts.

Discharge

The **discharge** of a stream is the volume of water flowing past a certain point in a given unit of time. This is usually measured in cubic meters per second or cubic feet per second. Dis-

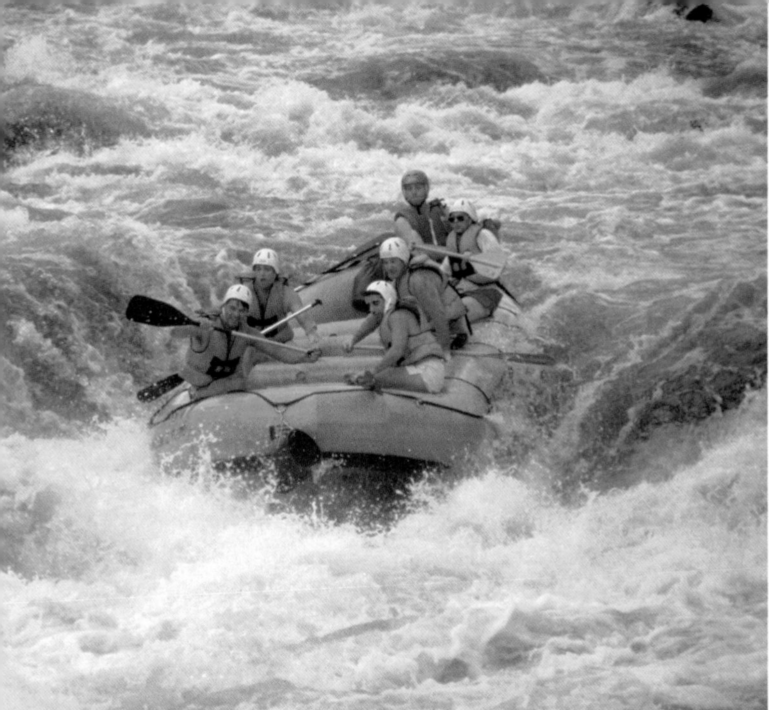

Figure 3.8 Rapids are common in mountain streams where the gradient is steep and the channel is rough and irregular. (Photo by Bruce Gaylord/Visuals Unlimited)

charge is determined by multiplying a stream's cross-sectional area by its velocity:

$$\text{discharge (m}^3\text{/second)} = \text{channel width (meters)} \times \text{channel depth (meters)} \times \text{velocity (meters/second)}$$

The largest river in North America, the Mississippi, discharges an average of 17,300 cubic meters (611,000 cubic feet) per second. Although this is a huge quantity of water, it is nevertheless dwarfed by the mighty Amazon in South America, the world's largest river. Fed by a vast rainy region that is nearly three-fourths the size of the conterminous United States, the Amazon discharges 12 times more water than the Mississippi.

Did You Know?

The Amazon River is responsible for about 20 percent of all the water reaching the ocean via rivers. Its nearest rival, Africa's Congo River, delivers about 4 percent of the total.

The discharges of most rivers are far from constant. This is true because of such variables as rainfall and snowmelt. In areas with seasonal variations in precipitation, streamflow will tend to be highest during the wet season, or during spring snowmelt, and lowest during the dry season or during periods when high temperatures increase the water losses through evapotranspiration. However, not all channels maintain a continuous flow of water. Streams that exhibit flow only during wet periods are referred to as *intermittent streams.* In arid climates, many streams carry water only occasionally after a heavy rainstorm and are called *ephemeral streams.*

Changes from Upstream to Downstream

One useful way of studying a stream is to examine its *profile.* A profile is simply a cross-sectional view of a stream from its source area (called the *head* or *headwaters*) to its *mouth,* the point downstream where the river empties into another water body. By examining Figure 3.9, you can see that the most

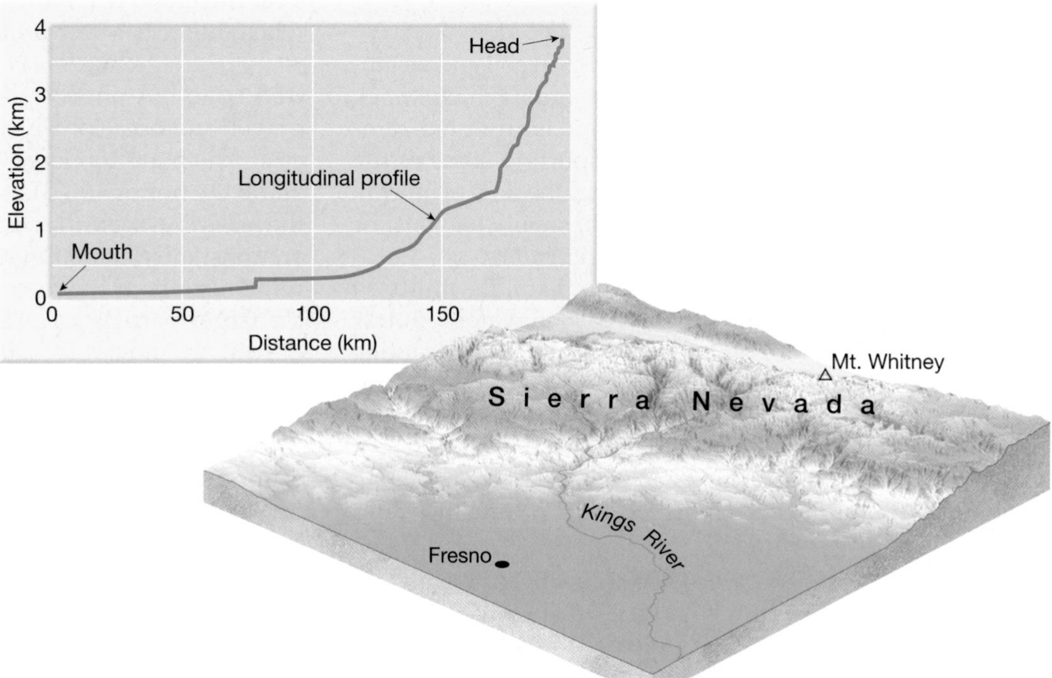

Figure 3.9 Longitudinal profile of California's Kings River. It originates in the Sierra Nevada and flows westward into the San Joaquin Valley. A longitudinal profile is a cross-section along the length of a stream. Note the concave upward curve of the profile, with a steeper gradient upstream and a gentler gradient downstream.

obvious feature of a typical profile is a constantly decreasing gradient from the head to the mouth. Although many local irregularities may exist, the overall profile is a relatively smooth concave-upward curve.

The profile shows that the gradient decreases downstream. To see how other factors change in a downstream direction, observations and measurements must be made. When data are collected from several gaging stations along a river, they show that in a humid region, discharge increases from the head toward the mouth. This should come as no surprise because as we move downstream, more and more tributaries contribute water to the main channel. Furthermore, in most humid regions, additional water is added from the groundwater supply. Thus, as you move downstream, the stream's width, depth, and velocity change in response to the increased volume of water carried by the stream.

Streams that begin in mountainous areas where precipitation is abundant and then flow through arid regions may experience the opposite situation. Here, discharge may actually decrease downstream because of water loss due to evaporation, infiltration into the streambed, and removal by irrigation. The Colorado River in the southwestern United States is such an example.

The Work of Running Water

Streams are Earth's most important erosional agent. Not only do they have the ability to downcut and widen their channels, but streams also have the capacity to transport the enormous quantities of sediment that are delivered to them by sheet flow, mass wasting, and groundwater. Eventually, much of this material is dropped by the water to create a variety of depositional features.

Erosion

A stream's ability to accumulate and transport soil and weathered rock is aided by the work of raindrops, which knock sediment particles loose. When the ground is saturated, rainwater begins to flow downslope, transporting some of the material it has dislodged. On barren slopes the flow of muddy water, called *sheet flow,* will often erode small channels, or *rills,* which in time may evolve into larger *gullies.*

Once the surface flow reaches a stream, its ability to erode is greatly enhanced by the increase in water volume. When the flow of water is sufficiently strong, it can dislodge particles from the channel and lift them into the moving water. In this manner, the force of running water swiftly erodes poorly consolidated materials on the bed and sides of a stream channel. On occasion, the banks of the channel may be undercut, dumping even more loose debris into the water to be carried downstream.

In addition to eroding unconsolidated materials, the hydraulic force of streamflow can also cut a channel into solid bedrock. A stream's ability to erode bedrock is greatly enhanced by the particles it carries. These particles can be any size, from large boulders in fast-flowing waters to sand and gravel-size particles in somewhat slower flows. Just as the particles of grit on sandpaper can wear away a piece of wood, so too can the sand and gravel carried by a stream abrade a bedrock channel. Moreover, pebbles caught in swirling eddies can act like "drills" and bore circular *potholes* into the channel floor.

Transportation

All streams regardless of size transport some rock material. Streams also sort the solid sediment they transport because finer, lighter material is carried more rapidly than coarser, heavier rock debris. Depending on the nature of the rock material, the stream load consists of material (1) in solution (**dissolved load**), (2) in suspension (**suspended load**), and (3) sliding or rolling along the bottom (**bed load**).

Dissolved Load Most of the *dissolved load* is brought to a stream by groundwater and is dispersed throughout the flow. The quantity of material carried in solution is highly variable and is most abundant in humid areas where limestone and other soluble rock forms the bedrock. Usually, the amount of dissolved load is small and therefore is expressed as parts of dissolved material per million parts of water (parts per million, or ppm). Although some rivers may have a dissolved load of 1000 ppm or more, the average figure for the world's rivers is estimated at 115 to 120 ppm.

Suspended Load Most large rivers carry the largest part of their load in *suspension.* Indeed, the visible cloud of sediment suspended in the water is the most obvious portion of a stream's load. Usually, only fine particles consisting of silt and clay can be carried this way, but during a flood, sand and even gravel-size particles are transported as well. Also, during a flood, the total quantity of material carried in suspension increases dramatically, as can be verified by anyone whose home has been a site for the deposition of this material.

Bed Load A portion of a stream's load consists of sand, gravel, and occasionally large boulders. These coarser particles, which are too large to be carried in suspension, move along the bottom (bed) of the stream channel and constitute the *bed load.* Unlike the suspended and dissolved loads, which are constantly in motion, the bed load is in motion only intermittently, when the force of the water is sufficient to move the larger particles. The smaller particles, mainly sand and gravel, are moved by *saltation,* which resembles a series of jumps or skips. Larger particles either roll or slide along the bottom, depending on their shape.

Competence and Capacity Streams vary in their ability to carry a load. Their ability is determined by two criteria. First, the **competence** of a stream measures the maximum size of particles it is capable of transporting. The stream's velocity determines its competence. If the velocity of a stream doubles, its competence increases four times; if the velocity triples, its competence increases nine times; and so forth. This explains how large boulders that seem immovable can be transported during a flood, which greatly increases a stream's velocity.

Second, the **capacity** of a stream is the maximum load it can carry. The capacity of a stream is directly related to its discharge. The greater the volume of water flowing in a stream, the greater is its capacity for hauling sediment.

By now it should be clear why the greatest erosion and transportation of sediment occur during floods (Figure 3.10). The increase in discharge results in a greater capacity, and the increase in velocity results in greater competence. With rising velocity, the water becomes more turbulent, and larger and larger particles are set in motion. In just a few days or perhaps a few hours, a stream in flood stage can erode and transport more sediment than it does during months of normal flow.

Deposition

Whenever a stream slows down, the situation reverses. As its velocity decreases, its competence is reduced and sediment begins to drop out, largest particles first. Each particle size has a *critical settling velocity.* As streamflow drops below the critical setting velocity of a certain particle size, sediment in that category begins to settle out. Thus, stream transport provides a mechanism by which solid particles of various sizes are separated. This process, called **sorting,** explains why particles of similar size are deposited together.

The material deposited by a stream is called **alluvium,** the general term for any stream-deposited sediment. Many different depositional features are composed of alluvium. Some occur within stream channels, some occur on the valley floor adjacent to the channel, and some exist at the mouth of the stream. We will consider the nature of these features in the next section.

Figure 3.10 During floods, both capacity and competence increase. Therefore, the greatest erosion and sediment transport occur during these high-water periods. Here, we see the sediment-filled floodwaters of Kenya's Mara River. (Photo by Tim Davis/Stone)

Did You Know?

In the United States, hydroelectric power plants contribute about 5 percent of the country's energy needs. Most of this energy is produced at large dams.

Stream Channels

A basic characteristic of streamflow that distinguishes it from overland flow is that it is usually confined to a channel. A stream channel can be thought of as an open conduit that consists of the streambed and banks that act to confine the flow except during floods.

Although somewhat oversimplified, we can divide stream channels into two types. *Bedrock channels* are those in which the streams are actively cutting into solid rock. In contrast, when the bed and banks are composed mainly of unconsolidated sediment, the channel is called an *alluvial channel.*

Bedrock Channels

In their headwaters, where the gradient is steep, most rivers cut into bedrock. These streams typically transport coarse particles that actively abrade the bedrock channel. Potholes are often visible evidence of the erosional forces at work.

Bedrock channels typically alternate between relatively gently sloping segments where alluvium tends to accumulate, and steeper segments where bedrock is exposed. These steeper areas may contain rapids or occasionally a waterfall (see Figure 3.8 on p. 73). The channel pattern exhibited by streams cutting into bedrock is controlled by the underlying geologic structure. Even when flowing over rather uniform bedrock, streams tend to exhibit a winding or irregular pattern rather than flowing in a straight channel. Anyone who has gone on a white-water rafting trip has observed the steep, winding nature of a stream flowing in a bedrock channel.

Alluvial Channels

Many stream channels are composed of loosely consolidated sediment (alluvium) and therefore can undergo major changes in shape because the sediments are continually being eroded, transported, and redeposited. The major factors affecting the shapes of these channels is the average size of the sediment being transported, the channel gradient, and the discharge.

Alluvial channel patterns reflect a stream's ability to transport its load at a uniform rate, while expending the least amount of energy. Thus, the size and type of sediment being

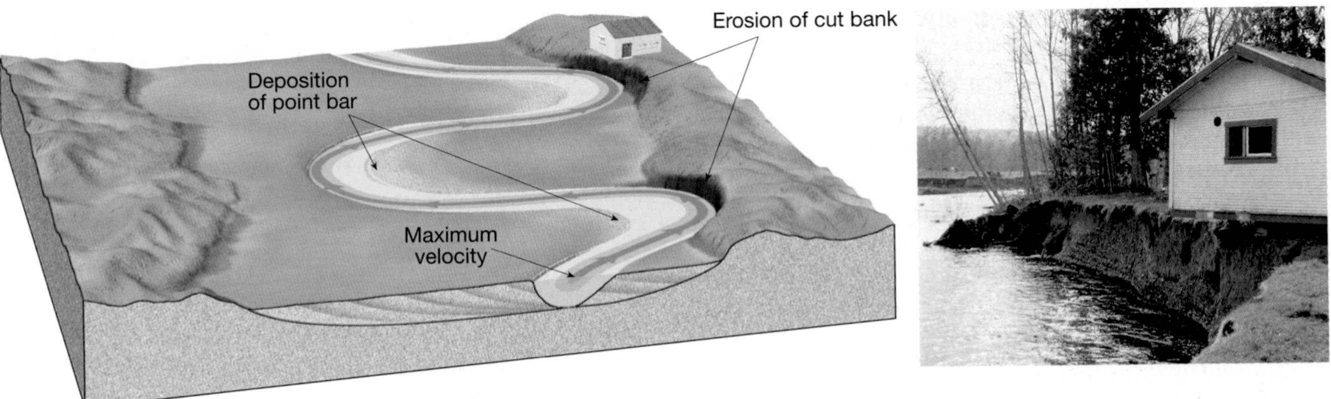

Figure 3.11 When a stream meanders, its zone of maximum speed shifts toward the outer bank. Because the outside of a meander is a zone of active erosion, it is called the *cut bank*. The house in the photo is located next to a cut bank. A point bar is deposited when water on the inside of a meander slows. (Photo by P. A. Glancy/U.S. Geological Survey)

carried help determine the nature of the stream channel. Two common types of alluvial channels are *meandering channels* and *braided channels.*

Meandering Channels Streams that transport much of their load in suspension generally move in sweeping bends called **meanders.** These streams flow in relatively deep, smooth channels and transport mainly mud (silt and clay). The lower Mississippi River exhibits a channel of this type.

Because of the cohesiveness of consolidated mud, the banks of stream channels carrying fine particles tend to resist erosion. As a consequence, most of the erosion in such channels occurs on the outside of the meander, where velocity and turbulence are greatest. In time, the outside bank is undermined, especially during periods of high water. Because the outside of a meander is a zone of active erosion, it is often referred to as the *cut bank* (Figure 3.11). The debris acquired by the stream at the cut bank moves downstream with the coarser material generally being deposited as *point bars* in zones of decreased velocity on the insides of meanders. In this manner, meanders migrate laterally by eroding the outside of the bends and depositing on the inside.

In addition to migrating laterally, the bends in a channel also migrate down the valley. This occurs because erosion is more effective on the downstream (downslope) side of the meander. The downstream migration of a meander is sometimes slowed when it reaches a more resistant material. This allows the next meander upstream to "catch up" and overtake it as shown in Figure 3.12. The neck of land between the meanders is gradually narrowed. Eventually, the river may erode through the narrow neck of land to the next loop (Figure 3.12). The new, shorter channel segment is called a **cutoff** and, because of its shape, the abandoned bend is called an **oxbow lake** (Figure 3.13).

Braided Channels Some streams consist of a complex network of converging and diverging channels that thread their way among numerous islands or gravel bars (Figure 3.14). Because these channels have an interwoven appearance, they are said to be **braided.** Braided channels form where a large

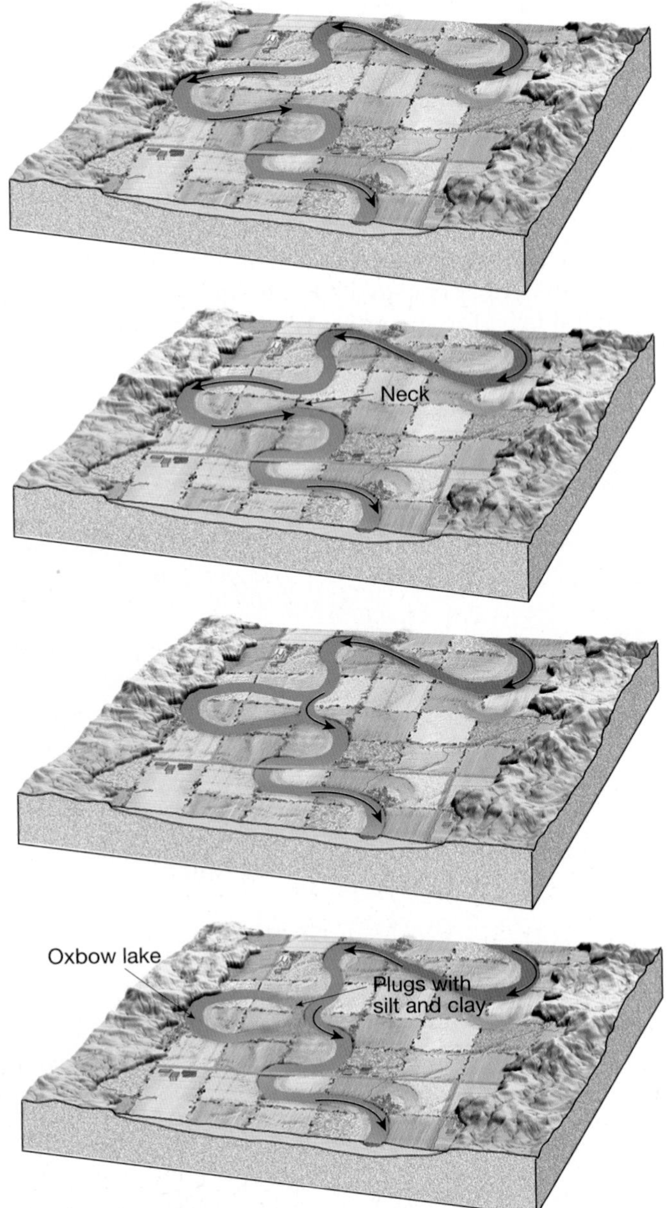

Figure 3.12 Formation of a cutoff and oxbow lake.

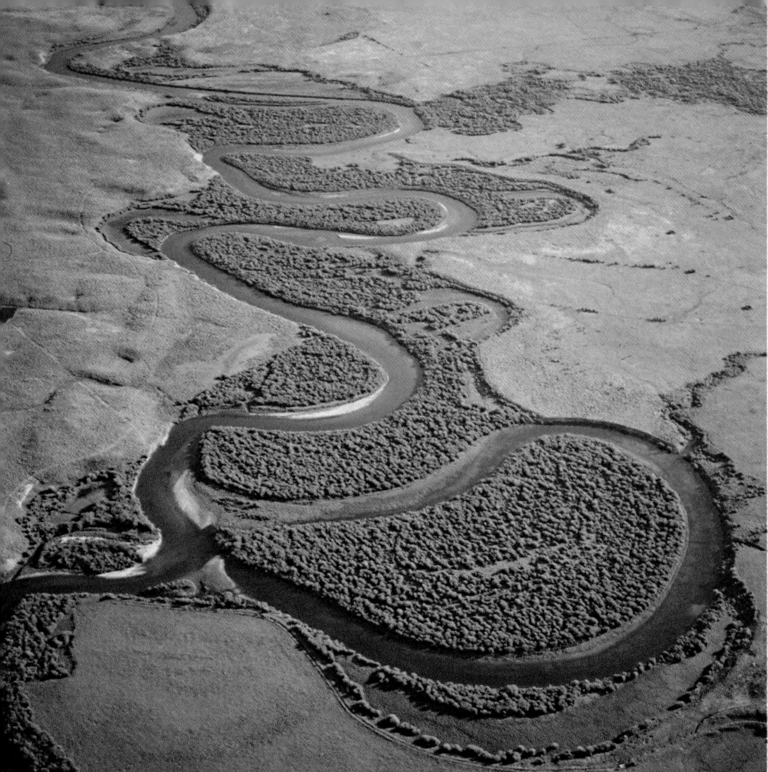

Figure 3.13 Oxbow lakes occupy abandoned meanders. As they fill with sediment, oxbow lakes gradually become swampy meander scars. Aerial view of an oxbow lake created by the meandering Green River near Bronx, Wyoming. (Photo by Michael Collier)

proportion of the stream's load consists of coarse material (sand and gravel) and the stream has a highly variable discharge. Because the bank material is readily erodable, braided channels are wide and shallow.

One circumstance in which braided streams form is at the end of a glacier where there is a large seasonal variation in discharge. Here, large amounts of ice-eroded sediment are dumped into the meltwater streams flowing away from the glacier. When flow is sluggish, the stream is unable to move all of the sediment and therefore deposits the coarsest material as bars that force the flow to split and follow several paths. Usually the laterally shifting channels completely rework most of the surface sediments each year, thereby transforming the entire streambed. In some braided streams, however, the bars have built up to form islands that are anchored by vegetation.

Figure 3.14 Braided stream choked with sediment near the terminus of a melting glacier. (Photo by Bradford Washburn)

In summary, meandering channels develop where the load consists largely of fine-grained particles that are transported as suspended load in a deep, smooth channel. By contrast, wide, shallow braided channels develop where coarse-grained alluvium is transported as bedload.

Base Level and Stream Erosion

Streams cannot endlessly erode their channels deeper and deeper. There is a lower limit to how deep a stream can erode, and that limit is called **base level.** Most often, a stream's base level occurs where a stream enters the ocean, a lake, or another stream.

Two general types of base level are recognized. Sea level is considered the *ultimate base level,* because it is the lowest level to which stream erosion could lower the land. *Temporary,* or *local, base levels* include lakes, resistant layers of rock, and main streams that act as base level for their tributaries. For example, when a stream enters a lake, its velocity quickly approaches zero and its ability to erode ceases. Thus, the lake prevents the stream from eroding below its level at any point upstream from the lake. However, because the outlet of the lake can cut downward and drain the lake, the lake is only a temporary hindrance to the stream's ability to downcut its channel. In a similar manner, the layer of resistant rock at the lip of the waterfall in Figure 3.15 acts as a temporary base level. Until the ledge of hard rock is eliminated, it will limit the amount of downcutting upstream.

Any change in base level will cause a corresponding readjustment of stream activities. When a dam is built along a stream, the reservoir that forms behind it raises the base level of the stream (Figure 3.16). Upstream from the dam, the gradient is reduced, lowering the stream's velocity and, hence, its sediment-transporting ability. The stream, now having too little energy to transport its entire load, will deposit sediment. This builds up its channel. Deposition will be the dominant process until the stream's gradient increases sufficiently to transport its load.

Shaping Stream Valleys

Streams, with the aid of weathering and mass wasting, shape the landscape through which they flow. As a result, streams continuously modify the valleys that they occupy.

A **stream valley** consists not only of the channel but also the surrounding terrain that directly contributes water to the stream. Thus, it includes the valley bottom, which is the lower, flatter area that is partially or totally occupied by the stream channel, and the sloping valley walls that rise above the valley bottom on both sides. Most stream valleys are much broader at the top than is the width of their channel at the bottom. This would not be the case if the only agent responsible for eroding valleys were the streams flowing through them. The sides of most valleys are shaped by a

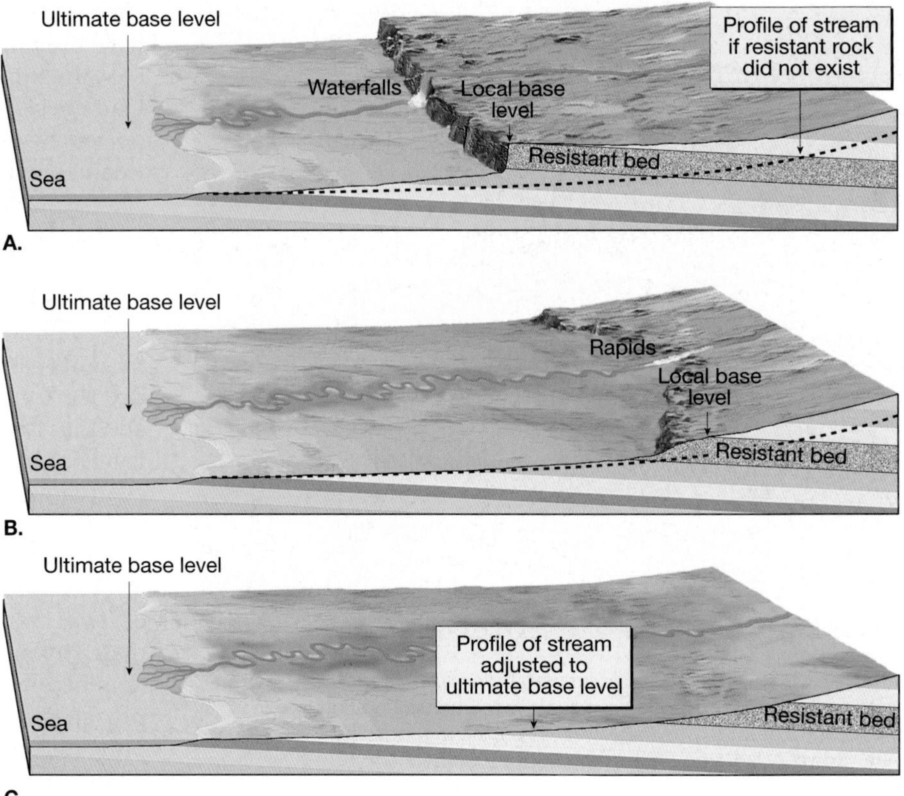

Figure 3.15 A resistant layer of rock can act as a local (temporary) base level. Because the durable layer is eroded more slowly, it limits the amount of downcutting upstream.

combination of weathering, overland flow, and mass wasting. In some arid regions, where weathering is slow and where rock is particularly resistant, narrow valleys having nearly vertical walls are common.

Stream valleys can be divided into two general types—narrow V-shaped valleys and wide valleys with flat floors—with many gradations between.

Valley Deepening

When a stream's gradient is steep and the channel is well above base level, downcutting is the dominant activity. Abrasion caused by bed-load sliding and rolling along the bottom, and the hydraulic power of fast-moving water, slowly lower the streambed. The result is usually a V-shaped valley

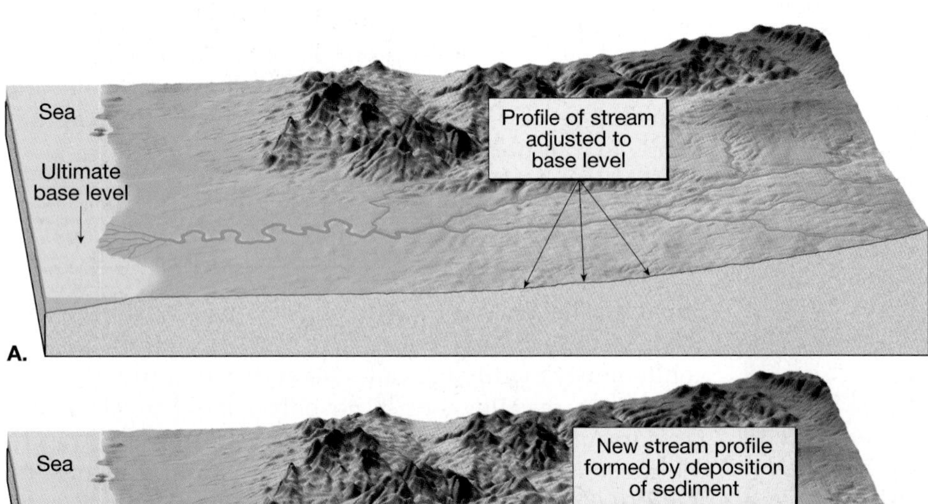

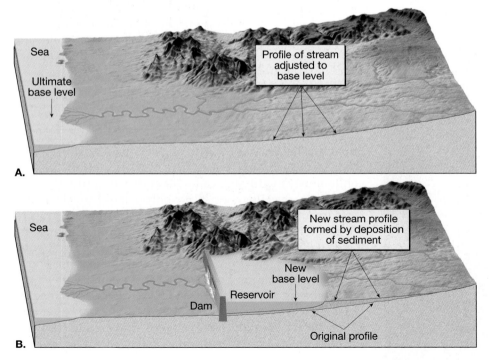

Figure 3.16 When a dam is built and a reservoir forms, the stream's base level is raised. This reduces the stream's velocity and leads to deposition and a reduction of the gradient upstream from the reservoir.

with steep sides. A classic example of a V-shaped valley is located in the section of Yellowstone River shown in Figure 3.17.

The most prominent features of a V-shaped valley are *rapids* and *waterfalls*. Both occur where the stream's gradient increases significantly, a situation usually caused by variations in the erodability of the bedrock into which a stream channel is cutting. Resistant beds create rapids by acting as a temporary base level upstream while allowing downcutting to continue downstream. In time, erosion usually eliminates the resistant rock. Waterfalls are places where the stream makes an abrupt vertical drop.

Valley Widening

Once a stream has cut its channel closer to base level, downward erosion becomes less dominant. At this point, the stream's channel takes on a meandering pattern, and more of the stream's energy is directed from side to side. The result is a widening of the valley as the river cuts away first at one bank and then at the other (Figure 3.18). The continuous lateral erosion caused by shifting of the stream's meanders produces an increasingly broader, flat valley floor covered with alluvium. This feature, called a **floodplain,** is appropriately named because when a river overflows its banks during flood stage, it inundates the floodplain.

Over time, the floodplain will widen to the point that the stream is only actively eroding the valley walls in a few places. In fact, in large rivers such as the lower Mississippi River valley, the distance from one valley wall to another can exceed 160 kilometers (100 miles).

Did You Know?

The world's highest uninterrupted waterfall is Angel Falls on Venezuela's Churun River. Named for American aviator Jimmie Angel, who first sighted the falls from the air in 1933, the river plunges 979 meters (3212 feet).

Figure 3.17 V-shaped valley of the Yellowstone River. The rapids and waterfalls indicate that the river is vigorously downcutting. (Photo by Art Wolfe, Inc.)

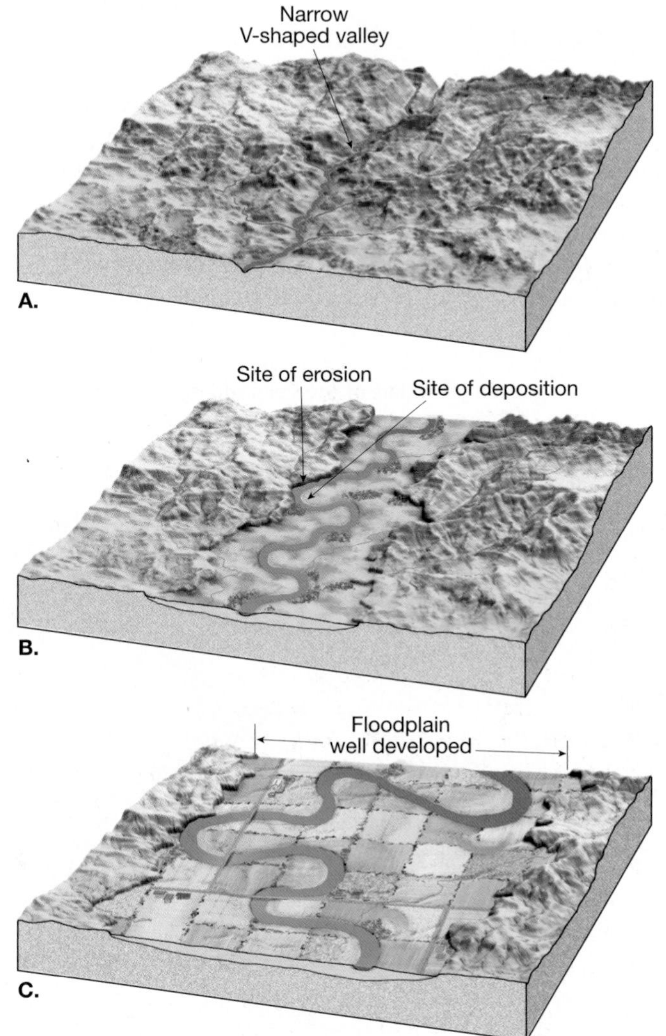

Figure 3.18 Stream eroding its floodplain.

Changing Base Level and Incised Meanders

We usually expect a stream with a highly meandering course to be on a floodplain in a wide valley. However, certain rivers exhibit meandering channels that flow in steep, narrow valleys. Such meanders are called **incised** (*incisum* = to cut into) **meanders** (Figure 3.19). How do such features form?

Originally, the meanders probably developed on the floodplain of a stream that was relatively near base level. Then, a change in base level caused the stream to begin downcutting. One of two events could have occurred. Either base level dropped or the land upon which the river flowed was uplifted.

An example of the first circumstance happened during the Ice Age when large quantities of water were withdrawn from the ocean and locked up in glaciers on land. The result was that sea level (ultimate base level) dropped, causing rivers flowing into the ocean to begin to downcut.

Regional uplift of the land, the second cause for incised meanders, is exemplified by the Colorado Plateau in the southwestern United States. Here, as the plateau was gradually uplifted, numerous meandering rivers adjusted to being higher above base level by downcutting (Figure 3.19).

Depositional Landforms

As indicated earlier, whenever a stream's velocity slows, it begins to deposit some of the sediment it is carrying. Also recall that streams continually pick up sediment in one part of their channels and redeposit it downstream. These channel deposits are most often composed of sand and gravel and are commonly referred to as **bars.** Such features, however, are only temporary, for the material will be picked up again and eventually carried to the ocean. In addition to sand and gravel bars, streams also create other depositional features that have somewhat longer life spans. These include deltas and natural levees.

Figure 3.19 Incised meanders of the Colorado River in Canyonlands National Park, Utah. Here, as the Colorado Plateau was gradually uplifted, the meandering river adjusted to being higher above base level by downcutting. (Photo by Michael Collier)

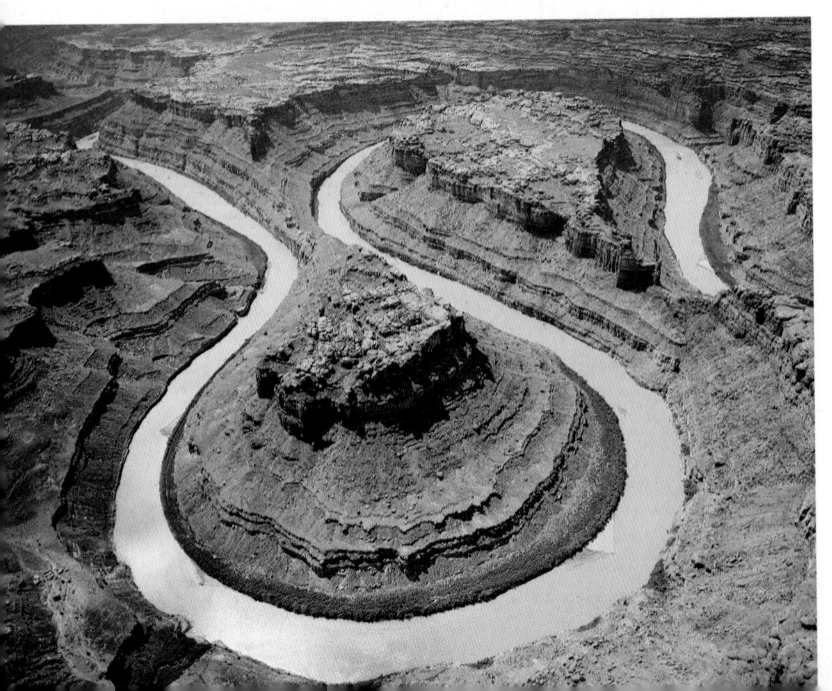

Deltas

When a stream enters the relatively still waters of an ocean or lake, its velocity drops abruptly, and the resulting deposits form a **delta** (Figure 3.20). As the delta grows outward, the stream's gradient continually lessens. This circumstance eventually causes the channel to become choked with sediment deposited from the slowing water. As a consequence, the river seeks a shorter, higher-gradient route to base level, as illustrated in Figure 3.20B. This illustration shows the main channel dividing into several smaller ones, called **distributaries.** Most deltas are characterized by these shifting channels that act in an opposite way to that of tributaries.

Rather than carrying water into the main channel, distributaries carry water away from the main channel. After numerous shifts of the channel, a delta may grow into a roughly triangular shape like the Greek letter delta (Δ), for which it is named. Note, however, that many deltas do not exhibit the idealized shape. Differences in the configurations of shorelines and variations in the nature and strength of wave activity result in many shapes. Many large rivers have deltas extending over thousands of square kilometers. The delta of the Mississippi River is one example. It resulted from the accumulation of huge quantities of sediment derived from the vast region drained by the river and its tributaries. Today, New Orleans rests where there was ocean less than 5000 years ago. Figure 3.21 shows that portion of the Mississippi delta that has been built over the past 5000 to 6000 years. As you can see, the delta is actually a series of seven coalescing subdeltas. Each formed when the river left its existing channel in favor of a shorter, more direct path to the Gulf of Mexico. The individual subdeltas interfinger and partially cover one another to produce a very complex structure. The present subdelta, called a *bird-foot* delta because of the configuration of its distributaries, has been built by the Mississippi in the past 500 years.

Natural Levees

Some rivers occupy valleys with broad floodplains and build **natural levees** that parallel their channels on both banks (Figure 3.22). Natural levees are built by successive floods over many years. When a stream overflows its banks, its velocity immediately diminishes, leaving coarse sediment deposited in strips bordering the channel. As the water spreads out over the valley, a lesser amount of fine sediment is deposited over the valley floor. This uneven distribution of material produces the very gentle slope of the natural levee.

The natural levees of the lower Mississippi rise 6 meters (20 feet) above the floodplain. The area behind the levee is characteristically poorly drained for the obvious reason that water cannot flow up the levee and into the river. Marshes called **backswamps** result. A tributary stream that cannot enter a river

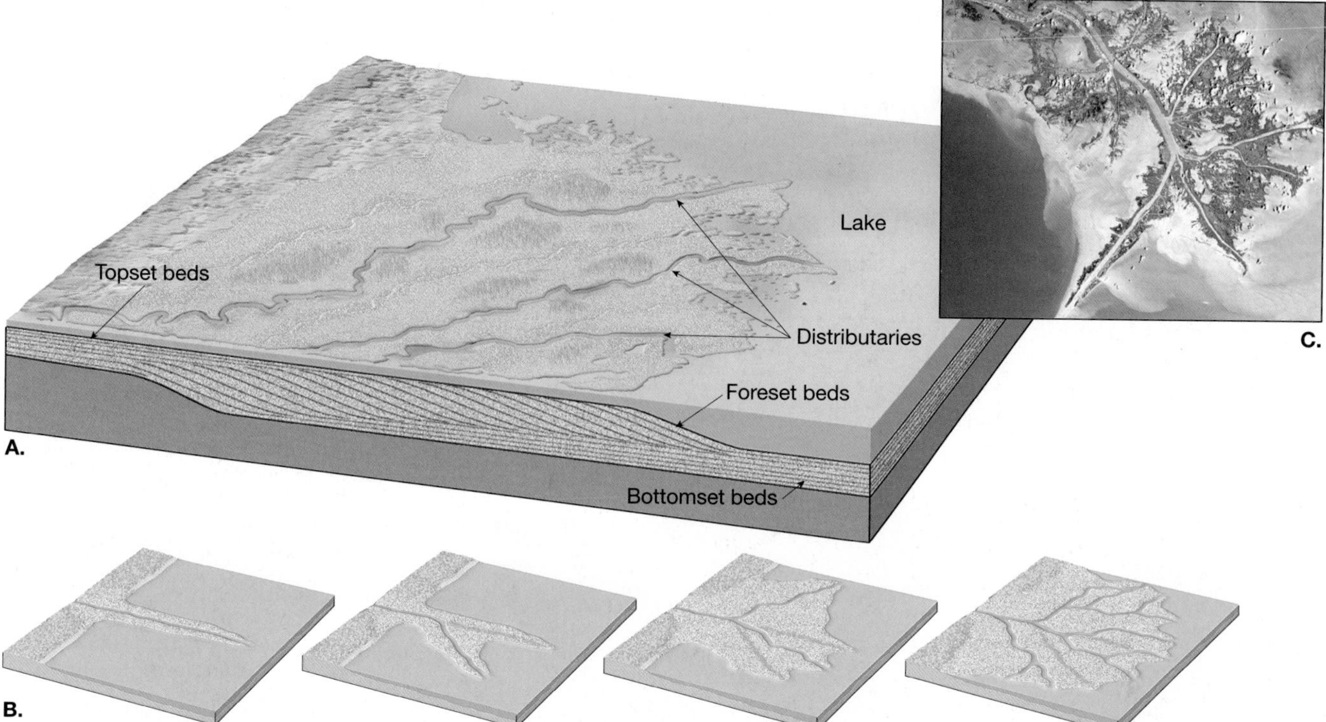

Figure 3.20 **A.** Structure of a simple delta that forms in the relatively quiet waters of a lake. **B.** Growth of a simple delta. As a stream extends its channel, the gradient is reduced. During flood stage, the river is frequently diverted to a higher-gradient route, forming a new distributary. Old, abandoned distributaries are gradually invaded by aquatic vegetation and fill with sediment. (After Ward's Natural Science Establishment, Inc., Rochester, NY) **C.** Satellite image of a portion of the delta of the Mississippi River in May 2001. For the past 500 years or so, the main flow of the river has been along its present course, extending southeast from New Orleans. During that span, the delta advanced into the Gulf of Mexico at a rate of about 10 kilometers (6 miles) per century. Notice the numerous distributaries. (NASA/GSFC/METI/ERSDAC/JAROS, and U.S./Japan ASTER Science Team)

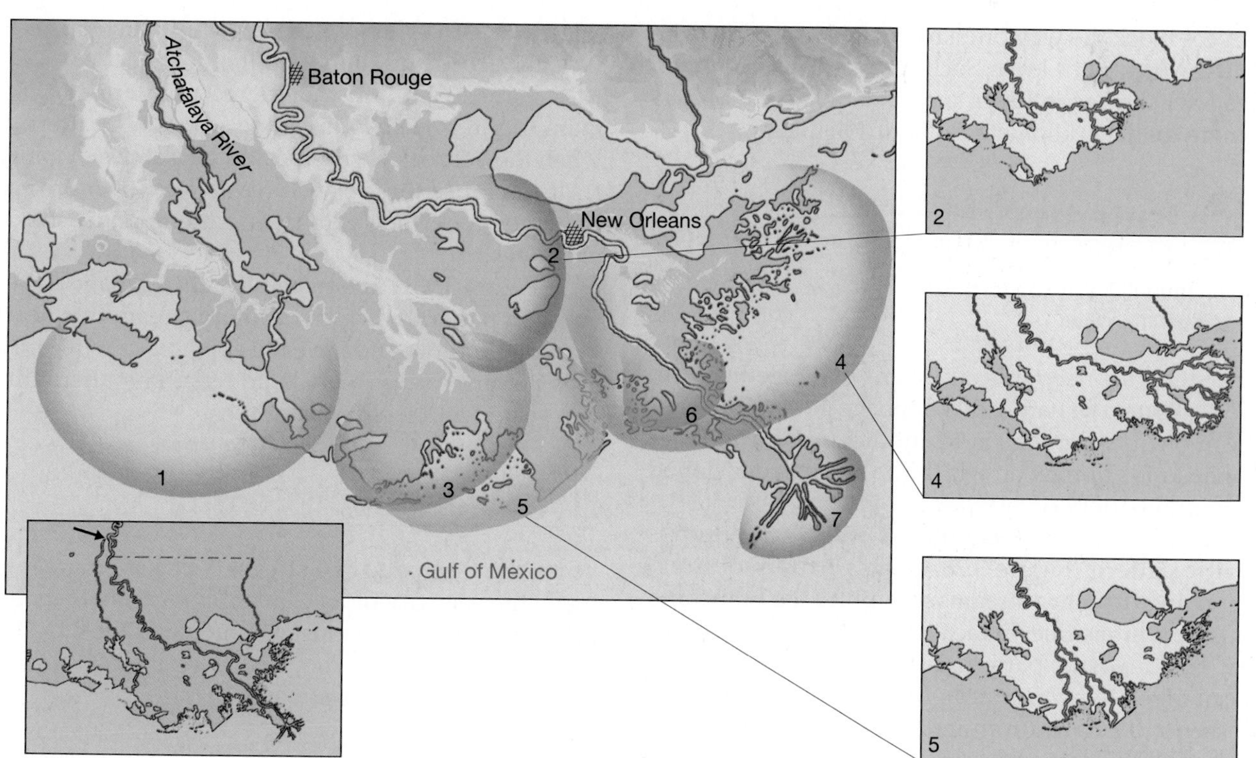

Figure 3.21 During the past 5000 to 6000 years, the Mississippi River has built a series of seven coalescing subdeltas. The numbers indicate the order in which the subdeltas were deposited. The present birdfoot delta (number 7) represents the activity of the past 500 years. Without ongoing human efforts, the present course will shift and follow the path of the Atchafalaya River. The inset on the left shows the point where the Mississippi may someday break through (arrow) and the shorter path it would take to the Gulf of Mexico. (After C. R. Kolb and J. R. Van Lopik)

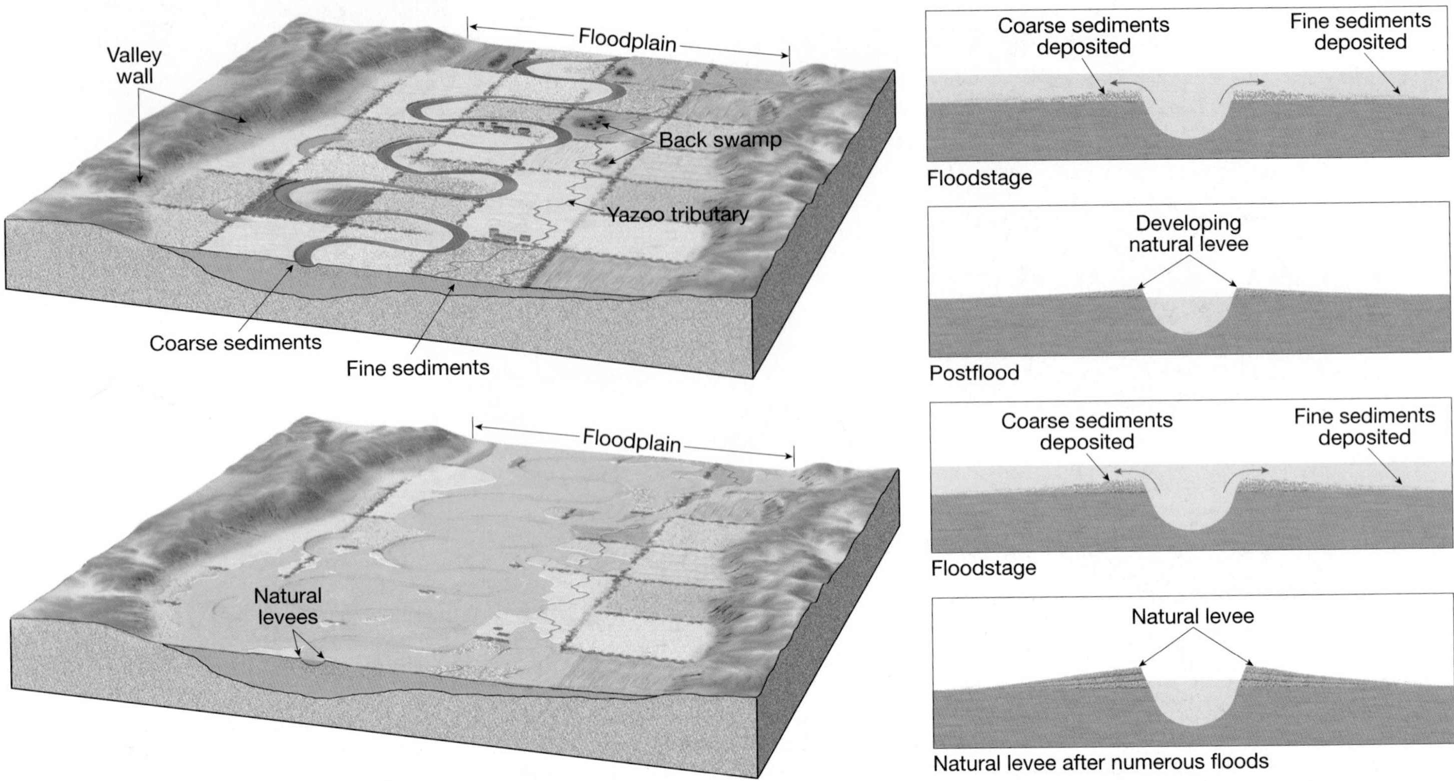

Figure 3.22 Natural levees are gently sloping landforms created by repeated floods. The diagrams on the right show the sequence of development. Because the ground next to the stream channel is higher than the adjacent floodplain, back swamps and yazoo tributaries may develop.

because levees block the way often has to flow parallel to the river until it can breach the levee. Such streams are called **yazoo tributaries** after the Yazoo River, which parallels the Mississippi for more than 300 kilometers (about 190 miles).

Drainage Patterns

Sculpturing Earth's Surface
▼ Running Water

Drainage systems are networks of streams that together form distinctive patterns. The nature of a drainage pattern can vary greatly from one type of terrain to another, primarily in response to the kinds of rock on which the streams developed or the structural pattern of faults and folds.

The most commonly encountered drainage pattern is the **dendritic pattern** (Figure 3.23A). This pattern of irregularly branching tributary streams resembles the branching pattern of a deciduous tree. In fact, the word *dendritic* means "treelike." The dendritic pattern forms where the underlying material is relatively uniform. Because the surface material is essentially uniform in its resistance to erosion, it does not control the pattern of streamflow. Rather, the pattern is determined chiefly by the direction of slope of the land.

When streams diverge from a central area like spokes from the hub of a wheel, the pattern is said to be **radial** (Figure 3.23B). This pattern typically develops on isolated volcanic cones and domal uplifts.

Figure 3.23C illustrates a **rectangular** pattern, in which many right-angle bends can be seen. This pattern develops when the bedrock is crisscrossed by a series of joints and/or faults. Because these structures are eroded more easily than unbroken rock, their geometric pattern guides the directions of valleys.

Figure 3.23D illustrates a **trellis** drainage pattern, a rectangular pattern in which tributary streams are nearly parallel to one another and have the appearance of a garden trellis. This pattern forms in areas underlain by alternating bands of resistant and less-resistant rock.

Floods and Flood Control

When the discharge of a stream becomes so great that it exceeds the capacity of its channel, it overflows its banks as a **flood.** Floods are the most common and most destructive of all geologic hazards. They are, nevertheless, simply part of the *natural* behavior of streams.

Causes of Floods

Rivers flood because of the weather. Rapid melting of snow in the sping and/or major storms that bring heavy rains over a large region cause most floods. Exceptional rains caused

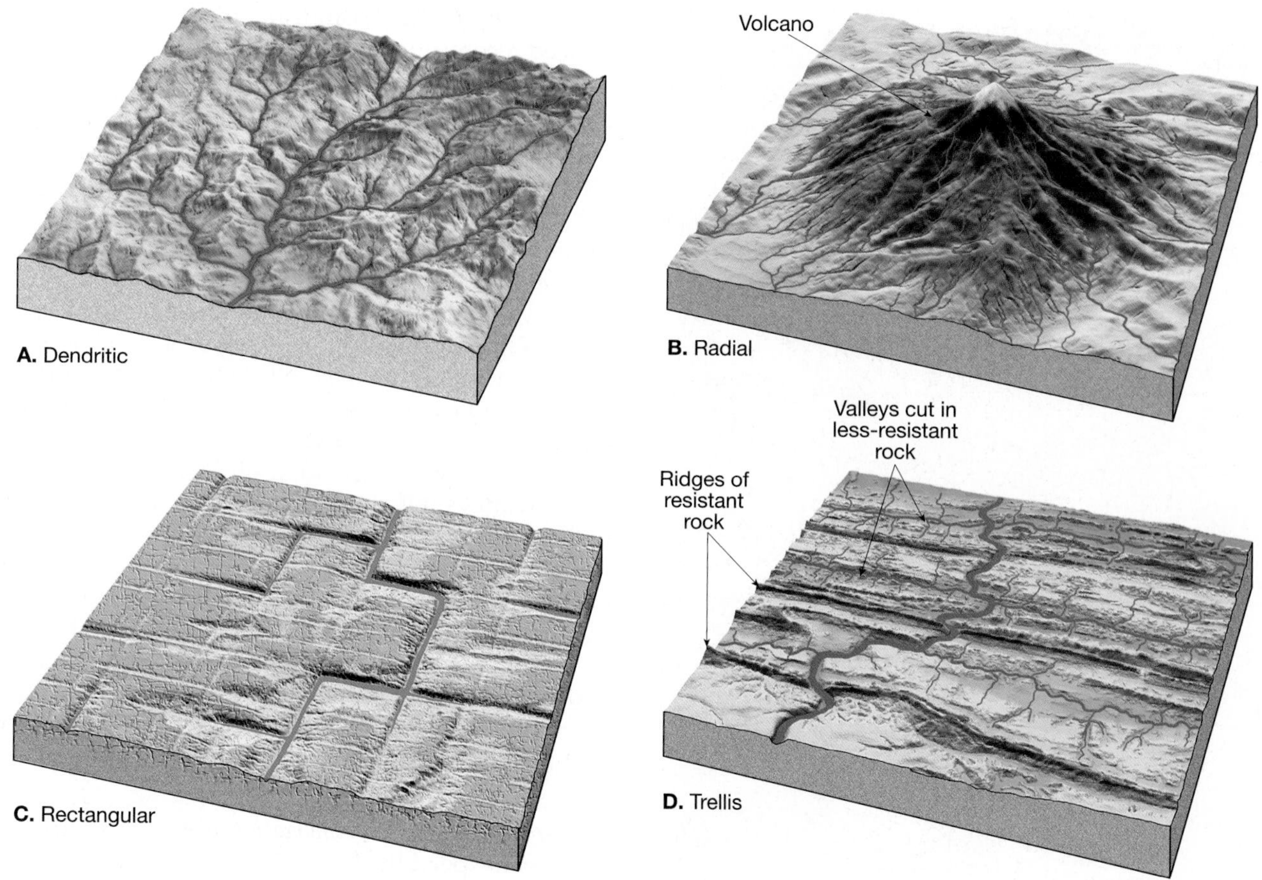

Figure 3.23 Drainage patterns. **A.** Dendritic **B.** Radial **C.** Rectangular **D.** Trellis

the devastating floods in the upper Mississippi River Valley during the summer of 1993 (Figure 3.24).

Unlike the extensive regional floods just mentioned, *flash floods* are more limited in extent. Flash floods occur with little warning and can be deadly because they produce a rapid rise in water levels and can have a devastating flow velocity. Several factors influence flash flooding. Among them are rainfall intensity and duration, surface conditions, and topography. Urban areas are susceptible to flash floods because a high percentage of the surface area is composed of impervious roofs, streets, and parking lots, where runoff is very rapid. Mountainous areas are susceptible because steep slopes can funnel runoff into narrow canyons.

Did You Know?

Humans have covered an amazing amount of land with buildings, parking lots, and roads. A recent study indicated that the area of such impervious surfaces in the United States (excluding Alaska and Hawaii) amounts to more than 112,600 square kilometers (nearly 44,000 square miles), which is slightly less than the area of the state of Ohio.

Human interference with the stream system can worsen or even cause floods. A prime example is the failure of a dam or an artificial levee. These structures are built for flood protection. They are designed to contain floods of a certain magnitude. If a larger flood occurs, the dam or levee is overtopped. If the dam or levee fails or is washed out, the water behind it is released to become a flash flood. The bursting of a dam in 1889 on the Little Conemaugh River caused the devastating Johnstown, Pennsylvania, flood that took some 3000 lives. A second dam failure occured there in 1977 and caused 77 fatalities.

Flood Control

Several strategies have been devised to eliminate or lessen the catastrophic effects of floods. Engineering efforts include the construction of artificial levees, the building of flood-control dams, and river channelization.

Artificial Levees *Artificial levees* are earthen mounds built on the banks of a river to increase the volume of water the channel can hold. These most common of stream-containment structures have been used since ancient times and continue to be used today.

Artificial levees are usually easy to distinguish from natural levees because their slopes are much steeper. When a

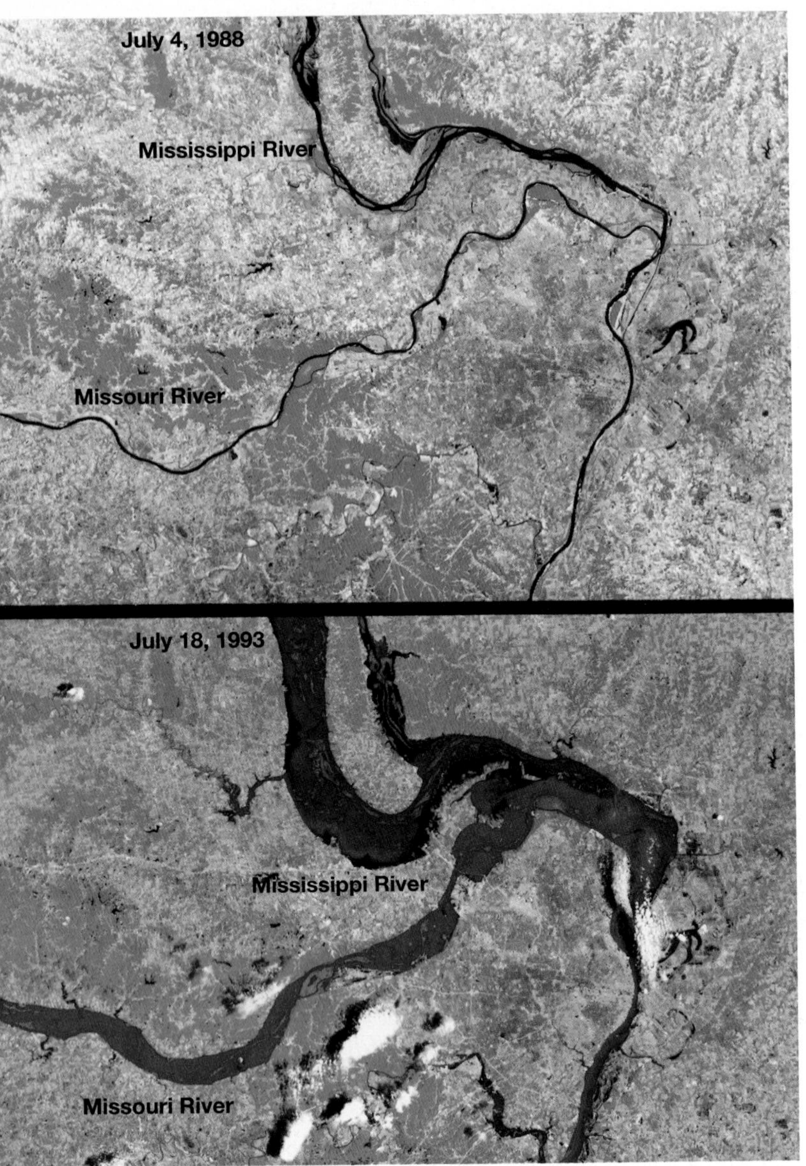

Figure 3.24 Satellite views of the Missouri River flowing into the Mississippi River. St. Louis is just south of their confluence. The upper image shows the rivers during a drought that occurred in summer 1988. The lower image depicts the peak of the record-breaking 1993 flood. Exceptional rains produced the wettest spring and early summer of the twentieth century in the upper Mississippi River basin. In all, nearly 14 million acres were inundated, displacing at least 50,000 people. (Photos courtesy of Spaceimaging.com)

river is confined by levees during periods of high water, it frequently deposits material in its channel as the discharge diminishes. This is sediment that otherwise would have been dropped on the floodplain. Thus, each time there is a high flow, deposits are left on the river bed and the bottom of the channel is built up. With the buildup of the bed, less water is required to overflow the original levee. As a result, the height of the levee may have to be raised periodically to protect the floodplain. Moreover, many artificial levees are not built to withstand periods of extreme flooding. For example, levee failures were numerous in the Midwest during summer 1993, when the upper Mississippi and many of its tributaries experienced record floods (Figure 3.25).

Flood-Control Dams *Flood-control dams* are built to store floodwater and then let it out slowly. This lowers the flood crest by spreading it out over a longer time span. Ever since the 1920s, thousands of dams have been built on nearly every major river in the United States. Many dams have significant non-flood-related functions such as providing water for irrigated agriculture and for hydroelectric power generation. Many reservoirs are also major regional recreational facilities.

Although dams may reduce flooding and provide other benefits, building these structures also has significant costs and consequences. For example, reservoirs created by dams may cover fertile farmland, useful forests, historic sites, and scenic valleys. Of course, dams trap sediment. Therefore, deltas and floodplains downstream erode because they are no longer replenished with silt during floods. Large dams can also cause significant ecologic damage to river environments that took thousands of years to establish.

Building a dam is not a permanent solution to flooding. Sedimentation behind a dam means that the volume of its reservoir will gradually diminish, reducing the effectiveness of this flood-control measure.

Channelization *Channelization* involves altering a stream channel to speed the flow of water to prevent it from reaching flood height. This may simply involve clearing a channel of obstructions or dredging a channel to make it wider and deeper.

A more radical alteration involves straightening a channel by creating *artificial cutoffs.* The idea is that by shortening the stream, the gradient and hence the velocity are increased. By increasing velocity, the larger discharge associated with flooding can be dispersed more rapidly.

Beginning in the early 1930s, the U.S. Army Corps of Engineers created many artificial cutoffs on the Mississippi for the purpose of increasing the efficiency of the channel and reducing the threat of flooding. In all, the river has been shortened more than 240 kilometers (150 miles). The program has been somewhat successful in reducing the height of the river in flood stage. However, because the river's tendency toward meandering still exists, preventing the river from returning to its previous course has been difficult.

A Nonstructural Approach All of the flood-control measures described so far have involved structural solutions aimed at "controlling" a river. These solutions are expensive and often give people residing on the floodplain a false sense of security.

Today, many scientists and engineers advocate a nonstructural approach to flood control. They suggest that an alternative to artificial levees, dams, and channelization is sound floodplain management. By identifying high-risk areas, appropriate zoning regulations can be implemented to minimize development and promote more appropriate land use.

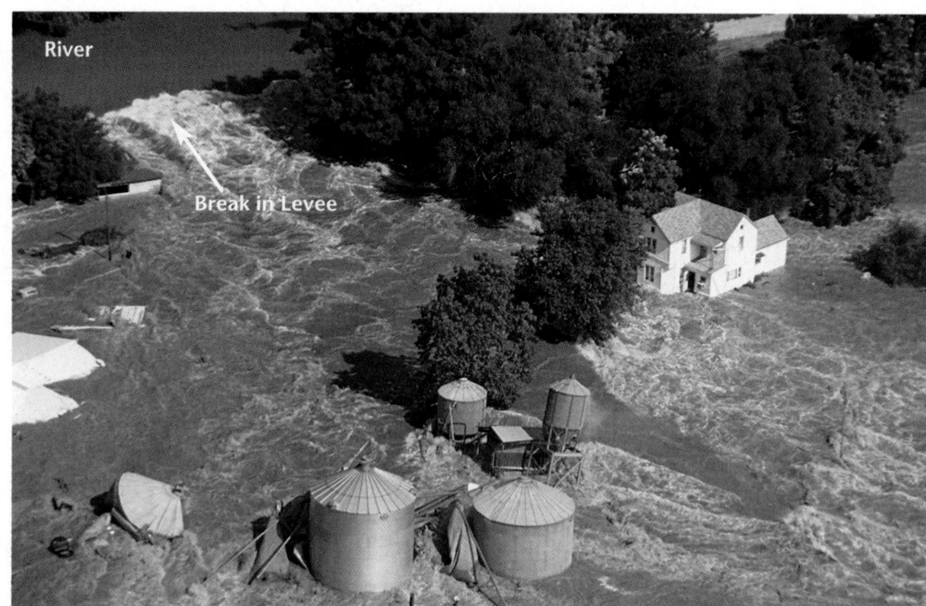

Figure 3.25 Water rushes through a break in an artificial levee in Monroe County, Illinois. During the record-breaking 1993 Midwest floods, many artificial levees could not withstand the force of the floodwaters. Sections of many weakened structures were overtopped or simply collapsed. (Photo by James A. Finley/AP/Wide World Photos)

Groundwater: Water Beneath the Surface

Sculpturing Earth's Surface
▼ Groundwater

Groundwater is one of our most important and widely available resources. Yet people's perceptions of the subsurface environment from which it comes are often unclear and incorrect. The reason is that groundwater is hidden from view except in caves and mines, and the impressions people gain from these subsurface openings are often misleading. Observations on the land surface give an impression that Earth is "solid." This view is not changed very much when we enter a cave and see water flowing in a channel that appears to have been cut into solid rock.

Because of such observations, many people believe that groundwater occurs only in underground "rivers." But actual rivers underground are extremely rare. In reality, most of the subsurface environment is not 'solid' at all. Rather, it includes countless tiny *pore spaces* between grains of soil and sediment plus narrow joints and fractures in bedrock. Together, these spaces add up to an immense volume. It is in these tiny openings that groundwater collects and moves.

The Importance of Groundwater

Considering the entire hydrosphere, or all of Earth's water, only about six-tenths of 1 percent occurs underground. Nevertheless, this small percentage, stored in the rocks and sediments beneath Earth's surface, is a vast quantity. When the oceans are excluded and only sources of freshwater are considered, the significance of groundwater becomes more apparent.

Figure 3.26 contains estimates of the distribution of freshwater in the hydrosphere. Clearly, the largest volume occurs as glacial ice. Second in rank is groundwater, with slightly more than 14 percent of the total. However, when ice is excluded and just liquid water is considered, more than 94 percent is groundwater. Without question, *groundwater represents the largest reservoir of freshwater that is readily available to humans.* Its value in terms of economics and human well-being is incalculable.

Worldwide, wells and springs provide water for cities, crops, livestock, and industry. In the United States, groundwater is the source of about 40 percent of the water used for all purposes (except hydroelectric power generation and power plant cooling). Groundwater is the drinking water for more than 50 percent of the population, is 40 percent of the water for irrigation, and provides more than 25 percent of industry's needs. In some areas, however, overuse of this basic resource has caused serious problems, including streamflow

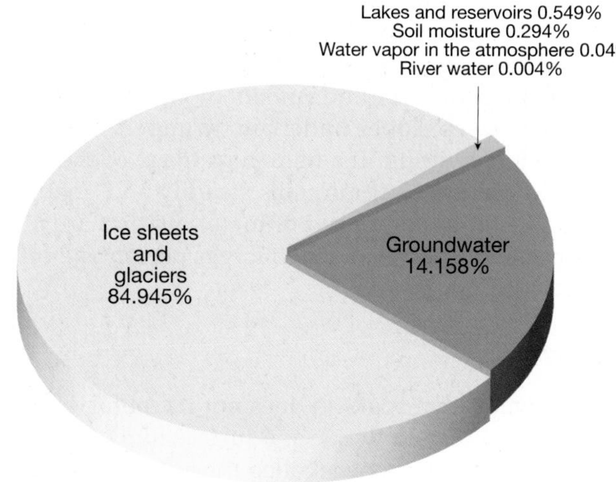

Figure 3.26 Estimated distribution of freshwater in the hydrosphere, according to the U.S. Geological Survey.

Did You Know?

Although most caves and sinkholes are associated with regions underlain by limestone, these features can also form in gypsum and rock salt (halite) because these rocks are highly soluble and readily dissolved.

depletion, land subsidence, and increased pumping costs. In addition, groundwater contamination resulting from human activities is a real and growing threat in many places.

Groundwater's Geologic Roles

Geologically, groundwater is important as an erosional agent. The dissolving action of groundwater slowly removes rock, allowing surface depressions known as sinkholes to form as well as creating subterranean caverns (Figure 3.27). Groundwater is also an equalizer of streamflow. Much of the water that flows in rivers is not direct runoff from rain and snowmelt. Rather, a large percentage of precipitation soaks in and then moves slowly underground to stream channels. Groundwater is thus a form of storage that sustains streams during periods when rain does not fall. When we see water flowing in a river during a dry period, it is water from rain that fell at some earlier time and was stored underground.

A.

B.

Figure 3.27 **A.** A view of the interior of Three Fingers Cave, Lincoln County, New Mexico. The dissolving action of groundwater created the cavern. Later, groundwater deposited the limestone decorations. (Photo by Harris Photographic/Tom Stack and Associates) **B.** Groundwater was responsible for creating these sinkholes in a limestone plateau north of Jajce, Bosnia and Herzegovina. (Photo by Jerome Wyckoff)

Distribution and Movement of Groundwater

Sculpturing Earth's Surface
▼ Groundwater

When rain falls, some of the water runs off, some returns to the atmosphere by evaporation and transpiration, and the remainder soaks into the ground. This last path is the primary source of practically all groundwater. The amount of water that takes each of these paths, however, varies greatly from time to time and place to place. Influential factors include steepness of the slope, nature of the surface material, intensity of the rainfall, and type and amount of vegetation. Heavy rains falling on steep slopes underlain by impervious materials will obviously result in a high percentage of the water running off. Conversely, if rain falls steadily and gently on more gradual slopes composed of materials that are more easily penetrated by water, a much larger percentage of the water soaks into the ground.

Distribution

Some of the water that soaks in does not travel far, because it is held by molecular attraction as a surface film on soil particles. This near-surface zone is called the *belt of soil moisture.* It is crisscrossed by roots, voids left by decayed roots, and animal and worm burrows that enhance the infiltration of rainwater into the soil. Soil water is used by plants for life functions and transpiration. Some of this water also evaporates directly back into the atmosphere.

Water that is not held as soil moisture penetrates downward until it reaches a zone where all the open spaces in sediment and rock are completely filled with water. This is the

zone of saturation. Water within it is called **groundwater.** The upper limit of this zone is known as the **water table.** The area above the water table where the soil, sediment, and rock are not saturated is called the **unsaturated zone** (Figure 3.28). Although a considerable amount of water can be present in the unsaturated zone, this water cannot be pumped by wells because it clings too tightly to rock and soil particles. By contrast, below the water table, the water pressure is great enough to allow water to enter wells, thus permitting groundwater to be withdrawn for use. We will examine wells more closely later in the chapter.

The water table is rarely level, as we might expect a table to be. Instead, its shape is usually a subdued replica of the surface, reaching its highest elevations beneath hills and decreasing in height toward valleys (Figure 3.28). The water table of a wetland (swamp) is right at the surface. Lakes and streams generally occupy areas low enough that the water table is above the land surface.

Several factors contribute to the irregular surface of the water table. One important influence is the fact that groundwater moves very slowly. Because of this, water tends to "pile up" beneath high areas between stream valleys. If rainfall were to cease completely, these water "hills" would slowly subside and gradually approach the level of the adjacent valleys. However, new supplies of rainwater are usually added often enough to prevent this. Nevertheless, in times of extended drought, the water table may drop enough to dry up shallow wells. Other causes for the uneven water table are variations in rainfall and permeability of Earth materials from place to place.

Factors Influencing the Storage and Movement of Groundwater

The nature of subsurface materials strongly influences the rate of groundwater movement and the amount of groundwater that can be stored. Two factors are especially important—porosity and permeability.

Porosity Water soaks into the ground because bedrock, sediment, and soil contain countless voids or openings. These openings are similar to those of a sponge and are often called pore spaces. The quantity of groundwater that can be stored depends on the **porosity** of the material, which is the percentage of the total volume of rock or sediment that consists of pore spaces. Voids most often are spaces between sedimentary particles, but also common are joints, faults, cavities formed by the dissolving of soluble rock such as limestone, and vesicles (voids left by gases escaping from lava).

Variations in porosity can be great. Sediment is commonly quite porous, and open spaces may occupy 10 percent to 50 percent of the sediment's total volume. Pore space depends on the size and shape of the grains; how they are packed together; the degree of sorting; and in sedimentary rocks, the amount of cementing material. Most igneous and metamorphic rocks, as well as some sedimentary rocks, are composed of tightly interlocking crystals so the voids between grains may be negligible. In these rocks, fractures must provide the voids.

Permeability Porosity alone cannot measure a material's capacity to yield groundwater. Rock or sediment may be very

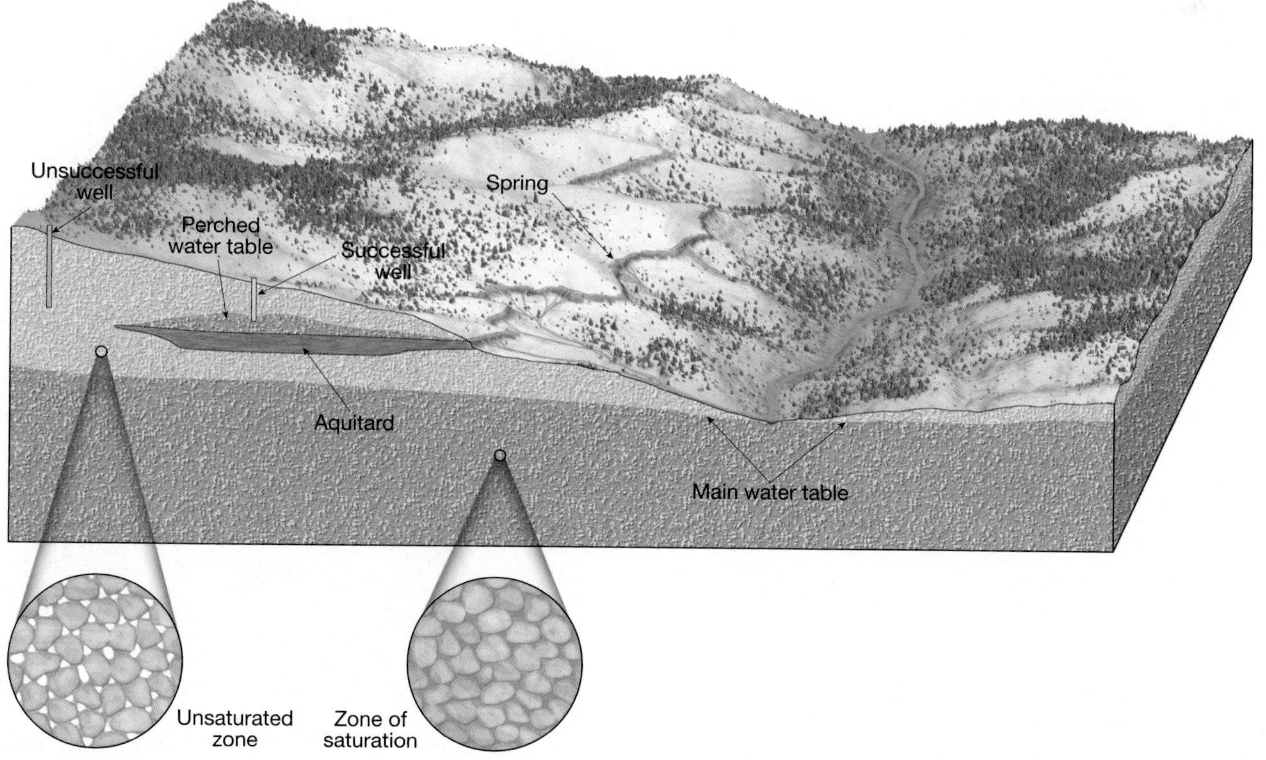

Figure 3.28 This diagram illustrates the relative positions of many features associated with subsurface water.

Did You Know?

Because of its high porosity, excellent permeability, and great size, the Ogallala Formation, the largest aquifer in the United States, accumulated huge amounts of groundwater—enough freshwater to fill Lake Huron.

porous and still prohibit water from moving through it. The **permeability** of a material indicates its ability to *transmit* a fluid. Groundwater moves by twisting and turning through interconnected small openings. The smaller the pore spaces, the slower the groundwater moves. If the spaces between particles are too small, water cannot move at all. For example, clay's ability to store water can be great, owing to its high porosity, but its pore spaces are so small that water is unable to move through it. Thus, we say that clay is *impermeable.*

Aquitards and Aquifers Impermeable layers such as clay that hinder or prevent water movement are termed **aquitards** (*aqua* = water, *tard* = slow). In contrast, larger particles, such as sand or gravel, have larger pore spaces. Therefore, the water moves with relative ease. Permeable rock strata or sediments that transmit groundwater freely are called **aquifers** ("water carriers"). Aquifers are important because they are the water-bearing layers sought after by well drillers.

Groundwater Movement

The movement of most groundwater is exceedingly slow, from pore to pore. A typical rate is a few centimeters per day. The energy that makes the water move is provided by the force of gravity. In response to gravity, water moves from areas where the water table is high to zones where the water table is lower. This means that water usually gravitates toward a stream channel, lake, or spring. Although some water takes the most direct path down the slope of the water table, much of the water follows long, curving paths toward the zone of discharge.

Figure 3.29 shows how water percolates into a stream from all possible directions. Some paths clearly turn upward, apparently against the force of gravity, and enter through the bottom of the channel. This is easily explained: The deeper you go into the zone of saturation, the greater the water pressure. Thus, the looping curves followed by water in the saturated zone may be thought of as a compromise between the downward pull of gravity and the tendency of water to move toward areas of reduced pressure.

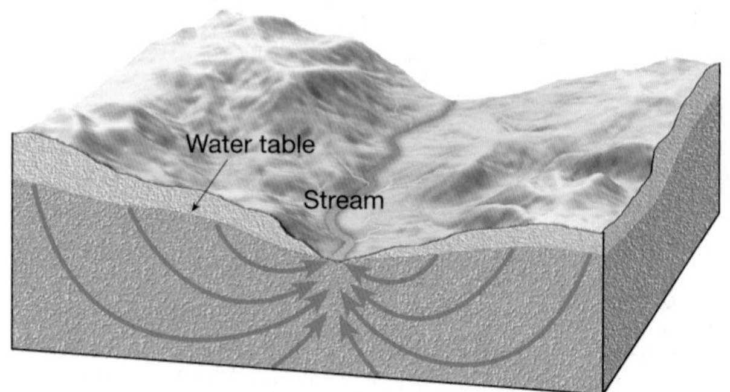

Figure 3.29 Arrows indicate groundwater movement through uniformly permeable material. The looping curves may be thought of as a compromise between the downward pull of gravity and the tendency of water to move toward areas of reduced pressure.

Springs

Sculpturing Earth's Surface
▼ Groundwater

Springs have aroused the curiosity and wonder of people for thousands of years. The fact that springs were (and to some people still are) rather mysterious phenomena is not difficult to understand, for here is water flowing freely from the ground in all kinds of weather in seemingly inexhaustible supply but with no obvious source. Today, we know that the source of springs is water from the zone of saturation and that the ultimate source of this water is precipitation.

Whenever the water table intersects Earth's surface, a natural outflow of groundwater results, which we call a **spring.** Springs such as the one pictured in Figure 3.30 form when an aquitard blocks the downward movement of groundwater and forces it to move laterally. Where the per-

Figure 3.30 Spring in Arizona's Marble Canyon. (Photo by Michael Collier)

meable bed (aquifer) outcrops, a spring or several springs result.

Another situation that can produce a spring is illustrated in Figure 3.28 (p. 87). Here an aquitard is situated above the main water table. As water percolates downward, a portion of it is intercepted by the aquitard, thereby creating a localized zone of saturation and a *perched water table.* Springs, however, are not confined to places where a perched water table creates a flow at the surface. Many geologic situations lead to the formation of springs because subsurface conditions vary greatly from place to place.

Figure 3.31 A wintertime eruption of Old Faithful, one of the world's most famous geysers. It emits as much as 45,000 liters (almost 12,000 gallons) of hot water and steam about once each hour. (Photo by Marc Muench/David Muench Photography, Inc.)

Hot Springs

By definition, the water in **hot springs** is 6° to 9°C (10° to 15°F) warmer than the average annual air temperature for the localities where they occur. In the United States alone, there are well over 1000 such springs.

Temperatures in deep mines and oil wells usually rise with increasing depth, an average of about 2°C per 100 meters (1°F per 100 feet). Therefore, when groundwater circulates at great depths, it becomes heated. If it rises to the surface, the water may emerge as a hot spring. The water of some hot springs in the eastern United States is heated in this manner. The great majority (more than 95 percent) of the hot springs (and geysers) in the United States are found in the West. The reason for such a distribution is that the source of heat for most hot springs is cooling igneous rock, and it is in the West that igneous activity has occurred most recently.

Geysers

Geysers are intermittent hot springs or fountains in which columns of water are ejected with great force at various intervals, often rising 30 to 60 meters (100 to 200 feet) into the air. After the jet of water ceases, a column of steam rushes out, often with a thunderous roar. Perhaps the most famous geyser in the world is Old Faithful in Yellowstone National Park, which erupts about once each hour (Figure 3.31). Geysers are also found in other parts of the world, notably New Zealand and Iceland. In fact, the Icelandic word *geysa,* to gush, gives us the name *geyser.*

Geysers occur where extensive underground chambers exist within hot igneous rocks. As relatively cool groundwater enters the chambers, it is heated by the surrounding rock. At the bottom of the chamber, the water is under great pressure because of the weight of the overlying water. This great pressure prevents the water from boiling at the normal surface temperature of 100°C (212°F). For example, at the bottom of a 300-meter (1000-foot) water-filled chamber, water must attain a temperature of nearly 230°C (450°F) before it will boil. The heating causes the water to expand, with the result that some is forced out at the surface. This loss of water reduces the pressure on the remaining water in the chamber, which lowers the boiling point. A portion of the water deep within the chamber quickly turns to steam and causes the geyser to erupt. Following the eruption, cool groundwater again seeps into the chamber, and the cycle begins anew.

Did You Know?

The largest geyser, if we use the word *large* to mean "tall," is Yellowstone's Steamboat Geyser. During a major eruption, it can spew plumes of water 90 meters (300 feet) high for up to 40 minutes. This is followed by a steam phase that features bursts of hot mist that can extend 150 meters (500 feet) into the sky.

Wells

Sculpturing Earth's Surface
▼ Groundwater

The most common method for removing groundwater is the **well,** a hole bored into the zone of saturation (see Figure 3.28, p. 87). Wells serve as small reservoirs into which groundwater migrates and from which it can be pumped to the surface. The use of wells dates back many centuries and continues to be an important method of obtaining water. By far the single greatest use of this water in the United States is irrigation for agriculture. More than 65 percent of the groundwater used each year is for this purpose. Industrial uses rank

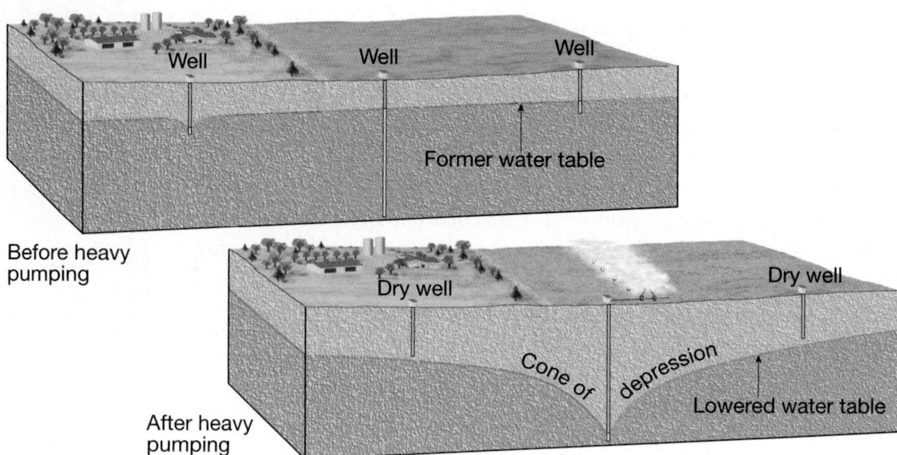

Figure 3.32 A cone of depression in the water table often forms around a pumping well. If heavy pumping lowers the water table, some wells may be left dry.

a distant second, followed by the amount used by homes in cities and rural areas.

The water-table level may fluctuate considerably during the course of a year, dropping during dry seasons and rising following periods of precipitation. Therefore, to ensure a continuous supply of water, a well must penetrate below the water table. Whenever a substantial amount of water is withdrawn from a well, the water table around the well is lowered. This effect, termed **drawdown,** decreases with increasing distance from the well. The result is a depression in the water table, roughly conical in shape, known as a **cone of depression** (Figure 3.32). For most small domestic wells, the cone of depression is negligible. However, when wells are used for irrigation or for industrial purposes, the withdrawal of water can be great enough to create a very wide and steep cone of depression that may substantially lower the water table in an area and cause nearby shallow wells to become dry. Figure 3.32 illustrates this situation.

Artesian Wells

Sculpturing Earth's Surface
▼ Groundwater

In most wells, water cannot rise on its own. If water is first encountered at 30 meters (100 feet) depth, it remains at that level, fluctuating perhaps a meter or two with seasonal wet and dry periods. However, in some wells, water rises, sometimes overflowing at the surface.

The term **artesian** is applied to any situation in which groundwater rises in a well above the level where it was initially encountered. For such a situation to occur, two conditions must exist (Figure 3.33): (1) Water must be confined to an aquifer that is inclined so that one end is exposed at the surface, where it can receive water; and (2) aquitards both above and below the aquifer must be present to prevent the water from escaping. When such a layer is tapped, the pres-

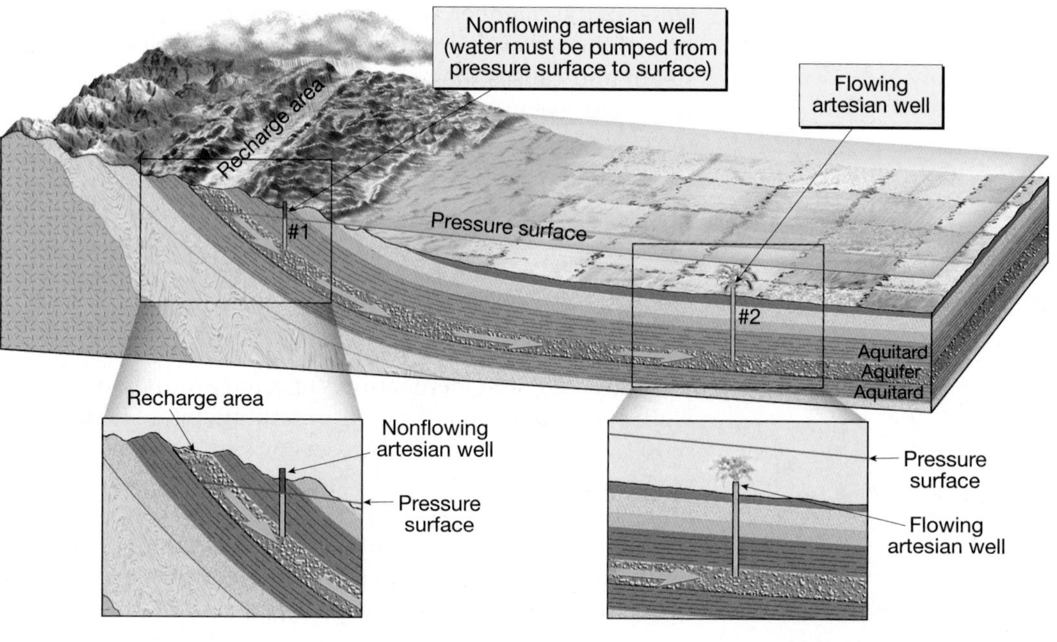

Figure 3.33 Artesian systems occur when an inclined aquifer is surrounded by impermeable beds.

sure created by the weight of the water above will force the water to rise. If there were no friction, the water in the well would rise to the level of the water at the top of the aquifer. However, friction reduces the height of this pressure surface. The greater the distance from the recharge area (area where water enters the inclined aquifer), the greater the friction and the smaller the rise of water.

In Figure 3.33, Well 1 is a *nonflowing artesian well,* because at this location the pressure surface is below ground level. When the pressure surface is above the ground and a well is drilled into the aquifer, a *flowing artesian well* is created (Well 2, Figure 3.33).

Artesian systems act as "natural pipelines," transmitting water from remote areas of recharge great distances to the points of discharge. In this manner, water that fell in central Wisconsin years ago is now taken from the ground and used by communities many kilometers to the south in Illinois. In South Dakota, such a system brings water from the western Black Hills eastward across the state.

On a different scale, city water systems may be considered examples of artificial artesian systems (Figure 3.34). The water tower, into which water is pumped, may be considered the area of recharge, the pipes the confined aquifer, and the faucets in homes the flowing artesian wells.

Environmental Problems of Groundwater

As with many of our valuable natural resources, groundwater is being exploited at an increasing rate. In some areas, overuse threatens the groundwater supply. In other places, groundwater withdrawal has caused the ground and everything resting upon it to sink. Still other localities are concerned with the possible contamination of their groundwater supply.

Treating Groundwater as a Nonrenewable Resource

Many natural systems tend to establish a condition of equilibrium. The groundwater system is no exception. The water table's height reflects a balance between the rate of water added by precipitation and the rate of water removed by discharge and withdrawal. An imbalance will either raise or lower the water table. A long-term drop in the water table can occur if there is either a decrease in recharge due to prolonged drought or an increase in groundwater discharge or withdrawal.

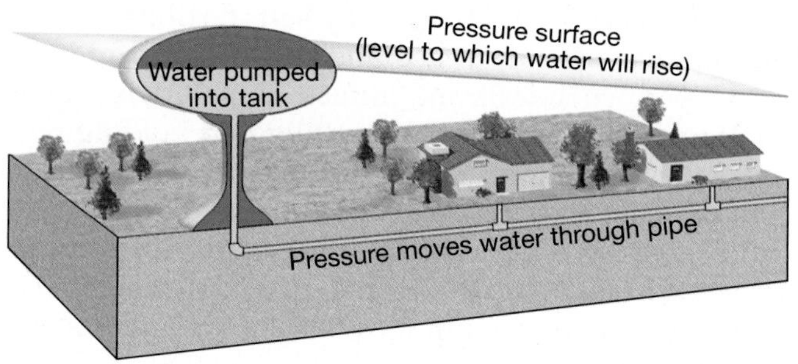

Figure 3.34 City water systems can be considered artificial artesian systems.

For many people, groundwater appears to be an endlessly renewable resource, for it is continually replenished by rainfall and melting snow. But in some regions, groundwater has been and continues to be treated as a *nonrenewable* resource because the amount of water available to recharge the aquifer is significantly less than the amount being withdrawn.

The High Plains, a relatively dry region that extends from the western Dakotas to western Texas, provides one example of an extensive agricultural economy that is largely dependent on irrigation (Figure 3.35A). Nearly 170,000 wells are now being used to irrigate more than 65,000 square kilometers (16 million acres) of land (Figure 3.35B). In the southern part of this region, which includes the Texas panhandle, the natural recharge of the aquifer is very slow and the problem of declining groundwater levels is acute. In fact, in years of average or below-average precipitation, recharge is negligible because all or nearly all of the meager rainfall is returned to the atmosphere by evaporation and transpiration.

Therefore, where intense irrigation has been practiced for an extended period, depletion of groundwater can be severe. Declines in the water table at rates as great as 1 meter per year have led to an overall drop of between 15 and 60 meters (50 and 200 feet) in some areas. Under these circumstances, it can be said that the groundwater is literally being "mined." Even if pumping were to cease immediately, it would take thousands of years for the groundwater to be fully replenished.

Groundwater depletion has been a concern in the High Plains and other areas of the West for many years, but it is worth pointing out that the problem is not confined to this part of the country. Increased demands on groundwater resources have overstressed aquifers in many areas, not just in arid and semiarid regions.

Land Subsidence Caused by Groundwater Withdrawal

As you will see later in this chapter, surface subsidence can result from natural processes related to groundwater. However, the ground may also sink when water is pumped from wells faster than natural recharge processes can replace it. This effect is particularly pronounced in areas underlain by thick layers of loose sediments. As water is withdrawn, the water pressure drops and the weight of the overburden is transferred to the sediment. The greater pressure packs the sediment grains more tightly together and the ground subsides.

Many areas can be used to illustrate such land subsidence. A classic example in the United States occurred in the San Joaquin Valley of California (Figure 3.36). Other well-known cases of land subsidence resulting from groundwater pumping in the United States include Las Vegas, Nevada;

Figure 3.35 **A.** The High Plains extend from the western Dakotas south to Texas. Despite being a land of little rain, this is an important agricultural region. The reason is a vast endowment of groundwater that makes irrigation possible through most of the region. The source of most of this water is the Ogallala formation, the largest aquifer in the United States. **B.** In some agricultural regions, water is pumped from the ground faster than it is replenished. In such instances, groundwater is being treated as a nonrenewable resource. This aerial view shows circular crop fields irrigated by center pivot irrigation systems in semiarid eastern Colorado. (Photo by James L. Amos/Corbis/Bettmann) **C.** Groundwater provides more than 190 billion liters (50 billion gallons) per day in support of the agricultural economy of the United States. (Photo by Michael Collier)

New Orleans and Baton Rouge, Louisiana; and the Houston-Galveston area of Texas. In the low-lying coastal area between Houston and Galveston, land subsidence ranges from 1.5 to 3 meters (5 to 10 feet). The result is that about 78 square kilometers (30 square miles) are permanently flooded.

Outside the United States, one of the most spectacular examples of subsidence occurred in Mexico City, which is built on a former lake bed. In the first half of the twentieth century, thousands of wells were sunk into the water-saturated sediments beneath the city. As water was withdrawn, portions of the city subsided by as much as 6 meters (20 feet) or more.

Groundwater Contamination

The pollution of groundwater is a serious matter, particularly in areas where aquifers provide a large part of the water supply. One common source of groundwater pollution is sewage emanating from an ever increasing number of septic tanks. Other sources are inadequate or broken sewer systems and farm wastes.

If sewage water, which is contaminated with bacteria, enters the groundwater system, it may become purified through natural processes. The harmful bacteria may be mechanically filtered by the sediment through which the water percolates, destroyed by chemical oxidation, and/or assimilated by other organisms. For purification to occur, however, the aquifer must be of the correct composition. For example, extremely permeable aquifers (such as highly fractured crystalline rock, coarse gravel, or cavernous limestone) have such large openings that contaminated groundwater may travel long distances without being cleansed. In this case, the water flows too rapidly and is not in contact with the surrounding material long enough for purification to occur. This is the problem at Well 1 in Figure 3.37A.

Conversely, when the aquifer consists of sand or permeable sandstone, the water can sometimes be purified after traveling only a few dozen meters through it. The openings between sand grains are large enough to permit water movement, yet the movement of the water is slow enough to allow ample time for its purification (Well 2, Figure 3.37B).

Other sources and types of contamination also threaten groundwater supplies (Figure 3.38). These include widely used substances such as highway salt, fertilizers that are spread across the land surface, and pesticides. In addition, a wide array of chemicals and industrial materials may leak from pipelines, storage tanks, landfills, and holding ponds. Some of these pollutants are classified as *hazardous,* meaning that they are either flammable, corrosive, explosive, or toxic. As rainwater oozes through the refuse, it may dissolve a variety of potential contaminants. If the leached material reaches the water table, it will mix with the groundwater and contaminate the supply.

Because groundwater movement is usually slow, polluted water may go undetected for a long time. In fact, contamination is sometimes discovered only after drinking water

Figure 3.36 The shaded area on the map shows California's San Joaquin Valley. This important agricultural region relies heavily on irrigation. The marks on the utility pole in the photo indicate the level of the surrounding land in preceding years. Between 1925 and 1975, this part of the San Joaquin Valley subsided almost 9 meters (30 feet) because of the withdrawal of groundwater and the resulting compaction of sediments. (Photo courtesy of U.S. Geological Survey, U.S. Department of the Interior)

has been affected and people become ill. By this time, the volume of polluted water could be quite large, and even if the source of contamination is removed immediately, the problem is not solved. Although the sources of groundwater contamination are numerous, the solutions are relatively few.

Once the source of the problem has been identified and eliminated, the most common practice is simply to abandon the water supply and allow the pollutants to be flushed away gradually. This is the least costly and easiest solution, but the aquifer must remain unused for many years. To accelerate this process, polluted water is sometimes pumped out and treated. Following removal of the tainted water, the aquifer is allowed to recharge naturally or, in some cases, the treated water or other fresh water is pumped back in. This process is costly, time consuming, and possibly risky because there is no way to be certain that all of the contamination has been removed. Clearly, the most effective solution to groundwater contamination is prevention.

The Geologic Work of Groundwater

Groundwater dissolves rock. This fact is key to understanding how caverns and sinkholes form. Because soluble rocks, especially limestone, underlie millions of square kilometers of Earth's surface, it is here that groundwater carries on its important role as an erosional agent. Limestone is nearly insoluble in pure water but is quite easily dissolved by water containing small quantities of carbonic acid, and most groundwater contains this acid. It forms because rainwater readily dissolves carbon dioxide from the air and from decaying plants. Therefore, when groundwater comes in contact

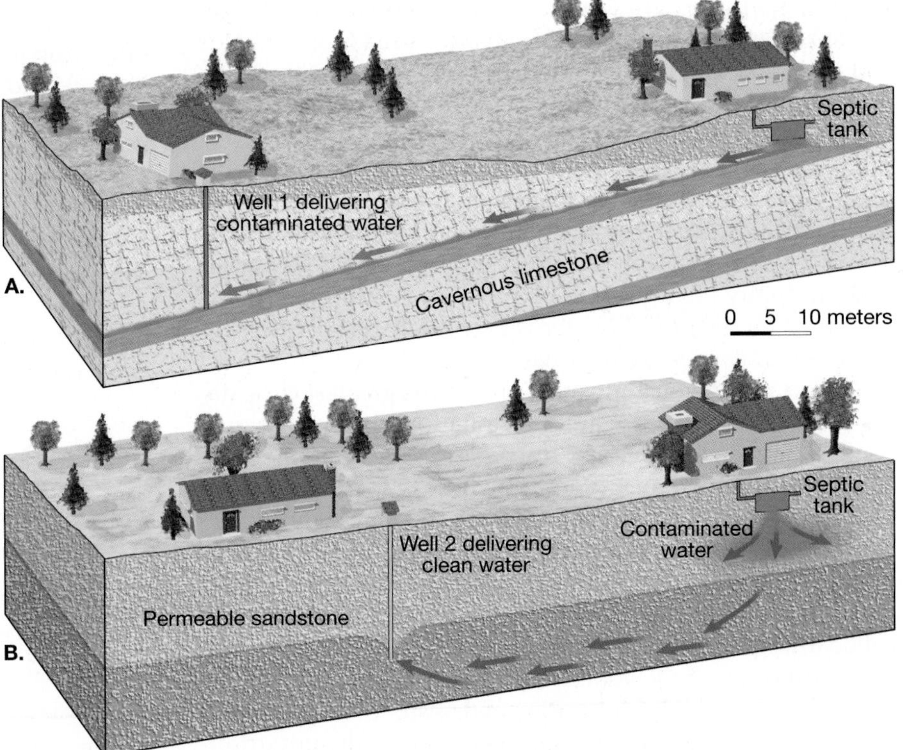

Figure 3.37 **A.** Although the contaminated water has traveled more than 100 meters (330 feet) before reaching Well 1, the water moves too rapidly through the cavernous limestone to be purified. **B.** As the discharge from the septic tank percolates through the permeable sandstone, it is purified in a relatively short distance.

A.

B.

Figure 3.38 Sometimes agricultural chemicals **(A)** and materials leached from landfills **(B)** find their way into the groundwater. These are two of the potential sources of groundwater contamination. (Photo A by Roy Morsch/The Stock Market; Photo B by F. Rossotto/The Stock Market)

with limestone, the carbonic acid reacts with calcite (calcium carbonate) in the rocks to form calcium bicarbonate, a soluble material that is then carried away in solution.

Caverns

The most spectacular results of groundwater's erosional handiwork are limestone **caverns.** In the United States alone, about 17,000 caves have been discovered. Although most are relatively small, some have spectacular dimensions. Carlsbad Caverns in southeastern New Mexico and Mammoth Cave in Kentucky are famous examples. One chamber in Carlsbad Caverns has an area equivalent to 14 football fields and enough height to accommodate the U.S. Capitol Building. At Mammoth Cave, the total length of interconnected caverns extends for more than 540 kilometers (340 miles).

Most caverns are created at or just below the water table in the zone of saturation. Acidic groundwater follows lines of weakness in the rock, such as joints and bedding planes. As time passes, the dissolving process slowly creates cavities and gradually enlarges them into caverns. Material that is dissolved by the groundwater is eventually discharged into streams and carried to the ocean.

Certainly, the features that arouse the greatest curiosity for most cavern visitors are the stone formations that give some caverns a wonderland appearance. These are not erosional features, like the caverns in which they reside, but depositional features. They are created by the seemingly endless dripping of water over great spans of time. The calcium carbonate that is left behind produces the limestone we call travertine. These cave deposits, however, are also commonly called *dripstone,* an obvious reference to their mode of origin.

Although the formation of caverns takes place in the zone of saturation, the deposition of dripstone is not possible until the caverns are above the water table, in the unsaturated zone. This commonly occurs as nearby streams cut their valleys deeper, lowering the water table as the elevation of the rivers drops. As soon as the chamber is filled with air, the conditions are right for the decoration phase of cavern building to begin.

Of the various dripstone features found in caverns, perhaps the most familiar are **stalactites.** These icicle-like pendants hang from the ceiling of the cavern and form where water seeps through cracks above. When water reaches air in the cave, some of the dissolved carbon dioxide escapes from the drop and calcite begins to precipitate. Deposition occurs as a ring around the edge of the water drop. As drop after drop follows, each leaves an infinitesimal trace of calcite behind, and a hollow limestone tube is created. Water then moves through the tube, remains suspended momentarily at the end, contributes a tiny ring of calcite, and falls to the cavern floor.

The stalactite just described is appropriately called a *soda straw* (Figure 3.39A). Often the hollow tube of the soda straw becomes plugged or its supply of water increases. In either case, the water is forced to flow and deposit along the outside of the tube. As deposition continues, the stalactite takes on the more common conical shape.

Did You Know?

America's largest bat colonies are found in caves. For example, Braken Cave in central Texas is the summer home of 20 million Mexican free-tail bats. They spend their days in total darkness more than 3 kilometers (2 miles) inside the cave. Each night, they leave the cave to feed, consuming more than 200,000 kilograms (220 tons) of insects.

A.

B.

Figure 3.39 **A.** "Live" soda-straw stalactites, Lehman Caves, Great Basin National Park, Nevada. (Photo by Tom Bean) **B.** Stalagmites grow upward from the cavern floor, Chinese Theater, Carlsbad Caverns National Park, New Mexico. (Photo by David Muench Photography, Inc.)

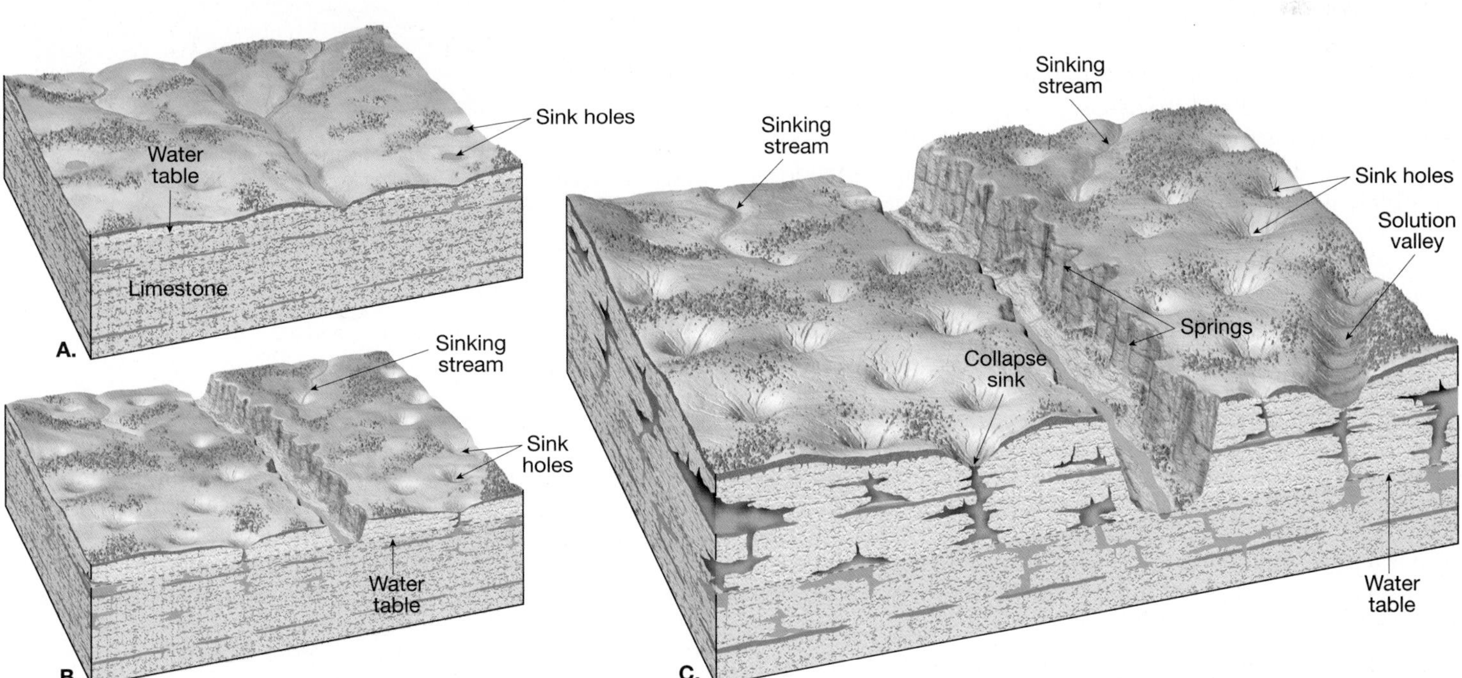

Figure 3.40 Development of a karst landscape. **A.** During early stages, groundwater percolates through limestone along joints and bedding planes. Solution activity creates and enlarges caverns at and below the water table. **B.** In this view, sinkholes are well developed and surface streams are funneled below ground. **C.** With the passage of time, caverns grow larger and the number and size of sinkholes increase. Collapse of caverns and coalescence of sinkholes form larger, flat-floored depressions. Eventually, solution activity may remove most of the limestone from the area, leaving only isolated remnants.

Formations that develop on the floor of a cavern and reach upward toward the ceiling are called **stalagmites** (Figure 3.39B). The water supplying the calcite for stalagmite growth falls from the ceiling and splatters over the surface. As a result, stalagmites do not have a central tube and are usually more massive in appearance and more rounded on their upper ends than are stalactites. Given enough time, a downward-growing stalactite and an upward-growing stalagmite may join to form a *column*.

Karst Topography

Many areas of the world have landscapes that, to a large extent, have been shaped by the dissolving power of groundwater. Such areas are said to exhibit **karst topography,** named for the *Krs* region in the border area between Slovenia (formerly a part of Yugoslavia) and Italy where such topography is strikingly developed. In the United States, karst landscapes occur in many areas that are underlain by limestone, including portions of Kentucky, Tennessee, Alabama, southern Indiana, and central and northern Florida (Figure 3.40). Generally, arid and semiarid areas do not develop karst topography because there is insufficient groundwater. When solution features exist in such regions, they are likely to be remnants of a time when rainier conditions prevailed.

Karst areas typically have irregular terrain punctuated with many depressions called **sinkholes** or, simply, **sinks** (see Figure 3.27B, p. 86). In the limestone areas of Florida, Kentucky, and southern Indiana, literally tens of thousands of these depressions vary in depth from just a meter or two to a maximum of more than 50 meters.

Sinkholes commonly form in one of two ways. Some develop gradually over many years without any physical disturbance to the rock. In these situations, the limestone immediately below the soil is dissolved by downward-seeping rainwater that is freshly charged with carbon dioxide. These depressions are usually not deep and are characterized by relatively gentle slopes. By contrast, sinkholes can also form suddenly and without warning when the roof of a cavern collapses under its own weight. Typically, the depressions created in this manner are steep sided and deep. When they form in populous areas, they may represent a serious geologic hazard. Such a situation is clearly the case in Figure 3.41.

In addition to a surface pockmarked by sinkholes, karst regions characteristically show a striking lack of surface drainage (streams). Following a rainfall, runoff is quickly funneled belowground through sinks. It then flows through caverns until it finally reaches the water table. Where streams do exist at the surface, their paths are usually short. The names of such streams often give a clue to their fate. In the Mammoth Cave area of Kentucky, for example, there is Sinking Creek, Little Sinking Creek, and Sinking Branch. Some sinkholes become plugged with clay and debris, creating small lakes or ponds.

Figure 3.41 This small sinkhole formed suddenly in 1991 when the roof of a cavern collapsed, destroying this home in Frostproof, Florida. (Photo by *St. Petersburg Times*/Liaison Agency, Inc.)

The Chapter in Review

1. Weathering, mass wasting, and erosion are responsible for transforming solid rock into sediment. They are called external processes because they occur at or near Earth's surface and are powered by energy from the Sun. By contrast, internal processes, such as volcanism and mountain building, derive their energy from Earth's interior.

2. Mass wasting is the downslope movement of rock and soil under the direct influence of gravity. Gravity is the controlling force of mass wasting, but water often influences mass wasting by saturating the pore spaces and destroying the cohesion between particles. Oversteepening of slopes can trigger mass wasting.

3. The *water cycle* describes the continuous interchange of water among the oceans, atmosphere, and continents. Powered by energy from the Sun, it is a global system in which the atmosphere provides the link between the oceans and continents. The processes involved in the water cycle include *precipitation, evaporation, infiltration* (the movement of water into rocks or soil through cracks and pore spaces), *runoff* (water that flows over the land rather than infiltrating into the ground), and *transpiration* (the release of water vapor to the atmosphere by plants).

4. The factors that determine a stream's *velocity* are *gradient* (slope of the stream channel); *shape, size,* and *roughness* of the channel; and the stream's *discharge* (amount of water passing a given point per unit of time). Most often, the gradient and roughness of a stream decrease downstream, whereas width, depth, discharge, and velocity increase.

5. Streams transport their load of sediment in solution (*dissolved load*), in suspension (*suspended load*), and along the bottom of the channel (*bed load*). Much of the dissolved load is contributed by groundwater. Most streams carry the greatest part of their load in suspension. The bed load moves only intermittently and is usually the smallest portion of a stream's load.

6. A stream's ability to transport solid particles is described using two criteria: *capacity* (the maximum load of solid particles a stream can carry) and *competence* (the maximum particle size a stream can transport). Competence increases as the square of stream velocity, so if velocity doubles, the water's force increases fourfold.

7. Streams deposit sediment when velocity slows and competence is reduced. This results in *sorting*, the process by which like-sized particles are deposited together. Stream deposits are called *alluvium* and may occur as channel deposits called *bars;* as floodplain deposits, which include *natural levees;* and as *deltas* at the mouths of streams.

8. Stream channels are of two basic types: *bedrock channels* and *alluvial channels.* Bedrock channels are most common in headwaters regions where gradients are steep. Rapids and waterfalls are common features. Two types of alluvial channels are *meandering channels* and *braided channels.*

9. The two general types of *base level* (the lowest point to which a stream may erode its channel) are (1) *ultimate base level* and (2) *temporary,* or *local base level.* Any change in base level will cause a stream to adjust and establish a new balance. Lowering base level will cause a stream to downcut, whereas raising base level results in deposition of material in the channel.

10. When a stream has cut its channel closer to base level, its energy is directed from side to side, and erosion produces a flat valley floor, or *floodplain.* Streams that flow upon floodplains often move in sweeping bends called *meanders.* Widespread meandering may result in shorter channel segments, called *cutoffs,* and/or abandoned bends, called *oxbow lakes.*

11. Common *drainage patterns* produced by streams include (1) *dendritic,* (2) *radial,* (3) *rectangular,* and (4) *trellis.*

12. *Floods* are triggered by heavy rains and/or snowmelt. Sometimes human interference can worsen or even cause floods. Flood-control measures include the building of *artificial levees* and dams, as well as *channelization,* which could involve creating *artificial cutoffs.* Many scientists and engineers advocate a nonstructural approach to flood control that involves more appropriate land use.

13. As a resource, *groundwater* represents the largest reservoir of freshwater that is readily available to humans. Geologically, the dissolving action of groundwater produces caves and sinkholes. Groundwater is also an equalizer of streamflow.

14. Groundwater is water that occupies the pore spaces in sediment and rock in a zone beneath the surface, called the *zone of saturation.* The upper limit of this zone is the *water table.* The *unsaturated zone* is above the water table where the soil, sediment, and rock are not saturated. The quantity of water that can be stored depends on the *porosity* (the volume of open spaces) of the material. The *permeability* (the ability to transmit a fluid through interconnected pore spaces) of a material is a key factor affecting the movement of groundwater.

15. *Springs* occur whenever the water table intersects the land surface and a natural flow of groundwater results. *Wells,* openings bored into the zone of saturation, withdraw groundwater and create roughly conical depressions in the water table, known as *cones of depression. Artesian wells* occur when water rises above the level where it was initially encountered.

16. When groundwater circulates at great depths, it becomes heated. If it rises, the water may emerge as a *hot spring. Geysers* occur when groundwater is heated in underground chambers, and expands, and some water quickly changes to steam, causing the geyser to erupt. The source of heat for most hot springs and geysers is hot igneous rock.

17. Some of the current environmental problems involving groundwater include (1) *overuse* by intense irrigation, (2) *land subsidence* caused by groundwater withdrawal, and (3) *contamination* by pollutants.

18. Most *caverns* form in limestone at or below the water table when acidic groundwater dissolves this soluble rock. *Karst topography* exhibits an irregular terrain punctuated with many depressions, called *sinkholes.*

Key Terms

alluvium (p. 75)
angle of repose (p. 68)
aquifer (p. 88)
aquitard (p. 88)
artesian well (p. 90)
backswamp (p. 80)
bar (p. 80)
base level (p. 77)
bed load (p. 74)
braided channels (p. 76)
capacity (p. 75)
cavern (p. 94)
competence (p. 74)
cone of depression (p. 90)
cutoff (p. 76)
delta (p. 80)
dendritic pattern (p. 82)
discharge (p. 72)
dissolved load (p. 74)
distributary (p. 80)
divide (p. 71)
drainage basin (p. 71)
drawdown (p. 90)
erosion (p. 66)
external process (p. 66)
flood (p. 82)
floodplain (p. 79)
geyser (p. 89)
gradient (p. 72)
groundwater (p. 87)
hot spring (p. 89)
incised meander (p. 80)
infiltration (p. 69)
internal process (p. 66)
karst topography (p. 96)
laminar flow (p. 72)

mass wasting (p. 66)
meander (p. 76)
natural levee (p. 80)
oxbow lake (p. 76)
permeability (p. 88)
porosity (p. 87)
radial pattern (p. 82)
rectangular pattern (p. 82)
runoff (p. 69)
sinkhole (sink) (p. 96)
sorting (p. 75)
spring (p. 88)
stalactite (p. 90)
stalagmite (p. 96)
stream valley (p. 77)
suspended load (p. 74)
transpiration (p. 69)
trellis pattern (p. 82)
turbulent flow (p. 72)
unsaturated zone (p. 87)
water cycle (p. 69)
water table (p. 87)
weathering (p. 66)
well (p. 89)
yazoo tributary (p. 82)
zone of saturation (p. 87)

Questions for Review

1. Describe the role of external processes in the rock cycle.
2. What is the controlling force of mass wasting? What other factors can influence or trigger mass-wasting processes?
3. What role does mass wasting play in sculpting Earth's landscapes?
4. Describe the movement of water through the water cycle. Once precipitation has fallen on land, what paths might it take?
5. What are the three main parts (zones) of a river system?
6. A stream starts out 2000 meters above sea level and travels 250 kilometers to the ocean. What is its average gradient in meters per kilometer?
7. Suppose that the stream mentioned in Question 6 developed extensive meanders so that its course was lengthened to 500 kilometers. Calculate its new gradient. How does meandering affect gradient?
8. When the discharge of a stream increases, what happens to the stream's velocity?
9. In what three ways does a stream transport its load?
10. If you collect a jar of water from a stream, which part of its load will settle to the bottom of the jar? Which portion will remain in the water? Which portion of a stream's load would probably not be present in your sample?
11. Differentiate between competency and capacity.
12. Are bedrock channels more likely to be found near the head or near the mouth of a stream?
13. Describe a situation that might cause a stream channel to become braided.
14. Define base level. Name the main river in your area. For what streams does it act as base level? What is base level for the Mississippi River? The Missouri River?
15. Describe two situations that would trigger the formation of incised meanders.
16. Briefly describe the formation of a natural levee. How is this feature related to backswamps and yazoo tributaries?
17. List and briefly describe three basic flood-control strategies. What are some drawbacks of each?
18. Each of the following statements refers to a particular drainage pattern. Identify the pattern.
    a. Streams diverge from a central high area, such as a volcano.
    b. Streams form a branching, "treelike" pattern.
    c. Bedrock is crisscrossed by joints and faults.
19. What percentage of freshwater is groundwater? (See Figure 3.26, p. 85). If glacial ice is excluded and only liquid freshwater is considered, about what percentage is groundwater?
20. Geologically, groundwater is important as an erosional agent. Name another significant geologic role of groundwater.
21. Define groundwater and relate it to the water table.
22. How do porosity and permeability differ?
23. What is the source of heat for most hot springs and geysers? How is this reflected in the distribution of these features?
24. What is meant by the term *artesian*? Under what circumstances do artesian wells form?
25. Which problem is associated with the pumping of groundwater for irrigation in the southern part of the High Plains?
26. Briefly explain what happened in California's San Joaquin Valley, Mexico City, and other places as the result of excessive groundwater withdrawal.
27. Which aquifer would be most effective in purifying polluted groundwater: one composed mainly of coarse gravel, one consisting of sand, or one of cavernous limestone?
28. Differentiate between stalactites and stalagmites. How do these features form?
29. If you were to explore an area that exhibited karst topography, what features might you find? This area would probably be underlain by what rock type?

Online Study Guide

The *Foundations of Earth Science* Web site uses the resources and flexibility of the Internet to aid in your study of the topics in this chapter. Written and developed by Earth science instructors, this site will help improve your understanding of Earth science. Visit **http://www.prenhall .com/lutgens** and click on the cover of *Foundations of Earth Science 5e* to find:

- Online review quizzes.
- Critical thinking exercises.
- Links to chapter-specific Web resources.
- Internet-wide key-term searches.

http://www.prenhall.com/lutgens

GEODe: Earth Science

GEODe: Earth Science makes studying more effective by reinforcing key concepts using animation, video, narration, interactive exercises, and practice quizzes. A copy is included with every copy of *Foundations of Earth Science 5e.*

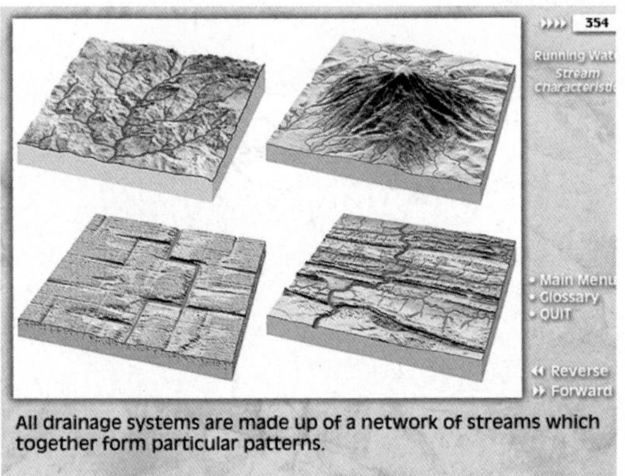
354
• Main Menu
• Glossary
• QUIT
Reverse
Forward
All drainage systems are made up of a network of streams which together form particular patterns.

The single greatest use of groundwater in the United States is for irrigation.
395
• Main Menu
• Glossary
• QUIT
Reverse
Forward
Groundwater is a very important source of drinking water and water used for irrigation and industry.

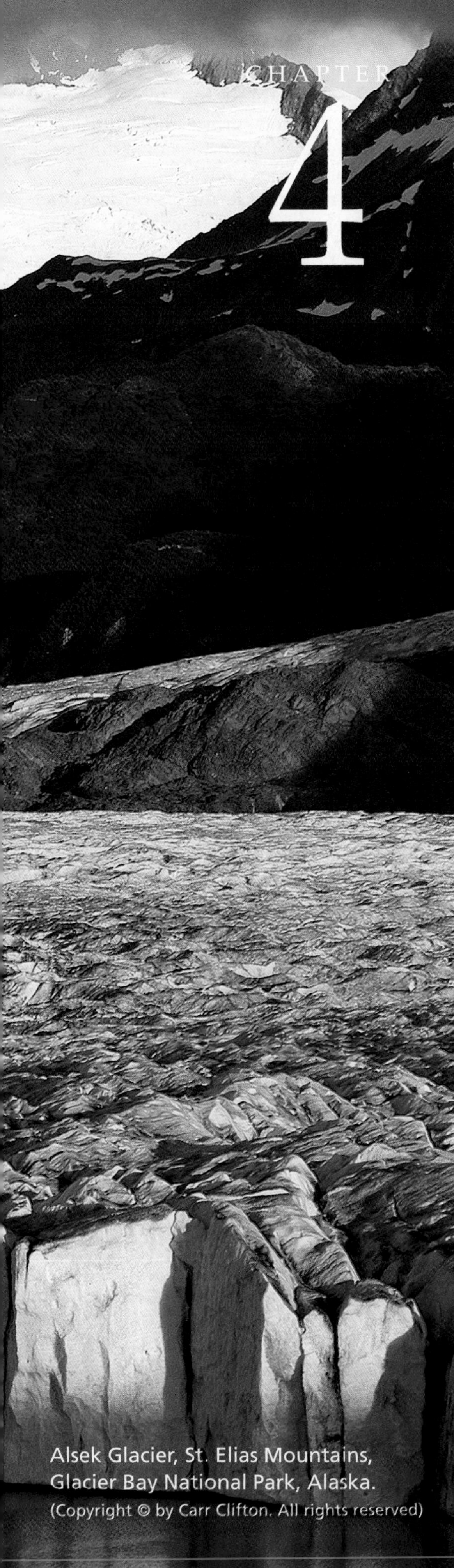

Alsek Glacier, St. Elias Mountains, Glacier Bay National Park, Alaska.

Glacial and Arid Landscapes

FOCUS ON LEARNING

To assist you in learning the important concepts in this chapter, you will find it helpful to focus on the following questions:

1. What is a glacier? Where on Earth are the two general types of glaciers located today?

2. How do glaciers move, and what are the various processes of glacial erosion?

3. What materials make up the features created by glacial deposition? What are the most widespread features?

4. What is some evidence for the Ice Age? What indirect effects did Ice Age glaciers have on the land and sea?

5. What are the causes of deserts in both the lower and middle latitudes?

6. What are the roles of water and wind in arid climates?

7. How did many of the landscapes in the dry Basin and Range region of the United States evolve?

8. What are the ways that wind erodes?

9. What are some depositional features produced by wind?

Like the running water and groundwater that were the focus of Chapter 3, glaciers and wind are significant erosional processes. They are responsible for creating many different landforms and are part of an important link in the rock cycle in which the products of weathering are transported and deposited as sediment.

Climate has a strong influence on the nature and intensity of Earth's external processes. This fact is dramatically illustrated in this chapter. The existence and extent of glaciers are largely controlled by Earth's changing climate. Another excellent example of the strong link between climate and geology is seen when we examine the development of arid landscapes.

Today, glaciers cover nearly 10 percent of Earth's land surface; however, in the recent geologic past, ice sheets were three times more extensive, covering vast areas with ice thousands of meters thick. Many regions still bear the mark of these glaciers. The first part of this chapter will examine glaciers and the erosional and depositional features they create. The second part will explore dry lands and the geologic work of wind. Because desert and near-desert conditions prevail over an area as large as that affected by the massive glaciers of the Ice Age, the nature of such landscapes is indeed worth investigating.

Glaciers: A Part of Two Basic Cycles

Sculpturing Earth's Surface
▼ Glaciers

Many present-day landscapes were modified by the widespread glaciers of the most recent Ice Age and still strongly reflect the handiwork of ice. The basic features of such diverse places as the Alps, Cape Cod, and Yosemite Valley were fashioned by now vanished masses of glacial ice. Moreover, Long Island, New York; the Great Lakes; and the fiords of Norway and Alaska all owe their existence to glaciers. Glaciers, of course, are not just a phenomenon of the geologic past. As you will see, they are still sculpturing and depositing in many regions today.

Glaciers are a part of two of Earth's basic cycles—the water cycle and the rock cycle. In the water cycle, when precipitation falls at high elevations or high latitudes, the water may not immediately make its way back toward the sea. Instead, it may become part of a glacier. Although the ice will eventually melt, allowing the water to continue its path to the sea, water can be stored as glacial ice for tens, hundreds, or even thousands of years. During the time that the water is part of a glacier, the moving mass of ice can do enormous amounts of work—scouring the land surface and acquiring, transporting, and depositing great quantities of sediment. This activity is a basic part of the rock cycle.

A **glacier** is a thick ice mass that forms over hundreds or thousands of years. It originates on land from the accumulation, compaction, and recrystallization of snow. A glacier appears to be motionless, but it is not—glaciers move very slowly. Like running water, groundwater, wind, and waves, glaciers are dynamic erosional agents that accumulate, transport, and deposit sediment. Although glaciers are found in many parts of the world today, most are located in remote areas, either near Earth's poles or in high mountains.

Did You Know?

Approximately 160,000 glaciers presently occupy Earth's polar regions and high mountain environments. Today's glaciers cover only about one-third the area that was covered at the height of the most recent Ice Age.

Valley (Alpine) Glaciers

Literally thousands of relatively small glaciers exist in lofty mountain areas, where they usually follow valleys originally occupied by streams. Unlike the rivers that previously flowed in these valleys, the glaciers advance slowly, perhaps only a few centimeters per day. Because of their location, these moving ice masses are termed **valley glaciers** or **alpine glaciers** (Figure 4.1). Each glacier is a stream of ice, bounded by

Figure 4.1 Aerial view of Chigmit Mountains in Lake Clark National Park, Alaska. Valley glaciers continue to modify this landscape. (Photo by Michael Collier)

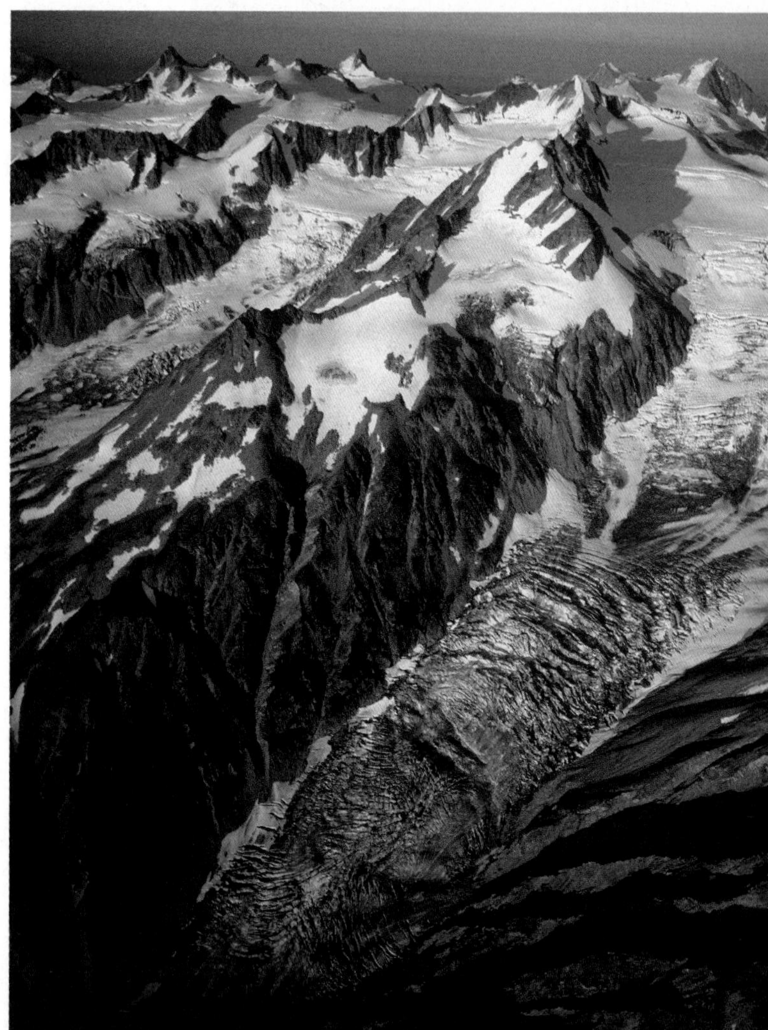

Did You Know?

The total volume of all valley glaciers is estimated to be about 210,000 cubic kilometers (50,400 cubic miles). This is comparable to the combined volume of all the world's fresh water and saline lakes.

precipitous rock walls, that flows downvalley from a snow accumulation center near its head. Like rivers, valley glaciers can be long or short, wide or narrow, single or with branching tributaries. The widths of alpine glaciers are generally small compared to their lengths. In length, some extend for just a fraction of a kilometer, whereas others go on for many tens of kilometers. The west branch of the Hubbard Glacier, for example, runs through 112 kilometers (70 miles) of mountainous terrain in Alaska and Canada's Yukon Territory.

Ice Sheets

In contrast to valley glaciers, **ice sheets** exist on a much larger scale. These enormous masses flow out in all directions from one or more centers and completely obscure all but the highest areas of underlying terrain. Although many ice sheets have existed in the past, just two achieve this status at present (Figure 4.2). In the Northern Hemisphere, Greenland is covered by an imposing ice sheet averaging nearly 1500 meters (5000 feet) thick. It occupies 1.7 million square kilometers (more than 660 thousand square miles) or about 80 percent of this large island.

In the Southern Hemisphere, the huge Antarctic Ice Sheet attains a maximum thickness of nearly 4300 meters (14,000 feet) and covers an area of more than 13.9 million square kilometers (5.4 million square miles). Because of the proportions of these huge features, they are often called *continental ice sheets.* Indeed, the combined areas of present-day continental ice sheets represent almost 10 percent of Earth's land area.

Along portions of the Antarctic coast, glacial ice flows into the adjacent ocean, creating features called **ice shelves.** These are large, relatively flat masses of floating ice that extend seaward from the coast but remain attached to the land along one or more sides. The shelves are thickest on their landward sides, and they become thinner seaward. They are sustained by ice from the adjacent ice sheet as well as being nourished by snowfall and the freezing of seawater to their bases. Antarctica's ice shelves extend over approximately 1.4 million square kilometers (0.6 million square miles). The Ross and Filchner ice shelves are the largest, with the Ross Ice Shelf alone covering an area approximately the size of Texas (Figure 4.2).

Did You Know?

The Antarctic ice sheet represents 80 percent of the world's ice and nearly two-thirds of Earth's fresh water.

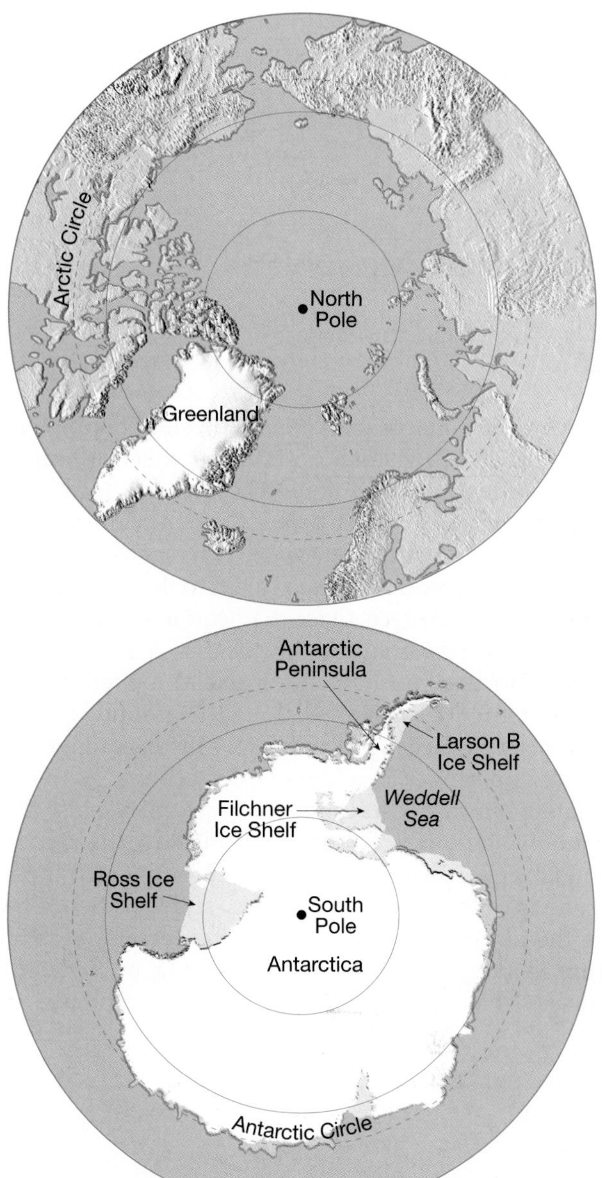

Figure 4.2 The only present-day continental ice sheets are those covering Greenland and Antarctica. Their combined areas represent almost 10 percent of Earth's land area. Greenland's ice sheet occupies 1.7 million square kilometers, or about 80 percent of the island. The area of the Antarctic Ice Sheet is almost 14 million square kilometers. Ice shelves occupy an additional 1.4 million square kilometers adjacent to the Antarctic Ice Sheet.

Other Types of Glaciers

In addition to valley glaciers and ice sheets, other types of glaciers have also been identified. Covering some uplands and plateaus are masses of glacial ice called **ice caps.** Like ice sheets, ice caps completely bury the underlying landscape but are much smaller than the continental-scale features. Ice caps occur in many places, including Iceland and several of the large islands in the Arctic Ocean. Another type, known as **piedmont glaciers,** occupy broad lowlands at the bases of steep mountains and form when one or more valley glaciers emerge from the confining walls of mountain valleys. The advancing ice spreads out to form a broad sheet. The size of individual piedmont glaciers varies greatly. Among the largest is the broad Malaspina Glacier along the coast of southern Alaska. It covers more than 5000 square kilometers (2000 square miles) of the flat coastal plain at the foot of the lofty St. Elias range (Figure 4.3).

How Glaciers Move

Sculpturing Earth's Surface
▼ Glaciers

The movement of glacial ice is generally referred to as *flow.* The fact that glacial movement is described in this way seems paradoxical—how can a solid flow? The way in which ice flows is complex and of two basic types. One mechanism involves plastic movement within the ice. Ice behaves as a brittle solid until the pressure upon it is equivalent to the weight of about 50 meters (165 feet) of ice. Once that load is surpassed, ice behaves as a plastic material and flow begins. A second and often equally important mechanism of glacial movement occurs when the entire ice mass slips along the ground. The lowest portions of most glaciers probably move by this sliding process.

The uppermost 50 meters (165 feet) of a glacier is appropriately referred to as the *zone of fracture.* Because there is not enough overlying ice to cause plastic flow, this upper part of the glacier consists of brittle ice. Consequently, the ice in this zone is carried along piggyback-style by the ice below. When the glacier moves over irregular terrain, the zone of fracture is subjected to tension, resulting in cracks called **crevasses.** These gaping cracks, which often make travel across glaciers dangerous, may extend to depths of 50 meters (165 feet). Below this depth, plastic flow seals them off.

Observing and Measuring Movement

Unlike streamflow, glacial movement is not obvious. If we could watch a valley glacier move, we would see that like the water in a river, all of the ice does not move downstream at the same rate. Flow is greatest in the center of the glacier because of the drag created by the walls and floor of the valley.

Early in the nineteenth century, the first experiments involving the movement of glaciers were designed and carried out in the Alps. Markers were placed in a straight line across an alpine glacier. The position of the line was marked on the valley walls so that if the ice moved, the change in position could be detected. The positions of the markers were noted periodically, revealing the movement just described. Although most glaciers move too slowly for direct visual detection, the experiments succeeded in demonstrating that movement nevertheless occurs. The experiment illustrated in Figure 4.4 was carried out at Switzerland's Rhône Glacier later in the nineteenth century. It not only traced the movement of markers within the ice but also mapped the position of the glacier's terminus.

How rapidly does glacial ice move? Average rates vary considerably from one glacier to another. Some move so slowly that trees and other vegetation may become well established in the debris that accumulates on the glacier's surface. Others advance up to several meters each day. Movement of some glaciers is characterized by periods of extremely rapid advance followed by periods during which movement is practically nonexistent.

Budget of a Glacier: Accumulation versus Wastage

Snow is the raw material from which glacial ice originates. Therefore, glaciers form in areas where more snow falls in winter than can melt during the summer. Glaciers are constantly gaining and losing ice. Snow accumulation and ice formation occur in the **zone of accumulation** (Figure 4.5). Here, the addition of snow thickens the glacier and promotes movement. Beyond this area of ice formation is the **zone of wastage,** where there is a net loss to the glacier as all of the

Did You Know?

Glaciers are found on all continents except Australia. Surprisingly, tropical Africa has a small area of glacial ice atop its highest mountain, Mount Kilimanjaro. However, studies show that the mountain has lost more than 80 percent of its ice since 1912 and that all of it may be gone by 2020.

Figure 4.3 Malaspina Glacier in southern Alaska is a classic piedmont glacier. Piedmont glaciers occur where valley glaciers exit a mountain range onto broad lowlands, are no longer laterally confined, and spread to become wide lobes. Malaspina Glacier is actually a compound glacier, formed by the merger of several valley glaciers, the most prominent of which seen here are Agassiz Glacier (left) and Seward Glacier (right). In total, Malaspina Glacier is up to 65 kilometers (40 miles) wide and extends up to 45 kilometers (28 miles) from the mountain front nearly to the sea. (NASA/JPL)

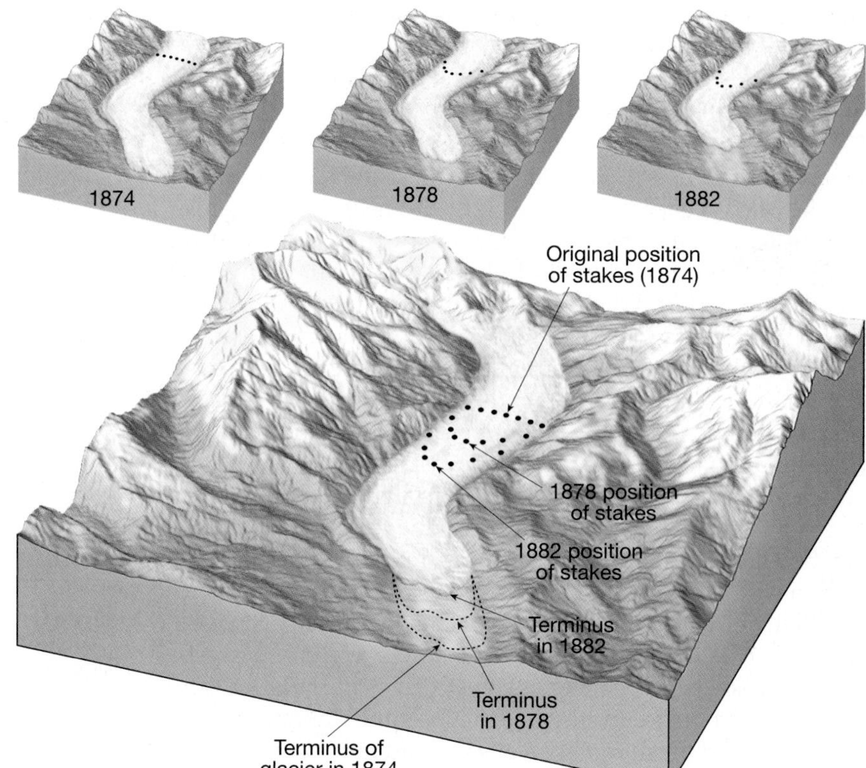

Figure 4.4 Ice movement and changes in the terminus of Rhône Glacier, Switzerland. In this classic study of a valley glacier, the movement of stakes clearly shows that glacial ice moves and that movement along the sides of the glacier is slower than movement in the center. Also notice that even though the ice front was retreating, the ice within the glacier was advancing.

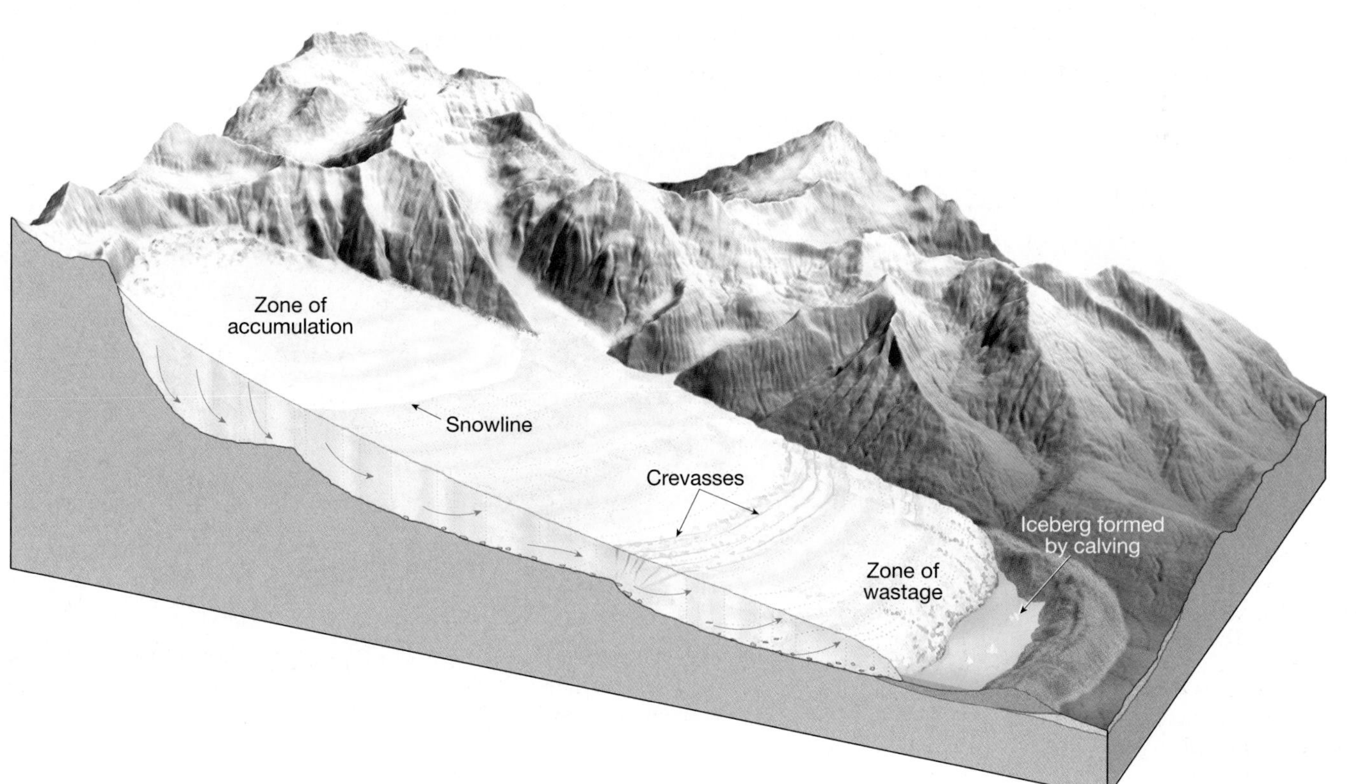

Figure 4.5 The snowline separates the zone of accumulation and the zone of wastage. Above the snowline, more snow falls each winter than melts each summer. Below the snowline, the snow from the previous winter completely melts as does some of the underlying ice. Whether the margin of a glacier advances, retreats, or remains stationary depends on the balance between accumulation and wastage. When a glacier moves across irregular terrain, crevasses form in the brittle upper portion.

snow from the previous winter melts, as does some of the glacial ice (Figure 4.5).

In addition to melting, glaciers also waste as large pieces of ice break off the front of a glacier in a process called *calving*. Calving creates *icebergs* when glaciers reach the sea (Figure 4.6). Because icebergs are just slightly less dense than seawater, they float very low in the water, with nearly 90 percent of their mass submerged. The margins of the Greenland Ice Sheet produce thousands of icebergs each year. Many drift southward and find their way into the North Atlantic, where they are a hazard to navigation.

Whether the margin of a glacier is advancing, retreating, or remaining stationary depends on the *budget of the glacier.* That is, it depends on the balance or lack of balance between accumulation on the one hand and wastage on the other. If ice accumulation exceeds wastage, the glacial front advances until the two factors balance. At this point, the terminus of the glacier becomes stationary. If a warming trend increases wastage and/or if a drop in snowfall decreases accumulation, the ice front will retreat. As the terminus of the glacier retreats, the extent of the zone of wastage diminishes. Therefore, in time a new balance will be reached between accumulation and wastage, and the ice front will again become stationary.

Whether the margin of a glacier is advancing, retreating, or stationary, the ice within the glacier continues to flow forward. In the case of a receding glacier, the ice simply does not flow forward rapidly enough to offset wastage. This point is illustrated in Figure 4.4. While the line of stakes within the Rhône Glacier continued to move downvalley, the terminus of the glacier slowly retreated upvalley.

Did You Know?

If the Antarctic ice sheet were to melt at a suitable uniform rate, it could feed the Mississippi River for more than 50,000 years or it could maintain the flow of the Amazon River for approximately 5000 years.

Glacial Erosion

Sculpturing Earth's Surface
▼ Glaciers

Glaciers erode tremendous volumes of rock. For anyone who has observed the terminus of an alpine glacier, the evidence of its erosive force is plain. You can witness firsthand the re-

Figure 4.6 Icebergs are created when large pieces calve from the front of a glacier after it reaches a water body. Here, ice is calving from the terminus of Alaska's Hubbard Glacier in Wrangell–St. Elias National Park, Alaska. (Photo by Tom & Susan Bean, Inc.) As the inset in the lower left illustrates, only about 10 percent of an iceberg protrudes above the waterline. Some icebergs can be very large. The satellite image in the upper right shows a huge iceberg that broke from an Antarctic ice shelf. Its area (3000 square kilometers; 1200 square miles) is about the same as the state of Rhode Island. (NASA)

lease of rock fragments of various sizes from the ice as it melts. All signs lead to the conclusion that the ice has scraped, scoured, and torn rock debris from the floor and walls of the valley and carried it downvalley. Indeed, as a transporter of sediment, ice has no equal.

Once rock debris is acquired by the glacier, it cannot settle out as does the load carried by a stream or by the wind. Consequently, glaciers can carry huge blocks that no other erosional agent could possibly budge. Although today's glaciers are of limited importance as erosional agents, many landscapes that were modified by the widespread glaciers of the recent Ice Age still reflect to a high degree the work of ice.

How Glaciers Erode

Glaciers erode land primarily in two ways—by plucking and abrasion. First, as a glacier flows over a fractured bedrock surface, it loosens and lifts blocks of rock and incorporates them into the ice. This process, known as **plucking,** occurs when meltwater penetrates the cracks and joints along the rock floor of the glacier and freezes. As the water expands, it exerts tremendous leverage that pries the rock loose. In this manner, sediment of all sizes becomes part of the glacier's load.

The second major erosional process is **abrasion.** As the ice and its load of rock fragments slide over bedrock, they function like sandpaper to smooth and polish the surface below. The pulverized rock produced by the glacial "grist mill" is appropriately called **rock flour.** So much rock flour may be produced that meltwater streams leaving a glacier often have the grayish appearance of skim milk—visible evidence of the grinding power of the ice.

When the ice at the bottom of a glacier contains large rock fragments, long scratches and grooves called **glacial striations** may be gouged into the bedrock (Figure 4.7A). These linear scratches provide clues to the direction of ice flow. By mapping the striations over large areas, patterns of glacial flow can often be reconstructed.

In contrast, not all abrasive action produces striations. The rock surfaces over which the glacier moves may become highly polished by the ice and its load of finer particles. The broad expanses of smoothly polished granite in California's Yosemite National Park provide an excellent example (Figure 4.7B).

As is the case with other agents of erosion, the rate of glacial erosion is highly variable. This differential erosion by ice is largely controlled by four factors: (1) rate of glacial movement; (2) thickness of the ice; (3) shape, abundance, and hardness of the rock fragments contained in the ice at the base of the glacier; and (4) the erodibility of the surface beneath the glacier. Variations in any or all of these factors from time to time and/or from place to place mean that the features, effects, and degree of landscape modification in glaciated regions can vary greatly.

Landforms Created by Glacial Erosion

Although the erosional accomplishments of ice sheets can be tremendous, landforms carved by these huge ice masses usually do not inspire the same awe as do the erosional features created by valley glaciers. In regions where the erosional effects of ice sheets are significant, glacially scoured surfaces and subdued terrain are the rule. By contrast, in mountainous areas, erosion by valley glaciers produces many truly spectacular features. Much of the rugged mountain scenery so celebrated for its majestic beauty is the product of erosion by valley glaciers.

A. B.

Figure 4.7 **A.** Glacial abrasion created the scratches and grooves in this bedrock. Glacier Bay National Park, Alaska. (Photo © by Carr Clifton) **B.** Glacially polished granite in California's Yosemite National Park. (Photo by E. J. Tarbuck)

Take a moment to study Figure 4.8, which shows a mountain setting before, during, and after glaciation. You will refer to this figure often in the following discussion.

Glaciated Valleys Prior to glaciation, alpine valleys are characteristically V-shaped because streams are well above base level and are therefore downcutting (Figure 4.8A). However, in mountainous regions that have been glaciated, the valleys are no longer narrow. As a glacier moves down a valley once occupied by a stream, the ice modifies it in three ways: The glacier widens, deepens, and straightens the valley, so that what was once a narrow V-shaped valley is transformed into a U-shaped **glacial trough** (Figures 4.8C and 4.9A).

The amount of glacial erosion depends in part on the thickness of the ice. Consequently, main glaciers, also called *trunk glaciers,* cut their valleys deeper than do their smaller tributary glaciers. Thus, after the ice has receded, the valleys of tributary glaciers are left standing above the main glacial trough and are termed **hanging valleys.** Rivers flowing through hanging

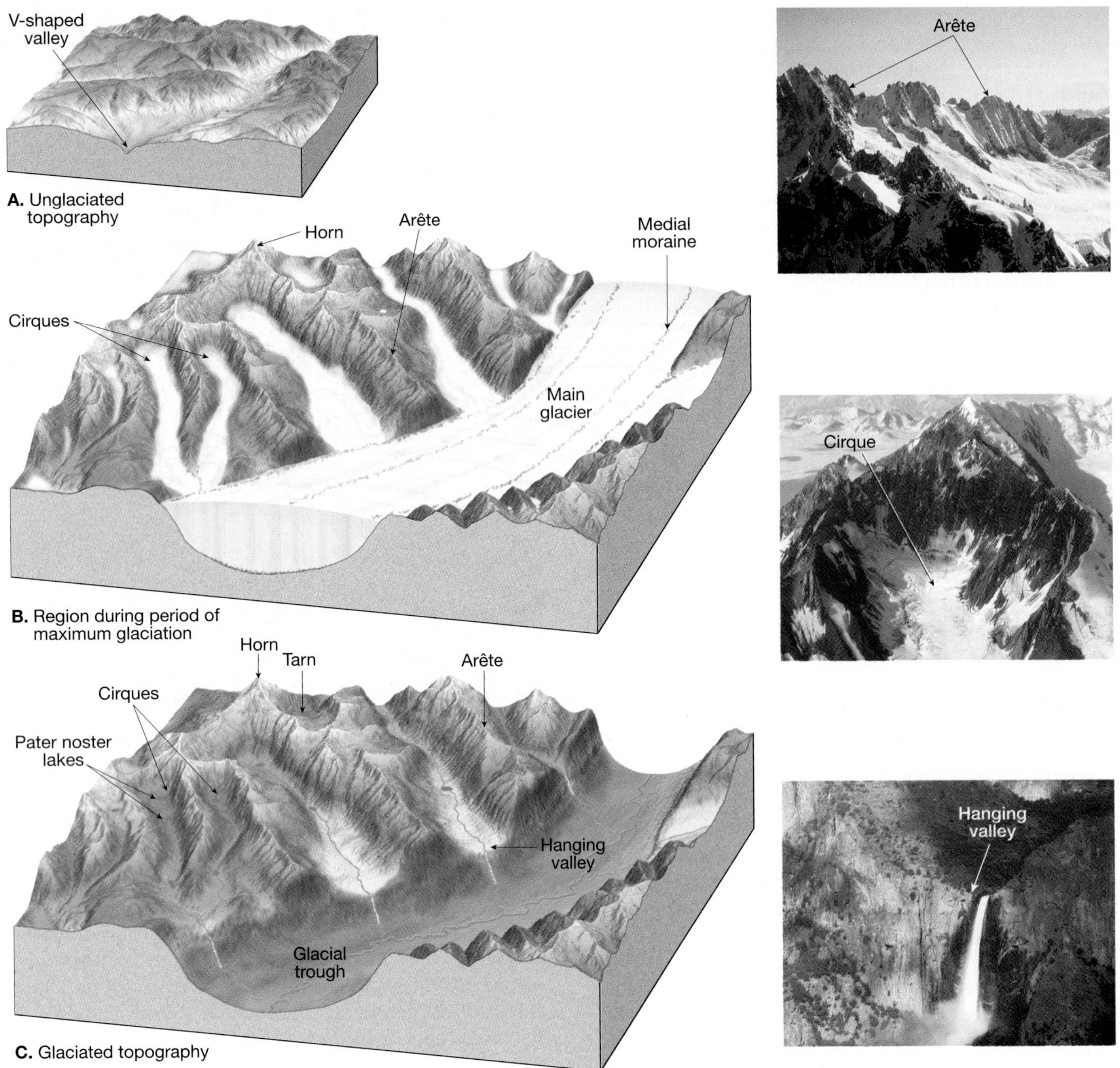

Figure 4.8 These diagrams of a hypothetical area show the development of erosional landforms created by alpine glaciers. The unglaciated landscape in part **A** is modified by valley glaciers in part **B.** After the ice recedes, in part **C,** the terrain looks very different than it did before glaciation. (Arête photo from James E. Patterson Collection; Cirque photo by Marli G. Miller; Hanging Valley photo by Marc Muench)

valleys may produce spectacular waterfalls, such as those in Yosemite National Park, California (Figure 4.8).

Cirques At the head of a glacial valley is a characteristic and often imposing feature associated with an alpine glacier—a **cirque.** As Figure 4.8 illustrates, these hollowed-out, bowl-shaped depressions have precipitous walls on three sides but are open on the downvalley side. The cirque is the focal point of the glacier's growth because it is the area of snow accumulation and ice formation. Cirques begin as irregularities in the mountainside that are subsequently enlarged by frost wedging and plucking along the sides and bottom of the glacier. The glacier, in turn, acts as a conveyor belt that carries the debris away. After the glacier has melted away, the cirque basin is often occupied by a small lake called a *tarn.*

Arêtes and Horns The Alps, Northern Rockies, and many other mountain landscapes carved by valley glaciers reveal more than glacial troughs and cirques. In addition, sinuous, sharp-edged ridges called **arêtes** and sharp, pyramid-like peaks called **horns** project above the surroundings (Figure 4.8C). Both features can originate from the same basic process—the enlargement of cirques produced by plucking and frost action. Cirques around a single high mountain create the spires of rock called horns. As the cirques enlarge and

A.

B.

Figure 4.9 **A.** Prior to glaciation, a mountain valley is typically narrow and V-shaped. During glaciation, an alpine glacier widens, deepens, and straightens the valley, creating the U-shaped glacial trough seen here. The string of lakes is called pater noster lakes. This valley is in Glacier National Park, Montana. (Photo by John Montagne). **B.** Like other fiords, this one at Tracy Arm, Alaska, is a drowned glacial trough. The fiord continues to get larger as the glacier in the background retreats and the sea takes its place. (Photo by Tom & Susan Bean, Inc.)

converge, an isolated horn is produced. A famous example is the Matterhorn in the Swiss Alps (Figure 4.10).

Arêtes can form in a similar manner except that the cirques are not clustered around a point but rather exist on opposite sides of a divide. As the cirques grow, the divide separating them is reduced to a very narrow, knifelike partition. An arête may also be created when the area separating two parallel glacial valleys is narrowed as the glaciers scour and widen their valleys.

Fiords **Fiords** are deep, often spectacular, steep-sided inlets of the sea that exist in many high-latitude areas of the world where mountains are adjacent to the ocean (Figure 4.9B). Norway, British Columbia, Greenland, New Zealand, Chile, and Alaska all have coastlines characterized by fiords. They are glacial troughs that became submerged as the ice left the valley and sea level rose following the Ice Age.

The depth of some fiords exceeds 1000 meters (3300 feet). However, the great depths of these flooded troughs are only partly explained by the post–Ice Age rise in sea level. Unlike the situation governing the downward erosional work of rivers, sea level does not act as base level for glaciers. As a consequence, glaciers are capable of eroding their beds far below the surface of the sea. For example, a valley glacier 300 meters (1000 feet) thick can carve its valley floor more than 250 meters (800 feet) below sea level before downward erosion ceases and the ice begins to float.

Figure 4.10 Horns are sharp, pyramid-like peaks that are fashioned by alpine glaciers. This example is the famous Matterhorn in the Swiss Alps. (Photo by Gavriel Jecan/Art Wolfe, Inc.)

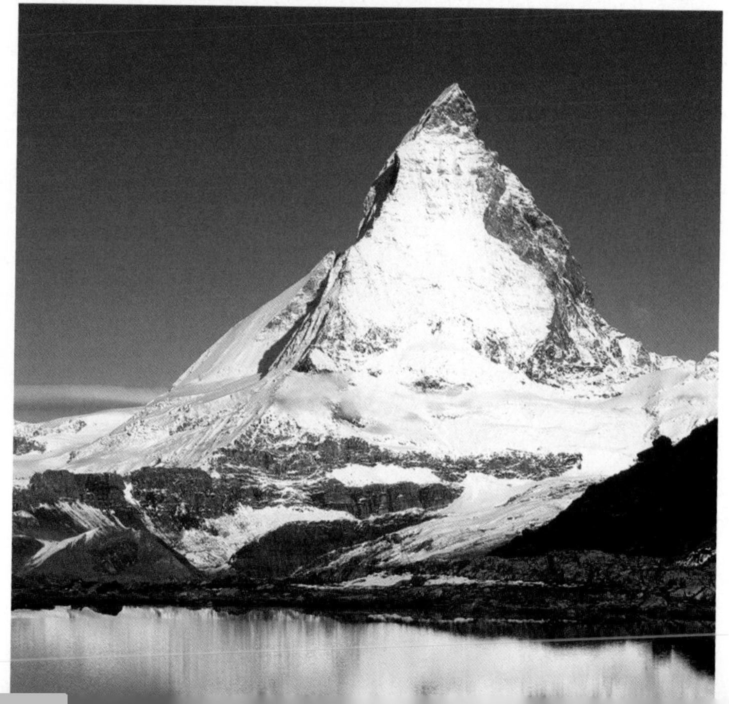

Did You Know?

If all of Earth's glaciers melted, sea level would rise approximately 70 meters (230 feet) worldwide.

Glacial Deposits

Sculpturing Earth's Surface
▼ Glaciers

Glaciers pick up and transport a huge load of debris as they slowly advance across the land. These materials are ultimately deposited when the ice melts. Glacial sediment can play a truly significant role in forming the physical landscape in regions where it is deposited. For example, in many areas once covered by the ice sheets of the recent Ice Age, the bedrock is rarely exposed because glacial deposits that are tens or even hundreds of meters thick completely bury the terrain. The general effect of these deposits is to level the topography. Indeed, many of today's familiar rural landscapes—rocky pastures in New England, wheat fields in the Dakotas, rolling farmland in the Midwest—resulted directly from glacial deposition.

Figure 4.11 Glacial till is an unsorted mixture of many different sediment sizes. A close examination often reveals cobbles that have been scratched as they were dragged along by the glacier. (Photos by E. J. Tarbuck)

Types of Glacial Drift

Long before the theory of an extensive Ice Age was proposed, much of the soil and rock debris covering portions of Europe was recognized as coming from elsewhere. At the time, these foreign materials were believed to have "drifted" into their present positions by floating ice during an ancient flood. As a consequence, the term *drift* was applied to this sediment. Although rooted in a concept that was not correct, this term was so well established by the time the true glacial origin of the debris became widely recognized that it remained in the glacial vocabulary. Today, **glacial drift** is an all-embracing term for sediments of glacial origin, no matter how, where, or in what form they were deposited.

Glacial drift is divided into two distinct types: (1) materials deposited directly by the glacier, which are known as *till,* and (2) sediments laid down by glacial meltwater, called *stratified drift.* Here is the difference: **Till** is deposited as glacial ice melts and drops its load of rock fragments. Unlike moving water and wind, ice cannot sort the sediment it carries; therefore, deposits of till are characteristically unsorted mixtures of many particle sizes (Figure 4.11). **Stratified drift** is sorted according to the size and weight of the fragments. Because ice is not capable of such sorting activity, these sediments are not deposited directly by the glacier. Rather, they reflect the sorting action of glacial meltwater.

Some deposits of stratified drift are made by streams coming directly from the glacier. Other stratified deposits involve sediment that was originally laid down as till and later picked up, transported, and redeposited by meltwater beyond the margin of the ice. Accumulations of stratified drift often consist largely of sand and gravel, because the meltwater is not capable of moving larger material and because the finer rock flour remains suspended and is commonly carried far from the glacier. An indication that stratified drift consists primarily of sand and gravel can be seen in many areas where these deposits are actively mined as aggregate for roadwork and other construction projects.

Boulders found in the till or lying free on the surface are called **glacial erratics** if they are different from the bedrock below (Figure 4.12). Of course, this means that they must have been derived from a source outside the area where they are found. Although the locality of origin for most erratics is unknown, the origin of some can be determined. There-

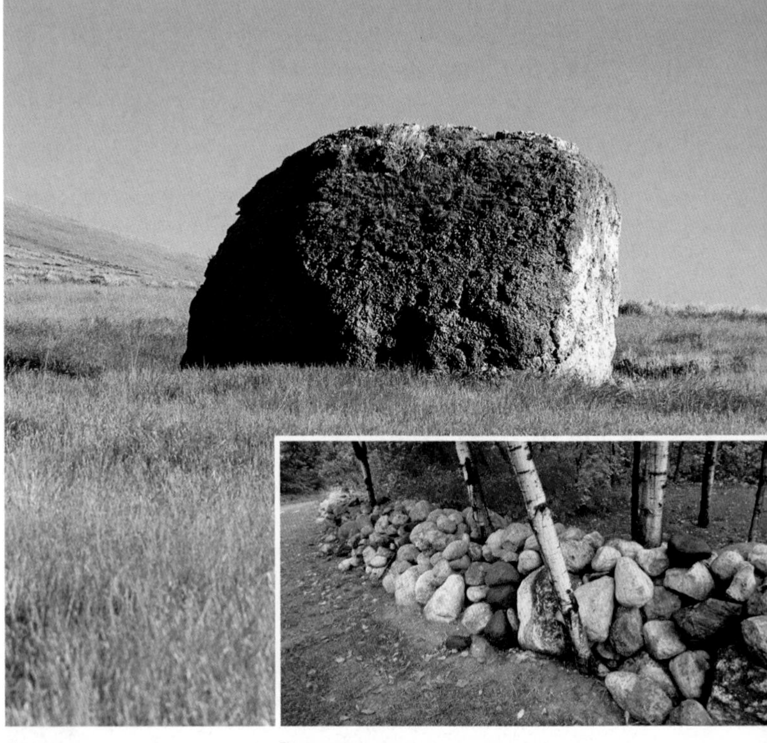

Figure 4.12 **A.** A large, glacial erratic, central Wyoming. (Photo by Yva Momatiuk and John Eastcott/Photo Researchers, Inc.) **B.** Stone wall made of glacial erratics near West Bend, Wisconsin. (Photo by Tom Bean)

fore, by studying glacial erratics as well as the mineral composition of the till, geologists can sometimes trace the path of a lobe of ice. In portions of New England as well as other areas, erratics can be seen dotting pastures and farm fields. In some places, these rocks are cleared from fields and piled to make stone walls and fences (Figure 4.12 inset).

Moraines, Outwash Plains, and Kettles

Perhaps the most widespread features created by glacial deposition are *moraines,* which are simply layers or ridges of till. Several types of moraines are identified; some are common only to mountain valleys, and others are associated with areas affected by either ice sheets or valley glaciers. Lateral and medial moraines fall in the first category, whereas end moraines and ground moraines are in the second.

Lateral and Medial Moraines The sides of a valley glacier accumulate large quantities of debris from the valley walls. When the glacier wastes away, these materials are left as ridges, called **lateral moraines,** along the sides of the valley (Figure 4.13). **Medial moraines** are formed when two advancing valley glaciers come together to form a single ice stream. The till that was once carried along the edges of each glacier joins to form a single dark stripe of debris within the newly enlarged glacier. The creation of these dark stripes within the ice stream is one obvious proof that glacial ice moves, because the medial moraine could not form if the ice did not flow downvalley (Figure 4.13). It is common to see several dark debris stripes within a large alpine glacier because one will form each time a tributary glacier joins the main valley.

End Moraines and Ground Moraines An **end moraine** is a ridge of till that forms at the terminus of a glacier. These relatively common landforms are deposited when a state of equilibrium is attained between wastage and ice accumulation. That is, the end moraine forms when the ice is melting near the end of the glacier at a rate equal to the forward advance of the glacier from its region of nourishment. Although the terminus of the glacier is stationary, the ice continues to flow forward, delivering a continuous supply of sediment in the same manner a conveyor belt delivers goods to the end of a production line. As the ice melts, the till is dropped and the end moraine grows. The longer the ice front remains stable, the larger the ridge of till will become.

Eventually, the time comes when wastage exceeds nourishment. At this point, the front of the glacier begins to recede in the direction from which it originally advanced. As the ice front retreats, however, the conveyor-belt action of the glacier continues to provide fresh supplies of sediment to the terminus. In this manner, a large quantity of till is deposited as the ice melts away, creating a rock-strewn, undulating plain. This gently rolling layer of till deposited as the ice front recedes is termed **ground moraine.** Ground moraine has a leveling effect, filling in low spots and clogging old stream channels, often leading to a derangement of the existing drainage system. In areas where this layer of till is still relatively fresh, such as the northern Great Lakes region, poorly drained swampy lands are quite common.

Periodically, a glacier will retreat to a point where wastage and nourishment once again balance. When this happens, the ice front stabilizes and a new end moraine forms.

The pattern of end moraine formation and ground moraine deposition may be repeated many times before the glacier has completely vanished. Such a pattern is illustrated by Figure 4.14. The very first end moraine to form marks the farthest advance of the glacier and is called the *terminal end moraine.* Those end moraines that form as the ice front occasionally stabilizes during retreat are termed *recessional end moraines.* Terminal and recessional moraines are essentially alike; the only difference between them is their relative positions.

End moraines deposited by the most recent major stage of Ice Age glaciation are prominent features in many parts of the Midwest and Northeast. In Wisconsin, the wooded, hilly terrain of the Kettle Moraine near Milwaukee is a particularly

Figure 4.13 Lateral moraines form from the accumulation of debris along the sides of a valley glacier. Medial moraines form when the lateral moraines of merging valley glaciers join. Medial moraines could not form if the ice did not advance downvalley. Therefore, these dark stripes are proof that glacial ice moves. Saint Elias National Park, Alaska. (Photo by Tom Bean)

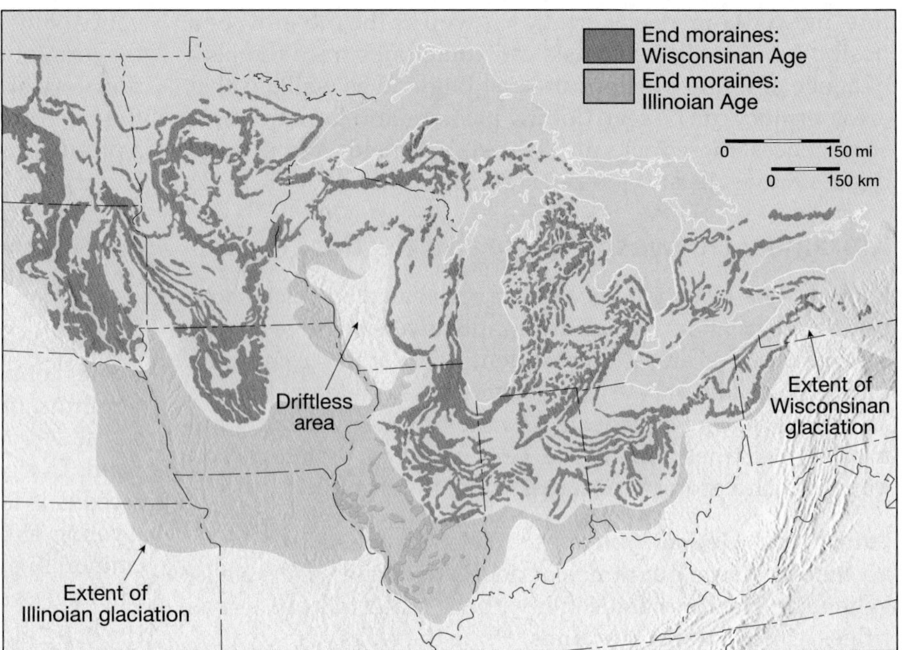

Figure 4.14 End moraines of the Great Lakes region. Those deposited during the most recent stage of glaciation (a span called the Wisconsinan stage) are most prominent.

picturesque example. A well-known example in the Northeast is Long Island. This linear strip of glacial sediment that extends northeastward from New York City is part of an end moraine complex that stretches from eastern Pennsylvania to Cape Cod, Massachusetts.

Figure 4.15 represents a hypothetical area during and following glaciation. It shows the end moraines that were just described as well as the depositional features that are discussed in the sections that follow. This figure depicts landscape features similar to those that might be encountered if you were traveling in the upper Midwest or New England. As you read about other glacial deposits, you will be referred to this figure again.

Outwash Plains and Valley Trains At the same time that an end moraine is forming, meltwater emerges from the ice in rapidly moving streams that are often choked with suspended material and carry a substantial bed load. As the water leaves the glacier, it rapidly loses velocity and much of its bed load is dropped. In this way, a broad, ramplike accumulation of stratified drift is built adjacent to the downstream edge of most end moraines. When the feature is formed in association with an ice sheet, it is termed an **outwash plain,** and when it is confined to a mountain valley, it is commonly referred to as a **valley train** (Figure 4.15).

Kettles Often, end moraines, outwash plains, and valley trains are pockmarked with basins or depressions known as **kettles** (Figure 4.15). Kettles form when blocks of stagnant ice become buried in drift and eventually melt, leaving pits in the glacial sediment. Most kettles do not exceed 2 kilometers (1.25 miles) in diameter and the typical depth of most kettles is less than 10 meters (33 feet). Water often fills the depression and forms a pond or lake. One well-known example is Walden Pond near Concord, Massachusetts. It is here that Henry David Thoreau lived alone for two years in the 1840s and about which he wrote *Walden,* his American literature classic.

Drumlins, Eskers, and Kames

Moraines are not the only landforms deposited by glaciers. Some landscapes are characterized by numerous elongate parallel hills made of till. Other areas exhibit conical hills and relatively narrow winding ridges composed mainly of stratified drift.

Drumlins **Drumlins** are streamlined asymmetrical hills composed of till (Figure 4.15). They range in height from 15 to 60 meters (50 to 200 feet) and average 0.4 to 0.8 kilometer (0.25 to 0.50 mile) in length. The steep side of the hill faces the direction from which the ice advanced, while the gentler slope points in the direction the ice moved. Drumlins are not found singly but rather occur in clusters, called *drumlin fields.* One such cluster, east of Rochester, New York, is estimated to contain about 10,000 drumlins. Their streamlined shape indicates that they were molded in the zone of flow within an active glacier. It is thought that drumlins originate when glaciers advance over previously deposited drift and reshape the material.

Eskers and Kames In some areas that were once occupied by glaciers, sinuous ridges composed largely of sand and gravel may be found. These ridges, called **eskers,** are deposits made by streams flowing in tunnels beneath the ice, near the terminus of a glacier (Figure 4.15). They may be several meters high and extend for many kilometers. In some areas, they are mined for sand and gravel and for this reason eskers are disappearing in some localities.

Kames are steep-sided hills that, like eskers, are composed largely of stratified drift (Figure 4.15). Kames originate when glacial meltwater washes sediment into openings and

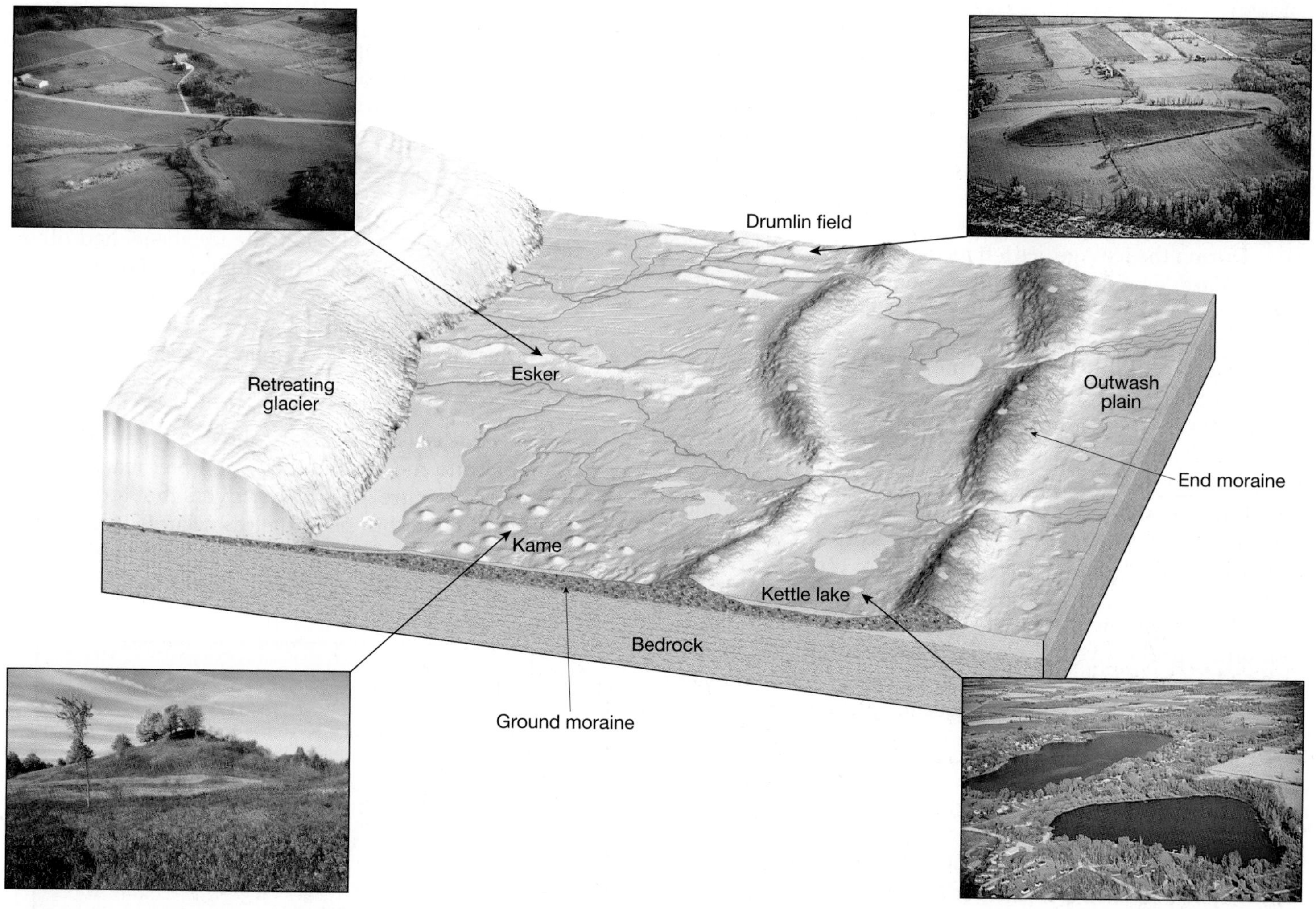

Figure 4.15 This hypothetical area illustrates many common depositional landforms. The outermost end moraine marks the limit of glacial advance and is called the *terminal end moraine.* End moraines that form as the ice front occasionally becomes stationary during retreat are called *recessional end moraines.* (Drumlin photo courtesy of Ward's Natural Science Establishment; Kame, Esker, and Kettle photos by Richard P. Jacobs/JLM Visuals)

Did You Know?

Studies have shown that the retreat of glaciers may reduce the stability of faults and hasten earthquake activity. When the weight of the glacier is removed, the ground rebounds. In tectonically active areas experiencing postglacial rebound, earthquakes may occur sooner and/or be stronger than if the ice were present.

depressions in the stagnant wasting terminus of a glacier. When the ice eventually melts away, the stratified drift is left behind as mounds or hills.

Glaciers of the Ice Age

We have mentioned the Ice Age, a time when ice sheets and alpine glaciers were far more extensive than they are today. There was a time when the most popular explanation for drift was that the material had been brought by icebergs or perhaps simply swept across the landscape by a catastrophic flood. During the nineteenth century, however, field investigations by many scientists provided convincing proof that an extensive Ice Age was responsible for these deposits and for many other features.

By the beginning of the twentieth century, geologists had largely determined the extent of Ice Age glaciation. Further, they discovered that many glaciated regions did not have just one layer of drift, but several. Close examination of these older deposits showed well-developed zones of chemical weathering and soil formation as well as the remains of plants that require warm temperatures. The evidence was clear: There had not been just one glacial advance but several, each separated by extended periods when climates were as warm as or warmer than at present. The Ice Age had not simply been a time when the ice advanced over the land, lingered for a while, and then receded. Rather, it was a complex period characterized by a number of advances and withdrawals of glacial ice.

The glacial record on land is punctuated by many erosional gaps. This makes it difficult to reconstruct the episodes of the Ice Age. But sediment on the ocean floor provides an uninterrupted record of climate cycles for this period. Studies of cores drilled from these seafloor sediments show that glacial/interglacial cycles have occurred about every 100,000 years. About 20 such cycles of cooling and warming were identified for the span we call the Ice Age.

During the Ice Age, ice left its imprint on almost 30 percent of Earth's land area, including about 10 million square kilometers (nearly 4 million square miles) of North America, 5 million square kilometers (2 million square miles) of Europe, and 4 million square kilometers (1.6 million square miles) of Siberia (Figure 4.16). The amount of glacial ice in the Northern Hemisphere was roughly twice that in the Southern Hemisphere. The primary reason is that the Southern Hemisphere has little land in the middle latitudes, and, therefore, the southern polar ice could not spread far beyond the margins of Antarctica. By contrast, North America and Eurasia provided great expanses of land for the spread of ice sheets.

We now know that the Ice Age began between 2 million and 3 million years ago. This means that most of the major glacial episodes occurred during a division of the geologic time scale called the **Pleistocene epoch.** Although the Pleistocene is commonly used as a synonym for the Ice Age, this epoch does not encompass it all. The Antarctic Ice Sheet, for example, formed at least 30 million years ago.

Did You Know?

Antarctica's ice sheet weighs so much that it depresses Earth's crust by an estimated 900 meters (3000 feet) or more.

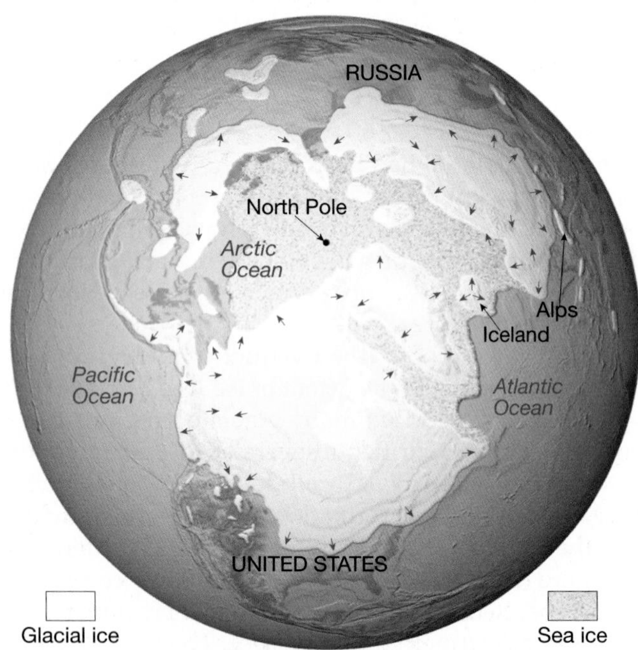

Figure 4.16 Maximum extent of glaciation in the Northern Hemisphere during the Ice Age. Sea ice (the ice surrounding the North Pole in the Arctic Ocean) is *not* glacial ice. Rather, it is the frozen upper portion of the ocean.

Other Effects of Ice Age Glaciers

In addition to the massive erosional and depositional work carried on by Pleistocene glaciers, the ice sheets had other, sometimes profound, effects on the landscape. For example, as the ice advanced and retreated, animals and plants were forced to migrate. This led to stresses that some organisms could not tolerate. Furthermore, many present-day stream courses bear little resemblance to their preglacial routes. The Missouri River once flowed northward toward Hudson Bay in Canada. The Mississippi River followed a path through central Illinois, and the head of the Ohio River reached only as far as Indiana (Figure 4.17). Other rivers that today carry only a trickle of water but nevertheless occupy broad channels are a testament to the fact that they once carried torrents of glacial meltwater.

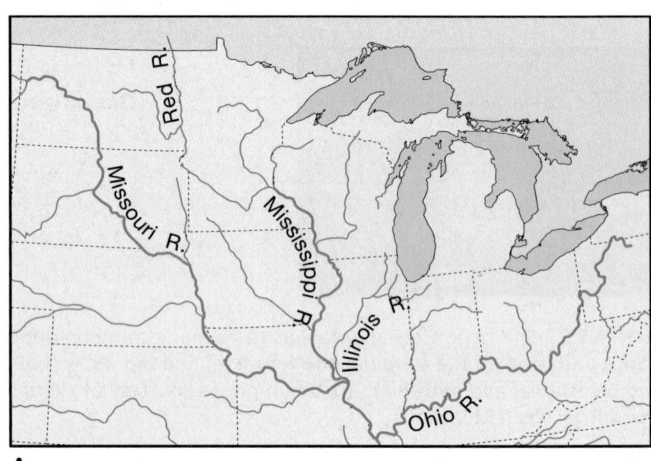

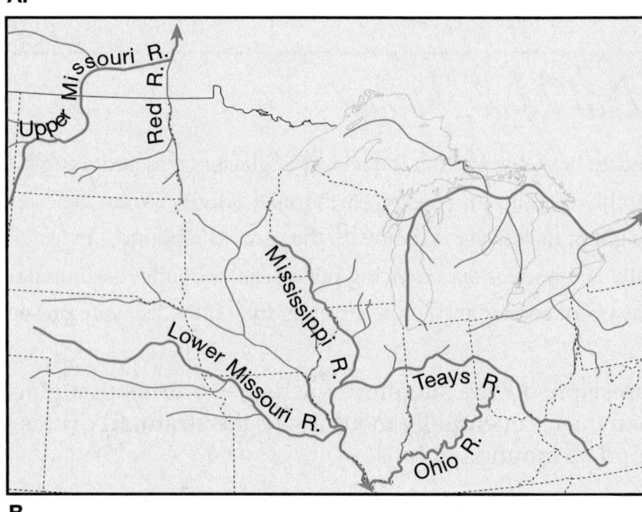

Figure 4.17 **A.** This map shows the Great Lakes and the familiar present-day pattern of rivers in the central United States. Pleistocene ice sheets played a major role in creating this pattern. **B.** Reconstruction of drainage systems in the central United States prior to the Ice Age. The pattern was very different from today's, and there were no Great Lakes.

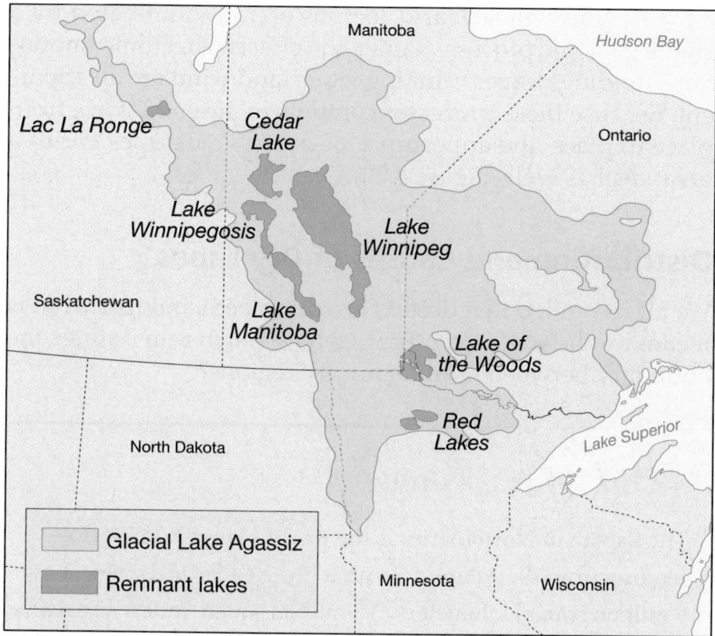

Figure 4.18 Map showing the extent of glacial Lake Agassiz. This proglacial lake came into existence about 12,000 years ago and lasted for about 4500 years. It was an immense feature—bigger than all of the present-day Great Lakes combined. The modern-day remnants of this water body are still major landscape features.

In areas that were centers of ice accumulation, such as Scandinavia and northern Canada, the land has been slowly rising for the past several thousand years. The land had downwarped under the tremendous weight of 3-kilometer-thick (almost 2-mile-thick) masses of ice. Following the removal of this immense load, the crust has been adjusting by gradually rebounding upward ever since.

Ice sheets and alpine glaciers can act as dams to create lakes by trapping glacial meltwater and blocking the flow of rivers. Some of these lakes are relatively small and short-lived. Others can be large and exist for hundreds or thousands of years.

Figure 4.18 is a map of Lake Agassiz—the largest lake to form during the Ice Age in North America. With the retreat of the ice sheet came enormous volumes of meltwater. The Great Plains generally slope upward to the west. As the terminus of the ice sheet receded northeastward, meltwater was trapped between the ice on one side and the sloping land on the other, causing Lake Agassiz to deepen and spread across the landscape. It came into existence about 12,000 years ago and lasted for about 4500 years. Such water bodies are termed *proglacial lakes*, referring to their position just beyond the outer limits of a glacier or ice sheet.

A far-reaching effect of the Ice Age was the worldwide change in sea level that accompanied each advance and retreat of the ice sheets. The snow that nourishes glaciers ultimately comes from moisture evaporated from the oceans. Therefore, when the ice sheets increased in size, sea level fell and the shoreline moved seaward (Figure 4.19). Estimates suggest that sea level was as much as 100 meters (330 feet) lower than it is today. Consequently, the Atlantic Coast of the United States was located more than 100 kilometers (60 miles) to the east of New York City. Moreover, France and Britain were joined where the English Channel is today. Alaska and Siberia were connected across the Bering Strait, and Southeast Asia was tied by dry land to the islands of Indonesia.

The formation and growth of ice sheets was an obvious response to significant changes in climate. But the existence of the glaciers themselves triggered climatic changes in the regions beyond their margins. In arid and semiarid areas on all continents, temperatures were lower, which meant evaporation rates were also lower. At the same time, precipitation was moderate. This cooler, wetter climate resulted in the formation of many lakes called **pluvial lakes** (from the Latin term *pluvia* meaning "rain"). In North America, pluvial lakes were concentrated in the vast Basin and Range region of Nevada and Utah. Although most are now gone, a few remnants remain, the largest being Utah's Great Salt Lake.

Figure 4.19 This map of a portion of North America shows the present-day coastline compared to the coastline that existed during the last Ice-Age maximum (18,000 years ago) and the coastline that would exist if present ice sheets in Greenland and Antarctica melted. (After R. H. Dott Jr. and R. L. Battan)

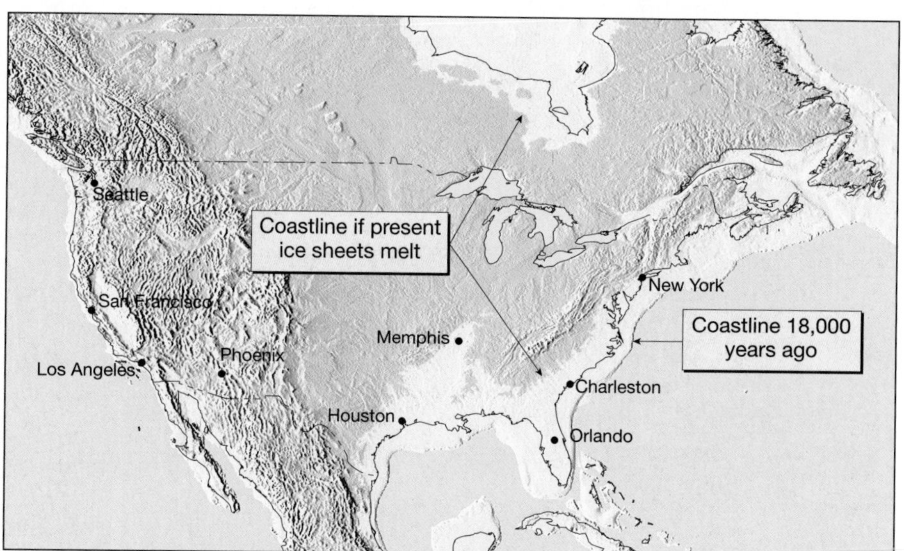

Deserts

Sculpturing Earth's Surface
▼ Deserts

The dry regions of the world encompass about 42 million square kilometers (more than 16 million square miles), a surprising 30 percent of Earth's land surface. No other climate group covers so large a land area. The word *desert* literally means "deserted" or "unoccupied." For many dry regions, this is a very appropriate description. Yet where water is available in deserts, plants and animals thrive. Nevertheless, the world's dry regions are among the least familiar land areas on Earth outside of the polar realm.

Desert landscapes frequently appear stark. Their profiles are not softened by a carpet of soil and abundant plant life. Instead, barren rocky outcrops with steep, angular slopes are common. Some rocks are tinted orange and red; others in different locations are gray and brown and streaked with black. For many visitors, desert scenery exhibits a striking beauty; to others, the terrain seems bleak. No matter which feeling is elicited, it is clear that deserts are very different from the more humid places where most people live.

As you will see, arid regions are not dominated by a single geologic process. Rather, the effects of tectonic (mountain-building) forces, running water, and wind are all apparent. Because these processes combine in different ways from place to place, the appearance of desert landscapes varies a great deal as well (Figure 4.20).

Distribution and Causes of Dry Lands

We all recognize that deserts are dry places, but just what is meant by the word *dry*? That is, how much rain defines the boundary between humid and dry regions?

Did You Know?

The Sahara of North Africa is the world's largest desert. Extending from the Atlantic Ocean to the Red Sea, it covers about 9 million square kilometers (3.5 million square miles) an area about the size of the United States. By comparison, the largest desert in the United States, Nevada's Great Basin Desert, has an area that is less than 5 percent as large as the Sahara.

Figure 4.20 Arizona's Organ Pipe Cactus National Monument is in the Sonoran Desert. The Ajo Mountains are in the background. The appearance of desert landscapes varies a great deal from place to place. (Photo by Jeff Lepore/Photo Researchers, Inc.)

Sometimes, it is arbitrarily defined by a single rainfall figure—for example, 25 centimeters (10 inches) per year of precipitation. However, the concept of dryness is relative; it refers to *any situation in which a water deficiency exists.* Climatologists define *dry climate* as one in which yearly precipitation is less than the potential loss of water by evaporation.

Within these water-deficient regions, two climatic types are commonly recognized: **desert,** or arid, and **steppe,** or semiarid. The two categories have many features in common; their differences are primarily a matter of degree. The steppe is a marginal and more humid variant of the desert and represents a transition zone that surrounds the desert and separates it from bordering humid climates. The world map showing the distribution of desert and steppe regions reveals that dry lands are concentrated in the subtropics and in the middle latitudes (Figure 4.21).

Deserts in places such as Africa, Arabia, and Australia are primarily the result of the prevailing global distribution of air pressure and winds (Figure 4.22). Coinciding with dry regions in the lower latitudes are zones of high air pressure known as the *subtropical highs.* These pressure systems are characterized by subsiding air currents. When air sinks, it is compressed and warmed. Such conditions are just the opposite of what is needed to produce clouds and precipitation. Consequently, these regions are known for their clear skies, sunshine, and ongoing drought.

Middle-latitude deserts and steppes exist principally because they are sheltered in the deep interiors of large landmasses. They are far removed from the ocean, which is the ultimate source of moisture for cloud formation and precipitation. In addition, the presence of high mountains across the paths of prevailing winds further acts to separate these areas from water-bearing, maritime air masses. In North America, the Coast Ranges, Sierra Nevada, and Cascades are the foremost mountain barriers to moisture from the Pacific (Figure 4.23). In their rain shadow lies the dry and expansive Basin and Range region of the American West.

Middle-latitude deserts provide an example of how mountain-building processes affect climate. Without mountains, wetter climates would prevail where dry regions exist today.

Did You Know?

All deserts are not hot. Cold temperatures are experienced in middle-latitude deserts. For example, at Ulan Bator in Mongolia's Gobi Desert, the average *high* temperature on January days is only −19°C (−2°F)!

The Role of Water in Arid Climates

Permanent streams are normal in humid regions, but practically all desert streambeds are dry most of the time (Figure 4.24A). Deserts have **ephemeral streams,** which means that they carry water only in response to specific episodes of rainfall. A typical ephemeral stream may flow only a few days or perhaps just a few hours during the year. In some years, the channel might carry no water at all.

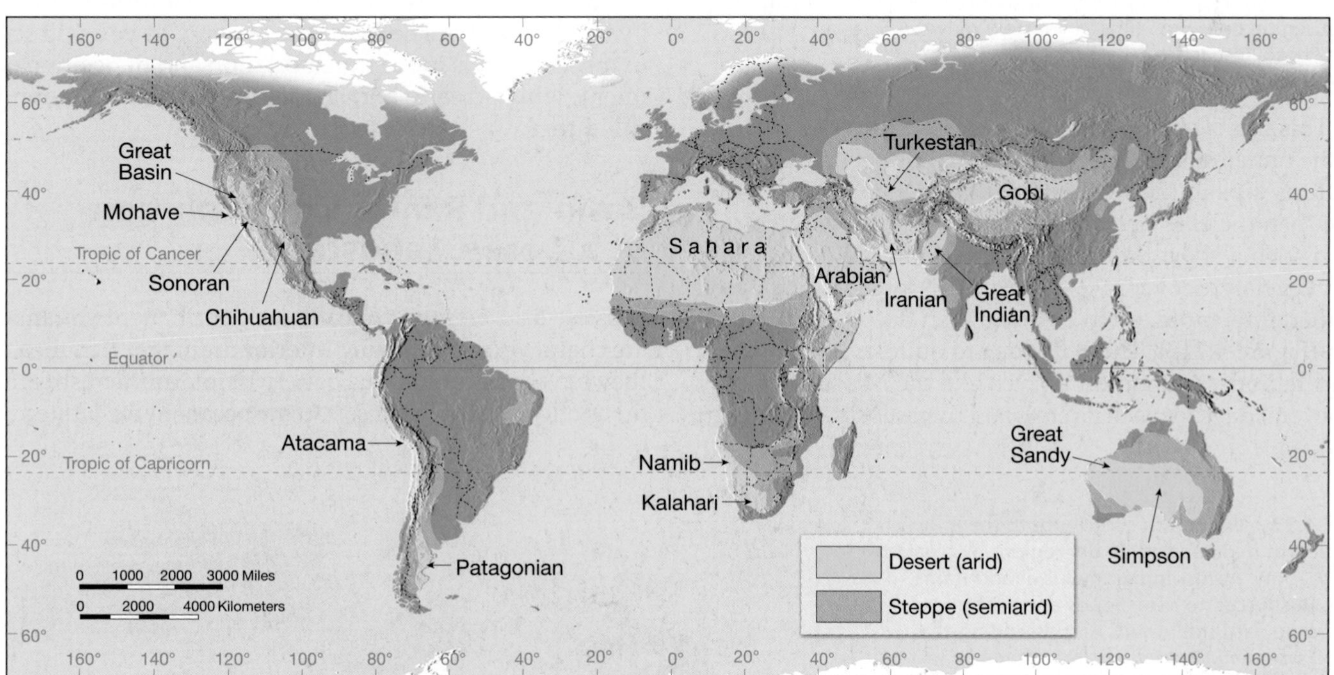

Figure 4.21 Arid and semiarid climates cover about 30 percent of Earth's land surface. No other climate group covers so large an area. Lower latitude deserts, such as the Sahara, Kalahari, and Great Sandy, are associated with the dry subsiding air masses of Earth's subtropical high pressure belts. Middle latitude deserts, such as the Gobi and Great Basin, are in the deep interior of continents and in the rain shadow of mountains.

Figure 4.22 In this view of Earth from space, North Africa's Sahara Desert, the adjacent Arabian Desert, and the Kalahari and Namib deserts in southern Africa are clearly visible as tan-colored, cloud-free zones. These low-latitude deserts are dominated by the dry, subsiding air associated with pressure belts known as the *subtropical highs.* By contrast, the band of clouds that extends across central Africa and the adjacent oceans coincides with the equatorial low-pressure belt, the rainiest region on Earth. (Image courtesy of NASA/Science Source/Photo Researchers, Inc.)

This fact is obvious even to the casual traveler who notices the numerous bridges with no streams beneath them, or numerous dips in the road where dry channels cross. However, when the rare heavy showers do occur, so much rain falls in such a short time that it cannot all soak in. Because desert vegetative cover is sparse, runoff is largely unhindered and, therefore, rapid, often creating flash floods along valley floors (Figure 4.24B). These floods are quite unlike floods in humid regions. A flood on a river like the Mississippi may take several days to reach its crest and then subside. But desert floods arrive suddenly and subside quickly. Because much surface material in a desert is not anchored by vegetation, the amount of erosional work that occurs during a single short-lived rain event is impressive.

In the dry western United States, different names are used for ephemeral streams, including *wash* and *arroyo.* In other parts of the world, a dry desert stream may be a *wadi* (Arabia and North Africa), a *donga* (South America), or a *nullah* (India).

Humid regions are notable for their well-developed drainage systems. But in arid regions, streams usually lack an extensive system of tributaries. In fact, a basic characteristic of desert streams is that they are small and die out before reaching the sea. Because the water table is usually far below the surface, few desert streams can draw upon it as streams do in humid regions. Without a steady supply of water, the combination of evaporation and soaking in soon depletes the stream.

The few permanent streams that do cross arid regions, such as the Colorado and Nile rivers, originate *outside* the desert, often in mountains where water is more plentiful. Here the supply of water must be great to compensate for the losses that occur as the stream crosses the desert. For example, after the Nile leaves its headwaters in the lakes and mountains of central Africa, it crosses almost 3000 kilometers (nearly 1900 miles) of the Sahara Desert *without a single tributary.*

It should be emphasized that *running water, although infrequent, nevertheless is responsible for most of the erosional work in deserts.* This is contrary to a common belief that wind is the most important erosional agent sculpturing desert landscapes. Although wind erosion is indeed more significant in dry areas than elsewhere, most desert landforms are carved by running water. The main role of wind, as you will see later in this chapter, is in the transportation and deposition of sediment, which creates and shapes the ridges and mounds we call dunes.

Basin and Range: The Evolution of a Desert Landscape

Because arid regions typically lack permanent streams, they are characterized as having **interior drainage.** This means that they have a discontinuous pattern of intermittent streams that do not flow out of the desert to the ocean. In the United States,

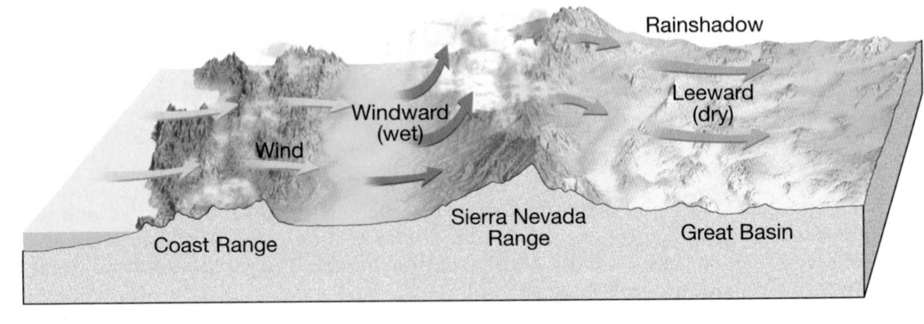

Figure 4.23 Mountains contribute to the formation of middle-latitude dry regions by creating a rain shadow. As moving air meets a mountain barrier, it is forced to rise. Clouds and precipitation on the windward side result. Air descending the leeward side is much drier. The mountains effectively cut the leeward side off from the source of moisture, creating a rain shadow. The Great Basin desert is a rain-shadow desert that covers nearly all of Nevada and portions of adjacent states.

A.

B.

Figure 4.24 **A.** Most of the time, desert stream channels are dry. **B.** This photo shows an ephemeral stream shortly after a heavy shower. Although such floods are short-lived, large amounts of erosion occur. (Photos by E. J. Tarbuck)

the dry Basin and Range region provides an excellent example. The region includes southern Oregon, all of Nevada, western Utah, southeastern California, southern Arizona, and southern New Mexico. The name Basin and Range is an apt description for this almost 800,000-square-kilometer (312,000-square-mile) region, as it is characterized by more than 200 relatively small mountain ranges that rise 900 to 1500 meters (3000 to 5000 feet) above the basins that separate them.

In this region, as in others like it around the world, most erosion occurs without reference to the ocean (ultimate base level), because the interior drainage never reaches the sea. Even where permanent streams flow to the ocean, few tributaries exist, and only a narrow strip of land adjacent to the stream has sea level as its ultimate level of land reduction.

The block models in Figure 4.25 depict how the landscape has evolved in the Basin and Range region. During and following uplift of the mountains, running water carves the elevated masses and deposits large quantities of sediment in the basin. In this early stage, relief (difference in elevation between high and low points in an area) is greatest. As erosion lowers the mountains and sediment fills the basins, elevation differences diminish.

When the occasional torrents of water produced by sporadic rains or periods of snowmelt move down the mountain canyons, they are heavily loaded with sediment. Emerging from the confines of the canyon, the runoff spreads over the gentler slopes at the base of the mountains and quickly loses velocity. Consequently, most of its sediment load is dumped within a short distance. The result is a cone of debris known as an **alluvial fan** at the mouth of a canyon (Figure 4.26). Over the years, a fan enlarges, eventually coalescing with fans from adjacent canyons to produce an apron of sediment (*bajada*) along the mountain front.

On the rare occasions of abundant rainfall, streams may flow across the alluvial fans to the center of the basin, converting the basin floor into a shallow **playa lake.** Playa lakes last only a few days or weeks, before evaporation and infiltration remove the water. The dry, flat lake bed that remains is termed a *playa.*

Playas occasionally become encrusted with salts that are left behind when the water in which they were dissolved evaporates. A case in point is the sodium borate (better known as borax) mined from ancient playa lake deposits in Death Valley, California.

With the ongoing erosion of the mountain mass and the accompanying sedimentation, the local relief continues to diminish. Eventually, nearly the entire mountain mass is gone. Thus, by the late stages of erosion, the mountain areas are reduced to a few large bedrock knobs (called *inselbergs*) projecting above the sediment-filled basin.

Each of the stages of landscape evolution in an arid climate depicted in Figure 4.25 can be observed in the Basin and Range region. Recently uplifted mountains in an early stage of erosion are found in southern Oregon and northern Nevada. Death Valley, California, and southern Nevada fit into the more advanced middle stage, whereas the late stage, with its inselbergs, can be seen in southern Arizona.

Wind Erosion

Sculpturing Earth's Surface
▼ Deserts

Moving air, like moving water, is turbulent and able to pick up loose debris and transport it to other locations. Just as in a river, the velocity of wind increases with height above the surface. Also like a river, wind transports fine particles in suspension while heavier ones are carried as bed load (Figure 4.27). However, the transport of sediment by wind differs

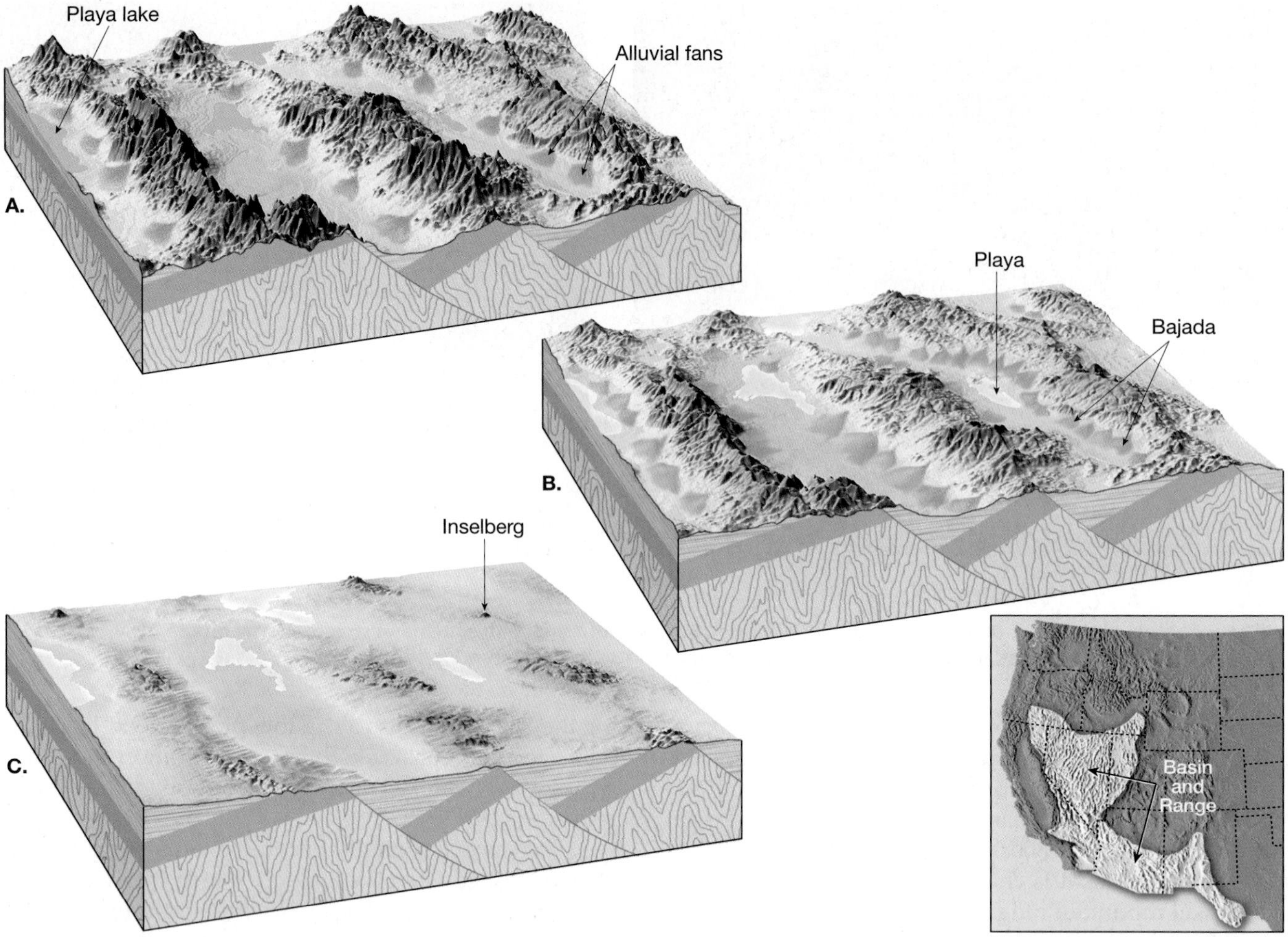

Figure 4.25 Stages of landscape evolution in a mountainous desert such as the Basin and Range region of the West. As erosion of the mountains and deposition in the basins continue, relief diminishes. **A.** Early stage. **B.** Middle stage. **C.** Late stage.

Figure 4.26 Alluvial fans develop where the gradient of a stream changes abruptly from steep to flat. Such a situation exists in Death Valley, California, where streams emerge from the mountains into a flat basin. As a result, Death Valley has many large alluvial fans. (Photo by Michael Collier)

A.

B.

Figure 4.27 **A.** This satellite image from July 16, 2003, captured Saharan dust streaming northeastward from North Africa across the Mediterranean Sea toward Italy. (NASA) **B.** The bed load carried by winds consists of sand grains, many of which move by bouncing along the surface. Sand never travels far from the surface, even when winds are very strong. (Photo by Stephen Trimble)

from that by running water in two significant ways. First, wind's lower density compared to water renders it less capable of picking up and transporting coarse materials. Second, because wind is not confined to channels, it can spread sediment over large areas, as well as high into the atmosphere.

Compared to running water and glaciers, wind is a relatively insignificant erosional agent. Recall that even in deserts, most erosion is performed by intermittent running water, not by the wind. Wind erosion is more effective in arid lands than in humid areas because in humid places moisture binds particles together and vegetation anchors the soil. For wind to be an effective erosional force, dryness and scanty vegetation are important prerequisites. When such circumstances exist, wind may pick up, transport, and deposit great quantities of fine sediment. During the 1930s, parts of the Great Plains experienced vast dust storms (Figure 4.28). The plowing under of the natural vegetative cover for farming, followed by severe drought, exposed the land to wind erosion and led to the area being labeled the Dust Bowl.

Deflation, Blowouts, and Desert Pavement

One way that wind erodes is by **deflation,** the lifting and removal of loose material. Wind can suspend only fine sediment such as clay and silt. Larger grains of sand are rolled or skipped along the surface (a process called *saltation*) and comprise the bed load. Particles larger than sand are usually not transported by wind. Deflation sometimes is difficult to notice because the entire surface is being lowered at the same time, but it can be significant. In portions of the 1930s Dust Bowl, the land was lowered by as much as 1 meter (3 feet) in only a few years (Figure 4.29).

The most noticeable result of deflation in some places is shallow depressions called **blowouts.** In the Great Plains region, from Texas north to Montana, thousands of blowouts can be seen. They range from small dimples less than 1-meter (3-feet) deep and 3-meters (10-feet) wide to depressions that are more than 45 meters (150 feet) deep and several kilometers across.

In portions of many deserts, the surface is characterized by a layer of coarse pebbles and cobbles that are too large to be moved by the wind. This stony veneer, called **desert pavement,** may form as deflation lowers the surface

Figure 4.28 Dust blackens the sky on May 21, 1937, near Elkhart, Kansas. It was because of storms like this that portions of the Great Plains were called the Dust Bowl in the 1930s. (Photo reproduced from the collection of the Library of Congress)

Figure 4.29 This photo was taken north of Granville, North Dakota, in July 1936, during a prolonged drought. Strong winds removed the soil that was not anchored by vegetation. The mounds are 1.2 meters (4 feet) high and show the level of the land prior to deflation. (Photo courtesy of the State Historical Society of North Dakota 0278-01)

by removing sand and silt from poorly sorted materials. As Figure 4.30 A illustrates, the concentration of larger particles at the surface gradually increases as the finer particles are blown away. Eventually, a continuous cover of coarse particles remains.

Studies have shown that the process depicted in Figure 4.30A is not an adequate explanation for all environments in which desert pavement exists. As a result, an alternate explanation was formulated and is illustrated in Figure 4.30B. This hypothesis suggests that pavement develops on a surface that initially consists of coarse pebbles. Over time, protruding cobbles trap fine wind-blown grains that settle and sift downward through the spaces between the larger surface stones. The process is aided by infiltrating rainwater.

Once desert pavement becomes established, a process that might take hundreds of years, the surface is effectively protected from further deflation if left undisturbed. However, as the layer is only one or two stones thick, the passage of vehicles or animals can dislodge the pavement and expose the fine-grained material below. If this happens, the surface is no longer protected from deflation.

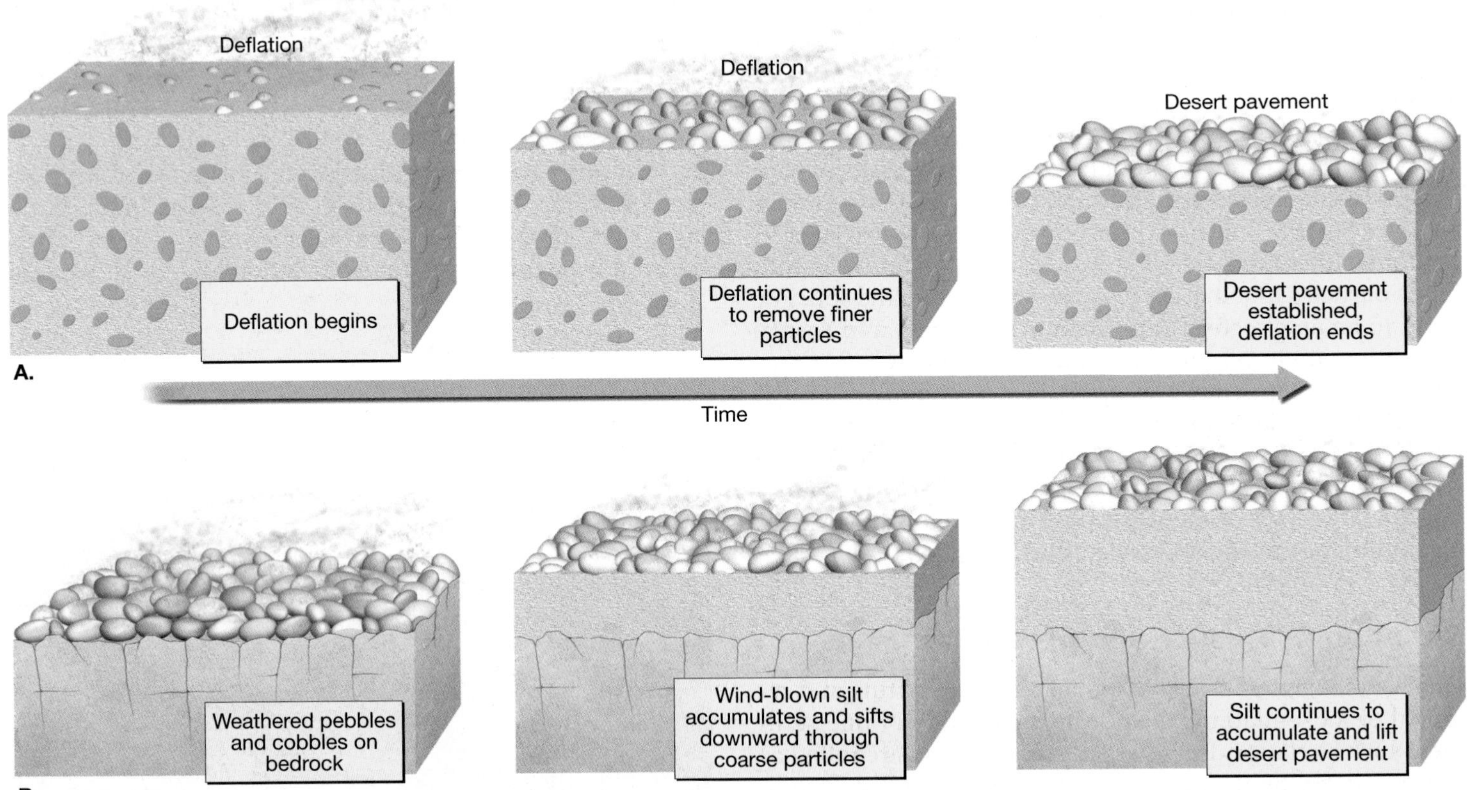

Figure 4.30 Two models of desert pavement formation. **A.** This model portrays an area with poorly sorted surface deposits. Coarse particles gradually become concentrated into a tightly packed layer as deflation lowers the surface by removing sand and silt. Here desert pavement is the result of wind erosion. **B.** This model shows the formation of desert pavement on a surface initially covered with coarse pebbles and cobbles. Wind-blown dust accumulates at the surface and gradually sifts downward through spaces between coarse particles. Infiltrating rainwater aids the process. This depositional process raises the surface and produces a layer of coarse pebbles and cobbles underlain by a substantial layer of fine sediment.

Did You Know?

Deserts do not necessarily consist of mile after mile of drifting sand dunes. Surprisingly, sand accumulations represent only a small percentage of the total desert area. In the Sahara, dunes cover only one-tenth of its area. The sandiest of all deserts, the Arabian, is one-third sand covered.

Wind Abrasion

Like glaciers and streams, wind erodes in part by *abrasion*. In dry regions as well as along some beaches, windblown sand will cut and polish exposed rock surfaces. Abrasion is often given credit for accomplishments beyond its actual capabilities. Such features as balanced rocks that stand high atop narrow pedestals and intricate detailing on tall pinnacles are not the results of abrasion by windblown sand. Sand is seldom lifted more than a meter above the surface, so the wind's sandblasting effect is obviously limited in vertical extent. But in areas prone to such activity, telephone poles have actually been cut through near their bases. For this reason, collars are often fitted on the poles to protect them from being "sawed" down.

Wind Deposits

Sculpturing Earth's Surface
▼ Deserts

Although wind is relatively unimportant in producing erosional landforms, significant depositional landforms are created by the wind in some regions. Accumulations of windblown sediment are particularly conspicuous in the world's dry lands and along many sandy coasts. Wind deposits are of two distinctive types: (1) extensive blankets of silt, called *loess*, that once were carried in suspension, and (2) mounds and ridges of sand from the wind's bed load, which we call *dunes*.

Loess

In some parts of the world the surface topography is mantled with deposits of windblown silt, called **loess.** Dust storms deposited this material over thousands of years. When loess is breached by streams or road cuts, it tends to maintain vertical cliffs and lacks any visible layers, as you can see in Figure 4.31A.

The distribution of loess worldwide indicates two primary sources for this sediment: deserts and glacial deposits of stratified drift. The thickest and most extensive loess deposits occur in western and northern China. They were blown there from the extensive desert basins of central Asia. Accumulations of 30 meters (100 feet) are not uncommon, and

Figure 4.31 **A.** This vertical loess bluff near the Mississippi River in southern Illinois is about 3 meters (10 feet) high. (Photo by James E. Patterson) **B.** This satellite image from November 1, 2006, shows streamers of windblown dust moving southward into the Gulf of Alaska. It illustrates a process similar to the one that created many loess deposits in the American Midwest during the Ice Age. Fine silt is produced by the grinding action of glaciers and then transported beyond the margin of the ice by running water and deposited. Later, the fine silt is picked up by strong winds and deposited as loess. (NASA)

A.

B.

thicknesses of more than 100 meters (300 feet) have been measured. It is this fine, buff-colored sediment that gives the Yellow River (Huang Ho) its name.

In the United States, deposits of loess are significant in many areas, including South Dakota, Nebraska, Iowa, Missouri, and Illinois, as well as portions of the Columbia Plateau in the Pacific Northwest. Unlike the deposits in China, which originated in deserts, the loess in the United States and Europe is an indirect product of glaciation. Its source is deposits of stratified drift. During the retreat of the ice sheets, many river valleys were choked with glacial sediment. Strong winds sweeping across the barren floodplains picked up the finer sediment and dropped it as a blanket on areas adjacent to the valleys (Figure 4.31B).

Sand Dunes

Like running water, wind releases its load of sediment when its velocity falls and the energy available for transport diminishes. Thus, sand begins to accumulate wherever an obstruction across the path of the wind slows its movement. Unlike deposits of loess, which form blanketlike layers over broad areas, winds commonly deposit sand in mounds or ridges called **dunes** (Figure 4.32).

As moving air encounters an object, such as a clump of vegetation or a rock, the wind sweeps around and over it, leaving a shadow of more slowly moving air behind the obstacle as well as a smaller zone of quieter air just in front of the obstacle. Some of the sand grains moving with the wind come to rest in these *wind shadows.* As the accumulation of sand continues, it forms an increasingly efficient wind barrier to trap even more sand. If there is a sufficient supply of sand and the wind blows steadily long enough, the mound of sand grows into a dune.

Many dunes have an asymmetrical profile, with the leeward (sheltered) slope being steep and the windward slope more gently inclined. Sand is rolled up the gentle slope on the windward side by the force of the wind. Just beyond the crest of the dune, the wind velocity is reduced and the sand accumulates. As more sand collects, the slope steepens and eventually some of it slides under the pull of gravity. In this

Did You Know?

The highest dunes in the world are located along the southwest coast of Africa in the Namib Desert. In places, these huge dunes reach heights of 300 to 350 meters (1000 to 1167 feet). The dunes at Great Sand Dunes National Park in southern Colorado are the highest in North America, rising more than 210 meters (700 feet) above the surrounding terrain.

A.

B.

Figure 4.32 **A.** Dunes composed of gypsum sand at White Sands National Monument in southeastern New Mexico. These dunes are gradually migrating from the background (top) toward the foreground (bottom) of the photo. Strong winds move sand up the more gentle windward slopes. As sand accumulates near the dune crest, the slope steepens and some of the sand slides down the steeper *slip face.* **B.** Sand sliding down the steep slip face of a dune in White Sands National Monument, New Mexico. (Photos by Michael Collier)

A.

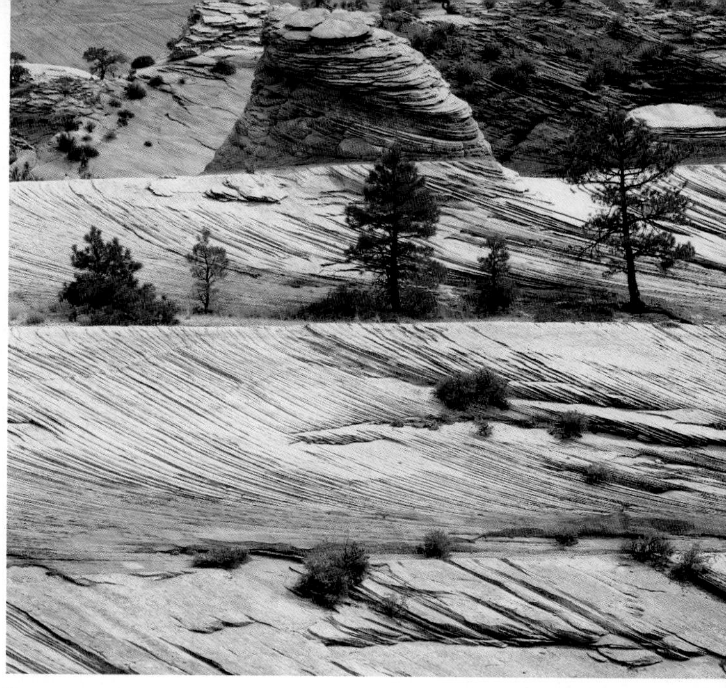

B.

Figure 4.33 **A.** The cutaway section of this sand dune shows cross bedding. (Photo by John S. Shelton) **B.** Cross beds are an obvious characteristic of the Navajo Sandstone, here exposed in Zion National Park, Utah. When dunes are buried and become part of the sedimentary record, the cross-bedded structure is preserved. (Photo by David Muench Photography, Inc.)

way, the leeward slope of the dune, called the **slip face,** maintains a relatively steep angle. Continued sand accumulation, coupled with periodic slides down the slip face, results in the slow migration of the dune in the direction of air movement (Figure 4.32B).

As sand is deposited on the slip face, it forms layers inclined in the direction the wind is blowing. These sloping layers are called **cross beds.** When the dunes are eventually buried under layers of sediment and become part of the sedimentary rock record, their asymmetrical shape is destroyed, but the cross beds remain as a testimony to their origin. Nowhere is cross bedding more prominent than in the sandstone walls of Zion Canyon in southern Utah (Figure 4.33).

The Chapter in Review

1. A *glacier* is a thick mass of ice originating on the land from the compaction and recrystallization of snow. It shows evidence of past or present movement. Today, *valley,* or *alpine glaciers,* are found in mountain areas where they usually follow valleys originally occupied by streams. *Ice sheets* exist on a much larger scale, covering most of Greenland and Antarctica.

2. Glaciers move in part by flowing. On the surface of a glacier, ice is brittle. However, below about 50 meters (165 feet), pressure is great and ice behaves like a *plastic material* and *flows.* A second important mechanism of glacial movement consists of the whole ice mass *slipping* along the ground.

3. Glaciers form in areas where more snow falls in winter than melts during summer. Snow accumulation and ice formation occur in the *zone of accumulation.* Beyond this area is the *zone of wastage,* where there is a net loss to the glacier. The *glacial budget* is the balance, or lack of balance, between accumulation at the upper end of the glacier and loss at the lower end.

4. Glaciers erode land by *plucking* (lifting pieces of bedrock out of place) and abrasion (grinding and scraping of a rock surface). Erosional features produced by valley glaciers include *glacial troughs, hanging valleys, cirques, arêtes, horns,* and *fiords.*

5. Any sediment of glacial origin is called *drift.* The two distinct types of glacial drift are (a) *till,* which is material deposited directly by the ice, and (b) *stratified drift,* which is sediment laid down by meltwater from a glacier.

6. The most widespread features created by glacial deposition are layers or ridges of till, called *moraines.* Associated with valley glaciers are *lateral moraines,* formed along the sides of the valley, and *medial moraines,* formed between two valley glaciers that have joined. *End moraines,* which mark the former position of the front of a glacier, and *ground moraines,* undulating layers of till deposited as the ice front retreats, are common to both valley glaciers and ice sheets.

7. Perhaps the most convincing evidence for several glacial advances during the *Ice Age* is the widespread existence of *multiple layers of drift* on land and an uninterrupted record of climate cycles preserved in *seafloor sediments.* In addition to massive erosional and depositional work, other effects of Ice Age glaciers included the *forced migration* of animals, *changes in river courses, adjustment of the crust* by rebounding after the removal of the immense load of ice, and *climate changes* caused by the existence of the glaciers themselves. In the sea, the most far-reaching effect of the Ice Age was the *worldwide change in sea level* that accompanied each advance and retreat of the ice sheets.

8. Deserts in the lower latitudes coincide with zones of high air pressure known as *subtropical highs.* Middle-latitude deserts exist because of their positions in the deep interiors of large continents, far removed from oceans. Mountains also act to shield these regions from marine air masses.

9. Practically all desert streams are dry most of the time and are said to be *ephemeral.* Nevertheless, *running water is responsible for most of the erosional work in a desert.* Although wind erosion is more significant in dry areas than elsewhere, the main role of wind in a desert is the transportation and deposition of sediment.

10. Many of the landscapes of the Basin and Range region of the western United States are characterized by *interior drainage* with streams eroding uplifted mountain blocks and depositing the sediment in interior basins. *Alluvial fans, playas, playa lakes,* and *inselbergs* are features often associated with these landscapes.

11. For wind erosion to be effective, dryness and scant vegetation are essential. *Deflation,* the lifting and removal of loose material, often produces shallow depressions called *blowouts* and can also lower the surface by removing sand and silt. *Abrasion,* the "sandblasting" effect of wind, is often given too much credit for producing desert features. However, abrasion does cut and polish rock near the surface.

12. Wind deposits are of two distinct types: (1) extensive *blankets of silt,* called *loess,* that were carried by wind in *suspension,* and (2) *mounds and ridges of sand,* called *dunes,* formed from sediment that was carried as part of the wind's *bed load.*

Key Terms

abrasion (p. 107)
alluvial fan (p. 119)
alpine glacier (p. 102)
arête (p. 108)
blowout (p. 121)
cirque (p. 108)
crevasses (p. 104)
cross beds (p. 125)
deflation (p. 121)
desert (p. 117)
desert pavement (p. 121)
drumlin (p. 112)
dune (p. 124)
end moraine (p. 111)
ephemeral stream (p. 117)
esker (p. 112)
fiord (p. 109)
glacial drift (p. 110)
glacial erratic (p. 110)
glacial striations (p. 107)
glacial trough (p. 107)
glacier (p. 102)
ground moraine (p. 111)
hanging valley (p. 107)
horn (p. 108)
ice cap (p. 103)
ice sheet (p. 102)
ice shelf (p. 103)
interior drainage (p. 118)
kame (p. 112)
kettle (p. 112)
lateral moraine (p. 111)
loess (p. 123)
medial moraine (p. 111)
outwash plain (p. 112)
piedmont glacier (p. 103)
playa lake (p. 119)
Pleistocene epoch (p. 114)
plucking (p. 107)
pluvial lake (p. 115)
rock flour (p. 107)
slip face (p. 125)
steppe (p. 117)
stratified drift (p. 110)
till (p. 110)
valley glacier (p. 102)
valley train (p. 112)
zone of accumulation (p. 104)
zone of wastage (p. 106)

Questions for Review

1. What is a glacier? What percentage of Earth's land area do glaciers cover?
2. Contrast valley glaciers and ice sheets.
3. Describe how glaciers fit into the hydrologic cycle. What role do they play in the rock cycle?
4. Describe the two components of glacial flow. At what rates do glaciers move? In a valley glacier, does all of the ice move at the same rate? Explain.
5. Why do crevasses form in the upper portion of a glacier but not below a depth of about 50 meters (165 feet)?
6. Under what circumstances will the terminus of a glacier advance? Retreat? Remain stationary?
7. Describe two basic processes by which glaciers erode.
8. How does the shape of a glaciated mountain valley differ from that of an unglaciated mountain valley?
9. List and describe the erosional features you might expect to see in an area where valley glaciers exist or have recently existed.
10. What is glacial drift? What is the difference between till and stratified drift? What general effect do glacial deposits have on the landscape?
11. List and briefly describe the four basic moraine types. What do all moraines have in common?
12. List and briefly describe four depositional features other than moraines.
13. How does a kettle form?
14. About what percentage of Earth's land surface was covered at some time by Pleistocene glaciers? How does this compare to the area presently covered by ice sheets and glaciers? (Check your answer with that for Question 1.)

15. List three indirect effects of Ice Age glaciers.
16. How extensive are Earth's dry regions?
17. Most desert streams are said to be ephemeral. What does that mean?
18. What is the most important erosional agent in deserts?
19. Describe the features and characteristics associated with each of the stages in the evolution of a mountainous desert region such as the Basin and Range region in the American West.
20. Why is wind erosion relatively more important in dry regions than in humid areas?
21. List two types of wind erosion and the features that may result from each.
22. Although sand dunes are the best-known wind deposits, accumulations of loess are very significant in some parts of the world. What is loess? Where are such deposits found? What are the origins of this sediment?
23. How do sand dunes migrate?

Online Study Guide

The *Foundations of Earth Science* Web site uses the resources and flexibility of the Internet to aid in your study of the topics in this chapter. Written and developed by Earth science instructors, this site will help improve your understanding of Earth science. Visit **http://www.prenhall.com/lutgens** and click on the cover of *Foundations of Earth Science 5e* to find:

- Online review quizzes.
- Critical thinking exercises.
- Links to chapter-specific Web resources.
- Internet-wide key-term searches.

http://www.prenhall.com/lutgens

GEODe: Earth Science

GEODe: Earth Science makes studying more effective by reinforcing key concepts using animation, video, narration, interactive exercises, and practice quizzes. A copy is included with every copy of *Foundations of Earth Science 5e.*

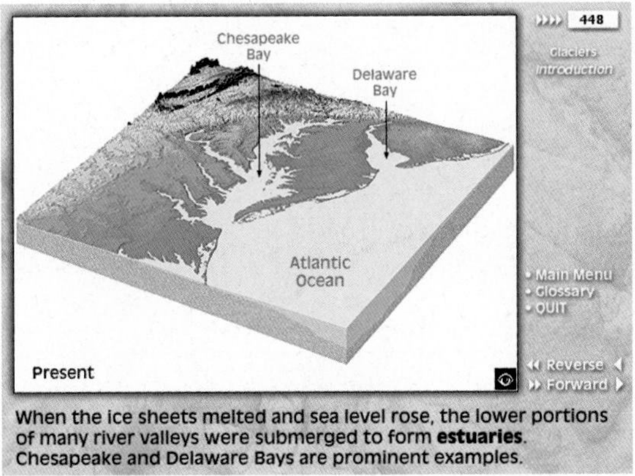

When the ice sheets melted and sea level rose, the lower portions of many river valleys were submerged to form **estuaries**. Chesapeake and Delaware Bays are prominent examples.

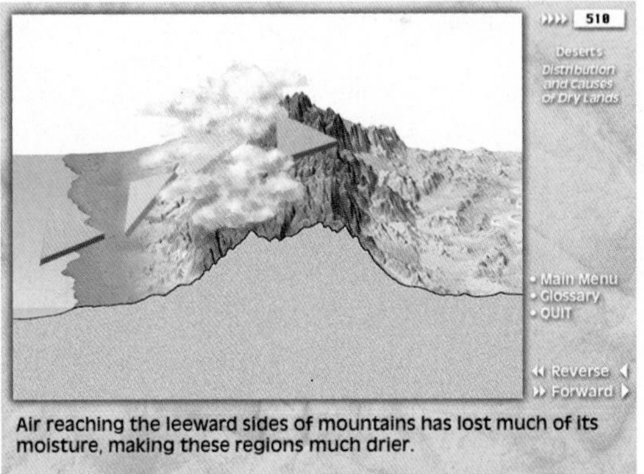

Air reaching the leeward sides of mountains has lost much of its moisture, making these regions much drier.

UNIT III FORCES WITHIN

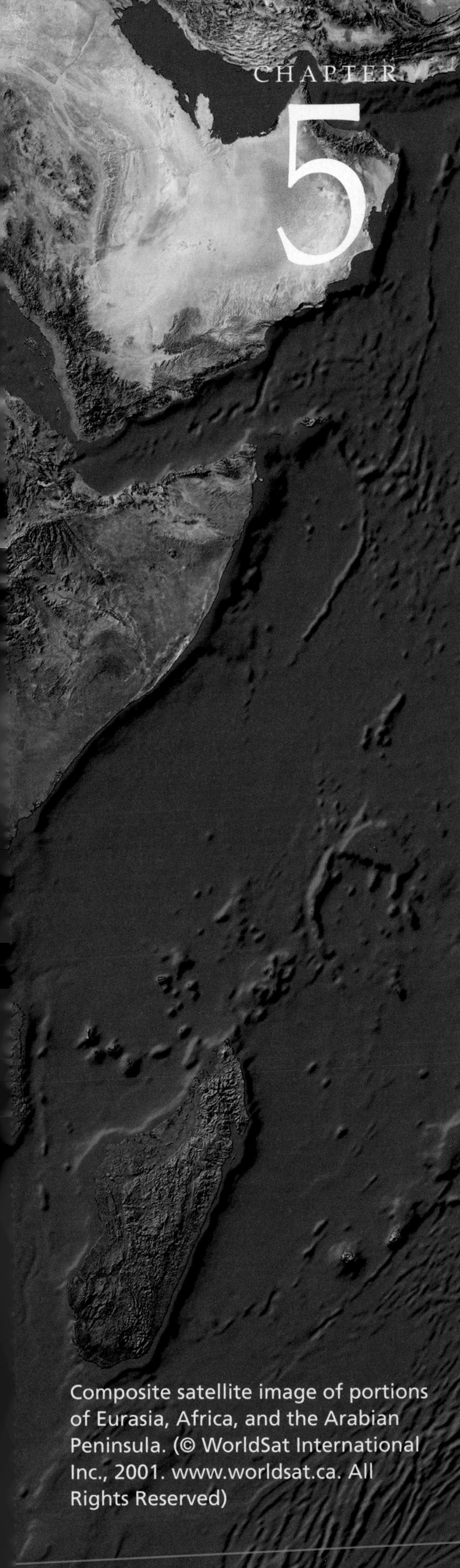

Composite satellite image of portions of Eurasia, Africa, and the Arabian Peninsula. (© WorldSat International Inc., 2001. www.worldsat.ca. All Rights Reserved)

Plate Tectonics: A Scientific Theory Unfolds

FOCUS ON LEARNING

To assist you in learning the important concepts in this chapter, you will find it helpful to focus on the following questions:

1. What lines of evidence were used to support the continental drift hypothesis?
2. What was one of the main objections to the continental drift hypothesis?
3. What is the theory of plate tectonics?
4. In what major way does the plate tectonics theory depart from the continental drift hypothesis?
5. What are the three types of plate boundaries?
6. What is the evidence used to support the plate tectonics theory?
7. What models have been proposed to explain the driving mechanism for plate motion?

A century ago, most geologists believed that the geographic positions of the ocean basins and continents were fixed. During the past few decades, however, vast amounts of new data have dramatically changed our understanding of the nature and workings of our planet. Earth scientists now realize that the continents gradually migrate across the globe. Where landmasses split apart, new ocean basins are created between the diverging blocks. Meanwhile, older portions of the seafloor are carried back into the mantle in regions where trenches occur in the deep ocean floor. Because of these movements, blocks of continental crust eventually collide and form Earth's great mountain ranges (Figure 5.1). In short, a revolutionary new model of Earth's tectonic* processes has emerged.

This profound reversal of scientific understanding has been appropriately described as a *scientific revolution.* The revolution began as a relatively straightforward proposal by Alfred Wegener, called *continental drift.* After many years of heated debate, Wegener's hypothesis of drifting continents was rejected by the vast majority of the scientific community. The concept of a mobile Earth was particularly distasteful to North American geologists, perhaps because much of the supporting evidence had been gathered from the southern continents, with which most North American geologists were unfamiliar.

During the 1950s and 1960s, new kinds of evidence began to rekindle interest in this nearly abandoned proposal. By 1968, these new developments led to the unfolding of a far more encompassing explanation, which incorporated aspects of continental drift and seafloor spreading—a theory known as *plate tectonics.*

In this chapter, we will examine the events that led to this dramatic reversal of scientific opinion in an attempt to provide some insight into how science works. We will also briefly trace the development of the concept of continental drift, examine why it was first rejected, and consider the evidence that finally led to the acceptance of the theory of plate tectonics.

**Tectonics* refers to the deformation of Earth's crust and results in the formation of structural features such as mountains.

Continental Drift: An Idea Before Its Time

The idea that continents, particularly South America and Africa, fit together like pieces of a jigsaw puzzle originated with improved world maps. However, little significance was given to this idea until 1915, when Alfred Wegener, a German meteorologist and geophysicist, published *The Origin of Continents and Oceans.* In this book, Wegener set forth his radical hypothesis of **continental drift.***

Wegener suggested that a supercontinent he called **Pangaea** (meaning "all land") once existed (Figure 5.2). He

*Wegener's ideas were actually preceded by those of an American geologist, F. B. Taylor, who in 1910 published a paper on continental drift. Taylor's paper provided little supporting evidence for continental drift, which may have been the reason that it had a relatively small impact on the scientific community.

Figure 5.1 Climbers (left) camping on a shear rock face of a mountain known as K7 in Pakistan's Karakoram, a part of the Himalayas. These mountains formed as India collided with Eurasia. (Photo by Jimmy Chin/National Geographic/Getty)

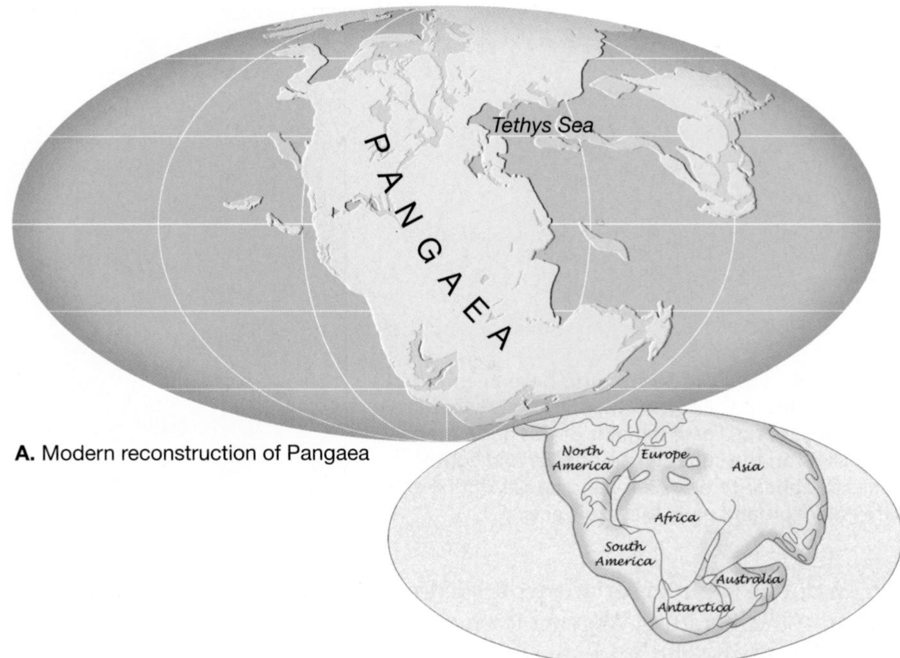

Figure 5.2 Reconstruction of Pangaea as it is thought to have appeared 200 million years ago. **A.** Modern reconstruction. **B.** Reconstruction done by Wegener in 1915.

further hypothesized that about 200 million years ago, this supercontinent began breaking into smaller continents, which then "drifted" to their present positions.

Wegener and others collected substantial evidence to support these claims. The fit of South America and Africa and the geographic distribution of fossils, rock structures, and ancient climates all seemed to support the idea that these now separate landmasses were once joined. Let us examine their evidence.

Evidence: The Continental Jigsaw Puzzle

Like a few others before him, Wegener first suspected that the continents might have been joined when he noticed the remarkable similarity between the coastlines on opposite sides of the South Atlantic. However, his use of present-day shorelines to make a fit of the continents was challenged immediately by other Earth scientists. These opponents correctly argued that shorelines are continually modified by erosional processes, and even if continental displacement had taken place, a good fit today would be unlikely. Wegener appeared to be aware of this problem, and, in fact, his original jigsaw fit of the continents was crude.

Scientists have determined that a much better approximation of the true outer boundary of the continents is the seaward edge of the continental shelf, which lies submerged several hundred meters below sea level. In the early 1960s, scientists produced a map that attempted to fit the edges of the continental shelves at a depth of 900 meters (3000 feet). The remarkable fit that was obtained is shown in Figure 5.3. The few places where the continents overlap are regions where streams have deposited large quantities of sediment, thus enlarging the continental shelves. The overall fit was even better than researchers suspected it would be.

Evidence: Fossils Match Across the Seas

Although the seed for Wegener's hypothesis came from the remarkable similarities of the continental margins on opposite sides of the Atlantic, he thought the idea of a mobile Earth was improbable. Not until he learned that identical fossil organisms were known from rocks in both South America and

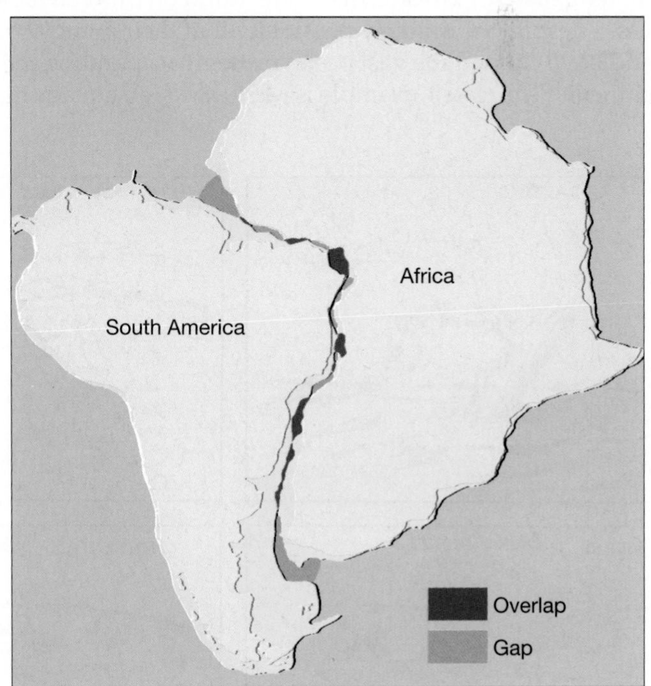

Figure 5.3 This figure shows the best fit of South America and Africa along the continental slope at a depth of 500 fathoms (about 900 meters [3000 feet]). The areas where continental blocks overlap appear in brown. (After A. G. Smith, "Continental Drift," in *Understanding the Earth*, edited by I. G. Gass.)

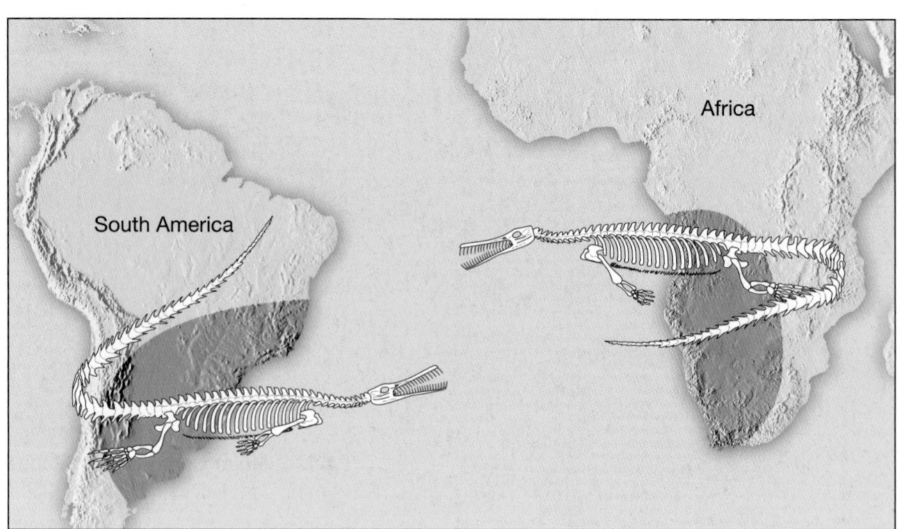

Figure 5.4 Fossils of *Mesosaurus* have been found on both sides of the South Atlantic and nowhere else in the world. Fossil remains of this and other organisms on the continents of Africa and South America appear to link these landmasses during the late Paleozoic and early Mesozoic eras.

Africa did he begin to seriously pursue this idea. Through a review of the literature, Wegener learned that most paleontologists (scientists who study the fossilized remains of organisms) were in agreement that some type of land connection was needed to explain the existence of identical fossils of Mesozoic life forms on widely separated landmasses. (Just as modern life forms native to North America are quite different from those of Africa, one would expect that during the Mesozoic era, organisms on widely separated continents would be quite distinct.)

Mesosaurus To add credibility to his argument for the existence of a supercontinent, Wegener cited documented cases of several fossil organisms that were found on different landmasses despite the unlikely possibility that their living forms could have crossed the vast ocean presently separating these continents. The classic example is *Mesosaurus,* an aquatic fish-catching reptile whose fossil remains are found only in black shales of the Permian period (about 260 million years ago) in eastern South America and southern Africa (Figure 5.4). If *Mesosaurus* had been able to make the long journey across the vast South Atlantic Ocean, its remains should be more widely distributed. As this is not the case, Wegener argued that South America and Africa must have been joined during that period of Earth's history.

How did scientists during Wegener's era explain the existence of identical fossil organisms in places separated by thousands of kilometers of open ocean? Transoceanic land bridges (isthmian links) were the most widely accepted explanations of such migrations (Figure 5.5). We know, for example, that during the recent Ice Age, the lowering of sea level allowed animals to cross the narrow Bering Strait between Asia and North America. Was it possible that land bridges once connected Africa and South America but later subsided

Figure 5.5 These sketches by John Holden illustrate various explanations for the occurrence of similar species on landmasses that are presently separated by vast oceans. (Reprinted with permission of John Holden)

below sea level? Modern maps of the seafloor substantiate Wegener's contention that land bridges of this magnitude had not existed. If they had, remnants would still lie below sea level.

Present-Day Organisms In a later edition of his book, Wegener also cited the distribution of present-day organisms as evidence to support the drifting of continents. For example, modern organisms with similar ancestries clearly had to evolve in isolation during the last few tens of millions of years. Most obvious of these are the Australian marsupials (such as kangaroos), which have a direct fossil link to the marsupial opossums found in the Americas. After the breakup of Pangaea, the Australian marsupials followed a different evolutionary path than related life forms in the Americas.

Evidence: Rock Types and Structures Match

Anyone who has worked a picture puzzle knows that in addition to the pieces fitting together, the picture must be continuous as well. The "picture" that must match in the "continental drift puzzle" is one of rock types and mountain belts found on the continents. If the continents were once together, the rocks found in a particular region on one continent should closely match in age and type those found in adjacent positions on the adjoining continent. Wegener found evidence of 2.2-billion-year-old igneous rocks in Brazil that closely resembled similarly aged rocks in Africa.

Similar evidence exists in the form of mountain belts that terminate at one coastline, only to reappear on landmasses across the ocean. For instance, the mountain belt that includes the Appalachians trends northeastward through the eastern United States and disappears off the coast of Newfoundland. Mountains of comparable age and structure are found in Greenland, the British Isles, and Scandinavia. When these landmasses are reassembled, as in Figure 5.6, the mountain chains form a nearly continuous belt.

Wegener must have been convinced that the similarities in rock structure on both sides of the Atlantic linked these landmasses when he said, "It is just as if we were to refit the torn pieces of a newspaper by matching their edges and then check whether the lines of print run smoothly across. If they do, there is nothing left but to conclude that the pieces were in fact joined in this way."*

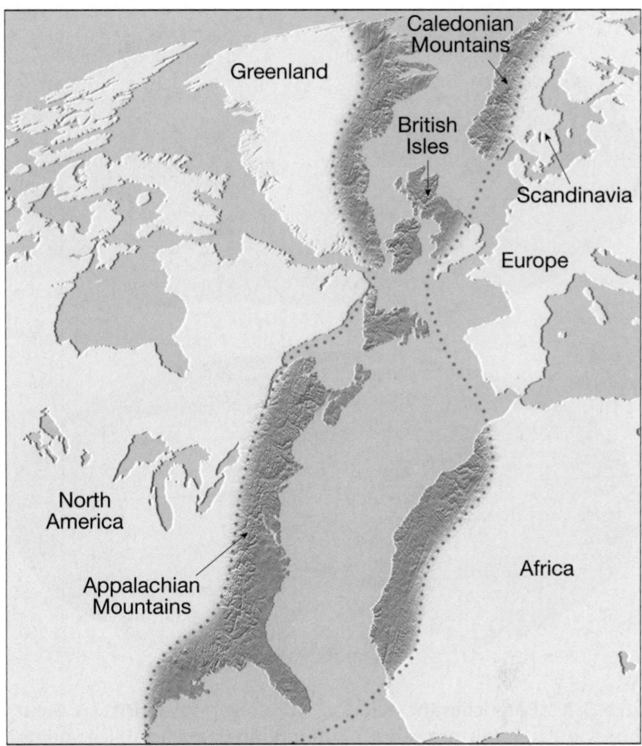

Figure 5.6 Matching mountain ranges across the North Atlantic. The Appalachian Mountains trend along the eastern flank of North America and disappear off the coast of Newfoundland. Mountains of comparable age and structure are found in Greenland, the British Isles, and Scandinavia. When these landmasses are placed in their predrift locations, these ancient mountain chains form a nearly continuous belt. These folded mountain belts formed roughly 300 million years ago as the landmasses collided during the formation of the supercontinent of Pangaea.

Evidence: Ancient Climates

Because Alfred Wegener was a meteorologist by profession, he was keenly interested in obtaining paleoclimatic (*paleo* = ancient, *climatic* = climate) data to support continental drift. His efforts were rewarded when he found evidence for apparently dramatic global climatic changes during the geologic past. In particular, he learned of ancient glacial deposits that indicated that near the end of the Paleozoic era (about 300 million years ago), ice sheets covered extensive areas of the Southern Hemisphere and India (Figure 5.7A). Layers of glacially transported sediments of the same age were found in southern Africa and South America, as well as in India and Australia. Much of the land area containing evidence of this late Paleozoic glaciation presently lies within 30 degrees of the equator in subtropical or tropical climates.

Could Earth have gone through a period of sufficient cooling to have generated extensive ice sheets in areas that are presently tropical? Wegener rejected this explanation because during the late Paleozoic, large tropical swamps existed in the Northern Hemisphere. These swamps, with their lush vegetation, eventually became the major coal fields of the eastern United States, Europe, and Siberia.

*Alfred Wegener, *The Origin of Continents and Oceans.* Translated from the 4th revised German edition of 1929 by J. Birman (London: Methuen, 1966).

Did You Know?

Alfred Wegener, best known for his continental drift hypothesis, also wrote numerous scientific papers on weather and climate. Following his interest in meteorology, Wegener made four extended trips to the Greenland Ice Sheet in order to study its harsh winter weather. In November 1930, while making a month-long trek across the ice sheet, Wegener and a companion perished.

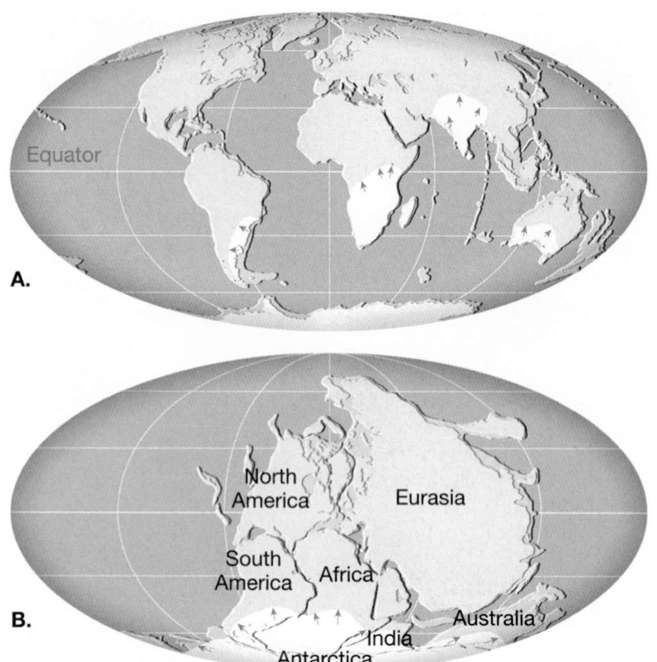

Figure 5.7 Paleoclimatic evidence for continental drift. **A.** Near the end of the Paleozoic era (about 300 million years ago), ice sheets covered extensive areas of the Southern Hemisphere and India. Arrows show the direction of ice movement that can be inferred from glacial grooves in the bedrock. **B.** Shown are the continents restored to their former position with the South Pole located roughly between Antarctica and Africa. This configuration accounts for the conditions necessary to generate a vast glacial ice sheet and also explains the direction of ice movements that radiated away from the South Pole.

Fossils from these coal fields indicate that the tree ferns that produced the coal deposits had large fronds, which are indicative of tropical settings. Furthermore, unlike trees in colder climates, these trees lacked growth rings, a characteristic of tropical plants that grow in regions having minimal fluctuations in temperature.

Wegener suggested that a more plausible explanation for the late Paleozoic glaciation was provided by the supercontinent of Pangaea. In this configuration, the southern continents are joined together and located near the South Pole (Figure 5.7B). This would account for the conditions necessary to generate extensive expanses of glacial ice over much of the Southern Hemisphere. At the same time, this geography would place today's northern landmasses nearer the equator and account for their vast coal deposits. Wegener was so convinced that his explanation was correct that he wrote, "This evidence is so compelling that by comparison all other criteria must take a back seat."

How does a glacier develop in hot, arid central Australia? How do land animals migrate across wide expanses of open water? As compelling as this evidence may have been, 50 years passed before most of the scientific community accepted the concept of continental drift and the logical conclusions to which it led.

The Great Debate

Wegener's proposal did not attract much open criticism until 1924, when his book was translated into English. From this time on, until his death in 1930, his drift hypothesis encountered a great deal of hostile criticism. To quote the respected American geologist T. C. Chamberlin, "Wegener's hypothesis . . . takes considerable liberty with our globe, and is less bound by restrictions or tied down by awkward, ugly facts than most of its rival theories. Its appeal seems to lie in the fact that it plays a game in which there are few restrictive rules and no sharply drawn code of conduct."

One of the main objections to Wegener's hypothesis stemmed from his inability to provide a mechanism that was capable of moving the continents across the globe. Wegener proposed that the tidal influence of the Moon was strong enough to give the continents a westward motion. However, the prominent physicist Harold Jeffreys quickly countered with the argument that tidal friction of the magnitude needed to displace the continents would bring Earth's rotation to a halt in a matter of a few years.

Wegener also proposed that the larger and sturdier continents broke through the oceanic crust, much like ice breakers cut through ice. However, no evidence existed to suggest that the ocean floor was weak enough to permit passage of the continents without their being appreciably deformed in the process.

Although most of Wegener's contemporaries opposed his views, even to the point of open ridicule, a few considered his ideas plausible. For those geologists who continued the search for additional evidence, the exciting concept of continents in motion held their interest. Others viewed continental drift as a solution to previously unexplainable observations.

Plate Tectonics: The New Paradigm

Forces Within
▼ Plate Tectonics

Following World War II, oceanographers equipped with new marine tools and ample funding from the U.S. Office of Naval Research embarked on an unprecedented period of oceanographic exploration. Over the next two decades, a much better picture of large expanses of the seafloor slowly and painstak-

Did You Know?

A group of scientists proposed an interesting although incorrect explanation for the cause of continental drift. This proposal suggested that early in Earth's history, our planet was only about half its current diameter and completely covered by continental crust. Through time, Earth expanded, causing the continents to split into their current configurations, while new seafloor "filled in" the spaces as they drifted apart.

ingly began to emerge. From this work came the discovery of a global **oceanic ridge system** that winds through all of the major oceans in a manner similar to the seams on a baseball.

In other parts of the ocean, new discoveries were also being made. Earthquake studies conducted in the western Pacific demonstrated that tectonic activity was occuring at great depths beneath deep-ocean trenches. Of equal importance was the fact that dredging of the seafloor did not bring up any oceanic crust that was older than 180 million years. Further, sediment accumulations in the deep-ocean basins were found to be thin, not the thousands of meters that were predicted.

By 1968, these developments, among others, led to the unfolding of a far more encompassing theory than continental drift, known as **plate tectonics.** The implications of plate tectonics are so far-reaching that this theory is today the framework within which to view most geologic processes.

Earth's Major Plates

According to the plate tectonics model, the uppermost mantle, along with the overlying crust, behaves as a strong, rigid layer, known as the **lithosphere,** which is broken into pieces called **plates** (Figure 5.8). Lithospheric plates are thinnest in the oceans where their thickness may vary from as little as a few kilometers at the oceanic ridges to 100 kilometers (62 miles) in the deep-ocean basins. By contrast, continental lithosphere is generally 100 to 150 kilometers (62 to 93 miles) thick but may be more than 250 kilometers (155 miles) thick below older portions of landmasses. The lithosphere overlies a weaker region in the mantle known as the **asthenosphere** (weak sphere). The temperature/pressure regime in the upper asthenosphere is such that the rocks there are near their melting temperatures. This results in a very weak zone that permits the lithosphere to be effectively detached from the layers below. Thus, the weak rock within the upper asthenosphere allows Earth's rigid outer shell to move.

As shown in Figure 5.9, seven major lithospheric plates are recognized. They are the North American, South American, Pacific, African, Eurasian, Australian-Indian, and Antarctic plates. The largest is the Pacific plate, which encompasses a significant portion of the Pacific Ocean basin. Notice from Figure 5.9 that most of the large plates include an entire continent plus a large area of ocean floor (for example, the South American plate). This is a major departure from Wegener's continental drift hypothesis, which proposed that the continents moved through the ocean floor, not with it. Note also that no plate is defined entirely by the margins of a continent.

Intermediate-sized plates include the Caribbean, Nazca, Philippine, Arabian, Cocos, Scotia, and Juan de Fuca plates. In addition, there are more than a dozen smaller plates that have been identified but are not shown in Figure 5.9.

One of the main tenets of the plate tectonic theory is that plates move as coherent units relative to all other plates. As plates move, the distance between two locations on the same plate—New York and Denver, for example—remains relatively constant, whereas the distance between sites on different plates, such as New York and London, gradually changes. (It has been recently shown that plates can suffer *some* internal deformation, particularly to the oceanic lithosphere.)

Lithospheric plates move relative to one another at a very slow but continuous rate that averages about 5 centimeters (2 inches) per year. This movement is ultimately driven by the unequal distribution of heat within Earth. Hot material found deep in the mantle moves slowly upward and serves as one part of our planet's internal convection system. Concurrently, cooler, denser slabs of oceanic lithosphere descend into the mantle, setting Earth's rigid outer shell into motion. Ultimately, the titanic, grinding movements of Earth's lithospheric plates

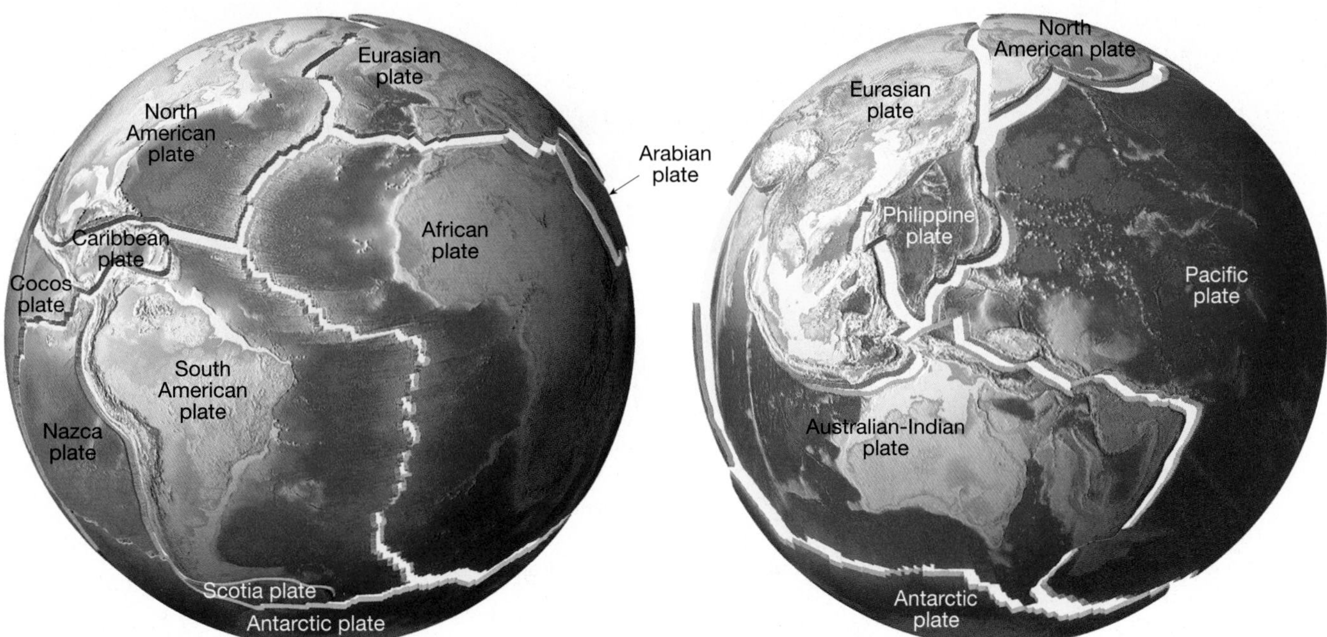

Figure 5.8 Illustration of some of Earth's lithospheric plates.

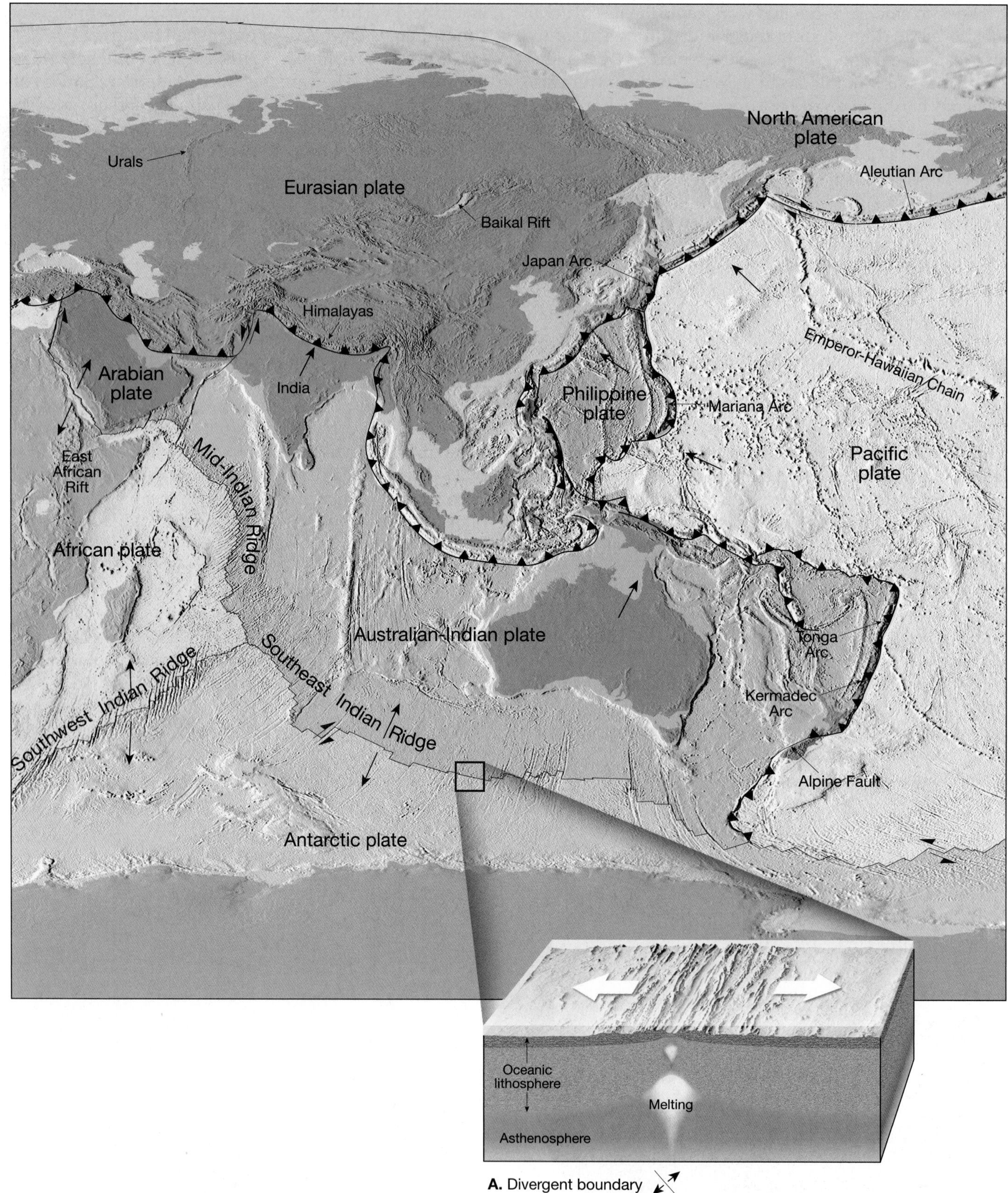

Figure 5.9 A mosaic of rigid plates constitutes Earth's outer shell. (After W. B. Hamilton and others)

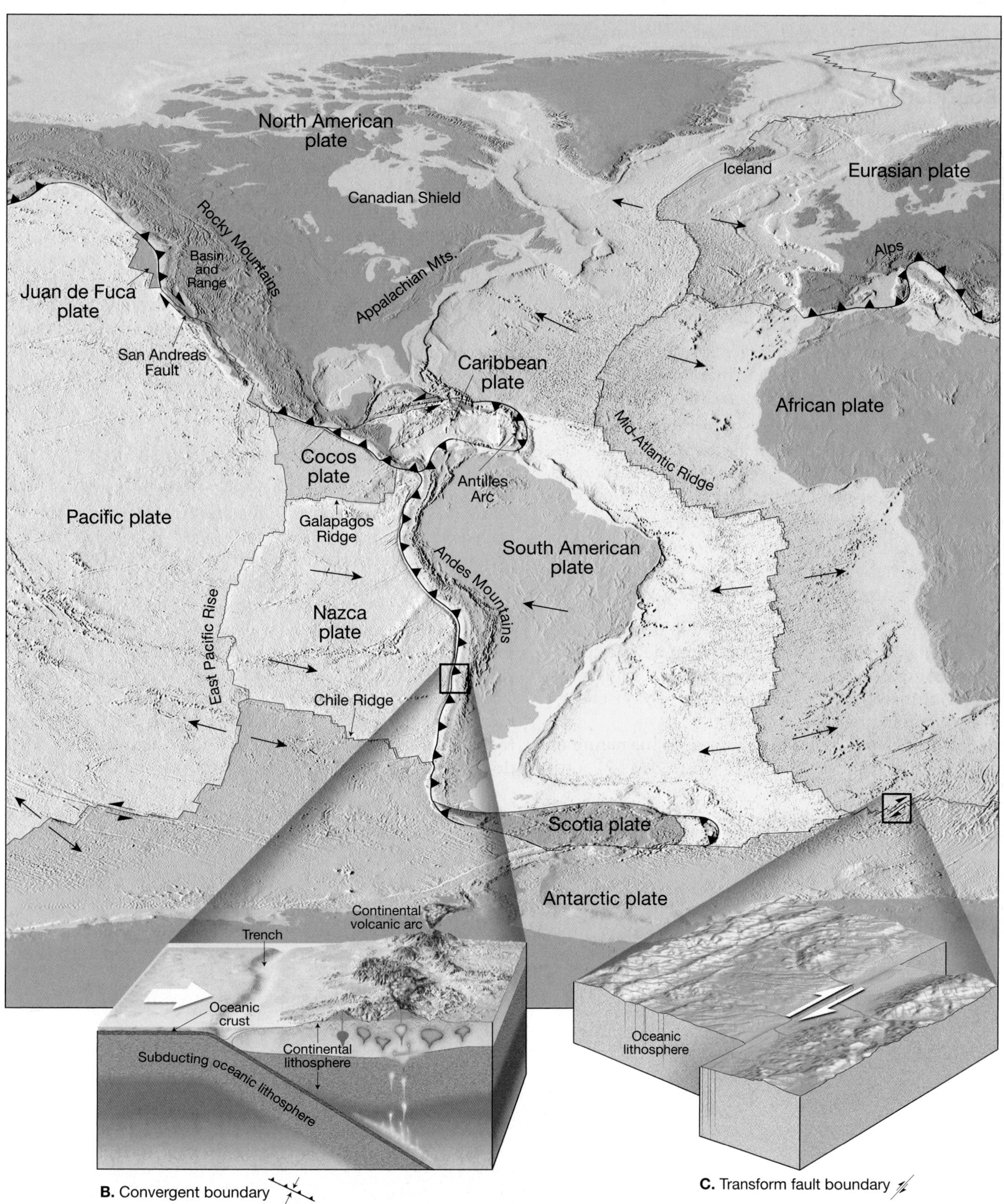
North American plate
Canadian Shield
Iceland
Eurasian plate
Rocky Mountains
Basin and Range
Appalachian Mts.
Alps
Juan de Fuca plate
San Andreas Fault
Caribbean plate
African plate
Cocos plate
Antilles Arc
Mid-Atlantic Ridge
Pacific plate
Galapagos Ridge
South American plate
Andes Mountains
Nazca plate
East Pacific Rise
Chile Ridge
Scotia plate
Antarctic plate
Continental volcanic arc
Trench
Oceanic crust
Continental lithosphere
Subducting oceanic lithosphere
Oceanic lithosphere

B. Convergent boundary

C. Transform fault boundary

generate earthquakes, create volcanoes, and deform large masses of rock into mountains.

Plate Boundaries

Lithospheric plates move as coherent units relative to all other plates. Although the interiors of plates may experience some deformation, all major interactions among individual plates (and, therefore, most deformation) occur along their *boundaries.* In fact, plate boundaries were first established by plotting the locations of earthquakes. Moreover, plates are bounded by three distinct types of boundaries, which are differentiated by the type of movement they exhibit. These boundaries are depicted at the bottom of Figure 5.9 and are briefly described here:

1. **Divergent boundaries** *(constructive margins)*—where two plates move apart, resulting in upwelling of material from the mantle to create new seafloor (Figure 5.9A).
2. **Convergent** *(destructive margins)*—where two plates move together, resulting in oceanic lithosphere descending beneath an overriding plate, eventually to be reabsorbed into the mantle, or possibly in the collision of two continental blocks to create a mountain system (Figure 5.9B).
3. **Transform fault boundaries** *(conservative margins)*—where two plates grind past each other without the production or destruction of lithosphere (Figure 5.9C).

Each plate is bounded by a combination of these three types of plate margins. For example, the Juan de Fuca plate has a divergent zone on the west, a convergent boundary on the east, and numerous transform faults, which offset segments of the oceanic ridge (Figure 5.9).

The following sections briefly summarize the nature of the three types of plate boundaries.

Divergent Boundaries

Forces Within
▼ Plate Tectonics

Most **divergent boundaries** are located along the crests of oceanic ridges and can be thought of as *constructive plate margins* since this is where new oceanic lithosphere is generated (Figure 5.10). As the plates move away from the ridge axis, the fractures that form are filled with molten rock that wells up from the hot mantle below. This magma gradually, cools to produce new slivers of seafloor. In a continuous manner, adjacent plates spread apart and new oceanic lithosphere forms between them. As we will see later, divergent boundaries are not confined to the ocean floor but can also form on the continents.

Oceanic Ridges and Seafloor Spreading

Along well-developed divergent plate boundaries, the seafloor is elevated, forming the *oceanic ridge.* The interconnected oceanic ridge system is the longest topographic feature on Earth's surface, exceeding 70,000 kilometers (43,000 miles) in length. Representing 20 percent of Earth's surface, the oceanic ridge system winds through all major ocean basins like the seam on a baseball (see Figure 5.7, p. 134). Although the crest of the oceanic ridge is commonly 2 to 3 kilometers (1 to 2 miles) higher than the adjacent ocean basins, the term "ridge" may be misleading as this feature is not narrow but has widths from 1000 to 4000 kilometers (600 to 2500 miles). In addition, along the axis of some ridge segments is a deep downfaulted structure called a **rift valley.**

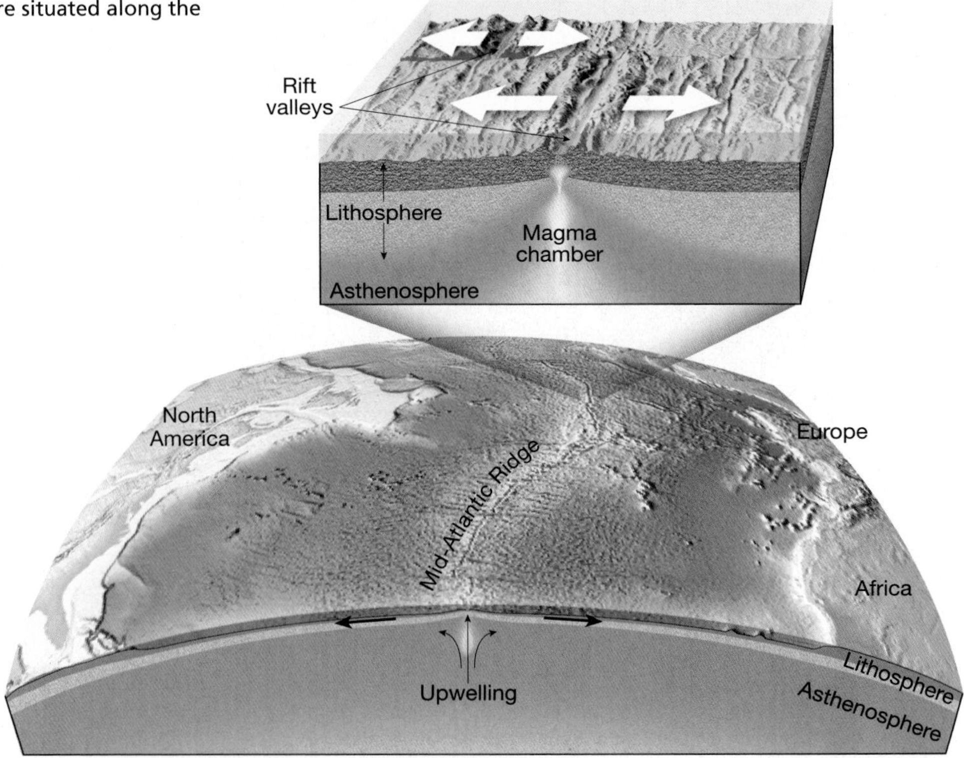

Figure 5.10 Most divergent plate boundaries are situated along the crests of oceanic ridges.

The mechanism that operates along the oceanic ridge system to create new seafloor is appropriately called **seafloor spreading.** Typical rates of spreading average around 5 centimeters (2 inches) per year. This is roughly the same rate at which human fingernails grow. Comparatively slow spreading rates of 2 centimeters per year are found along the Mid-Atlantic Ridge, whereas spreading rates exceeding 15 centimeters (6 inches) have been measured along sections of the East Pacific Rise. Although these rates of lithospheric production are slow on a human time scale, they are nevertheless rapid enough to have generated all of Earth's ocean basins within the last 200 million years. In fact, none of the ocean floor that has been dated exceeds 180 million years in age.

The primary reason for the elevated position of the oceanic ridge is that newly created oceanic crust is hot and occupies more volume, which makes it less dense than cooler rocks. As new lithosphere is formed along the oceanic ridge, it is slowly yet continually displaced away from the zone of upwelling along the ridge axis. Thus, it begins to cool and contract, thereby increasing in density. This thermal contraction accounts for the greater ocean depths that exist away from the ridge crest.

It takes about 80 million years before the cooling and contracting cease completely. By this time, rock that was once part of the elevated oceanic ridge system is located in the deep-ocean basin, where it is buried by substantial accumulations of sediment. In addition, cooling causes the mantle rocks below the oceanic crust to strengthen, thereby adding to the plate's thickness. Stated another way, the thickness of oceanic lithosphere is age-dependent. The older (cooler) it is, the greater its thickness.

Continental Rifting

Divergent plate boundaries can also develop within a continent, in which case the landmass may split into two or more smaller segments, as Alfred Wegener had proposed for the breakup of Pangaea (Figure 5.11). The splitting of a continent

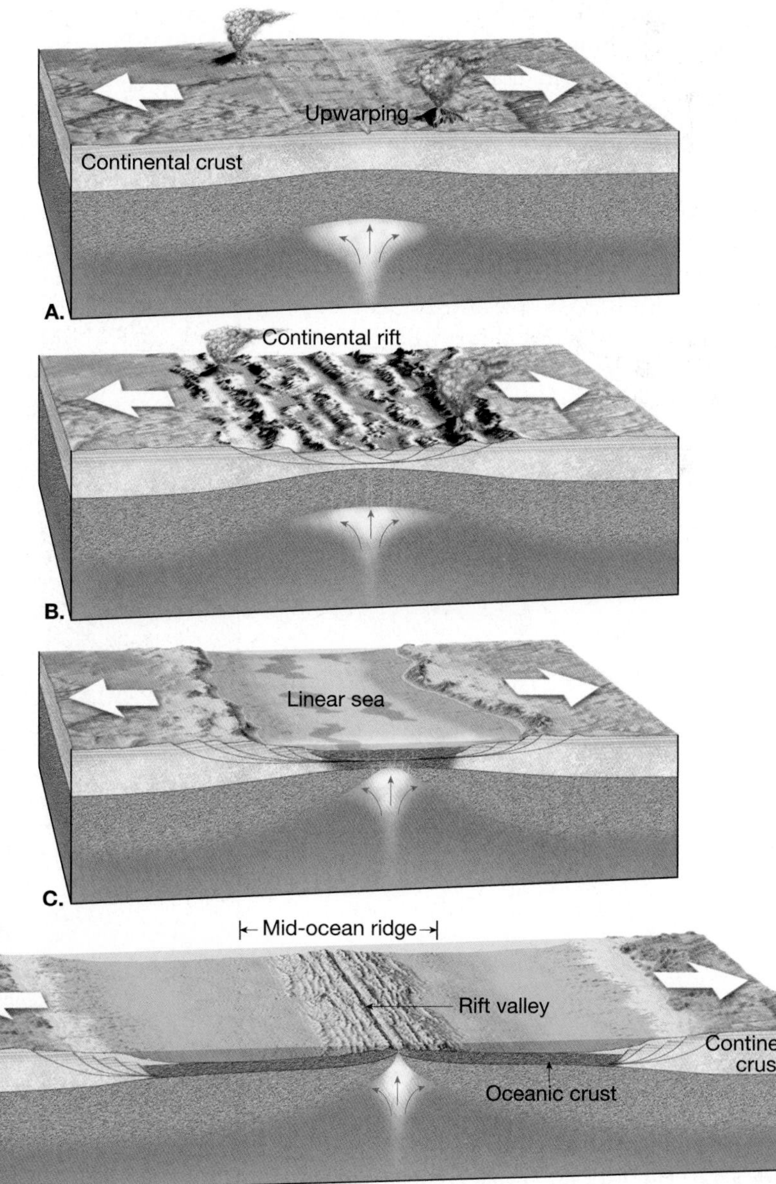

Figure 5.11 Continental rifting and the formation of a new ocean basin. **A.** Continental rifting is thought to occur where tensional forces stretch and thin the crust. As a result, molten rock ascends from the asthenosphere and initiates volcanic activity at the surface. **B.** As the crust is pulled apart, large slabs of rock sink, generating a rift valley. **C.** Further spreading generates a narrow sea. **D.** Eventually, an expansive ocean basin and ridge system are created.

Did You Know?

An observer on another planet would notice, after only a few million years, that all of the continents and ocean basins on Earth are indeed moving. The Moon, on the other hand, is tectonically dead, so it would look virtually unchanged millions of years into the future.

Did You Know?

The remains of some of the earliest humans, *Homo habilis* and *Homo erectus,* were discovered by Louis and Mary Leakey in the East African Rift. Scientists consider this region to be the "birthplace" of the human race.

is thought to begin with the formation of an elongated depression called a *continental rift.* A modern example of a continental rift is the East African Rift (Figure 5.12). Whether this rift will develop into a full-fledged spreading center and eventually split the continent of Africa is a matter of much speculation.

Nevertheless, the East African Rift represents the initial stage in the breakup of a continent. Tensional forces have stretched and thinned the continental crust. As a result, molten rock ascends from the asthensophere and initiates volcanic activity at the surface (Figure 5.11A). Large volcanic mountains such as Mount Kilimanjaro and Mount Kenya exemplify the extensive volcanic activity that accompanies continental rifting. Research suggests that if tensional forces are maintained, the rift valley will lengthen and deepen, eventually extending out to the margin of the continent, splitting it in two (Figure 5.11C). At this point, the rift becomes a narrow sea with an outlet to the ocean, similar to the Red Sea. The Red Sea formed when the Arabian Peninsula rifted from Africa, an event that began about 20 million years ago. If spreading continues, the Red Sea will grow wider and develop an elevated oceanic ridge similar to the Mid-Atlantic Ridge (Figure 5.11D).

Not all rift valleys develop into full-fledged spreading centers. Running through the central United States is an aborted rift zone that extends from Lake Superior into central Kansas. This once active rift valley is filled with rock that was extruded onto the crust more than a billion years ago. Why one rift valley develops into a full-fledged oceanic spreading center while others are abandoned is not yet known.

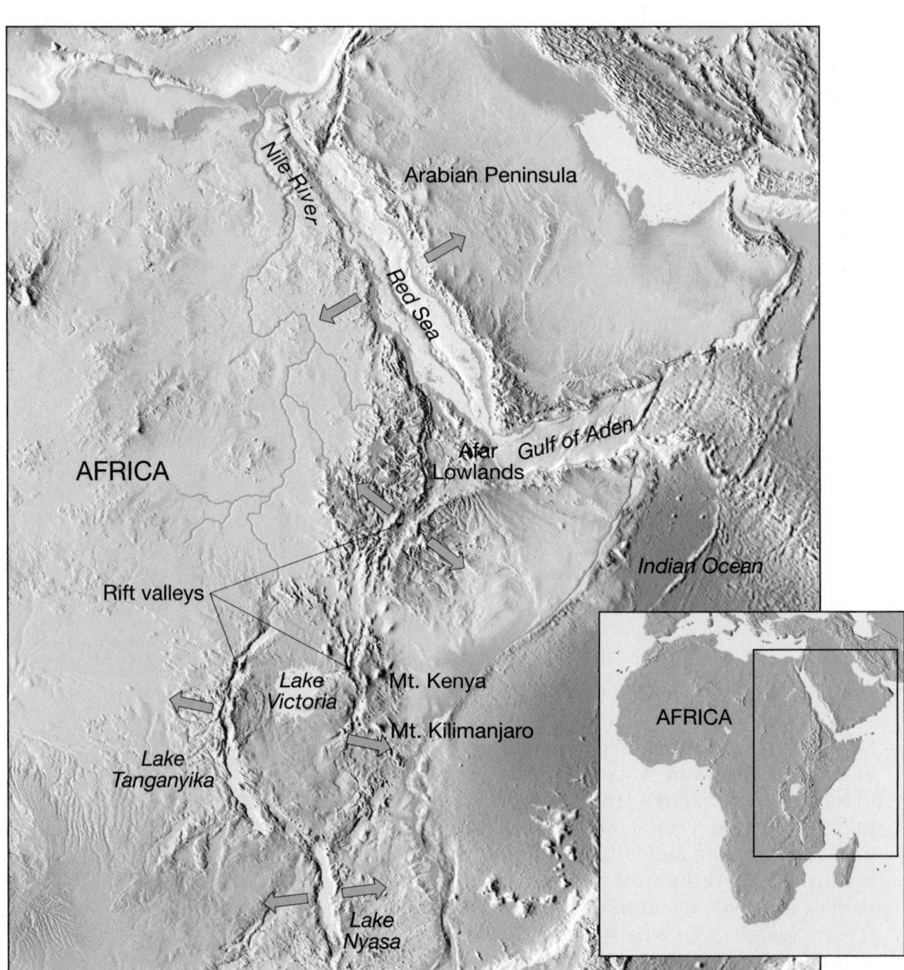

Figure 5.12 East African rift valleys and associated features.

Convergent Boundaries

Forces Within
▼ Plate Tectonics

Although new lithosphere is constantly being produced at the oceanic ridges, our planet is not growing larger—its total surface area remains constant. To balance the addition of newly created lithosphere, older, denser portions of oceanic lithosphere descend into the mantle along **convergent boundaries.** Because lithosphere is "destroyed" at convergent boundaries, they are also called *destructive plate margins.*

Convergent plate margins occur where two plates move toward each other and the leading edge of one is bent downward, allowing it to slide beneath the other. The surface expression produced by the descending plate is a **deep-ocean trench,** such as the Peru-Chile trench (Figure 5.13). Trenches formed in this manner may be thousands of kilometers long, 8 to 12 kilometers (5 to 7 miles) deep, and between 50 and 100 kilometers (31 to 62 miles) wide.

Convergent boundaries are also called **subduction zones,** because they are sites where lithosphere is descending (being subducted) into the mantle. Subduction occurs because the density of the descending lithospheric plate is greater than the density of the underlying asthenosphere. In general, oceanic lithosphere is more dense than the underlying asthenosphere whereas continental lithosphere is less dense and resists subduction. As a consequence, it is always the oceanic lithosphere that is subducted.

Slabs of oceanic lithosphere descend into the mantle at angles that vary from a few degrees to nearly vertical (90 degrees), but average about 45 degrees. The angle at which oceanic lithosphere descends depends largely on its density. For example, when a spreading center is located near a subduction zone, the lithosphere is young and, therefore, warm and buoyant. Hence, the angle of descent is small, as it is along parts of the Peru-Chile trench. Low dip angles usually result in considerable interaction between the descending slab and the overriding plate. Consequently, these regions experience great earthquakes.

As oceanic lithosphere ages (gets farther from the spreading center), it gradually cools, which causes it to thicken and increase in density. Once oceanic lithosphere is about 15 million years old, it becomes more dense than the supporting asthenosphere and will sink when given the opportunity. In parts of the western Pacific, some oceanic lithosphere is more than 180 million years old. This is the thickest and densest in today's oceans. The subducting slabs in this region typically plunge into the mantle at angles approaching 90 degrees.

Although all convergent zones have the same basic characteristics, they are highly variable features. Each is controlled by the type of crustal material involved and the tectonic setting. Convergent boundaries can form between two oceanic plates, one oceanic and one continental plate, or two continental plates. All three situations are illustrated in Figure 5.14.

Oceanic-Continental Convergence

Whenever the leading edge of a plate capped with continental crust converges with a slab of oceanic lithosphere, the buoyant continental block remains "floating," while the denser oceanic slab sinks into the mantle (Figure 5.14A). When a descending oceanic slab reaches a depth of about 100

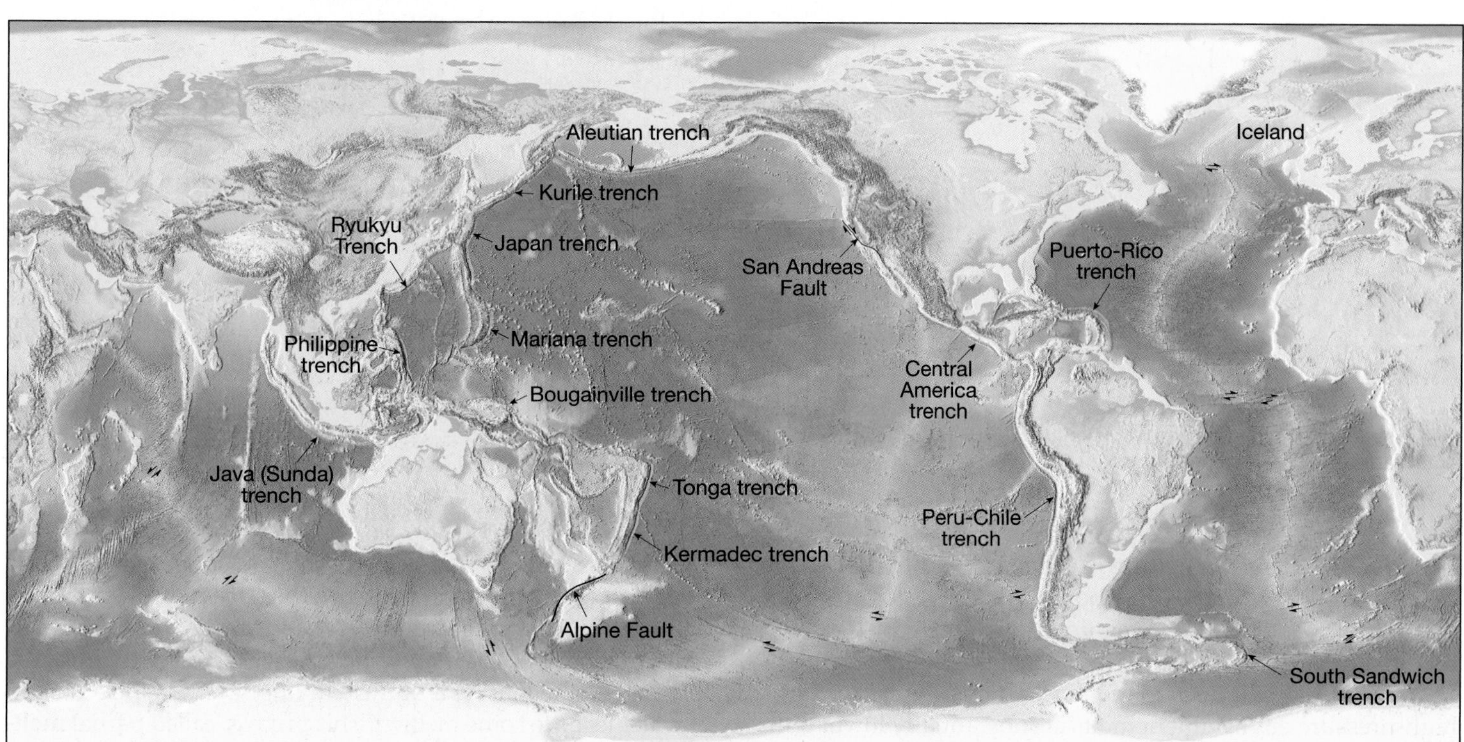

Figure 5.13 Distribution of the world's oceanic trenches, ridge system, and transform faults. Where transform faults offset ridge segments, they permit the ridge to change direction (curve) as can be seen in the Atlantic Ocean.

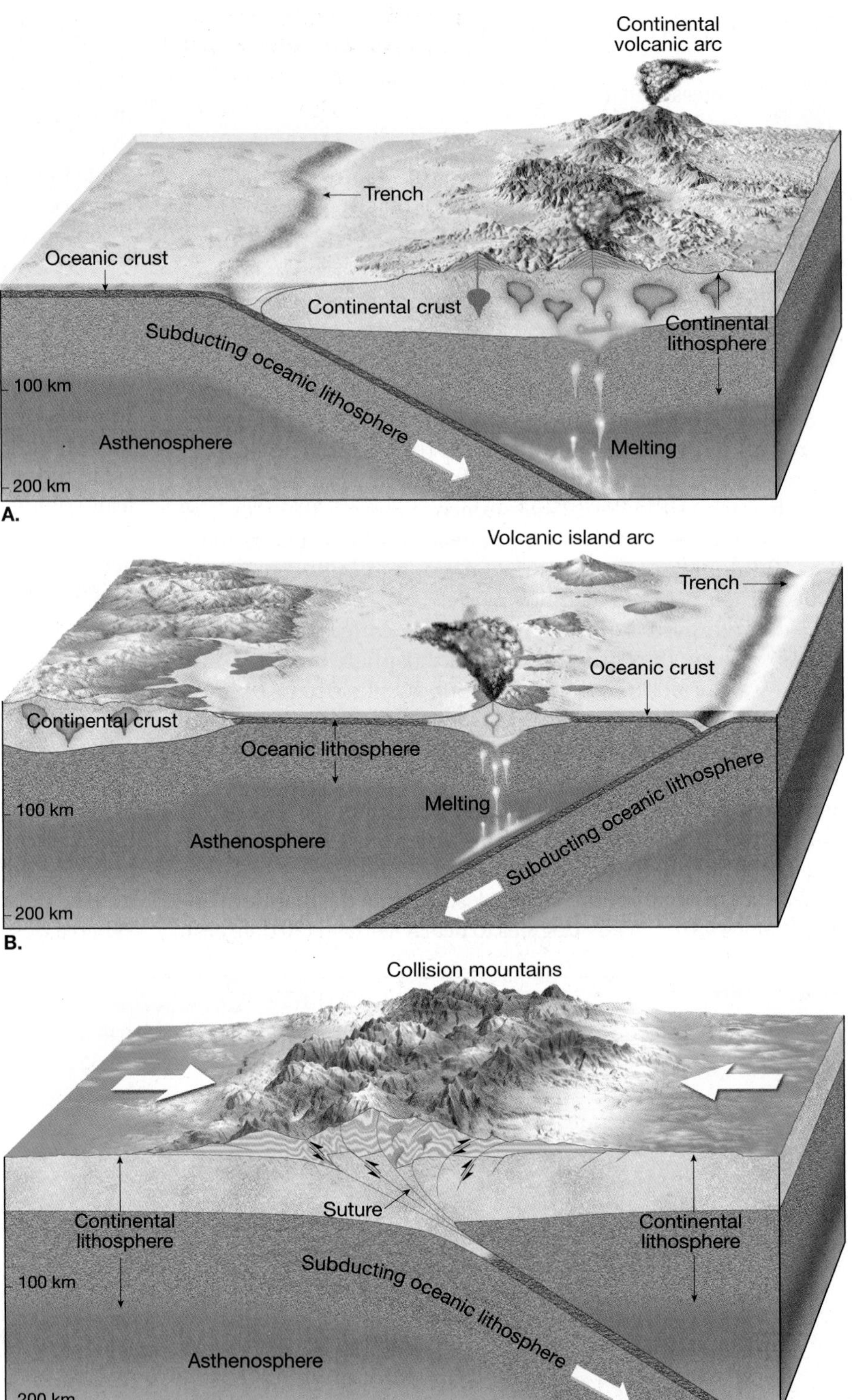

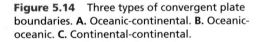

Figure 5.14 Three types of convergent plate boundaries. **A.** Oceanic-continental. **B.** Oceanic-oceanic. **C.** Continental-continental.

kilometers (62 miles), melting is triggered within the wedge of hot asthenosphere that lies above it. But how does the subduction of a cool slab of oceanic lithosphere cause mantle rock to melt? The answer lies in the fact that volatiles (mainly water) act as salt does to melt ice. That is, "wet" rock in a high-pressure environment melts at substantially lower temperatures than does "dry" rock of the same composition.

Sediments and oceanic crust contain a large amount of water, which is carried to great depths by a subducting plate. As the plate plunges downward, water is "squeezed" from the pore spaces as confining pressure increases. At even greater depths, heat and pressure drive water from hydrated (water-rich) minerals such as the *amphiboles*. At a depth of roughly 100 kilometers, the mantle is sufficiently hot that the introduction of water leads to some melting. This process, called **partial melting,** generates as little as 10 percent molten material, which is intermixed with unmelted mantle rock. Being less dense than the surrounding mantle, this hot mobile material gradually rises toward the surface as a teardrop-shaped structure. Depending on the environment, these mantle-derived magmas may ascend through the crust and give rise to a volcanic eruption. However, much of this molten rock never reaches the surface; rather, it solidifies at depth where it acts to thicken the crust.

Partial melting of mantle rock generates molten rock that has a *basaltic composition* similar to the lava that erupts on the Island of Hawaii. In a continental setting, however, basaltic magma typically melts and assimilates some of the crustal rocks through which it ascends. The result is the formation of a silica-rich (SiO_2) magma. When silica-rich magmas reach the surface, they often erupt explosively, generating large columns of volcanic ash and gases. A classic example of such an eruption was the 1980 eruption of Mount St. Helens.

The volcanoes of the towering Andes are the product of magma generated by the subduction of the Nazca plate beneath the South American continent. Mountains such as the Andes, which are produced in part by volcanic activity associated with the subduction of oceanic lithosphere, are called **continental volcanic arcs.** The Cascade Range of Washington, Oregon, and California is another volcanic arc that consists of several well-known volcanic mountains, including Mount Rainier, Mount Shasta, and Mount St. Helens. This active volcanic arc also extends into Canada, where it includes Mount Garibaldi, Mount Silverthrone, and others.

Oceanic-Oceanic Convergence

An oceanic-oceanic convergent boundary has many features in common with oceanic-continental plate margins. The differences are mainly attributable to the nature of the crust capping the overriding plate. Where two oceanic slabs converge, one descends beneath the other, initiating volcanic activity by the same mechanism that operates at oceanic-continental plate boundaries. Water squeezed from the subducting slab of oceanic lithosphere triggers melting in the hot wedge of mantle rock that lies above. In this setting, volcanoes grow up from the ocean floor, rather than upon a continental platform. When subduction is sustained, it will eventually build a chain of volcanic structures that emerge as islands. The volcanic islands are spaced about 80 kilometers (50 miles) apart and are built upon submerged ridges of volcanic material a few hundred kilometers wide. This newly formed land consisting of an arc-shaped chain of small volcanic islands is called a **volcanic island arc,** or simply an **island arc** (Figure 5.14B).

The Aleutian, Mariana, and Tonga islands are examples of volcanic island arcs. Island arcs such as these are generally located 100 to 300 kilometers (62 to 190 miles) from a deep-ocean trench. Located adjacent to the island arcs just mentioned are the Aleutian trench, the Mariana trench, and the Tonga trench (see Figure 5.13, p. 141).

Most volcanic island arcs are located in the western Pacific. Only two volcanic island arcs are located in the Atlantic—the Lesser Antilles arc adjacent to the Caribbean Sea, and the Sandwich Islands in the South Atlantic. The Lesser Antilles are a product of the subduction of the Atlantic beneath the Caribbean plate. Located within this arc is the island of Martinique, where Mount Pelée erupted in 1902, destroying the town of St. Pierre and killing an estimated 28,000 people, as well as the island of Montserrat.

Relatively young island arcs are fairly simple structures that are underlain by deformed oceanic crust that is generally less than 20 kilometers (12 miles) thick. Examples include the arcs of Tonga, the Aleutians, and the Lesser Antilles. By contrast, older island arcs are more complex and are underlain by crust that ranges in thickness from 20 to 35 kilometers (12 to 22 miles). Examples include the Japanese and Indonesian arcs, which are built on material generated by earlier episodes of subduction or sometimes on a small piece of continental crust.

Continental-Continental Convergence

As you saw earlier, when an oceanic plate is subducted beneath continental lithosphere, an Andean-type volcanic arc develops along the margin of the continent. However, if the subducting plate also contains continental lithosphere, continued subduction eventually brings the two continental blocks together (Figure 5.15A). Whereas oceanic lithosphere is relatively dense and sinks into the asthenosphere, continental lithosphere is buoyant, which prevents it from being subducted to any great depth. The result is a collision between the two continental fragments (Figure 5.15C).

Such a collision occurred when the subcontinent of India "rammed" into Asia, producing the Himalayas—the most spectacular mountain range on Earth (see Figure 5.1, p. 130). During this collision, the continental crust buckled and fractured and was generally shortened and thickened. In addition to the Himalayas, several other major mountain systems, including the Alps, Appalachians, and Urals, formed during continental collisions.

Prior to a continental collision, the landmasses involved are separated by an ocean basin. As the continental blocks converge, the intervening seafloor is subducted beneath one of the plates. Subduction initiates partial melting in the overlying mantle, which in turn results in the growth of a volcanic arc. Depending on the location of the subduction zone, the volcanic arc could develop on either of the converging landmasses, or if the subduction zone developed several hundred kilometers seaward from the coast, a volcanic island arc would form. Eventually, as the intervening seafloor is consumed, these continental masses collide. This folds and deforms the accumulation of sediments and sedimentary rocks along the continental margin as if they had been placed in a gigantic vise. The result is the formation of a new mountain range composed of deformed and metamorphosed sedimentary rocks, fragments of the volcanic arc, and often slivers of oceanic crust.

Transform Fault Boundaries

Forces Within
▼ Plate Tectonics

The third type of plate boundary is the **transform fault,** where plates slide horizontally past one another without the production or destruction of lithosphere (*conservative plate margins*). The nature of transform faults was discovered in 1965

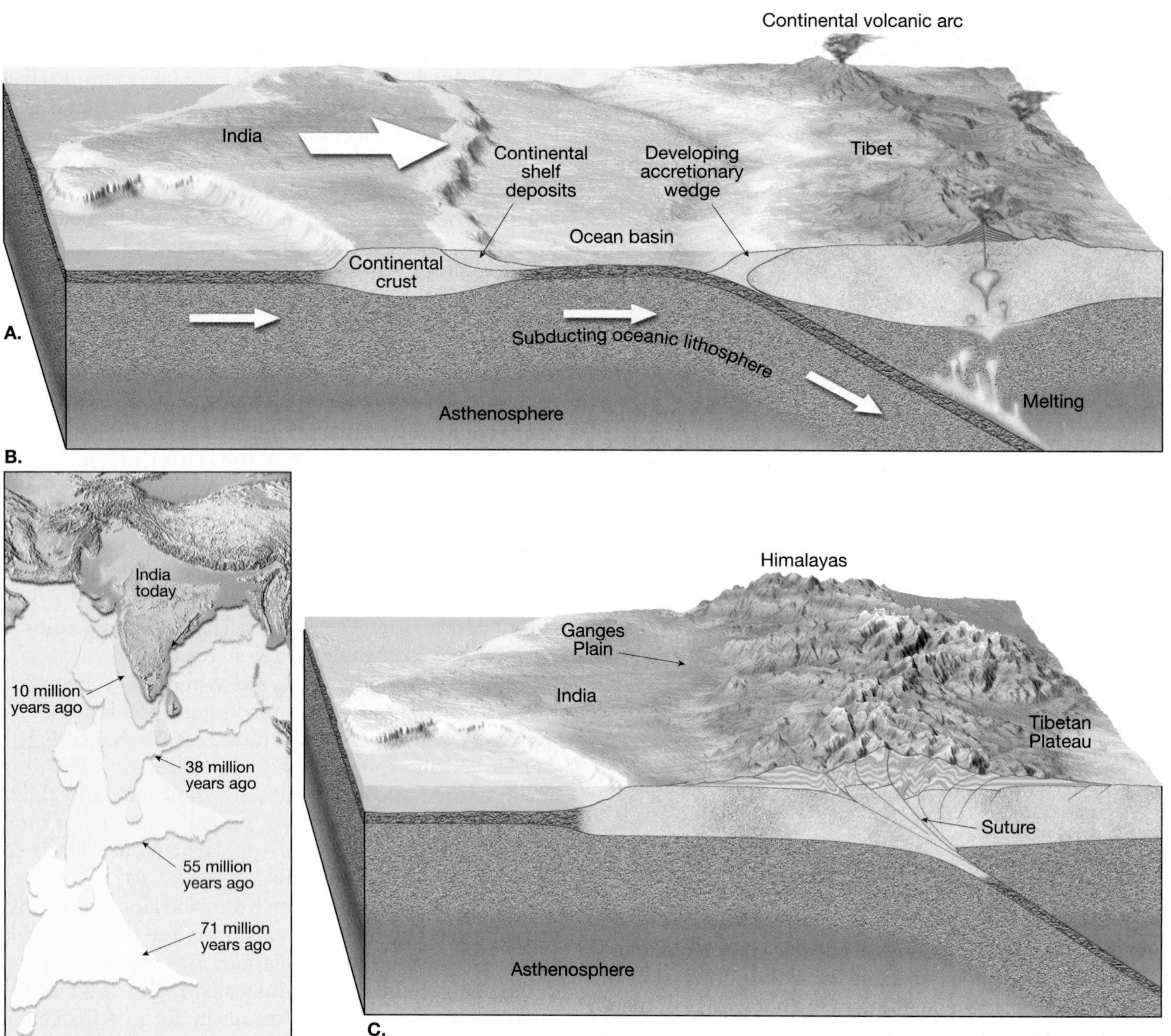

Figure 5.15 The ongoing collision of India and Asia, starting about 45 million years ago, produced the majestic Himalayas. **A.** Converging plates generated a subduction zone, while partial melting triggered by the subducting oceanic slab produced a continental volcanic arc. Sediments scraped from the subducting plate were added to the accretionary wedge. **B.** Position of India in relation to Eurasia at various times. (Modified after Peter Molnar) **C.** The two landmasses eventually collided, deforming and elevating the sediments that had been deposited along their continental margins.

by J. Tuzo Wilson of the University of Toronto. Wilson suggested that these large faults connect the global active belts (convergent boundaries, divergent boundaries, and other transform faults) into a continuous network that divides Earth's outer shell into several rigid plates. Thus, Wilson became the first to suggest that Earth was made of individual plates, while at the same time identifying the faults along which relative motion between the plates is made possible.

Most transform faults join two segments of an oceanic ridge (Figure 5.16). Here, they are part of prominent linear breaks in the oceanic crust known as **fracture zones,** which include both the active transform faults as well as their inactive extensions into the plate interior. These fracture zones are present approximately every 100 kilometers along the trend of a ridge axis. As shown in Figure 5.16, active transform faults lie *only between* the two offset ridge segments. Here seafloor produced at one ridge axis moves in the opposite direction as seafloor produced at an opposing ridge segment. Thus, between the ridge segments, these adjacent slabs of oceanic crust are grinding past each other along the fault. Beyond the ridge crests are the inactive zones, where the fractures are preserved as linear topographic scars. The trend of these fracture zones roughly parallels the direction of plate motion at the time of their formation. Thus, these structures can be used to map the direction of plate motion in the geologic past.

In another role, transform faults provide the means by which the oceanic crust created at ridge crests can be transported to a site of destruction: the deep-ocean trenches (Figure 5.17). Notice that the Juan de Fuca plate moves in a southeasterly direction, eventually being subducted under

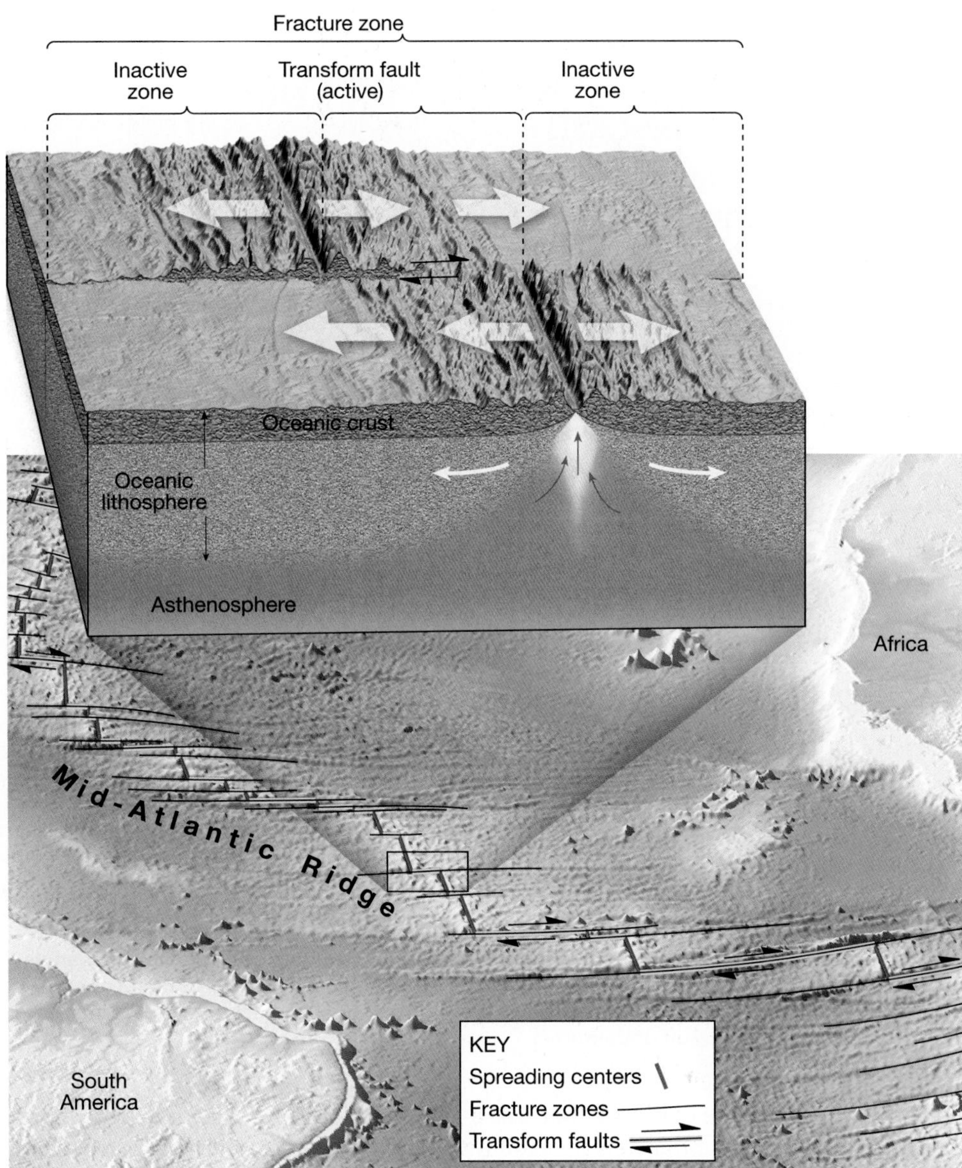

Figure 5.16 Illustration of a transform fault joining segments of the Mid-Atlanic Ridge.

Did You Know?

The Great Alpine Fault of New Zealand is a transform fault that runs down South Island, marking the line where two plates meet. The northwestern part of South Island sits on the Australian plate, while the rest of the island lies on the Pacific plate. Like its sister fault, California's San Andreas, displacement has been measured in hundreds of miles.

the West Coast of the United States. The southern end of this plate is bounded by the Mendocino fault. This transform fault boundary connects the Juan de Fuca ridge to the Cascadia subduction zone. Therefore, it facilitates the movement of the crustal material created at the ridge crest to its destination beneath the North American continent.

Although most transform faults are located within the ocean basins, a few cut through continental crust. Two examples are the earthquake-prone San Andreas Fault of California and the Alpine Fault of New Zealand. Notice in Figure 5.17 that the San Andreas Fault connects a spreading center located in the Gulf of California to the Cascadia subduction zone and the Mendocino fault located along the northwest coast of the United States. Along the San Andreas Fault, the Pacific plate is moving toward the northwest, past the North American plate. If this movement continues, that part of California west of the fault zone, including the Baja Peninsula, will eventually become an island off the West Coast of the United States and Canada. It could eventually reach Alaska. However, a more immediate concern is the earthquake activity triggered by movements along this fault system.

Testing the Plate Tectonics Model

With the development of the theory of plate tectonics, researchers from all of the Earth sciences began testing this new model of how Earth works. Some of the evidence supporting continental drift and seafloor spreading has already been

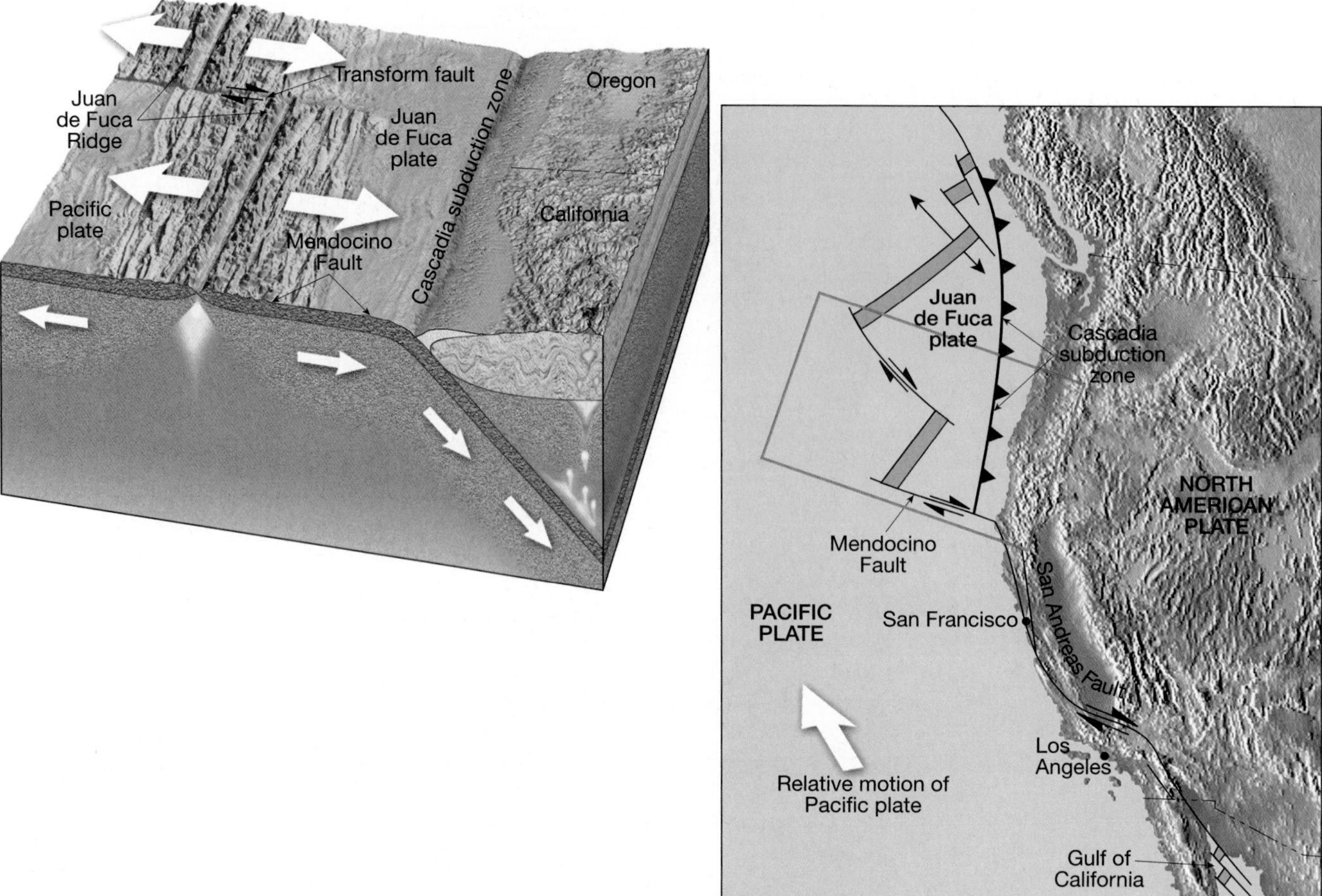

Figure 5.17 The Mendocino transform fault permits seafloor generated at the Juan de Fuca ridge to move southeastward past the Pacific plate and beneath the North American plate. Thus, this transform fault connects a divergent boundary to a subduction zone. Furthermore, the San Andreas Fault, also a transform fault, connects two spreading centers: the Juan de Fuca Ridge and a divergent zone located in the Gulf of California.

presented. Some of the evidence that was instrumental in solidifying the support for this new idea follows. Note that much of this evidence was not new; rather, it was new interpretations of already existing data that swayed the tide of opinion.

Evidence: Ocean Drilling

Some of the most convincing evidence confirming seafloor spreading has come from drilling directly into the ocean floor. From 1968 until 1983, the source of these important data was the Deep Sea Drilling Project, an international program sponsored by several major oceanographic institutions and the National Science Foundation. The primary goal was to gather firsthand information about the age of the ocean basins and the processes that formed them. To accomplish this, a new drilling ship, the *Glomar Challenger,* was built.

Operations began in August 1968 in the South Atlantic. At several sites, holes were drilled through the entire thickness of sediments to the basaltic rock below. An important objective was to gather samples of sediment from just above the igneous crust as a means of dating the seafloor at each site.* Because sedimentation begins immediately after the oceanic crust forms, remains of microorganisms found in the oldest sediments—those resting directly on the crust—can be used to date the ocean floor at that site.

When the oldest sediment from each drill site was plotted against its distance from the ridge crest, the plot demonstrated that the age of the sediment increased with increasing distance from the ridge. This finding supported the seafloor-spreading hypothesis, which predicted that the youngest oceanic crust would be found at the ridge crest and the oldest oceanic crust would be at the continental margins.

The data from the Deep Sea Drilling Project also reinforced the idea that the ocean basins are geologically youthful because no seafloor with an age in excess of 180 million years was found. By comparison, continental crust that exceeds 4 billion years in age has been dated.

The thickness of ocean-floor sediments provided additional verification of seafloor spreading. Drill cores from the *Glomar Challenger* revealed that sediments are almost entirely absent on the ridge crest and the sediment thickens with increasing distance from the ridge. Because the ridge crest is younger than the areas farther away from it, this pattern of sediment distribution should be expected if the seafloor-spreading hypothesis is correct.

*Radiometric dates of the ocean crust itself are unreliable because of the alteration of basalt by seawater.

The Ocean Drilling Program succeeded the Deep Sea Drilling Project and, like its predecessor, was a major international program. The more technologically advanced drilling ship, the *JOIDES Resolution*, continued the work of the *Glomar Challenger*.* The *JOIDES Resolution* can drill in water depths as great as 8200 meters (27,000 feet) and contains onboard laboratories equipped with a large and sophisticated array of seagoing scientific research equipment (Figure 5.18).

In October 2003, the *JOIDES Resolution* became part of a new program, the Integrated Ocean Drilling Program (IODP). This new international effort does not rely on just one drilling ship but uses multiple vessels for exploration. One of the new additions is the massive 210-meter-long (690-foot-long) *Chikyu,* which began operations in 2006.

Evidence: Hot Spots

Mapping of seamounts (submarine volcanoes) in the Pacific Ocean revealed several linear chains of volcanic structures. One of the most studied chains consists of at least 129 volcanoes and extends from the Hawaiian Islands to Midway Island and continues northward toward the Aleutian trench, a distance of nearly 6000 kilometers (3700 miles) (Figure 5.19). This nearly continuous string of volcanic islands and seamounts is called the Hawaiian Island–Emperor Seamount chain. Radiometric dating of these structures showed that the volcanoes increase in age with increasing distance from Hawaii. Hawaii, the youngest volcanic island in the chain, began rising from the seafloor less than a million years ago, whereas Midway Island is 27 million years old, and Suiko Seamount, near the Aleutian trench, is more than 60 million years old (Figure 5.19).

*JOIDES stands for Joint Oceanographic Institutions for Deep Earth Sampling.

Taking a closer look at the Hawaiian Islands, we see a similar increase in age from the volcanically active island of Hawaii, at the southeastern end of the chain, to the inactive volcanoes that make up the island of Kauai in the northwest (Figure 5.19).

Researchers are in agreement that a rising plume of mantle material is located beneath the island of Hawaii. As the ascending **mantle plume** enters the low-pressure environment at the base of the lithosphere, melting occurs. The surface manifestation of this activity is a **hot spot.** Hot spots are areas of volcanism, high heat flow, and crustal uplifting that are a few hundred kilometers across. As the Pacific plate moved over this hot spot, successive volcanic structures were built. As shown in Figure 5.19, the age of each volcano indicates the time when it was situated over the relatively stationary mantle plume. These chains of volcanic structures, known as *hot spot tracks,* trace the direction of plate motion.

Kauai is the oldest of the large islands in the Hawaiian chain. Five million years ago, when it was positioned over the hot spot, Kauai was the only Hawaiian Island (Figure 5.19). Visible evidence of the age of Kauai can be seen by examining its extinct volcanoes, which have been eroded into jagged peaks and vast canyons. By contrast, the relatively young island of Hawaii exhibits young lava flows, and two of Hawaii's volcanoes, Mauna Loa and Kilauea, remain active.

Notice in Figure 5.19 that the Hawaiian Island–Emperor Seamount chain bends. This particular bend in the track occurred about 50 million years ago when the motion of the Pacific plate changed from nearly due north to a northwesterly path. Similarly, hot spots found on the floor of the Atlantic have increased our understanding of the migration of landmasses following the breakup of Pangaea.

Research suggests that at least some mantle plumes originate at great depth, perhaps at the mantle-core boundary. Others, however, may have a much shallower origin. Of

Figure 5.18 The *JOIDES Resolution,* the drilling ship of the Ocean Drilling Program. This modern drilling ship replaced the *Glomar Challenger* in the important work of sampling the floor of the world's oceans. (Photo courtesy of Ocean Drilling Program)

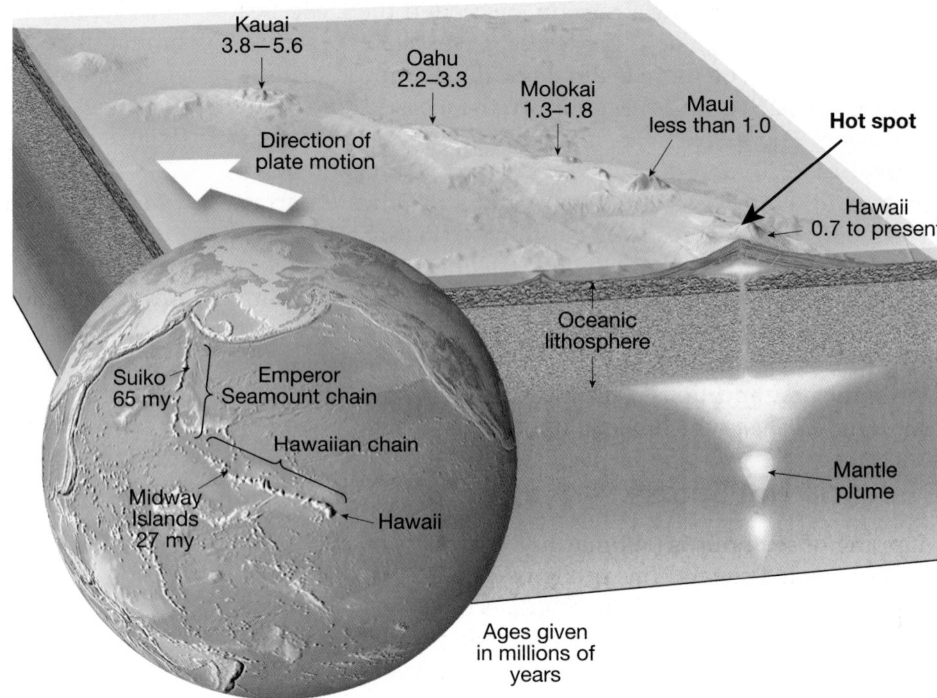

Figure 5.19 The chain of islands and seamounts that extends from Hawaii to the Aleutian trench results from the movement of the Pacific plate over an apparently stationary hot spot. Radiometric dating of the Hawaiian Islands shows that the volcanic activity decreases in age toward the island of Hawaii.

Did You Know?

Olympus Mons is a huge volcano on Mars that strongly resembles the Hawaiian shield volcanoes. Rising 25 kilometers (16 miles) above the surrounding plains, Olympus Mons owes its massiveness to the fact that plate tectonics does not operate on Mars. Consequently, instead of being carried away from the hot spot by plate motion, as occurred with the Hawaiian volcanoes, Olympus Mons remained fixed and grew to a gigantic size.

the 40 or so hot spots that have been identified, more than a dozen are located near spreading centers. For example, the mantle plume located beneath Iceland is responsible for the large accumulation of volcanic rocks found along the northern section of the Mid-Atlantic Ridge.

The existence of mantle plumes and their association with hot spots is well documented. Most mantle plumes are long-lived features that appear to maintain relatively fixed positions within the mantle. However, recent evidence has shown that some hot spots may slowly migrate. If this is the case, models of past plate motion that are based on a fixed hot spot frame of reference will need to be reevaluated.

Evidence: Paleomagnetism

Anyone who has used a compass to find direction knows that Earth's magnetic field has north and south magnetic poles. Today these magnetic poles align closely, but not exactly, with the geographic poles. (The geographic poles, or true north and south poles, are where Earth's rotational axis intersects the surface.) Earth's magnetic field is similar to that produced by a simple bar magnet. Invisible lines of force pass through the planet and extend from one magnetic pole to the other as shown in Figure 5.20. A compass needle, itself a small magnet free to rotate on an axis, becomes aligned with the magnetic lines of force and points to the magnetic poles.

Unlike the pull of gravity, we cannot feel Earth's magnetic field, yet its presence is revealed because it deflects a compass needle. In a similar manner, certain rocks contain minerals that serve as "fossil compasses." These iron-rich minerals, such as *magnetite,* are abundant in lava flows of

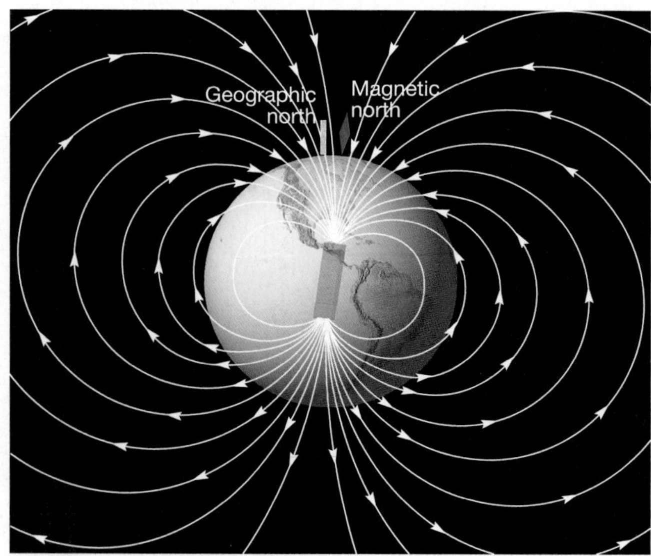

Figure 5.20 Earth's magnetic field consists of lines of force much like those a giant bar magnet would produce if placed at the center of Earth.

Did You Know?

When all of the continents were joined to form Pangaea, the rest of Earth's surface was covered with a huge ocean called *Panthalassa* (*pan* = all, *thalassa* = sea). Today, all that remains of Panthalassa is the Pacific Ocean, which has been decreasing in size since the breakup of Pangaea.

basaltic composition.* When heated above a temperature known as the **Curie point,** these magnetic minerals lose their magnetism. However, when these iron-rich grains cool below their Curie point (about 585°C for magnetite), they gradually become magnetized in the direction of the existing magnetic lines of force. Once the minerals solidify, the magnetism they possess will usually remain "frozen" in this position. They behave much like a compass needle: They "point" toward the position of the magnetic poles at the time of their formation. If the rock is moved, the rock magnetism will retain its original alignment. Rocks that formed thousands or millions of years ago and contain a record of the direction of the magnetic poles at the time of their formation are said to possess **fossil magnetism,** or **paleomagnetism.**

Apparent Polar Wandering A study of rock magnetism conducted during the 1950s in Europe led to an interesting discovery. The magnetic alignment in the iron-rich minerals in lava flows of different ages indicated that many different peleomagnetic poles once existed. A plot of the apparent positions of the magnetic north pole with respect to Europe revealed that during the past 500 million years, the location of the pole had gradually wandered from a location near Hawaii northward through eastern Siberia and finally to its present location (Figure 5.21A). This was strong evidence that either the magnetic poles had migrated through time, an idea known as *polar wandering,* or that the lava flows moved—in other words, Europe had drifted in relation to the poles.

Although the magnetic poles are known to move in an erratic path around the geographic poles, studies of paleomagnetism from numerous locations show that the positions of the magnetic poles, averaged over thousands of years, correspond closely to the positions of the geographic poles. Therefore, a more acceptable explanation for the apparent polar wandering paths was provided by Wegener's hypothesis. If the magnetic poles remain stationary, their *apparent movement* is produced by continental drift.

The latter idea was further supported by comparing the latitude of Europe as determined from fossil magnetism with evidence obtained from paleoclimatic studies. Recall that during the Pennsylvanian period (about 300 million years ago), coal-producing swamps covered much of Europe. During this same time period, paleomagnetic evidence places Europe near the equator—a fact consistent with the tropical environment indicated by these coal deposits.

Further evidence for continental drift came a few years later when a polar-wandering path was constructed for North America (Figure 5.21A). It turned out that paths for North America and Europe had similar shapes but were separated by about 30 degrees of longitude. At the time these rocks

*Some sediments and sedimentary rocks contain enough iron-bearing mineral grains to acquire a measurable amount of magnetization.

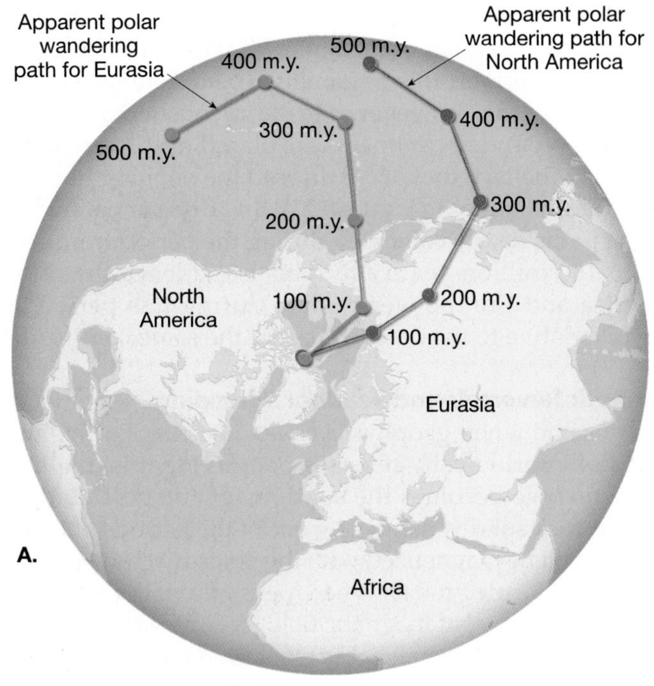

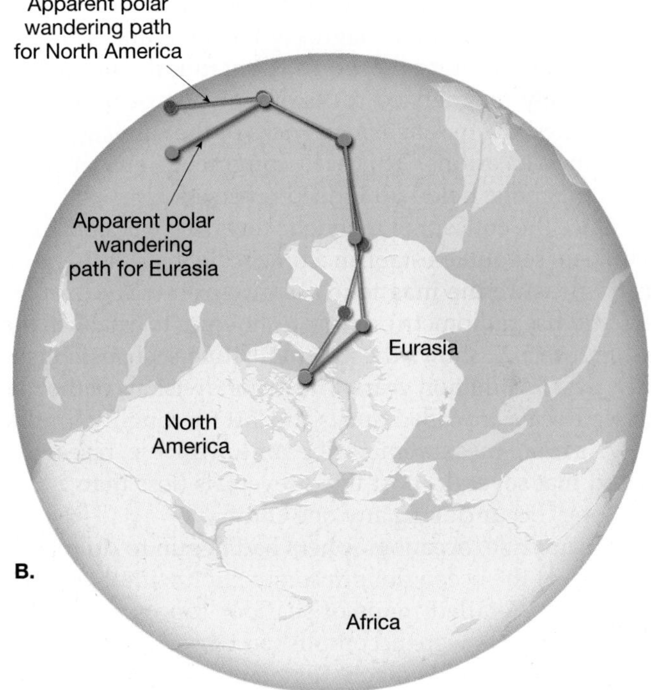

Figure 5.21 Simplified apparent polar wandering paths as established from North American and Eurasian paleomagnetic data. **A.** The more westerly path determined from North American data is thought to have been caused by the westward drift of North America by about 24 degrees from Eurasia. **B.** The positions of the wandering paths when the landmasses are reassembled in their predrift locations.

crystallized, could there have been two magnetic north poles that migrated parallel to each other? Investigators found no evidence to support this possibility. The differences in these migration paths, however, can be reconciled if the two presently separated continents are placed next to each other, as we now believe they were prior to the opening of the Atlantic Ocean. Notice in Figure 5.21B that these apparent wandering paths nearly coincided during the period from about 400 to 160 million years ago. This is evidence that North America and Europe were joined during this period and moved relative to the poles as part of the same continent.

Magnetic Reversals and Seafloor Spreading Another discovery came when geophysicists learned that Earth's magnetic field periodically reverses polarity; that is, the north magnetic pole becomes the south magnetic pole, and vice versa. A rock solidifying during one of the periods of reverse polarity will be magnetized with the polarity opposite that of rocks being formed today. When rocks exhibit the same magnetism as the present magnetic field, they are said to possess **normal polarity,** whereas rocks exhibiting the opposite magnetism are said to have **reverse polarity.**

Evidence for **magnetic reversals** was obtained when investigators measured the magnetism of lavas and sediments of various ages around the world. They found that normally and reversely magnetized rocks of a given age in one location matched the magnetism of rocks of the same age found in all other locations. This was convincing evidence that Earth's magnetic field had indeed reversed.

Once the concept of magnetic reversals was confirmed, researchers set out to establish a time scale for them. The task was to measure the magnetic polarity of hundreds of lava flows and use radiometric dating techniques to establish their ages. Figure 5.22 shows the **magnetic time scale** established for the past few million years. The major divisions of the magnetic time scale are called *chrons* and last for roughly 1 million years. As more measurements became available, researchers realized that several, short-lived reversals (less than 200,000 years long) occur during any one chron.

Meanwhile, oceanographers had begun to do magnetic surveys of the ocean floor in conjunction with their efforts to construct detailed maps of seafloor topography. These magnetic surveys were accomplished by towing very sensitive instruments called **magnetometers** behind research vessels. The goal of these geophysical surveys was to map variations in the strength of Earth's magnetic field that arise from differences in the magnetic properties of the underlying crustal rocks.

The first comprehensive study of this type was carried out off the Pacific Coast of North America and had an unexpected outcome. Researchers discovered alternating strips of high- and low-intensity magnetism as shown in Figure 5.23.

This relatively simple pattern of magnetic variation defied explanation until 1963, when Fred Vine and D. H. Matthews demonstrated that the high- and low-intensity stripes supported the concept of seafloor spreading. Vine and Matthews suggested that the stripes of high-intensity magnetism are regions where the paleomagnetism of the ocean crust

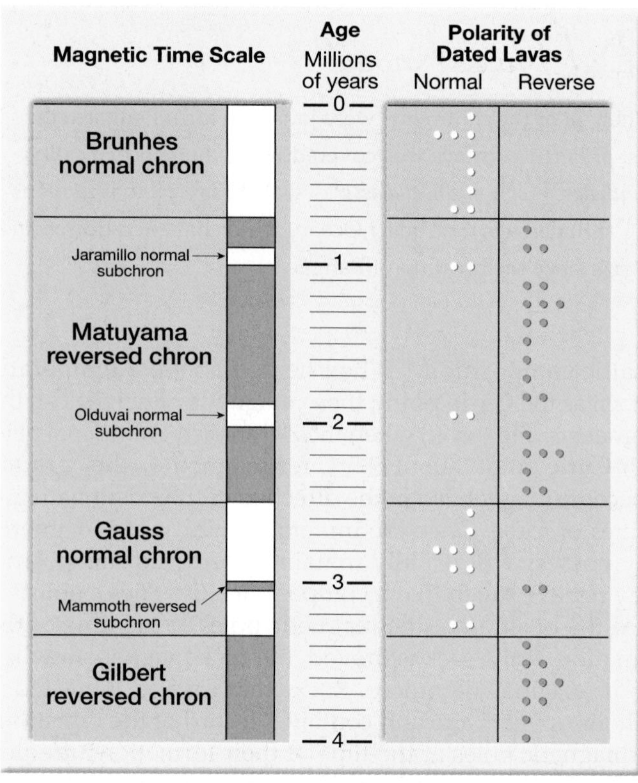

Figure 5.22 Time scale of Earth's magnetic field in the recent past. This time scale was developed by establishing the magnetic polarity for lava flows of known age. (Data from Allen Cox and G. B. Dalrymple)

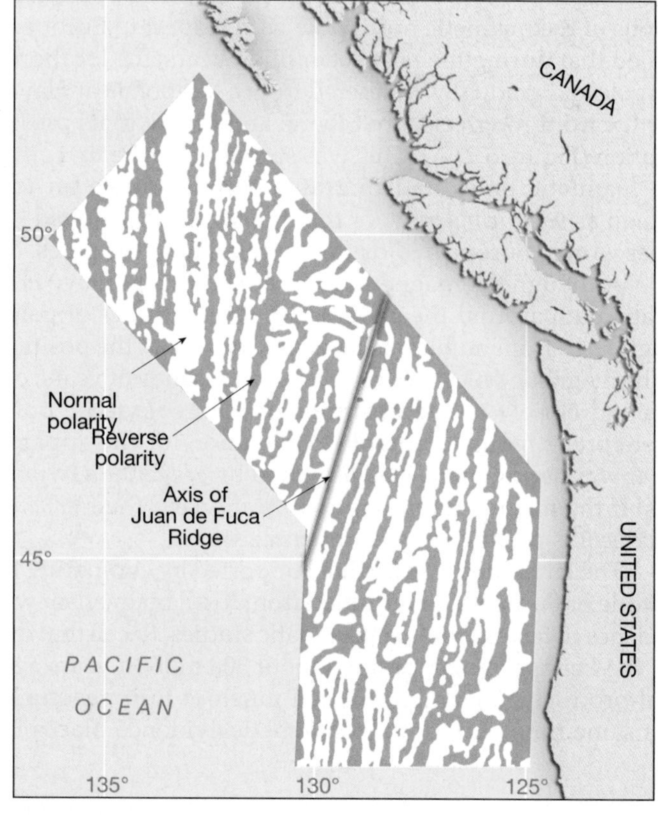

Figure 5.23 Pattern of alternating stripes of high- and low-intensity magnetism discovered off the Pacific Coast of North America.

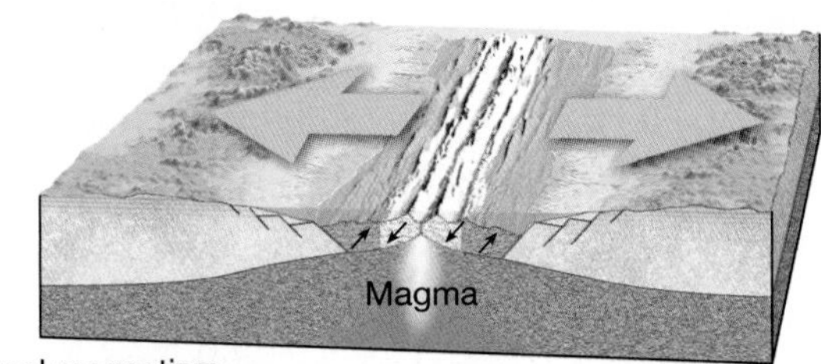

A. Period of normal magnetism

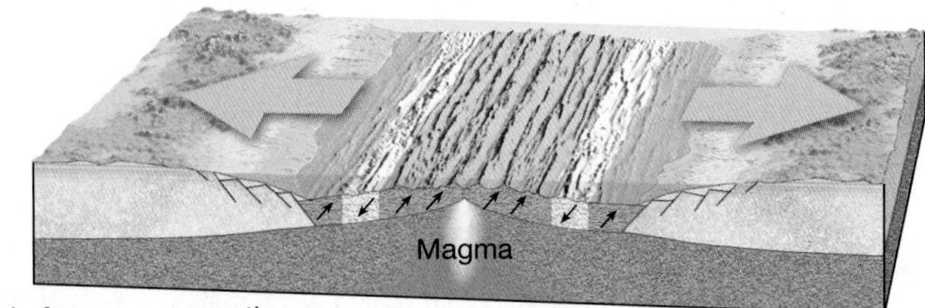

B. Period of reverse magnetism

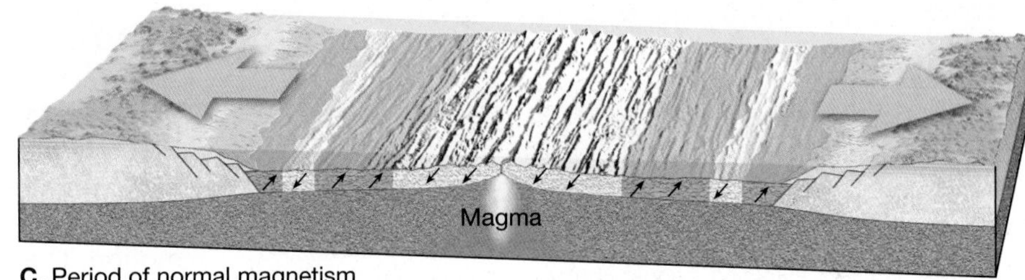

C. Period of normal magnetism

Figure 5.24 As new basalt is added to the ocean floor at mid-ocean ridges, it is magnetized according to Earth's existing magnetic field. Hence, it behaves much like a tape recorder as it records each reversal of the planet's magnetic field.

exhibits normal polarity (Figure 5.23). Consequently, these rocks *enhance* (reinforce) Earth's magnetic field. Conversely, the low-intensity stripes are regions where the ocean crust is polarized in the reverse direction and, therefore, *weakens* the existing magnetic field. But how do parallel stripes of normally and reversely magnetized rock become distributed across the ocean floor?

Vine and Matthews reasoned that as magma solidifies along narrow rifts at the crest of an oceanic ridge, it is magnetized with the polarity of the existing magnetic field (Figure 5.24). Because of seafloor spreading, this stripe of magnetized crust would gradually increase in width. When Earth's magnetic field reverses polarity, any newly formed seafloor (having the opposite polarity) would form in the middle of the old stripe. Gradually, the two parts of the old stripe are carried in opposite directions away from the ridge crest. Subsequent reversals would build a pattern of normal and reverse stripes as shown in Figure 5.24. Because new rock is added in equal amounts to both trailing edges of the spreading ocean floor, we should expect that the pattern of stripes (size and polarity) found on one side of an oceanic ridge to be a mirror image of the other side. A few years later, a survey across the Mid-Atlantic Ridge just south of Iceland revealed a pattern of magnetic stripes exhibiting a remarkable degree of symmetry to the ridge axis (Figure 5.25).

Because the dates of magnetic reversals going back nearly 200 million years have been established, the rate at which spreading occurs at the various ridges can be determined accurately. In the Pacific Ocean, for example, the magnetic stripes are much wider for corresponding time intervals than those of the Atlantic Ocean. Hence, we conclude that a faster spreading rate exists for the spreading center of the Pacific as compared to the Atlantic. When we apply numerical dates to these magnetic events, we find that the spreading rate for the North Atlantic Ridge is only 2 centimeters (1 inch) per year. The rate is somewhat faster for the South Atlantic. The spreading rates for the East Pacific Rise generally range between 6 and 12 centimeters (2.5 to 5 inches) per year, with a maximum rate of about 20 centimeters (8 inches) a year. Thus, we have a magnetic tape recorder that records changes in Earth's magnetic field. This recorder also permits us to determine the rate of seafloor spreading.

The Breakup of Pangaea

Wegener used evidence from fossils, rock types, and ancient climates to create a jigsaw-puzzle fit of the continents, thereby creating his supercontinent of Pangaea. In a similar manner, but employing modern tools not available to Wegener, geologists have recreated the steps in the breakup of this supercontinent, an event that began nearly 200 million years ago. From this work, the dates when individual crustal fragments separated from one another and their relative motions have been well established (Figure 5.26).

An important consequence of the breakup of Pangaea was the creation of a "new" ocean basin: the Atlantic. As you can see in Figure 5.26, splitting of the supercontinent did not occur simultaneously along the margins of the Atlantic. The first split developed between North America and Africa.

Here, the continental crust was highly fractured, providing pathways for huge quantities of fluid lavas to reach the

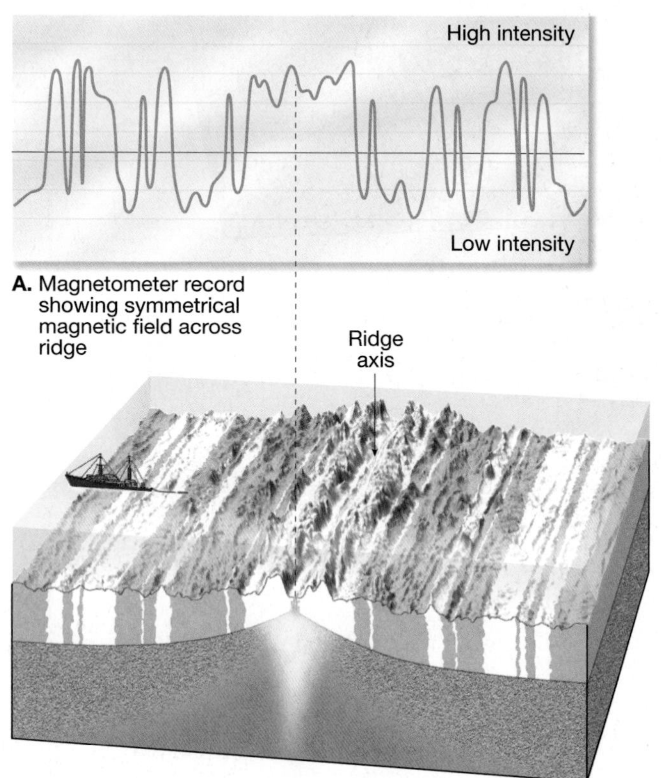

Figure 5.25 The ocean floor as a magnetic tape recorder. **A.** Schematic representation of magnetic intensities recorded as a magnetometer is towed across a segment of the oceanic ridge. **B.** Notice the symmetrical stripes of low- and high-intensity magnetism that parallel the ridge crest. Vine and Matthews suggested that the stripes of high-intensity magnetism occur where normally magnetized oceanic basalts enhance the existing magnetic field. Conversely, the low-intensity stripes are regions where the crust is polarized in the reverse direction, which weakens the existing magnetic field.

surface. Today, these lavas are represented by weathered igneous rocks found along the eastern seaboard of the United States—primarily buried beneath the sedimentary rocks that form the continental shelf. Radiometric dating of these solidified lavas indicates that rifting began in various stages between 180 million and 165 million years ago. This time span represents the "birth date" for this section of the North Atlantic.

By 130 million years ago, the South Atlantic began to open near the tip of what is now South Africa. As this zone of rifting migrated northward, it gradually opened the South Atlantic (compare Figures 5.26B and C). Continued breakup of the southern landmass led to the separation of Africa and Antarctica and sent India on a northward journey. By the early Cenozoic, about 50 million years ago, Australia had separated from Antarctica, and the South Atlantic had emerged as a full-fledged ocean (Figure 5.26D).

A modern map (Figure 5.26F) shows that India eventually collided with Asia, an event that began about 45 million years ago and created the Himalayas as well as the Tibetan Highlands. About the same time, the separation of Greenland from Eurasia completed the breakup of the northern landmass. During the past 20 million years or so of Earth's history, Arabia rifted from Africa to form the Red Sea, and Baja California separated from Mexico to form the Gulf of California (Figure 5.26E). Meanwhile, the Panama Arc joined North America and South America to produce our globe's familiar, modern appearance.

Did You Know?

Researchers have estimated that the continents join to form supercontinents roughly every 500 million years. Since it has been about 200 million years since Pangaea broke up, we have only 300 million years to wait before the next supercontinent is completed.

What Drives Plate Motions?

The plate tectonics theory *describes* plate motion and the role that this motion plays in generating and/or modifying the major features of Earth's crust. Therefore, acceptance of plate tectonics does not rely on knowing exactly what drives plate motion. This is fortunate, because none of the models yet proposed can account for all major facets of plate tectonics. Nevertheless, researchers generally agree on the following:

1. Convective flow in the rocky 2900-kilometer-thick (1800-mile-thick) mantle—in which warm, buoyant rock rises and cooler, dense material sinks under its own weight—is the underlying driving force for plate movement.
2. Mantle convection and plate tectonics are part of the same system. Subducting oceanic plates drive the cold downward-moving portion of convective flow while shallow upwelling of hot rock along the oceanic ridge and buoyant mantle plumes are the upward-flowing arm of the convective mechanism.
3. The slow movements of Earth's plates and mantle are ultimately driven by the unequal distribution of heat within Earth's interior.

What is not known with any high degree of certainty is the precise nature of this convective flow.

Forces That Drive Plate Motion

Several mechanisms contribute to plate motion; these include *slab pull, ridge push,* and *slab suction.* There is general agreement that the subduction of cold, dense slabs of oceanic lithosphere is the main driving force of plate motion (Figure 5.27). As these slabs sink into the asthenosphere, they "pull" the trailing plate along. This phenomenon, called **slab pull,** results because old slabs of oceanic lithosphere are more dense than the underlying asthenosphere and hence "sink like a rock."

Another important driving force is called **ridge push** (Figure 5.27). This gravity-driven mechanism results from

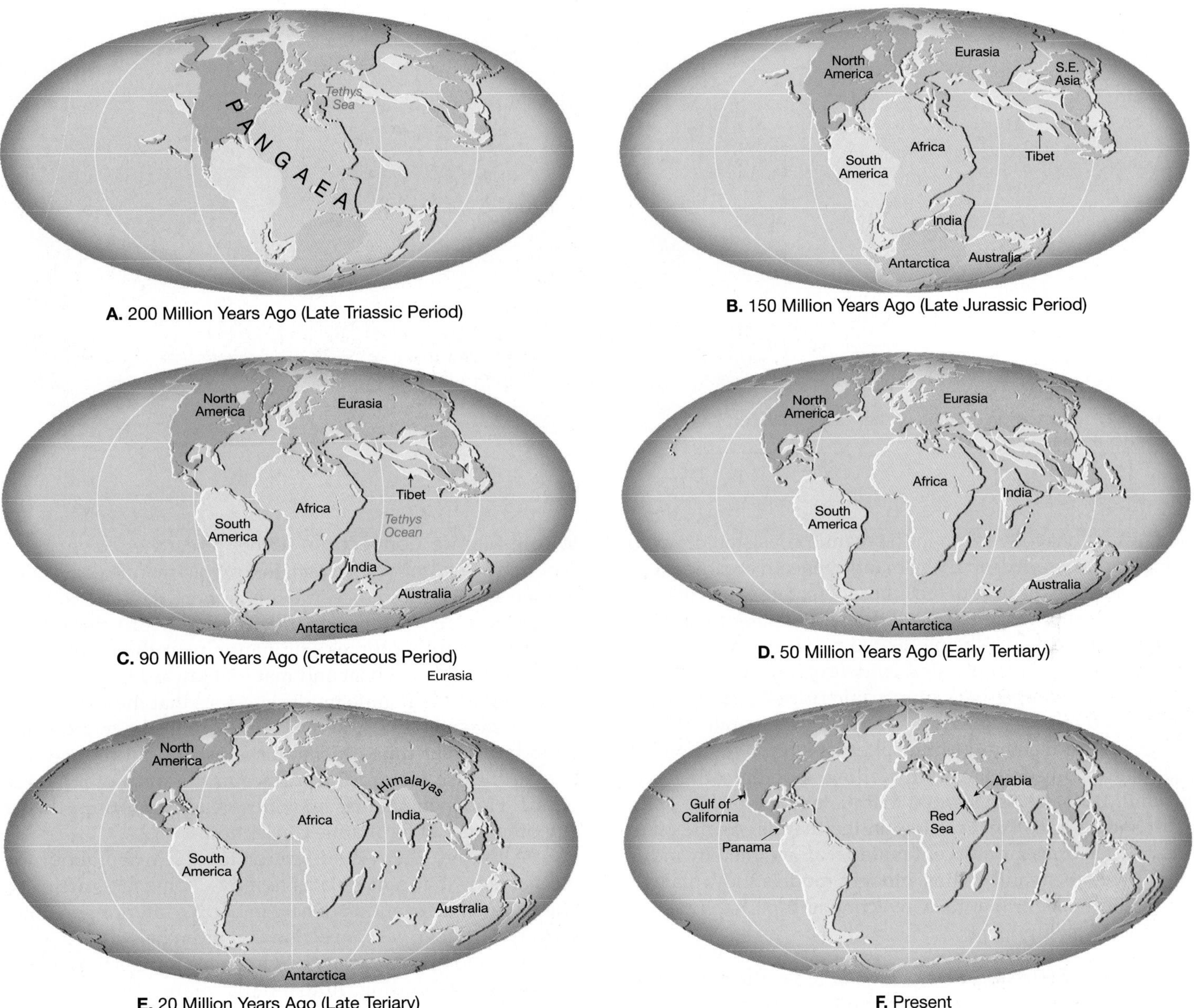

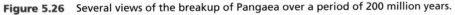

Figure 5.26 Several views of the breakup of Pangaea over a period of 200 million years.

the elevated position of the oceanic ridge, which causes slabs of lithosphere to "slide" down the flanks of the ridge. Ridge push appears to contribute far less to plate motions than does slab pull. The primary evidence for this comes from known spreading rates on ridge segments having different elevations. For example, despite its greater average height above the seafloor, spreading rates along the Mid-Atlantic Ridge are considerably less than spreading rates along the less steep East Pacific Rise. The fact that fast-moving plates are being subducted along a large percentage of their margins also supports the theory that slab pull has more significant impact than ridge push. Examples of fast-moving plates include the Pacific, Nazca, and Cocos plates, all of which have spreading rates that exceed 10 centimeters (4 inches) per year.

Yet another driving force arises from the drag of a subducting slab on the adjacent mantle. The result is an induced mantle circulation that pulls both the subducting and overriding plates toward the trench. Because this mantle flow tends to "suck" in nearby plates (similar to pulling the plug in a bathtub), it is called **slab suction** (Figure 5.27). Even if a subducting slab becomes detached from the overlying plate, its descent will continue to create flow in the mantle and hence will continue to drive plate motion.

Models of Plate-Mantle Convection

Any model of mantle-plate convection must be consistent with observed physical and chemical properties of the mantle. When seafloor spreading was first proposed, geologists

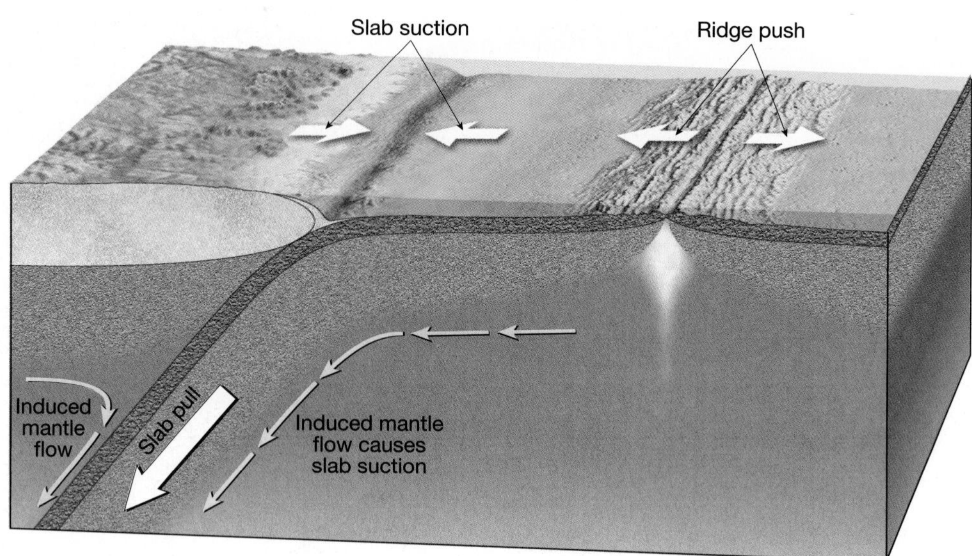

Figure 5.27 Illustration of some of the forces that act on plates.

suggested that convection in the mantle consisted of upcurrents coming from deep in the mantle beneath oceanic ridges. Upon reaching the base of the lithosphere, these currents were thought to spread laterally and drag the plates along. Thus, plates were viewed as being carried passively by the flow in the mantle. Based on physical evidence, however, it became clear that upwelling beneath oceanic ridges is shallow and not related to deep convection in the lower mantle. It is the horizontal movement of lithospheric plates away from the ridge that causes mantle upwelling, not the other way around. We have also learned that plate motion is the dominant source of convective flow in the mantle. As the plates move, they drag the adjacent material along, thereby inducing flow in the mantle. Thus, modern models show plates as an integral part of mantle convection and perhaps even its most active component.

Layering at 660 Kilometers One of the earliest proposals regarding mantle convection came to be called the "layer cake" model. As shown in Figure 5.28A, this model has two zones of convection—a thin convective layer above 660 kilometers (410 miles) and a thick one located below. This model successfully explains why the basaltic lavas that erupt along oceanic ridges have a somewhat different composition than those that erupt in Hawaii as a result of hot-spot activity. The mid-ocean ridge basalts come from the upper convective layer, which is well mixed, whereas the mantle plume that feeds the Hawaiian volcanoes taps a deeper, more primitive magma source that resides in the lower convective layer.

Despite evidence that supports this model, data gathered from the study of earthquake waves have shown that at least some subducting slabs of cold, oceanic lithosphere penetrate the 660-kilometer boundary and descend deep into the mantle. The subducting lithosphere should serve to mix the upper and lower layers. As a result, the layered mantle structure proposed in this model would be destroyed.

Whole-Mantle Convection Other researchers favor some type of *whole-mantle convection.* In a whole-mantle convection model, slabs of cold, oceanic lithosphere descend to great depths and stir the entire mantle (Figure 5.28B). Simultaneously, hot mantle plumes originating near the mantle-core boundary transport heat and material toward the surface.

One whole-mantle model suggests that the subducting slabs of oceanic lithosphere accumulate at the mantle-core boundary. Over time, this material is thought to melt and buoyantly rise toward the surface in the form of a mantle plume. However, there is no widespread support for this or any other model for mantle convection.

Although there is still much to be learned about the mechanisms that cause plates to move, some facts are clear. The unequal distribution of heat in Earth's interior generates some type of thermal convection that ultimately drives plate-mantle motion. The primary driving force for this flow is provided by descending lithospheric plates that serve to transport cold material into the mantle. Moreover, mantle plumes that are generated at the core-mantle boundary carry heat from the core upward into the mantle.

Did You Know?

Because plate tectonic processes are powered by heat from Earth's interior, the forces that drive plate motion will cease sometime in the distant future. The work of external forces (wind, water, and ice), however, will continue to erode Earth's surface. Eventually, landmasses will be nearly flat. What a different world it will be—an Earth with no earthquakes, no volcanoes, and no mountains.

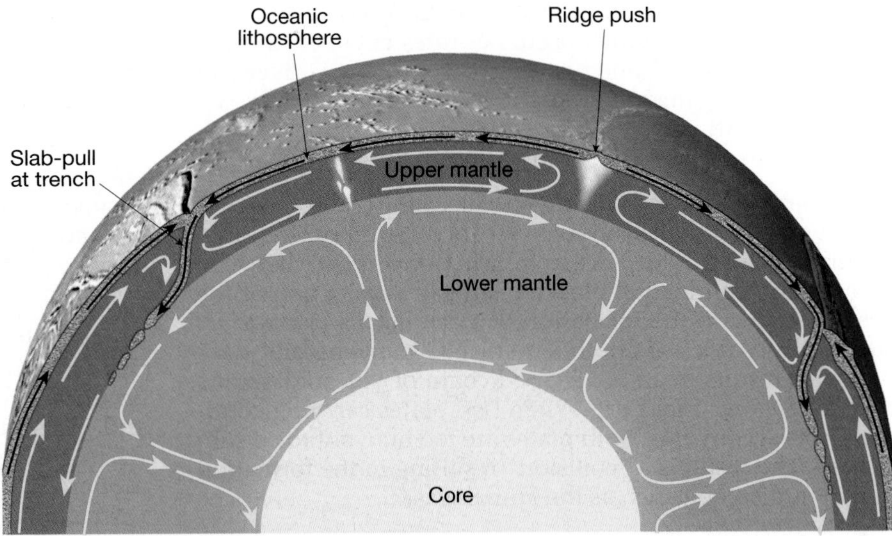

A. Layering at 660 kilometers

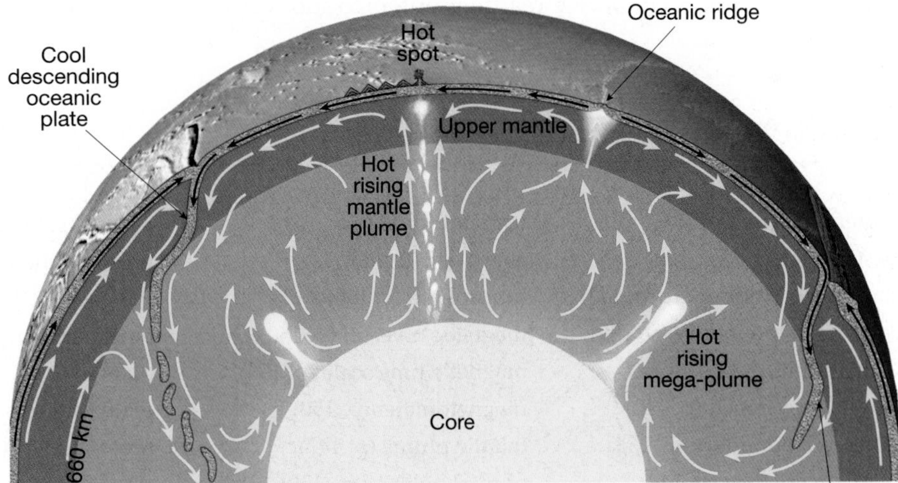

B. Whole mantle convection

Cool descending oceanic plate

Figure 5.28 Proposed models for mantle convection. **A.** This model consists of two convection layers—a thin, convective layer above 660 kilometers (410 miles) and a thick one below. **B.** In this whole-mantle convection model, cold, oceanic lithosphere descends into the lowermost mantle while hot mantle plumes transport heat toward the surface.

The Chapter in Review

1. In the early 1900s, *Alfred Wegener* set forth his *continental drift* hypothesis. One of its major tenets was that a supercontinent called *Pangaea* began breaking apart into smaller continents about 200 million years ago. The smaller continental fragments then *drifted* to their present positions. To support the claim that the now-separate continents were once joined, Wegener and others used the *fit of South America and Africa, the distribution of ancient climates, fossil evidence,* and *rock structures.*

2. One of the main objections to the continental drift hypothesis was the inability of its supporters to provide an acceptable mechanism for the movement of continents.

3. The theory of *plate tectonics,* a far more encompassing theory than continental drift, holds that Earth's rigid outer shell, called the *lithosphere,* consists of seven large and numerous smaller segments called *plates* that are in motion relative to one another. Most of Earth's *seismic activity, volcanism,* and *mountain building* occur along the dynamic margins of these plates.

4. A major departure of the plate tectonics theory from the continental drift hypothesis is that large plates contain both continental and ocean crust and the entire plate moves. By contrast, in continental drift, Wegener proposed that the sturdier continents "drifted" by breaking through the oceanic crust, much like ice breakers cut through ice.

5. *Divergent plate boundaries* occur where plates move apart, resulting in upwelling of material from the mantle to create new seafloor. Most divergent boundaries occur along the axis of the oceanic ridge system and are associated with seafloor spreading, which occurs at rates between about 2 and 15 centimeters (1 and 6 inches) per year. New divergent boundaries may form within a continent (for example, the East African rift valleys), where they may fragment a landmass and develop a new ocean basin.

6. *Convergent plate boundaries* occur where plates move together, resulting in the subduction of oceanic lithosphere into the mantle along a deep-ocean trench. Convergence between oceanic and continental blocks results in subduction of the oceanic slab and the formation of a *continental volcanic arc* such as the Andes of South America. Oceanic-oceanic convergence results in an arc-shaped chain of volcanic islands called a *volcanic island arc.* When two plates carrying continental crust converge, both plates are too buoyant to be subducted. The result is a "collision" resulting in the formation of a mountain belt such as the Himalayas.

7. *Transform fault boundaries* occur where plates grind past each other without the production or destruction of lithosphere. Most transform faults join two segments of a mid-ocean ridge. Others connect spreading centers to subduction zones and thus facilitate the transport of oceanic crust created at a ridge crest to its site of destruction, at a deep-ocean trench. Still others, like the San Andreas Fault, cut through continental crust.

8. The theory of plate tectonics is supported by the ages and thickness of *sediments* from the floors of the deep-ocean basins and the existence of island chains that formed over *hot spots* and provide a frame of reference for tracing the direction of plate motion.

9. Two basic models for mantle convection are currently being evaluated. Mechanisms that contribute to this convective flow are slab pull, ridge push, and slab suction. *Slab pull* occurs where cold, dense oceanic lithosphere is subducted and pulls the trailing lithosphere along. *Ridge push* results when gravity sets the elevated slabs astride oceanic ridges in motion. Hot, buoyant *mantle plumes* are considered the upward flowing arms of mantle convection. One model suggests that mantle convection occurs in two layers separated at a depth of 660 kilometers (410 miles). Another model proposes whole-mantle convection that stirs the entire 2900-kilometer-thick (1800-mile-thick) rocky mantle.

Key Terms

asthenosphere (p. 135)
continental drift (p. 130)
continental volcanic arc (p. 143)
convergent boundary (p. 139)
Curie point (p. 149)
deep-ocean trench (p. 140)
divergent boundary (p. 138)
fossil magnetism (p. 149)
fracture zone (p. 144)
hot spot (p. 147)
island arc (p. 143)
lithosphere (p. 135)
magnetic reversal (p. 150)
magnetic time scale (p. 150)
magnetometer (p. 150)
mantle plume (p. 147)
normal polarity (p. 150)
oceanic ridge system (p. 135)
paleomagnetism (p. 149)
Pangaea (p. 130)
partial melting (p. 142)
plate (p. 135)
plate tectonics (p. 135)
reverse polarity (p. 150)
ridge push (p. 152)
rift valley (p. 139)
seafloor spreading (p. 139)
slab pull (p. 152)
slab suction (p. 153)
subduction zone (p. 140)
transform fault boundary (p. 145)
volcanic island arc (p. 143)

Questions for Review

1. Who is credited with developing the continental drift hypothesis?
2. What was probably the first evidence that led some to suspect the continents were once connected?
3. What was Pangaea?
4. List the evidence that Wegener and his supporters gathered to substantiate the continental drift hypothesis.
5. Explain why the discovery of the fossil remains of *Mesosaurus* in both South America and Africa, but nowhere else, supports the continental drift hypothesis.
6. Early in this century, what was the prevailing view of how land animals migrated across vast expanses of ocean?
7. How did Wegener account for the existence of glaciers in the southern landmasses, while at the same time areas in North America, Europe, and Siberia supported lush tropical swamps?
8. On what basis were plate boundaries first established?
9. What are the three major types of plate boundaries? Describe the relative plate motion at each of these boundaries.
10. What is seafloor spreading? Where is active seafloor spreading occurring today?
11. What is a subduction zone? With what type of plate boundary is it associated?
12. Where is lithosphere being consumed? Why must the production and destruction of lithosphere be going on at approximately the same rate?
13. Briefly describe how the Himalaya Mountains formed.

14. Differentiate between transform faults and the other two types of plate boundaries.
15. Some predict that California will sink into the ocean. Is this idea consistent with the theory of plate tectonics?
16. Define the term *paleomagnetism.*
17. How does the continental drift hypothesis account for the apparent wandering of Earth's magnetic poles?
18. What is the age of the oldest sediments recovered by deep-ocean drilling? How do the ages of these sediments compare to the ages of the oldest continental rocks?
19. How do hot spots and the plate tectonics theory account for the fact that the Hawaiian Islands vary in age?
20. Briefly describe the three mechanisms that drive plate motion.
21. With which of the three types of plate boundaries are the following places or features associated: Himalayas, Aleutian Islands, Red Sea, Andes Mountains, San Andreas Fault, Iceland, Japan, Mount St. Helens?

Online Study Guide

The *Foundations of Earth Science* Web site uses the resources and flexibility of the Internet to aid in your study of the topics in this chapter. Written and developed by Earth science instructors, this site will help improve your understanding of Earth science. Visit **http://www.prenhall.com/lutgens** and click on the cover of *Foundations of Earth Science 5e* to find:

- Online review quizzes.
- Critical thinking exercises.
- Links to chapter-specific Web resources.
- Internet-wide key-term searches.

http://www.prenhall.com/lutgens

GEODe: Earth Science

GEODe: Earth Science makes studying more effective by reinforcing key concepts using animation, video, narration, interactive exercises, and practice quizzes. A copy is included with every copy of *Foundations of Earth Science 5e.*

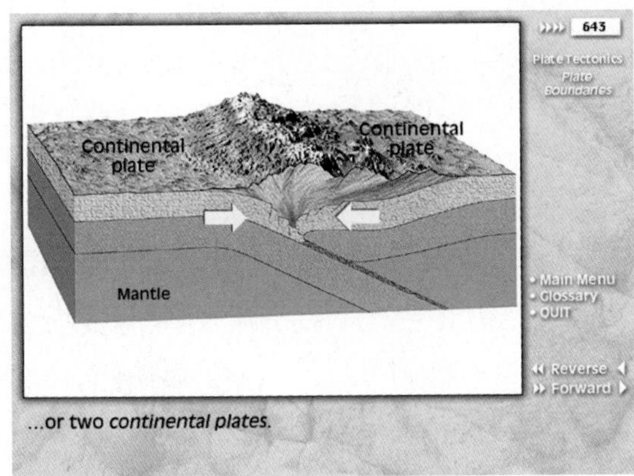

...or two *continental plates.*

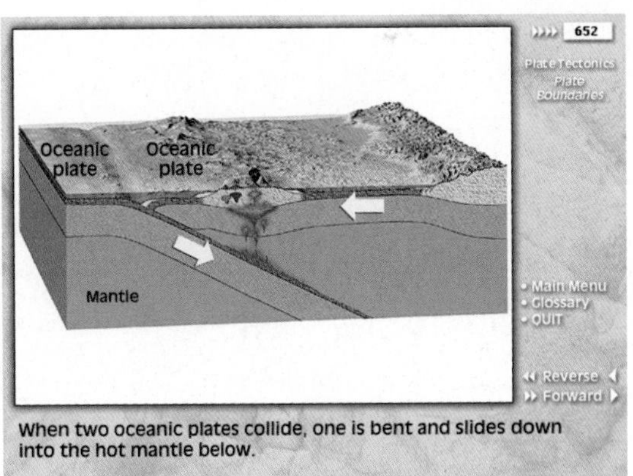

When two oceanic plates collide, one is bent and slides down into the hot mantle below.

CHAPTER 6

Hikers camping in Pakistan's Charakusa Valley with the bold peaks of the Karakoram Range in the background. (Photo by Jimmy Chin/National Geographic/Getty)

Restless Earth: Earthquakes, Geologic Structures, and Mountain Building

FOCUS ON LEARNING

To assist you in learning the important concepts in this chapter, you will find it helpful to focus on the following questions:

1. What is an earthquake?
2. What are the types of earthquake waves?
3. How is the epicenter of an earthquake determined?
4. Where are Earth's principal earthquake zones?
5. How is earthquake magnitude expressed by the Richter scale?
6. What key factors determine the destructiveness of an earthquake?
7. What are the four major zones of Earth's interior?
8. How do continental crust and oceanic crust differ?
9. What are the two basic types of rock deformation?
10. How is mountain building associated with the various kinds of convergent plate boundaries?

On October 17, 1989, at 4:04 P.M. Pacific daylight time, millions of television viewers around the world were settling in to watch the third game of the World Series. Instead, they saw their TV sets go black as tremors hit San Francisco's Candlestick Park. Although the earthquake was centered in a remote section of the Santa Cruz Mountains, 100 kilometers (62 miles) to the south, major damage occurred in the Marina District of San Francisco.

The most tragic result of the violent shaking was the collapse of some double-decked sections of Interstate 880, also known as the Nimitz Freeway. The ground motions caused the upper deck to sway, shattering the concrete support columns along a mile-long section of the freeway. The upper deck then collapsed onto the lower roadway, flattening cars as if they were aluminum cans. This earthquake, named the Loma Prieta quake for its point of origin, claimed 67 lives.

In mid-January 1994, less than five years after the Loma Prieta earthquake devastated portions of the San Francisco Bay area, a major earthquake struck the Northridge area of Los Angeles. Although not the fabled "Big One," this moderate 6.7 magnitude earthquake left 57 dead, more than 5000 injured, and tens of thousands of households without water and electricity. The damage exceeded $40 billion and was attributed to an apparently unknown fault that ruptured 18 kilometers (11 miles) beneath Northridge.

The Northridge earthquake, which occurred northwest of downtown Los Angeles, began at 4:31 A.M. and lasted roughly 40 seconds. During this brief period, the quake terrorized the entire Los Angeles area. The three-story Northridge Meadows apartment complex, where 16 people died, was devastated. Most of the deaths resulted when sections of the upper floors collapsed onto the first-floor units. In addition, nearly 300 schools were seriously damaged and a dozen major roadways buckled. Among these were two of California's major arteries—the Golden State Freeway (Interstate 5), where an overpass collapsed, completely blocking the roadway, and the Santa Monica Freeway. Fortunately, these roadways had practically no traffic at this early morning hour.

In nearby Granada Hills, broken gas lines were set ablaze while the streets flooded from broken water mains. In Los Angeles County alone, more than 100 calls were made to the fire department. Seventy homes burned in the Sylmar area. A 64-car freight train derailed, including some cars carrying hazardous cargo. Despite the huge economic losses, it is remarkable that the destruction was not greater. Unques-

Figure 6.1 Destruction caused by a major earthquake that struck northwestern Turkey on August 17, 1999. More than 17,000 people perished. (Photo by Yann Arthus-Bertrand/Peter Arnold, Inc.)

tionably, the upgrading of structures to meet the requirements of building codes developed for this earthquake-prone area helped minimize what could have been a much greater human tragedy.

What Is an Earthquake?

Forces Within
▼ Earthquakes

An **earthquake** is the vibration of Earth produced by a rapid release of energy (Figure 6.1). This energy radiates in all directions from its source, the **focus,** in the form of waves. The waves are like those produced when a stone is dropped into a calm pond (Figure 6.2). Just as the impact of the stone sets water waves in motion, an earthquake generates seismic waves that radiate throughout Earth. Even though the energy dissipates rapidly with increasing distance from the focus, sensitive instruments worldwide record the event.

More than 30,000 earthquakes that are strong enough to be felt occur worldwide annually. Fortunately, most are minor tremors and do very little damage. Only about 75 significant earthquakes take place each year, and many of these occur in remote regions. Occasionally, however, a large earthquake occurs near a major population center. Under these conditions, an earthquake is among the most destructive natural forces on Earth.

Did You Know?

Literally thousands of earthquakes occur daily! Fortunately, the majority of them are too small to be felt by people, and the majority of larger ones occur in remote regions. Their existence is known only because of sensitive seismographs.

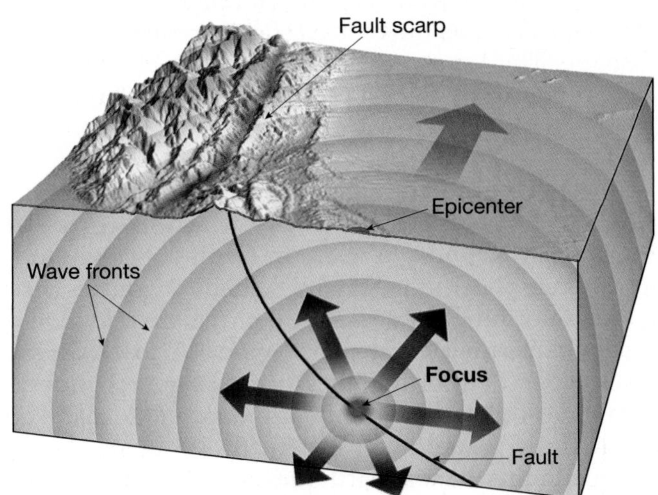

Figure 6.2 The focus of an earthquake is located at depth. The surface location directly above it is called the epicenter.

The shaking of the ground, coupled with the liquefaction of some soils, wreaks havoc on buildings and other structures. In addition, when a quake occurs in a populated area, power and gas lines are often ruptured, causing numerous fires. In the 1906 San Francisco earthquake, much of the damage was caused by fires, which quickly became uncontrollable when broken water mains left firefighters with only trickles of water (Figure 6.3).

Earthquakes and Faults

The tremendous energy released by atomic explosions or by volcanic eruptions can produce an earthquake, but these events are relatively weak and infrequent. What mechanism produces a destructive earthquake? Ample evidence exists that Earth is not a static planet. We know that Earth's crust has been uplifted at times, because we have found numerous ancient wave-cut benches many meters above the level of the highest tides. Other regions exhibit evidence of extensive subsidence. In addition to these vertical displacements, offsets in fence lines, roads, and other structures indicate that horizontal movement is common (Figure 6.4). These movements are usually associated with large fractures in Earth's crust called *faults.*

Typically, earthquakes occur along preexisting faults that formed in the distant past along zones of weakness in Earth's crust. Some are very large and can generate major earthquakes. One example is the San Andreas Fault, which is a transform fault boundary that separates two great sections of Earth's lithosphere: the North American plate and the Pacific plate. Other faults are small and produce only minor earthquakes.

Most faults are not perfectly straight or continuous; instead, they consist of numerous branches and smaller fractures that display kinks and offsets. Such a pattern is displayed in Figure 6.30 (p. 182), which shows that the San Andreas Fault is actually a system that consists of several large faults and innumerable small fractures (not shown).

It is also clear that most faults are locked, except for brief, abrupt movements that accompany an earthquake rupture. The primary reason faults are locked is that the confining pressure exerted by the overlying crust is enormous and essentially squeezes the fractures in the crust shut. Nevertheless, even faults that have been inactive for thousands of years

Did You Know?

Humans have inadvertently triggered earthquakes. In 1962, Denver began experiencing frequent tremors. The earthquakes were located near an army waste-disposal well used to inject waste into the ground. Investigators concluded that the pressurized fluids made their way along a buried fault surface, which reduced friction and triggered fault slippage and earthquakes. Sure enough, when the pumping halted, so did the tremors.

Figure 6.3 San Francisco in flames after the 1906 earthquake. (Reproduced from the collection of the Library of Congress) Inset photo shows fire triggered when a gas line ruptured during the Northridge earthquake in southern California in 1994. (AFP/Getty Images)

can rupture again if the stresses acting on the region increase sufficiently.

Discovering the Cause of Earthquakes

The actual mechanism of earthquake generation eluded geologists until H. F. Reid of Johns Hopkins University conducted a study following the great 1906 San Francisco earthquake. The earthquake was accompanied by horizontal surface displacements of several meters along the northern portion of the San Andreas Fault. Field investigations determined that during this single earthquake, the Pacific plate lurched as much as 4.7 meters (15 feet) northward past the adjacent North American plate.

Figure 6.4 Slippage along a fault produced an offset in this orange grove located east of Calexico, California. (Photo by John S. Shelton)

The mechanism for earthquake formation that Reid deduced from this information is illustrated in Figure 6.5. Part A of the figure shows an existing fault, or break in the rock. In part B, tectonic forces ever so slowly deform the crustal rocks on both sides of the fault, as demonstrated by the bent features. Under these conditions, rocks are bending and storing elastic energy, much like a wooden stick does if bent. Eventually, the frictional resistance holding the rocks in place is overcome. As slippage occurs at the weakest point (the focus), displacement will exert stress farther along the fault, where additional slippage will occur releasing the built-up strain (Figure 6.5C). This slippage allows the deformed rock to "snap back." The vibrations we know as an earthquake occur as the rock elastically returns to its original shape. The "springing back" of the rock was termed **elastic rebound** by Reid, because the rock behaves elastically, much like a stretched rubber band does when it is released.

Most of the motion along faults can be satisfactorily explained by the plate tectonics theory, which states that large slabs of Earth's lithosphere are in continual slow motion. These mobile plates interact with neighboring plates, straining and deforming the rocks at their margins. In fact, it is along faults associated with plate boundaries that most earthquakes occur. Furthermore, earthquakes are repetitive: As soon as one is over, the continuous motion of the plates adds strain to the rocks until they eventually fail again.

In summary, most earthquakes are produced by the rapid release of elastic energy stored in rock that has been subjected to great stress. Once the strength of the rock is exceeded, it suddenly ruptures, causing the vibrations of an

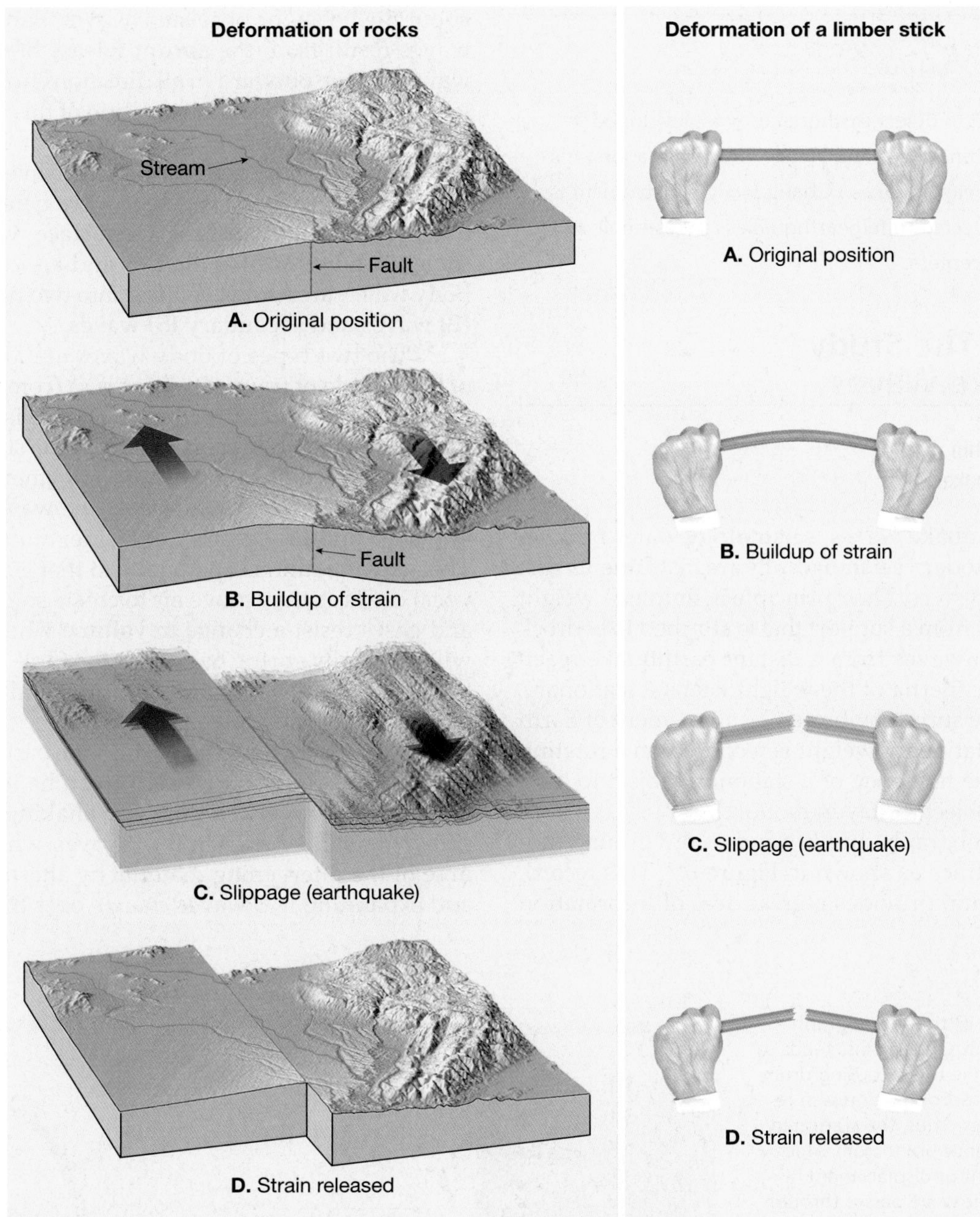

Figure 6.5 Elastic rebound. As rock is deformed, it bends, storing elastic energy. Once the rock is strained beyond its breaking point, it ruptures, releasing the stored-up energy in the form of earthquake waves.

earthquake. Earthquakes most often occur along existing faults whenever the frictional forces on the fault surfaces are overcome.

Foreshocks and Aftershocks

The intense vibrations of the 1906 San Francisco earthquake lasted about 40 seconds. Although most of the displacement along the fault occurred in this rather short period, additional movements along this and other nearby faults lasted for several days following the main quake. The adjustments that follow a major earthquake often generate smaller earthquakes called **aftershocks.** Although these aftershocks are usually much weaker than the main earthquake, they can sometimes destroy already badly weakened structures. This occurred during a 1988 earthquake in Armenia. A large aftershock of magnitude 5.8 collapsed many structures that had been weakened by the main tremor.

In addition, small earthquakes called **foreshocks** often precede a major earthquake by days or, in some cases, by as much as several years. Monitoring of these foreshocks has been used as a means of predicting forthcoming major earthquakes, with mixed success. We will consider the topic of earthquake prediction in a later section of this chapter.

Did You Know?

The first instrument to detect earthquakes was developed in about A.D. 132 in China by Chang Heng. (China has a long history of devastating earthquakes.) Chang Heng's instrument is thought to have detected unfelt earthquakes and estimated the direction to the epicenters.

Seismology: The Study of Earthquake Waves

Forces Within
▼ Earthquakes

The study of earthquake waves, **seismology,** dates back almost 2000 years. Modern **seismographs** are instruments that record earthquake waves. Their principle is simple. A weight is freely suspended from a support that is attached to bedrock (Figure 6.6). When waves from a distant earthquake reach the instrument, the inertia of the weight keeps it stationary, while Earth and the support vibrate. The movement of Earth in relation to the stationary weight is recorded on a rotating drum. (*Inertia* is the tendency of a stationary object to hold still, or a moving object to stay in motion.)

Modern seismographs amplify and record ground motion, producing a trace as shown in Figure 6.7. This record, called a **seismogram,** provides a great deal of information about the behavior of seismic waves. Simply stated, seismic waves result from the abrupt release of stored energy. The waves radiate outward in all directions from the focus, as you saw in Figure 6.2. The transmission of this energy can be compared to the shaking of gelatin in a bowl. Seismograms reveal that two main types of seismic waves are generated by the slippage of a rock mass. One wave type travels around the outer layer of Earth. These are **surface waves.** Other waves travel through Earth's interior and are called **body waves.** Body waves are further divided into two types called **primary (P) waves** and **secondary (S) waves.**

The two types of body waves are divided on the basis of their mode of travel. P waves *push* (compress) and *pull* (expand) rocks in the direction the wave is traveling (Figure 6.8A). Imagine holding someone by the shoulders and shaking him or her. This push-pull movement is how P waves move through Earth materials. The wave energy does not move up and down or side to side, but through the rocks. This wave motion is analogous to that generated by human vocal cords as they move air to create sound. Solids, liquids, and gases resist a change in volume when compressed and will elastically spring back once the force is removed. Therefore, P waves, which are compressional waves, can travel through all of these materials.

Conversely, S waves "shake" the particles at right angles to their direction of travel. This can be illustrated by tying one end of a rope to a post and shaking the other end, as shown in Figure 6.8C. Unlike P waves, which change the volume of the intervening material by alternately compressing and expanding it, S waves change only the shape of the ma-

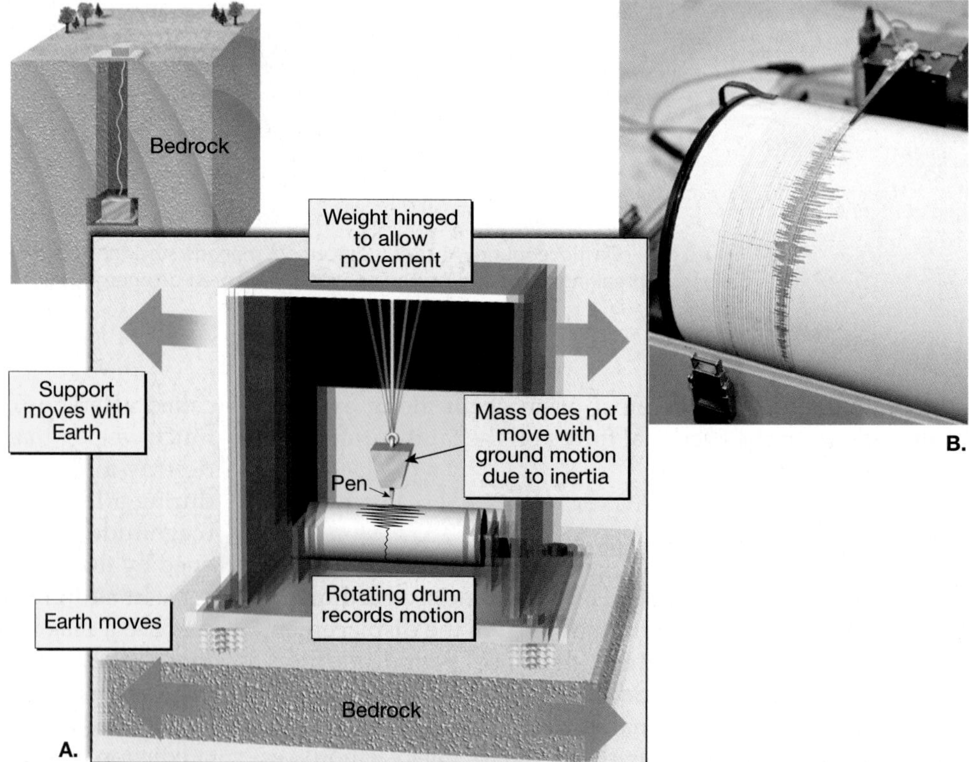

Figure 6.6 Principle of the seismograph. **A.** The inertia of the suspended mass tends to keep it motionless, while the recording drum, which is anchored to bedrock, vibrates in response to seismic waves. Thus, the stationary mass provides a reference point from which to measure the amount of displacement occurring as the seismic wave passes through the ground. **B.** Seismograph recording earthquake tremors. (Photo courtesy of Zephyr/Photo Researchers, Inc.)

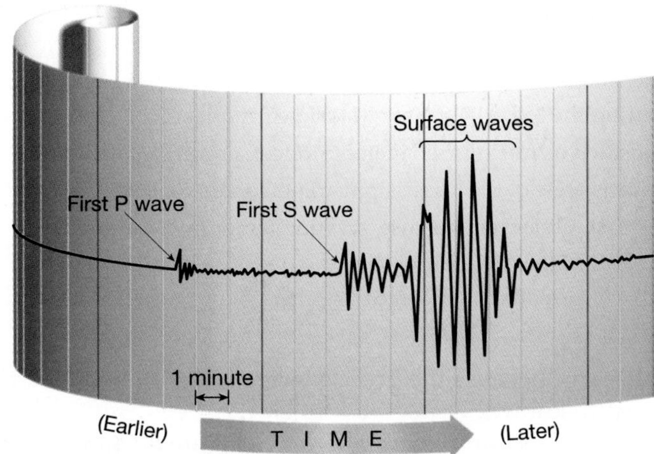

Figure 6.7 Typical seismic record. Note the time interval between the arrival of the first P wave and the arrival of the first S wave.

Did You Know?

Some of the most interesting uses of seismographs involve the reconstruction of unfortunate events, such as airline crashes, pipeline explosions, and mining disasters. For example, a seismologist helped with the investigation of Pan Am Flight 103, which was brought down in 1988 over Lockerbie, Scotland, by a terrorist's bomb. Nearby seismographs recorded six separate impacts, indicating that the plane had broken into that many large pieces before it crashed.

terial that transmits them. Because fluids (gases and liquids) do not resist changes in shape, they cannot transmit S waves.

The motion of surface waves is somewhat more complex. As surface waves travel along the ground, they cause the ground and anything resting on it to move, much like ocean swells toss a ship. In addition to their up-and-down motion, surface waves have a side-to-side motion similar to an S wave oriented in a horizontal plane. This latter motion is particularly damaging to the foundations of structures.

By observing a "typical" seismic record, as shown in Figure 6.7, you can see a major difference between these seismic waves: P waves arrive at the recording station *before* S waves, which arrive before surface waves. This is a consequence of

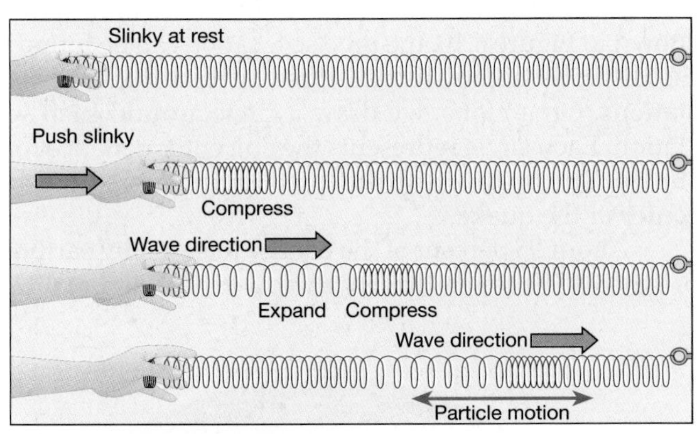

A. P waves generated using a slinky

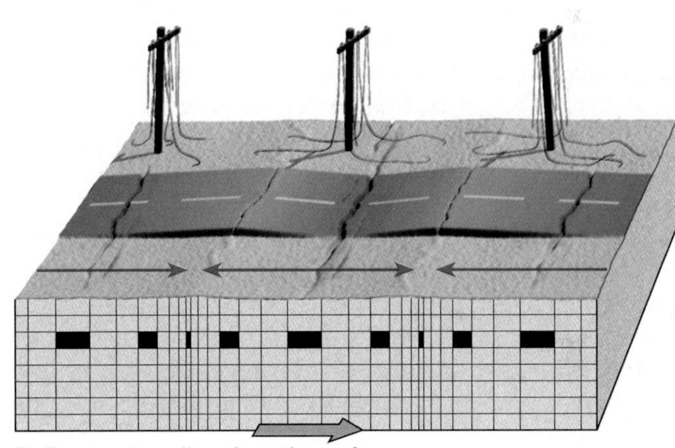

B. P waves traveling along the surface

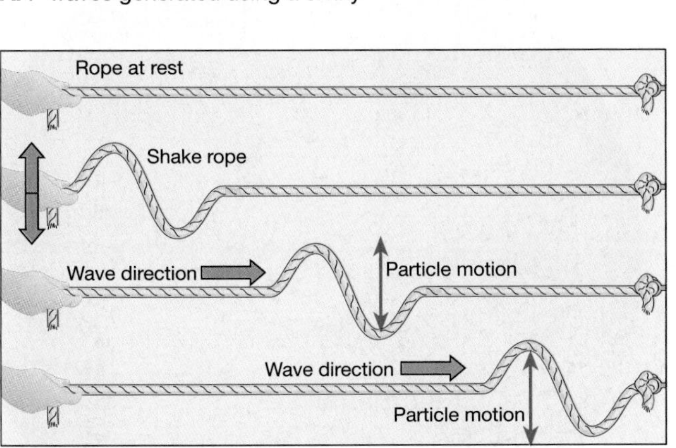

C. S waves generated using a rope

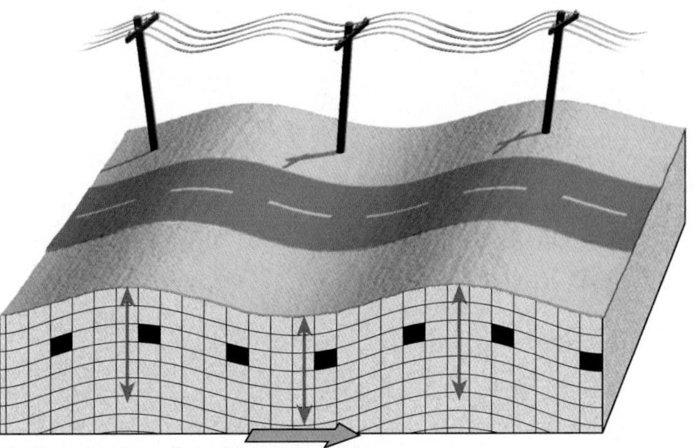

D. S waves traveling along the surface

Figure 6.8 Types of seismic waves and their characteristic motion. (Note that during a strong earthquake, ground shaking consists of a combination of various kinds of seismic waves.) **A.** As illustrated by a slinky, P waves are compressional waves that alternately compress and expand the material through which they pass. **B.** The back-and-forth motion produced as compressional waves travel along the surface can cause the ground to buckle and fracture and may cause power lines to break. **C.** S waves cause material to oscillate at right angles to the direction of wave motion. **D.** Because S waves can travel in any plane, they produce up-and-down and sideways shaking of the ground.

their speeds. For purposes of illustration, the velocity of P waves through granite within the crust is about 6 kilometers (4 miles) per second. Under the same conditions, S waves travel only 3.5 kilometers (2 miles) per second. Differences in density and elastic properties of the transmitting material greatly influence the velocities of these waves. Nevertheless, in any solid material, P waves travel about 1.7 times faster than S waves, and surface waves are about 10 percent slower than S waves.

As you will see, seismic waves allow us to determine the location and magnitude of earthquakes. In addition, seismic waves provide us with a tool for probing Earth's interior.

Did You Know?

During the 1811–1812 New Madrid earthquake, the ground subsided as much as 4.5 meters (15 feet) and created Lake St. Francis west of the Mississippi and enlarged Reelfoot Lake to the east. Other regions rose, creating temporary waterfalls in the bed of the Mississippi River.

Locating an Earthquake

Forces Within
▼ Earthquakes

Recall that the *focus* is the place within Earth where the earthquake waves originate. The **epicenter** is the location *on the surface directly above the focus* (see Figure 6.2, p. 161).

The difference in velocities of P and S waves provides a method for locating the epicenter. The principle used is analogous to a race between two autos, one faster than the other. The P wave always wins the race, arriving ahead of the S wave. But the greater the distance of the race, the greater will be the *difference* in the arrival times at the finish line (the seismic station). Therefore, the greater the interval measured on a seismogram between the arrival of the first P wave and the first S wave, the greater is the distance to the earthquake source.

The system for locating earthquake epicenters was developed by using seismograms from earthquakes whose epicenters could be easily pinpointed from physical evidence. From the seismograms, travel-time graphs were constructed (Figure 6.9). The first travel-time graphs were greatly improved when seismograms became available from nuclear explosions, because the precise location and time of detonation were known.

Using the sample seismogram in Figure 6.7 and the travel-time curve in Figure 6.9, we can determine the distance separating the recording station from the earthquake in two steps: (1) using the seismogram, determine the time interval between the arrival of the first P wave and the first S wave, and (2) using the travel-time graph, find the P-S interval on the vertical axis and use that information to determine the distance to the epicenter on the horizontal axis. From this information, we can determine that this earthquake occurred 3400 kilometers (2100 miles) from the recording instrument.

Now we know the distance, but which way? Its location could be in any direction from the seismic station. As shown in Figure 6.10, the precise location can be found when the distance is known from three or more different seismic stations. On a globe, we draw a circle around each seismic station. Each circle represents the epicenter distance for each station. The point where the three circles intersect is the epicenter of the quake.

About 95 percent of the energy released by earthquakes originates in a few relatively narrow zones (Figure 6.11). The

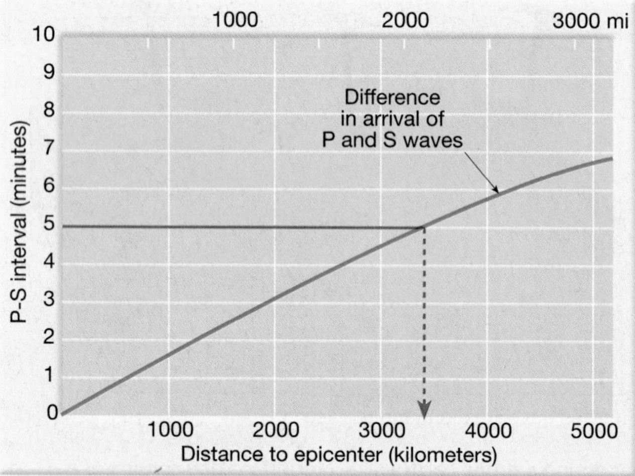

Figure 6.9 A travel-time graph is used to determine the distance to the epicenter. The difference in arrival times of the first P and S waves in the example is 5 minutes. Thus, the epicenter is roughly 3400 kilometers (2100 miles) away.

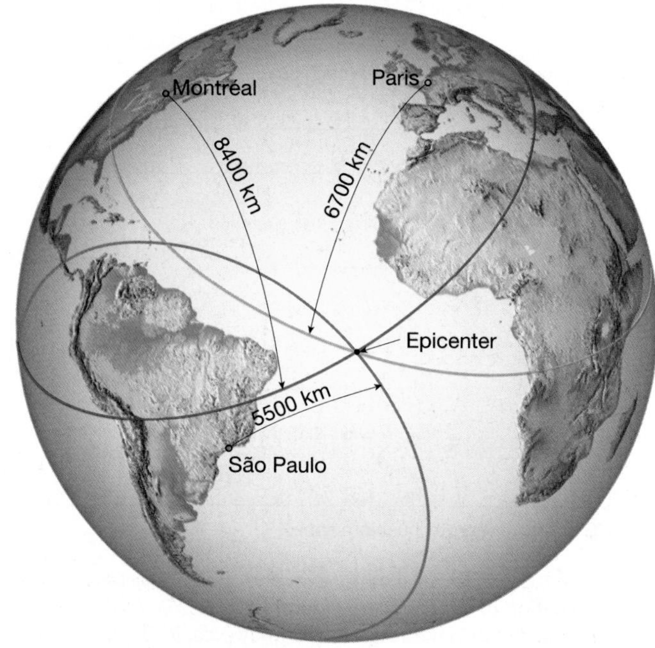

Figure 6.10 An earthquake epicenter is located using the distances obtained from three seismic stations.

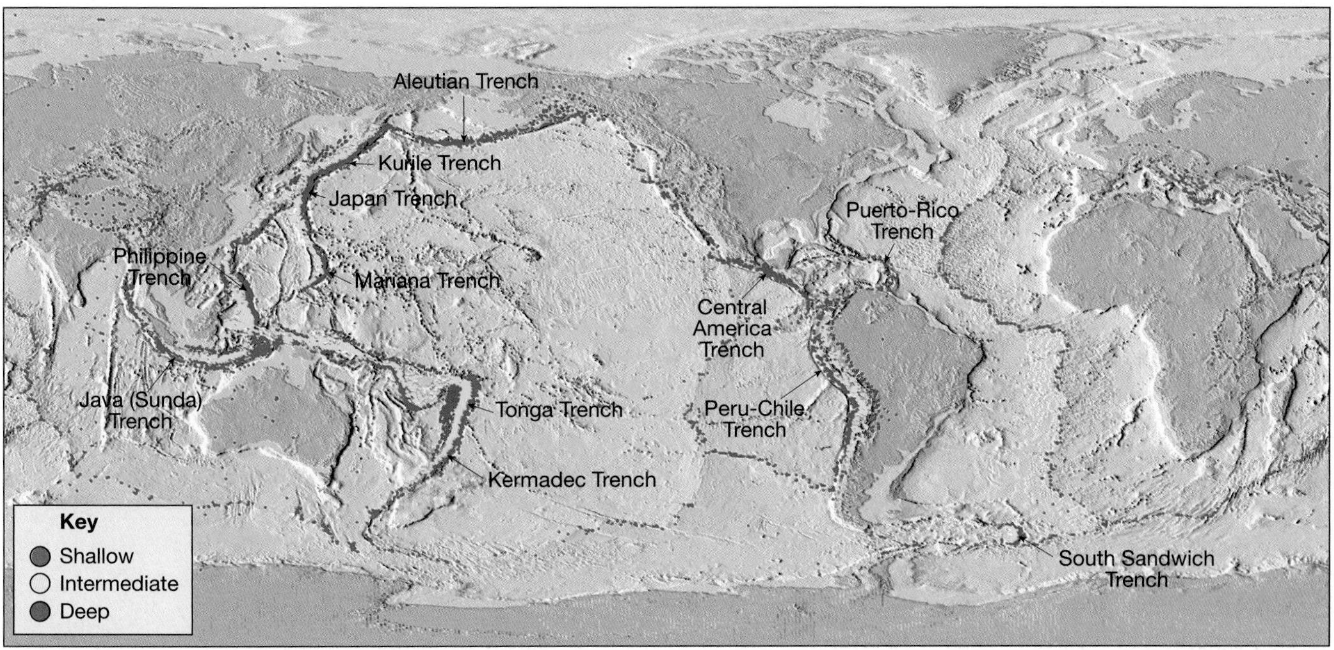

Figure 6.11 Distribution of the 14229 earthquakes with magnitudes equal to or greater than 5.0 for a 10-year period. (Data from National Geophysical Data Center/NOAA)

greatest energy is released along a path around the outer edge of the Pacific Ocean known as the *circum-Pacific belt.* Included in this zone are regions of great seismic activity, such as Japan, the Philippines, Chile, and numerous volcanic island chains, as exemplified by Alaska's Aleutian Islands.

Figure 6.11 reveals another continuous belt that extends for thousands of kilometers through the world's oceans. This zone coincides with the oceanic ridge system, an area of frequent but low-intensity seismic activity. By comparing this figure with Figure 5.9 (pp. 136–137), you can see a close correlation between the location of earthquake epicenters and plate boundaries.

Did You Know?

The greatest loss of life attributed to an earthquake occurred in the Hwang River of northern China. The Chinese made their homes here by carving numerous structures into a powdery, windblown material called *loess.* Early on the morning of January 23, 1556, an earthquake struck the region, collapsing the cavelike structures and killing about 830,000 people. Sadly, in 1920, a similar event collapsed loess structures in much the same way, taking about 200,000 lives.

Measuring the Size of Earthquakes

Seismologists have employed a variety of methods to obtain two fundamentally different measures that describe the size of an earthquake—intensity and magnitude. The first of these to be used was **intensity**—a measure of the degree of earthquake shaking at a given locale based on the amount of damage. With the development of seismographs, it became clear that a quantitative measure of an earthquake based on seismic records rather than uncertain personal estimates of damage was desirable. The measurement that was developed, called **magnitude,** relies on calculations that use data provided by seismic records (and other techniques) to estimate the amount of energy released at the source of the earthquake.

Intensity Scales

Until a little more than a century ago, historical records provided the only accounts of the severity of earthquake shaking and destruction. Using these descriptions—which were compiled without any established standards for reporting—made accurate comparisons of earthquake sizes difficult, at best.

In order to standardize the study of earthquake severity, scientists developed various intensity scales that considered damage done to buildings, as well as individual descriptions of the event, and secondary effects such as landslides and the extent of ground rupture. By 1902, Guiseppe Mercalli had developed a relatively reliable intensity scale, which in a modified form is still used today. The **Modified Mercalli Intensity Scale** shown in Table 6.1 was developed using California buildings as its standard, but it is appropriate for use throughout most of the United States and Canada to estimate the strength of an earthquake. For example, if some well-built wood structures and most masonry buildings are destroyed by an earthquake, a region would be assigned an intensity of X on the Mercalli scale.

Despite their usefulness in providing seismologists with a tool to compare earthquake severity, particularly in regions where there are no seismographs, intensity scales have

Table 6.1 Modified Mercalli Intensity Scale

I	Not felt except by a very few under especially favorable circumstances.
II	Felt only by a few persons at rest, especially on upper floors of buildings.
III	Felt quite noticeably indoors, especially on upper floors of buildings, but many people do not recognize it as an earthquake.
IV	During the day felt indoors by many, outdoors by few. Sensation like heavy truck striking building.
V	Felt by nearly everyone, many awakened. Disturbances of trees, poles, and other tall objects sometimes noticed.
VI	Felt by all; many frightened and run outdoors. Some heavy furniture moved; few instances of fallen plaster or damaged chimneys. Damage slight.
VII	Everybody runs outdoors. Damage negligible in buildings of good design and construction; slight-to-moderate in well-built ordinary structures; considerable in poorly built or badly designed structures.
VIII	Damage slight in specially designed structures; considerable in ordinary substantial buildings with partial collapse; great in poorly built structures. (Fall of chimneys, factory stacks, columns, monuments, walls.)
IX	Damage considerable in specially designed structures. Buildings shifted off foundations. Ground cracked conspicuously.
X	Some well-built wooden structures destroyed. Most masonry and frame structures destroyed. Ground badly cracked.
XI	Few, if any, (masonry) structures remain standing. Bridges destroyed. Broad fissures in ground.
XII	Damage total. Waves seen on ground surfaces. Objects thrown upward into air.

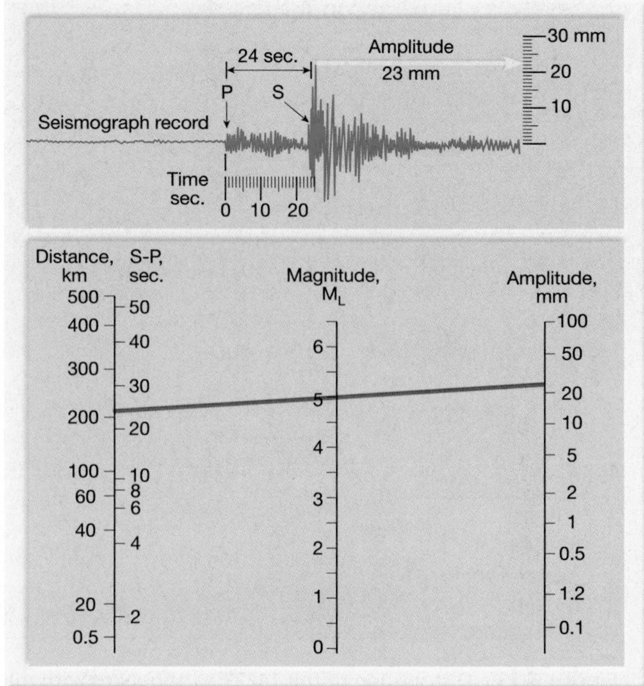

Figure 6.12 Illustration showing how the Richter magnitude of an earthquake can be determined graphically using a seismograph record from a Wood-Anderson instrument. First, measure the height (amplitude) of the largest wave on the seismogram (23mm) and then the distance to the focus using the time interval between S and P waves (24 seconds). Next, draw a line between the distance scale (left) and the wave amplitude scale (right). By doing this, you should obtain the Richter magnitude (M_L) of 5. (Data from California Institute of Technology)

severe drawbacks. In particular, intensity scales are based on effects (largely destructive) of earthquakes that depend not only on the severity of ground shaking but also on factors such as population density, building design, and the nature of surface materials.

Magnitude Scales

In order to compare earthquakes across the globe, a measure was needed that did not rely on parameters that vary considerably from one part of the world to another, such as construction practices. As a consequence, a number of magnitude scales were developed.

Richter Magnitude In 1935, Charles Richter of the California Institute of Technology developed the first magnitude scale using seismic records to estimate the relative sizes of earthquakes. As shown in Figure 6.12 (top), the **Richter scale** is based on the amplitude of the largest seismic wave (P, S, or surface wave) recorded on a seismogram. Because seismic waves weaken as the distance between the earthquake focus and the seismograph increases (in a manner similar to light), Richter developed a method that accounted for the decrease in wave amplitude with increased distance. Theoretically, as long as the same, or equivalent, instruments were used, monitoring stations at various locations would obtain the same Richter magnitude for every recorded earthquake. (Richter selected the Wood-Anderson seismograph as the standard recording device.)

Although the Richter scale has no upper limit, the largest magnitude recorded on a Wood-Anderson seismograph was 8.9. These great shocks release approximately 10^{26} ergs of energy—roughly equivalent to the detonation of 1 billion tons of TNT. Conversely, earthquakes with a Richter magnitude of less than 2.0 are not felt by humans. With the development of more sensitive instruments, tremors of a magnitude of minus 2 were recorded. Table 6.2 shows how Richter magnitudes and their effects are related.

Did You Know?

It is a commonly held belief that moderate earthquakes decrease the chances of a major earthquake in the same region, but this is not the case. When you compare the amount of energy released by earthquakes of different magnitudes, it turns out that thousands of moderate tremors would be needed to release the huge amount of energy released during one "great" earthquake.

Table 6.2 Earthquake magnitude and expected world incidence

Richter Magnitudes	Effects Near Epicenter	Estimated Number per Year
<2.0	Generally not felt, but recorded.	600,000
2.0–2.9	Potentially perceptible.	300,000
3.0–3.9	Felt by some.	49,000
4.0–4.9	Felt by most.	6200
5.0–5.9	Damaging shocks.	800
6.0–6.9	Destructive in populous regions.	266
7.0–7.9	Major earthquakes. Inflict serious damage.	18
≥8.0	Great earthquakes. Destroy communities near epicenter.	1–2

Source: Earthquake Information Bulletin and others.

Earthquakes vary enormously in strength, and great earthquakes produce wave amplitudes that are thousands of times larger than those generated by weak tremors. To accommodate this wide variation, Richter used a *logarithmic scale* to express magnitude, where a *tenfold* increase in wave amplitude corresponds to an increase of 1 on the magnitude scale. Thus, the amount of ground shaking for a 5-magnitude earthquake is 10 times greater than that produced by an earthquake having a Richter magnitude of 4.

In addition, each unit of Richter magnitude equates to roughly a *32-fold energy increase.* Thus, an earthquake with a magnitude of 6.5 releases 32 times more energy than one with a magnitude of 5.5, and roughly 1000 times more energy than a 4.5-magnitude quake. A major earthquake with a magnitude of 8.5 releases millions of times more energy than the smallest earthquakes felt by humans.

Richter's original goal was modest in that he attempted to rank only the earthquakes of southern California (shallow-focus earthquakes) into groups of large, medium, and small magnitude. Hence, Richter magnitude was designed to study nearby (or local) earthquakes and is denoted by the symbol (M_L)—where *M* is for *magnitude* and *L* is for *local.*

The convenience of describing the size of an earthquake by a single number that could be calculated quickly from seismograms makes the Richter scale a powerful tool. Further, unlike intensity scales that could be applied only to populated areas of the globe, Richter magnitudes could be assigned to earthquakes in more remote regions and even to events that occurred in the ocean basins. Richter's method was adapted to a number of different seismographs located throughout the world. In time, seismologists modified Richter's work and developed new magnitude scales.

Moment Magnitude Seismologists have recently been employing a more precise measure called **moment magnitude** (M_W), which can be calculated using several techniques. In one method, the moment magnitude is calculated from field studies using a combination of factors that include the average amount of displacement along the fault, the area of the rupture surface, and the shear strength of the faulted rock—a measure of how much energy a rock can store before it suddenly slips and releases this energy in the form of an earthquake (and heat).

The moment magnitude can also be readily calculated from seismograms by examining very long period seismic waves. The values obtained have been calibrated so that small- and moderate-sized earthquakes have moment magnitudes that are roughly equivalent to Richter magnitudes. However, moment magnitudes are much better for describing very large earthquakes. For example, on the moment magnitude scale, the 1906 San Francisco earthquake, which had a Richter magnitude of 8.3, would be demoted to 7.9 on the moment magnitude scale, whereas the 1964 Alaskan earthquake with an 8.3 Richter magnitude would be increased to 9.2. The strongest earthquake on record is the 1960 Chilean earthquake with a moment magnitude of 9.5 (Table 6.3).

Moment magnitude has gained wide acceptance among seismologists and engineers because (1) it is the only magnitude scale that estimates adequately the size of very large earthquakes; (2) it is a measure that can be derived mathematically from the size of the rupture surface and the amount of displacement and better reflects the total energy released during an earthquake; and (3) it can be verified by two independent methods—field studies that are based on measurements of fault displacement and seismographic methods using long-period waves.

Destruction from Earthquakes

The most violent earthquake to jar North America this century—the Good Friday Alaskan earthquake—occurred in 1964. Felt throughout that state, the earthquake had a moment magnitude (M_W) of 9.2 and reportedly lasted three to four minutes. This brief event left 131 people dead, thousands homeless, and the economy of the state badly disrupted (Figure 6.13). Had the schools and business districts been open on this holiday, the toll surely would have been higher. Within 24 hours of the initial shock, 28 aftershocks were recorded, 10 of which exceeded a magnitude of 6 on the Richter scale.

Did You Know?

In Japanese folklore, earthquakes were caused by a giant catfish that lived beneath the ground. Whenever the catfish flopped about, the ground shook. Meanwhile, people in Russia's Kamchatka Peninsula believed that earthquakes occurred when a mighty dog, Kozei, shook freshly fallen snow from its coat.

Table 6.3 Some notable earthquakes

Year	Location	Deaths (est.)	Magnitude	Comments
1556	Shensi, China	830,000	Unknown	Possibly the greatest natural disaster.
1755	Lisbon, Portugal	70,000	Unknown	Tsunami damage extensive.
*1811–1812	New Madrid, Missouri	Few	Unknown	Three major earthquakes.
*1886	Charleston, South Carolina	60	Unknown	Greatest historical earthquake in the eastern United States.
*1906	San Francisco, California	1500	8.3	Fires caused extensive damage.
1908	Messina, Italy	120,000	Unknown	
1923	Tokyo, Japan	143,000	7.9	Fire caused extensive destruction.
1960	Southern Chile	5700	9.5	The largest-magnitude earthquake ever recorded.
*1964	Alaska	131	9.2	Greatest North American earthquake.
1970	Peru	66,000	7.8	Great rockslide.
*1971	San Fernando, California	65	6.5	Damage exceeded $1 billion.
1975	Liaoning Province, China	1328	7.5	First major earthquake to be predicted.
1976	Tangshan, China	240,000	7.6	Not predicted.
1985	Mexico City	9500	8.1	Major damage occurred 400 km from epicenter.
1988	Armenia	25,000	6.9	Poor construction practices.
*1989	San Francisco Bay area	62	7.1	Damages exceeded $6 billion.
1990	Iran	50,000	7.3	Landslides and poor construction practices caused great damage.
1993	Latur, India	10,000	6.4	Located in stable continental interior.
*1994	Northridge, California	51	6.7	Damages in excess of $15 billion.
1995	Kobe, Japan	5472	6.9	Damages estimated to exceed $100 billion.
1999	Izmit, Turkey	17,127	7.4	Nearly 44,000 injured and more than 250,000 displaced.
1999	Chi-Chi, Taiwan	2300	7.6	Severe destruction; 8700 injuries.
2001	Bhuj, India	25,000+	7.9	Millions homeless.
2003	Bam, Iran	41,000+	6.6	Ancient city with poor construction.
2004	Indian Ocean	230,000	9.0	Devastating tsunami damage.
2005	Pakistan/Kashmir	83,000	7.6	Many landslides, 4 million homeless.

*U.S. earthquakes.
Source: U.S. National Oceanic and Atmospheric Administration

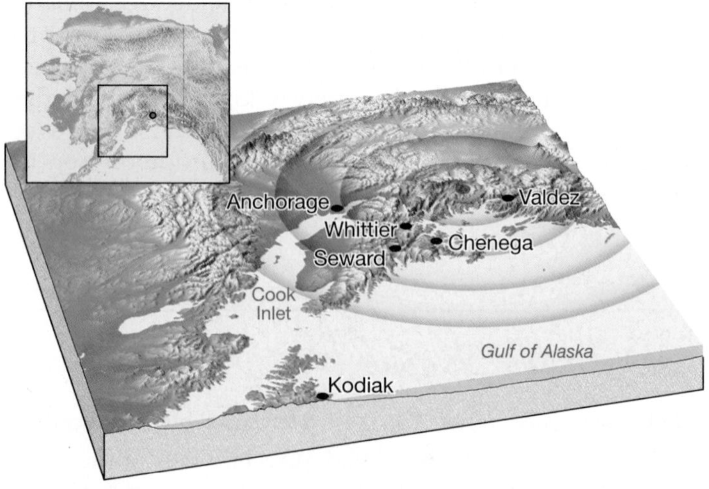

Figure 6.13 Region most affected by the Good Friday earthquake of 1964. Note the location of the epicenter (red dot). (After U.S. Geological Survey)

Structural Damage from Seismic Vibrations

The 1964 Alaskan earthquake provided geologists with new insights into the role of *ground shaking* as a destructive force. As the energy released by an earthquake travels along Earth's surface, it causes the ground to vibrate in a complex manner by moving up and down as well as from side to side. The amount of structural damage attributable to the vibrations depends on several factors, including (1) the intensity and (2) the duration of vibrations, (3) the nature of the material upon which the structure rests, and (4) the design of the structure.

All of the multistory structures in Anchorage were damaged by the vibrations. The more flexible wood-frame residential buildings fared best. A striking example of how construction variations affect earthquake damage is shown in Figure 6.14. You can see that the steel-frame building on the left withstood the vibrations, whereas the poorly designed JCPenney building was badly damaged. Engineers have learned that unreinforced masonary buildings are the most serious safety threats in earthquakes.

Figure 6.14 Damage to the five-story JCPenney building, Anchorage, Alaska. Very little structural damage was incurred by the adjacent building. (Courtesy of NOAA)

Most large structures in Anchorage were damaged, even though they were built to the earthquake provisions of the Uniform Building Code. Perhaps some of that destruction can be attributed to the unusually long duration of this earthquake. Most earthquakes consist of tremors that last less than one minute. The 1994 Northridge earthquake was felt for about 40 seconds, and the strong vibrations of the 1989 Loma Prieta earthquake lasted less than 15 seconds. But the Alaska quake reverberated for three or four minutes.

Amplification of Seismic Waves Although the region within 20 to 50 kilometers (12 to 31 miles) of the epicenter will experience about the same intensity of ground shaking, the destruction varies considerably within this area. This difference is mainly attributable to the nature of the ground on which the structures are built. Soft sediments, for example, generally amplify the vibrations more than solid bedrock. Thus, the buildings located in Anchorage, which were situated on unconsolidated sediments, experienced heavy structural damage. By contrast, most of the town of Whittier, although much nearer the epicenter, rests on a firm foundation of granite and hence suffered much less damage. However, Whittier was damaged by a tsunami (described in the next section).

Liquefaction In areas where unconsolidated materials are saturated with water, earthquake vibrations can generate a phenomenon known as **liquefaction** (*liqueo* = to be fluid, *facio* = to make). Under these conditions, what had been a stable soil turns into a mobile fluid that is not capable of supporting buildings or other structures (Figure 6.15). As a result, underground objects such as storage tanks and sewer lines may literally float toward the surface of their newly liquefied environment. Buildings and other structures may settle and collapse. During the 1989 Loma Prieta earthquake, in San Francisco's Marina District, foundations failed and geysers of sand and water shot from the ground, indicating that liquefaction had occurred (Figure 6.16).

Figure 6.15 Effects of liquefaction. This tilted building rests on unconsolidated sediment that imitated quicksand during a major 1985 Mexican earthquake. (Photo by James L. Beck)

A.

B.

Figure 6.16 Liquefaction. **A.** These "mud volcanoes" were produced by the Loma Prieta earthquake of 1989. They formed when geysers of sand and water shot from the ground, an indication that liquefaction occurred. (Photo by Richard Hilton, courtesy of Dennis Fox) **B.** Students experiencing the nature of liquefaction. (Photo by M. Miller)

What Is a Tsunami?

Large undersea earthquakes occasionally set in motion massive waves of water called **seismic sea waves,** or **tsunami*** (*tsu* = harbor, *nami* = waves). These destructive waves often are called "tidal waves" by the media. However, this name is inappropriate, because these waves are not created by the tidal effect of the Moon or Sun. Most tsunami result from vertical displacement along a fault located on the ocean floor, or from a large underwater landslide triggered by an earthquake (Figure 6.17).

Once formed, a tsunami resembles the ripples created when a pebble is dropped into a pond. In contrast to ripples, a tsunami advances across the ocean at speeds between 500 and 950 kilometers (300 and 600 miles) per hour. Despite this, a tsunami in the open ocean can pass undetected because its height is usually less than 1 meter and the distance between wave crests is great, ranging from 100 to 700 kilometers (62 to 435 miles). Upon entering shallower coastal water, these destructive waves are slowed and the water begins to pile up to heights that occasionally exceed 30 meters (100 feet), as shown in Figure 6.17. As the crest of a tsunami approaches shore, it appears as a rapid rise in sea level with a turbulent and chaotic surface.

The first warning of an approaching tsunami is usually a rapid withdrawal of water from beaches. Some residents living near the Pacific basin have learned to heed this warning and move to higher ground, because about 5 to 30 minutes later the retreat of water is followed by a surge capable of extending hundreds of meters inland. In a successive fashion, each surge is followed by a rapid oceanward retreat of the water.

*Seismic sea waves were given the name *tsunami* by the Japanese, who have suffered a great deal from them. The term *tsunami* is now used worldwide.

Did You Know?

During an earthquake near Port Royal, Jamaica, the water-saturated sand on which the city was built vigorously shook. As a result, the sand particles lost contact with one another, giving the mixture the consistency of a thick milk shake. Anything supported by the ground, such as buildings and people, either floated or sank. One eyewitness stated: "[W]hole streets with inhabitants were swallowed up. . . . Some were swallowed quite down, and cast up again by great quantities of water; others went down and were never more seen."

Did You Know?

Although there are no historical records of a tsunami produced by a meteorite impact in the oceans, such events have occurred. Geologic evidence indicates that the most recent occurrence was a megatsunami that devastated portions of the Australian coast about the year 1500. A meteorite impact 65 million years ago near Mexico's Yucatan Peninsula created one of the largest impact-induced tsunami ever. The giant wave swept hundreds of kilometers inland around the shore of the Gulf of Mexico.

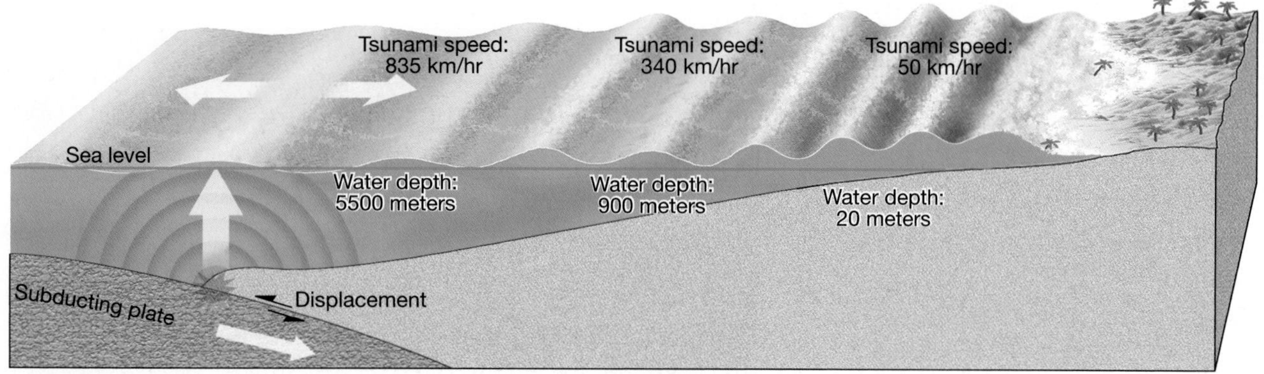

Figure 6.17 Schematic drawing of a tsunami generated by displacement of the ocean floor. The speed of a wave column correlates with ocean depth. As shown, waves moving in deep water advance at speeds in excess of 800 kilometers per hour. Speed gradually slows to 50 kilometers per hour at depths of 20 meters. Decreasing depth slows the movement of the wave column. As waves slow in shallow water, they grow in height until they topple and rush onto shore with tremendous force. The size and spacing of these swells are not to scale.

Tsunami Damage from the 2004 Indonesian Earthquake A massive undersea earthquake of moment magnitude 9.0 occurred near the island of Sumatra on December 26, 2004, and sent waves of water racing across the Indian Ocean and Bay of Bengal (Figure 6.18A). This tsunami was one of the deadliest natural disasters of any kind in modern times, claiming more than 230,000 lives. As water surged several kilometers inland, cars and trucks were flung around like toys in a bathtub, and fishing boats were rammed into homes. In some locations, the backwash of water dragged bodies and huge amounts of debris out to sea.

The destruction was indiscriminate, destroying luxury resorts and poor fishing hamlets on the Indian Ocean coast (Figure 6.18B). Devastation was most severe along the southeast coast of Sri Lanka, in the Indonesian province of Aceh, in the Indian state of Tamil Nada, and on Thailand's resort island of Phuket. Damages were reported as far away as the Somalia coast of Africa, 4100 kilometers (2500 miles) west of the earthquake epicenter.

The killer waves generated by this massive quake achieved heights as great as 10 meters (33 feet) and struck many unprepared areas within three hours of the event. Although the Pacific basin contains deep-sea buoys and tide gauges that can spot tsunami waves at sea, the Indian Ocean does not. (The deep-sea buoys have pressure sensors that detect changes in pressure as the earthquake's energy travels through the ocean, and tide gauges measure the rise and fall in sea level.) The rarity of tsunami in the Indian Ocean also contributed to the lack of preparedness for such an event. It should come as no surprise that the countries of India, Indonesia, and Thailand have announced plans to establish a tsunami warning system for the Indian Ocean.

Tsunami Warning System In 1946, a large tsunami struck the Hawaiian Islands without warning. A wave more than 15 meters (50 feet) high left several coastal villages in shambles. This destruction motivated the U.S. Coast and Geodetic Survey to establish a tsunami warning system for coastal areas of the Pacific. From seismic observatories throughout the region, large earthquakes are reported to the Tsunami Warning Center in Honolulu. Scientists at the center use tidal gauges to determine whether a tsunami has formed. Within an hour, a warning is issued. Although tsunami travel very rapidly, there is sufficient time to evacuate all but the region nearest the epicenter. For example, a tsunami generated near the Aleutian Islands would take five hours to reach Hawaii, and one generated near the coast of Chile would travel 15 hours before reaching Hawaii (Figure 6.19).

Fire

The 1906 earthquake in San Francisco reminds us of the formidable threat of fire (see Figure 6.3, p. 162). The central city contained mostly large, older wooden structures and brick buildings. Although many of the unreinforced brick buildings were extensively damaged by vibrations, the greatest destruction was caused by fires that started when gas and electrical lines were severed. The fires raged out of control for three days and devastated more than 500 blocks of the city. The problem was compounded by the initial ground shaking, which broke the city's water lines into hundreds of unconnected pieces.

The fire was finally contained when buildings were dynamited along a wide boulevard to provide a *fire break*. Only a few deaths were attributed to the San Francisco fire, but this is not always the case. An earthquake that rocked Japan in 1923 triggered an estimated 250 fires, which devastated the city of Yokohama and destroyed more than half the homes in Tokyo. More than 100,000 deaths were attributed to the fires, which were driven by unusually high winds.

Landslides and Ground Subsidence

In the 1964 Alaskan earthquake, the greatest damage to structures was from landslides and ground subsidence triggered by the vibrations. At Valdez and Seward, the violent shaking caused river-delta materials to experience liquefaction; the subsequent slumping carried both waterfronts away. Because it could happen again in another quake, the entire town of Valdez was relocated about 7 kilometers (4 miles) away on

A.

B.

Figure 6.18 A massive earthquake of moment magnitude 9.0 off the Indonesian island of Sumatra sent a tsunami racing across the Indian Ocean and Bay of Bengal on December 26, 2004. **A.** Unsuspecting foreign tourists, who at first walked on the sand after the water receded, now rush toward shore as the first of six tsunami start to roll toward Hat Rai Lay Beach near Krabi in southern Thailand. (AFP/Getty Images, Inc.) **B.** The Indonesian town of Banda Aceh, ten days after the tsunami. (Photo by Berbar Halin/Sipa Press)

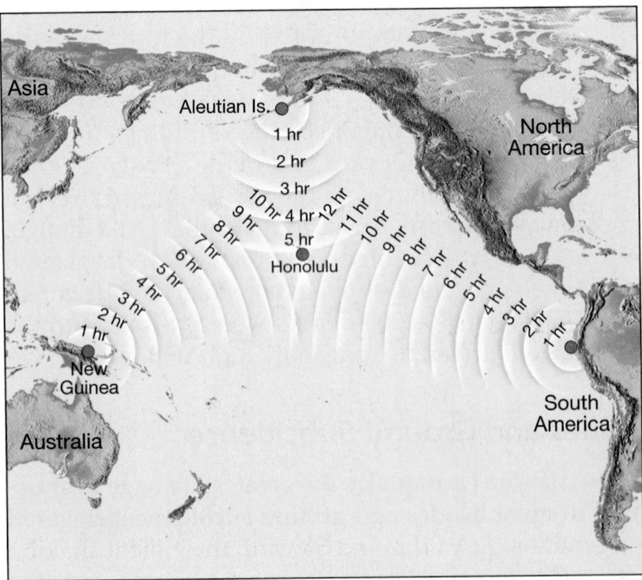

Figure 6.19 Tsunami travel times to Honolulu, Hawaii, from selected locations throughout the Pacific. (Data from NOAA)

more stable ground. In Valdez, 31 people on a dock died when it slid into the sea.

Most of the damage in Anchorage was attributed to landslides caused by the shaking and lurching ground. Many homes were destroyed in Turnagain Heights when a layer of clay lost its strength and more than 200 acres of land slid toward the ocean (Figure 6.20). A portion of this landslide was left in its natural condition as a reminder of this destructive event. The site was named "Earthquake Park." Downtown Anchorage was also disrupted as sections of the main business district dropped by as much as 3 meters (10 feet).

Did You Know?

The tsunami generated in the 1964 Alaskan earthquake heavily damaged communities along the Gulf of Alaska and killed 107 people. By contrast, only nine people died in Anchorage as a direct result of the vibrations. Furthermore, despite a one-hour warning, 12 persons perished in Crescent City, California, from the fifth and most destructive tsunami wave generated by this quake.

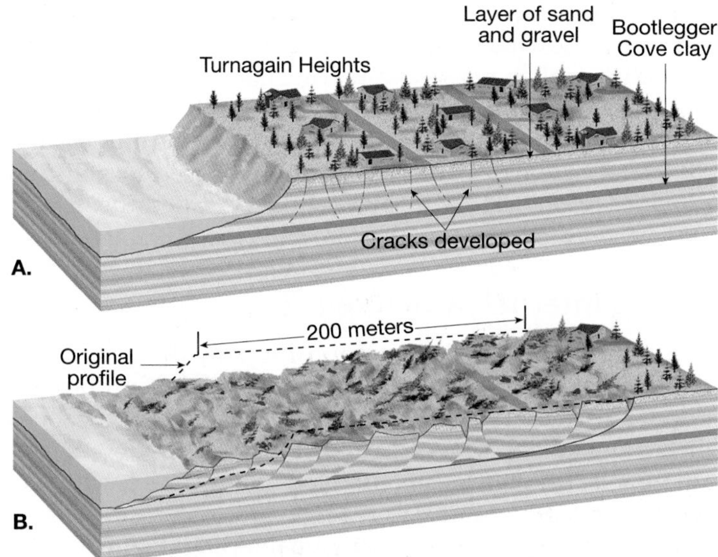

Figure 6.20 Turnagain Heights slide caused by the 1964 Alaskan earthquake. **A.** Vibrations from the earthquake caused cracks to appear near the edge of the bluff. **B.** Within seconds, blocks of land began to slide toward the sea on a weak layer of clay. In less than 5 minutes, as much as 200 meters (650 feet) of the Turnagain Heights bluff area had been destroyed. **C.** Photo of a small portion of the Turnagain Heights slide. (Photo courtesy of U.S. Geological Survey)

Probing Earth's Interior: "Seeing" Seismic Waves

Forces Within
▼ Earthquakes

The best way to learn about Earth's interior is to dig or drill a hole and examine it directly. Unfortunately, this is only possible at shallow depths. The deepest a drilling rig has ever penetrated is only 12.3 kilometers (8 miles), which is about 1/500 of the way to Earth's center! Even this was an extraordinary accomplishment because temperature and pressure increase so rapidly with depth.

Fortunately, many earthquakes are large enough that their seismic waves travel all the way through Earth and can be recorded on the other side (Figure 6.21). This means that the seismic waves act like medical x-rays used to take images of a person's insides. There are about 100 to 200 earthquakes each year that are large enough (about $M_W > 6$) to be well recorded by seismographs all around the globe. These large earthquakes provide the means to "see" into our planet and have been the source of most of the data that allowed us to figure out the nature of Earth's interior.

Interpreting the waves recorded on seismograms in order to identify Earth structure is complicated by the fact that seismic waves usually do not travel along straight paths. Instead, seismic waves are reflected, refracted, and diffracted as they pass through our planet. They reflect off boundaries between different layers, they refract (or bend) when passing from one layer to another layer, and they diffract around any obstacles they encounter (Figure 6.21). These different wave behaviors have been used to identify the boundaries that exist within Earth.

One of the most noticeable behaviors of seismic waves is that they follow strongly curved paths (Figure 6.21). This occurs because the velocity of seismic waves generally increases with depth. In addition, seismic waves travel faster when rock is stiffer or less compressible. These properties of stiffness and compressibility are then used to interpret the composition and temperature of the rock. For instance, when rock is hotter, it becomes less stiff (imagine taking a frozen chocolate bar and then heating it up!), and waves travel more slowly. Waves also travel at different speeds through rocks of different compositions. Thus, the speed that seismic waves travel can help determine both the kind of rock that is inside Earth and how hot it is.

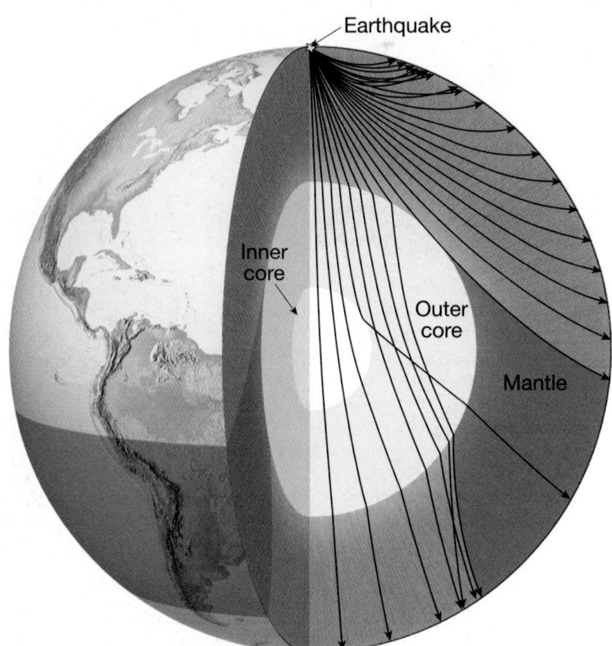

Figure 6.21 Slice through Earth's interior showing some of the ray paths that seismic waves from an earthquake would take. Notice that in the mantle, the rays follow curved (refracting) paths rather than straight paths because the seismic velocity of rocks increases with depth, a result of increasing pressure with depth.

Formation of Earth's Layered Structure

As material accumulated to form Earth (and for a short period afterward), the high-velocity impact of nebular debris and the decay of radioactive elements caused the temperature of our planet to steadily increase. During this time of intense heating, Earth became hot enough that iron and nickel began to melt. Melting produced liquid blobs of heavy metal that sank toward the center of the planet. This process occurred rapidly on the scale of geologic time and produced Earth's dense iron-rich core.

The early period of heating resulted in another process of chemical differentiation, whereby melting formed buoyant masses of molten rock that rose toward the surface, where they solidified to produce a primitive crust. These rocky materials were rich in oxygen and "oxygen-seeking" elements, particularly silicon and aluminum, along with lesser amounts of calcium, sodium, potassium, iron, and magnesium. In addition, some heavy metals such as gold, lead, and uranium, which have low melting points or were highly soluble in the ascending molten masses, were scavenged from Earth's interior and concentrated in the developing crust. This early period of chemical segregation established the three basic divisions of Earth's interior—the iron-rich *core;* the thin *primitive crust;* and Earth's largest layer, called the *mantle,* which is located between the core and crust (Figure 6.22).

Earth's Internal Structure

In addition to these three compositionally distinct layers, Earth can be divided into layers based on physical properties. The physical properties used to define such zones include whether the layer is solid or liquid and how weak or strong it is. Knowledge of both types of layers is essential to our understanding of basic geologic processes, such as volcanism, earthquakes, and mountain building (Figure 6.22).

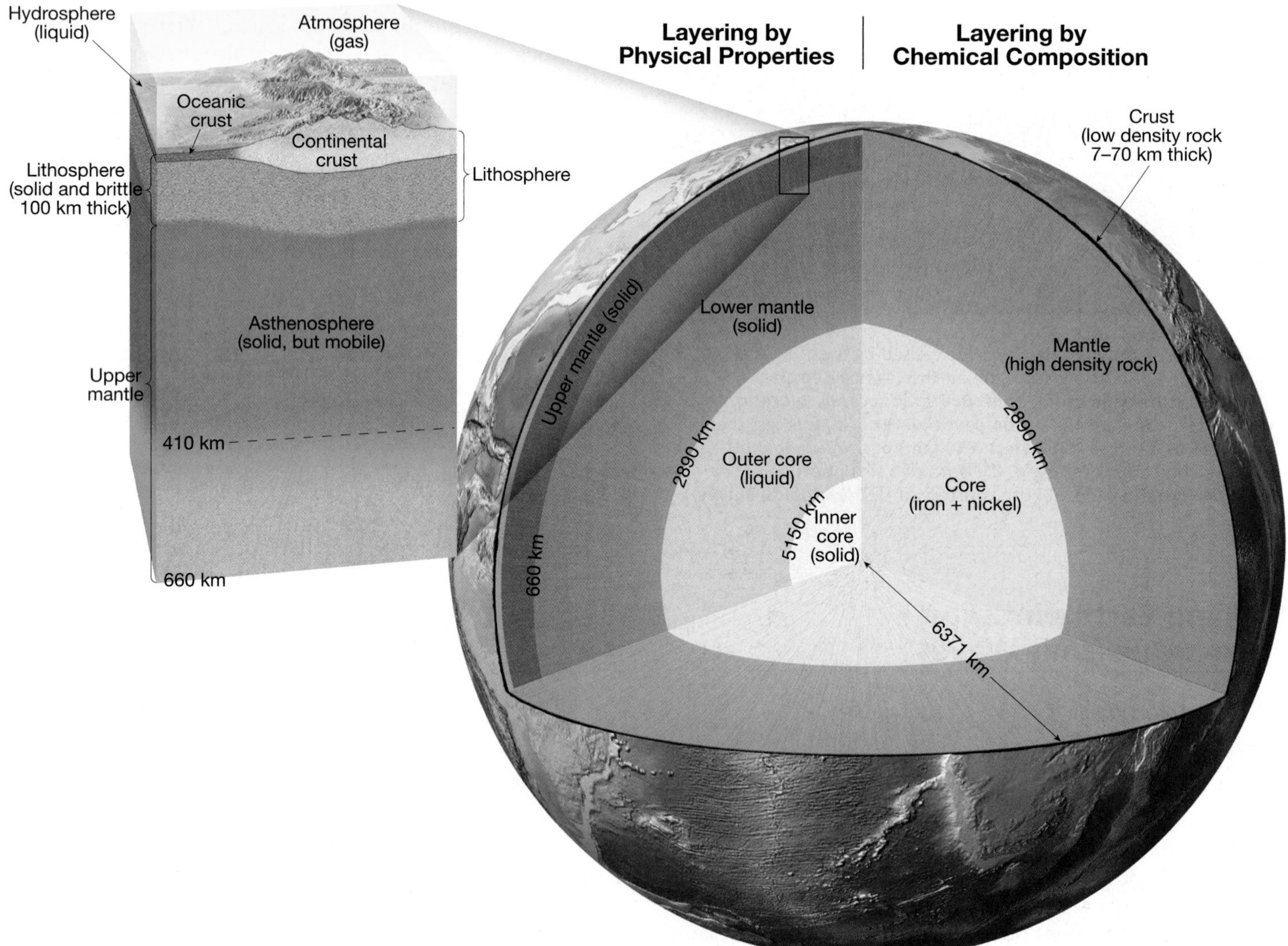

Figure 6.22 Views of Earth's layered structure. The right side of the large cross section shows that Earth's interior is divided into three different layers based on compositional differences—the crust, mantle, and core. The left side of the large cross section depicts the five main layers of Earth's interior based on physical properties and hence mechanical strength—the lithosphere, asthenosphere, lower mantle, outer core, and inner core. The block diagram to the left of the cross section shows an enlarged view of the upper portion of Earth's interior.

Earth's Crust The **crust,** Earth's relatively thin, rocky outer skin, is of two types—continental crust and oceanic crust. Both share the word *crust,* but the similarity ends there. The oceanic crust is roughly 7 kilometers (4 miles) thick and composed of the dark igneous rock *basalt.* By contrast, the continental crust averages 35 to 40 kilometers (22 to 25 miles) thick but may exceed 70 kilometers (40 miles) in some mountainous regions such as the Rockies and Himalayas. Unlike the oceanic crust, which has a relatively homogeneous chemical composition, the continental crust consists of many rock types. Although the upper crust has an average composition of a *granitic rock* called *granodiorite,* it varies considerably from place to place.

Continental rocks have an average density of about 2.7 grams per cubic centimeter, and some are 4 billion years old. The rocks of the oceanic crust are younger (180 million years or less) and denser (about 3.0 grams per cubic centimeter) than continental rocks.*

Earth's Mantle More than 82 percent of Earth's volume is contained in the **mantle,** a solid, rocky shell that extends to a depth of about 2900 kilometers (1800 miles). The boundary between the crust and mantle represents a marked change in chemical composition. The dominant rock type in the uppermost mantle is *peridotite,* which is richer in the metals magnesium and iron than the minerals found in either the continental or oceanic crust.

The upper mantle extends from the crust-mantle boundary down to a depth of about 660 kilometers (410 miles). The upper mantle can be divided into three different parts. The top portion of the upper mantle is part of the stiff *lithosphere,* and beneath that is the weaker *asthenosphere.* The bottom part of the upper mantle is called the *transition zone.*

The **lithosphere** (sphere of rock) consists of the entire crust and uppermost mantle and forms Earth's relatively cool, rigid outer shell. Averaging about 100 kilometers (62 miles) in thickness, the lithosphere is more than 250 kilometers (155 miles) thick below the oldest portions of the continents (Figure 6.22). Beneath this stiff layer to a depth of about 350 kilometers (217 miles) lies a soft, comparatively weak layer known as the **asthenosphere** ("weak sphere"). The top portion of the asthenosphere has a temperature/pressure regime that results in a small amount of melting. Within this very weak zone, the lithosphere is mechanically detached from the layer below. The result is that the lithosphere is able to move independently of the asthenosphere, a fact we will consider in the next chapter.

It is important to emphasize that the strength of various Earth materials is a function of both their composition and the temperature and pressure of their environment. The entire lithosphere does *not* behave like a brittle solid similar to rocks found on the surface. Rather, the rocks of the lithosphere get progressively hotter and weaker (more easily deformed) with increasing depth. At the depth of the uppermost asthenosphere, the rocks are close enough to their melting temperature that they are very easily deformed, and some melting may actually occur. Thus, the uppermost asthenosphere is weak because it is near its melting point, just as hot wax is weaker than cold wax.

From 660 kilometers (410 miles) deep to the top of the core, at a depth of 2900 kilometers (1800 miles), is the **lower mantle.** Because of an increase in pressure (caused by the weight of the rock above), the mantle gradually strengthens with depth. Despite their strength however, the rocks within the lower mantle are very hot and capable of very gradual flow.

Earth's Core The composition of the **core** is thought to be an iron-nickel alloy with minor amounts of oxygen, silicon, and sulfur—elements that readily form compounds with iron. At the extreme pressure found in the core, this iron-rich material has an average density of nearly 11 grams per cubic centimeter and approaches 14 times the density of water at Earth's center.

The core is divided into two regions that exhibit very different mechanical strengths. The **outer core** is a *liquid layer* 2270 kilometers (1410 miles) thick. It is the movement of metallic iron within this zone that generates Earth's magnetic field. The **inner core** is a sphere with a radius of 1216 kilometers (754 miles). Despite its higher temperature, the iron in the inner core is *solid* due to the immense pressures that exist in the center of the planet.

Geologic Structures

Earth is a dynamic planet. Evidence for the enormous forces that operate within it are seen in the Northern Rockies, where massive rock units have been thrust great distances over others. On a smaller scale, crustal movements of a few meters occur along faults during large earthquakes. The result of these tectonic activities is Earth's major mountain belts.

Before examining mountain building, we must first look at the nature of rock deformation and the structures that result.

Rock Deformation

Every body of rock, no matter how strong, has a point at which it will fracture or flow. **Deformation** (*de* = out, *forma* = form) is a general term that refers to all changes in the original shape and/or size of a rock body. Most crustal deformation occurs along plate margins. Recall from Chapter 5 that the lithosphere consists of large segments (plates) that move relative to one another. Plate motions and the interactions along plate boundaries generate forces that cause rock to deform.

When rocks are subjected to forces (stresses) greater than their own strength, they begin to deform, usually by folding, flowing, or fracturing (Figure 6.23). It is easy to visualize how rocks break, because we normally think of them as being brittle. But how can rock masses be bent into intricate folds without being broken during the process? To answer

*Liquid water has a density of 1 gram per cubic centimeter; therefore, the density of basalt is three times that of water.

Figure 6.23 Folded sedimentary layers exposed on the face of Mount Kidd, Alberta, Canada. (Photo by Peter French/DRK Photo)

this question, geologists performed laboratory experiments in which rocks were subjected to forces under conditions that simulated those existing at various depths within the crust.

Although each rock type deforms somewhat differently, the general characteristics of rock deformation were determined from these experiments. Geologists discovered that when stress is gradually applied, rocks first respond by deforming elastically. Changes that result from *elastic deformation* are recoverable; that is, like a rubber band, the rock will return to nearly its original size and shape when the force is removed. (Recall that the energy for most earthquakes comes from stored elastic energy that is released as rock snaps back to its original shape.) Once the elastic limit (strength) of a rock is surpassed, it either flows (*ductile deformation*) or fractures (*brittle deformation*).

The factors that influence the strength of a rock and how it will deform include temperature, confining pressure, rock type, and time. Rocks near the surface, where temperatures and confining pressures are low, tend to behave like a brittle solid and fracture once their strength is exceeded. This type of deformation is called **brittle failure** or **brittle deformation.** From our everyday experience, we know that glass objects, wooden pencils, china plates, and even our bones exhibit brittle failure once their strength is surpassed. By contrast, at depth, where temperatures and confining pressures are high, rocks exhibit *ductile* behavior. **Ductile deformation** is a type of solid-state flow that produces a change in the size and shape of an object without fracturing. Ordinary objects that display ductile behavior include modeling clay, bee's wax, caramel candy, and most metals. For example, a copper penny placed on a railroad track will be flattened and deformed (without breaking) by the force applied by a passing train.

Ductile deformation of a rock—strongly aided by high temperature and high confining pressure—is somewhat similar to the deformation of a penny flattened by a train. Rocks that display evidence of ductile flow were usually deformed at great depth and may exhibit contorted folds that give the impression that the strength of the rock was akin to soft putty (Figure 6.23).

Folds

During mountain building, flat-lying sedimentary and volcanic rocks are often bent into a series of wavelike undulations called **folds.** Folds in sedimentary strata are much like those that would form if you were to hold the ends of a sheet of paper and then push them together. In nature, folds come in a wide variety of sizes and configurations. Some folds are broad flexures in which rock units hundreds of meters thick have been slightly warped. Others are very tight microscopic structures found in metamorphic rocks. Size differences notwithstanding, most folds are the result of *compressional forces* that result in the shortening and thickening of the crust.

Types of Folds The two most common types of folds are anticlines and synclines (Figure 6.24). An **anticline** is most commonly formed by the upfolding, or arching, of rock layers.* Often found in association with anticlines are downfolds, or troughs, called **synclines.** Notice in Figure 6.24 that the limb of an anticline is also a limb of the adjacent syncline.

Depending on their orientation, these basic folds are described as *symmetrical* when the limbs are mirror images of each other and *asymmetrical* when they are not. An asymmetrical fold is said to be *overturned* if one limb is tilted beyond the vertical (Figure 6.24). An overturned fold can also lie on its side so that a plane extending through the axis of the fold would be horizontal. These *recombent* folds are common in mountainous regions such as the Alps.

Domes and Basins Broad upwarps in basement rock may deform the overlying cover of sedimentary strata and generate large folds. When this upwarping produces a circu-

*By strict definition, an *anticline* is a structure in which the oldest strata are found in the center. This most typically occurs when strata are upfolded. Further, a syncline is strictly defined as a structure in which the youngest strata are found in the center. This occurs most commonly when strata are downfolded.

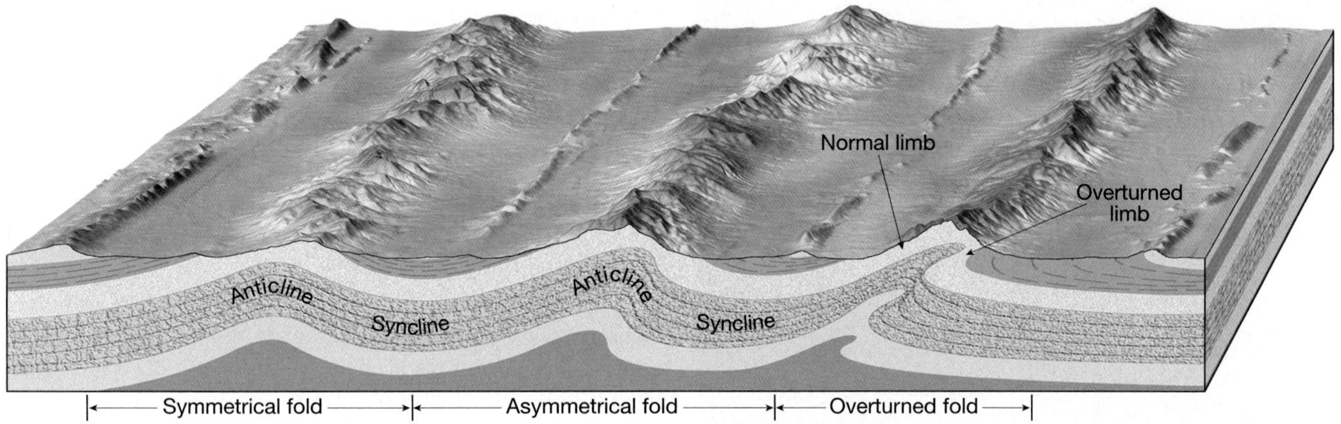

Figure 6.24 Block diagram of principal types of folded strata. The upfolded or arched structures are anticlines. The downfolds, or troughs, are synclines. Notice that the limb of an anticline is also the limb of the adjacent syncline.

lar or elongated structure, the feature is called a **dome.** Downwarped structures having a similar shape are termed **basins.**

The Black Hills of western South Dakota is a large domed structure thought to be generated by upwarping. Erosion has stripped away the highest portions of the upwarped sedimentary beds, exposing older igneous and metamorphic rocks in the center (Figure 6.25). Remnants of these once continuous sedimentary layers are visible, flanking the crystalline core of these mountains.

Several large basins exist in the United States. The basins of Michigan and Illinois have very gently sloping beds similar to saucers. These basins are thought to be the result of large accumulations of sediment, whose weight caused the crust to subside.

Because large basins usually contain sedimentary beds sloping at very low angles, they are usually identified by the age of the rocks comprising them. The youngest rocks are found near the center, and the oldest rocks are at the flanks. This is just the opposite order of a domed structure, such as the Black Hills, where the oldest rocks form the core.

Faults

Faults are fractures in the crust along which appreciable displacement has taken place. Occasionally, small faults can be recognized in road cuts where sedimentary beds have been offset a few meters, as shown in Figure 6.26. Faults of this scale usually occur as single discrete breaks. By contrast, large faults, like the San Andreas Fault in California, have displacements of hundreds of kilometers and consist of many interconnecting fault surfaces. These *fault zones* can be several kilometers wide and are often easier to identify from high-altitude photographs than at ground level.

Dip-Slip Faults Faults in which the movement is primarily parallel to the inclination, or dip, of the fault surface are called **dip-slip faults.** Vertical displacements along dip-slip faults may produce long, low cliffs called **fault scarps.** Fault scarps, such as the one shown in Figure 6.27, are produced by displacements that generate earthquakes.

It has become common practice to call the rock surface that is immediately above the fault the *hanging wall* and to call the rock surface below, the *footwall.* This nomenclature arose from prospectors and miners who excavated shafts and

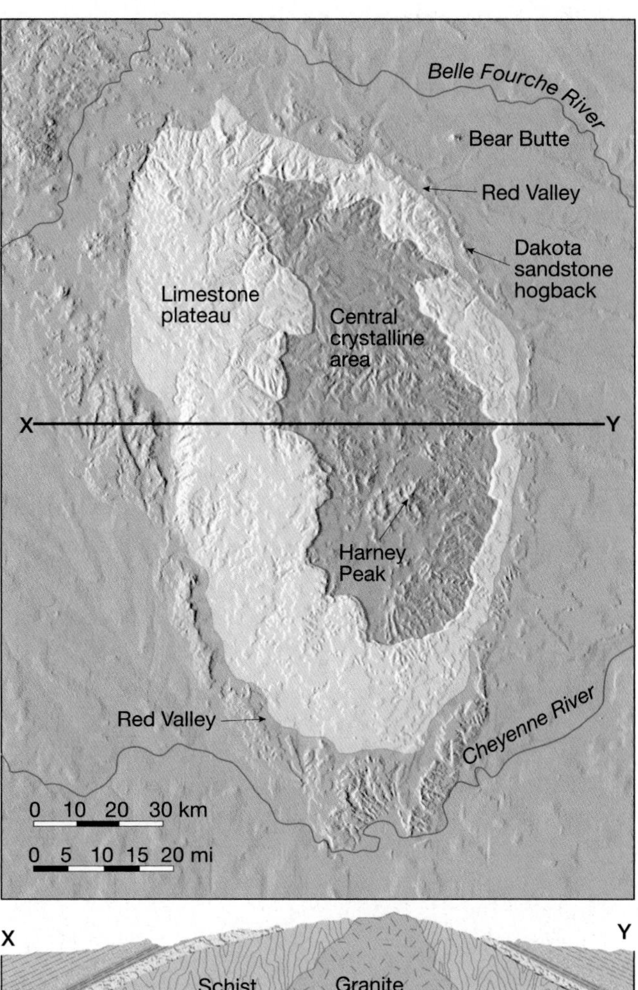

Figure 6.25 The Black Hills of South Dakota, an example of a domal structure in which the resistant igneous and metamorphic central core has been exposed by erosion. Map view (above), and cross-sectional view from point X to point Y (below).

Figure 6.26 Faulting caused the displacement of these beds. Arrows show relative motion of rock units. Exposure located along a road cut near Kanab, Utah. (Photo by Tom Bean/DRK Photo)

tunnels along fault zones because they are frequently sites of ore deposits. In these tunnels, the miners would walk on the rocks below the mineralized fault zone (the footwall) and hang their lanterns on the rocks above (the hanging wall).

Dip-slip faults are classified as **normal faults** when the hanging wall block moves down relative to the footwall block (Figure 6.28A). Most normal faults have steep dips of about 60 degrees, which tend to flatten out with depth. However, some dip-slip faults have much lower dips, with some approaching horizontal. Because of the downward motion of the hanging wall block, normal faults accommodate lengthening, or extension, of the crust.

Most normal faults are small, having displacements of only a meter or so, like the one shown in the road cut in Figure 6.26, p. 180. Others extend for tens of kilometers where they may sinuously trace the boundary of a mountain front. In the western United States, large-scale normal faults like these are associated with structures called **fault-block mountains** (Figure 6.29).

Examples of fault-block mountains include the Teton Range of Wyoming and the Sierra Nevada of California. Both are faulted along their eastern flanks, which were uplifted as the blocks tilted downward to the west. These precipitous mountain fronts were produced over a period of 5 to 10 million years by many irregularly spaced episodes of faulting. Each event was responsible for just a few meters of displacement.

Normal faulting is prevalent at spreading centers where plate divergence occurs. A central block called a **graben** is bounded by normal faults and drops as the plates separate (Figure 6.29). These grabens produce an elongated valley bounded by relatively uplifted structures called **horsts.** The Rift Valley of East Africa is made up of several large grabens, above which tilted horsts produce a linear mountainous topography. (See Figure 5.12, p. 140.) This valley, nearly 6000 kilometers (3700 miles) long, contains the excavation sites of some of the earliest human fossils.

An excellent example of horst and graben topography is found in the Basin and Range Province, a region that encompasses Nevada and portions of surrounding states (Figure 6.29). The crust has been elongated and broken to create more than 200 relatively small mountain ranges. Averaging about 80 kilometers (50 miles) in length, the ranges rise 900 to 1500 meters (3000 to 5000 feet) above the adjacent downfaulted basins. Notice in Figure 6.29 that slopes of the normal faults in the Basin and Range Province decrease with depth and join together to form a nearly horizontal fault called a *detachment fault.* These detachment faults extend for several kilometers below the surface. They form a major boundary between the rocks below, which exhibit ductile deformation, and the rocks above, which demonstrate brittle deformation via faulting.

Fault motion provides geologists with a method of determining the nature of the forces at work within Earth. Normal faults indicate the existence of tensional forces that pull the crust apart. This "pulling apart" can be accomplished either by uplifting that causes the surface to stretch and break or by opposing horizontal forces.

Reverse faults and **thrust faults** are dip-slip faults in which the hanging wall block moves up relative to the footwall block (Figure 6.28B, C). Reverse faults have dips greater than 45 degrees, and thrust faults have dips less than 45 degrees. Because the hanging wall block moves up and over the footwall block, reverse and thrust faults accommodate shortening of the crust.

Most high-angle reverse faults are small and accommodate local displacements in regions dominated by other types of faulting. Thrust faults, on the other hand, exist at all scales. In mountainous regions such as the Alps, Northern Rockies, Himalayas, and Appalachians, thrust faults have displaced

Did You Know?

People have actually observed the creation of a fault scarp—and have lived to tell about it. In Idaho, a large earthquake in 1983 created a 3-meter (10-foot) fault scarp that was witnessed by several people, many of whom were knocked off their feet. More often, though, fault scarps are noticed only after they form.

Figure 6.27 A fault scarp located near Joshua Tree National Park, California. (Photo by Alan P. Trujillo/APT Photos)

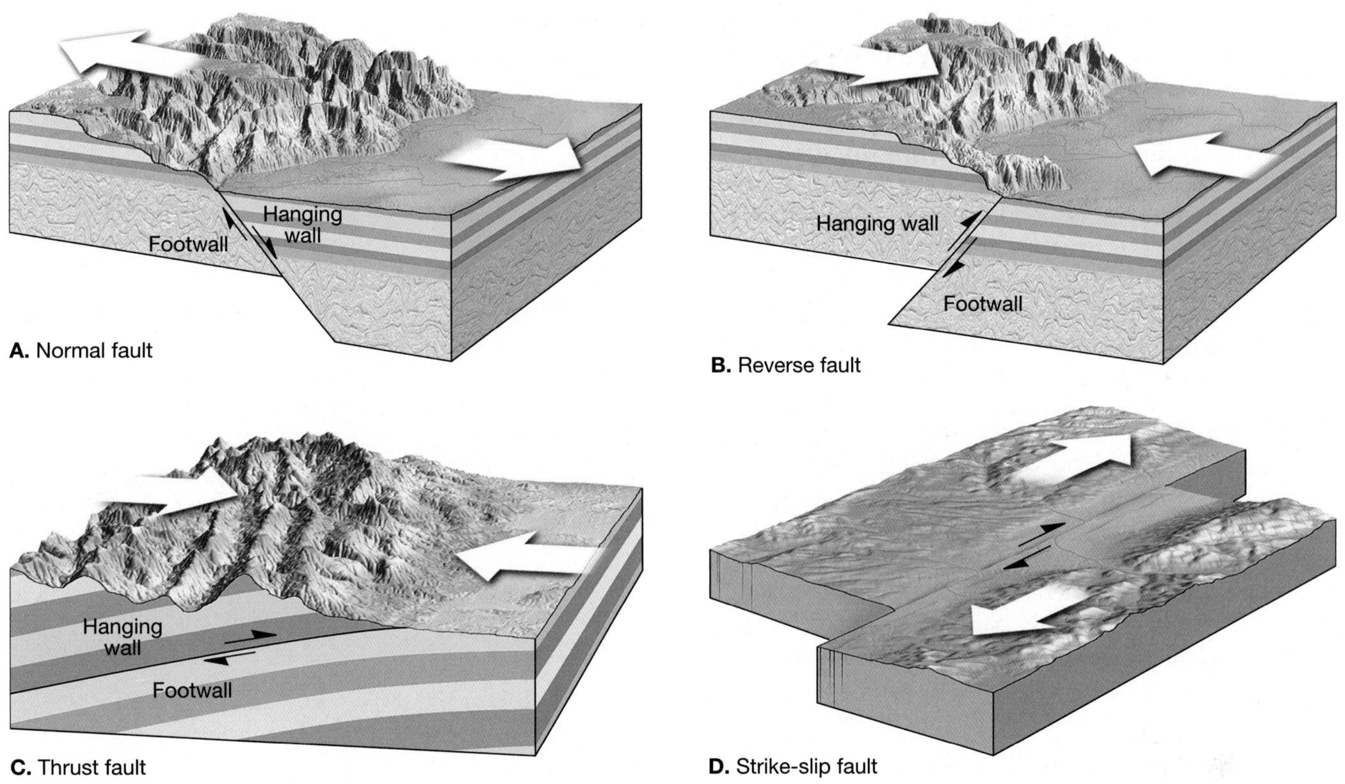

Figure 6.28 Block diagrams of four types of faults. **A.** Normal fault. **B.** Reverse fault. **C.** Thrust fault. **D.** Strike-slip fault.

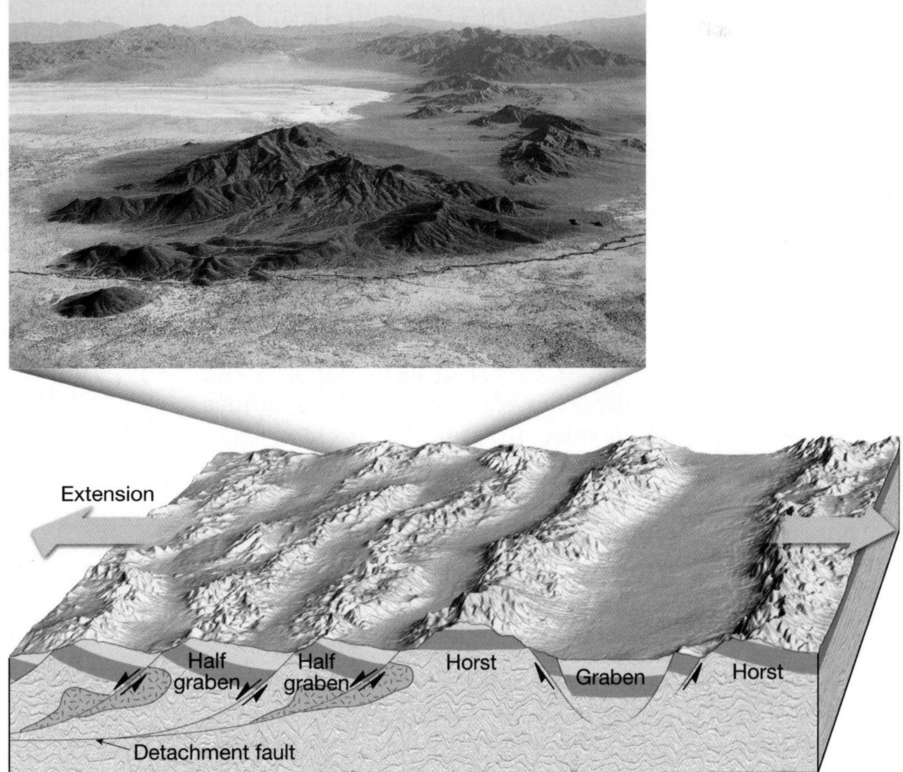

Figure 6.29 Normal faulting in the Basin and Range Province. Here, tensional stresses have elongated and fractured the crust into numerous blocks. Movement along these fractures has tilted the blocks, producing parallel mountain ranges called *fault-block mountains.* The downfaulted blocks (grabens) form basins, whereas the upfaulted blocks (horsts) are eroded to form rugged mountainous topography. In addition, numerous tilted blocks (half-grabens) form both basins and mountains. (Photo by Michael Collier)

strata as far as 50 kilometers (31 miles) over adjacent rock units. The result of this large-scale movement is that older strata end up overlying younger rocks.

Whereas normal faults occur in tensional environments, reverse and thrust faults result from strong compressional stresses. In these settings, crustal blocks are displaced *toward* one another, with the hanging wall being displaced upward relative to the footwall. Thrust faulting is most pronounced in subduction zones and other convergent boundaries where plates are colliding. Compressional forces generally produce folds as well as faults and result in a thickening and shortening of the material involved.

Strike-slip Faults Faults in which the dominant displacement is horizontal and parallel to the trend, or strike, of the fault surface are called **strike-slip faults** (Figure 6.28D). Because of their large size and linear nature, many strike-slip faults produce a trace that is visible over a great distance. Rather than a single fracture along which movement takes place, large strike-slip faults consist of a zone of roughly parallel fractures. The zone may be up to several kilometers wide. The most recent movement, however, is often along a strand only a few meters wide, which may offset features such as stream channels. Furthermore, crushed and broken rocks produced during faulting are more easily eroded, often producing linear valleys or troughs that mark the locations of strike-slip faults.

The earliest scientific records of strike-slip faulting were made following surface ruptures that produced large earthquakes. One of the most noteworthy of these was the great San Francisco earthquake of 1906. During this strong earthquake, structures such as fences that were built across the San Andreas Fault were displaced as much as 4.7 meters (15 feet). Because the movement along the San Andreas causes the crustal block on the opposite side of the fault to move to the right as you face the fault, it is called a *right-lateral* strike-slip fault. The Great Glen fault in Scotland is a well-known example of a *left-lateral* strike-slip fault, which exhibits the opposite sense of displacement.

Many major strike-slip faults cut through the lithosphere and accommodate motion between two large crustal plates. This special kind of strike-slip fault is called a **transform fault.** One of the best-known transform faults is California's San Andreas Fault (Figure 6.30). This plate-bounding fault can be traced for about 950 kilometers (600 miles) from the Gulf of California to a point along the Pacific Coast north of San Francisco, where it heads out to sea. Ever since its formation, about 29 million years ago, displacement along the San Andreas Fault has exceeded 560 kilometers (350 miles). This movement has accommodated the northward displacement of southwestern California and the Baja Peninsula of Mexico in relation to the remainder of North America.

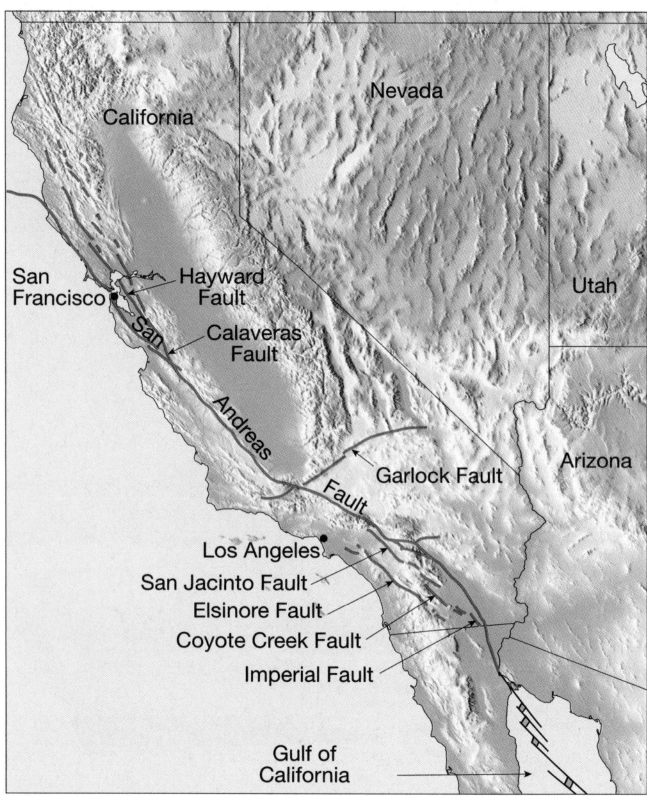

Figure 6.30 Map showing the extent of the San Andreas Fault system.

Mountain Building

Like other people, geologists have been inspired more by Earth's mountains than by any other landforms (Figure 6.31). Through extensive scientific exploration over the past 150 years, much has been learned about the internal processes that generate these often spectacular terrains. The name for the processes that collectively produce a mountain belt is **orogenesis,** (*oros* = mountain, *genesis* = to come into being). The rocks comprising mountains provide striking visual evidence of the enormous compressional forces that have deformed large sections of Earth's crust and subsequently elevated them to their present positions. Although folding is often the most conspicuous sign of these forces, thrust faulting, metamorphism, and igneous activity are always present in varying degrees.

Mountain building has occurred during the recent geologic past in several locations around the world. These young mountainous belts include the American Cordillera, which runs along the western margin of the Americas from Cape Horn to Alaska and includes the Andes and Rocky mountains; the Alpine-Himalaya chain, which extends from the Mediterranean through Iran to northern India and into Indochina; and the mountainous terrains of the western Pacific, which include volcanic island arcs such as Japan, the

Did You Know?

New Zealander Edmund Hillary and Tenzing Norgay of Nepal were the first to reach the summit of Mount Everest on May 29, 1953. Not one to rest on his laurels, Hillary later went on to lead the first crossing of Antarctica.

Figure 6.31 This peak is part of the Karakoram Range in Pakistan. The Karakoram are part of the Himalayan system. (Photo by Art Wolfe)

Philippines, and Sumatra. Most of these young mountain belts have come into existence within the past 100 million years. Some, including the Himalayas, began their growth as recently as 45 million years ago.

In addition to these relatively young mountain belts, several chains of older mountains exist on Earth as well. Although these older structures are deeply eroded and topographically less prominent, they clearly possess the same structural features found in younger mountains. Typical of this older group are the Appalachians in the eastern United States and the Urals in Russia.

Over the years, several hypotheses have been put forward regarding the formation of Earth's major mountain belts. One early proposal suggested that mountains are simply wrinkles in Earth's crust, produced as the planet cooled from its original semimolten state. As Earth lost heat, it contracted and shrank. In response to this process, the crust was deformed similar to when the peel of an orange wrinkles as the fruit dries out. However, neither this nor any other early hypothesis was able to withstand careful scrutiny and had to be discarded.

Mountain Building at Subduction Zones

With the development of the theory of plate tectonics, a model for orogenesis with excellent explanatory power has emerged. According to this model, most mountain building occurs at convergent plate boundaries. Here, the subduction of oceanic lithosphere triggers partial melting of mantle rock, providing a source of magma that intrudes the crustal rocks that form the margin of the overlying plate. In addition, colliding plates provide the tectonic forces that fold, fault, and metamorphose the thick accumulations of sediments that have been deposited along the flanks of landmasses. Together, these processes thicken and shorten the continental crust, thereby elevating rocks that may have formed near the ocean floor to lofty heights.

To unravel the events that produce mountains, researchers examine ancient mountain structures as well as sites where orogenesis is currently active. Of particular interest are active subduction zones, where lithospheric plates are converging. The subduction of oceanic lithosphere generates Earth's strongest earthquakes and most explosive volcanic eruptions, as well as playing a pivotal role in generating many of Earth's mountain belts.

The subduction of oceanic lithosphere gives rise to two different types of mountain belts. Where *oceanic lithosphere* subducts beneath an *oceanic plate,* an *island arc* and related tectonic features develop. Subduction beneath a *continental block,* on the other hand, results in the formation of a *continental volcanic arc* along the margin of the adjacent landmass. Plate boundaries that generate continental volcanic arcs are often referred to as *Andean-type plate margins.*

Island Arcs Island arcs form where two oceanic plates converge and one is subducted beneath the other (Figure 6.32). This activity results in partial melting of the mantle wedge located above the subducting plate and eventually leads to the growth of a volcanic island arc on the ocean floor. Because they are associated with subducting oceanic lithosphere,

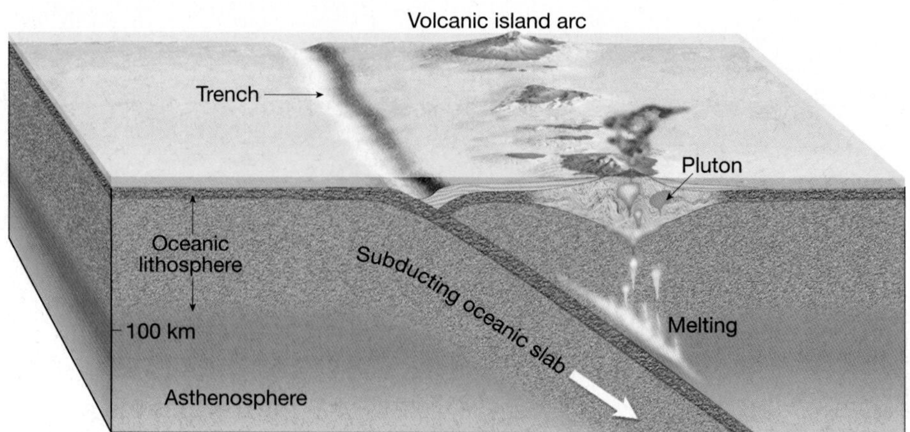

Figure 6.32 The development of a volcanic island arc by the convergence of two oceanic plates. Continuous subduction along these convergent zones results in the development of thick units of continental-type crust.

island arcs are typically found on the margins of an ocean basin, such as the Pacific—where the majority of volcanic island arcs are found. Examples of active island arcs include the Mariana, New Hebrides, Tonga, and Aleutian arcs.

Island arcs represent what are perhaps the simplest mountain belts. These structures result from the steady subduction of oceanic lithosphere, which may last for 100 million years or more. Somewhat sporadic volcanic activity, the emplacement of igneous bodies at depth, and the accumulation of sediment that is scraped from the subducting plate gradually increase the volume of crustal material capping the upper plate. Some mature volcanic island arcs, such as Japan, appear to have been built upon a preexisting fragment of crustal material.

The continued development of a mature volcanic island arc can result in the formation of mountainous topography consisting of belts of igneous and metamorphic rocks. This activity, however, is viewed as just one phase in the development of a major mountain belt. As you will see later, some volcanic arcs are carried by a subducting plate to the margin of a large continental block, where they become involved in a major mountain-building episode.

Mountain Building Along Andean-Type Margins Mountain building along continental margins involves the convergence of an oceanic plate and a plate whose leading edge contains continental crust. Exemplified by the Andes Mountains, an *Andean-type convergent zone* results in the formation of a continental volcanic arc and related tectonic features inland of the continental margin.

The first stage in the development of an idealized Andean-type mountain belt occurs prior to the formation of the subduction zone. During this period, the continental margin is a **passive continental margin;** that is, it is not a plate boundary but a part of the same plate as the adjoining oceanic crust. The East Coast of North America provides a present-day example of a passive continental margin. Here, as at other passive continental margins surrounding the Atlantic, deposition of sediment on the continental shelf is producing a thick wedge of shallow-water sandstones, limestones, and shales (Figure 6.33A). Beyond the continental shelf, turbidity currents are depositing sediments on the continental slope and rise (see Chapter 9).

At some point, the continental margin becomes active. A subduction zone forms and the deformation process begins (Figure 6.33B). A good place to examine an **active continental margin** is the west coast of South America, where the Nazca plate is being subducted beneath the South American plate along the Peru-Chile trench. This subduction zone probably formed prior to the breakup of the supercontinent of Pangaea.

In an idealized Andean-type subduction, convergence of the continental block and the subducting oceanic plate leads to deformation and metamorphism of the continental margin. Once the oceanic plate descends to about 100 kilometers (62 miles), partial melting of mantle rock above the subducting slab generates magma that migrates upward (Figure 6.33B). Thick continental crust greatly impedes the ascent of magma. Consequently, a high percentage of the magma that intrudes the crust never reaches the surface—instead, it crystallizes at depth to form plutons. Eventually, uplifting and erosion exhume these igneous bodies and associated metamorphic rocks. Once they are exposed at the surface, these massive structures are called *batholiths* (Figure 6.33C). Composed of numerous plutons, batholiths form the core of the Sierra Nevada in California and are prevalent in the Peruvian Andes.

During the development of this continental volcanic arc, sediment derived from the land and scraped from the subducting plate is plastered against the landward side of the

Did You Know?

The lake that is the home of the legendary Loch Ness Monster is located along the Great Glen fault that divides Northern Scotland. Strike-slip movement along this fault zone shattered a wide zone of rock, which glacial erosion subsequently removed, to produce the elongated valley in which Loch Ness is situated. The broken rock was easily eroded, a fact that accounts for the great depth of Loch Ness—as much as 180 meters (600 feet) below sea level.

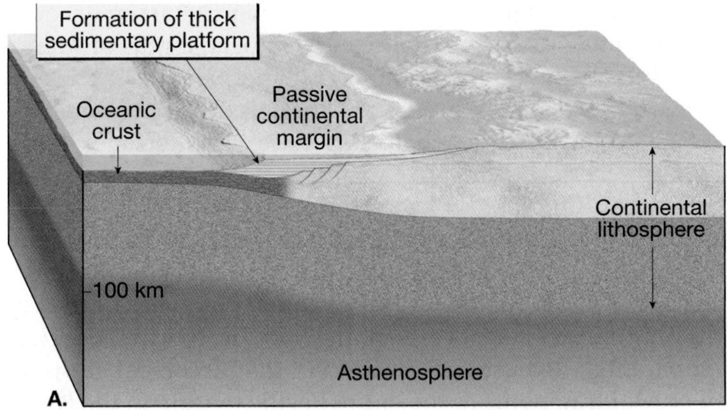

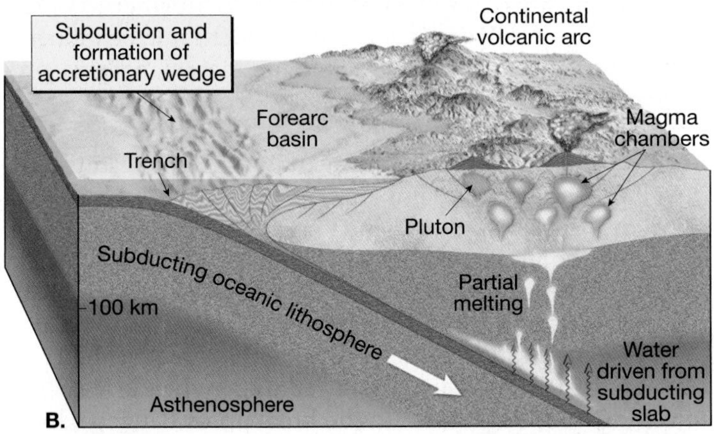

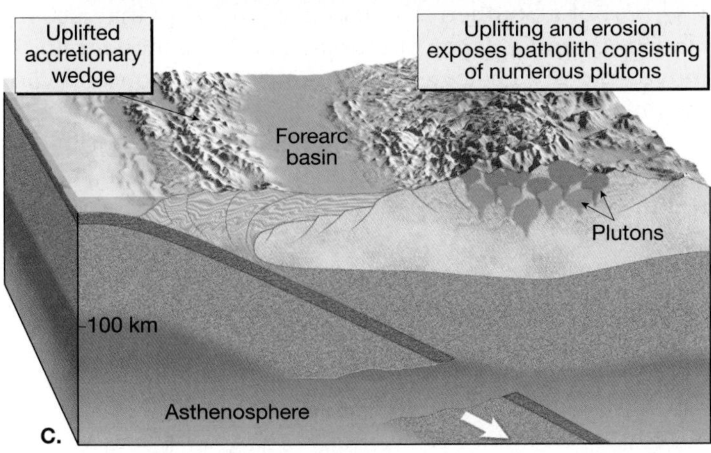

Figure 6.33 Mountain building along an Andean-type subduction zone. **A.** Passive continental margin with an extensive platform of sediments. **B.** Plate convergence generates a subduction zone, and partial melting produces a volcanic arc. Continued convergence and igneous activity further deform and thicken the crust, elevating the mountain belt, while an accretionary wedge develops. **C.** Subduction ends and is followed by a period of uplift and erosion.

trench like piles of dirt in front of a bulldozer. This chaotic accumulation of sedimentary and metamorphic rocks with occasional scraps of ocean crust is called an **accretionary wedge** (Figure 6.33B). Prolonged subduction can build an accretionary wedge that is large enough to stand above sea level (Figure 6.33C).

Andean-type mountain belts are composed of two roughly parallel zones. The volcanic arc develops on the continental block. It consists of volcanoes and large intrusive bodies intermixed with high-temperature metamorphic rocks. The seaward segment is the accretionary wedge. It consists of folded and faulted sedimentary and metamorphic rocks (Figure 6.33C).

One of the best examples of an inactive Andean-type orogenic belt is found in the western United States. It includes the Sierra Nevada and the Coast Ranges in California. These parallel mountain belts were produced by the subduction of a portion of the Pacific Basin under the western edge of the North American plate. The Sierra Nevada batholith is a remnant of a portion of the continental volcanic arc that was produced by several surges of magma over tens of millions of years. Subsequent uplifting and erosion have removed most of the evidence of past volcanic activity and exposed a core of crystalline, igneous, and associated metamorphic rocks.

In the trench region, sediments scraped from the subducting plate, plus those provided by the eroding continental volcanic arc, were intensely folded and faulted into an accretionary wedge. This chaotic mixture of rocks presently constitutes the Franciscan Formation of California's Coast Ranges. Uplifting of the Coast Ranges took place only recently, as evidenced by the young unconsolidated sediments that still mantle portions of these highlands.

Collisional Mountain Ranges

As you have seen, when a slab of oceanic lithosphere subducts beneath a continental margin, an Andean-type mountain belt develops. If the subducting plate also contains a slab of continental lithosphere, continued subduction eventually carries the continental block to the trench. Oceanic lithosphere is relatively dense and readily subducts, but continental crust is composed of low-density material that is too buoyant to undergo subduction. Consequently, the arrival of the continental block at the trench results in a collision with the overriding continent. The result is crustal shortening and thickening to produce a mountain belt.

Mountain belts can develop as a result of the collision and merger of an island arc or some other small crustal fragment with a continental block, as well as from the collision and joining of two or more continents.

Terranes and Mountain Building The process of collision and accretion (joining together) of comparatively small crustal fragments to a continental margin has generated many of the mountainous regions rimming the Pacific. Geologists refer to these accreted crustal blocks as *terranes*. Simply, the term **terrane** refers to any crustal fragment that has a geologic history distinct from that of adjoining terranes. Terranes come in various shapes and sizes.

Did You Know?

The highest city in the world, La Paz, Bolivia, is located 3630 meters (11,910 feet) above sea level, more than twice as high as Denver, Colorado—the highest major city in the United States.

What is the nature of these crustal fragments, and from where do they originate? Research suggests that prior to their accretion to a continental block, some of the fragments may have been *microcontinents* similar to the present-day island of Madagascar, located east of Africa in the Indian Ocean. Many others were island arcs similar to Japan, the Philippines, and the Aleutian Islands. Still others are submerged crustal fragments, such as *oceanic plateaus,* which were created by massive outpourings of basaltic lavas associated with hot-spot activity (Figure 6.34).

The widely accepted view is that as oceanic plates move, they carry embedded oceanic plateaus, volcanic island arcs, and microcontinents to an Andean-type subduction zone. When an oceanic plate contains a chain of small seamounts, these structures are generally subducted along with the descending oceanic slab. However, thick units of oceanic crust, such as the Ontong Java Plateau, or a mature island arc composed of abundant "light" igneous rocks may render the oceanic lithosphere too buoyant to subduct. In these situations, a collision between the crustal fragment and the continent occurs.

The sequence of events that occurs when a mature island arc reaches an Andean-type margin is shown in Figure 6.35. Because of its buoyancy, a mature island arc will not subduct beneath the continental plate. Instead, the upper portions of these thickened zones are peeled from the descending plate and thrust in relatively thin sheets upon the adjacent continental block. In some settings, continued subduction may carry another crustal fragment to the continental margin. When this fragment collides with the continental margin, it displaces the accreted island arc farther inland, adding to the zone of deformation and to the thickness and lateral extent of the continental margin.

The idea that mountain building occurs in association with the accretion of crustal fragments to a continental mass arose principally from studies conducted in the North American Cordillera (Figure 6.36). It was determined that some mountainous areas, principally those in the orogenic belts of Alaska and British Columbia, contain fossil and paleomagnetic evidence indicating that these strata once lay nearer the equator.

It is now assumed that many of the other terranes found in the North American Cordillera were once scattered throughout the eastern Pacific, much as we find island arcs and oceanic plateaus distributed in the western Pacific today (Figure 6.34). Since before the breakup of Pangaea, the eastern portion of the Pacific basin (Farallon plate) has been subducting under the western margin of North America. Apparently, this activity resulted in the piecemeal addition of crustal fragments to the entire Pacific margin of the continent—from Mexico's Baja Peninsula to northern Alaska (Figure 6.36). In a like manner, many modern microcontinents will eventually be accreted to active continental margins, producing new orogenic belts.

Continental Collisions Continental collisions result in the development of mountains that are characterized by shortened and thickened crust. Thicknesses of 50 kilometers (30 miles) are common, and some regions have crustal thicknesses in excess of 70 kilometers (40 miles). In these settings, crustal thickening is achieved through folding and faulting.

The mountain-building episode that created the Himalayas began roughly 45 million years ago when India began to collide with Asia. Prior to the breakup of Pangaea, India was part of a Southern Hemisphere landmass that also included Australia. Upon splitting from that continent, India moved rapidly, geologically speaking, a few thousand kilometers in a northward direction.

The subduction zone that facilitated India's northward migration was located near the southern margin of Asia (Figure 6.37). Ongoing subduction along Asia's margin created an Andean-type plate margin that contained a

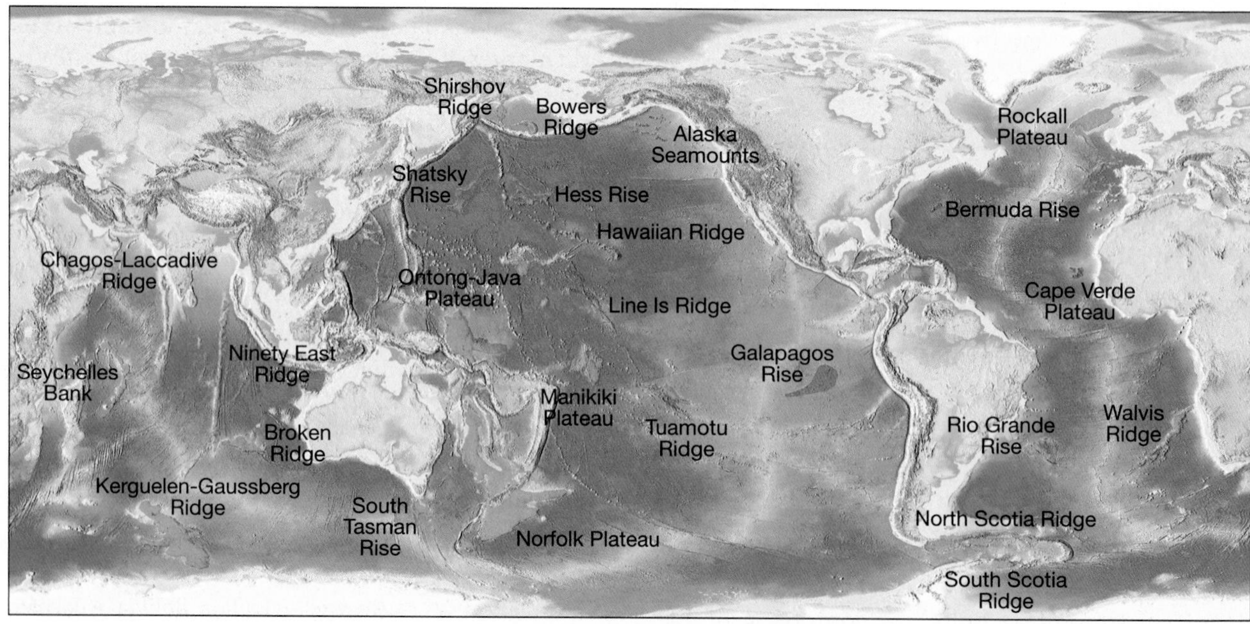

Figure 6.34 Distribution of present-day oceanic plateaus and other submerged crustal fragments. (Data from Ben-Avraham and others)

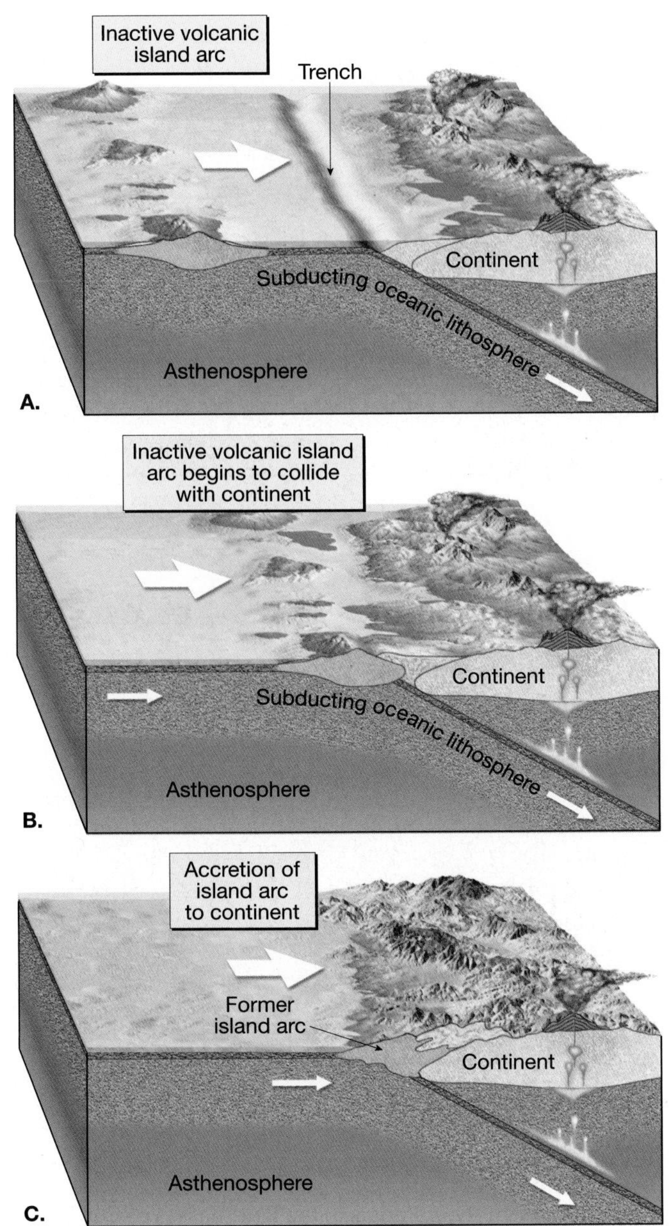

Figure 6.35 Sequence of events showing the collision and accretion of an island arc to a continental margin.

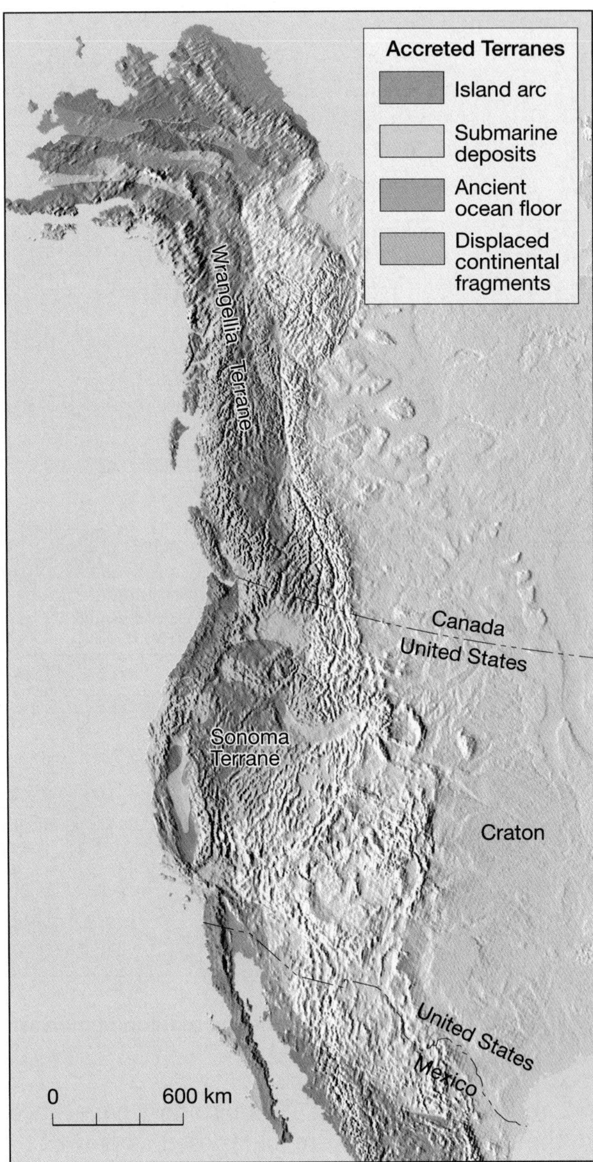

Figure 6.36 Map showing terranes thought to have been added to western North America during the past 200 million years. (Redrawn after D. R. Hutchinson and others)

well-developed volcanic arc and accretionary wedge. Eventually, the intervening ocean basin was consumed at the subduction zone and India collided with the Eurasian plate. The tectonic forces involved in the collision were immense and caused the more deformable materials located on the seaward edges of these landmasses to be highly folded and faulted (Figure 6.37). The shortening and thickening of the crust elevated great quantities of crustal material, thereby generating the spectacular Himalayan mountains (see Figure 6.31, p. 183).

In addition to uplift, crustal thickening caused lower layers to become deeply buried and to experience elevated temperatures and pressures. Partial melting within the deepest and most deformed region of the developing mountain belt produced magma bodies that intruded and further deformed the overlying rocks. It is in such environments where the metamorphic and igneous core of a major mountain belt is generated.

A similar but much older collision is believed to have taken place when the European continent collided with the Asian continent to produce the Ural Mountains, which extend in a north-south direction through Russia. Prior to the discovery of plate tectonics, geologists had difficulty explaining the existence of mountain ranges such as the Urals, which are located deep within continental interiors. How could thousands of meters of marine sediment be deposited and then become highly deformed while situated in the middle of a large stable landmass?

Other mountain ranges showing evidence of continental collisions are the Alps and the Appalachians. The Appalachians resulted from collisions among North America,

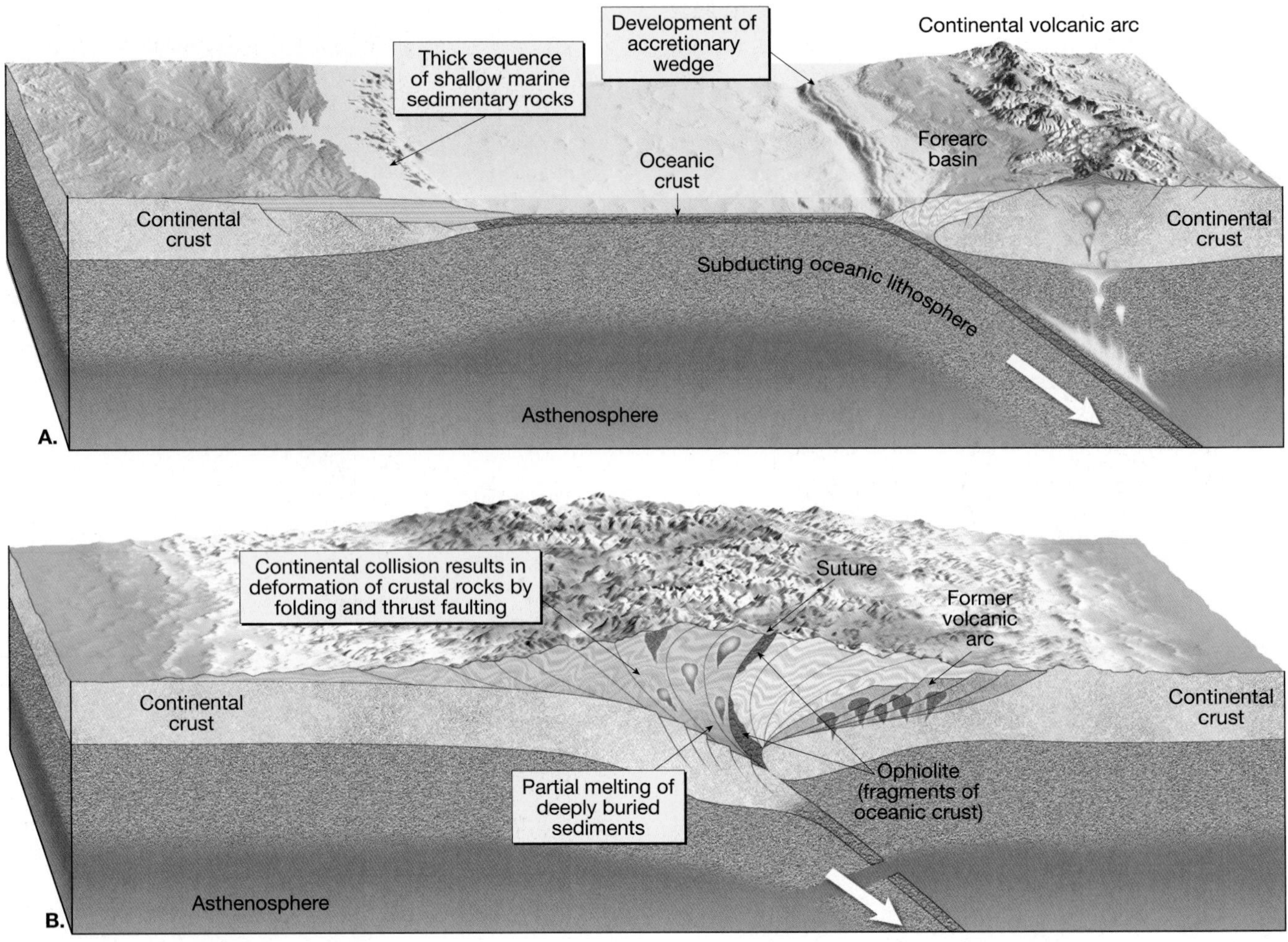

Figure 6.37 Illustration showing the collision of India with the Eurasian plate producing the spectacular Himalayas.

Europe, and northern Africa. Although they have since separated, these landmasses were juxtaposed as part of the supercontinent of Pangaea less than 200 million years ago. Detailed studies in the southern Appalachians indicate that the formation of this mountain belt was more complex than once thought. Rather than forming during a single continental collision, the Appalachians resulted from several distinct episodes of mountain building occurring over a period of nearly 300 million years.

Did You Know?

During the assembly of Pangaea, the landmasses of Europe and Siberia collided to produce the Ural Mountains. Long before the discovery of plate tectonics, this extensively eroded mountain chain was regarded as the boundary between Europe and Asia.

The Chapter in Review

1. *Earthquakes* are vibrations produced by the rapid release of energy from rocks that rupture when subjected to stresses beyond their limit. The movements that produce earthquakes occur along large fractures, called *faults,* which are associated with plate boundaries.

2. Two main groups of *seismic waves* are generated during an earthquake: (1) *surface waves,* which travel along the outer layer of Earth; and (2) *body waves,* which travel through Earth's interior. Body waves are further divided into *primary waves* (P waves), which push (compress) and pull (expand) rocks in the direction the wave is traveling, and *secondary waves* (S waves), which "shake" the particles in rock at right angles to their direction of travel. The P waves can travel through solids, liquids, and gases; the S waves travel only through solids. In any solid material, P waves travel about 1.7 times faster than S waves.

3. The location on Earth's surface directly above the focus of an earthquake is the *epicenter.* An epicenter is determined using the difference in velocities of P and S waves.

4. *There is a close correlation between earthquake epicenters and plate boundaries.* The principal earthquake epicenter zones are along the margin of the Pacific Ocean, known as the *circum-Pacific belt,* and through the world's oceans along the *oceanic ridge system.*

5. Seismologists use two fundamentally different measures to describe the size of an earthquake—intensity and magnitude. *Intensity* is a measure of the degree of ground shaking at a given locale based on the amount of damage. The *Modified Mercalli Intensity Scale* uses damages to buildings in California to estimate the intensity of ground shaking for a local earthquake. *Magnitude* is calculated from seismic records and estimates the amount of energy released at the source of an earthquake. Using the *Richter scale,* the magnitude of an earthquake is estimated by measuring the *amplitude* (maximum displacement) of the largest seismic wave recorded. A logarithmic scale is used to express magnitude, in which a tenfold increase in ground shaking corresponds to an increase of 1 on the magnitude scale. *Moment magnitude* is currently used to estimate the size of moderate and large earthquakes. It is calculated using the average displacement of the fault, the area of the fault surface, and the sheer strength of the faulted rock.

6. The most obvious factors that determine the amount of destruction accompanying an earthquake are the *magnitude* of the earthquake and the *proximity* of the quake to a populated area. *Structural damage* attributable to earthquake vibrations depends on several factors, including (1) *intensity,* (2) *duration* of the vibrations, (3) *nature of the material* upon which the structure rests, and (4) the *design* of the structure. Secondary effects of earthquakes include *tsunamis, landslides, ground subsidence,* and *fire.*

7. As indicated by the behavior of P and S waves as they travel through Earth, the four major zones of Earth's interior are: (1) *crust* (the very thin outer layer, 5 to 40 kilometers); (2) *mantle* (a rocky layer located below the crust with a thickness of 2900 kilometers); (3) *outer core* (a layer about 2270 kilometers thick, which exhibits the characteristics of a mobile liquid); and (4) *inner core* (a solid metallic sphere with a radius of about 1216 kilometers).

8. *Deformation* refers to changes in the shape and/or volume of a rock body. Rocks deform differently depending on the environment (temperature and confining pressure), the composition of the rock, and the length of time stress is maintained. Rocks first respond by deforming *elastically* and will return to their original shape when the stress is removed. Once their elastic limit (strength) is surpassed, rocks either deform by ductile flow or they fracture. *Ductile deformation* is a solid-state flow that results in a change in size and shape of rocks without fracturing. Ductile deformation occurs in a high-temperature/high-pressure environment. In a near-surface environment, most rocks deform by *brittle failure.*

9. Among the most basic geologic structures associated with rock deformation are *folds* (flat-lying sedimentary and volcanic rocks bent into a series of wavelike undulations). The two most common types of folds are *anticlines,* formed by the upfolding, or arching, of rock layers, and *synclines,* which are downfolds. Most folds are the result of horizontal *compressional stresses. Domes* (upwarped structures) and *basins* (downwarped structures) are circular or somewhat elongated folds formed by vertical displacements of strata.

10. *Faults* are fractures in the crust along which appreciable displacement has occurred. Faults in which the movement is primarily vertical are called *dip-slip faults.* Dip-slip faults include both *normal* and *reverse faults.* Low-angle reverse faults are called *thrust faults.* Normal faults indicate *tensional stresses* that pull the crust apart. Along spreading centers, divergence can cause a central block called a *graben,* bounded by normal faults, to drop as the plates separate.

11. Subduction of oceanic lithosphere under a continental block gives rise to an *Andean-type plate margin* that is characterized by a continental volcanic arc and associated igneous plutons. In addition, sediment derived from the land, as well as material scraped from the subducting plate, becomes plastered against the landward side of the trench, forming an *accretionary wedge.* An excellent example of an inactive Andean-type mountain belt is found in the western United States and includes the Sierra Nevada and the Coast Range in California.

12. Mountain belts can develop as a result of the collision and merger of an island arc, oceanic plateau, or some other small crustal fragment to a continental block. Many of the mountain belts of the North American Cordillera were generated in this manner.

13. Continued subduction of oceanic lithosphere beneath an Andean-type continental margin will eventually close an ocean basin. The result will be a *continental collision* and the development of compressional mountains that are characterized by shortened and thickened crust as exhibited by the Himalayas. The development of a major mountain belt is often complex, involving two or more distinct episodes of mountain building. Continental collisions have generated many mountain belts, including the Alps, Urals, and Appalachians.

Key Terms

accretionary wedge (p. 185)
active continental margin (p. 184)
aftershock (p. 163)
anticline (p. 178)
asthenosphere (p. 177)
basin (p. 179)
body wave (p. 164)
brittle failure or brittle deformation (p. 178)
core (p. 177)
crust (p. 177)
deformation (p. 177)
dip-slip fault (p. 179)
dome (p. 179)
ductile deformation (p. 178)
earthquake (p. 161)
elastic rebound (p. 162)
epicenter (p. 166)
fault (p. 179)
fault-block mountains (p. 180)
fault scarp (p. 179)

focus (p. 161)
fold (p. 178)
foreshock (p. 163)
graben (p. 180)
horst (p. 180)
inner core (p. 177)
intensity (p. 167)
liquefaction (p. 171)
lithosphere (p. 177)
lower mantle (p. 177)
magnitude (p. 167)
mantle (p. 177)
Modified Mercalli Intensity Scale (p. 167)
moment magnitude (p. 169)
normal fault (p. 180)
orogenesis (p. 182)
outer core (p. 177)
passive continental margin (p. 184)
primary (P) wave (p. 164)
reverse fault (p. 180)
Richter scale (p. 168)
secondary (S) wave (p. 164)
seismic sea wave (tsunami) (p. 172)
seismogram (p. 164)
seismograph (p. 164)
seismology (p. 164)
strike-slip fault (p. 182)
surface wave (p. 164)
syncline (p. 178)
terrane (p. 185)
thrust fault (p. 180)
transform fault (p. 182)

Questions for Review

1. What is an earthquake? Under what circumstances do earthquakes occur?
2. How are faults, foci, and epicenters related?
3. Who was first to explain the actual mechanism by which earthquakes are generated?
4. Explain what is meant by elastic rebound.
5. Faults that are experiencing no active creep may be considered "safe." Rebut or defend this statement.
6. Describe the principle of a seismograph.
7. List the major differences between P and S waves.
8. Most strong earthquakes occur in a zone on the globe known as the______.
9. An earthquake measuring 7 on the Richter scale releases about______times more energy than an earthquake with a magnitude of 6.
10. In addition to the destruction created directly by seismic vibrations, list three other types of destruction associated with earthquakes.
11. What is a tsunami? How is one generated?
12. Contrast the physical makeup of the asthenosphere and the lithosphere.
13. What is rock deformation?
14. How is brittle deformation different from ductile deformation?
15. Distinguish between anticlines and synclines, domes and basins, anticlines and domes.
16. Contrast the movements that occur along normal and reverse faults. What type of force is indicated by each fault?
17. How are reverse faults different from thrust faults? In what way are they the same?
18. The San Andreas Fault is an excellent example of a ______fault.
19. In the plate tectonics model, which type of plate boundary is most directly associated with mountain building?
20. What is an accretionary wedge? Briefly describe its formation.
21. What is a passive continental margin? Give an example. Give an example of an active continental margin.
22. How does the plate tectonics theory help explain the existence of fossil marine life in rocks atop the Ural Mountains?
23. Define the term *terrane.* How is it different from the term *terrain?*

Online Study Guide

The *Foundations of Earth Science* Web site uses the resources and flexibility of the Internet to aid in your study of the topics in this chapter. Written and developed by Earth science instructors, this site will help improve your understanding of Earth science. Visit **http://www.prenhall.com/lutgens** and click on the cover of *Foundations of Earth Science 5e* to find:

- Online review quizzes.
- Critical thinking exercises.
- Links to chapter-specific Web resources.
- Internet-wide key-term searches.

http://www.prenhall.com/lutgens

GEODe: Earth Science

GEODe: Earth Science makes studying more effective by reinforcing key concepts using animation, video, narration, interactive exercises, and practice quizzes. A copy is included with every copy of *Foundations of Earth Science 5e.*

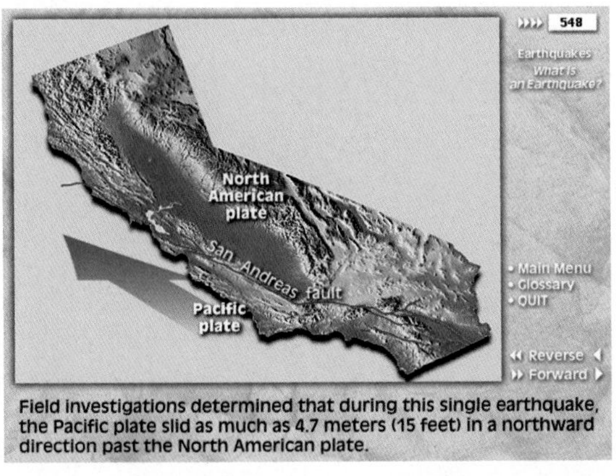

Field investigations determined that during this single earthquake, the Pacific plate slid as much as 4.7 meters (15 feet) in a northward direction past the North American plate.

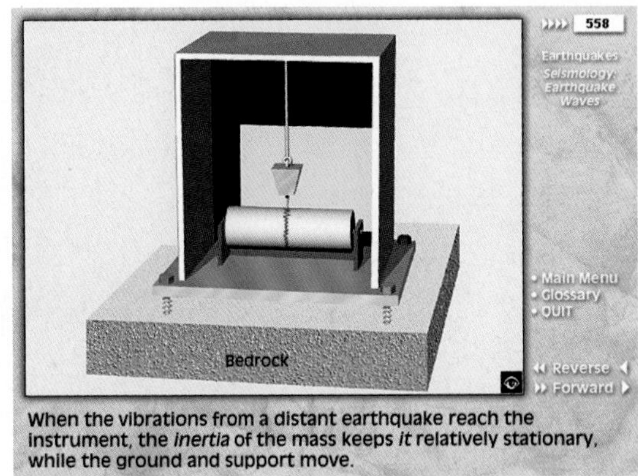

When the vibrations from a distant earthquake reach the instrument, the *inertia* of the mass keeps *it* relatively stationary, while the ground and support move.

CHAPTER

7

Fires Within: Igneous Activity

A recent eruption of Italy's Mount Etna. (Photo by Art Wolfe)

FOCUS ON LEARNING

To assist you in learning the important concepts in this chapter, you will find it helpful to focus on the following questions:

1. What primary factors determine the nature of volcanic eruptions? How do these factors affect a magma's viscosity?
2. What materials are associated with a volcanic eruption?
3. What are the eruptive patterns and characteristic shapes of the three groups of volcanoes generally recognized by volcanologists?
4. What are some other Earth features formed by volcanic activity?
5. What criteria are used to classify intrusive igneous bodies? What are some of these features?
6. What is the relation between volcanic activity and plate tectonics?

On Sunday, May 18, 1980, one of the largest volcanic eruptions to occur in North America in historic times transformed a picturesque volcano into a decapitated remnant. On this date in southwestern Washington State, Mount St. Helens erupted with tremendous force (Figure 7.1). The blast blew out the entire north flank of the volcano, leaving a gaping hole. In one brief moment, a prominent volcano whose summit had been more than 2900 meters (9500 feet) above sea level was lowered by more than 400 meters (1300 feet).

The event devastated a wide swath of timber-rich land on the north side of the mountain. Trees within a 400-square-kilometer (150-square-mile) area lay intertwined and flattened, stripped of their branches and appearing from the air like toothpicks strewn about (Figure 7.2). The accompanying mudflows carried ash, trees, and water-saturated rock debris 29 kilometers (18 miles) down the Toutle River. The eruption claimed 59 lives, some dying from the intense heat and the suffocating cloud of ash and gases.

The eruption ejected nearly a cubic kilometer of ash and rock debris. Following the devastating explosion, Mount St. Helens continued to emit great quantities of hot gases and ash, some of which were propelled more than 18,000 meters (11 miles) into the stratosphere. During the next few days, this very fine-grained material was carried around Earth by strong upper-air winds. Measurable deposits were reported in Oklahoma and Minnesota, with crop damage into central Montana. Meanwhile, ash fallout in the immediate vicinity exceeded 2 meters (7 feet) in depth. The air over Yakima, Washington (130 kilometers [80 miles] to the east), was so filled with ash that residents experienced midnight-like darkness at noon.

Mount St. Helens is one of 15 large volcanoes and innumerable smaller ones that comprise the Cascade Range, which extends from British Columbia to northern California. Eight of the largest volcanoes have been active in the past few hundred years. Of the remaining seven active volcanoes, the most likely to erupt again are Mount Baker and Mount Rainier in Washington, Mount Shasta and Lassen Peak in California, and Mount Hood in Oregon.

Not all volcanic eruptions are as violent as the 1980 Mount St. Helens event. Some volcanoes, such as Hawaii's Kilauea volcano, generate relatively quiet outpourings of fluid lavas. These "gentle" eruptions are not without some fiery displays; occasionally, fountains of incandescent lava spray hundreds of meters into the air. Such events, however, typically pose minimal threat to human life and property.

Testimony to the quiet nature of Kilauea's eruptions is the fact that the Hawaiian Volcanoes Observatory has operated on its summit since 1912—despite the fact that Kilauea has had more than 50 eruptive phases since record keeping began in 1823. Further, the longest and largest of Kilauea's eruptions began in 1983 and remains active at the time of this writing, although it has received only modest media attention.

Why do volcanoes such as Mount St. Helens erupt explosively, whereas others such as Kilauea are relatively quiet? Why do volcanoes occur in chains such as the Aleutian Islands or the Cascade Range? Why do some volcanoes form on the ocean floor, while others occur on the continents? This chapter will deal with these and other questions as we explore the nature and movement of magma and lava.

The Nature of Volcanic Eruptions

Forces Within
▼ Igneous Activity

Volcanic activity is commonly perceived as a process that produces a picturesque, cone-shaped structure that periodically erupts in a violent manner. Although some eruptions are very explosive, many others are not. What determines whether a volcano extrudes magma violently or "gently"? The primary factors include the magma's *composition*, its *temperature*, and the amount of *dissolved gases* it contains. To varying degrees, these factors affect the magma's **viscosity.** The more viscous ("thicker") the material, the greater its resistance to flow. For example, syrup is more viscous than water. The viscosity of magma associated with an explosive eruption may be thousands of times greater than magma that is extruded in a quiescent manner.

Factors Affecting Viscosity

The effect of temperature on viscosity is easily seen. Just as heating syrup makes it more fluid (less viscous), the mobility of lava is strongly influenced by temperature. As a lava flow cools and begins to congeal, its mobility decreases and eventually the flow halts.

A more significant influence on volcanic behavior is the chemical composition of magmas. This was discussed in Chapter 2, along with the classification of igneous rocks. Recall that a major difference among various igneous rocks is their silica (SiO_2) content. The same is true of the magmas from which rocks form. Magmas that produce basaltic rocks contain about 50 percent silica, whereas magmas that produce granitic rocks contain more than 70 percent silica (Table 7.1).

A magma's viscosity is directly related to its silica content. In general, the more silica in magma, the greater is its viscosity. The flow of magma is impeded because silicate structures link together into long chains, even before crystallization begins. Consequently, because of their high silica content, granitic lavas are very viscous and tend to form comparatively short, thick flows. By contrast, basaltic lavas, which contain less silica, tend to be more fluid and have been known to travel distances of 150 kilometers (90 miles) or more before congealing (Figure 7.3).

In Hawaiian eruptions, the magmas are hot and basaltic, so they are extruded with ease. By contrast, highly viscous magmas are more difficult to force through a vent. On occasion, the vent may become plugged with viscous magma, which can result in a buildup of gases that produces an explosive eruption. However, a viscous magma is not explosive by itself. It is the gas content that puts the "bang" into a violent eruption.

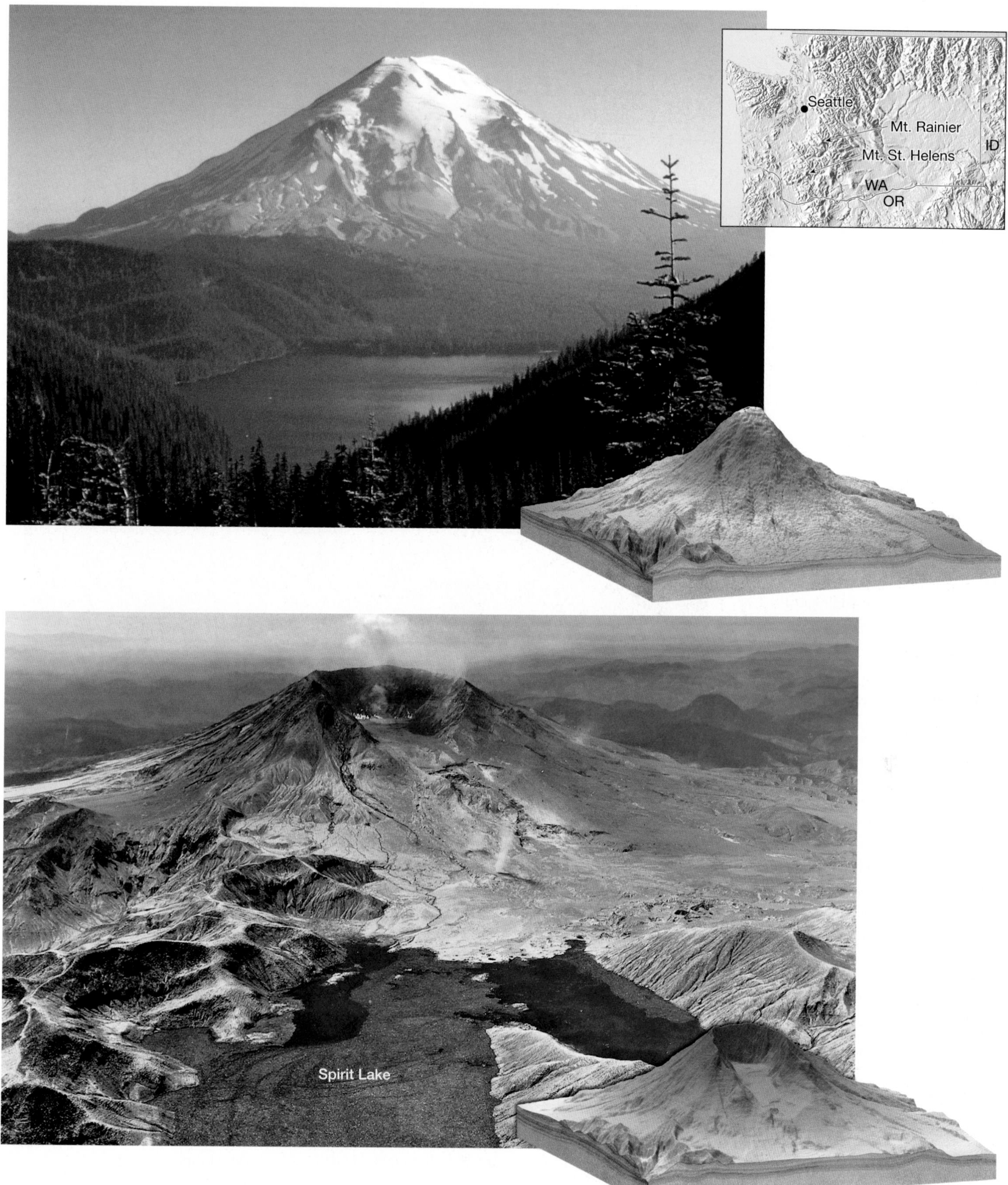

Figure 7.1 Before-and-after photographs show the transformation of Mount St. Helens caused by the May 18, 1980, eruption. The dark area in the "after" photo is debris-filled Spirit Lake, partially visible in the "before" photo. (Photos courtesy of U.S. Geological Survey)

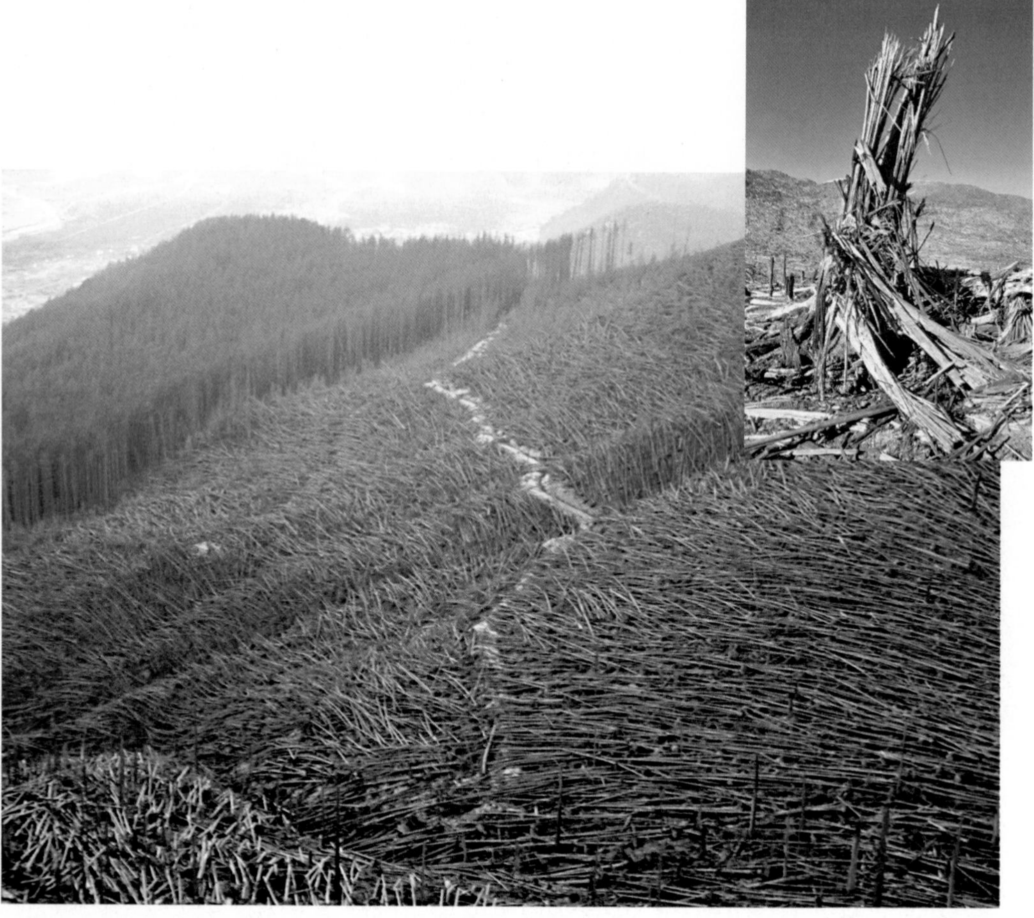

Figure 7.2 Douglas fir trees were snapped off or uprooted by the lateral blast of Mount St. Helens on May 18, 1980. (Photo by Lyn Topinka/AP Photo/Courtesy of USGS; inset photo by John Burnley/Photo Researchers, Inc.)

Importance of Dissolved Gases in Magma

Volatiles, the gaseous components of magma, provide the force that extrudes molten rock from a volcanic vent. These gases are mostly water vapor and carbon dioxide. As magma moves into a near-surface environment, such as within a volcano, the confining pressure in the uppermost portion of the magma body is greatly reduced. The reduction in confining pressure allows dissolved gases to be released suddenly, just as opening a soda bottle allows dissolved carbon dioxide gas bubbles to escape.

At high temperatures and low near-surface pressures, these gases will expand to occupy hundreds of times their original volume. Very fluid basaltic magmas allow the expanding gases to bubble upward and escape from the vent with relative ease. As they escape, the gases will often propel incandescent lava hundreds of meters into the air, producing lava fountains. Although spectacular, such fountains are mostly harmless and not generally associated with major explosive events that cause great loss of life and property

Did You Know?

Chile, Peru, and Ecuador boast the highest volcanoes in the world. Dozens of cones exceed 6000 meters (20,000 feet). Two volcanoes located in Ecuador, Chimborazo and Cotopaxi, were once considered the world's highest mountains. That distinction remained until the Himalayas were surveyed in the nineteenth century.

Table 7.1 Variations in properties among magmas of differing compositions

Property	Granitic Magma	Basaltic Magma	Andesitic Magma
Silica content	Least (about 70%)	Intermediate (about 50%)	Most (about 60%)
Viscosity	Least ("thinnest")	Intermediate	Greatest ("thickest")
Tendency to form lavas	Highest	Intermediate	Least
Tendency to form pyroclastics	Least	Intermediate	Greatest
Melting temperature	Highest	Intermediate	Lowest

Figure 7.3 A river of basaltic lava from the January 17, 2002, eruption of Mount Nyiragongo destroyed many homes in Goma, Congo. (AFP Photo/Marco Longari/Getty Images, Inc./Agence France Presse)

Did You Know?

The eruption of Tambora, Indonesia, in 1815 is the largest-known volcanic event in modern history. About 20 times more ash and rock were explosively ejected during this eruption than was emitted during the 1980 Mount St. Helens event. The sound of the explosion was heard an incredible 4800 kilometers (3000 miles) away, about the distance across the United States.

(Figure 7.4). Rather, eruptions of fluid basaltic lavas, such as those that occur in Hawaii, are relatively quiescent.

At the other extreme, highly viscous magmas impede the upward migration of expanding gases. The gases collect in bubbles and pockets that increase in size until they explosively eject the semimolten rock from the volcano. The result is an eruption such as that of Mount St. Helens, or Mount Pinatubo in the Philippines.

To summarize, the viscosity of magma, plus the quantity of dissolved gases and the ease with which they can escape, determines the nature of a volcanic eruption. We can now understand the "gentle" volcanic eruptions of hot, fluid lavas in Hawaii and the explosive, violent eruptions of viscous lavas from volcanoes such as Mount St. Helens.

What Is Extruded During Eruptions?

Forces Within
▼ Igneous Activity

Lava may appear to be the primary material extruded from a volcano, but this is not always the case. Just as often, explosive eruptions eject huge quantities of broken rock, lava "bombs," fine ash, and dust. Moreover, all volcanic eruptions emit large amounts of gas. This section will examine each of these materials associated with a volcanic eruption.

Lava Flows

The low silica content of hot basaltic lavas usually makes them very fluid. They flow in thin, broad sheets or streamlike ribbons. On the island of Hawaii, such lavas have been clocked at speeds of 30 kilometers (20 miles) per hour down steep slopes. These velocities are rare, however, and flow rates of 10 to 300 meters (30 to 1000 feet) per hour are more common.

Figure 7.4 Pyroclastic material and lava ejected from a flank eruption at Kilauea volcano, Hawaii. (Photo by Douglas Peebles)

In contrast, the movement of silica-rich (granitic) lava is on occasion too slow to be perceptible.

Two types of lava flows are known by their Hawaiian names. The most common of these, **aa** (pronounced ah-ah) **flows,** have surfaces of rough jagged blocks with dangerously sharp edges and spiny projections (Figure 7.5A). Crossing an aa flow can be a trying and miserable experience. By contrast, **pahoehoe** (pronounced pah-hoy-hoy) **flows** exhibit smooth surfaces that often resemble the twisted braids of ropes (Figure 7.5B). *Pahoehoe* means "on which one can walk."

Although aa and pahoehoe lavas can erupt from the same vent, pahoehoe flows have higher temperatures than aa flows. Consequently, pahoehoe flows are more fluid than lavas that form aa flows. In addition, pahoehoe lavas sometimes transform into aa flows. (Note, however, that aa flows cannot revert to pahoehoe.)

One of the factors thought to facilitate the change from pahoehoe to aa is cooling that occurs as the flow moves from the vent. Cooling promotes bubble formation, which, as stated previously, increases viscosity. Escaping gas bubbles produce numerous voids and sharp spines in the surface of the congealing lava. As the molten interior advances, the outer crust is broken further, transforming a rather smooth surface into an advancing mass of rough, clinkery rubble (Figure 7.5A).

The lava associated with the Mexican volcano Parícutin that buried the city of San Juan Parangaricutiro was of the aa type (see Figure 7.13, p. 204). One of the flows from Parícutin moved an average of only about a meter per day, but it advanced continuously for more than three months.

Pahoehoe lavas have also been known to move at agonizingly slow rates. In 1990, near the Hawaiian village of Kalapana, the villagers had to watch for weeks as a pahoehoe flow crept toward their homes at only a few meters per hour. Although most residents were able to escape injury, they were unable to stop the flow and their houses were eventually incinerated.

Gases

Magmas contain varied amounts of dissolved gases (*volatiles*) held in the molten rock by confining pressure, just as carbon dioxide is held in soft drinks. As with soft drinks, as soon as the pressure is reduced, the gases begin to escape. Obtaining gas samples from an erupting volcano is difficult and dangerous, so geologists can often only estimate the amount of gas originally contained within the magma.

The gaseous portion of most magmas is believed to make up 1 to 5 percent of the total weight, with most of this being water vapor. Although the percentage may be small, the actual quantity of emitted gas can exceed thousands of tons daily.

The composition of volcanic gases is important to scientists because these gases are thought to be the original source of the water for the oceans. Further, volcanic eruptions have contributed significantly to gases that make up the atmosphere. Analysis of samples taken during Hawaiian eruptions indicates that the gases are about 70 percent water vapor, 15 percent carbon dioxide, 5 percent nitrogen, 5 percent sulfur, and lesser amounts of chlorine, hydrogen, and argon. Sulfur compounds are easily recognized by their pungent odor and by the ease with which they form sulfuric acid—a natural source of air pollution.

Pyroclastic Materials

When basaltic lava is extruded, dissolved gases escape quite freely and continually. These gases propel incandescent blobs of lava to great heights. Some of this ejected material may land

A.

B.

Figure 7.5 **A.** Typical slow-moving aa lava flow. **B.** Typical pahoehoe (ropy) lava flow, Kilauea, Hawaii. (Photo by J. D. Griggs, U.S. Geological Survey, Denver)

near the vent and build a cone-shaped structure, whereas smaller particles will be carried great distances by the wind. By contrast, viscous (rhyolitic) magmas are highly charged with gases, and upon release they expand greatly as they blow pulverized rock, lava, and glass fragments from the vent. The particles produced in both these situations are referred to as **pyroclastic materials** (*pyro* = fire, *clast* = fragment). These ejected fragments range in size from very fine dust and sand-sized volcanic ash (less than 2 millimeters) to pieces that weigh several tons.

Ash and *dust* particles are produced from gas-laden viscous magma during an explosive eruption (see Figure 7.15, p. 206). As magma moves up in the vent, the gases rapidly expand, generating a froth of melt that might resemble the froth that flows from a just opened bottle of champagne. As the hot gases expand explosively, the froth is blown into very fine glassy fragments. When the hot ash falls, the glassy shards often fuse to form a rock called *welded tuff*. Sheets of this material, as well as ash deposits that later consolidate, cover vast portions of the western United States.

Also common are pyroclasts that range in size from small beads to walnuts termed *lapilli* ("little stones"). These ejecta are commonly called *cinders* (2 to 64 millimeters). Particles larger than 64 millimeters (2.5 inches) in diameter are called *blocks* when they are made of hardened lava and *bombs* when they are ejected as incandescent lava. Because bombs are semimolten upon ejection, they often take on a streamlined shape as they hurtle through the air (Figure 7.6). Because of their size, bombs and blocks usually fall on the slopes of the volcano; however, they are occasionally propelled far from the volcano by the force of escaping gases.

Did You Know?

Lava bombs 6 meters (20 feet) long and weighing over 200 tons were thrown 600 meters (2000 feet, about the length of 7 football fields) from the vent during an eruption of the Japanese volcano Asama.

Volcanic Structures and Eruptive Styles

Forces Within
▼ Igneous Activity

The popular image of a volcano is that of a solitary, graceful, snowcapped cone, such as Mount Hood in Oregon or Japan's Fujiyama. These picturesque, conical mountains are produced by volcanic activity that occurred intermittently over thousands, or even hundreds of thousands, of years. However, many volcanoes do not fit this image. Some volcanoes are only 30 meters (100 feet) high and formed during a single eruptive phase that may have lasted only a few days.

Volcanic landforms come in a wide variety of shapes and sizes, and each structure has a unique eruptive history. Nevertheless, volcanologists have been able to classify volcanic

Figure 7.6 Volcanic bombs forming during an eruption of Hawaii's Kilauea volcano. Ejected lava fragments take on a streamlined shape as they sail through the air. The bomb in the insert is about 10 centimeters (4 inches) long. (Photo by Arthur Roy/National Audubon Society/Photo Researchers, Inc.; inset photo by E. J. Tarbuck)

landforms and determine their eruptive patterns. This section will consider the general anatomy of a volcano and look at three major volcanic types: *shield volcanoes, cinder cones,* and *composite cones.* This discussion will be followed by an overview of other significant volcanic landforms.

Anatomy of a Volcano

Volcanic activity frequently begins when a fissure (crack) develops in the crust as magma moves forcefully toward the surface. As the gas-rich magma moves up this linear fissure, its path is usually localized into a circular **conduit,** or **pipe,** that terminates at a surface opening called a **vent** (Figure 7.7). Successive eruptions of lava, pyroclastic material, or frequently a combination of both often separated by long periods of inactivity eventually build the structure we call a **volcano.**

Located at the summit of most volcanoes is a somewhat funnel-shaped depression, called a **crater** (*crater* = a bowl). Volcanoes that are built primarily by the ejection of pyroclastic materials typically have craters that form by gradual accumulation of volcanic debris on the surrounding rim. Other craters form during explosive eruptions as the rapidly ejected particles erode the crater walls. Craters also form when the summit area of a volcano collapses following an eruption (Figure 7.8).

Some volcanoes have very large circular depressions called *calderas.* Whereas most craters have diameters ranging from a few tens to a few hundreds of meters, the diameters of calderas are typically greater than one kilometer and in rare cases can exceed 50 kilometers (31 miles). We will consider the formation of various types of calderas later in this chapter.

During early stages of growth, most volcanic discharges come from a central summit vent. As a volcano matures, material also tends to be emitted from fissures that develop along the flanks, or base, of the volcano. Continued activity from a flank eruption may produce a small **parasitic cone** (*parasitus* = one who eats at the table of another). Mount Etna in Italy, for example, has more than 200 secondary vents, some of which have built cones. Many of these vents, however, emit only gases and are appropriately called **fumaroles** (*fumus* = smoke).

The form of a particular volcano is largely determined by the composition of the contributing magma. As you will see, fluid Hawaiian-type lavas tend to produce broad structures with gentle slopes, whereas more viscous silica-rich lavas (and some gas-rich basaltic lavas) tend to generate cones with moderate to steep slopes.

Shield Volcanoes

Shield volcanoes are produced by the accumulation of fluid basaltic lavas and exhibit the shape of a broad slightly domed structure that resembles a warrior's shield (Figure 7.9). Most (but not all) shield volcanoes have grown up from the deep-ocean floor to form islands or seamounts. For example, the islands of the Hawaiian chain, Iceland, and the Galapagos are either a single shield volcano or the coalescence of several shields. Extensive study of the Hawaiian Islands confirms that each shield was built from a myriad of basaltic lava flows averaging a few meters thick. Also, these islands consist of only about 1 percent pyroclastic ejecta.

Mauna Loa: Earth's Largest Volcano Mauna Loa is one of five overlapping shield volcanoes that together comprise the Big Island of Hawaii (Figure 7.9). From its base on the floor of the Pacific Ocean to its summit, Mauna Loa is more than 9 kilometers (6 miles) high, exceeding the height of Mount Everest. This massive pile of basaltic rock has a volume of 40,000 cubic kilometers (9600 cubic miles) that was extruded over a period of nearly a million years. For comparison, the

Did You Know?

Kilauea volcano located on the Big Island of Hawaii is the most continuously active volcano on Earth. Eruptions have regularly led to lava flowing into the ocean at a rate of roughly 5 meters (16 feet) per second for years on end.

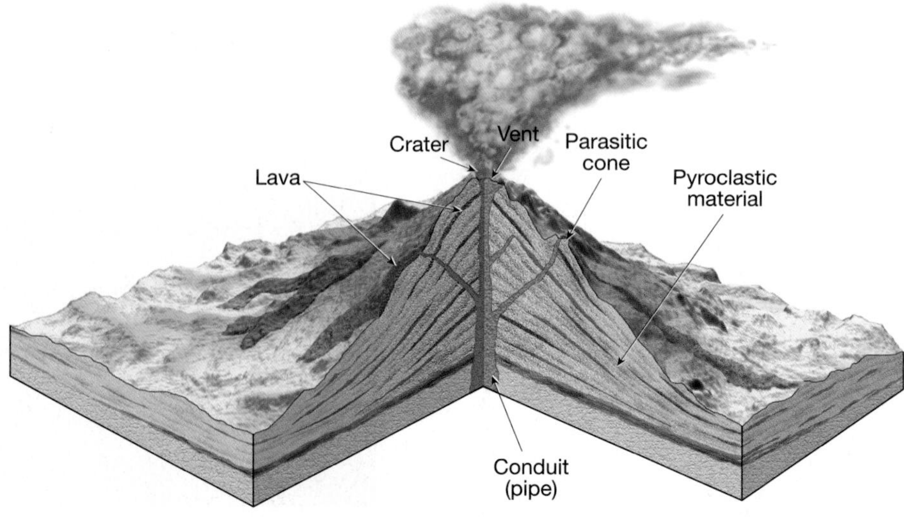

Figure 7.7 Anatomy of a "typical" composite cone (see also Figures 7.9 and 7.12 for a comparison with a shield and cinder cone, respectively).

Figure 7.8 The crater of Mount Vesuvius, Italy. The city of Naples is located northwest of Vesuvius, while Pompeii, the Roman town that was buried in an eruption in A.D. 79, is located southeast of the volcano. (Courtesy of NASA; inset image by Krafft/Photo Researchers, Inc.)

volume of material comprising Mauna Loa is roughly 200 times greater than the amount comprising a large composite cone such as Mount Rainier (Figure 7.10). Most shields, however, are more modest in size. For example, the classic Icelandic shield, Skjalbreidur, rises to a height of only about 600 meters (2000 feet) and is 10 kilometers (6 miles) across its base.

Young shields, particularly those located in Iceland, emit very fluid lava from a central summit vent and have sides with gentle slopes that vary from 1 to 5 degrees. Mature shields, as exemplified by Mauna Loa, have steeper flanks, while their summits are comparatively flat. During the mature stage, lavas are discharged from the summit vents, as well as from rift zones that develop along the slopes. Most lava is discharged as the fluid pahoehoe type, but as these flows cool downslope, many change into a clinkery aa flow. Once an eruption is well established, a large fraction of the lava (perhaps 80 percent) flows through a well-developed system of lava tubes (tunnels). This greatly increases the distance lava can travel before it solidifies.

Another feature common to a mature, active shield volcano is a large steep-walled caldera that occupies its summit. Mauna Loa's summit caldera measures 2.6 by 4.5 kilometers (1.6 by 2.8 miles) and has a depth that averages about 150 meters (500 feet). Calderas on large shield volcanoes form when the roof above the magma chamber collapses. This can occur through the evacuation of a magma reservoir following a large eruption or as magma migrates to the flank of a volcano to feed a fissure eruption.

Kilauea, Hawaii: Eruption of a Shield Volcano Kilauea, the most active and intensely studied shield volcano in the world, is located on the island of Hawaii in the shadow of Mauna Loa. More than 50 eruptions have been witnessed here since record keeping began in 1823. Several months before each eruptive phase, Kilauea inflates as magma gradually migrates upward and accumulates in a central reservoir located a few kilometers below the summit. For up to 24 hours in advance of an eruption, swarms of small earthquakes warn of the impending activity.

Most of the activity on Kilauea during the past 50 years occurred along the flanks of the volcano in a region called the East Rift Zone. The longest and largest rift eruption ever

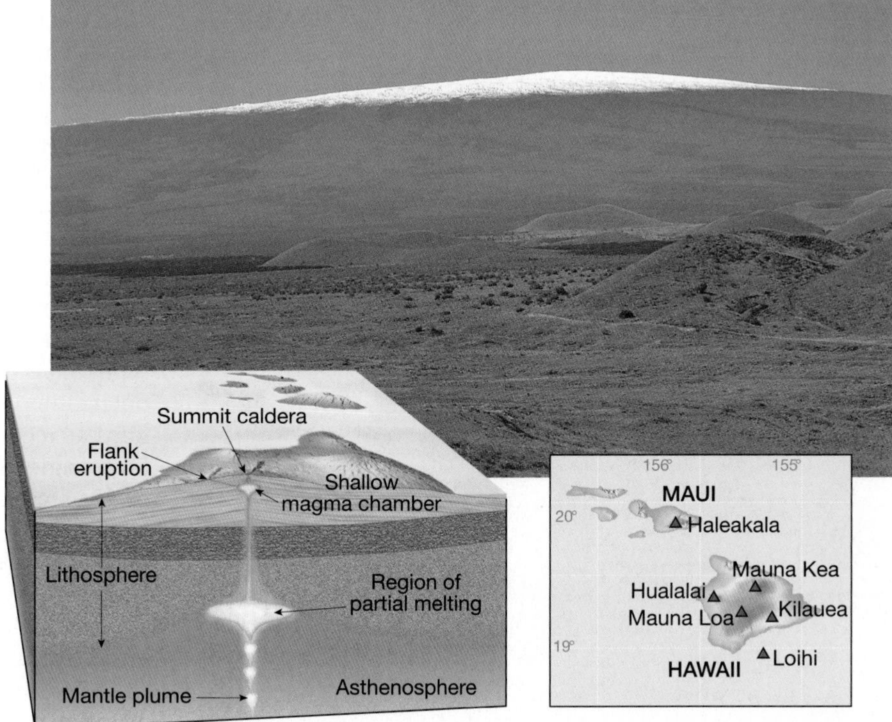

Figure 7.9 Mauna Loa is one of five shield volcanoes that together make up the island of Hawaii. Shield volcanoes are built primarily of fluid basaltic lava flows and contain only a small percentage of pyroclastic materials. (Photo by Greg Vaughn)

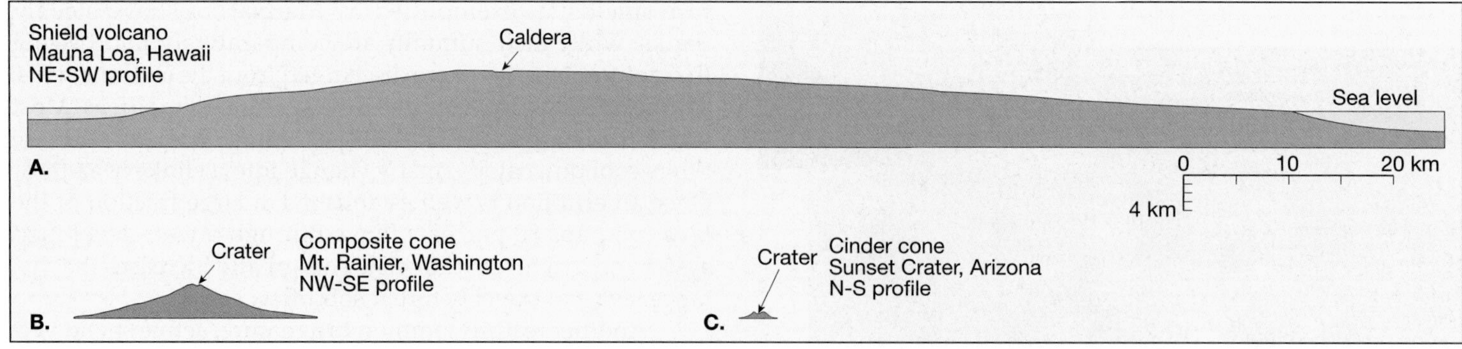

Figure 7.10 Profiles of volcanic landforms. **A.** Profile of Mauna Loa, Hawaii, the largest shield volcano in the Hawaiian chain. Note size comparison with Mount Rainier, Washington, a large composite cone. **B.** Profile of Mount Rainier, Washington. Note how it dwarfs a large cinder cone. **C.** Profile of Sunset Crater, Arizona, a relatively large steep-sided cinder cone.

recorded on Kilauea began in 1983 and continues to this day, with no signs of abating. The first discharge began along a 6-kilometer (4-mile) fissure where a 100-meter (330-foot) high "curtain of fire" formed as red-hot lava was ejected skyward. When the activity became localized, a cinder and spatter cone given the Hawaiian name *Puu Oo* was built. Over the next three years, the general eruptive pattern consisted of short periods (hours to days) when fountains of gas-rich lava sprayed skyward (Figure 7.11). Each event was followed by nearly a month of inactivity.

By the summer of 1986, a new vent opened up 3 kilometers (2 miles) downrift. Here, smooth-surfaced pahoehoe lava formed a lava lake. The lake occasionally overflowed, but more often lava escaped through tunnels to feed pahoehoe flows that moved down the southeastern flank of the volcano toward the sea. These flows destroyed nearly a hundred rural homes, covered a major roadway, and eventually reached the sea. Lava has been intermittently pouring into the ocean ever since, adding new land to the island of Hawaii.

Figure 7.11 Lava extruded along the East Rift Zone, Kilauea, Hawaii. (Photo by Greg Vaughn)

Did You Know?

According to legend, Pele, the Hawaiian goddess of volcanoes, makes her home at the summit of Kilauea volcano. Evidence for her existence is "Pele's hair"—thin, delicate strands of glass, which are soft and flexible and have a golden-brown color. This threadlike volcanic glass forms when blobs of hot lava are spattered and shredded by escaping gases.

Cinder Cones

As the name suggests, **cinder cones** (also called **scoria cones**) are built from ejected lava fragments that take on the appearance of cinders or clinkers as they begin to harden while in flight. These pyroclastic fragments range in size from fine ash to bombs that may exceed a meter in diameter. However, most of the volume of a cinder cone consists of pea- to walnut-sized fragments that are markedly vesicular (containing voids) and have a black to reddish-brown color. Although cinder cones are composed mostly of loose pyroclastic material, they sometimes extrude lava. On such occasions, the discharges come from vents located at or near the base rather than from the summit crater.

Cinder cones are the most abundant of the three major types of volcanoes. They have a very simple, distinctive shape determined by the slope that loose pyroclasic material maintains as it comes to rest (Figure 7.12). Because cinders have a high angle of repose (the steepest angle at which material remains stable), young cinder cones are steep sided, having slopes between 30 and 40 degrees. In addition, cinder cones have large, deep craters in relation to the overall size of the structure. Although relatively symmetrical, many cinder cones are elongated and higher on the side that was downwind during the eruptions.

Most cinder cones are produced by a single, short-lived eruptive event. One study found that half of all cinder cones

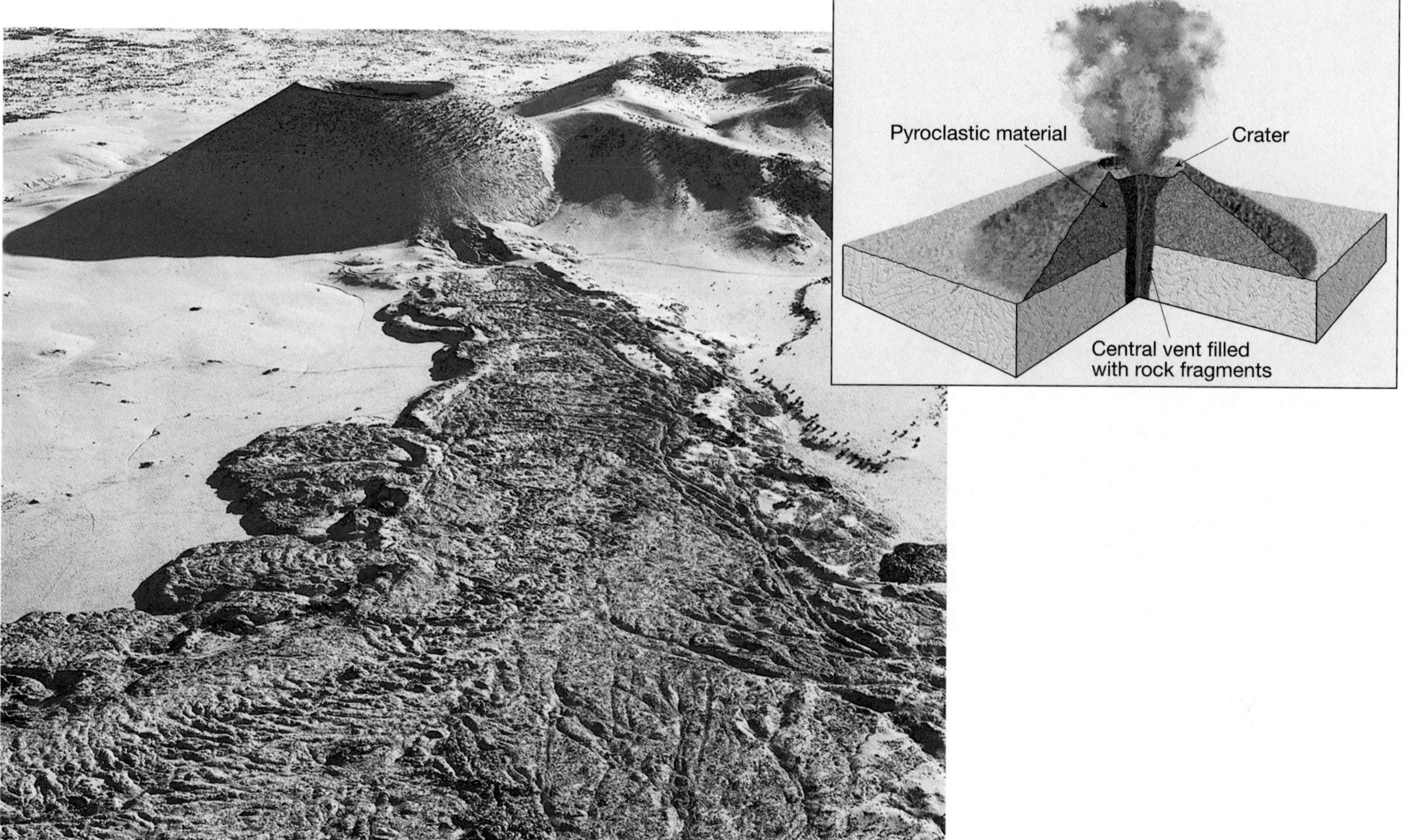

Figure 7.12 SP Crater, a cinder cone north of Flagstaff, Arizona. Cinder cones are built from ejected lava fragments and are usually less than 300 meters (1000 feet) high. (Photo by Michael Collier)

examined were constructed in less than one month, and that 95 percent formed in less than one year. However, in some cases, they remain active for several years. Parícutin, shown in Figure 7.13 had an eruptive cycle that spanned 9 years. Once the event ceases, the magma in the "plumbing" connecting the vent to the magma source solidifies, so the volcano never erupts again. As a consequence of this short life span, cinder cones are small, usually between 30 meters (100 feet) and 300 meters (1000 feet). A few rare examples exceed 700 meters (2300 feet) in height.

Cinder cones are found by the thousands all around the globe. Some are located in volcanic fields such as the one near Flagstaff, Arizona, which consists of about 600 cones. Others are parasitic cones that form on the flanks of larger volcanoes. Mount Etna, for example, has dozens of cinder cones like these.

Did You Know?

A short distance off the south coast of Hawaii is a submarine volcano called Loihi. Although very active, it has another 900 meters (3000 feet) to go before it breaks the surface and becomes another island in the Hawaiian chain.

Parícutin: Life of a Garden-Variety Cinder Cone One of the very few volcanoes studied by geologists from beginning to end is the cinder cone called Parícutin, located about 320 kilometers (200 miles) west of Mexico City. In 1943, its eruptive phase began in a cornfield owned by Dionisio Pulido, who witnessed the event as he prepared the field for planting.

For two weeks prior to the first eruption, numerous Earth tremors caused apprehension in the nearby village of Parícutin. Then on February 20, sulfurous gases began billowing from a small depression that had been in the cornfield for as long as people could remember. During the night, hot, glowing rock fragments were ejected from the vent, producing a spectacular fireworks display. Explosive discharges continued, throwing hot fragments and ash occasionally as high as 6000 meters (20,000 feet) above the crater rim. Larger fragments fell near the crater, some remaining incandescent as they rolled down the slope. These built an aesthetically pleasing cone, while finer ash fell over a much larger area, burning and eventually covering the village of Parícutin. In the first day, the cone grew to 40 meters (130 feet), and by the fifth day, it was more than 100 meters (330 feet) high. Within

Figure 7.13 The village of San Juan Parangaricutiro engulfed by lava from Parícutin, shown in the background. Only the church towers remain. (Photo by Tad Nichols)

the first year, more than 90 percent of the total ejecta had been discharged.

The first lava flow came from a fissure that opened just north of the cone, but after a few months, flows began to emerge from the base of the cone itself. In June 1944, a clinkery aa flow 10 meters (30 feet) thick moved over much of the village of San Juan Parangaricutiro, leaving only the church steeple exposed (Figure 7.13). After nine years of intermittent pyroclastic explosions and nearly continuous discharge to lava from vents at its base, the activity ceased almost as quickly as it had begun. Today, Parícutin is just another one of the scores of cinder cones dotting the landscape in this region of Mexico. Like the others, it will not erupt again.

Composite Cones

Earth's more picturesque yet potentially dangerous volcanoes are **composite cones** or **stratovolcanoes** (Figure 7.14). Most are located in a relatively narrow zone that rims the Pacific Ocean, appropriately called the *Ring of Fire* (see Figure 7.26, p. 213). This active zone includes a chain of continental volcanoes that are distributed along the west cost of South and North America, including the large cones of the Andes and the Cascade Range of the western United States and Canada. The latter group includes Mount St. Helens, Mount Rainier, and Mount Garibaldi. The most active regions in the Ring of Fire are located along curved belts of volcanic islands situated adjacent to the deep-ocean trenches of the northern and western Pacific. This nearly continous chain of volcanoes stretches from the Aleutian Islands to Japan and the Philippines and ends on North Island, New Zealand.

The classic composite cone is a large, nearly symmetrical structure composed of both lava and pyroclastic deposits. Just as shield volcanoes owe their shape to fluid basaltic lavas, composite cones reflect the nature of the erupted material. For the most part, composite cones are the product of gas-rich magma having an andesitic composition. Relative to shields, the silica-rich magmas typical of composite cones generate thick viscous lavas that travel short distances. In addition, composite cones may generate explosive eruptions that eject huge quantities of pyroclastic material.

The growth of a "typical" composite cone begins with both pyroclastic material and lava being emitted from a central vent. As the structure matures, lavas tend to flow from fissures that develop on the lower flanks of the cone. This activity may alternate with explosive eruptions that eject pyroclastic material from the summit crater. Sometimes both activities occur simultaneously.

Figure 7.14 These two volcanoes, Pomerape and Parinacota, exhibit the classic shape of a composite cone, Lauca National Park, Chile. (Photo by Michael Giannechini/Photo Researchers, Inc.)

A conical shape, with a steep summit area and more gradually sloping flanks, is typical of many large composite cones. This classic profile, which adorns calendars and postcards, is partially a consequence of the way viscous lavas and pyroclastic ejecta contribute to the growth of the cone. Coarse fragments ejected from the summit crater tend to accumulate near their source. Because of their high angle of repose, coarse materials contribute to the steep slopes of the summit area. Finer ejecta, on the other hand, are deposited as a thin layer over a large area. This acts to flatten the flank of the cone. In addition, during the early stages of growth, lavas tend to be more abundant and flow greater distances from the vent than do later lavas. This contributes to the cone's broad base. As the volcano matures, the short flows that come from the central vent serve to armor and strengthen the summit area. Consequently, steep summit slopes exceeding 40 degrees are sometimes possible. Two of the most perfect cones—Mount Mayon in the Philippines and Fujiyama in Japan—exhibit the classic form we expect of a composite cone, with its steep summit and gently sloping flanks.

Despite their symmetrical form, most composite cones have a complex history. Huge mounds of volcanic debris surrounding many cones provide evidence that, in the distant past, a large section of the volcano slid downslope as a massive landslide. Others developed horseshoe-shaped depressions at their summits as a result of explosive eruptions or as occurred during the 1980 eruption of Mount St. Helens, a combination of a landslide and the eruption of 0.6 cubic kilometer (0.1 cubic mile) of magma left a gaping void on the north side of the cone. Often, so much rebuilding has occurred since these eruptions that no trace of the amphitheater-shaped scar remains.

Living in the Shadow of a Composite Cone

More than 50 volcanoes have erupted in the United States in the past 200 years (see Figure 7.28, p. 217). Fortunately, the most explosive of these eruptions, except for Mount St. Helens in 1980, occurred in sparsely inhabited regions of Alaska. On a global scale, numerous destructive eruptions have occurred during the past few thousand years, a few of which may have influenced the course of human civilization.

Nuée Ardente: A Deadly Pyroclastic Flow

One of the most devastating phenomena associated with composite cones are **pyroclastic flows,** which consist of hot gases infused with incandescent ash and larger rock fragments. The most destructive of these fiery flows, called **nuée ardentes** (also referred to as *glowing avalanches*) are capable of racing down steep volcanic slopes at speeds that can approach 200 kilometers (125 miles) per hour (Figure 7.15).

The ground-hugging portion of a glowing avalanche is rich in particulate matter, which is suspended by jets of buoyant gases passing upward through the flow. Some of these

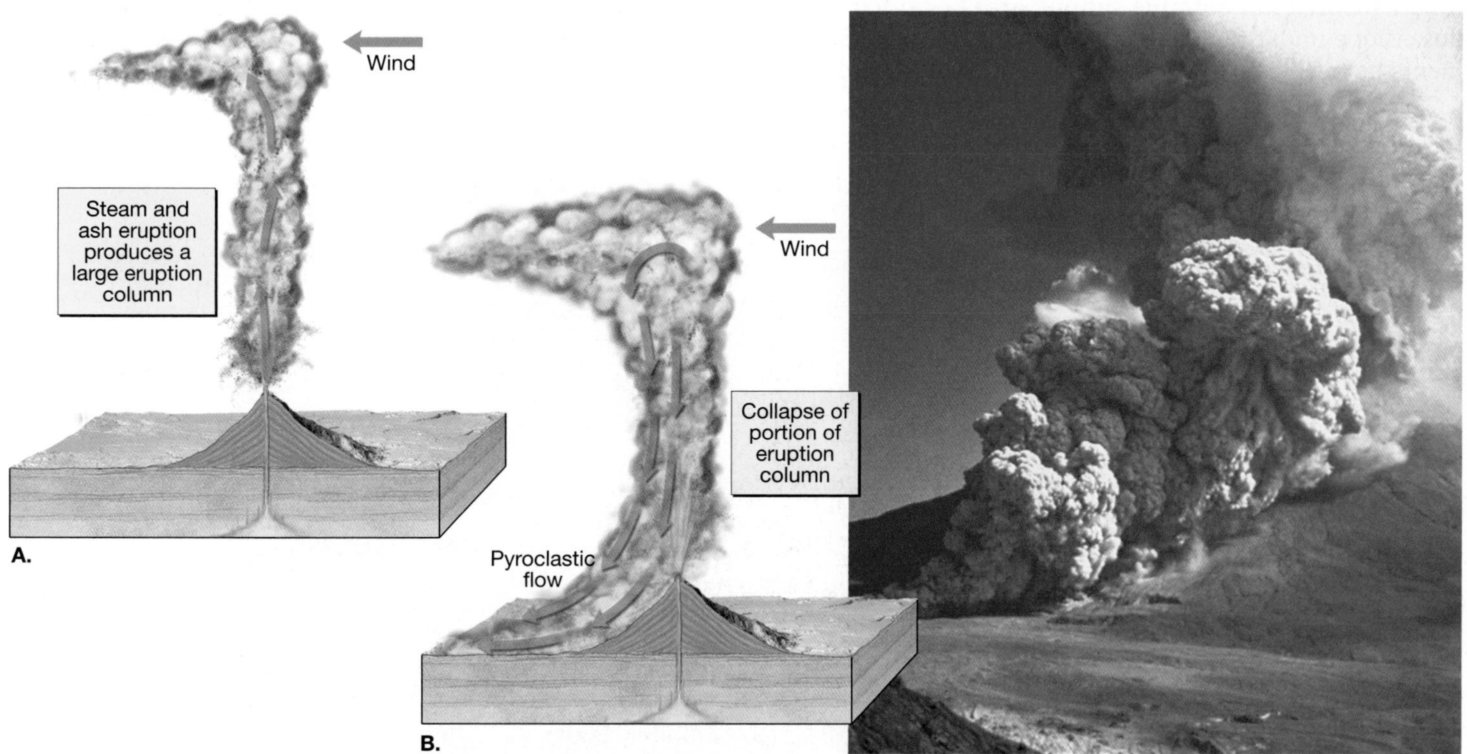

Figure 7.15 Parts **A** and **B** illustrate the process that creates a pyroclastic flow. The photo shows a hot pyroclastic flow called a nuée ardente racing down the slope of Mount St. Helens on August 7, 1980. Such flows can reach speeds in excess of 100 kilometers (62 miles) per hour. (Photo by Peter W. Lipman, U.S. Geological Survey, Denver)

Did You Know?

In 1902, when Mount Pelée erupted, the U.S. Senate was preparing to vote on the location of a canal to connect the Pacific and Atlantic oceans. The choice was between Panama and Nicaragua, a country that featured a smoking volcano on its postage stamp. Supporters of the Panamanian route used the potential threat of volcanic eruptions in Nicaragua to bolster their argument in favor of Panama. The U.S. Senate subsequently approved the construction of the Panama Canal by eight votes.

gases have escaped from newly erupted volcanic fragments. In addition, air that is overtaken and trapped by an advancing flow may be heated sufficiently to provide buoyancy to the particulate matter of the nuée ardente. Thus, these flows, which can include large rock fragments in addition to ash, travel downslope in a nearly frictionless environment. This helps to explain why some nuée ardente deposits are found more than 100 kilometers (60 miles) from their source.

The pull of gravity is the force that causes these heavier-than-air flows to sweep downslope much like a snow avalanche. Some pyroclastic flows result when a powerful eruption blasts material laterally out the side of a volcano. Probably more often, nuée ardentes form from the collapse of tall eruption columns that form over a volcano during an explosive event. Once gravity overcomes the initial upward thrust provided by the escaping gases, the ejecta begin to fall. Massive amounts of incandescent blocks, ash, and pumice fragments that fall onto the summit area begin to cascade downslope under the influence of gravity. The largest fragments have been observed bouncing down the flanks of a cone, while the finer materials travel rapidly as an expanding tongue-shaped cloud.

The Destruction of St. Pierre In 1902, an infamous nuée ardente from Mount Pelée, a small volcano on the Caribbean island of Martinique, destroyed the port town of St. Pierre. The destruction happened in moments and was so devastating that almost all of St. Pierre's 28,000 inhabitants were killed. Only one person on the outskirts of town—a prisoner protected in a dungeon—and a few people on ships in the harbor were spared (Figure 7.16).

Shortly after this calamitous eruption, scientists arrived on the scene. Although St. Pierre was mantled by only a thin layer of volcanic debris, they discovered that masonry walls nearly a meter thick were knocked over like dominoes; large trees were uprooted and cannons were torn from their mounts. A further reminder of the destructive force of this nuée ardente is preserved in the ruins of the mental hospital. One of the immense steel chairs that had been used to confine alcoholic patients can be seen today, contorted as though it were made of plastic.

Lahars: Mudflows on Active and Inactive Cones

In addition to violent eruptions, large composite cones may generate a type of mudflow referred to by its Indonesian name **lahar.** These destructive mudflows occur when volcanic debris becomes saturated with water and rapidly moves down steep volcanic slopes, generally following gullies and stream valleys. Some lahars are triggered when large volumes of ice and snow melt during an eruption. Others are generated when heavy rainfall saturates weathered volcanic deposits. Thus, lahars can occur even when a volcano is *not* erupting.

Figure 7.16 St. Pierre as it appeared shortly after the eruption of Mount Pelée, 1902. (Courtesy of the Library of Congress)

When Mount St. Helens erupted in 1980, several lahars formed. These flows and accompanying flood waters raced down the valleys of the north and south forks of the Toutle River at speeds exceeding 30 kilometers (20 miles) per hour. Water levels in the river rose to 4 meters (13 feet) above flood stage, destroying or severely damaging nearly all the homes and bridges along the impacted area (Figure 7.17). Fortunately, the area was not densely populated.

In 1985, deadly lahars were produced during a small eruption of Nevado del Ruiz, a 5300 meter (17,400 foot) volcano in the Andes Mountains of Colombia. Hot pyroclastic material melted ice and snow that capped the mountain (*Nevado* means "snow" in Spanish) and sent torrents of ash and debris down three major river valleys that flank the volcano. Reaching speeds of 100 kilometers (60 miles) per hour, these mudflows tragically took 25,000 lives.

Mount Rainier, Washington, is considered by many to be America's most dangerous volcano because, like Nevado del Ruiz, it has a thick year-round mantle of snow and ice. Adding to the risk is the fact that 100,000 people live in the valleys around Rainier, and many homes are built on lahars that flowed down the volcano hundreds or thousands of years ago. A future eruption, or perhaps just a period of heavy rainfall, may produce lahars that will likely take similar paths.

Other Volcanic Landforms

The most obvious volcanic structure is a cone. But other distinctive and important landforms are also associated with volcanic activity.

Calderas

Calderas (*caldaria* = a cooking pot) are large collapse depressions having a more or less circular form. Their diameters exceed one kilometer, and many are tens of kilometers across. (Those less than a kilometer across are called *collapse pits.*) Most calderas are formed by one of the following processes: (1) the collapse of the summit of a large composite volcano following an explosive eruption of silica-rich pumice and ash fragments (*Crater Lake–type calderas*); (2) the collapse of the top of a shield volcano caused by subterranean drainage from a central magma chamber (*Hawaiian-type calderas*); and (3) the collapse of a large area, caused by the discharge of colossal volumes of silica-rich pumice and ash along ring fractures (*Yellowstone-type calderas*).

Crater Lake–Type Calderas Crater Lake, Oregon, is located in a caldera that has a maximum diameter of 10 kilometers (6 miles) and is 1175 meters (3900 feet) deep. This caldera formed about 7000 years ago when a composite cone, later named Mount Mazama, violently extruded 50 to 70 cubic kilometers (12 to 17 cubic miles) of pyroclastic material (Figure 7.18). With the loss of support, 1500 meters (5000 feet) of the summit of this once prominent cone collapsed. After the collapse, rainwater filled the caldera (Figure 7.18). Later volcanic activity built a small cinder cone in the lake. Today, this cone, called Wizard Island, provides a mute reminder of past activity.

Figure 7.17 A house damaged by a lahar along the Toutle River, west-northwest of Mount St. Helens. The end section of the house was torn free and lodged against trees. (Photo by D. R. Crandell, U.S. Geological Survey)

Hawaiian-Type Calderas Although some calderas are produced by collapse following an explosive eruption, many are not. For example, Hawaii's active shield volcanoes, Mauna Loa and Kilauea, both have large calderas at their summits. Kilauea's measures 3.3 by 4.4 kilometers (about 2 by 3 miles) and is 150 meters (500 feet) deep. The walls of this caldera are steep sided, almost vertical, and as a result, it looks like a vast nearly flat-bottomed pit. Collapse was not initiated by the eruption of voluminous lavas directly from the caldera, although lava lakes occasionally formed on its floor. Instead, Kilauea's caldera formed by gradual subsidence as magma slowly drained laterally from the underlying magma chamber to the East Rift Zone, leaving the summit unsupported.

Yellowstone-Type Calderas Although the 1980 eruption of Mount St. Helens was spectacular, it pales by comparison to what happened 630,000 years ago in the region now occupied by Yellowstone National Park. Here, approximately 1000 cubic kilometers (240 cubic miles) of pyroclastic material erupted, eventually producing a caldera 70 kilometers (43 miles) across. This event produced showers of ash as far away as the Gulf of Mexico. Vestiges of this activity are the many hot springs and geysers in the region.

The formation of a large Yellowstone-type caldera begins when a silica-rich (rhyolitic) magma body is emplaced near the surface, upwarping the overlying rocks. Next, ring fractures develop in the roof, providing a pathway to the surface for the gas-rich magma. This initiates an explosive eruption of colossal proportions ejecting huge volumes (usually exceeding 100 cubic kilometers) of pyroclastic materials, mainly in the form of ash and pumice fragments. These materials typically form a pyroclastic flow that spreads across the landscape at speeds that may exceed 100 kilometers (60 miles) per hour, destroying most living things in its path. After coming to rest, the hot fragments of ash and pumice fuse together, forming a welded tuff that closely resembles a solidified lava flow. Finally, with the loss of support, the roof of the magma chamber collapses, generating a large caldera.

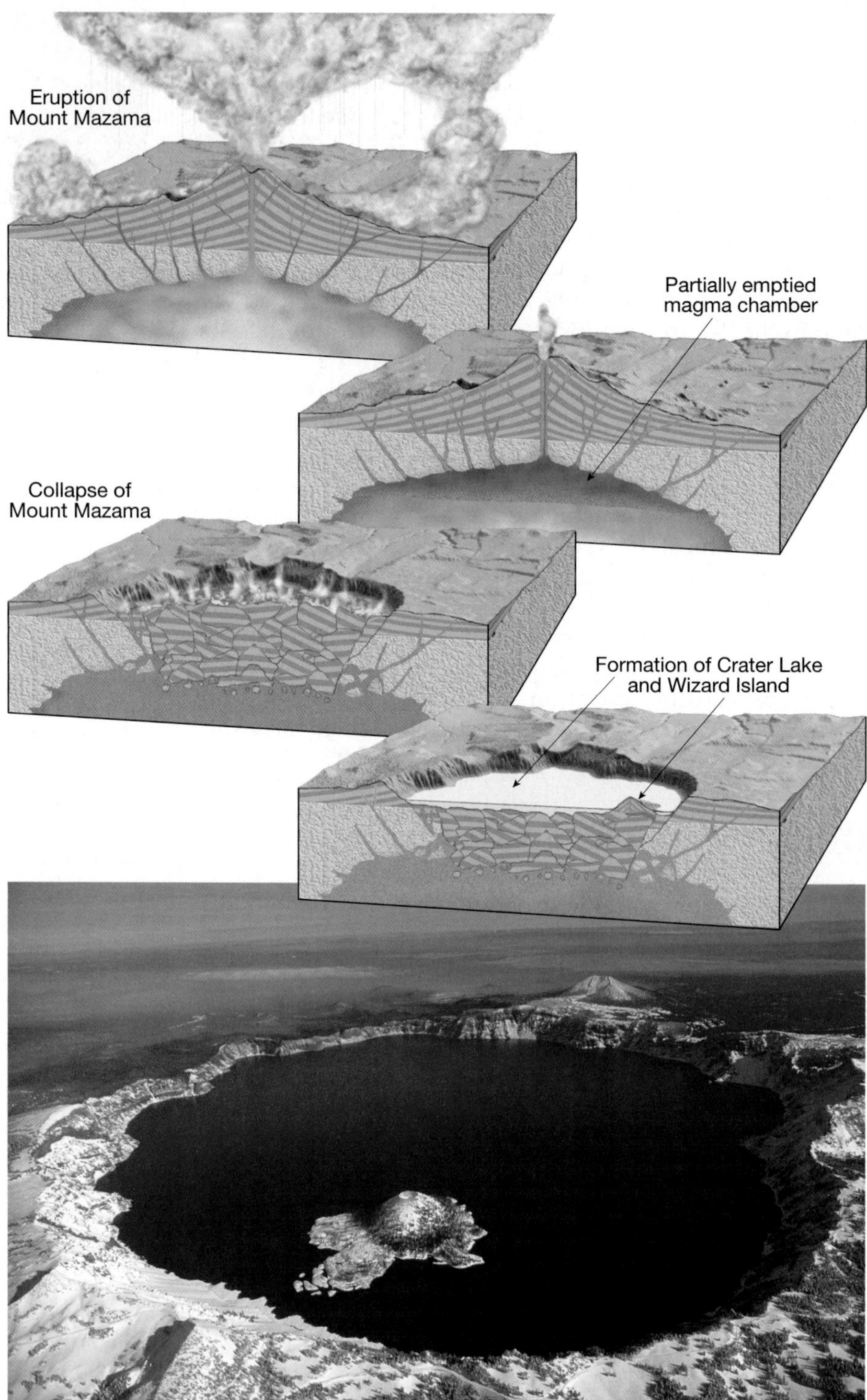

Figure 7.18 Sequence of events that formed Crater Lake, Oregon. About 7000 years ago, the summit of former Mount Mazama collapsed following a violent eruption that partly emptied the magma chamber. Subsequent eruptions produced the cinder cone called Wizard Island. Rainfall and groundwater contributed to form the lake. (After H. Williams, *The Ancient Volcanoes of Oregon.* Photo by Greg Vaughn/Tom Stock and Associates)

Did You Know?

Some calderas are so large that most people who visit geothermal features, such as those found in Yellowstone National Park, are unaware that they are standing within one of the largest volcanic depressions on Earth.

Calderas of the Yellowstone-type are the largest volcanic structures on Earth. Some geologists have compared their destructive force with that of the impact of a small asteroid. Fortunately, no eruption of this type has occurred in historic times. Other examples of large calderas located in the United States are California's Long Valley Caldera and Valles Caldera located west of Los Alamos, New Mexico.

Fissure Eruptions and Lava Plateaus

We think of volcanic eruptions as building a cone or shield from a central vent. But by far the greatest volume of volcanic material is extruded from fractures in the crust called **fissures** (*fissura* = to split). Rather than building a cone, these long, narrow cracks may emit a low-viscosity basaltic lava, blanketing a wide area.

The extensive Columbia Plateau in the northwestern United States was formed this way (Figure 7.19). Numerous **fissure eruptions** extruded very fluid basaltic lava (Figure 7.20). Successive flows, some 50 meters (165 feet) thick, buried the existing landscape as they built a lava plateau nearly a mile thick. The fluid nature of the lava is evident, because some remained molten long enough to flow 150 kilometers (90 miles) from its source. The term **flood basalts** appropriately describes these flows. Massive accumulations of basaltic lava, similar to those of the Columbia Plateau, occur worldwide. One of the largest is the Deccan Traps, a thick sequence of flat-lying basalt flows covering nearly 500,000 square kilometers (195,000 square miles) of west central India. When the Deccan Traps formed about 66 million years ago, nearly 2 million cubic kilometers (almost 500,000 cubic miles) of lava were extruded in less than 1 million years. Another huge deposit of flood basalts, called the Ontong Java Plateau, is found on the floor of the Pacific Ocean.

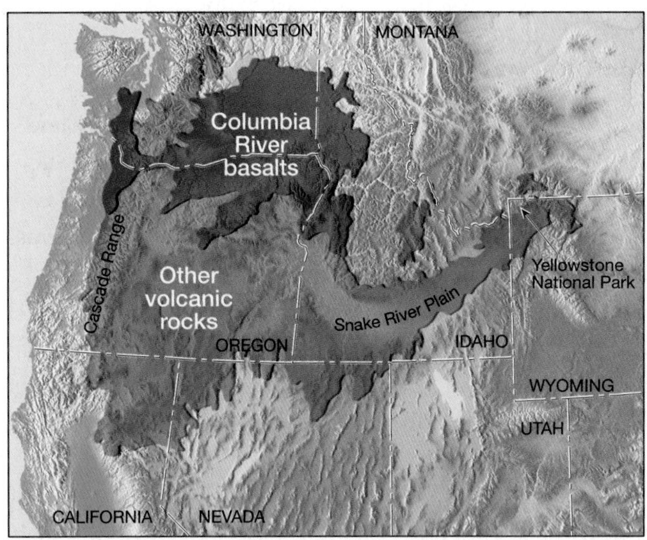

Figure 7.19 Volcanic areas in the northwestern United States. The Columbia River basalts cover an area of nearly 200,000 square kilometers (80,000 square miles). Activity here began about 17 million years ago as lava began to pour out of large fissures, eventually producing a basalt plateau with an average thickness of more than 1 kilometer. (After U.S. Geological Survey)

Volcanic Pipes and Necks

Most volcanoes are fed magma through short conduits, called *pipes,* that connect a magma chamber to the surface. In rare circumstances, pipes may extend tubelike to depths exceeding 200 kilometers (125 miles). When this occurs, the ultramafic magmas that migrate up these structures produce rocks that are thought to be samples of the mantle that have undergone very little alteration during their ascent. Geologists consider these unusually deep conduits to be "windows" into Earth, for they allow us to view rock normally found only at great depth.

The best-known volcanic pipes are the diamond-bearing structures of South Africa. The rocks filling the pipes here originated at depths of at least 150 kilometers (90 miles), where pressure is high enough to generate diamonds and other high-pressure minerals. The task of transporting essentially unaltered magma (along with diamond inclusions)

Did You Know?

Prehistoric caldera eruptions in Yellowstone National Park were about 1000 times larger than the Mount St. Helens episode, strong enough to decimate much of western North America. Volcanologists have compared their destructive forces with that of the impact of a small asteroid. Other caldera eruptions of similar scale occurred in the Jemez Mountains near Santa Fe, New Mexico, Long Valley near Bishop, California, and in Colorado's San Juan Mountains.

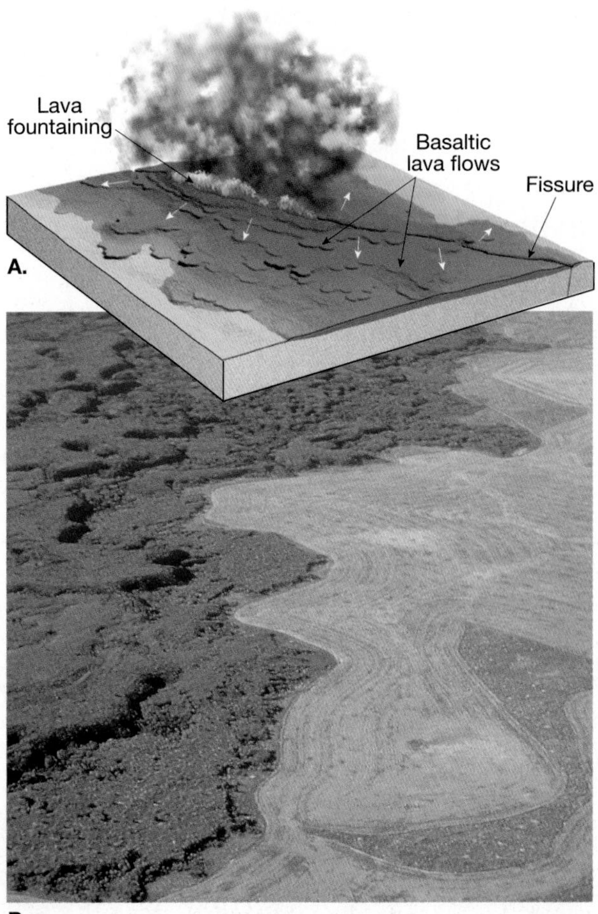

Figure 7.20 Basaltic fissure eruption. **A.** Lava fountaining from a fissure forming fluid lava flows called flood basalts. **B.** Photo of basalt flows (dark) near Idaho Falls. (Photo by John S. Shelton)

Did You Know?

In 1783, the deadly Laki fissure eruption took place in Iceland along a 24-kilometer (15-mile) rift. The eruption extruded 12 cubic kilometers (3 cubic miles) of fluid basaltic lava. In addition, sulfurous gases and ash were emitted that destroyed grasslands, killing most of Iceland's livestock. The eruption and ensuing famine killed 10,000 Icelanders—roughly one in five. Recently, a volcanologist who is an expert on the Laki eruption concluded that a similar eruption today would disrupt commercial aviation in the Northern Hemisphere for several months.

through 150 kilometers of solid rock is exceptional. This fact accounts for the scarcity of natural diamonds.

Volcanoes on land are continually being lowered by weathering and erosion. Cinder cones are easily eroded, because they are composed of unconsolidated materials. However, all volcanoes will eventually succumb to relentless erosion over geologic time. As erosion progresses, the rock occupying the volcanic pipe is often more resistant and may remain standing above the surrounding terrain long after most of the cone has vanished. Shiprock, New Mexico, is such a feature and is called a **volcanic neck** (Figure 7.21). This structure, higher than many skyscrapers, is but one of many such landforms that protrude conspicuously from the red desert landscapes of the American Southwest.

Intrusive Igneous Activity

Forces Within
▼ Igneous Activity

Although volcanic eruptions can be among the most violent and spectacular events in nature and, therefore, worthy of detailed study, most magma is emplaced at depth. Thus, an understanding of intrusive igneous activity is as important to geologists as the study of volcanic events.

The structures that result from the emplacement of igneous material at depth are called **plutons,** named for Pluto, the god of the lower world in classical mythology. Because all plutons form out of view beneath Earth's surface, they can be studied directly only after uplifting and erosion have exposed them. The challenge lies in reconstructing the events that generated these structures millions or even hundreds of millions of years ago.

Plutons are known to occur in a great variety of sizes and shapes. Some of the most common types are illustrated in Figure 7.22. Notice that some of these structures have a tabular (tabletop) shape; others are quite massive. Also, observe that some of these bodies cut across existing structures, such as layers of sedimentary rock; others form when magma is injected between sedimentary layers. Because of these differences, intrusive igneous bodies are generally classified according to their shape as either *tabular* (*tabula* = table) or *massive* and by their orientation with respect to the host rock. Plutons are said to be *discordant* (*discordare* = to disagree) if they cut across existing structures and *concordant* (*concordare* = to agree) if they form parallel to features such as sedimentary strata. As you can see in Figure 7.22, plutons are closely associated with volcanic activity.

Dikes

Dikes are tabular discordant bodies that are produced when magma is injected into fractures. The force exerted by the emplaced magma can be great enough to separate the walls of the fracture farther. Once crystallized, these sheetlike structures have thicknesses ranging from less than a centimeter to more than a kilometer. The largest have lengths of hundreds of kilometers. Most dikes, however, are a few meters thick and extend laterally for no more than a few kilometers.

Dikes are often found in groups that once served as vertically oriented pathways followed by molten rock that fed ancient lava flows. The parent pluton is generally not observable. Some dikes are found radiating, like spokes on a wheel,

Figure 7.21 Ship Rock, New Mexico, is a volcanic neck. This structure, which stands more than 420 meters (1380 feet) high, consists of igneous rock that crystallized in the vent of a volcano that has long since been eroded away. (Photo by Tom Bean/ DRK Photo)

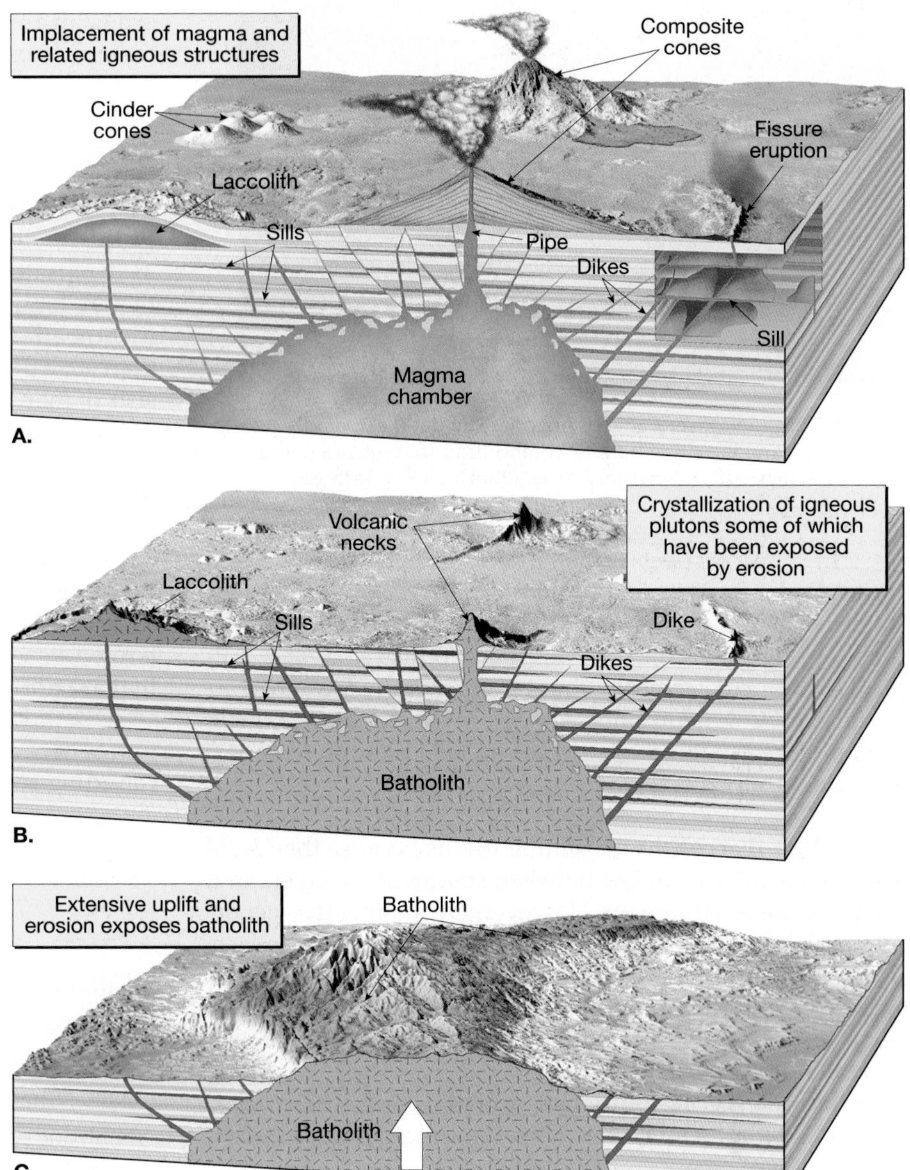

Figure 7.22 Basic igneous structures. **A.** This block diagram shows the relationship between volcanism and intrusive igneous activity. **B.** This view illustrates the basic intrusive igneous structures, some of which have been exposed by erosion long after their formation. **C.** After millions of years of uplifting and erosion, a batholith is exposed at the surface.

from an eroded volcanic neck. In these situations, the active ascent of magma is thought to have generated fissures in the volcanic cone out of which lava flowed.

Sills and Laccoliths

Sills and laccoliths are concordant plutons that form when magma is intruded in a near-surface environment. They differ in shape and usually differ in composition.

Sills **Sills** are tabular plutons formed when magma is injected along sedimentary bedding surfaces (Figure 7.23). Horizontal sills are the most common, although all orientations, even vertical, are known to exist. Because of their relatively uniform thickness and large areal extent, sills are likely the product of very fluid magmas. Magmas having a low silica content are more fluid, so most sills are composed of the rock basalt.

The emplacement of a sill requires that the overlying sedimentary rock be lifted to a height equal to the thickness of the sill. Although this is a formidable task, in shallow environments it often requires less energy than forcing the magma up the remaining distance to the surface. Consequently, sills form only at shallow depths, where the pressure exerted by the weight of overlying rock layers is low. Although sills are intruded between layers, they can be locally discordant. Large sills frequently cut across sedimentary layers and resume their concordant nature at a higher level.

One of the largest and most studied of all sills in the United States is the Palisades Sill. Exposed for 80 kilometers (50 miles) along the west bank of the Hudson River in southeastern New York and northeastern New Jersey, this sill is about 300 meters (1000 feet) thick. Because of its resistant nature, the Palisades Sill forms an imposing cliff that can be seen easily from the opposite side of the Hudson.

Figure 7.23 Salt River Canyon, Arizona. The dark, essentially horizontal band is a sill of basaltic composition that intruded into horizontal layers of sedimentary rock. (Photo by E. J. Tarbuck)

In many respects, sills closely resemble buried lava flows. Both are tabular and often exhibit columnar jointing (Figure 7.24). **Columnar joints** form as igneous rocks cool and develop shrinkage fractures that produce elongated, pillarlike columns. Further, because sills generally form in near-surface environments and may be only a few meters thick, the emplaced magma often cools quickly enough to generate a fine-grained texture.

Laccoliths **Laccoliths** are similar to sills because they form when magma is intruded between sedimentary layers in a near-surface environment. However, the magma that generates laccoliths is more viscous. This less fluid magma collects as a lens-shaped mass that arches the overlying strata upward (see Figure 7.22, p. 211). Consequently, a laccolith can occasionally be detected because of the dome-shaped bulge it creates at the surface.

Most large laccoliths are probably not much wider than a few kilometers. The Henry Mountains in southeastern Utah are largely composed of several laccoliths believed to have been fed by a much larger magma body emplaced nearby.

Batholiths

By far the largest intrusive igneous bodies are **batholiths** (*bathos* = depth, *lithos* = stone). Most often, batholiths occur in groups that form linear structures several hundreds of kilometers long and up to 100 kilometers (62 miles) wide. The Idaho batholith, for example, encompasses an area of more than 40,000 square kilometers (15,000 square miles) and consists of many plutons. Indirect evidence gathered from gravitational studies indicates that batholiths are also very thick, possibly extending dozens of kilometers into the crust. Based on the amount exposed by erosion, some batholiths are at least several kilometers thick.

By definition, a plutonic body must have a surface exposure greater than 100 square kilometers (40 square miles) to be considered a batholith. Smaller plutons of this type are termed *stocks.* Many stocks appear to be portions of batholiths that are not yet fully exposed.

Batholiths may compose the core of a mountain system where uplifting and erosion have removed the surrounding rock, thereby exposing the resistant igneous body.

Did You Know?

In the eastern United States, some exposed granitic intrusions have dome-shaped, nearly treeless summits, hence the name "summit balds." Examples include Cadillac Mountain in Maine, Chocurua in New Hampshire, Black Mountain in Vermont, and Stone Mountain in Georgia.

Figure 7.24 Columnar jointing in basalt, Giants Causeway National Park, Northern Ireland. These five-to-seven-sided columns are produced by contraction and fracturing that results as a lava flow or sill gradually cools. (Photo by Tom Till)

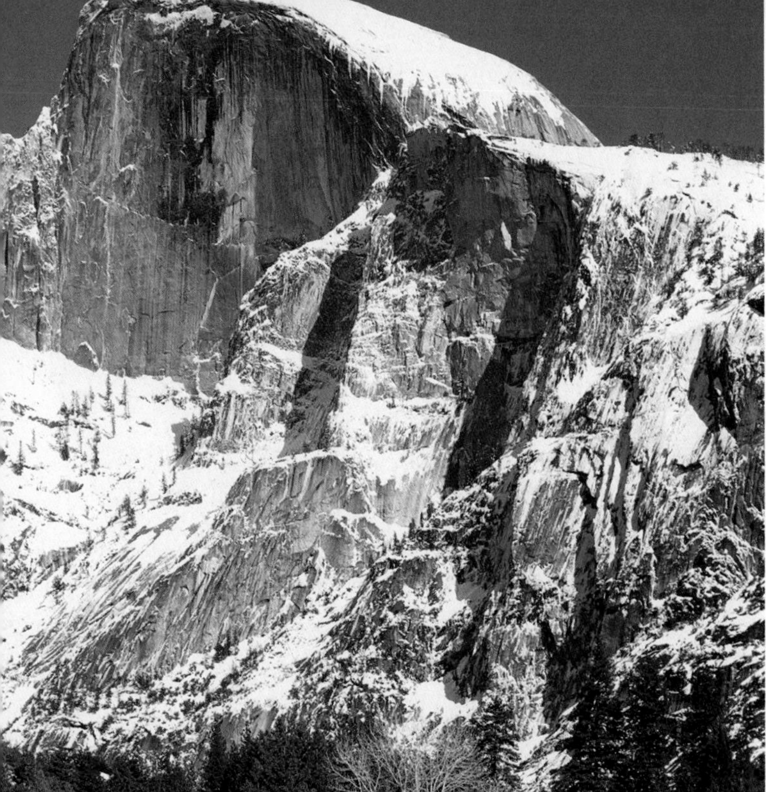

Figure 7.25 Half Dome in Yosemite National Park, California, is part of the Sierra Nevada Batholith, a huge structure that extends nearly 400 kilometers (250 miles). (Photo by Charles Gurche)

Some of the best-known structures in the Sierra Nevada, such as Half Dome, are carved from such a granitic mass (Figure 7.25).

Large expanses of granitic rock also occur in the stable interiors of the continents, such as the Canadian Shield of North America. These relatively flat exposures are the remains of ancient mountains that have long since been leveled by erosion. Thus, the rocks that make up the batholiths of youthful mountain ranges, such as the Sierra Nevada, were generated near the top of a magma chamber, whereas in shield areas, the roots of former mountains and, thus, the lower portions of batholiths, are exposed.

Plate Tectonics and Igneous Activity

Geologists have known for decades that the global distribution of volcanism is not random. Of the more than 800 active volcanoes that have been identified, most are located along the margins of the ocean basins—most notably within the circum-Pacific belt known as the *Ring of Fire* (Figure 7.26). This group of volcanoes consists mainly of composite cones that emit volatile-rich magma having an intermediate (andesitic) composition that occasionally produce awe-inspiring eruptions.

The volcanoes comprising a second group emit very fluid basaltic lavas and are confined to the deep-ocean basins, including well-known examples on Hawaii and Iceland. In addition, this group contains many active submarine volcanoes that dot the ocean floor; particularly notable are the innumerable small seamounts that lie along the axis of the oceanic ridge.

A third group includes those volcanic structures that are irregularly distributed in the interiors of the continents. Volcanism on continents is the most diverse, ranging from eruptions of very fluid basaltic lavas, like those that generated the Columbia River basalts to explosive eruptions of silica-rich rhyolitic magma as occurred in Yellowstone.

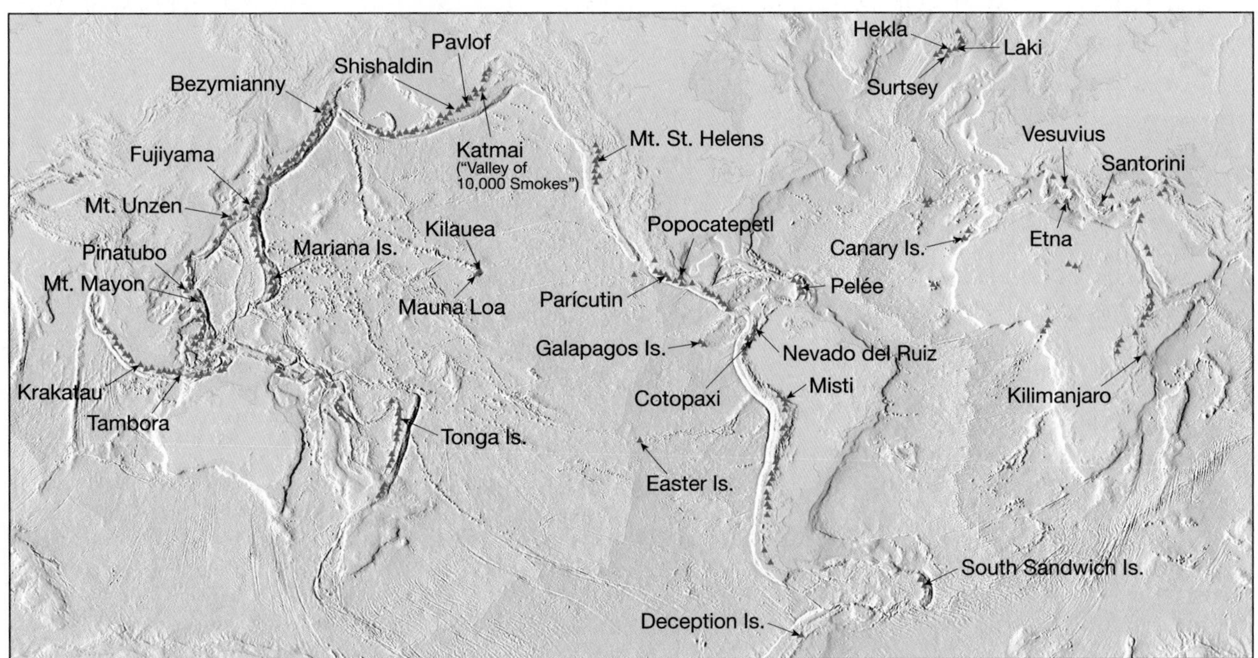

Figure 7.26 Locations of some of Earth's major volcanoes. Note the concentration of volcanoes encircling the Pacific basin, known as the Ring of Fire.

Mount Augustine, Alaska (Stephen Kaufman)

Kilauea, Hawaii
(J. D. Griggs/USGS)

Figure 7.27 Three zones of volcanism. Two of these zones are plate boundaries, and the third includes areas within the plates.

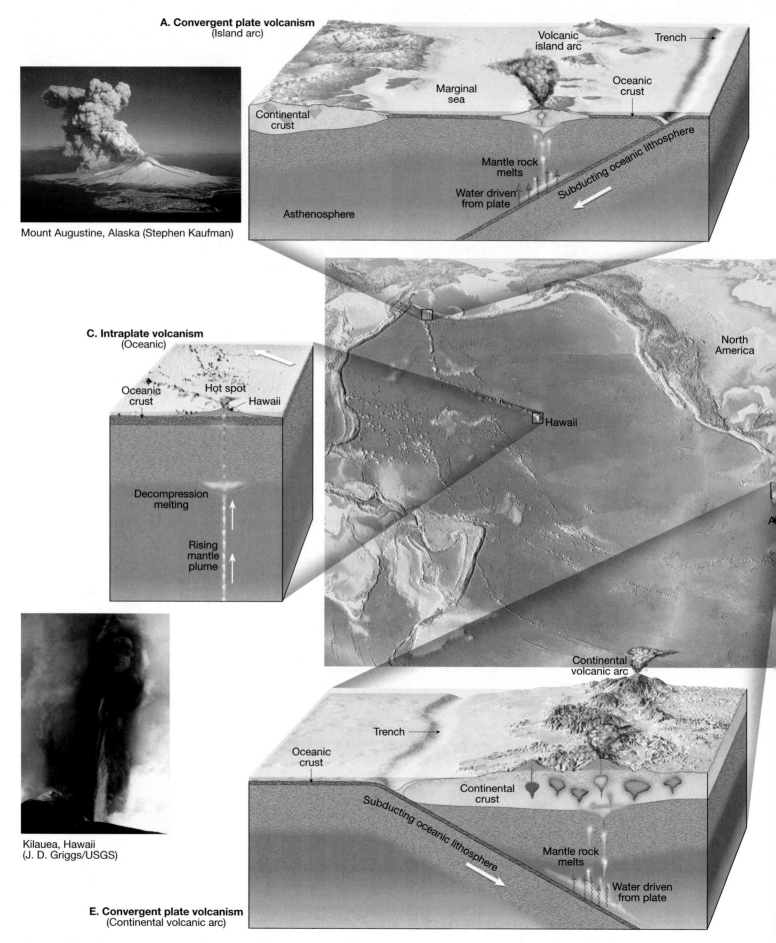

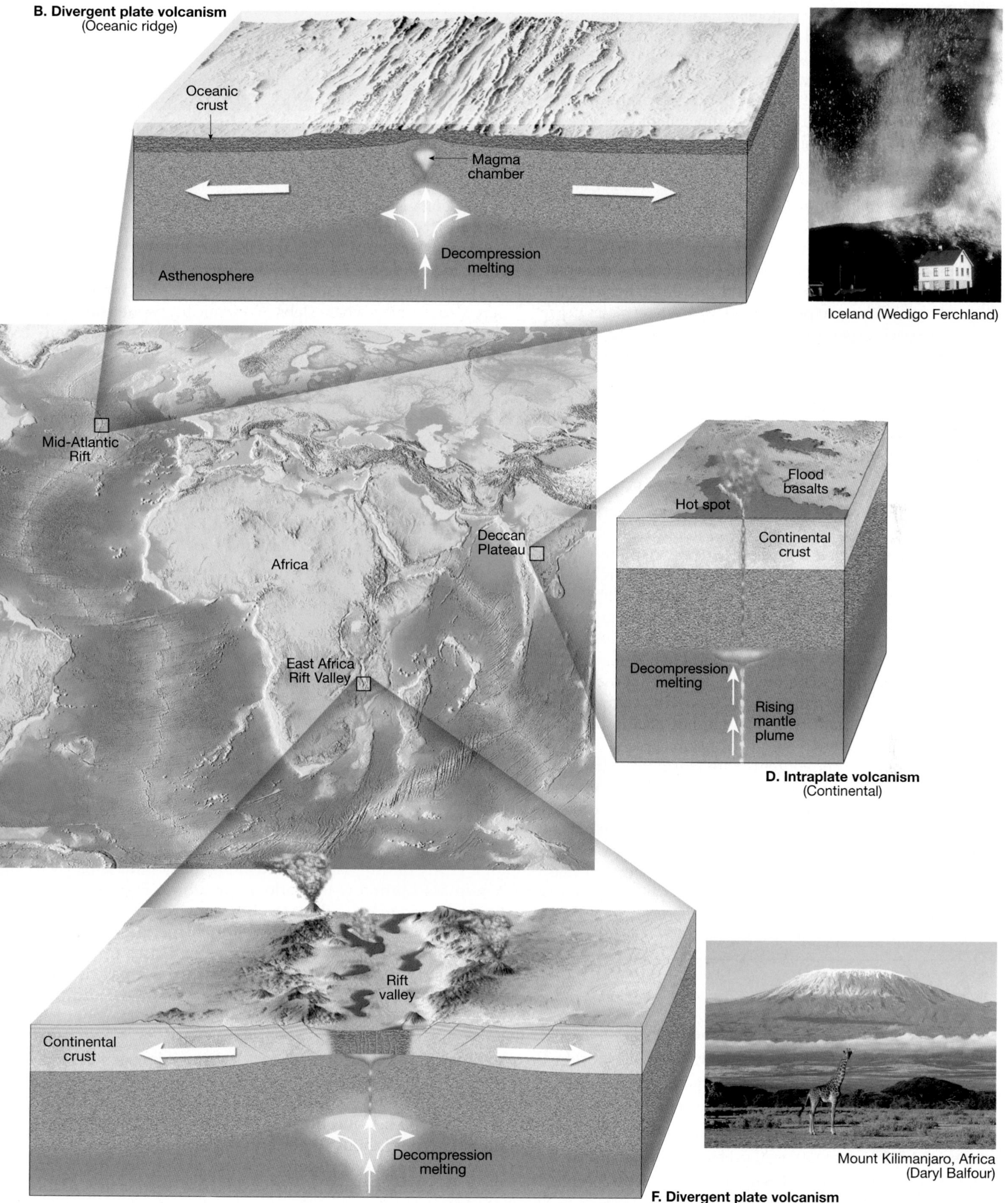
B. Divergent plate volcanism
(Oceanic ridge)
Oceanic crust
Magma chamber
Decompression melting
Asthenosphere
Iceland (Wedigo Ferchland)
Mid-Atlantic Rift
Africa
Deccan Plateau
East Africa Rift Valley
Flood basalts
Hot spot
Continental crust
Decompression melting
Rising mantle plume
D. Intraplate volcanism
(Continental)
Rift valley
Continental crust
Decompression melting
Mount Kilimanjaro, Africa
(Daryl Balfour)
F. Divergent plate volcanism
(Continental rifting)

Figure 7.27 (Continued)

Until the late 1960s, geologists had no explanation for the apparently haphazard distribution of continental volcanoes, nor were they able to account for the almost continuous chain of volcanoes that circles the margin of the Pacific basin. With the development of the theory of plate tectonics, the picture was greatly clarified. Recall that most magma originates in the upper mantle and that the mantle is essentially solid, *not molten* rock. The basic connection between plate tectonics and volcanism is that *plate motions provide the mechanisms by which mantle rocks melt to generate magma.*

We will examine three zones of igneous activity and their relationship to plate boundaries (Figure 7.27). These active areas are located (1) along convergent plate boundaries, where plates move toward each other and one sinks beneath the other; (2) along divergent plate boundaries, where plates move away from each other and new seafloor is created; and (3) areas within the plates proper that are not associated with any plate boundary.

Igneous Activity at Convergent Plate Boundaries

Recall that at convergent plate boundaries slabs of oceanic crust are bent as they descend into the mantle, generating an oceanic trench. As a slab sinks deeper into the mantle, the increase in temperature and pressure drives volatiles (mostly H_2O) from the oceanic crust. These mobile fluids migrate upward into the wedge-shaped piece of mantle located between the subducting slab and overriding plate. Once the sinking slab reaches a depth of about 100 kilometers (60 miles), these water-rich fluids reduce the melting point of hot mantle rock sufficiently to trigger some melting. The partial melting of mantle rock generates magma with a basaltic composition. After a sufficient quantity of magma has accumulated, it slowly migrates upward.

Volcanism at a convergent plate margin results in the development of a linear or slightly curved chain of volcanoes called a *volcanic arc.* These volcanic chains develop roughly parallel to the associated trench—at distances of 200 to 300 kilometers (125 to 190 miles). Volcanic arcs can be constructed on oceanic, or continental, lithosphere. Those that develop within the ocean and grow large enough for their tops to rise above the surface are labeled *island archipelagos* in most atlases. Geologists prefer the more descriptive term **volcanic island arcs,** or simply **island arcs** (Figure 7.27A). Several young volcanic island arcs of this type border the western Pacific basin, including the Aleutians, the Tongas, and the Marianas.

Volcanism associated with convergent plate boundaries may also develop where slabs of oceanic lithosphere are subducted under continental lithosphere to produce a **continental volcanic arc** (Figure 7.27C). The mechanisms that generate these magmas are essentially the same as those operating at island arcs. The major difference is that continental crust is much thicker and is composed of rocks having a higher silica content than oceanic crust. Hence, through the assimilation of silica-rich crustal rocks, a magma body may change composition as it rises through continental crust. Stated another way, magmas generated in the mantle may change from a comparatively dry, fluid basaltic magma to a viscous andesitic or rhyolitic magma having a high concentration of volatiles as it moves up through the continental crust. The chain of volcanoes found along the axis of the Andes Mountains is perhaps the best example of a mature continental volcanic arc.

Since the Pacific basin is essentially bordered by convergent plate boundaries (and associated subduction zones), it is easy to see why the irregular belt of explosive volcanoes we call the Ring of Fire formed in this region. The volcanoes of the Cascade Range in the northwestern United States, including Mount Hood, Mount Rainier, and Mount Shasta, are included in this group (Figure 7.28).

Did You Know?

Although tsunami are most often associated with displacement along a fault located on the seafloor, some result from the collapse of a volcanic cone. This was dramatically demonstrated during the 1883 eruption on Krakatau, when the northern half of a volcano plunged into the Sunda Strait, creating a tsunami that exceeded 30 meters (100 feet) in height. Although the island of Krakatau was uninhabited, an estimated 36,000 people were killed along the coastline of the Indonesian islands of Java and Sumatra.

Did You Know?

At 4392 meters (14,411 feet) in altitude, Washington's Mount Rainier is the tallest of the 15 great volcanoes that make up the backbone of the Cascade Range. Although considered active, its summit is covered by more than 25 alpine glaciers.

Igneous Activity at Divergent Plate Boundaries

The greatest volume of magma (perhaps 60 percent of Earth's total yearly output) is produced along the oceanic ridge system in association with seafloor spreading (Figure 7.27D). Below the ridge axis where the lithospheric plates are being continually pulled apart, the solid yet mobile mantle responds to the decrease in overburden and rises upward to fill in the rift. As rock rises, it experiences a decrease in confining pressure and undergoes melting without the addition of heat. This process, called *decompression melting,* generates large quantities of magma.

Partial melting of mantle rock at spreading centers produces basaltic magma. Because this newly formed magma is less dense than the mantle rock from which it was derived, it buoyantly rises. Collecting in reservoirs located just beneath the ridge crest, about 10 percent of this melt eventually migrates upward along fissures to erupt as flows on the ocean floor. This activity continuously adds new basaltic rock to the plate margins, temporarily welding them together, only to break again as spreading continues. Along some ridges, out-

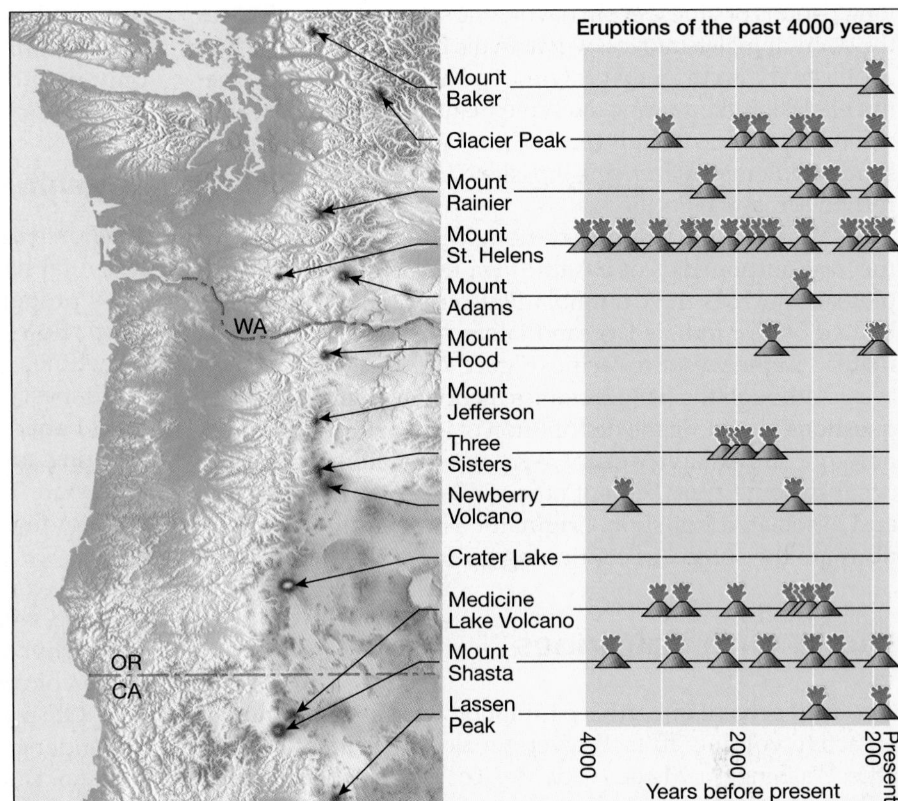

Figure 7.28 Of the 13 potentially active volcanoes in the Cascade Range, 11 have erupted in the past 4000 years and 7 in just the past 200 years. More than 100 eruptions, most of which were explosive, have occurred in the past 4000 years. Mount St. Helens is the most active volcano in the Cascades. Its eruptions have ranged from relatively quiet outflows of lava to explosive events much larger than that of May 18, 1980. Each eruption symbol in the diagram represents from one to several dozen eruptions closely spaced in time. (After U.S. Geological Survey)

pouring of lava build numerous small seamounts. At other locations, erupted lavas produce fluid flows that create more subdued topography.

Although most spreading centers are located along the axis of an oceanic ridge, some are not. In particular, the East African Rift is a site where continental crust is being ripped apart (Figure 7.27F). Vast outpourings of fluid basaltic lavas are common in this region. The East African Rift zone also contains numerous small volcanoes and even a few large composite cones, as exemplified by Mount Kilimanjaro.

Did You Know?

Iceland, which is the largest volcanic island in the world, has more than 20 active volcanoes and numerous geysers and hot springs. The Icelanders named their finest gusher of boiling water *Geysir,* a name used to describe similar features around the world, including "Old Faithful Geyser" in Yellowstone National Park.

Intraplate Igneous Activity

We know why igneous activity is initiated along plate boundaries, but why do eruptions occur in the interiors of plates? Hawaii's Kilauea is considered the world's most active volcano, yet it is situated thousands of kilometers from the nearest plate boundary in the middle of the vast Pacific plate. Other sites of **intraplate volcanism** (meaning "within the plate") include the Canary Islands, Yellowstone, and several volcanic centers that you may be surprised to learn are located in the Sahara Desert of northern Africa.

We now recognize that most intraplate volcanism occurs where a mass of hotter than normal mantle material called a **mantle plume** ascends toward the surface (Figure 7.27B). Although the depth at which (at least some) mantle plumes originate is still hotly debated, many appear to form deep within Earth at the core-mantle boundary. These plumes of solid yet mobile mantle rock rise toward the surface in a manner similar to the blobs that form within a lava lamp. (These are the gaudy lamps that contain two immiscible liquids in a glass container. As the base of the lamp is heated, the denser liquid at the bottom becomes buoyant and forms blobs that rise to the top.) Like the blobs in a lava lamp, a mantle plume has a bulbous head that draws out a narrow stalk beneath it as it rises. Once the plume head nears the top of the mantle, decompression melting generates basaltic magma that may eventually trigger volcanism at the surface. The result is a localized volcanic region a few hundred kilometers across called a **hot spot** (Figure 7.27E). More than 100 hot spots have been identified, and most have persisted for millions of years. The land surface around hot spots is often elevated, showing that it is buoyed up by a plume of warm low-density material. Furthermore, by measuring the heat flow in these regions, geologists have determined that the mantle beneath hot spots must be 100 to 150°C hotter than normal.

The volcanic activity on the island of Hawaii, with its outpourings of basaltic lava, is certainly the result of hot-spot volcanism (Figure 7.27 B). Where a mantle plume has persisted for long periods of time, a chain of volcanic structures may form as the overlying plate moves over it. In the Hawaiian Islands, hot-spot activity is currently centered on Kilauea. However, over the past 80 million years, the same mantle plume generated a chain of volcanic islands (and seamounts) that extend thousands of kilometers from the Big Island in a northwestward direction across the Pacific.

Mantle plumes are also thought to be responsible for the vast outpourings of basaltic lava that create large basalt plateaus such as the Columbia Plateau in the northwestern United States, India's Deccan Plateau, and the Ontong Java Plateau in the western Pacific (Figure 7.27 E).

Although the plate tectonics theory has answered many questions regarding the distribution of igneous activity, many new questions have arisen: Why does seafloor spreading occur in some areas but not others? How do mantle plumes and associated hot spots originate? These and other questions are the subject of continuing geologic research.

Living with Volcanoes

About 10 percent of Earth's population lives in the vicinity of an active volcano. In fact, several major cities including Seattle, Washington; Mexico City, Mexico; Tokyo, Japan; Naples, Italy; and Quito, Ecuador, are located on or near a volcano.

Until recently, the dominant view of Western societies was that humans possess the wherewithal to subdue volcanoes and other types of catastrophic natural hazards. It is now becoming increasingly apparent that volcanoes are not only very destructive but unpredictable as well. With this awareness, the new focus is on how to live with volcanoes.

Volcanic Hazards

As shown in Figure 7.29, volcanoes produce a wide variety of potential hazards that can kill people and wildlife and destroy property. Perhaps the greatest threats to life are pyroclastic flows. These hot mixtures of gas, ash, and pumice that sometimes exceed 800°C (1500°F) speed down the flanks of volcanoes, giving people little chance of surviving.

Lahars, which can occur even when a volcano is not erupting, are perhaps the next most dangerous volcanic phenomenon. These mixtures of volcanic debris and water can flow for tens of kilometers down a valley at speeds that may exceed 100 kilometers (62 miles) per hour. Lahars pose a potential hazard to many communities downstream from glacier-clad volcanoes such as Mount Rainier. Other potentially destructive mass-wasting events include the rapid collapse of the volcano's summit or flank.

Other obvious hazards include explosive eruptions that can endanger people and property at great distances from a volcano. During the past 15 years, at least 80 commercial jets have been damaged by inadvertently flying into clouds of volcanic ash. One of these was a near crash that occurred in 1989 when a Boeing 747, with more than 300 passengers

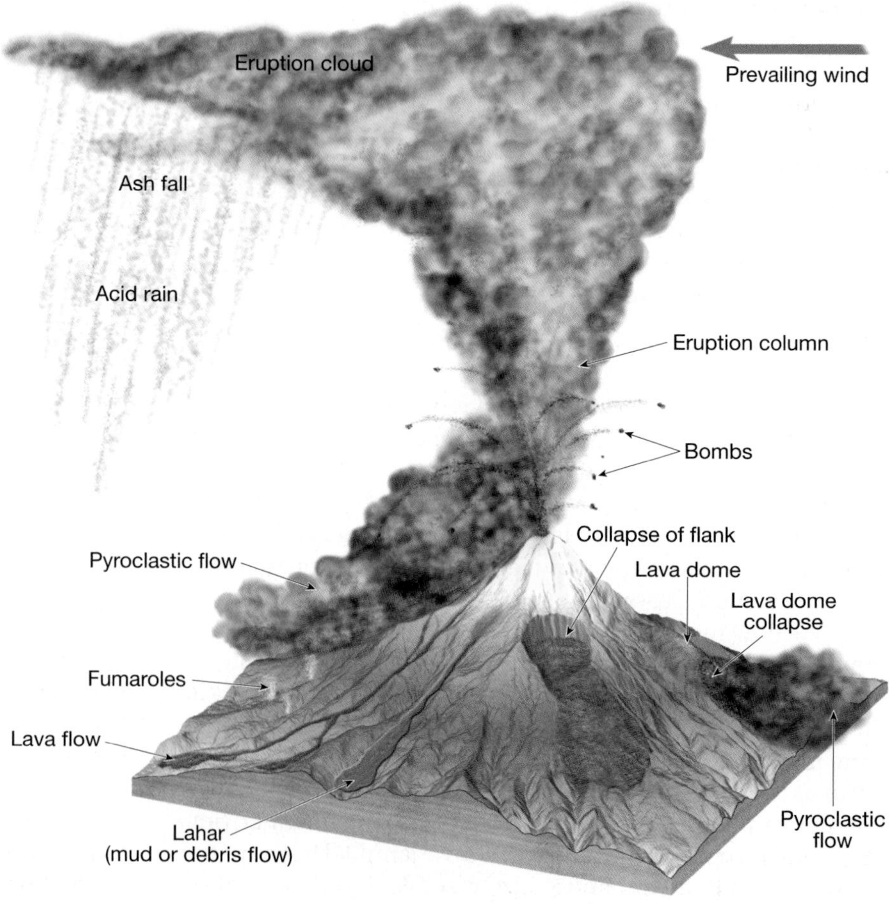

Figure 7.29 Simplified drawing showing a wide variety of natural hazards associated with volcanoes. (After U.S. Geological Survey)

aboard, encountered an ash cloud from Alaska's Redoubt volcano. All four engines stalled after they became clogged with ash. Fortunately, the engines were restarted at the last minute and the aircraft managed to land safely in Anchorage.

Monitoring Volcanic Activity

A number of volcano monitoring techniques are now employed; most of them aimed at detecting the movement of magma from a subterranean reservoir (typically several kilometers deep) toward the surface. The four most noticeable changes in a volcanic landscape caused by the migration of magma are (1) changes in the pattern of volcanic earthquakes; (2) expansion of a near-surface magma chamber, which leads to inflation of the volcano; (3) changes in the amount and/or composition of the gases that are released from a volcano; and (4) an increase in ground temperature caused by the emplacement of new magma.

Almost a third of all volcanoes that have erupted in historic times are now monitored seismically (through the use of instruments that detect earthquake tremors). In general, a sharp increase in seismic unrest followed by a period of relative quiet has shown to be a precursor for many volcanic eruptions. However, some large volcanic structures have exhibited lengthy periods of seismic unrest. For example, Rabaul Caldera in New Guinea recorded a strong increase in seismicity in 1981. This activity lasted 13 years and finally culminated with an eruption in 1994. Occasionally, a large earthquake has triggered a volcanic eruption, or at least disturbed the volcano's plumbing. Kilauea, for example, began to erupt after the Kalapana earthquake of 1975.

The roof of a volcano may rise as new magma accumulates in its interior—a phenomena that precedes many volcanic eruptions. Because the accessibility of many volcanoes is limited, remote sensing devices, including lasers, Doppler radar, and Earth-orbiting satellites, are often used to determine whether or not a volcano is swelling. The recent discovery of ground doming at Three Sisters volcanoes in Oregon was first detected using radar images obtained from satellites.

Volcanologists also frequently monitor the gases that are released from volcanoes in an effort to detect even minor changes in their amount and/or composition. Some volcanoes show an increase in sulfur dioxide (SO_2) emissions months or years prior to an eruption. On the other hand, a few days prior to the 1991 eruption of Mount Pinatubo, emission of carbon dioxide (CO_2) dropped dramatically.

Figure 7.30 Photo taken by Jeff Williams from the International Space Station of a steam-and-ash eruption emitted from Cleveland Volcano in the Aluetian Islands. This event was reported to the Alaska Volcano Observatory, which issued a warning to air traffic control. (NASA)

The development of remote sensing devices has greatly increased our ability to monitor volcanoes. These instruments and techniques are particularly useful for monitoring eruptions in progress. Photographic images and infrared (heat) sensors can detect lava flows and volcanic columns rising from a volcano (Figure 7.30). Furthermore, satellites can detect ground deformation as well as monitor SO_2 emissions.

The overriding goal of all monitoring techniques is to discover precursors that may warn of an imminent eruption. This is accomplished by first diagnosing the current condition of a volcano and then using this baseline data to predict its future behavior. Stated another way, a volcano must be observed over an extended period to recognize significant changes from its "resting state."

The Chapter in Review

1. The primary factors that determine the nature of volcanic eruptions include the magma's *temperature*, its *composition*, and the *amount of dissolved gases* it contains. As lava cools, it begins to congeal; and as *viscosity* increases, its mobility decreases. *The viscosity of magma is directly related to its silica content. Rhyolitic* lava, with its high silica content, is very viscous and forms short, thick flows. *Basaltic* lava, with a lower silica content, is more fluid and may travel a long distance before congealing. Dissolved gases provide the force that propels molten rock from the vent of a volcano.

2. The materials associated with a volcanic eruption include *lava flows* (*pahoehoe* and *aa* flows for basaltic lavas); *gases* (primarily in the form of *water vapor*); and *pyroclastic material* (pulverized rock and lava fragments blown from the volcano's vent, which include *ash, pumice, lapilli, cinders, blocks,* and *bombs*).

3. Successive eruptions of lava from a central vent result in a mountainous accumulation of material known as a *volcano*. Located at the summit of many volcanoes is a steep-walled depression called a *crater. Shield cones* are broad, slightly domed volcanoes built primarily of fluid, basaltic lava. *Cinder cones* have steep slopes composed of pyroclastic material. *Composite cones*, or *stratovolcanoes*, are large, nearly symmetrical structures built of interbedded lavas and pyroclastic deposits. Composite cones produce some of the most violent volcanic activity. Often associated with a violent eruption is a *nuée ardente*, a fiery cloud of hot gases infused with incandescent ash that races down steep volcanic slopes. Large composite cones may also generate a type of mudflow known as a *lahar.*

4. Most volcanoes are fed by *conduits* or *pipes.* As erosion progresses, the rock occupying the pipe is often more resistant and may remain standing above the surrounding terrain as a *volcanic neck.* The summits of some volcanoes have large, nearly circular depressions called *calderas* that result from collapse following an explosive eruption. Calderas also form on shield volcanos by subterranean drainage from a central magma chamber, and the largest calderas form by the discharge of colossal volumes of silica-rich pumice along ring fractures. Although volcanic eruptions from a central vent are the most familiar, by far the largest amounts of volcanic material are extruded from cracks in the crust called *fissures.* The term *flood basalt* describes the fluid, waterlike, basaltic lava flows that cover an extensive region in the northwestern United States known as the Columbia Plateau. When silica-rich magma is extruded, *pyroclastic flows* consisting largely of ash and pumice fragments usually result.

5. Intrusive igneous bodies are classified according to their *shape* and by their *orientation with respect to the host rock,* generally sedimentary rock. The two general shapes are *tabular* (tablelike) and *massive.* Intrusive igneous bodies that cut across existing sedimentary beds are said to be *discordant,* whereas those that form parallel to existing sedimentary beds are *concordant.*

6. *Dikes* are tabular, discordant igneous bodies produced when magma is injected into fractures that cut across rock layers. Tabular, concordant bodies called *sills* form when magma is injected along the bedding surfaces of sedimentary rocks. *Laccoliths* are similar to sills but form from less-fluid magma that collects as a lens-shaped mass that arches the overlying strata upward. *Batholiths,* the largest intrusive igneous bodies with surface exposures of more than 100 square kilometers (40 square miles), frequently make up the cores of mountains.

7. *Most active volcanoes are associated with plate boundaries.* Active areas of volcanism are found along oceanic ridges where seafloor spreading is occurring (*divergent plate boundaries*), in the vicinity of ocean trenches where one plate is being subducted beneath another (*convergent plate boundaries*), and in the interiors of plates themselves (*intraplate volcanism*). Rising plumes of hot mantle rock are the source of most intraplate volcanism

Key Terms

aa flow (p. 198)
batholith (p. 212)
caldera (p. 200)
cinder cone (p. 202)
columnar joint (p. 212)
composite cone (p. 204)
conduit (p. 200)
continental volcanic arc (p. 216)
crater (p. 200)
dike (p. 210)
fissure (p. 209)
fissure eruption (p. 209)
flood basalt (p. 209)
fumarole (p. 200)
hot spot (p. 217)
intraplate volcanism (p. 217)
island arc (p. 216)
laccolith (p. 212)
lahar (p. 207)
mantle plume (p. 217)
nuée ardente (p. 205)
pahoehoe flow (p. 198)
parasitic cone (p. 200)
pipe (p. 200)
pluton (p. 210)
pyroclastic flow (p. 205)
pyroclastic materials (p. 199)
scoria cone (p. 202)
shield volcano (p. 200)
sill (p. 211)
stratovolcano (p. 204)
vent (p. 200)
viscosity (p. 194)
volatiles (p. 196)
volcanic island arc (p. 216)
volcanic neck (p. 210)
volcano (p. 200)

Questions for Review

1. What is the difference between *magma* and *lava?*
2. What three factors determine the nature of a volcanic eruption? What role does each play?
3. Why is a volcano fed by highly viscous magma likely to be a greater threat than a volcano supplied with very fluid magma?
4. Describe *pahoehoe* and *aa* lava.
5. List the main gases released during a volcanic eruption.
6. Analysis of samples taken during Hawaiian eruptions on the island of Hawaii indicate that _____ was the most abundant gas released.
7. How do volcanic bombs differ from blocks of pyroclastic debris?
8. Compare a volcanic *crater* to a *caldera.*
9. Compare and contrast the three main types of volcanoes (size, shape, eruptive style, and so forth).
10. Name one example of each of the three types of volcanoes.
11. Compare the formation of Hawaii with that of Parícutin.
12. Describe the formation of Crater Lake. Compare it to the caldera formed during the eruption of Kilauea.
13. What are the largest volcanic structures on Earth?
14. What is Ship Rock, New Mexico, and how did it form?
15. How do the eruptions that created the Columbia Plateau differ from eruptions that create volcanic peaks?
16. Describe each of the four intrusive igneous features discussed in the text (*dike, sill, laccolith,* and *batholith*).

17. Why might a laccolith be detected at Earth's surface before being exposed by erosion?
18. What is the largest of all intrusive igneous bodies? Is it tabular or massive? Concordant or discordant?
19. Spreading-center volcanism is associated with which rock type? What causes rocks to melt in regions of spreading-center volcanism?
20. What is the Ring of Fire?
21. Are volcanic eruptions in the Ring of Fire generally quiet or violent? Name a volcano that would support your answer.
22. The Hawaiian Islands and Yellowstone National Park are thought to be associated with which of the three zones of volcanism?
23. Volcanic islands in the deep ocean are composed primarily of which igneous rock type?

Online Study Guide

The *Foundations of Earth Science* Web site uses the resources and flexibility of the Internet to aid in your study of the topics in this chapter. Written and developed by Earth science instructors, this site will help improve your understanding of Earth science. Visit **http://www.prenhall.com/lutgens** and click on the cover of *Foundations of Earth Science 5e* to find:

- Online review quizzes.
- Critical thinking exercises.
- Links to chapter-specific Web resources.
- Internet-wide key-term searches.

http://www.prenhall.com/lutgens

GEODe: Earth Science

GEODe: Earth Science makes studying more effective by reinforcing key concepts using animation, video, narration, interactive exercises, and practice quizzes. A copy is included with every copy of *Foundations of Earth Science 5e.*

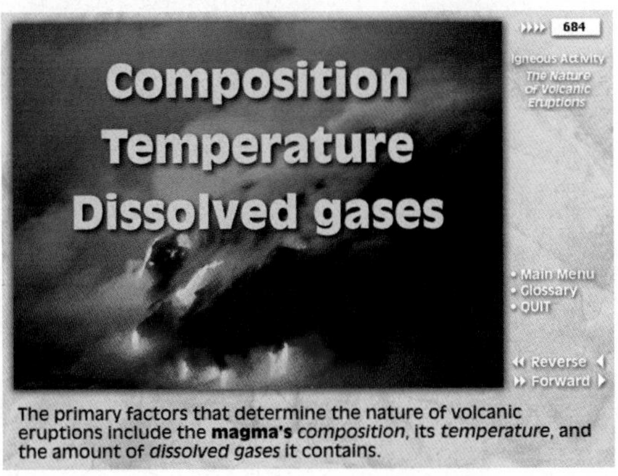

The primary factors that determine the nature of volcanic eruptions include the **magma's** *composition*, its *temperature*, and the amount of *dissolved gases* it contains.

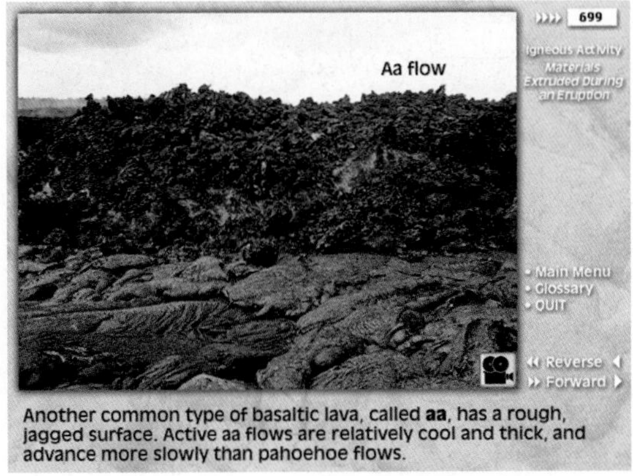

Another common type of basaltic lava, called **aa**, has a rough, jagged surface. Active aa flows are relatively cool and thick, and advance more slowly than pahoehoe flows.

UNIT IV DECIPHERING EARTH'S HISTORY

The strata exposed in the Grand Canyon contain clues to millions of years of Earth's history. (Photo by Marc Muench/Muench Photography, Inc.)

Geologic Time

FOCUS ON LEARNING

To assist you in learning the important concepts in this chapter, you will find it helpful to focus on the following questions:

1. What is the doctrine of uniformitarianism? How does it differ from catastrophism?
2. What are the two types of dates used by geologists to interpret Earth's history?
3. What are the basic laws, principles, and techniques used to establish relative dates?
4. What are fossils? What conditions favor the preservation of organisms as fossils?
5. How are fossils used to correlate rocks of similar ages that are in different places?
6. What is radioactivity, and how are radioactive isotopes used in radiometric dating?
7. What is the geologic time scale, and what are its principal subdivisions?
8. Why is it difficult to assign reliable numerical dates to samples of sedimentary rock?

In the eighteenth century, James Hutton recognized the immensity of Earth's history and the importance of time as a component in all geologic processes. In the nineteenth century, others effectively demonstrated that Earth had experienced many episodes of mountain building and erosion, which must have required great spans of geologic time. Although these pioneering scientists understood that Earth was very old, they had no way of knowing its true age. Was it tens of millions, hundreds of millions, or even billions of years old? Rather, a geologic time scale was developed that showed the sequence of events based on relative dating principles. What were these principles? What part did fossils play? With the discovery of radioactivity and the development of radiometric dating techniques, geologists can now assign fairly accurate dates to many of the events in Earth's history. What is radioactivity? Why is it a good "clock" for dating the geologic past? This chapter will answer these questions.

Geology Needs a Time Scale

In 1869, John Wesley Powell, who was later to head the U.S. Geological Survey, led a pioneering expedition down the Colorado River and through the Grand Canyon (Figure 8.1). Writing about the strata that were exposed by the downcutting of the river, Powell observed that "the canyons of this region would be a Book of Revelations in the rock-leaved Bible of geology." He was undoubtedly impressed with the millions of years of Earth's history exposed along the walls of the Grand Canyon (see photo, p. 222).

Powell realized that the evidence for an ancient Earth is concealed in its rocks. Like the pages in a long and complicated history book, rocks record the geologic events and changing life forms of the past. The book, however, is not complete. Many pages, especially in the early chapters, are missing. Others are tattered, torn, or smudged. Yet, enough of the book remains to allow much of the story to be deciphered.

Interpreting Earth's history is an important goal of geology. Like a modern-day sleuth, the geologist must interpret clues found preserved in the rocks. By studying rocks, especially sedimentary rocks, and the features they contain, geologists can unravel the complexities of the past.

Geologic events by themselves, however, have little meaning until they are put into a time perspective. Studying history, whether it be the Civil War or the Age of Dinosaurs, requires a calendar. Among geology's major contributions is a calendar called the *geologic time scale* and the discovery that Earth's history is exceedingly long.

The geologists who developed the geologic time scale revolutionized the way people think about time and perceive our planet. They learned that Earth is much older than anyone had previously imagined and that its surface and interior have been changed over and over again by the same geologic processes that operate today.

Some Historical Notes About Geology

The nature of our Earth—its materials and processes—has been a focus of study for centuries. However, the late 1700s is generally regarded as the beginning of modern geology. It was

A.

B.

Figure 8.1 **A.** Start of the expedition from Green River station. A drawing from Powell's 1875 book. **B.** Major John Wesley Powell, pioneering geologist and the second director of the U.S. Geological Survey. (Courtesy of the U.S. Geological Survey, Denver)

during this time that James Hutton published his important work, *Theory of the Earth.* Prior to this time, most explanations about Earth's history relied on supernatural explanations.

Catastrophism

In the mid-1600s, James Ussher, Anglican Archbishop of Armagh, Primate of All Ireland, published a work that had immediate and profound influence on people's view of Earth's age. A respected scholar of the Bible, Ussher constructed a chronology of human and Earth history in which he determined that Earth was only a few thousands of years old, having been created in 4004 B.C. Ussher's treatise earned widespread acceptance among Europe's scientific and religious leaders, and his chronology was soon printed in the margins of the Bible itself.

During the 1600s and 1700s, the doctrine of **catastrophism** strongly influenced people's thinking about Earth. Briefly stated, catastrophists believed that Earth's varied landscapes had been fashioned primarily by great catastrophes. Features such as mountains and canyons, which today we know take great periods of time to form, were explained as having been produced by sudden and often worldwide disasters of unknowable causes that no longer operate. This philosophy was an attempt to fit the rate of Earth's processes to the prevailing ideas on Earth's age.

The Birth of Modern Geology

When James Hutton, a Scottish physician and gentleman farmer, published his *Theory of the Earth* in the late 1700s, he put forth a principle that came to be known as the doctrine of **uniformitarianism.** Uniformitarianism is still a fundamental principle of modern geology. It simply states that *the physical, chemical, and biological laws that operate today have also operated in the geologic past.* This means that the forces and processes that we observe presently shaping our planet have been at work for a very long time. Thus, to understand ancient rocks, we must first understand present-day processes and their results. This idea is commonly expressed by saying, "The present is the key to the past."

Prior to Hutton's *Theory of the Earth,* no one had effectively demonstrated that geologic processes occur over extremely long periods of time. However, Hutton persuasively argued that processes that appear weak and slow acting could, over long spans of time, produce effects that were just as great as those resulting from sudden catastrophic events. Unlike his predecessors, Hutton cited verifiable observations to support his ideas.

For example, when he argued that mountains are sculpted and ultimately destroyed by weathering and the work of running water, and that their wastes are carried to the oceans by processes that can be observed, Hutton said, "We have a chain of facts which clearly demonstrates that the materials of the wasted mountains have traveled through the rivers"; and further, "There is not one step in all this progress that is not to be actually perceived." He then went on to summarize this thought by asking a question and immediately providing the answer: "What more can we require? Nothing but time."

Geology Today

The basic tenets of uniformitarianism are just as viable today as in Hutton's day. We realize more strongly than ever that the present gives us insight into the past and that the physical, chemical, and biological laws that govern geologic processes remain unchanging through time. However, we also understand that the doctrine should not be taken too literally. To say that geologic processes in the past were the same as those occurring today is not to suggest that they always had the same relative importance or that they operated at precisely the same rate. Moreover, some important geologic processes are not currently observable, but evidence that they occur is well established. For example, we know that Earth has experienced impacts from large meteorites even though we have no human witnesses. Such events altered Earth's crust, modified its climate, and strongly influenced life on the planet.

The acceptance of the concept of uniformitarianism, however, meant the acceptance of a very long history for Earth. Although Earth's processes vary in their intensity, they still take a long time to create or destroy major landscape features. For example, geologists have established that mountains once existed in portions of present-day Minnesota, Wisconsin, and Michigan. Today, the region consists of low hills and plains. Erosion (processes that wear land away) gradually destroyed these peaks. Estimates indicate that the North American continent is being lowered at a rate of about 3 centimeters (1 inch) per 1000 years. At this rate, it would take 100 million years for water, wind, and ice to lower mountains that were 3000 meters (10,000 feet) high.

But even 100 million years is a relatively short span on the time scale of Earth's history, for the rock record contains evidence that shows Earth has experienced *many cycles* of mountain building and erosion. Concerning the ever-changing nature of Earth through great expanses of geologic time, Hutton made a statement that was to become his most famous. In concluding his classic 1788 paper published in the *Transactions of the Royal Society of Edinburgh,* he stated, "The results, therefore, of our present enquiry is, that we find no vestige of a beginning—no prospect of an end."

Did You Know?

Early attempts at determining Earth's age proved to be unreliable. One method reasoned that if the rate at which sediment accumulates could be determined as well as the total thickness of sedimentary rock that had been deposited during Earth's history, an estimate of Earth's age could be made. All that was necessary was to divide the rate of sediment accumulation into the total thickness of sedimentary rock. This method was riddled with difficulties. Can you think of some?

It is important to remember that although many features of our physical landscape may seem to be unchanging over the decades we might observe them, they are nevertheless changing, but on time scales of hundreds, thousands, or even many millions of years.

Relative Dating—Key Principles

Deciphering Earth's History
▼ Relative Dating

The geologists who developed the geologic time scale revolutionized the way people think about time and perceive our planet. They learned that Earth is much older than anyone had previously imagined and that its surface and interior have been changed over and over again by the same geologic processes that operate today.

During the late 1800s and early 1900s, various attempts were made to determine Earth's age. Although some of the methods appeared promising at the time, none proved to be reliable. What these scientists were seeking was a **numerical date.** Such dates pinpoint the time in history when something took place—for example, the extinction of the dinosaurs about 65 million years ago. Today, our understanding of radioactivity allows us to accurately determine numerical dates for many rocks that represent important events in Earth's distant past. We will study radioactivity later in this chapter. Prior to the discovery of radioactivity, geologists had no accurate and dependable method of numerical dating and had to rely solely on relative dating.

Relative dating means placing rocks in their proper *sequence of formation*—which formed first, second, third, and so on. Relative dating cannot tell us how long ago something took place, only that it followed one event and preceded another. The relative dating techniques that were developed are valuable and still widely used. Numerical dating methods did not replace these techniques; they simply supplemented them. To establish a relative time scale, a few simple principles or rules had to be discovered and applied. Although they may seem obvious to us today, they were major breakthroughs in thinking at the time, and their discovery and acceptance were important scientific achievements.

Law of Superposition

Nicolaus Steno (1638–1686), a Danish anatomist, geologist, and priest, is credited with being the first to recognize a sequence of historical events in an outcrop of sedimentary rock layers. Working in the mountains of western Italy, Steno applied a very simple rule that became the most basic principle of relative dating—the **law of superposition.** The law simply states that in an undeformed sequence of sedimentary rocks, each bed is older than the one above it and younger than the one below. Although it may seem obvious that a rock layer could not be deposited unless it had something older beneath it for support, it was not until 1669 that Steno clearly stated the principle.

This rule also applies to other surface-deposited materials such as lava flows and beds of ash from volcanic eruptions. Applying the law of superposition to the beds exposed in the upper portion of the Grand Canyon (Figure 8.2), you can easily place the layers in their proper order. Among those that are shown, the sedimentary rocks in the Supai Group must be the oldest, followed in order by the Hermit Shale, Coconino Sandstone, Toroweap Formation, and Kaibab Limestone.

A.

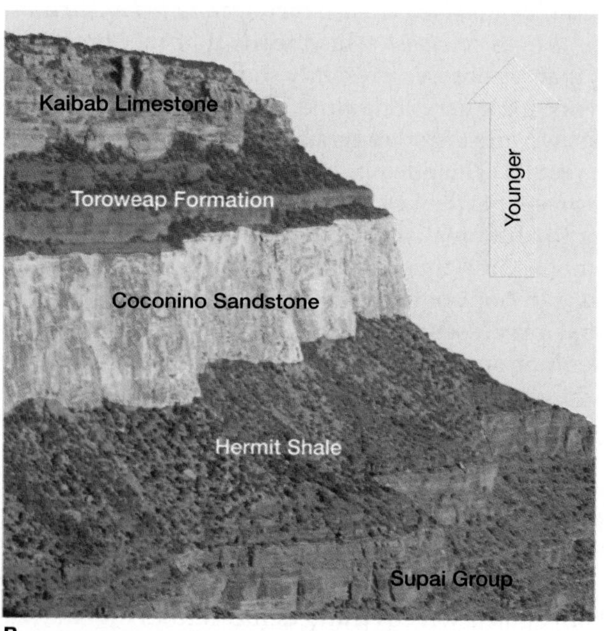

B.

Figure 8.2 Applying the law of superposition to these layers exposed in the upper portion of the Grand Canyon, the Supai Group is oldest and the Kaibab Limestone is youngest. (Photo by E. J. Tarbuck)

Figure 8.3 Most layers of sediment are deposited in a nearly horizontal position. Thus, when we see rock layers that are folded or tilted, we can assume that they must have been moved into that position by crustal disturbances *after* their deposition. These tilted layers are exposed in Hartland Quay, Devon, England. (Photo by Tom Bean/DRK Photo)

Principle of Original Horizontality

Steno is also credited with recognizing the **principle of original horizontality.** This principle simply states that most layers of sediment are deposited in a horizontal position. Thus, if we observe rock layers that are flat, it means they have not been disturbed and still have their *original* horizontality. The layers in the Grand Canyon illustrate this in the photo on p. 222 and in Figure 8.2. But if they are folded or inclined at a steep angle, they must have been moved into that position by crustal disturbances sometime *after* their deposition (Figure 8.3).

Principle of Cross-Cutting Relationships

When a fault cuts through other rocks, or when magma intrudes and crystallizes, we can assume that the fault or intrusion is younger than the rocks affected. For example, in Figure 8.4, the faults and dikes clearly must have occurred *after* the sedimentary layers were deposited.

This is the **principle of cross-cutting relationships.** By applying the cross-cutting principle, you can see that fault A occurred *after* the sandstone layer was deposited, because it "broke" the layer. However, fault A occurred *before* the conglomerate was laid down, because that layer is unbroken.

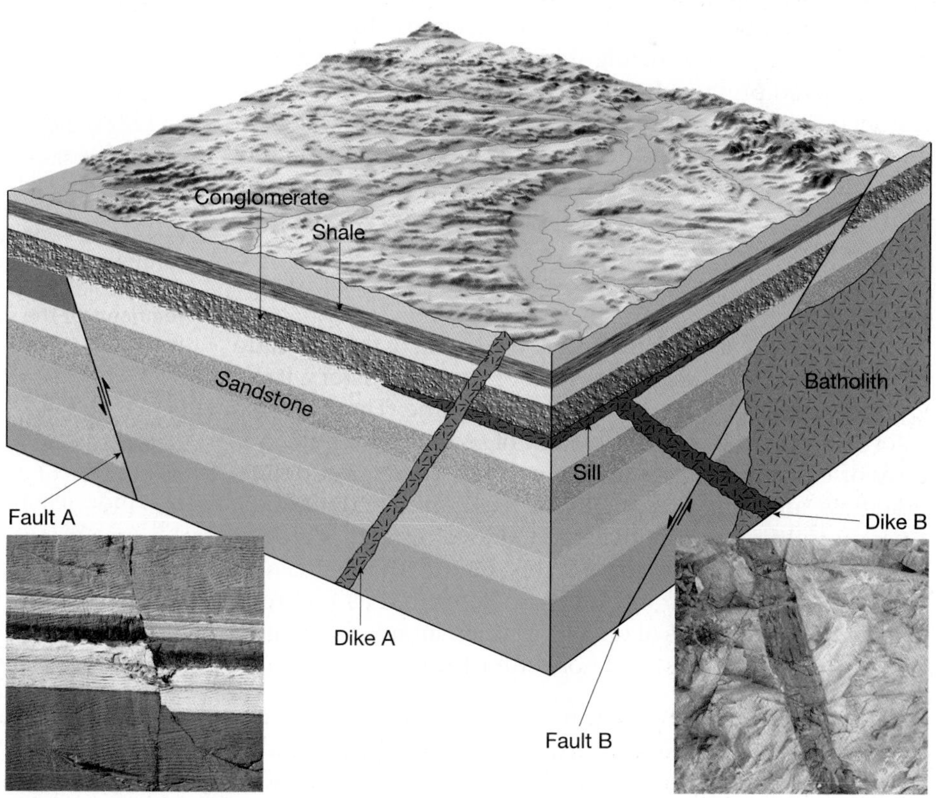

Figure 8.4 Cross-cutting relationships are an important principle used in relative dating. An intrusive rock body is younger than the rocks it intrudes. A fault is younger than the rock layers it cuts.

We can also state that dike B and its associated sill are older than dike A, because dike A cuts the sill. In the same manner, we know that the batholith was emplaced after movement occurred along fault B, but before dike B was formed. This is true because the batholith cuts across fault B, and dike B cuts across the batholith.

Inclusions

Sometimes inclusions can aid the relative dating process. **Inclusions** are fragments of one rock unit that have been enclosed within another. The basic principle is logical and straightforward. The rock mass adjacent to the one containing the inclusions must have been there first in order to provide the rock fragments. Therefore, the rock mass containing inclusions is the younger of the two. Figure 8.5 provides an example. In part A, we know that the intrusive igneous rock is younger because it contains inclusions of the surrounding rock. In part C, the inclusions of intrusive igneous rock in the adjacent sedimentary layer (above the line labeled "nonconformity") indicate that the sedimentary layer was deposited on top of an eroded igneous mass rather than being intruded from below by a mass of magma that later crystallized.

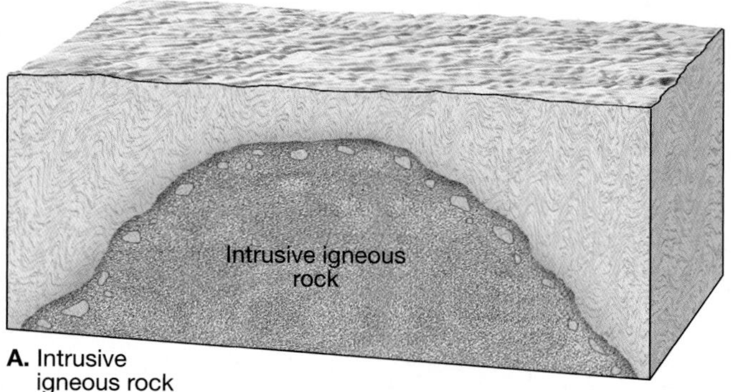

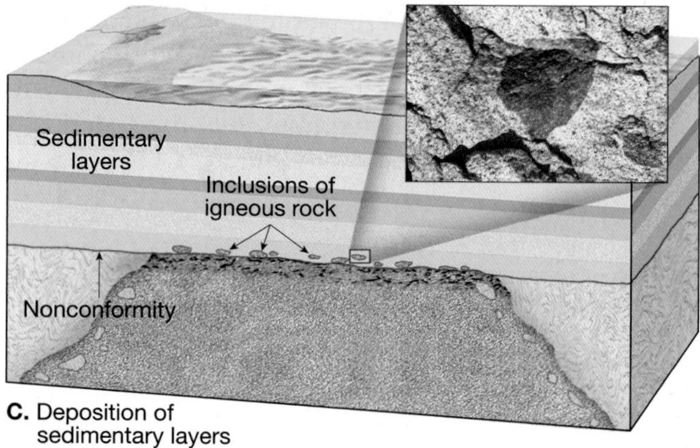

Figure 8.5 These diagrams illustrate two ways that inclusions can form, as well as a type of unconformity termed a *nonconformity.* In part **A,** the inclusions in the igneous mass represent unmelted remnants of the surrounding host rock that were broken off and incorporated at the time the magma was intruded. In part **C,** the igneous rock must be older than the overlying sedimentary beds because the sedimentary beds contain inclusions of the igneous rock. When older intrusive igneous rocks are overlain by younger sedimentary layers, a nonconformity is said to exist. The photo shows an inclusion of dark igneous rock in a lighter-colored and younger host rock. (Photo by Tom Bean)

Unconformities

When we observe layers of rock that have been deposited essentially without interruption, we call them **conformable.** Many areas exhibit conformable beds representing certain spans of geologic time. However, no place on Earth has a complete set of conformable strata.

Throughout Earth's history, the deposition of sediment has been interrupted again and again. All such breaks in the rock record are termed *unconformities.* An **unconformity** represents a long period during which deposition ceased, erosion removed previously formed rocks, and then deposition resumed. In each case, Earth's crust has undergone uplift and erosion, followed by subsidence and renewed sedimentation. Unconformities are important features because they represent significant geologic events in Earth's history. Moreover, their recognition helps us identify which intervals of time are not represented by strata and are missing from the geologic record.

The rocks exposed in the Grand Canyon of the Colorado River represent a tremendous span of geologic history. It is a wonderful place to take a trip through time. The canyon's colorful strata record a long history of sedimentation in a variety of environments—advancing seas, rivers and deltas, tidal flats, and sand dunes. But the record is not continuous. Unconformities represent vast amounts of time that are not accounted for in the canyon's layers. Figure 8.6 is a geologic cross section of the Grand Canyon. Refer to it as you read about three basic types of unconformities: angular unconformities, disconformities, and nonconformities.

Angular Unconformity The most easily recognized unconformity is **angular unconformity.** It consists of tilted or folded sedimentary rocks that are overlain by younger, more flat-lying strata. An angular unconformity indicates that during the pause in deposition, a period of deformation (folding or tilting) as well as erosion occurred. The steps in Figure 8.7 illustrate this process.

When James Hutton studied an angular unconformity in Scotland more than 200 years ago, it was clear to him that it represented a major episode of geologic activity (Figure 8.7E). He also appreciated the immense time span implied by such relationships. Later, when a companion wrote of their

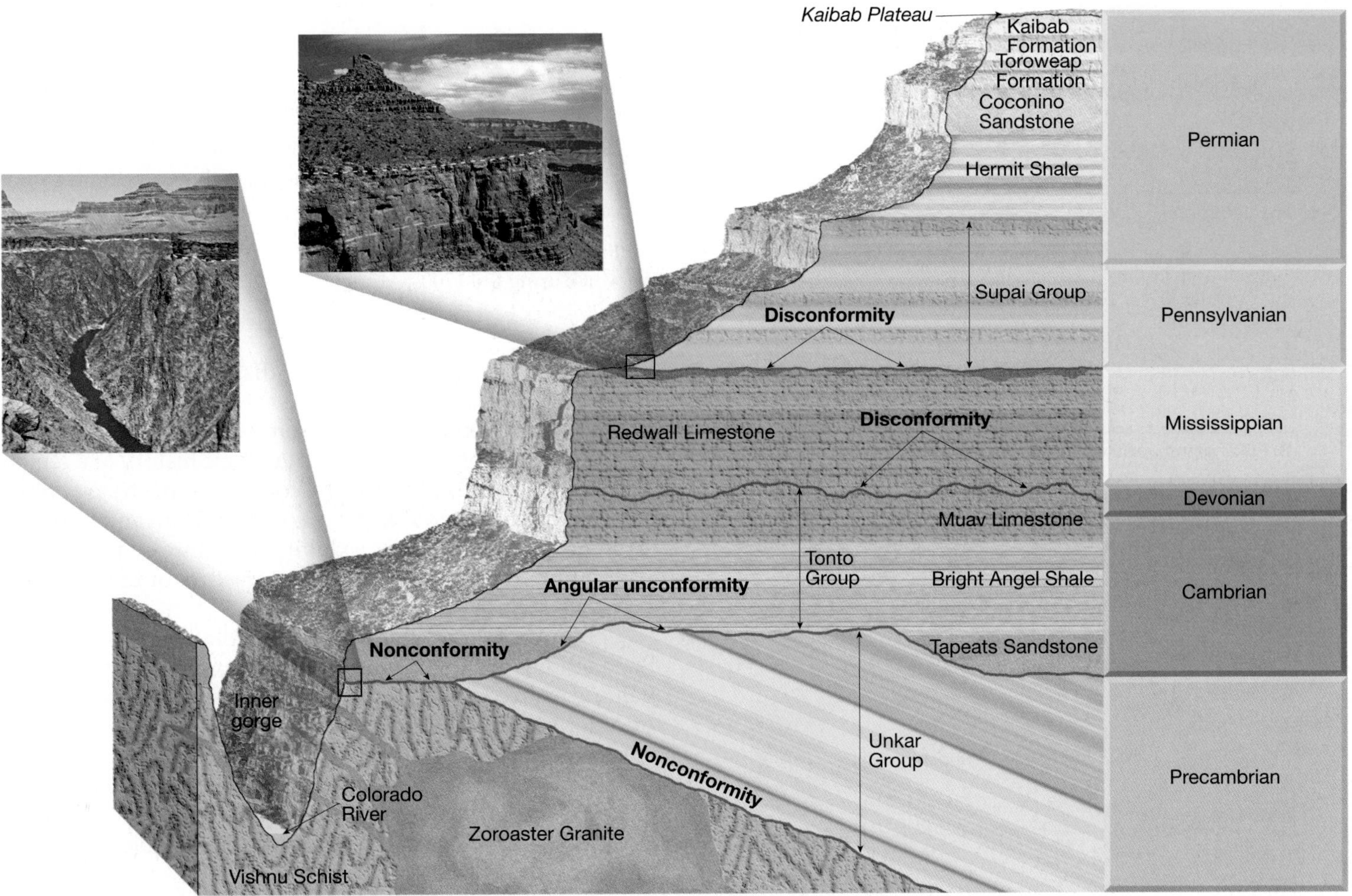

Figure 8.6 This cross section through the Grand Canyon illustrates the three basic types of unconformities. An angular unconformity can be seen between the tilted Precambrian Unkar Group and the Cambrian Tapeats Sandstone. Two disconformities are marked, above and below the Redwall Limestone. A nonconformity occurs between the igneous and metamorphic rocks exposed in the inner gorge and the sedimentary strata of the Unkar Group. A nonconformity, highlighted by a photo, also occurs between the rocks of the inner gorge and the Tapeats Sandstone.

visit to the site, he stated that, "The mind seemed to grow giddy by looking so far into the abyss of time."

Disconformity When contrasted with angular unconformities, **disconformities** are more common but usually far less conspicuous because the strata on either side are essentially parallel. Many disconformities are difficult to identify because the rocks above and below are similar and there is little evidence of erosion. Such a break often resembles an ordinary bedding plane. Other disconformities are easier to identify because the ancient erosion surface is cut deeply into the older rocks below.

Nonconformity The third basic type of unconformity is **nonconformity.** Here, the break separates older metamorphic or intrusive igneous rocks from younger sedimentary strata (see Figure 8.5). Just as angular unconformities and disconformities imply crustal movements, so, too, do nonconformities. Intrusive igneous masses and metamorphic rocks originate far below the surface. For a nonconformity to develop, a period of uplift and the erosion of overlying rocks must occur. Once exposed at the surface, the igneous or metamorphic rocks are subjected to weathering and erosion prior to subsidence and the renewal of sedimentation.

Using Relative Dating Principles

By applying the principles of relative dating to the hypothetical geologic cross section shown in Figure 8.8, the rocks and the events in Earth's history they represent can be placed into their proper sequence. The statements within the figure summarize the logic used to interpret the cross section. In this example, we establish a relative time scale for the rocks and events in the area of the cross section. Remember, we have no idea how many years of Earth's history are represented, nor do we know how this area compares to any other.

Correlation of Rock Layers

To develop a geologic time scale that applies to the entire Earth, rocks of similar age in different regions must be matched up. Such a task is referred to as **correlation.** Within a limited area, correlating the rocks of one locality with those

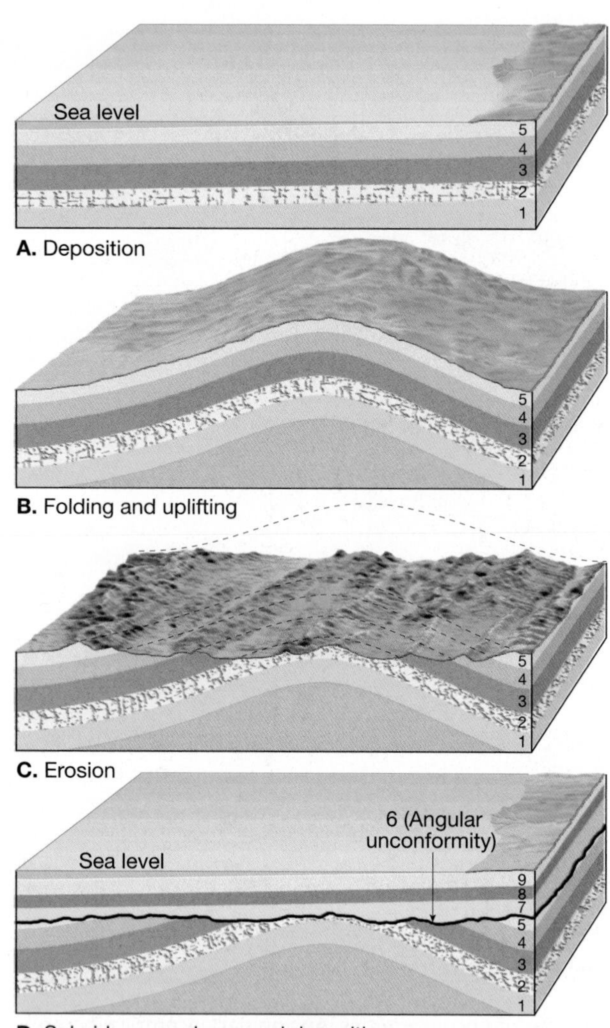

Figure 8.7 Formation of an angular unconformity. An angular unconformity represents a period during which deformation and erosion occurred. Part **E** shows an angular unconformity at Siccar Point, Scotland, that was first described by James Hutton more than 200 years ago. (Photo by Edward Hay)

of another may be done simply by walking along the outcropping edges. However, this may not be possible when a bed is not continuously exposed. Correlation over short distances is often achieved by noting the position of a rock layer in a sequence of strata. Or, a layer may be identified in another location if it consists of very distinctive or uncommon minerals.

By correlating the rocks from one place to another, a more comprehensive view of the geologic history of a region is possible. Figure 8.9 shows the correlation of strata at three sites on the Colorado Plateau in southern Utah and northern Arizona. No single locale exhibits the entire sequence, but correlation reveals a more complete picture of the sedimentary rock record.

Many geologic studies involve relatively small areas. Such studies are important in their own right, but their full value is realized only when the rocks are correlated with those of other regions. Although the methods just described are sufficient to trace a rock formation over relatively short distances, they are not adequate for matching rocks that are separated by great distances. When correlation between widely separated areas or between continents is the objective, geologists must rely on fossils.

Did You Know?

The word *fossil* comes from the Latin *fossilium,* which means "dug up from the ground." As originally used by medieval writers, a fossil was *any* stone, ore, or gem that came from an underground source. In fact, many early books on mineralogy are called books of fossils. The current meaning of fossil came about during the 1700s.

Fossils: Evidence of Past Life

Fossils, the remains or traces of prehistoric life, are important inclusions in sediment and sedimentary rocks. They are important basic tools for interpreting the geologic past. The scientific study of fossils is called **paleontology.** It is an interdisciplinary science that blends geology and biology in an attempt to understand all aspects of the succession of life over the vast expanse of geologic time. Knowing the nature of the life forms that existed at a particular time helps researchers understand past environmental conditions. Further, fossils are important time indicators and play a key role in correlating rocks of similar ages that are from different places.

Did You Know?

People frequently confuse paleontology and archaeology. Paleontologists study fossils and are concerned with *all* life forms in the geologic past. By contrast, archaeologists focus on the material remains of past human life. These remains include both the objects used by people long ago, called *artifacts,* and the buildings and other structures associated with where people lived, called *sites.*

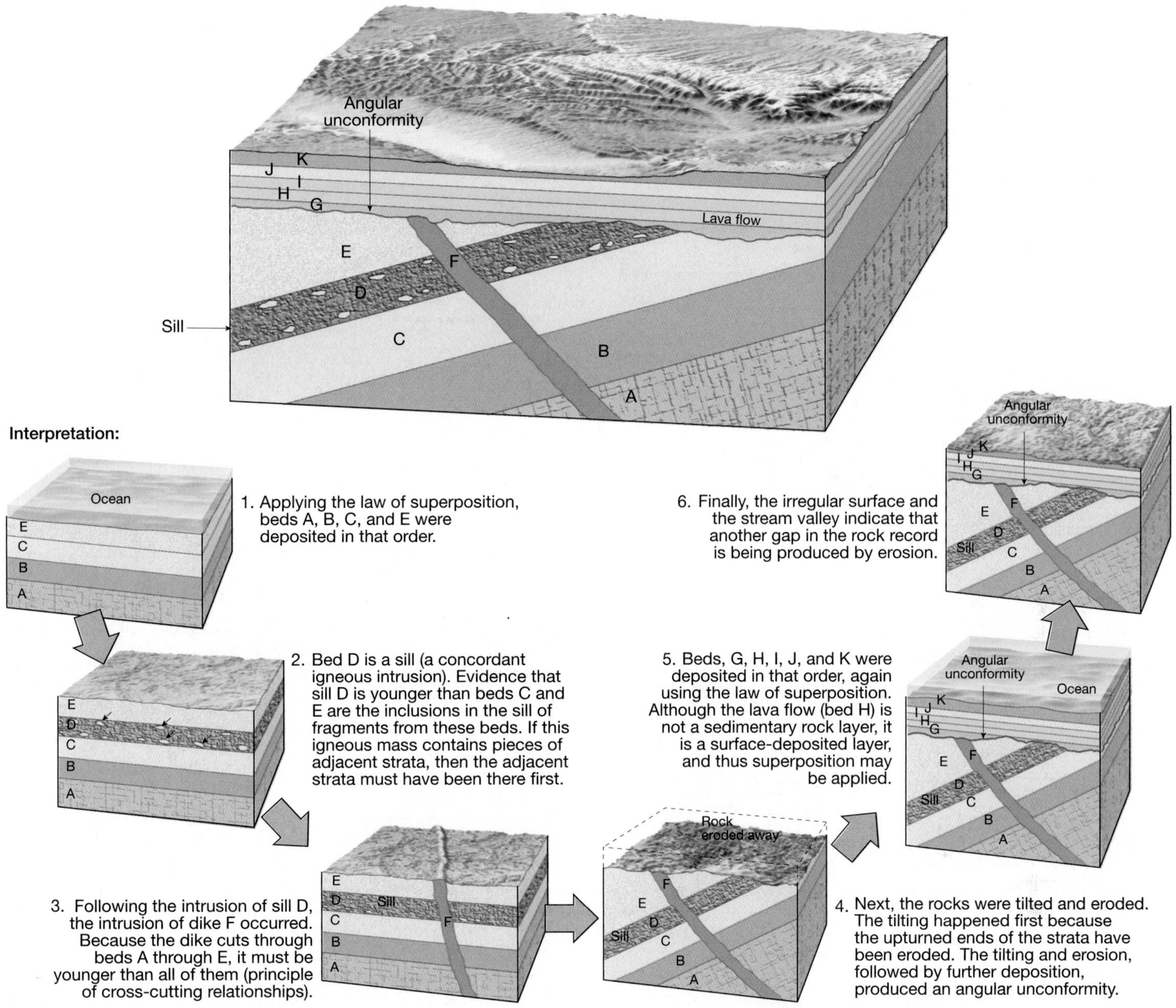

Figure 8.8 Applying relative dating principles to a geologic cross section of a hypothetical region.

Types of Fossils

Fossils are of many types. The remains of relatively recent organisms may not have been altered at all. Such objects as teeth, bones, and shells are common examples. Far less common are entire animals, flesh included, that have been preserved because of rather unusual circumstances. Remains of prehistoric elephants called mammoths that were frozen in the Arctic tundra of Siberia and Alaska are examples, as are the mummified remains of sloths preserved in a dry cave in Nevada.

Given enough time, the remains of an organism are likely to be modified. Fossils often become *petrified* (literally, "turned into stone"), meaning that the small internal cavities and pores of the original structure are filled with precipitated mineral matter (Figure 8.10A). In other instances *replacement* may occur. The cell walls and other solid material are removed and replaced with mineral matter. The microscopic details of the replaced structure are sometimes faithfully retained.

Molds and casts constitute another common class of fossils. When a shell or other structure is buried in sediment and then dissolved by underground water, a *mold* is created. The mold faithfully reflects only the shape and surface marking of the organism; it does not reveal any information concerning its internal structure. If these hollow spaces are subsequently filled with mineral matter, *casts* are created (Figure 8.10B).

A type of fossilization called *carbonization* is particularly effective in preserving leaves and delicate animal forms. It

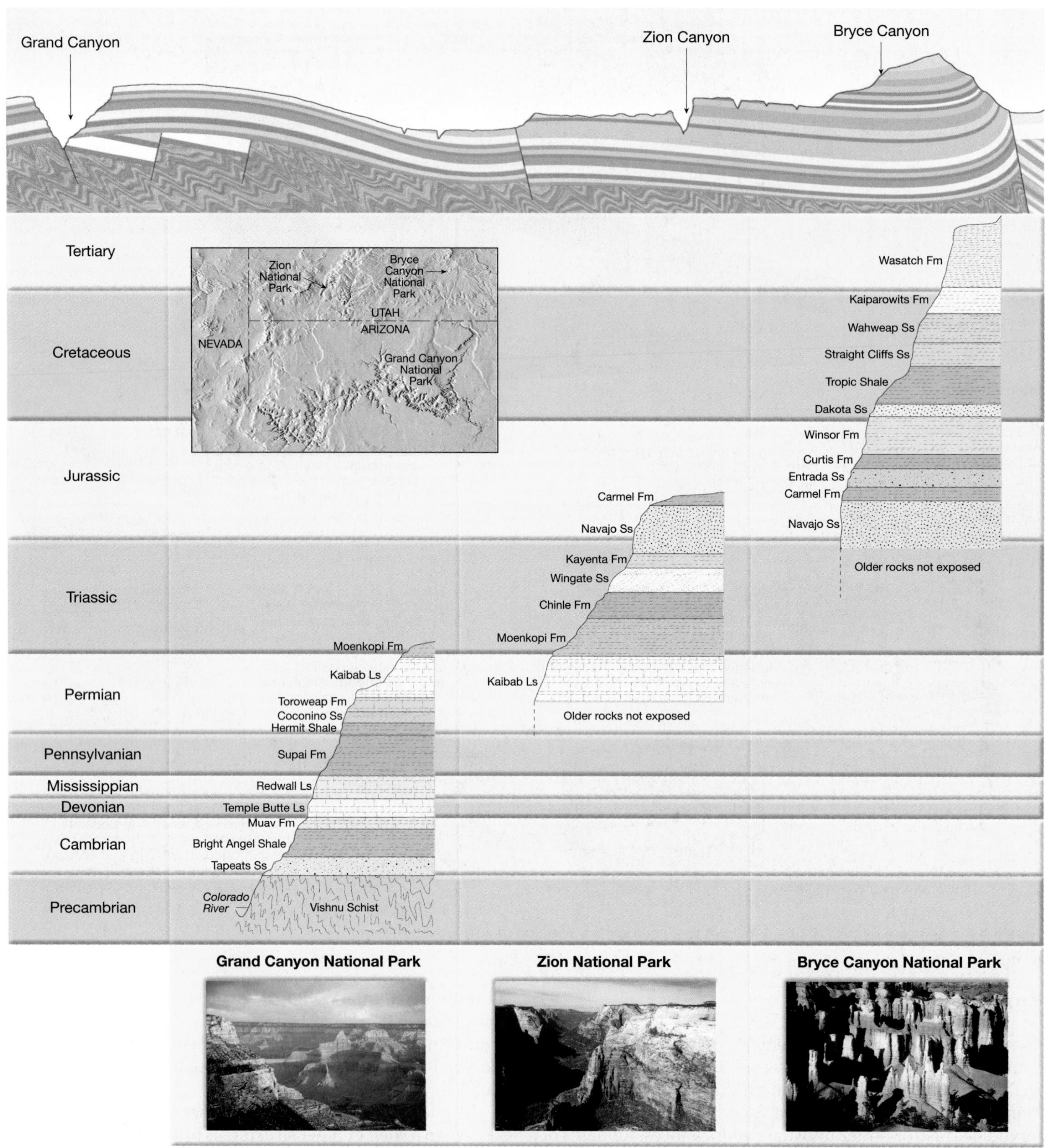

Figure 8.9 Correlation of strata at three locations on the Colorado Plateau allows for a more complete view of the extent of sedimentary rocks in the region. (After U.S. Geological Survey. Insert photos E. J. Tarbuck)

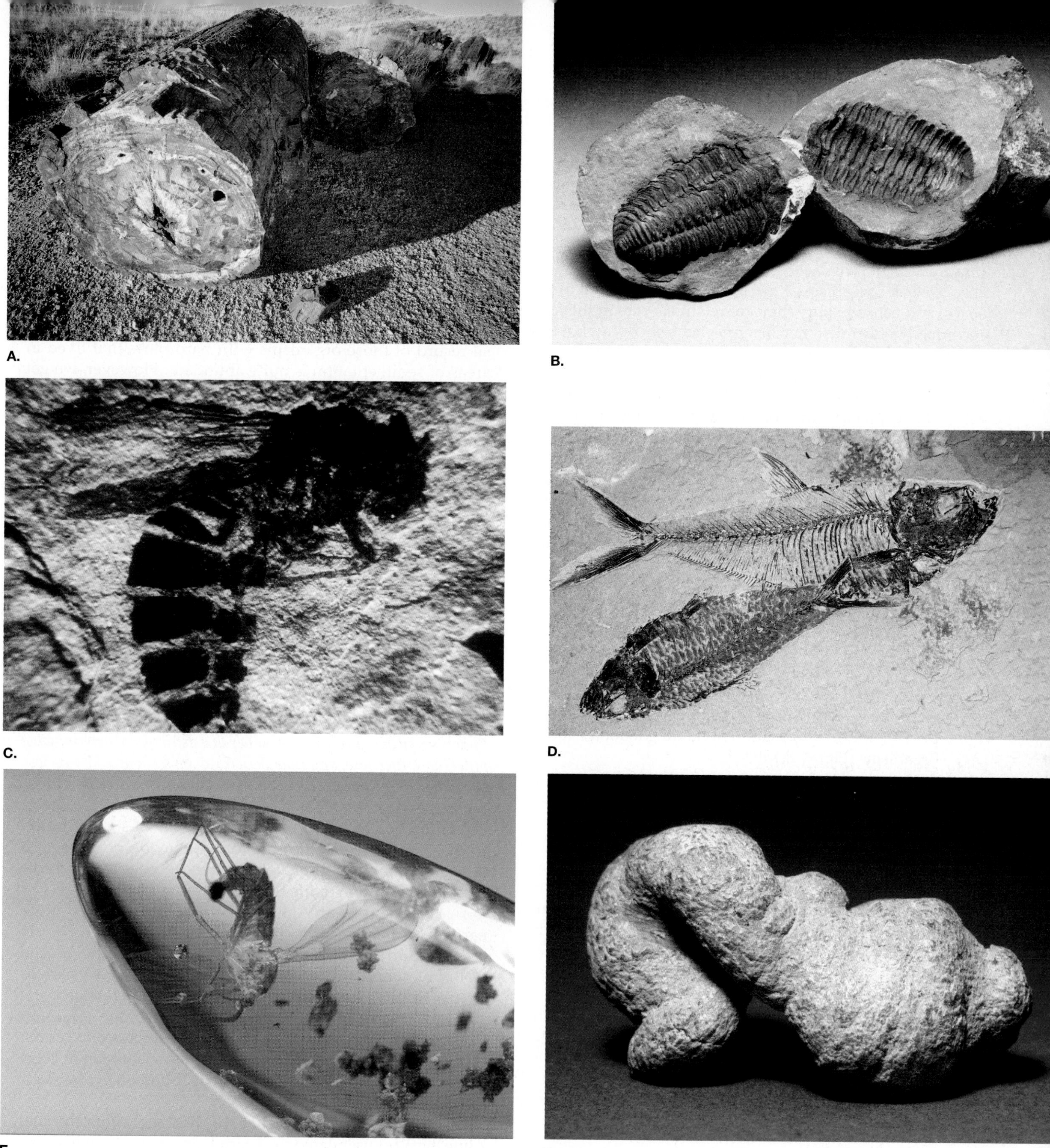

Figure 8.10 There are many types of fossilization. Six examples are shown here. **A.** Petrified wood in Petrified Forest National Park, Arizona. **B.** This trilobite photo illustrates mold and cast. **C.** A fossil bee preserved as a thin carbon film. **D.** Impressions are common fossils and often show considerable detail. **E.** Insect in amber. **F.** Coprolite or fossil dung. (Photo A by David Muench Photography, Inc.; Photos B, D, and F by E. J. Tarbuck; Photo C courtesy of Florissant Fossil Beds National Monument; Photo E by Breck P. Kent)

occurs when fine sediment encases the remains of an organism. As time passes, pressure squeezes out the liquid and gaseous components and leaves behind a thin residue of carbon (Figure 8.10C). Black shales deposited as organic-rich mud in oxygen-poor environments often contain abundant carbonized remains. If the film of carbon is lost from a fossil preserved in fine-grained sediment, a replica of the surface, called an *impression,* may still show considerable detail (Figure 8.10D).

Delicate organisms, such as insects, are difficult to preserve, and consequently they are relatively rare in the fossil record. Not only must they be protected from decay, but they must not be subjected to any pressure that would crush them. One way in which some insects have been preserved is in *amber,* the hardened resin of ancient trees. The fly in Figure 8.10E was preserved after being trapped in a drop of sticky resin. Resin sealed off the insect from the atmosphere and protected the remains from damage by water and air. As the resin hardened, a protective pressure-resistant case formed.

In addition to the fossils already mentioned, there are numerous other types, many of them only traces of prehistoric life. Examples of such indirect evidence include the following:

1. Tracks—animal footprints made in soft sediment that was later lithified.
2. Burrows—tubes in sediment, wood, or rock made by an animal. These holes may later become filled with mineral matter and preserved. Some of the oldest-known fossils are believed to be worm burrows.
3. Coprolites—fossil dung and stomach contents that can provide useful information pertaining to food habits of organisms (Figure 8.10F).
4. Gastroliths—highly polished stomach stones that were used in the grinding of food by some extinct reptiles.

Conditions Favoring Preservation

Only a tiny fraction of the organisms that have lived during the geologic past have been preserved as fossils. The remains of an animal or plant are normally destroyed. Under what circumstances are they preserved? Two special conditions appear to be necessary: rapid burial and the possession of hard parts.

When an organism perishes, its soft parts are usually quickly eaten by scavengers or decomposed by bacteria. Occasionally, however, the remains are buried by sediment. When this occurs, the remains are protected from the environment where destructive processes operate. Rapid burial, therefore, is an important condition favoring preservation.

In addition, animals and plants have a much better chance of being preserved as part of the fossil record if they have hard parts. Although traces and imprints of soft-bodied animals such as jellyfish, worms, and insects exist, they are not common. Flesh usually decays so rapidly that preservation is exceedingly unlikely. Hard parts such as shells, bones, and teeth predominate in the record of past life.

Because preservation is contingent on special conditions, the record of life in the geologic past is biased. The fossil record of those organisms with hard parts that lived in areas of sedimentation is quite abundant. However, we get only an occasional glimpse of the vast array of other life forms that did not meet the special conditions favoring preservation.

Did You Know?

Even when organisms die and their tissues decay, the organic compounds (hydrocarbons) of which they were made may survive in sediments. These are called *chemical fossils.* These hydrocarbons form oil and gas, but some residues can persist in the rock record and be analyzed to determine the kinds of organisms they are derived from.

Fossils and Correlation

The existence of fossils had been known for centuries, yet it was not until the late 1700s and early 1800s that their significance as geologic tools was made evident. During this period, an English engineer and canal builder, William Smith, discovered that each rock formation in the canals he worked on contained fossils unlike those in the beds either above or below. Further, he noted that sedimentary strata in widely separated areas could be identified—and correlated—by their distinctive fossil content.

Based on Smith's classic observations and the findings of many geologists who followed, one of the most important and basic principles in historical geology was formulated: *Fossil organisms succeed one another in a definite and determinable order, and, therefore, any time period can be recognized by its fossil content.* This has come to be known as the **principle of fossil succession.** In other words, when fossils are arranged according to their age by applying the law of superposition to the rocks in which they are found, they do not present a random or haphazard picture. To the contrary, fossils document the evolution of life through time.

For example, an Age of Trilobites is recognized quite early in the fossil record. Then, in succession, paleontologists recognize an Age of Fishes, an Age of Coal Swamps, an Age of Reptiles, and an Age of Mammals. These "ages" pertain to groups that were especially plentiful and characteristic during particular time periods. Within each of the ages, subdivisions are based on certain species of trilobites, and certain types of fish, reptiles, and so on. This same succession of dominant organisms, never out of order, is found on every major landmass.

Once fossils were recognized as time indicators, they became the most useful means of correlating rocks of similar ages in different regions. Geologists pay particular attention to certain fossils called **index fossils.** These fossils are widespread geographically and are limited to a short span of geologic time, so their presence provides an important method of matching rocks of the same age. Rock formations, however, do not always contain a specific index fossil. In such situations, groups of fossils, called *fossil assemblages,* are used to establish the age of the bed. Figure 8.11 illustrates how an assemblage of fossils can be used to date rocks more precisely than could be accomplished by the use of only one of the fossils.

In addition to being important and often essential tools for correlation, fossils are important environmental indicators. Although much can be deduced about past environments by studying the nature and characteristics of sedimentary rocks, a close examination of the fossils present can usually provide a great deal more information.

For example, when the remains of certain clam shells are found in limestone, the geologist can assume that the region was once covered by a shallow sea, because that is where clams live today. Also, by using what we know of living organisms, we can conclude that fossil animals with thick shells capable of withstanding pounding and surging waves must have inhabited shorelines. On the other hand, animals with thin, delicate shells probably indicate deep, calm offshore waters. Hence, by looking closely at the types of fossils, the approximate position of an ancient shoreline may be identified.

Further, fossils can indicate the former temperature of the water. Certain present-day corals require warm and shallow tropical seas like those around Florida and the Bahamas. When similar corals are found in ancient limestones, they indicate that a Florida-like marine environment must have existed when they were alive. These are just a few examples of how fossils can help unravel the complex story of Earth's history.

Did You Know?

Numerical dates of the fossil record show that life began in the ocean approximately 3.8 billion years ago.

Dating with Radioactivity

Deciphering Earth's History
▼ Radiometric Dating

In addition to establishing relative dates by using the principles described in the preceding sections, it is also possible to obtain reliable numerical dates for events in the geologic past. We know that Earth is about 4.5 billion years old and that the dinosaurs became extinct about 65 million years ago. Dates that are expressed in millions and billions of years truly stretch our imagination because our personal calendars involve time measured in hours, weeks, and years. Nevertheless, the vast expanse of geologic time is a reality, and it is radiometric dating that allows us to measure it accurately. In this section, you will learn about radioactivity and its application in radiometric dating.

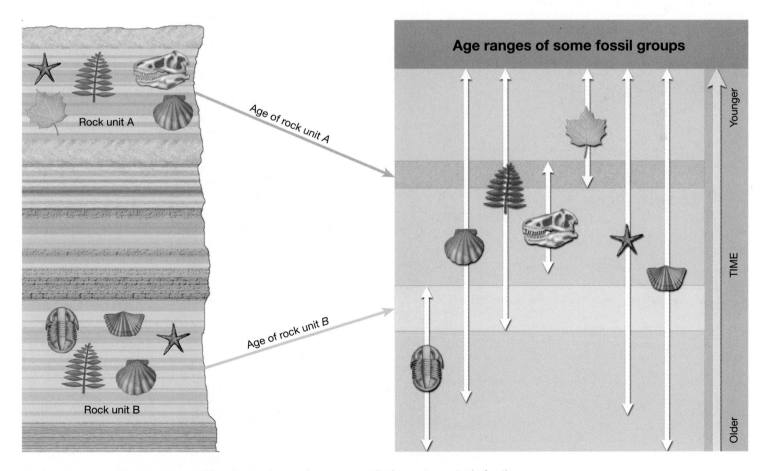

Figure 8.11 Overlapping ranges of fossils help date rocks more exactly than using a single fossil.

Reviewing Basic Atomic Structure

Recall from Chapter 1 that each atom has a *nucleus* containing protons and neutrons and that the nucleus is orbited by electrons. *Electrons* have a negative electrical charge, and *protons* have a positive charge. A *neutron* is actually a proton and an electron combined, so it has no charge (it is neutral).

The *atomic number* (the element's identifying number) is the number of protons in the nucleus. Every element has a different number of protons in the nucleus and thus a different atomic number (hydrogen = 1, oxygen = 8, uranium = 92, etc.). Atoms of the same element always have the same number of protons, so the atomic number is constant.

Practically all (99.9 percent) of an atom's mass is found in the nucleus, indicating that electrons have practically no mass at all. By adding together the number of protons and neutrons in the nucleus, the *mass number* of the atom is determined. The number of neutrons in the nucleus can vary. These variants, called *isotopes*, have different mass numbers.

To summarize with an example, uranium's nucleus always has 92 protons, so its atomic number always is 92. But its neutron population varies, so uranium has three isotopes: uranium-234 (number of protons + neutrons = 234), uranium-235, and uranium-238. All three isotopes are mixed in nature. They look the same and behave the same in chemical reactions.

Radioactivity

The forces that bind protons and neutrons together in the nucleus are usually strong. However, in some isotopes, the nuclei are unstable because the forces binding protons and neutrons together are not strong enough. As a result, the nuclei spontaneously break apart (decay), a process called **radioactivity.** What happens when unstable nuclei break apart?

Three common types of radioactive decay are illustrated in Figure 8.12 and are summarized as follows:

1. *Alpha particles* (α particles) may be emitted from the nucleus. An alpha particle consists of 2 protons and 2 neutrons. Consequently, the emission of an alpha particle means (a) the mass number of the isotope is reduced by 4, and (b) the atomic number is decreased by 2.
2. When a *beta particle* (β particle), or electron, is given off from a nucleus, the mass number remains unchanged, because electrons have practically no mass. However, because the electron has come from a neutron (remember, a neutron is a combination of a proton and an elec-

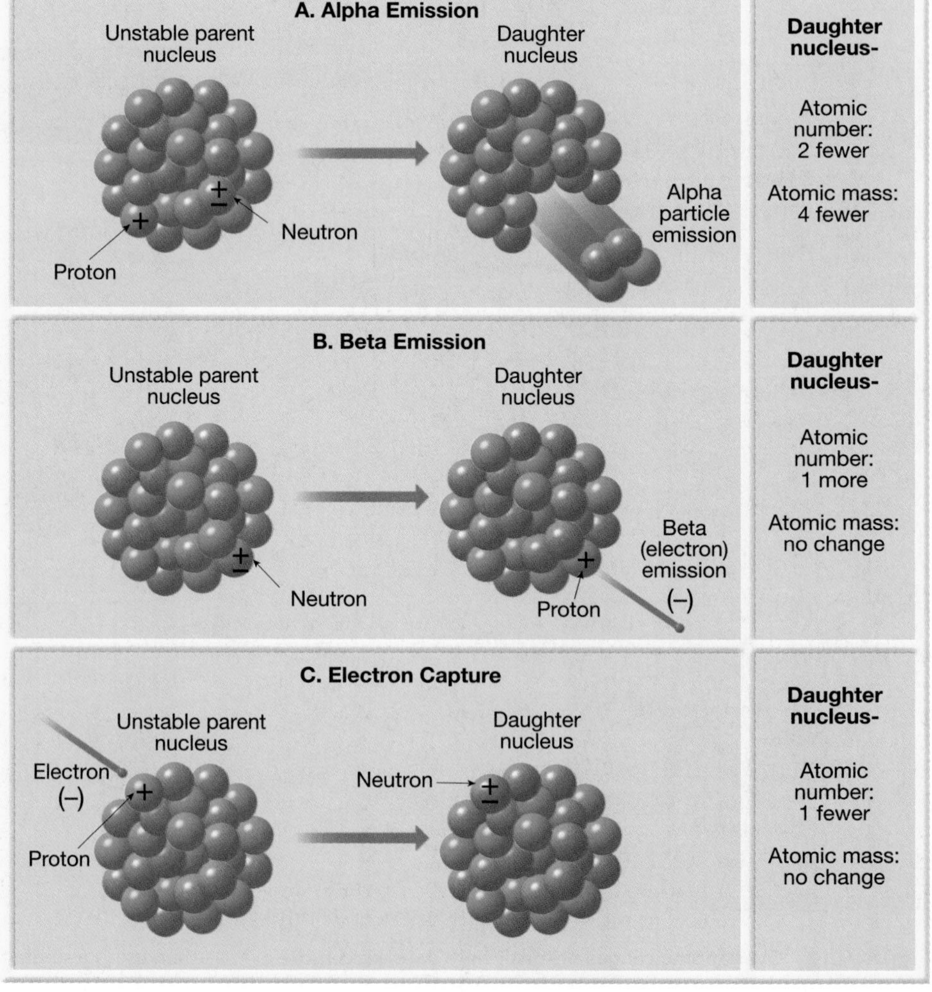

Figure 8.12 Common types of radioactive decay. Notice that in each case, the number of protons (atomic number) in the nucleus changes, thus producing a different element.

tron), the nucleus contains one more proton than before. Therefore, the atomic number increases by 1.

3. Sometimes an electron is captured by the nucleus. The electron combines with a proton and forms an additional neutron. As in the last example, the mass number remains unchanged. However, because the nucleus now contains one less proton, the atomic number decreases by 1.

An unstable (radioactive) isotope is referred to as the *parent.* The isotopes resulting from the decay of the parent are the *daughter products.* Figure 8.13 provides an example of radioactive decay. When the radioactive parent, uranium-238 (atomic number 92, mass number 238), decays, it follows a number of steps, emitting eight alpha particles and six beta particles before finally becoming the stable daughter product lead-206 (atomic number 82, mass number 206).

Certainly among the most important results of the discovery of radioactivity is that it provides a reliable method of calculating the ages of rocks and minerals that contain particular radioactive isotopes. The procedure is called **radiometric dating.** Why is radiometric dating reliable? The rates of decay for many isotopes have been precisely measured and do not vary under the physical conditions that exist in Earth's outer layers. Therefore, each radioactive isotope used for dating has been decaying at a fixed rate ever since the formation of the rocks in which it occurs, and the products of decay have been accumulating at a corresponding rate. For example, when uranium is incorporated into a mineral that crystallizes from magma, there is no lead (the stable daughter product) from previous decay. The radiometric "clock" starts at this point. As the uranium in this newly formed mineral disintegrates, atoms of the daughter product are trapped and measurable amounts of lead eventually accumulate.

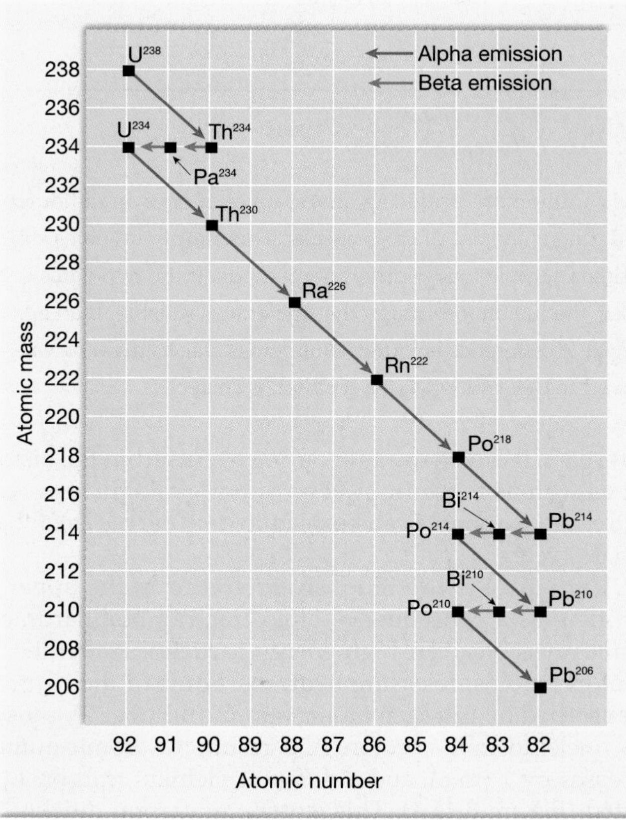

Figure 8.13 The most common isotope of uranium (U-238) is an example of a radioactive decay series. Before the stable end product (Pb-206) is reached, many different isotopes are produced as intermediate steps.

Half-Life

The time required for one-half of the nuclei in a sample to decay is called the **half-life** of the isotope. Half-life is a common way of expressing the rate of radioactive disintegration. Figure 8.14 illustrates what occurs when a radioactive parent decays directly into its stable daughter product. When the quantities of parent and daughter are equal (ratio 1:1), we know that one half-life has transpired. When one-quarter of the original parent atoms remain and three-quarters have decayed to the daughter product, the parent/daughter ratio is 1:3, and we know that two half-lives have passed. After three half-lives, the ratio of parent atoms to daughter atoms is 1:7 (one parent atom for every seven daughter atoms).

If the half-life of a radioactive isotope is known and the parent/daughter ratio can be determined, the age of the sample can be calculated. For example, assume that the half-life of a hypothetical unstable isotope is 1 million years and the parent/daughter ratio in a sample is 1:15. Such a ratio indicates that four half-lives have passed and that the sample must be 4 million years old.

Radiometric Dating

Notice that the *percentage* of radioactive atoms that decay during one half-life is always the same: 50 percent. However, the *actual number* of atoms that decay with the passing of each half-life continually decreases. As the percentage of radioactive parent atoms declines, the proportion of stable daughter atoms rises, with the increase in daughter atoms just matching the drop in parent atoms. This fact is the key to radiometric dating.

Of the many radioactive isotopes that exist in nature, five have proved particularly important in providing radiometric ages for ancient rocks (Table 8.1). Rubidium-87, uranium-238, and uranium-235 are used for dating rocks that are millions of years old, but potassium-40 is more versatile. Although the half-life of potassium-40 is 1.3 billion years, analytical techniques make possible the detection of tiny amounts of its stable daughter product, argon-40, in some rocks that are younger than 100,000 years. Another important reason for its frequent use is that potassium is abundant in many common minerals, particularly micas and feldspars.

It is important to realize that an accurate radiometric date can be obtained only if the mineral remained a closed system during the entire period since its formation. A correct date is not possible unless there was neither the addition nor loss of parent or daughter isotopes. This is not always the case. In fact, an important limitation of the potassium-argon method arises from the fact that argon is a gas, and it may leak from minerals, throwing off measurements.

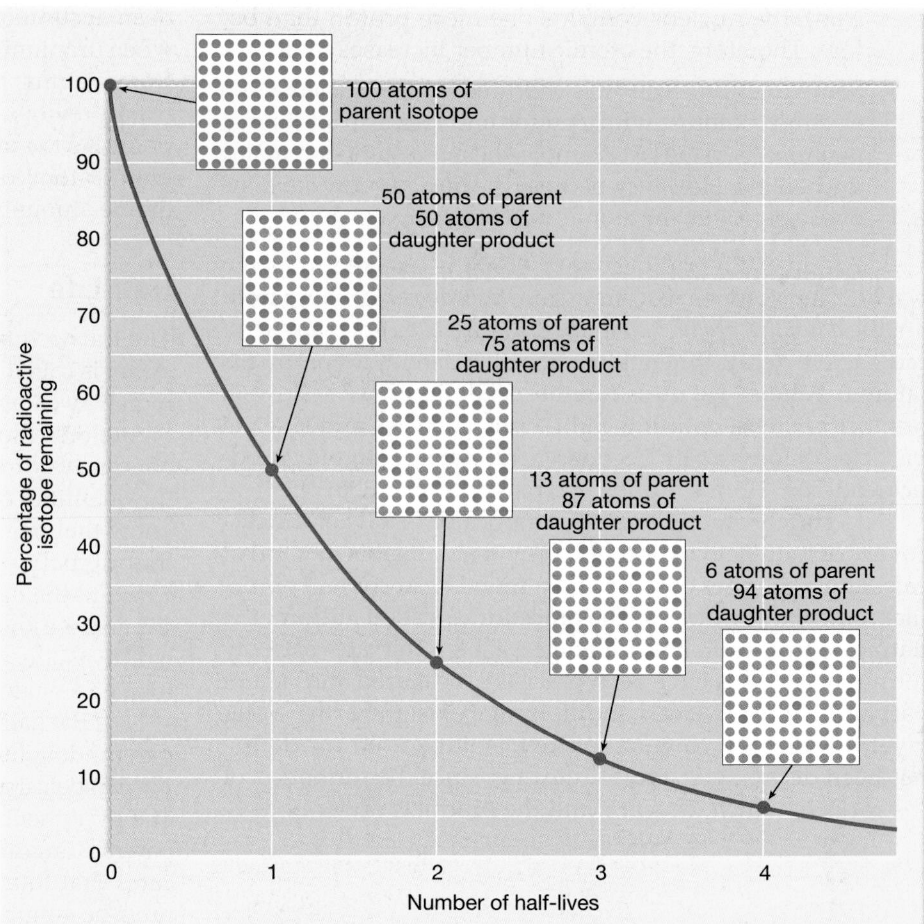

Figure 8.14 The radioactive decay curve shows change that is exponential. Half of the radioactive parent remains after one half-life. After a second half-life, one-quarter of the parent remains, and so forth.

Bear in mind that although the basic principle of radiometric dating is simple, the actual procedure is quite complex. The analysis that determines the quantities of parent and daughter must be painstakingly precise. In addition, some radioactive materials do not decay directly into the stable daughter product. As you saw in Figure 8.13, uranium-238 produces 13 intermediate unstable daughter products before the fourteenth and final daughter product, the stable isotope lead-206, is produced.

Dating with Carbon-14

To date very recent events, carbon-14 is used. Carbon-14 is the radioactive isotope of carbon. The process is often called **radiocarbon dating.** Because the half-life of carbon-14 is only 5,730 years, it can be used for dating events from the historic past as well as those from very recent geologic history. In some cases, carbon-14 can be used to date events as far back as 75,000 years.

Carbon-14 is continuously produced in the upper atmosphere as a consequence of cosmic-ray bombardment. Cosmic rays, which are high-energy particles, shatter the nuclei of gas atoms, releasing neutrons. Some of the neutrons are absorbed by nitrogen atoms (atomic number 7), causing their nuclei to emit a proton. As a result, the atomic number decreases by 1 (to 6), and a different element, carbon-14, is created (Figure 8.15A). This isotope of carbon quickly becomes incorporated into carbon dioxide, which circulates in the atmosphere and is absorbed by living matter. As a result, all organisms contain a small amount of carbon-14, including yourself.

Table 8.1 Radioactive isotopes frequently used in radiometric dating

Radioactive Parent	Stable Daughter Product	Currently Accepted Half-Life Values
Uranium-238	Lead-206	4.5 billion years
Uranium-235	Lead-207	713 million years
Thorium-232	Lead-208	14.1 billion years
Rubidium-87	Strontium-87	47.0 billion years
Potassium-40	Argon-40	1.3 billion years

Did You Know?

One common precaution against sources of error in radiometric dating is the use of cross checks. This simply involves subjecting a sample to two different methods. If the two dates agree, the likelihood is high that the date is reliable. If an appreciable difference is found, other cross checks must be employed to determine which, if either, is correct.

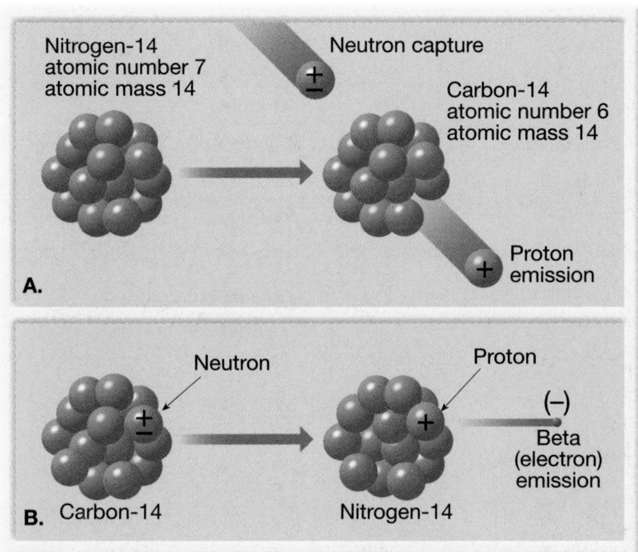

Figure 8.15 **A.** Production and **B.** decay of carbon-14. These sketches represent the nuclei of the respective atoms.

While an organism is alive, the decaying radiocarbon is continually replaced, and the proportions of carbon-14 and carbon-12 remain constant. Carbon-12 is the stable and most common isotope of carbon. However, when any plant or animal dies, the amount of carbon-14 gradually decreases as it decays to nitrogen-14 by beta emission (Figure 8.15B). By comparing the proportions of carbon-14 and carbon-12 in a sample, radiocarbon dates can be determined.

Although carbon-14 is useful in dating only the last small fraction of geologic time, it has become a valuable tool for anthropologists, archaeologists, and historians, as well as for geologists who study very recent Earth history. In fact, the development of radiocarbon dating was considered so important that the chemist who discovered this application, Willard F. Libby, received a Nobel prize.

Importance of Radiometric Dating

Radiometric dating methods have produced literally thousands of dates for events in Earth's history. Rocks exceeding 3.5 billion years in age are found on all of the continents. Earth's oldest rocks (so far) are gneisses from northern Canada near Great Slave Lake that have been dated at 4.03 billion years (b.y.). Rocks from western Greenland have been dated at 3.7 to 3.8 b.y., and rocks nearly as old are found in the Minnesota River Valley and northern Michigan (3.5 to 3.7 b.y.), in southern Africa (3.4 to 3.5 b.y.), and in western Australia (3.4 to 3.6 b.y.). It is important to point out that these ancient rocks are not from any sort of "primordial crust" but originated as lava flows, igneous intrusions, and sediments deposited in shallow water—an indication that Earth's history began *before* these rocks formed. Even older mineral grains have been dated. Tiny crystals of the mineral zircon having radiometric ages as old as 4.3 b.y. have been found in younger sedimentary rocks in western Australia. The source rocks for these tiny durable grains either no longer exist or have not yet been found.

Radiometric dating has vindicated the ideas of Hutton, Charles Darwin, and others who inferred that geologic time must be immense. Indeed, modern dating methods have proved that there has been enough time for the processes we observe to have accomplished tremendous tasks.

Did You Know?

Dating with carbon-14 is useful to archeologists and historians as well as geologists. For example, University of Arizona researchers recently used carbon-14 to determine the age of the Dead Sea Scrolls, considered among the great archeological discoveries of the twentieth century. Parchment from the scrolls dates between 150 B.C. and 5 B.C. Portions of the scrolls contain dates that match those determined by the carbon-14 measurements.

The Geologic Time Scale

Deciphering Earth's History
▼ Geologic Time Scale

Geologists have divided the whole of geologic history into units of varying magnitude. Together, they comprise the **geologic time scale** of Earth's history (Figure 8.16). The major units of the time scale were delineated during the nineteenth century, principally by scientists in Western Europe and Great Britain. Because radiometric dating was unavailable at that time, the entire time scale was created using methods of relative dating. It was only in the twentieth century that radiometric dating permitted numerical dates to be added.

Structure of the Time Scale

The geologic time scale subdivides the 4.5-billion-year history of Earth into many different units and provides a meaningful time frame within which the events of the geologic past are arranged. As shown in Figure 8.16, **eons** represent the greatest expanses of time. The eon that began about 542 million years ago is the **Phanerozoic,** a term derived from Greek words meaning "visible life." It is an appropriate description because the rocks and deposits of the Phanerozoic eon contain abundant fossils that document major evolutionary trends.

Did You Know?

Although movies and cartoons have depicted humans and dinosaurs living side by side, this was never the case. Dinosaurs flourished during the Mesozoic era and became extinct about 65 million years ago. Humans and their close ancestors did not appear on the scene until the late Cenozoic, more than 60 million years *after* the demise of dinosaurs.

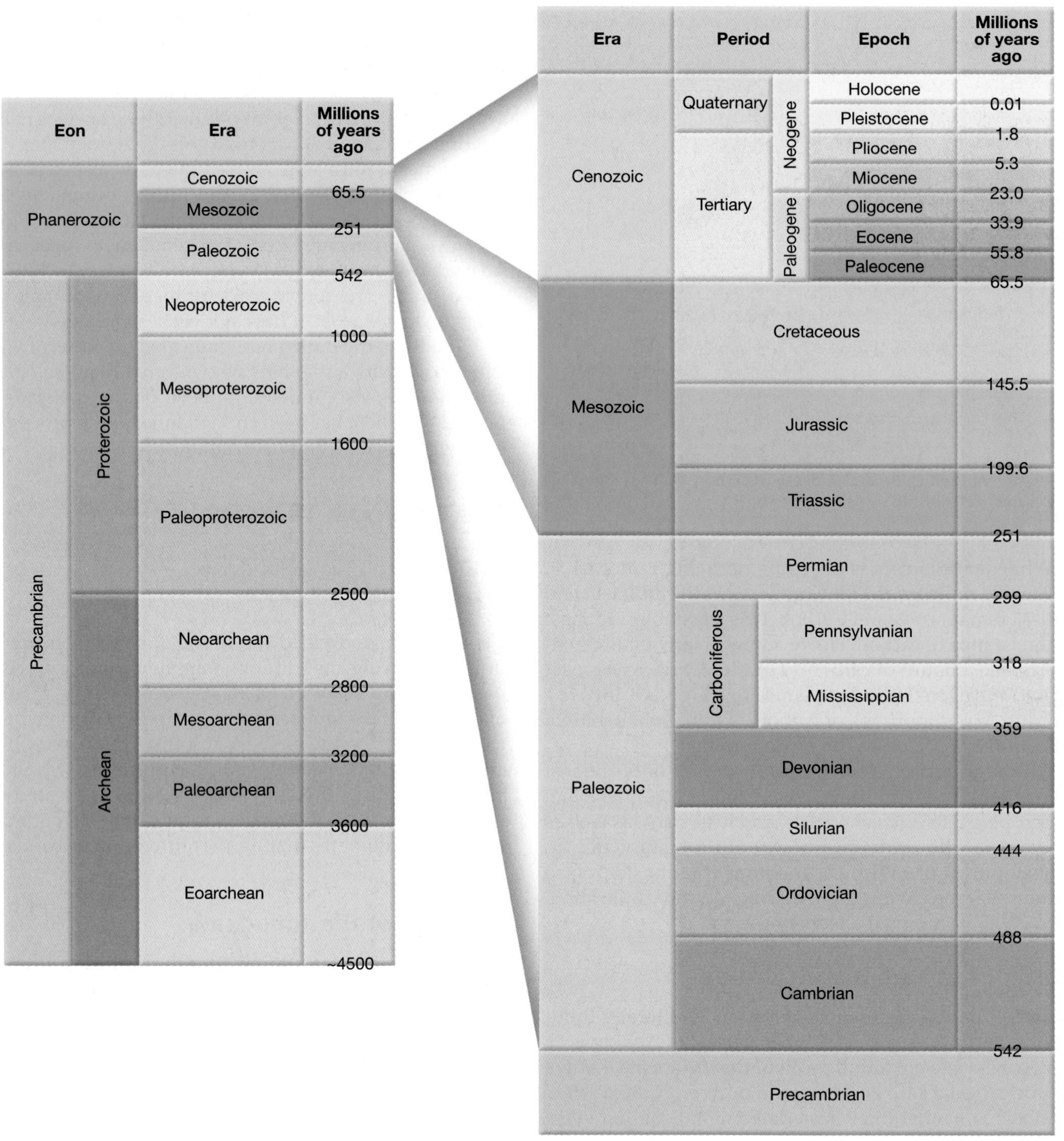

Eon
Era
Millions of years ago
Phanerozoic
Cenozoic
Mesozoic
Paleozoic
65.5
251
542
Precambrian
Proterozoic
Neoproterozoic
1000
Mesoproterozoic
1600
Paleoproterozoic
2500
Archean
Neoarchean
2800
Mesoarchean
3200
Paleoarchean
3600
Eoarchean
~4500
Era
Period
Epoch
Millions of years ago
Cenozoic
Quaternary
Tertiary
Neogene
Paleogene
Holocene
0.01
Pleistocene
1.8
Pliocene
5.3
Miocene
23.0
Oligocene
33.9
Eocene
55.8
Paleocene
65.5
Mesozoic
Cretaceous
145.5
Jurassic
199.6
Triassic
251
Paleozoic
Permian
299
Carboniferous
Pennsylvanian
318
Mississippian
359
Devonian
416
Silurian
444
Ordovician
488
Cambrian
542
Precambrian

Figure 8.16 The geologic time scale. Numbers on the time scale represent time in millions of years before the present. These dates were added long after the time scale had been established using relative dating techniques. The Precambrian accounts for more than 88 percent of geologic time. (Data from Geological Society of America)

Another glance at the time scale reveals that eons are divided into **eras.** The three eras within the Phanerozoic are the **Paleozoic** ("ancient life"), the **Mesozoic** ("middle life"), and the **Cenozoic** ("recent life"). As the names imply, these eras are bounded by profound worldwide changes in life forms.

Each era of the Phanerozoic eon is subdivided into units known as **periods.** The Paleozoic has seven, the Mesozoic three, and the Cenozoic two. Each of these dozen periods is characterized by a somewhat less profound change in life forms as compared with the eras.

Each of the periods is divided into still smaller units called **epochs.** As you can see in Figure 8.16, seven epochs have been named for the periods of the Cenozoic era. The epochs of other periods, however, are not usually referred to by specific names. Instead, the terms *early, middle,* and *late* are generally applied to the epochs of these earlier periods.

Precambrian Time

Notice that the detail of the geologic time scale does not begin until about 542 million years ago, the date for the beginning of the Cambrian period. The nearly 4 billion years prior to the Cambrian is divided into two eons, the *Archean* and the *Proterozoic.* It is also common for this vast expanse of time to simply be referred to as the **Precambrian.** Although it represents about 88 percent of Earth's history, the Precambrian is not divided into nearly as many smaller time units as is the Phanerozoic eon.

Why is the huge expanse of Precambrian time not divided into numerous eras, periods, and epochs? The reason is that Precambrian history is not known in great enough detail. The quantity of information geologists have deciphered about Earth's past is somewhat analogous to the detail of human history. The farther back we go, the less we know. Certainly, more data and information exist about the past 10 years than for the first decade of the twentieth century; the events of the nineteenth century have been documented much better than the events of the first century A.D., and so on. So it is with Earth's history. The more recent past has the freshest, least disturbed, and most observable record. The farther back in time the geologist goes, the more fragmented the record and clues become.

Difficulties in Dating the Geologic Time Scale

Although reasonably accurate numerical dates have been worked out for the periods of the geologic time scale (see Figure 8.16), the task is not without difficulty. The primary problem in assigning numerical dates is the fact that not all rocks can be dated by radiometric methods. For a radiometric date to be useful, all minerals in the rock must have formed at about the same time. For this reason, radioactive isotopes can be used to determine when minerals in an igneous rock crystallized and when pressure and heat created new minerals in a metamorphic rock.

However, samples of sedimentary rock can only rarely be dated directly by radiometric means. A sedimentary rock may include particles that contain radioactive isotopes, but the rock's age cannot be accurately determined because the grains that make up the rock are not the same age as the rock in which they occur. Rather, the sediments have been weathered from rocks of diverse ages.

Radiometric dates obtained from metamorphic rocks may also be difficult to interpret, because the age of a particular mineral in a metamorphic rock does not necessarily represent the time when the rock initially formed. Instead, the date may indicate any one of a number of subsequent metamorphic phases.

If samples of sedimentary rocks rarely yield reliable radiometric ages, how can numerical dates be assigned to sedimentary layers? Usually the geologist must relate them to datable igneous masses, as in Figure 8.17. In this example, radiometric dating has determined the ages of the volcanic ash bed within the Morrison Formation and the dike cutting the Mancos Shale and Mesaverde Formation. The sedimentary beds below the ash are obviously older than the ash, and all the layers above the ash are younger (principle of superposition). The dike is younger than the Mancos Shale and the Mesaverde Formation but older than the Wasatch Formation because the dike does not intrude the Tertiary rocks (cross-cutting relationships).

From this kind of evidence, geologists estimate that a part of the Morrison Formation was deposited about 160 million years ago, as indicated by the ash bed. Further, they conclude that the Tertiary period began after the intrusion of the dike, 66 million years ago. This is one example of literally thousands that illustrates how datable materials are used to "bracket" the various episodes in Earth's history within specific time periods. It shows the necessity of combining laboratory dating methods with field observations of rocks.

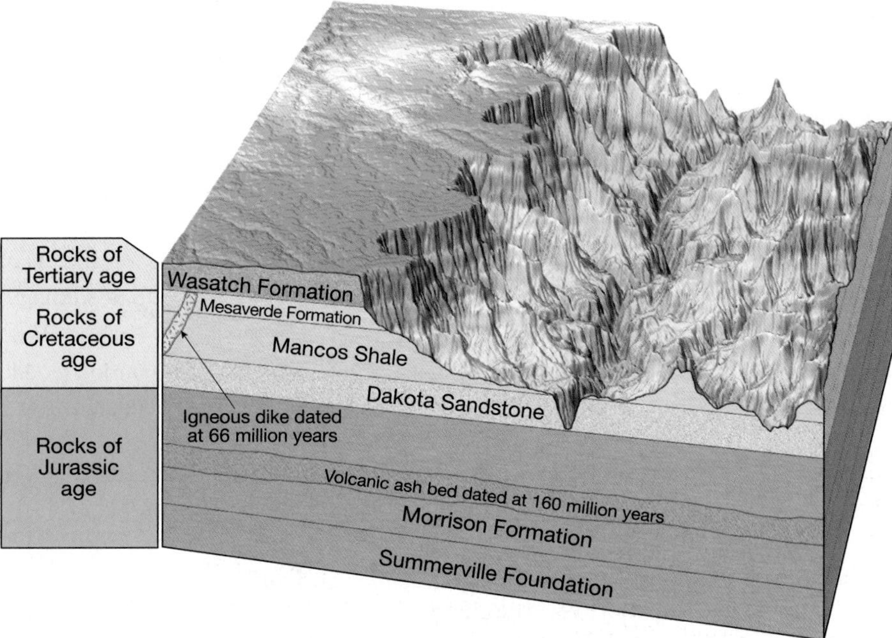

Figure 8.17 Numerical dates for sedimentary layers are usually determined by examining their relationship to igneous rocks. (After U.S. Geological Survey)

The Chapter in Review

1. During the seventeenth and eighteenth centuries, *catastrophism* influenced the formulation of explanations about Earth. Catastrophism states that Earth's landscapes have been developed primarily by great catastrophes. By contrast, *uniformitarianism,* one of the fundamental principles of modern geology advanced by *James Hutton* in the late 1700s, states that the physical, chemical, and biological laws that operate today have also operated in the geologic past. The idea is often summarized as "the present is the key to the past." Hutton argued that processes that appear to be slow acting could, over long spans of time, produce effects that were just as great as those resulting from sudden catastrophic events.

2. The two types of dates used by geologists to interpret Earth history are (1) *relative dates,* which put events in their *proper sequence of formation,* and (2) *numerical dates,* which pinpoint the *time in years* when an event took place.

3. Relative dates can be established using the *law of superposition, principle of original horizontality, principle of cross-cutting relationships, inclusions,* and *unconformities.*

4. *Correlation,* matching up two or more geologic phenomena of similar ages in different areas, is used to develop a geologic time scale that applies to the entire Earth.

5. *Fossils* are the remains or traces of prehistoric life. The special conditions that favor preservation are *rapid burial* and the possession of *hard parts* such as shells, bones, or teeth.

6. Fossils are used to *correlate* sedimentary rocks that are widely separated by using the rocks' distinctive fossil content and applying the *principle of fossil succession.* The principle of fossil succession states that fossil organisms succeed one another in a definite and determinable order, and, therefore, any time period can be recognized by its fossil content.

7. Each atom has a nucleus containing *protons* (positively charged particles) and *neutrons* (neutral particles). Orbiting the nucleus are negatively charged *electrons.* The *atomic number* of an atom is the number of protons in the nucleus. The *mass number* is the number of protons plus the number of neutrons in an atom's nucleus. *Isotopes* are variants of the same atom, but with a different number of neutrons and hence a different mass number.

8. *Radioactivity* is the spontaneous breaking apart (decay) of certain unstable atomic nuclei. Three common forms of radioactive decay are (1) emission of an alpha particle from the nucleus, (2) emission of a beta particle (or electron) from the nucleus, and (3) capture of an electron by the nucleus.

9. An unstable *radioactive isotope,* called the *parent,* will decay and form *daughter products.* The length of time for one-half of the nuclei of a radioactive isotope to decay is called the *half-life* of the isotope. If the half-life of the isotope is known, and the parent/daughter ratio can be measured, the age of a sample can be calculated.

10. The *geologic time scale* divides Earth's history into units of varying magnitude. It is commonly presented in chart form, with the oldest time and event at the bottom and the youngest at the top. The principal subdivisions of the geologic time scale, called *eons,* include the *Archean* and *Proterozoic* (together, these two eons are commonly referred to as the *Precambrian*), and, beginning about 542 million years ago, the *Phanerozoic.* The Phanerozoic (meaning "visible life") eon is divided into the following eras: *Paleozoic* ("ancient life"), *Mesozoic* ("middle life"), and *Cenozoic* ("recent life").

11. A significant problem in assigning numerical dates to units of time is that *not all rocks can be dated radiometrically.* A sedimentary rock may contain particles of many ages that have been weathered from different rocks that formed at various times. One way geologists assign numerical dates to sedimentary rocks is to relate them to datable igneous masses, such as dikes and volcanic ash beds.

Key Terms

angular unconformity (p. 228)
catastrophism (p. 225)
Cenozoic era (p. 240)
conformable (p. 228)
correlation (p. 229)
cross-cutting relationships, principle of (p. 227)
disconformity (p. 229)
eon (p. 239)
epoch (p. 240)
era (p. 240)
fossil (p. 230)
fossil succession, principle of (p. 234)
geologic time scale (p. 239)
half-life (p. 237)
inclusions (p. 228)
index fossil (p. 234)
Mesozoic era (p. 240)
nonconformity (p. 229)
numerical date (p. 226)
original horizontality, principle of (p. 227)
paleontology (p. 230)
Paleozoic era (p. 240)
period (p. 240)
Phanerozoic eon (p. 239)
Precambrian (p. 241)
radioactivity (p. 236)
radiocarbon dating (p. 238)
radiometric dating (p. 237)
relative dating (p. 226)
superposition, law of (p. 226)
unconformity (p. 228)
uniformitarianism (p. 225)

Questions for Review

1. Contrast the philosophies of *catastrophism* and *uniformitarianism.* How did the proponents of each perceive the age of Earth?
2. Distinguish between numerical and relative dating.
3. What is the law of superposition? How are cross-cutting relationships used in relative dating?

4. When you observe an outcrop of steeply inclined sedimentary layers, what principle allows you to assume that the beds became tilted *after* they were deposited?
5. Refer to Figure 8.4 (p. 227) and answer the following questions:
   a. Is fault A older or younger than the sandstone layer?
   b. Is dike A older or younger than the sandstone layer?
   c. Was the conglomerate deposited before or after fault A?
   d. Was the conglomerate deposited before or after fault B?
   e. Which fault is older, A or B?
   f. Is dike A older or younger than the batholith?
6. A mass of granite is in contact with a layer of sandstone. Using a principle described in this chapter, explain how you might determine whether the sandstone was deposited on top of the granite or the granite was intruded from below after the sandstone was deposited.
7. Distinguish among angular unconformity, disconformity, and nonconformity.
8. What is meant by the term *correlation?*
9. List and briefly describe at least five different types of fossils.
10. List two conditions that improve an organism's chances of being preserved as a fossil.
11. Why are fossils such useful tools in correlation?
12. If a radioactive isotope of thorium (atomic number 90, mass number 232) emits six alpha particles and four beta particles during the course of radioactive decay, what are the atomic number and mass number of the stable daughter product?
13. Why is radiometric dating the most reliable method of dating the geologic past?
14. Assume that a hypothetical radioactive isotope has a half-life of 10,000 years. If the ratio of radioactive parent to stable daughter product is 1:3, how old is the rock containing the radioactive material?
15. To make calculations easier, let us round the age of Earth to 5 billion years.
    a. What fraction of geologic time is represented by recorded history (assume 5000 years for the length of recorded history)?
    b. The first abundant fossil evidence does not appear until the beginning of the Cambrian period (542 million years ago). What percentage of geologic time is represented by abundant fossil evidence?
16. What subdivisions make up the geologic time scale? What is the primary basis for differentiating the eras?
17. Briefly describe the difficulties in assigning numerical dates to layers of sedimentary rock.

Online Study Guide

The *Foundations of Earth Science* Web site uses the resources and flexibility of the Internet to aid in your study of the topics in this chapter. Written and developed by Earth science instructors, this site will help improve your understanding of Earth science. Visit **http://www.prenhall.com/lutgens** and click on the cover of *Foundations of Earth Science 5e* to find:

- Online review quizzes.
- Critical thinking exercises.
- Links to chapter-specific Web resources.
- Internet-wide key-term searches.

http://www.prenhall.com/lutgens

GEODe: Earth Science

GEODe: Earth Science makes studying more effective by reinforcing key concepts using animation, video, narration, interactive exercises, and practice quizzes. A copy is included with every copy of *Foundations of Earth Science 5e.*

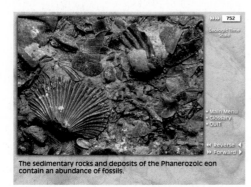

The sedimentary rocks and deposits of the Phanerozoic eon contain an abundance of fossils.

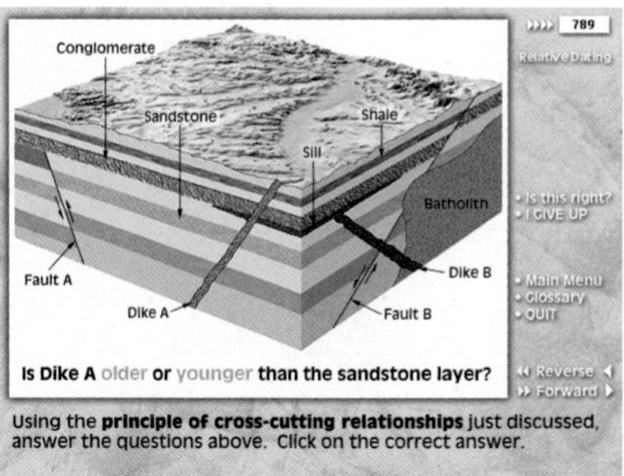

Using the **principle of cross-cutting relationships** just discussed, answer the questions above. Click on the correct answer.

UNIT V THE GLOBAL OCEAN

The *Kilo Moana* is a twin-hull oceanographic research ship. (AP Photo/U.S. Coast Guard, Marshalena Delaney)

CHAPTER

9

Oceans: The Last Frontier

FOCUS ON LEARNING

To assist you in learning the important concepts in this chapter, you will find it helpful to focus on the following questions:

1. What is oceanography?
2. What is the extent and distribution of the world's oceans?
3. What are the principal elements that contribute to the ocean's salinity? What are the sources for these elements?
4. How do temperature and salinity change with depth in the open ocean?
5. How does a passive continental margin differ from an active continental margin?
6. What are the features of the ocean basin floor? How are mid-ocean ridges related to seafloor spreading?
7. What are the various types of seafloor sediments? How can these sediments be used to study worldwide climate changes?

Calling Earth the "water planet" is certainly appropriate, because nearly 71 percent of its surface is covered by the global ocean. Although the ocean comprises a much greater percentage of Earth's surface than do the continents, it was only in the relatively recent past that the ocean became an important focus of study. Recently there has been a virtual explosion of data about the oceans, and with it, oceanography has grown dramatically. **Oceanography** is a composite science that draws on the methods and knowledge of biology, chemistry, physics, and geology to study all aspects of the world ocean.

The Vast World Ocean

A glance at a globe or a view of Earth from space reveals a planet dominated by the world ocean (Figure 9.1). It is for this reason that Earth is often referred to as the *blue planet.*

Geography of the Oceans

The area of Earth is about 510 million square kilometers (197 million square miles). Of this total, approximately 360 million square kilometers (140 million square miles), or 71 percent, is represented by oceans and marginal seas (seas around the ocean's margin, such as the Mediterranean Sea and the Caribbean Sea). Continents and islands comprise the remaining 29 percent, or 150 million square kilometers (58 million square miles).

By studying a globe or world map, it is readily apparent that the continents and oceans are not evenly divided between the Northern and Southern hemispheres (Figure 9.1). When we compute the percentages of land and water in the Northern Hemisphere, we find that nearly 61 percent of the surface is water, and about 39 percent is land. In the Southern Hemisphere, on the other hand, almost 81 percent of the surface is water, and only 19 percent is land. It is no wonder then that the Northern Hemisphere is called the *land hemisphere,* and the Southern Hemisphere the *water hemisphere.*

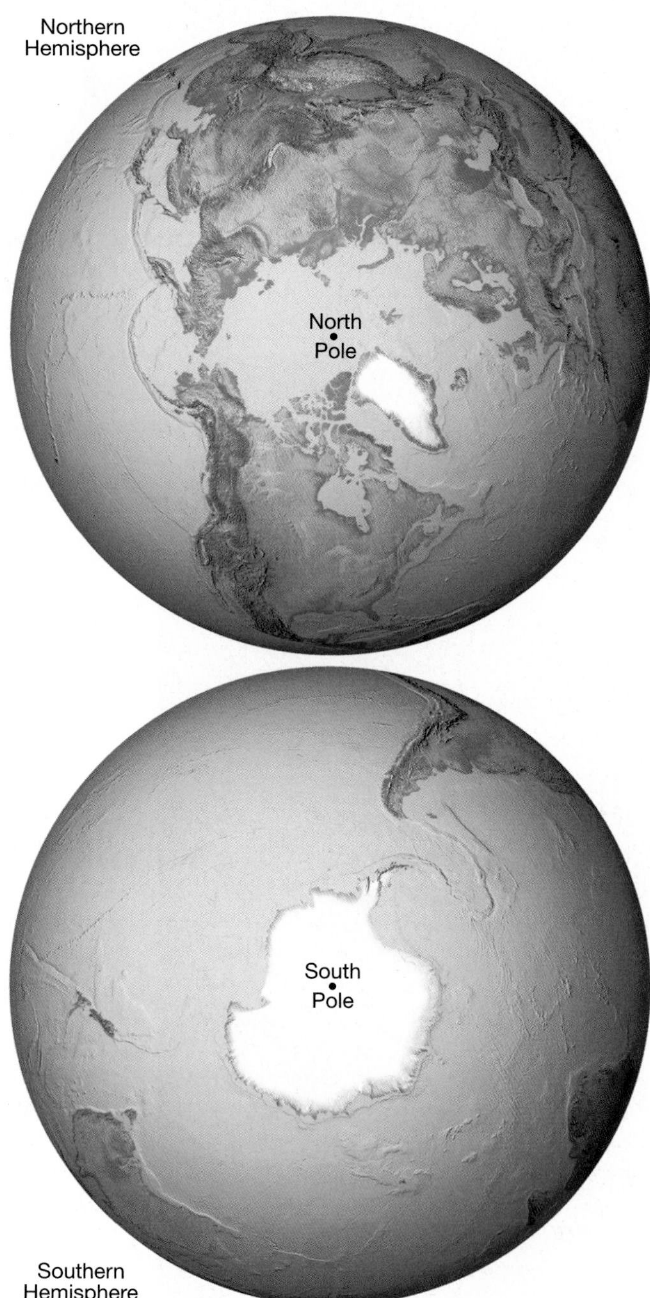

Figure 9.1 These views of Earth show the uneven distribution of land and water between the Northern and Southern hemispheres. Almost 81 percent of the Southern Hemisphere is covered by the oceans—20 percent more than the Northern Hemisphere.

Figure 9.2A shows the distribution of land and water in the Northern and Southern hemispheres by way of a graph. Between latitudes 45 degrees north and 70 degrees north, there is actually more land than water, whereas between 40 degrees south and 65 degrees south there is almost no land to interrupt the oceanic and atmospheric circulation.

The world ocean can be divided into four main ocean basins (Figure 9.2B):

1. The *Pacific Ocean,* which is the largest ocean (and the largest single geographic feature on the planet), covers more than half of the ocean surface area on Earth. In fact, the Pacific Ocean is so large that all of the continents could fit into the space occupied by it—with room left over! It is also the world's deepest ocean, with an average depth of 3940 meters (12,927 feet).
2. The *Atlantic Ocean* is about half the size of the Pacific Ocean but is not quite as deep. It is a relatively narrow ocean as compared to the Pacific and is bounded by almost parallel continental margins.
3. The *Indian Ocean* is slightly smaller than the Atlantic Ocean but has about the same average depth. Unlike the Pacific and Atlantic oceans, it is largely a Southern Hemisphere water body.
4. The *Arctic Ocean* is about 7 percent of the size of the Pacific Ocean and is only a little more than one-quarter as deep as the rest of the oceans.

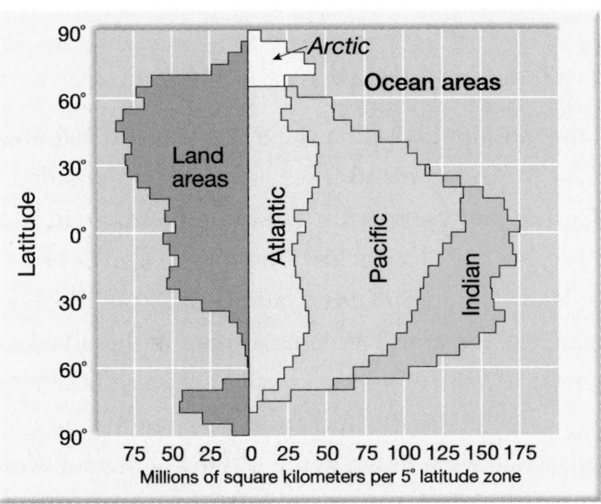

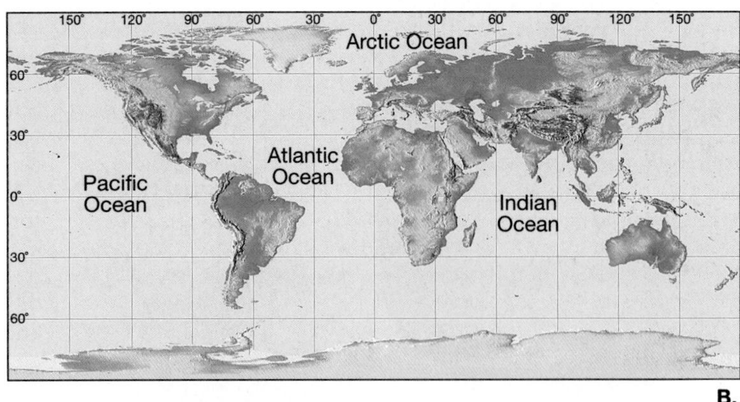

Figure 9.2 Distribution of land and water. **A.** The graph shows the amount of land and water in each 5-degree latitude belt. **B.** The world map provides a more familiar view.

Did You Know?

The Bering Sea is the most northerly marginal sea of the Pacific Ocean and connects to the Arctic Ocean through the Bering Straits. This water body is effectively cut off from the Pacific basins by the Aleutian Islands, which were created by volcanic activity associated with northward subduction of the Pacific basin.

Comparing the Oceans to the Continents

A major difference between continents and the ocean basins is their relative levels. The average elevation of the continents above sea level is about 840 meters (2756 feet), whereas the average depth of the oceans is nearly four and a half times this amount—3729 meters (12,234 feet). The volume of ocean water is so large that if Earth's solid mass were perfectly smooth (level) and spherical, the oceans would cover Earth's entire surface to a uniform depth of more than 2000 meters (1.2 miles)!

Composition of Seawater

What is the difference between pure water and seawater? One of the most obvious differences is that seawater contains dissolved substances that give it a distinctly salty taste. These dissolved substances are not simply sodium chloride (common table salt)— they include various other salts, metals, and even dissolved gases. In fact, every known naturally occurring element is found dissolved in at least trace amounts in seawater. Unfortunately, the salt content of seawater makes it unsuitable for drinking or for irrigating most crops and causes it to be highly corrosive to many materials. Yet, many parts of the ocean are teeming with life that is superbly adapted to the marine environment.

Salinity

Seawater consists of about 3.5 percent (by weight) dissolved mineral substances that are collectively termed *salts.* Although the percentage of dissolved components may seem small, the actual quantity is huge because the ocean is so vast.

Salinity is the total amount of solid material dissolved in water. More specifically, it is the ratio of the mass of dissolved substances to the mass of the water sample. Many common quantities are expressed in percent (%), which is really *parts per hundred.* Because the proportion of dissolved substances in seawater is such a small number, oceanographers typically express salinity in *parts per thousand* (‰). Thus,the average salinity of seawater is 3.5% or 35‰.

Figure 9.3 shows the principal elements that contribute to the ocean's salinity. Artificial seawater could be approximated by following the recipe in Table 9.1. This table shows that most of the salt in seawater is sodium chloride—common table salt. Sodium chloride together with the next four most abundant salts comprise more than 99 percent of all dissolved substances in the sea. Although only eight elements make up these five most abundant salts, seawater contains all of Earth's other naturally occurring elements. Despite their presence in minute quantities, many of these elements are very important in maintaining the necessary chemical environment for life in the sea.

Sources of Sea Salts

What are the primary sources for the vast quantities of dissolved substances in the ocean? Chemical weathering of rocks on the continents is one source. These dissolved materials are delivered to the oceans by streams at an estimated rate of more than 2.5 billion tons annually. The second major source of elements found in ocean water is Earth's interior. Through volcanic eruptions, large quantities of water and dissolved gases have been emitted during much of geolog-

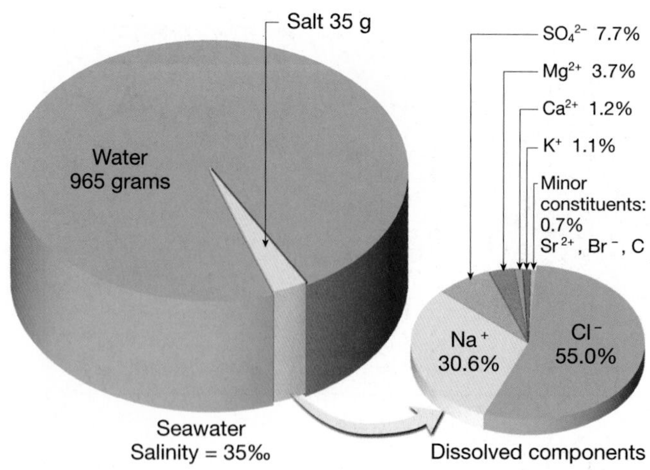

Figure 9.3 Relative proportions of water and dissolved components in seawater. Components shown by chemical symbol are chlorine (Cl^-), sodium (Na^+), sulfate (SO_4^{2-}), magnesium (Mg^{2+}), calcium (Ca^{2+}), potassium (K^+), strontium (Sr^{2+}), bromine (Br^-), and carbon (C).

ic time. This process, called **outgassing,** is the principal source of water in the oceans and in the atmosphere. Certain elements—notably chlorine, bromine, sulfur, and boron—were outgassed along with water and exist in the ocean in much greater abundance than could be explained by weathering of rocks alone.

Although rivers and volcanic activity continually contribute salts to the oceans, the salinity of seawater is not increasing. In fact, evidence suggests that the composition of seawater has been relatively stable for millions of years. Why doesn't the sea get saltier? The answer is that material is being removed just as rapidly as it is added. For example, some dissolved components are withdrawn from seawater by plants and animals as they build hard parts. Other components are removed when they chemically precipitate from the water as sediment. Still others are exchanged at oceanic ridges by hydrothermal activity. The net effect is that the overall makeup of seawater remains relatively constant through time.

Table 9.1 Recipe for artificial seawater.

To make seawater, combine:	Amount (grams)
Sodium chloride (NaCl)	23.48
Magnesium chloride ($MgCl_2$)	4.98
Sodium sulfate (Na_2SO_4)	3.92
Calcium chloride ($CaCl_2$)	1.10
Potassium chloride (KCl)	0.66
Sodium bicarbonate ($NaHCO_3$)	0.192
Potassium bromide (KBr)	0.096
Hydrogen borate (H_3BO_3)	0.026
Strontium chloride ($SrCl_2$)	0.024
Sodium fluoride (NaF)	0.003
Then add:	
Pure water (H_2O) to form 1000 grams of solution.	

Did You Know?

Some of the most highly saline water in the world is found in arid regions that have inland lakes, which are often called "seas" because they are so salty. The Great Salt Lake in Utah, for example, has a salinity of 280‰, and the Dead Sea on the border of Israel and Jordan has a salinity of 330‰. The water in the Dead Sea, therefore, contains 33 percent dissolved solids and is almost *10 times saltier than seawater.* As a result, these waters have such high density and are so buoyant that you can easily float while lying down in the water—with your arms and legs sticking up above water level!

Processes Affecting Seawater Salinity

Because the ocean is well mixed, the relative abundances of the major components in seawater are essentially constant, no matter where the ocean is sampled. Variations in salinity, therefore, are primarily a consequence of changes in the water content of the solution.

Various surface processes alter the amount of water in seawater, thereby affecting salinity (Figure 9.4). Processes that add large amounts of fresh water to seawater—and thereby decrease salinity—include precipitation, runoff from land, icebergs melting, and sea ice melting. Processes that remove large amounts of fresh water from seawater—and thereby increase seawater salinity—include evaporation and the formation of sea ice. High salinities, for example, are found where evaporation rates are high, as is the case in the dry subtropical regions (roughly between 25 and 35 degrees north or south latitude). Conversely, where large amounts of precipitation dilute ocean waters, as in the midlatitudes (between 35 and 60 degrees north or south latitude) and near the equator, lower salinities prevail.

Surface salinity in polar regions varies seasonally due to the formation and melting of sea ice. When seawater freezes in winter, sea salts do not become part of the ice. Therefore, the salinity of the remaining seawater increases. In summer when sea ice melts, the addition of the relatively fresh water dilutes the solution and salinity decreases.

Surface salinity variation in the open ocean normally ranges from 33‰ to 38‰. Some marginal seas, however, demonstrate extraordinary extremes. For example, in the restricted waters of the Middle East's Persian Gulf and Red Sea—where evaporation far exceeds precipitation—salinity may exceed 42‰. Conversely, very low salinities occur where large quantities of fresh water are supplied by rivers and precipitation. Such is the case for northern Europe's Baltic Sea, where salinity is often below 10‰.

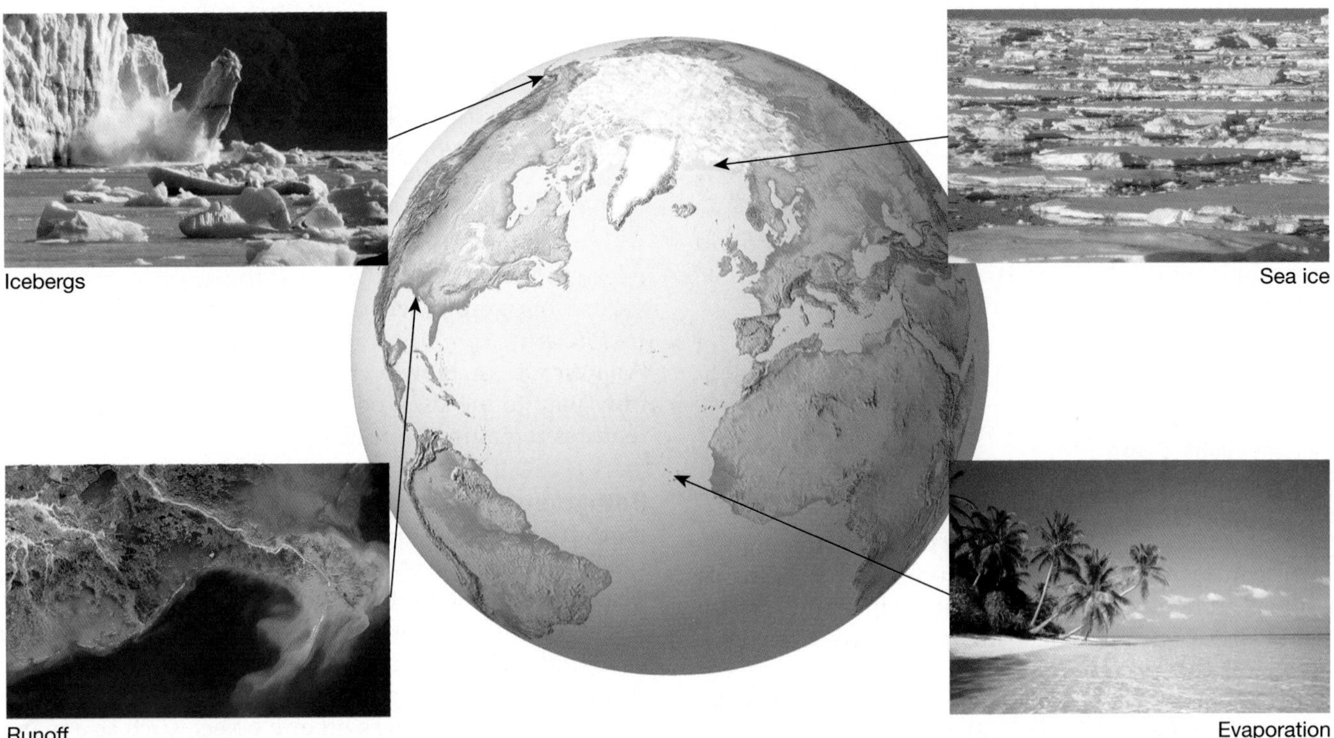

Figure 9.4 Processes affecting seawater salinity. Processes that *decrease* seawater salinity include precipitation, runoff, icebergs melting, and sea ice melting. Processes that *increase* seawater salinity include formation of sea ice and evaporation. (Photo credits: Top left, Tom & Susan Bean, Inc.; top right, Wolfgang Kaehler Photography; bottom left, NASA Headquarters; bottom right, Paul Steele/Corbis/The Stock Market)

The Ocean's Layered Structure

By sampling ocean waters, oceanographers have found that temperature and salinity change with depth. They recognize a general, three-layered structure in the open ocean: a shallow surface mixed zone, a transition zone, and a deep zone (Figure 9.5).

Because solar energy is received at the ocean surface, it is here that water temperatures are warmest. The mixing of these waters by waves as well as the turbulence from currents creates a mixed surface zone that has nearly uniform temperatures. The thickness and temperature of this layer vary, depending on latitude and season.

Below the Sun-warmed mixed zone, the temperature falls abruptly with depth. This layer of rapid temperature change, known as the **thermocline,** marks the transition between the warm surface layer and the deep zone of cold water (Figure 9.6). Below the thermocline, temperatures fall only a few more degrees. At depths greater than about 1500 meters (5000 feet), ocean-water temperatures are consistently below 4°C (39°F). Because the average ocean depth is approximately 3800 meters (12,500 feet), you can see that the temperature of most seawater is not much above freezing. In the high polar

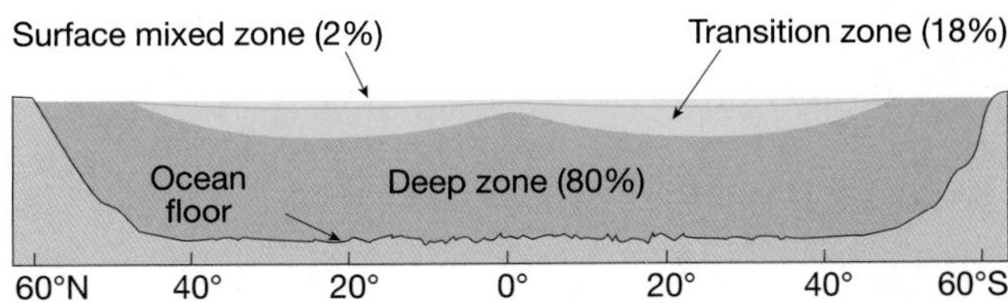

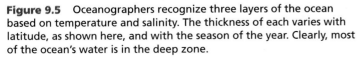

Figure 9.5 Oceanographers recognize three layers of the ocean based on temperature and salinity. The thickness of each varies with latitude, as shown here, and with the season of the year. Clearly, most of the ocean's water is in the deep zone.

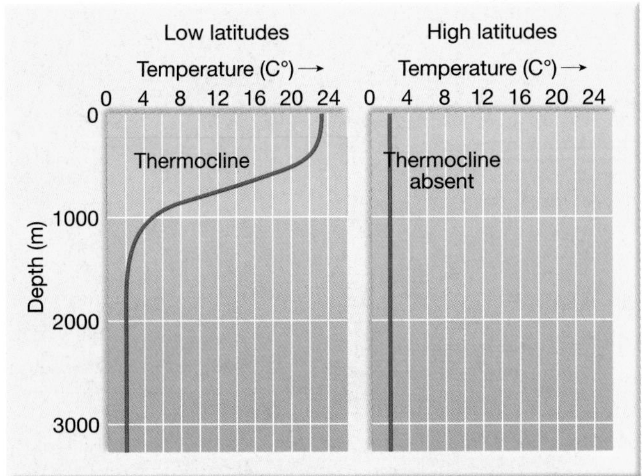

Figure 9.6 Variations in ocean water temperature with depth for low- and high-latitude regions. The layer of rapidly changing temperature, called the *thermocline,* is not present in the high latitudes.

latitudes, surface waters are cold and temperature changes with depth are slight. Consequently, the three-layered structure is not present (Figure 9.5).

Salinity variations with depth correspond to the general three-layered system described for temperatures. In the low and middle latitudes, a surface zone of higher salinity is created when freshwater is removed by evaporation. Below the surface zone, salinity decreases rapidly. This layer of rapid change, the **halocline,** corresponds closely to the thermocline. Below the halocline, salinity variations are small.

An Emerging Picture of the Ocean Floor

The Global Ocean
▼ Floor of the Ocean

If all the water were removed from the ocean basins, a great variety of features would be seen, including broad volcanic peaks, deep trenches, extensive plains, linear mountain chains, and large plateaus. In fact, the scenery would be nearly as diverse as that on the continents.

An understanding of seafloor features came with the development of techniques that measure the depth of the oceans. **Bathymetry** (*bathos* = depth, *metry* = measurement) is the measurement of ocean depths and the charting of the shape or topography of the ocean floor.

Mapping the Seafloor

The first understanding of the ocean floor's varied topography did not unfold until the historic three-and-a-half-year voyage of the HMS *Challenger* (Figure 9.7). From December 1872 to May 1876, the *Challenger* expedition made the first—and perhaps still most comprehensive—study of the global ocean ever attempted by one agency. The 127,500-kilometer (79,200-mile) trip took the ship and its crew of scientists to every ocean except the Arctic. Throughout the voyage, they sampled various ocean properties, including water depth, which was accomplished by laboriously lowering a long weighted line overboard. Not many years later, the knowledge gained by the *Challenger* of the ocean's great depth and varied topography was further expanded with the laying of transatlantic communication cables, especially in the North Atlantic Ocean. However, as long as a weighted line was the only way to measure ocean depths, knowledge of seafloor features remained limited.

Bathymetric Techniques Today, sound energy is used to measure water depths. The basic approach employs some type of **sonar,** an acronym for *so*und *na*vigation and *r*anging. The first devices that used sound to measure distances in water, called **echo sounders,** were developed early in the twentieth century. Echo sounders work by transmitting a sound wave (called a *ping*) into the water in order to produce an echo when it bounces off any object, such as a marine organism or the ocean floor (Figure 9.8A). A sensitive receiver intercepts the echo reflected from the bottom, and a clock precisely measures the travel time to fractions of a second. By knowing the speed of sound waves in water—about 1500 meters (4900 feet) per second—and the time required for the energy pulse to reach the ocean floor and return, depth can be calculated. The depths determined from continuous monitoring of these echoes are plotted so a profile of the ocean floor is obtained. By laboriously combining profiles from sev-

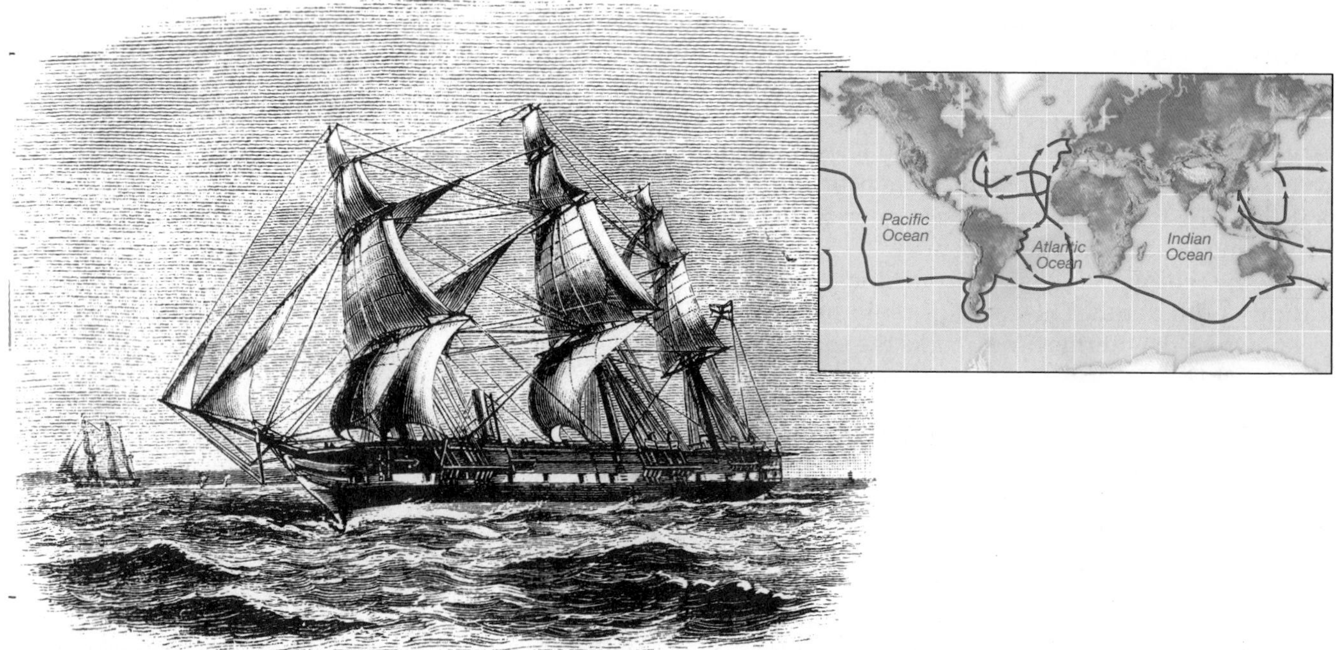

Figure 9.7 The first systematic bathymetric measurements of the ocean were made aboard the HMS *Challenger,* which departed England in December 1872 and returned in May 1876. (From C. W. Thompson and Sir John Murray, *Report on the Scientific Results of the Voyage of the H. M. S. Challenger,* Vol. 1, Great Britain: Challenger Office, 1895, Plate 1. Library of Congress)

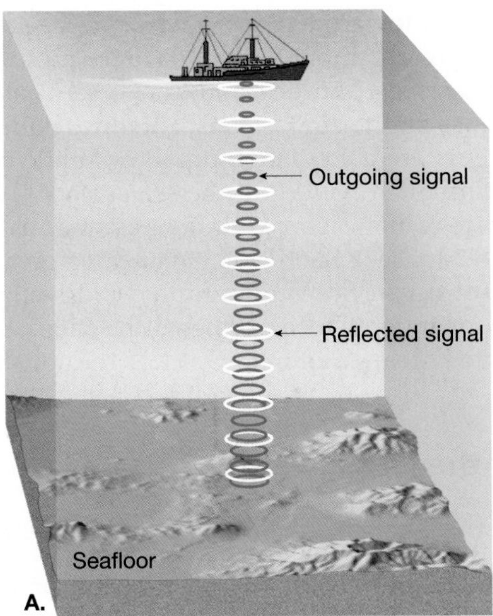

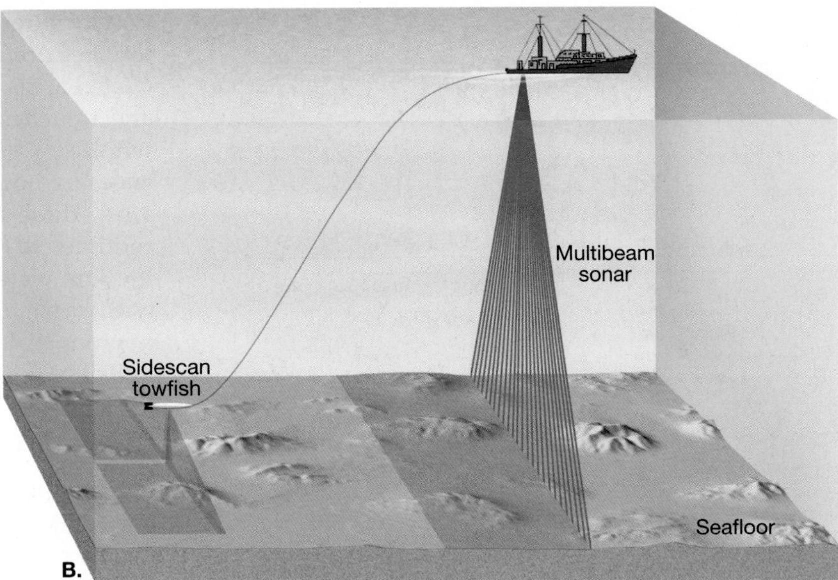

Figure 9.8 Various types of sonar. **A.** An echo sounder determines the water depth by measuring the time interval required for an acoustic wave to travel from a ship to the seafloor and back. The speed of sound in water is 1500 m/sec. Therefore, depth = $\frac{1}{2}$ (1500 m/sec × echo travel time). **B.** Modern multibeam sonar and sidescan sonar obtain an "image" of a narrow swath of seafloor every few seconds.

eral adjacent traverses, a chart of the seafloor can be produced.

Following World War II, the U.S. Navy developed *sidescan sonar* to look for mines and other explosive devices. These torpedo-shaped instruments can be towed behind a ship where they send out a fan of sound extending to either side of the ship's track. By combining swaths of sidescan sonar data, researchers produced the first photograph-like images of the seafloor. Although sidescan sonar provides valuable views of the seafloor, it does not provide bathymetric (water depth) data.

This problem is not present in the *high-resolution multibeam* instruments developed during the 1990s. These systems use hull-mounted sound sources that send out a fan of sound and then record reflections from the seafloor through a set of narrowly focused receivers aimed at different angles. Rather than obtaining the depth of a single point every few seconds, this technique makes it possible for a survey ship to map the features of the ocean floor along a strip tens of kilometers wide (Figure 9.8B). When a ship uses multibeam sonar to make a map of a section of seafloor, it travels through the area in a regularly spaced back-and-forth pattern known as, appropriately enough, "mowing the lawn." Furthermore, these systems can collect bathymetric data of such high resolution that they can distinguish depths that differ by less than a meter.

Despite their greater efficiency and enhanced detail, research vessels equipped with multibeam sonar travel at a mere 10 to 20 kilometers (6 to 12 miles) per hour. It would take at least 100 vessels outfitted with this equipment hundreds of years to map the entire seafloor. This explains why only about 5 percent of the seafloor has been mapped in detail—and why large areas of the seafloor have not yet been mapped with sonar at all.

Did You Know?

Ocean depths are often expressed in *fathoms.* One fathom equals 1.8 meters (6 feet), which is about the distance of a person's outstretched arms. The term is derived from how depth-sounding lines were brought back on board a vessel by hand. As the line was hauled in, a worker counted the number of arm lengths collected. By knowing the length of the person's outstretched arms, the amount of line taken in could be calculated. The length of one fathom was later standardized to 6 feet.

Seismic Reflection Profiles Marine geologists are also interested in viewing the rock structure beneath the sediments that blanket much of the seafloor. This can be accomplished by making a **seismic reflection profile.** To construct such a profile, strong low-frequency sounds are produced by explosions (depth charges) or air guns. These sound waves penetrate beneath the seafloor and reflect off the contacts between rock layers and fault zones, just like sonar reflects off the bottom of the sea. Figure 9.9 shows a seismic profile of a portion of the Madeira Abyssal Plain in the eastern Atlantic. Although the seafloor here is flat, notice the irregular ocean crust buried by a thick accumulation of sediments.

Viewing the Ocean Floor from Space

Another technological breakthrough that has led to an enhanced understanding of the seafloor involves measuring the shape of the ocean surface from space. After compensating for waves, tides, currents, and atmospheric effects, it was discovered that the ocean surface is not perfectly flat because

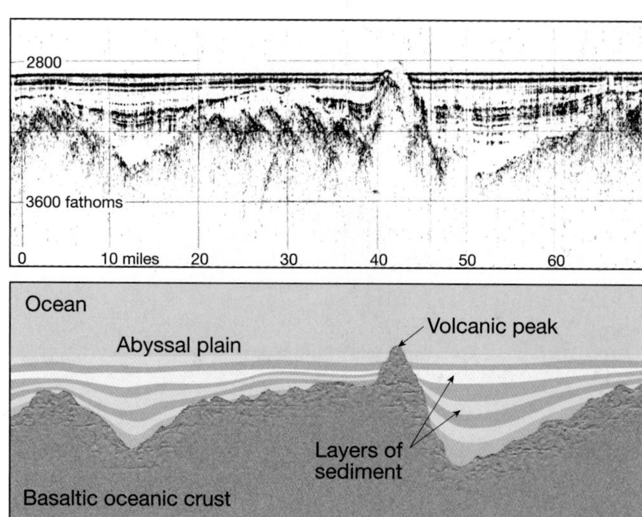

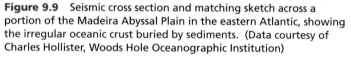

Figure 9.9 Seismic cross section and matching sketch across a portion of the Madeira Abyssal Plain in the eastern Atlantic, showing the irregular oceanic crust buried by sediments. (Data courtesy of Charles Hollister, Woods Hole Oceanographic Institution)

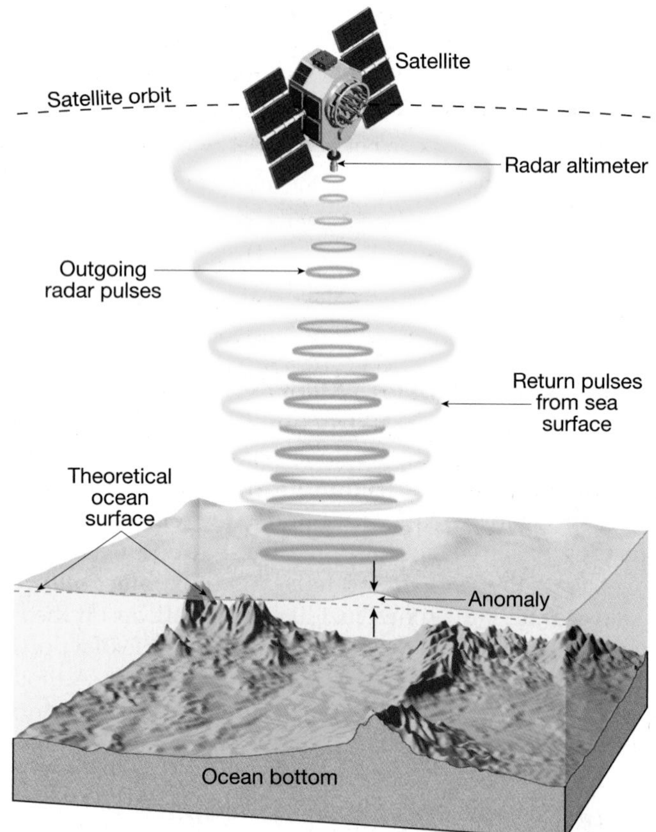

Figure 9.10 A satellite altimeter measures the variation in sea surface elevation, which is caused by gravitational attraction and mimics the shape of the seafloor. The sea surface anomaly is the difference between the measured and theoretical ocean surface.

gravity attracts water toward regions where massive seafloor features occur. Therefore, mountains and ridges produce elevated areas on the ocean surface, and, conversely, canyons and trenches cause slight depressions. Satellites equipped with *radar altimeters* are able to measure these subtle differences by bouncing microwaves off the sea surface (Figure 9.10). These devices can measure variations as small as 3 to 6 centimeters (1 to 2 inches). Such data have added greatly to the knowledge of ocean-floor topography. Cross-checked with traditional sonar depth measurements, the data are used to produce detailed ocean-floor maps, such as the one shown in Figure 9.11.

Provinces of the Ocean Floor

Oceanographers studying the topography of the ocean floor have delineated three major units: *continental margins,* the *ocean basin floor,* and the *oceanic (mid-ocean) ridge.* The map in Figure 9.12 on p. 256 outlines these provinces for the North Atlantic Ocean, and the profile at the bottom of the illustration shows the varied topography. Such profiles usually have their vertical dimension exaggerated many times—40 times in this case—to make topographic features more conspicuous. Vertical exaggeration, however, makes slopes shown in seafloor profiles appear to be *much* steeper than they actually are.

Continental Margins

The Global Ocean
▼ Floor of the Ocean

Two main types of **continental margins** have been identified—passive and active. Passive margins are found along most of the coastal areas that surround the Atlantic Ocean, including the east coasts of North and South America, as well as the coastal areas of Western Europe and Africa. Passive margins are *not* associated with plate boundaries and, therefore, experience little volcanism and few earthquakes. Here, weathered materials eroded from the adjacent landmass accumulate to form a thick, broad wedge of relatively undisturbed sediments.

By contrast, active continental margins occur where oceanic lithosphere is being subducted beneath the edge of a continent. The result is a relatively narrow margin, consisting of highly deformed sediments that were scraped from the descending lithospheric slab. Active continental margins are common around the Pacific Rim, where they parallel deep oceanic trenches.

Passive Continental Margins

The features comprising a **passive continental margin** include the continental shelf, the continental slope, and the continental rise (Figure 9.13 on p. 256).

Continental Shelf The **continental shelf** is a gently sloping submerged surface extending from the shoreline toward the deep-ocean basin. Because it is underlain by continental crust, it is clearly a flooded extension of the continents. The continental shelf varies greatly in width. Although almost nonexistent along some continents, the shelf may extend seaward as far as 1500 kilometers (930 miles) along others. On the average, the continental shelf is about 80 kilometers (50 miles) wide and 130 meters (423 feet) deep at its seaward edge. The average inclination of the continental shelf is only about one-tenth of 1 degree, a drop of only about 2 meters per kilometer (10 feet per mile). The slope is so slight that it would appear to an observer to be a horizontal surface.

Continental shelves represent only 7.5 percent of the total ocean area, but they have economic and political significance because they contain important mineral deposits, including large reservoirs of petroleum and natural gas, as well as huge sand and gravel deposits. The waters of the continental shelf also contain many important fishing grounds that are significant sources of food.

Although the continental shelf is relatively featureless, some areas are mantled by extensive glacial deposits and are thus quite rugged. In addition, some continental shelves are dissected by large valleys running from the coastline into deeper waters. Many of these *shelf valleys* are the seaward extensions of river valleys on the adjacent landmass. Such valleys appear to have been excavated during the Pleistocene epoch (Ice Age). During this time, great quantities of water were stored in vast ice sheets on the continents. This caused sea level to drop by 100 meters (330 feet) or more, exposing large areas of the continental shelves (see Figure 4.19, p. 115). Because of this drop in sea level, rivers extended their courses, and land-dwelling plants and animals inhabited the newly exposed portions of the continents.

Most continental shelves associated with passive margins, such as those along the East Coast of the United States, consist of thick accumulations of shallow-water sediments. These sediments are frequently several kilometers thick and are interbedded with limestones that formed during earlier periods of coral reef building, a process that occurs only in shallow water. Such evidence led researchers to conclude that these thick accumulations of sediment are produced along a gradually subsiding continental margin.

Continental Slope Marking the seaward edge of the continental shelf is the **continental slope,** a relatively steep zone (as compared with the shelf) that marks the boundary between continental crust and oceanic crust (Figure 9.13). Although the inclination of the continental slope varies greatly from place to place, it averages about 5 degrees and may exceed 25 degrees. Further, the continental slope is a relatively narrow feature, averaging only about 20 kilometers (12 miles) wide.

Continental Rise In regions where trenches do not exist, the steep continental slope merges into a more gradual incline known as the **continental rise.** Here, the slope drops to about one-third degree, or about 6 meters per kilometer (30 feet per mile). Whereas the width of the continental slope averages about 20 kilometers (12 miles), the continental rise may extend for hundreds of kilometers into the deep-ocean basin.

The continental rise consists of a thick accumulation of sediment that moved downslope from the continental shelf to the deep-ocean floor. The sediments are delivered to the base of the continental slope by *turbidity currents* that follow submarine canyons. (We will discuss these shortly.) When these muddy currents emerge from the mouth of a canyon onto the relatively flat ocean floor, they deposit sediment that forms a **deep-sea fan** (Figure 9.13). Deep-sea fans have the same basic shape as alluvial fans, which form at the foot of steep mountain slopes on land. As fans from adjacent submarine canyons grow, they coalesce to produce the continuous apron of sediment at the base of the continental slope that we call the continental rise.

Submarine Canyons and Turbidity Currents Deep, steep-sided valleys known as **submarine canyons** are cut into the continental slope and may extend across the entire continental rise to the deep-ocean basin (see Figure 9.14). Although some of these canyons appear to be the seaward extensions of river valleys, many others do not line up in this manner. Furthermore, these canyons extend to depths far below the maximum lowering of sea level during the Ice Age, so we cannot attribute their formation to stream erosion.

These submarine canyons have probably been excavated by turbidity currents. **Turbidity currents** are downslope movements of dense, sediment-laden water. They are created when sand and mud on the continental shelf and slope are dislodged and thrown into suspension. Because the mud-choked water is denser than normal seawater, it flows downslope as a mass, eroding and accumulating more sediment as it goes. The erosional work repeatedly carried on by these muddy torrents is thought to be the major force in the excavation of most submarine canyons.

Turbidity currents usually originate along the continental slope and continue across the continental rise, still cutting channels. Eventually, they lose momentum and come to rest along the ocean basin floor. As these currents slow, suspended sediments begin to settle out. First, the coarser sand is dropped, followed by successively finer accumulations of silt and then clay. These deposits, called *turbidites,* display a decrease in sediment grain size from bottom to top, a phenomenon known as *graded bedding.*

Turbidity currents are a very important mechanism of sediment transport in the ocean. By the action of turbidity currents, submarine canyons are excavated and sediments are carried to the deep-ocean floor.

Active Continental Margins

The continental slope along some coasts descends abruptly into a deep-ocean trench. In this situation, the landward wall of the trench and the continental slope are essentially the same feature. In such locations, the continental shelf is very narrow, if it exists at all.

Active continental margins are located primarily around the Pacific Ocean where oceanic lithosphere is being

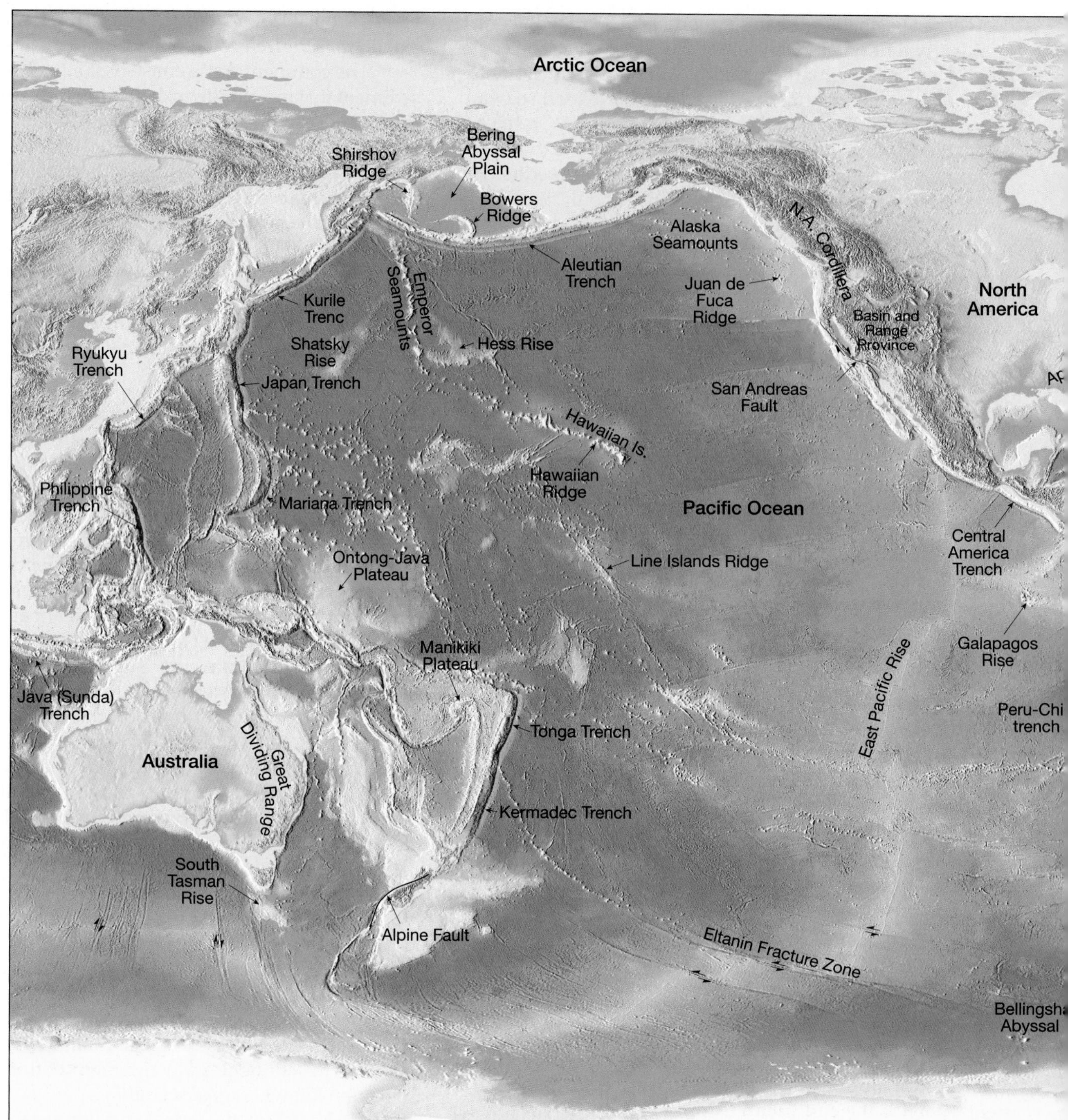

Figure 9.11 The topography of Earth's solid surface is shown on this map.

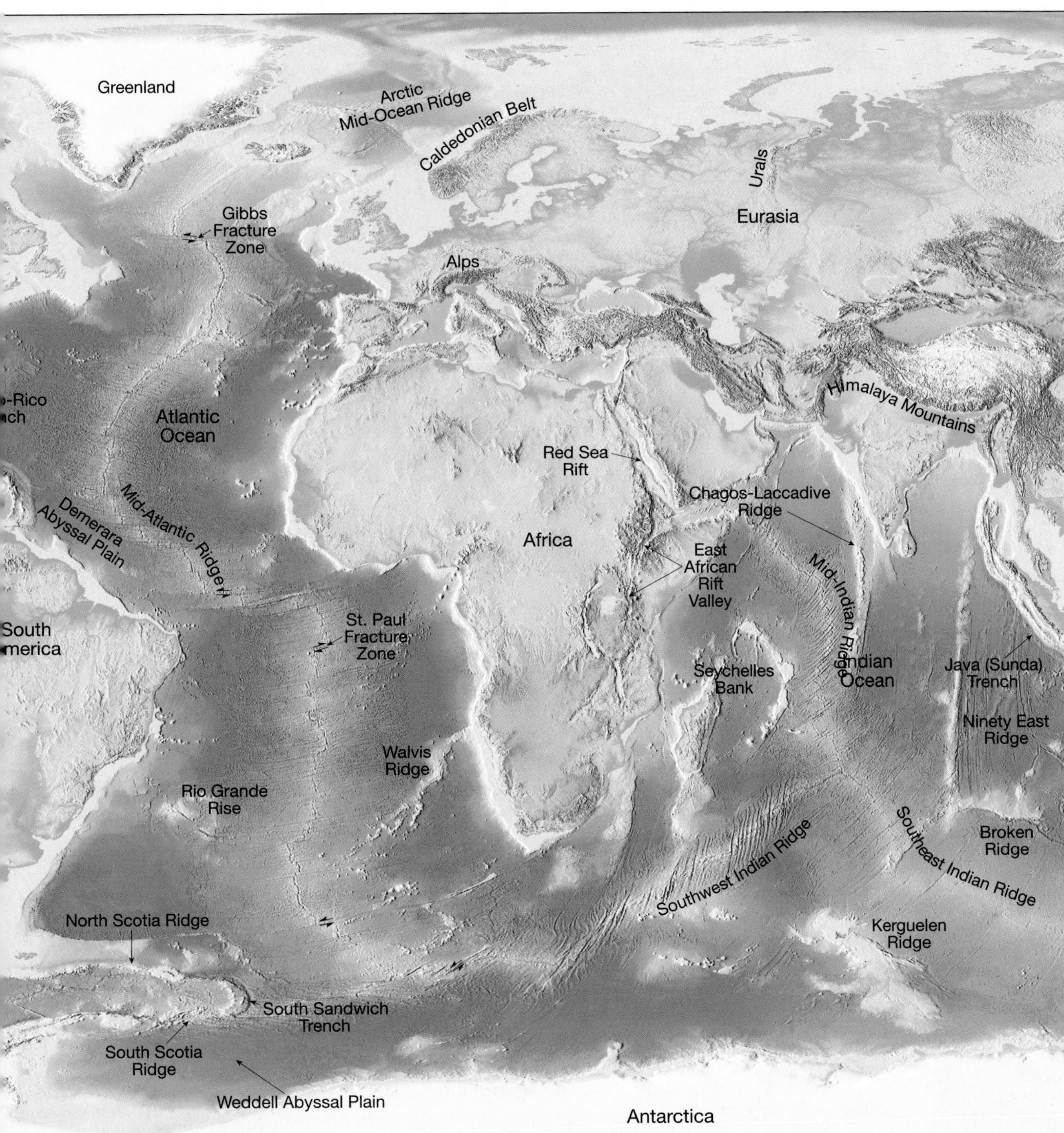

Figure 9.11 (Continued)

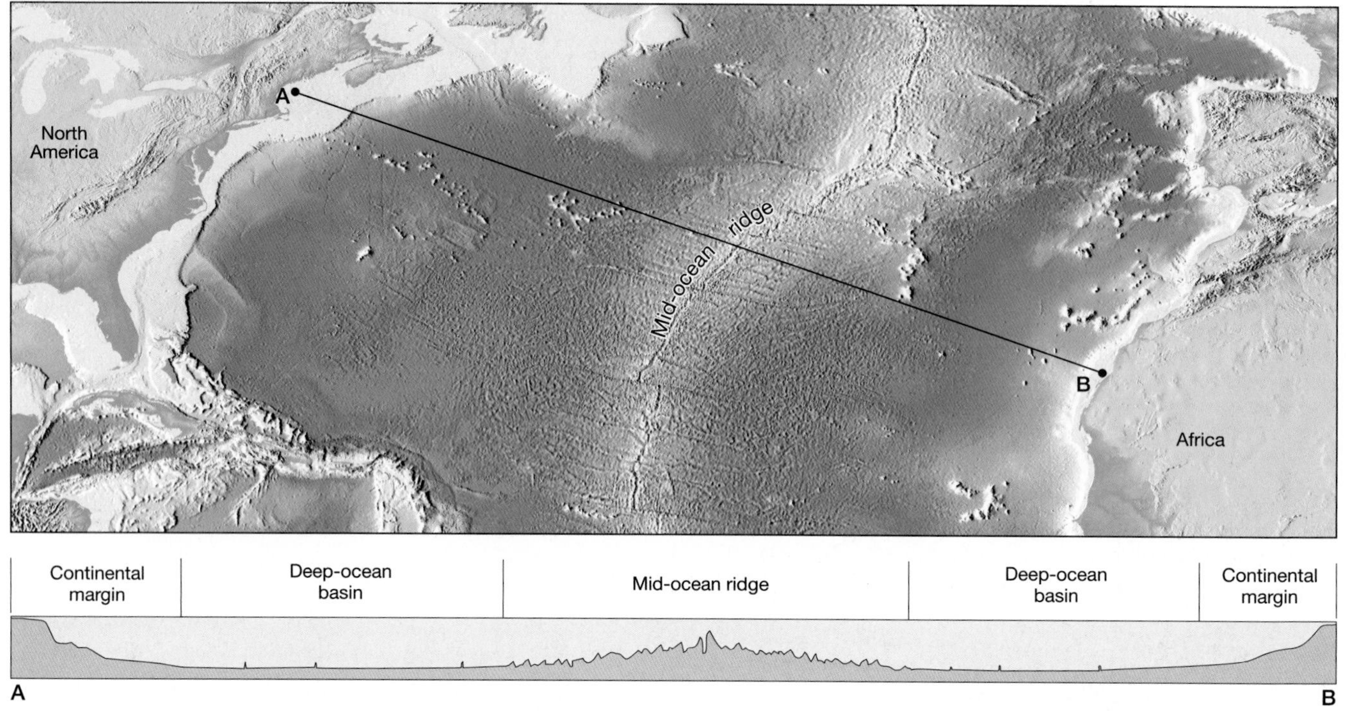

Figure 9.12 Map view (*top*) and corresponding profile view (*bottom*) showing the major topographic divisions of the North Atlantic Ocean. On the profile, the vertical scale has been expanded (exaggerated) by 40 times to make topographic features more conspicuous.

subducted beneath the leading edge of a continent (Figure 9.15). Here sediments from the ocean floor and pieces of oceanic crust are scraped from the descending oceanic plate and plastered against the edge of the overriding continent. This chaotic accumulation of deformed sediment and scraps of oceanic crust is called an *accretionary wedge.* Prolonged plate subduction, along with the accretion of sediments on the landward side of the trench, can produce a large accumulation of sediments along a continental margin.

Some subduction zones have little or no accumulation of sediments, indicating that ocean sediments are being carried into the mantle with the subducting plate. Here, the continental margin is very narrow, as the trench may lie a mere 50 kilometers (30 miles) offshore.

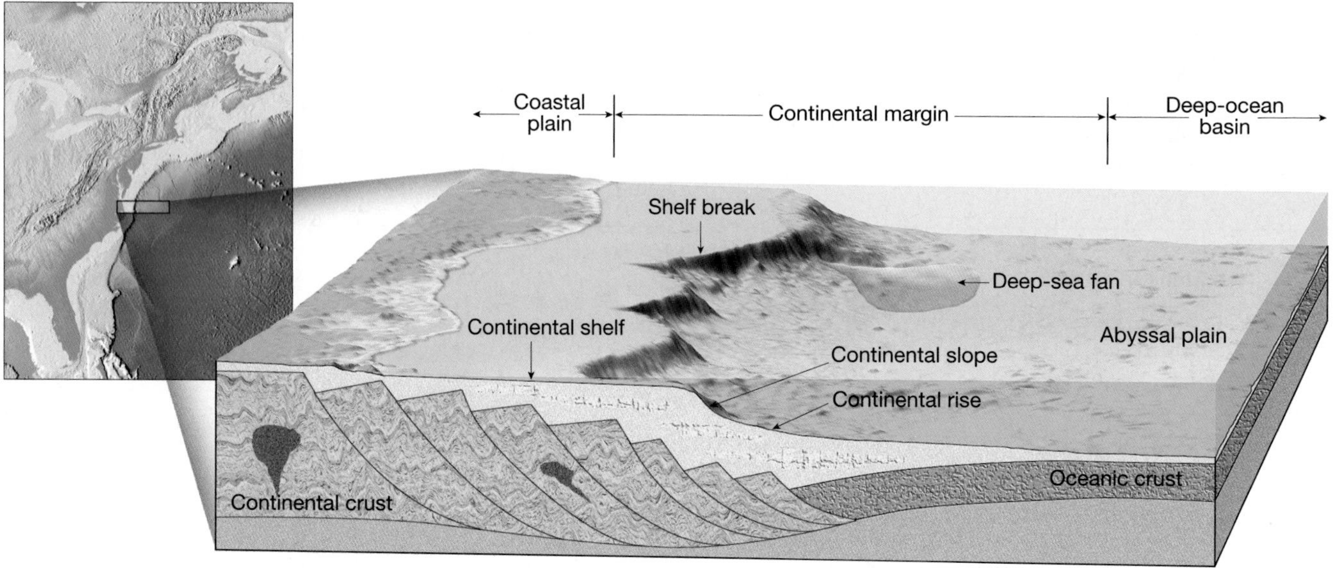

Figure 9.13 Schematic view showing the parts of a passive continental margin.

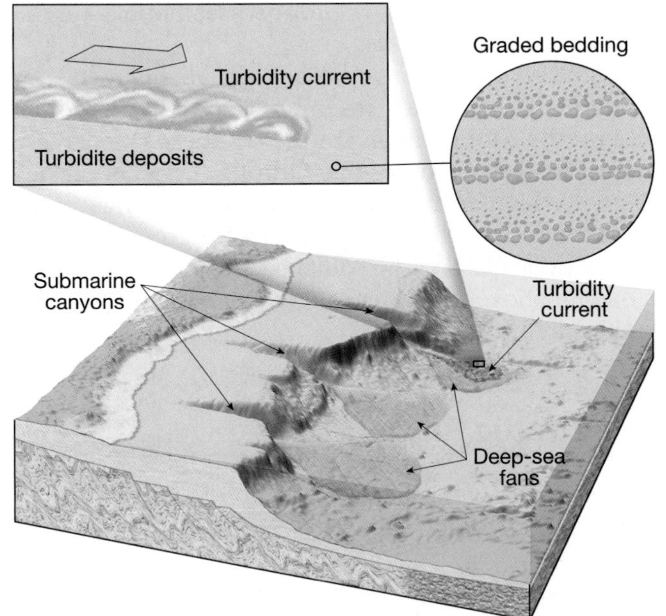

Figure 9.14 Turbidity currents are downslope movements of dense, sediment-laden water. They are created when sand and mud on the continental shelf and slope are dislodged and thrown into suspension. Because such mud-choked water is denser than normal seawater, it flows downslope, eroding and accumulating more sediment. Beds deposited by these currents are called *turbidites.* Each event produces a single bed characterized by a decrease in sediment size from bottom to top, a feature known as a *graded bed.*

Deep-Ocean Basin

The Global Ocean
▼ Floor of the Ocean

Between the continental margin and the mid-ocean ridge lies the deep-ocean basin (see Figure 9.11, pp. 254–255). The size of this region—almost 30 percent of Earth's surface—is roughly comparable to the percentage of the surface that presently projects above sea level as land. Here, we find *deep-ocean trenches,* which are extremely deep linear depressions in the ocean floor; remarkably flat regions, known as *abyssal plains;* broad volcanic peaks, called *seamounts;* and extensive areas of lava flows piled one atop the other, called *oceanic plateaus.*

Did You Know?

In January 1960, U.S. Navy Lt. Don Walsh and explorer Jacques Piccard descended to the bottom of the Challenger Deep region of the Mariana Trench. More than five hours after leaving the surface, they reached the bottom at 10,912 meters (35,800 feet)—a record depth of human descent that has never been surpassed.

Deep-Ocean Trenches

Deep-ocean trenches are long, relatively narrow troughs that are the deepest parts of the ocean. Most trenches are located along the margins of the Pacific Ocean, where many exceed 10,000 meters (33,000 feet) in depth (see Figure 9.11). A portion of one—the Challenger Deep in the Mariana Trench—has been measured at 11,022 meters (36,163 feet) below sea level, making it the deepest known part of the world ocean. Only two trenches are located in the Atlantic—the Puerto Rico Trench and the South Sandwich Trench (see Figure 9.11).

Although deep-ocean trenches represent only a very small portion of the area of the ocean floor, they are nevertheless significant geologic features. Trenches are sites of plate convergence where a moving lithospheric plate subducts and plunges back into the mantle. In addition to earthquakes being created as one plate "scrapes" beneath another, volcanic activity is also associated with these regions. The release of volatiles—especially water—from a descending plate triggers melting in the wedge of asthenosphere above it. This buoyant material slowly migrates upward and gives rise to

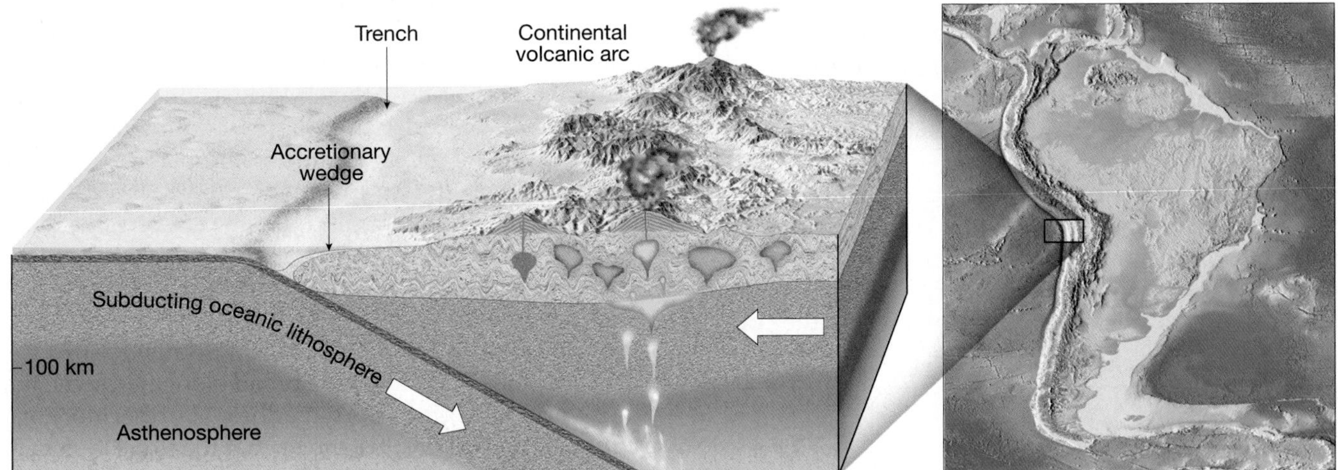

Figure 9.15 Active continental margin. Sediments from the ocean floor are scraped from the descending plate and added to the continental crust as an accretionary wedge.

volcanic activity at the surface. Thus, trenches are often paralleled by an arc-shaped row of active volcanoes called a **volcanic island arc.** Furthermore, **continental volcanic arcs,** such as those making up portions of the Andes, are located parallel to trenches that lie adjacent to continental margins (see Figure 9.15). The large number of trenches and associated volcanic activity along the margins of the Pacific Ocean cause the region to be known as the *Ring of Fire.*

Abyssal Plains

Abyssal plains are incredibly flat features; in fact, these regions are likely the most level places on Earth. The abyssal plain found off the coast of Argentina, for example, has less than 3 meters (10 feet) of relief over a distance exceeding 1300 kilometers (800 miles). The monotonous topography of abyssal plains is occasionally interrupted by the protruding summit of a buried volcanic structure.

Using seismic profilers, instruments whose signals penetrate far below the ocean floor, researchers have determined that abyssal plains consist of thick accumulations of sediment that have buried an otherwise rugged ocean floor (see Figure 9.9). The nature of the sediment indicates that these plains consist primarily of sediments transported far out to sea by turbidity currents.

Abyssal plains occur in all of the oceans. However, because the Atlantic Ocean has fewer trenches to act as traps for the sediments carried down the continental slope, it has more extensive abyssal plains than does the Pacific.

Seamounts, Guyots, and Oceanic Plateaus

Dotting the ocean floor are isolated volcanic peaks called **seamounts.** These features may rise hundreds of meters above the surrounding topography. Although these steep-sided conical peaks are found on the floors of all the oceans, the greatest number have been identified in the Pacific.

Some, like the Emperor Seamount chain that stretches north from the Hawaiian Islands to the Aleutian Trench, form in association with mantle plumes and volcanic hot spots. Others are born near oceanic ridges, divergent plate boundaries where the plates of the lithosphere move apart (see Chapter 5). If a volcano grows large enough before being carried from the zone that nourishes it, the structure emerges as an island. Examples in the Atlantic include the Azores, Ascension, Tristan da Cunha, and St. Helena.

During the time they exist as islands, some of these volcanoes are eroded to near sea level by running water and wave action. Over a span of millions of years, the islands gradually sink as the moving plate slowly carries them away from the elevated oceanic ridge or hot spot where they originated. These submerged, flat-topped seamounts are called **guyots** or **tablemounts.**

Mantle plumes have also generated several large **oceanic plateaus,** which resemble flood basalt provinces found on the continents. Examples of these extensive volcanic structures include the Ontong Java and Rockall plateaus, which formed from vast outpourings of fluid basaltic lavas onto the ocean floor (see Figure 6.34, p. 186). Hence, oceanic plateaus are composed mostly of pillow basalt and other mafic rocks that in some cases exceed 30 kilometers (19 miles) in thickness.

Did You Know?

It has been estimated that there are between 22,000 and 55,000 seamounts on the ocean floor—less than 2000 are active. The greatest number have been identified in the Pacific.

The Oceanic Ridge

The Global Ocean
▼ Floor of the Ocean

Along well-developed divergent plate boundaries, the seafloor is elevated, forming a broad linear swell called the **oceanic ridge,** or **mid-ocean ridge.** Our knowledge of the oceanic ridge system comes from soundings taken of the ocean floor, core samples obtained from deep-sea drilling, and visual inspection using deep-diving submersibles. We also have first-hand inspection of slices of ocean floor that have been displaced onto dry land along convergent plate boundaries. An elevated position, extensive faulting, and numerous volcanic structures characterize the oceanic ridge.

The interconnected oceanic ridge system is the longest topographic feature on Earth, exceeding 70,000 kilometers (43,000 miles) in length. This is quite evident as you follow the path of mid-ocean ridges in Figure 9.11, pp. 254–255. Representing 23 percent of Earth's surface, the oceanic ridge system winds through all major oceans in a manner similar to the seam on a baseball. The crest of this linear structure typically stands 2 to 3 kilometers (1 to 2 miles) above the adjacent deep-ocean basins and marks the plate margins where new oceanic crust is created.

Notice in Figure 9.16 that large sections of the oceanic ridge system have been named for their locations within the various ocean basins. Ideally, ridges run along the middle of ocean basins, where they are called *mid-ocean* ridges. This holds true for the Mid-Atlantic Ridge, which is positioned in the middle of the Atlantic, roughly paralleling the margins of the continents on either side (Figure 9.16A). This is also true for the Mid-Indian Ridge (Figure 9.16B), but note that the East Pacific Rise is displaced to the eastern side of the Pacific Ocean (Figure 9.16C).

The term *ridge* may be misleading, because these features are not narrow and steep as the term implies, but they have widths of from 1000 to 4000 kilometers (620 to 2500 miles) and the appearance of a broad elongated swell that exhibits various degrees of ruggedness. Furthermore, careful examination of Figure 9.11 shows that the ridge system is broken into segments that range from a few tens to hundreds of kilometers in length. Although each segment is offset from the adjacent segment, they are generally connected, one to the next, by a transform fault.

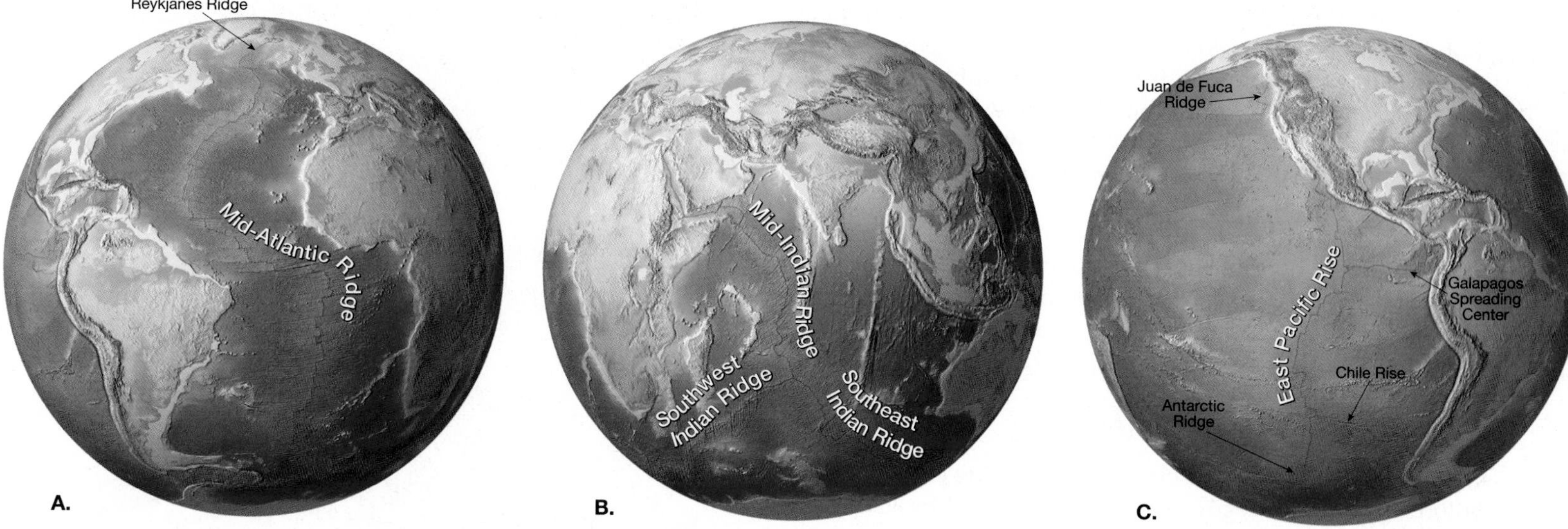

Figure 9.16 Distribution of the oceanic ridge system, which winds through all major ocean basins like a seam on a baseball.

Oceanic ridges are as high as some mountains found on the continents, and thus they are often described as mountainous in nature. However, the similarity ends there. Whereas most continental mountains form when compressional forces fold and metamorphose thick sequences of sedimentary rocks along convergent plate boundaries, oceanic ridges form where tensional forces fracture and pull the ocean crust apart. The oceanic ridge consists of layers and piles of newly formed basaltic rocks that have been faulted into elongated blocks that are buoyantly uplifted.

Along the axis of some segments of the oceanic ridge system are deep down-faulted structures called **rift valleys** (Figure 9.17). These features may exceed 50 kilometers (31 miles) in width and 2000 meters (6600 feet) in depth. Because they contain faulted and tilted blocks of oceanic crust, as well as volcanic cones that have grown upon the newly formed seafloor, rift valleys usually exhibit rugged topography. The name *rift valley* has been applied to these features because they are so strikingly similar to continental rift valleys such as the East African Rift.

Topographically, the outermost flanks of most ridges are relatively subdued (except for isolated volcanic peaks) and rise very gradually (slope less than 1 degree) toward the ridge axis. Approaching the ridge crest, the topography becomes more rugged as volcanic structures and faulted valleys that tend to parallel the ridge axis become more prominent. The most rugged topography is found on those ridges that exhibit large rift valleys.

Did You Know?

The Gulf of California, also known as the Sea of Cortez, formed over the past 6 million years by seafloor spreading. This (1200-kilometer-long (750-mile-long) basin is located between the west coast of mainland Mexico and the Baja Peninsula.

Seafloor Sediments

Except for steep areas of the continental slope and areas near the crest of the mid-ocean ridge, the ocean floor is covered with sediment. Part of this material has been deposited by turbidity currents, and the rest has slowly settled to the seafloor from above. The thickness of this carpet of debris varies greatly. In some trenches, which act as traps for

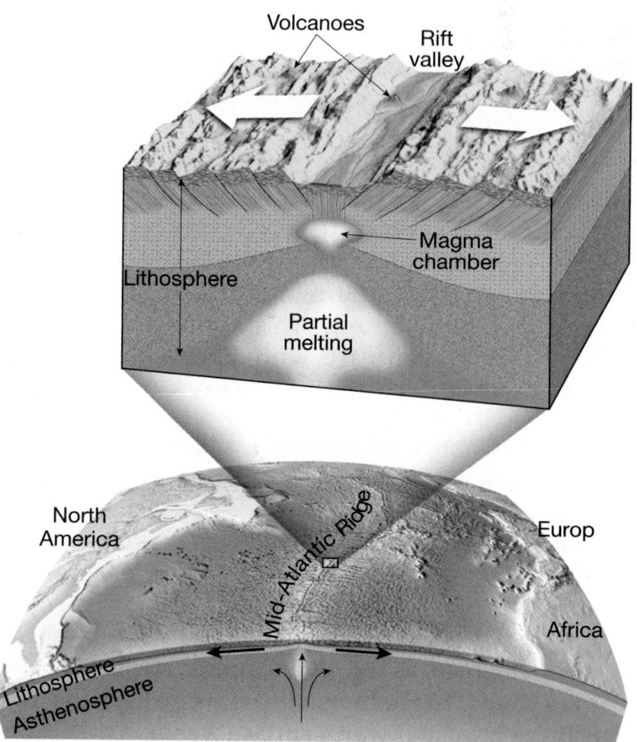

Figure 9.17 The axes of some segments of the oceanic ridge system contain deep down-faulted structures called *rift valleys.* Some may exceed 50 kilometers (31 miles) in width and 2000 meters (6600 feet) in depth.

sediments originating on the continental margin, accumulations may approach 10 kilometers (6 miles). In general, however, sediment accumulations are considerably less. In the Pacific Ocean, for example, uncompacted sediment measures about 600 meters (2000 feet) or less, whereas on the floor of the Atlantic, the thickness varies from 500 to 1000 meters (1600 to 3300 feet).

Although deposits of sand-sized particles are found on the deep-ocean floor, mud is the most common sediment covering this region. Muds also predominate on the continental shelf and slope, but the sediments in these areas are coarser overall because of greater quantities of sand.

Types of Seafloor Sediments

Seafloor sediments can be classified according to their origin into three broad categories: (1) **terrigenous** ("derived from land"), (2) **biogenous** ("derived from organisms"), and (3) **hydrogenous** ("derived from water"). Although each category is discussed separately, remember that all seafloor sediments are mixtures. No body of sediment comes entirely from a single source.

Terrigenous Sediment Terrigenous sediment consists primarily of mineral grains that were weathered from continental rocks and transported to the ocean. Larger particles (sand and gravel) usually settle rapidly near shore, whereas the very smallest particles take years to settle to the ocean floor and may be carried thousands of kilometers by ocean currents. As a consequence, virtually every area of the ocean receives some terrigenous sediment. The rate at which this sediment accumulates on the deep-ocean floor, though, is very slow. To form a 1-centimeter (0.4-inch) abyssal clay layer, for example, requires as much as 50,000 years. Conversely, on the continental margins near the mouths of large rivers, terrigenous sediment accumulates rapidly and forms thick deposits. In the Gulf of Mexico, for example, the sediment has reached a depth of many kilometers.

Because fine particles remain suspended in the water for a long time, there is ample opportunity for chemical reactions to occur. Because of this, the colors of deep-sea sediments are often red or brown. This results when iron in the particle or in the water reacts with dissolved oxygen in the water and produces a coating of iron oxide (rust).

Biogenous Sediment Biogenous sediment consists of shells and skeletons of marine animals and algae (Figure 9.18). This debris is produced mostly by microscopic organisms living in the sunlit waters near the ocean surface. Once these organisms die, their hard *tests* (*testa* = shell) continually "rain" down and accumulate on the seafloor.

The most common biogenous sediment is *calcareous* ($CaCO_3$) *ooze* which, as the name implies, has the consistency of thick mud. This sediment is produced from the tests of organisms that inhabit warm surface waters. When calcareous hard parts slowly sink through a cool layer of water, they begin to dissolve. This results because the deeper cold seawater is rich in carbon dioxide and is thus more acidic than warm water. In seawater deeper than about 4500 meters

Figure 9.18 Microscopic hard parts of radiolaria and foraminifera are examples of biogenous sediments. This photomicrograph has been enlarged hundreds of times. (Photo courtesy of Deep Sea Drilling Project, Scripps Institution of Oceanography, University of California, San Diego)

(15,000 feet), calcareous tests will completely dissolve before they reach bottom. Consequently, calcareous ooze does not accumulate at these greater depths.

Other biogenous sediments include *siliceous* (SiO_2) *ooze* and phosphate-rich material. The former is composed primarily of tests of diatoms (single-celled algae) and radiolaria (single-celled animals), whereas the latter is derived from the bones, teeth, and scales of fish and other marine organisms.

Hydrogenous Sediment Hydrogenous sediment consists of minerals that crystallize directly from seawater through various chemical reactions. For example, some limestones are formed when calcium carbonate precipitates directly from the water; however, most limestone is composed of biogenous sediment.

Some of the most common types of hydrogenous sediment include the following:

- *Manganese nodules* are rounded, hard lumps of manganese, iron, and other metals that precipitate in concentric layers around a central object (such as a volcanic pebble or a grain of sand). The nodules can be up to 20 centimeters (8 inches) in diameter and are often littered across large areas of the deep seafloor (Figure 9.19A).
- *Calcium carbonates* form by precipitation directly from seawater in warm climates. If this material is buried and hardened, it forms limestone. Most limestone, however, is composed of biogenous sediment.
- *Metal sulfides* are usually precipitated as coatings on rocks near black smokers associated with the crest of the mid-

Figure 9.19 **A.** Manganese nodules photographed at a depth of 2909 fathoms (5323 meters or 3.3 miles) beneath the *Robert Conrad*, south of Tahiti. (Photo courtesy of Lawrence Sullivan, Lamont-Doherty Earth Observatory/Columbia University) **B.** View from the submersible *Alvin* of a black smoker spewing hot, mineral-rich water along the East Pacific Rise. When heated solutions meet cold seawater, metal sulfides precipitate and form mounds of minerals around these hydrothermal vents. (Photo by Dudley Foster © Woods Hole Oceanographic Institution)

ocean ridge (Figure 9.19B). These deposits contain iron, nickel, copper, zinc, silver, and other metals in varying proportions.

- *Evaporites* form where evaporation rates are high and there is restricted open-ocean circulation. As water evaporates from such areas, the remaining seawater becomes saturated with dissolved minerals, which then begin to precipitate. Heavier than seawater, they sink to the bottom or form a characteristic white crust of evaporite minerals around the edges of these areas. Collectively termed *salts,* some evaporite minerals taste salty, such as *halite* (common table salt, NaCl), and some do not, such as the calcium sulfate minerals *anhydrite* ($CaSO_4$) and *gypsum* ($CaSO_4 \cdot 2H_2O$).

Distribution of Seafloor Sediments

Figure 9.20 shows the distribution of seafloor sediments. Coarse-grained terrigenous deposits dominate continental margin areas, whereas fine-grained terrigenous material

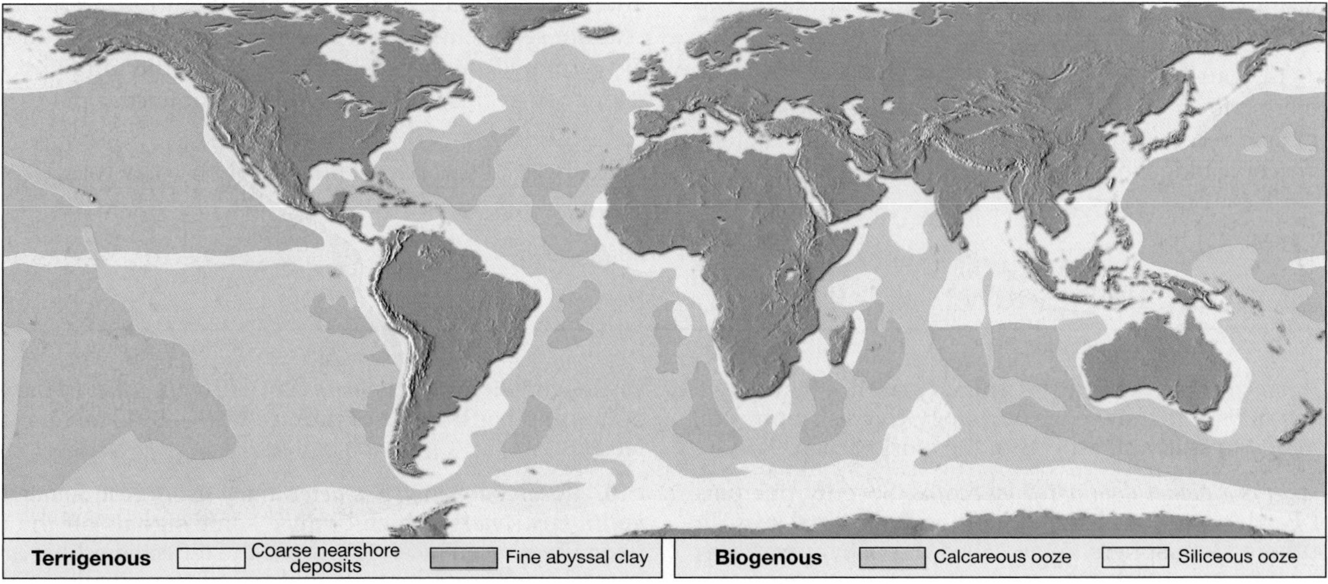

Figure 9.20 Distribution of marine sediment.

Figure 9.21 Scientists examine a sediment core aboard the *JOIDES Resolution*, the drilling ship of the Ocean Drilling Program. The seafloor represents a huge reservoir of data relating to global environmental change. (Photo courtesy of the Ocean Drilling Program, Science Operator, Texas A&M University)

(abyssal clay) is common in deeper areas of the ocean basins. However, deep-ocean deposits are dominated by calcareous oozes, which are found on the shallower portions of deep-ocean areas along the mid-ocean ridge. Siliceous oozes are found beneath areas of unusually high biologic productivity such as the Antarctic and the equatorial Pacific Ocean. Hydrogenous sediment comprises only a small proportion of deposits in the ocean.

There are a few places in the ocean where very little sediment accumulates. One such place is along the continental slope, where there is active erosion by turbidity and other deep-ocean currents. Another place where very little sediment can be found is along the oceanic ridge. Here, the seafloor along the crest of the ridge is so young (because of seafloor spreading), and the rates of sediment accumulation far from land are so slow that there has not been enough time for sediments to accumulate.

Various types of sediment accumulate on nearly all areas of the ocean floor in the same way dust accumulates in all parts of your home (which is why seafloor sediment is often referred to as "marine dust"). Even the deep-ocean floor far from land receives small amounts of windblown material, microscopic biogenous particles, and even space dust.

Seafloor Sediments and Climate Change

Reliable climate records go back only a couple of hundred years, at best. How do scientists learn about climates and climate change prior to that time? The answer is that they must reconstruct past climates from *indirect evidence;* that is, they must analyze phenomena that respond to and reflect changing atmospheric conditions. An interesting and important technique for analyzing Earth's climate history is the study of seafloor sediments.

Most seafloor sediments contain the remains of microscopic organisms that once lived near the sea surface (the ocean–atmosphere interface). When such near-surface organisms die, their tests slowly settle onto the ocean floor, where they can become buried and preserved over time. Thus, the deep-ocean floor has become a repository for sediment representing millions of years of Earth's history.

Seafloor sediments are useful recorders of worldwide climate change because the numbers and types of organisms living near the sea surface change with the climate:

> [W]e would expect that in any area of the ocean/atmosphere interface the average annual temperature of the surface water of the ocean would approximate that of the contiguous atmosphere. The temperature equilibrium established between surface seawater and the air above it should mean that . . . changes in climate should be reflected in changes in organisms living near the surface of the deep sea. . . . When we recall that the seafloor sediments in vast areas of the ocean consist mainly of shells of [oceanic] foraminifers, and that these animals are sensitive to variations in water temperature, the connection between such sediments and climatic change becomes obvious.*

In seeking to understand climate and other environmental changes, scientists are tapping the huge reservoir of data in seafloor sediments. Sediment cores gathered by drilling ships and other research vessels have provided invaluable data that have significantly expanded scientific knowledge and understanding of past climates (Figure 9.21). Analysis of seafloor sediments has revealed periods of Ice Ages and global warming, ocean circulation changes, the timing of major extinction events, and the movement of Earth's plates.

*Richard F. Flint, *Glacial and Quaternary Geology* (New York: Wiley, 1971), p. 718.

The Chapter in Review

1. *Oceanography* is an interdisciplinary science that draws on the methods and knowledge of biology, chemistry, physics, and geology to study all aspects of the world ocean.

2. *Earth is a planet dominated by oceans.* Seventy-one percent of Earth's surface area is oceans and marginal seas. In the Southern Hemisphere, often called the *water hemisphere,* about 81 percent of the surface is water. Of the three major oceans, Pacific, Atlantic, and Indian, the *Pacific Ocean* is the *largest,* contains slightly *more than half of the water* in the world ocean, and has the *greatest average depth*—3940 meters (12,927 feet).

3. *Ocean bathymetry is determined using echo sounders and multibeam sonars,* which bounce sonic signals off the ocean floor. Ship-based receivers record the reflected echoes and accurately measure the time interval of the signals. With this information, ocean depths are calculated and plotted to pro-

duce maps of ocean-floor topography. Recently, *satellite measurements* of the shape of the ocean surface have added data for mapping ocean-floor features.

4. *Salinity* is the proportion of dissolved salts to pure water, usually expressed in parts per thousand (‰). The average salinity in the open ocean ranges from 35‰ to 37‰. The principal elements that contribute to the ocean's salinity are *chlorine* (55‰) and *sodium* (31‰). The primary *sources for the elements in sea salt* are *chemical weathering* of rocks on the continents and *outgassing* through submarine volcanism. Outgassing is also considered to be the principal source of water in the oceans as well as in the atmosphere.

5. *Variations in seawater salinity* are primarily caused by changing the *water content.* Natural processes that add large amounts of fresh water to seawater and *decrease salinity* include *precipitation, runoff from land, icebergs melting,* and *sea ice melting.* Processes that remove large amounts of fresh water from seawater and *increase salinity* include the *formation of sea ice* and *evaporation.* Seawater salinity in the open ocean *ranges from 33‰ to 38‰*, with some marginal seas experiencing considerably more variation.

6. In most regions, open oceans exhibit a *three-layered temperature and salinity structure.* Ocean water temperatures are warmest at the surface because solar energy is absorbed there. Mixing can distribute this heat to a depth of about 450 meters (1500 feet) or more. Beneath the Sun-warmed zone of mixing, a layer of rapid temperature change, called the *thermocline,* occurs. Below the thermocline, in the deep zone, temperatures fall only a few more degrees. Salinity changes with depth correspond to the general three-layered temperature structure. In the low and middle latitudes, a surface zone of higher salinity is underlain by a layer of rapidly decreasing salinity, called the *halocline.* Below the halocline, salinity changes are small.

7. The zones that collectively make up a *passive continental margin* include the *continental shelf* (a gently sloping, submerged surface extending from the shoreline toward the deep-ocean basin), *continental slope* (the true edge of the continent, which has a steep slope that leads from the continental shelf into deep water), and in regions where trenches do not exist, the steep continental slope merges into a gradual incline known as the *continental rise.* The continental rise consists of sediments that have moved downslope from the continental shelf to the deep-ocean floor.

8. *Submarine canyons* are deep, steep-sided valleys that originate on the continental slope and may extend to the deep-ocean basin. Many submarine canyons have been excavated by *turbidity currents* (downslope movements of dense, sediment-laden water).

9. *Active continental margins* are located primarily around the margin of the Pacific Ocean in areas where the leading edge of a continent is overrunning oceanic lithosphere. Here, sediment scraped from the descending oceanic plate is plastered against the continent to form a collection of sediments called an *accretionary wedge.* An active continental margin generally has a narrow continental shelf, which grades into a deep-ocean trench.

10. The *deep-ocean basin* lies between the continental margin and the mid-ocean ridge system. The features of the deep-ocean basin include *deep-ocean trenches* (the deepest parts of the ocean, where moving crustal plates descend into the mantle), *abyssal plains* (very level regions consisting of thick accumulations of sediments that were deposited atop the low, rough portions of the ocean floor by turbidity currents), *seamounts* (isolated volcanic peaks on the ocean floor that originate near mid-ocean ridges or in association with volcanic hot spots), and *oceanic plateaus* (vast accumulations of basaltic lava flows).

11. The *oceanic (mid-ocean) ridge* winds through the middle of most ocean basins. Seafloor spreading occurs near the center of this broad feature, which is characterized by an elevated position, extensive faulting, and volcanic structures that have developed on newly formed oceanic crust. Most of the geologic activity associated with ridges occurs along a narrow region on the ridge crest, called the *rift valley,* where magma moves upward to create new slivers of oceanic crust.

12. *There are three broad categories of seafloor sediments. Terrigenous sediment* consists primarily of mineral grains that were weathered from continental rocks and transported to the ocean; *biogenous sediment* consists of shells and skeletons of marine animals and plants; and *hydrogenous sediment* includes minerals that crystallize directly from seawater through various chemical reactions. The global distribution of marine sediments is affected by proximity to source areas and water temperatures that favor the growth of certain marine organisms.

13. *Seafloor sediments are helpful when studying worldwide climate changes* because they often contain the remains of organisms that once lived near the sea surface. The numbers and types of these organisms change as the climate changes, and their remains in seafloor sediments record these changes.

Key Terms

abyssal plain (p. 258)
active continental margin (p. 253)
bathymetry (p. 250)
biogenous sediment (p. 260)
continental margin (p. 252)
continental rise (p. 253)
continental shelf (p. 252)
continental slope (p. 253)
continental volcanic arc (p. 258)
deep-ocean trench (p. 257)
deep-sea fan (p. 253)
echo sounder (p. 250)
guyot (p. 258)
halocline (p. 250)
hydrogenous sediment (p. 260)
oceanic plateau (p. 258)
oceanic (mid-ocean) ridge (p. 258)
oceanography (p. 246)
outgassing (p. 248)
passive continental margin (p. 252)
rift valley (p. 259)
salinity (p. 247)
seamount (p. 258)

seismic reflection profile (p. 251)
sonar (p. 250)
submarine canyon (p. 253)
tablemount (p. 258)
terrigenous sediment (p. 260)
thermocline (p. 249)
turbidity current (p. 253)
volcanic island arc (p. 258)

Questions for Review

1. How does the area of Earth's surface covered by the oceans compare with the area of the continents? Describe the distribution of land and water on Earth.
2. Name the four main ocean basins. Of the four:
   a. Which one is the largest in area? Which one is smallest?
   b. Which one is the deepest? Which one is shallowest?
   c. Which one is almost entirely within the Southern Hemisphere?
   d. Which one is exclusively in the Northern Hemisphere?
3. How does the average depth of the ocean compare to the average elevation of the continents?
4. What is meant by *salinity*? What is the average salinity of the ocean?
5. What are the six most abundant components (elements) dissolved in seawater? What is produced when the two most abundant elements are combined?
6. What are the two primary sources for the materials that comprise the dissolved components in seawater?
7. Describe the processes that affect seawater salinity. For each process, indicate whether water is added or removed and if it decreases or increases salinity. What physical conditions create high-salinity water in the Red Sea and low-salinity water in the Baltic Sea?
8. Assuming that the average speed of sound waves in water is 1500 meters per second, determine the water depth if the signal sent out by an echo sounder requires 6 seconds to strike bottom and return to the recorder (see Figure 9.8, p. 251).
9. Describe how satellites orbiting Earth can determine features on the seafloor without being able to directly observe them beneath several kilometers of seawater.
10. List the three subdivisions of a passive continental margin. Which subdivision is considered a flooded extension of the continent? Which has the steepest slope?
11. Describe the differences between active and passive continental margins. Be sure to include how various features relate to plate tectonics and give a geographic example of each type of margin.
12. Defend or rebut the statement "Most submarine canyons found on the continental slope and rise were formed during the Ice Age when rivers extended their valleys seaward."
13. Why are abyssal plains more extensive on the floor of the Atlantic than on the floor of the Pacific?
14. How are mid-ocean ridges and deep-ocean trenches related to plate tectonics?
15. Distinguish among the three basic types of seafloor sediment.
16. Why are seafloor sediments useful in studying climates of the past?

Online Study Guide

The *Foundations of Earth Science* Web site uses the resources and flexibility of the Internet to aid in your study of the topics in this chapter. Written and developed by Earth science instructors, this site will help improve your understanding of Earth science. Visit http://www.prenhall.com/lutgens and click on the cover of *Foundations of Earth Science 5e* to find:

- Online review quizzes.
- Critical thinking exercises.
- Links to chapter-specific Web resources.
- Internet-wide key-term searches.

http://www.prenhall.com/lutgens

GEODe: Earth Science

GEODe: Earth Science makes studying more effective by reinforcing key concepts using animation, video, narration, interactive exercises, and practice quizzes. A copy is included with every copy of *Foundations of Earth Science 5e.*

Passive margins are found along most of the coastal areas that surround the Atlantic Ocean.

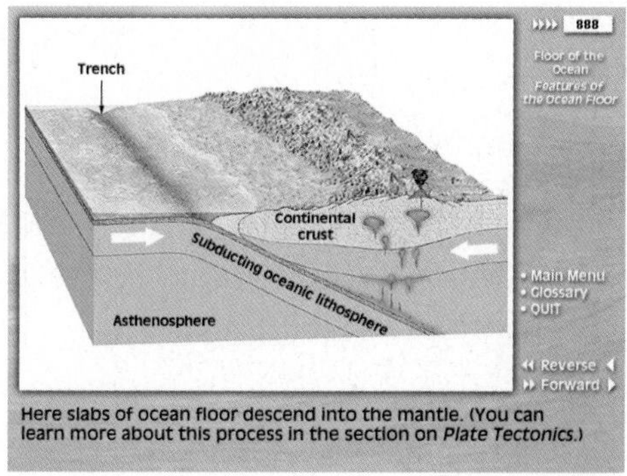

Here slabs of ocean floor descend into the mantle. (You can learn more about this process in the section on *Plate Tectonics*.)

The shore of Kaho'olawe, a small island south of Maui (in background) in the Hawaiian Islands. (Photo by David Muench)

The Restless Ocean

FOCUS ON LEARNING

To assist you in learning the important concepts in this chapter, you will find it helpful to focus on the following questions:

1. What forces create and influence surface ocean currents?
2. What two factors are most significant in creating a dense mass of ocean water?
3. What factors determine the height, length, and period of a wave?
4. What are some typical shoreline features produced by wave erosion and from sediment deposited by beach drift and longshore currents?
5. What is the difference between a submergent and an emergent coast?
6. How are tides produced?

The restless waters of the ocean are constantly in motion, powered by many different forces. Winds, for example, generate surface currents, which influence coastal climate and provide nutrients that affect the abundance of algae and other marine life in surface waters. Winds also produce waves that carry energy from storms to distant shores, where their impact erodes the land (Figure 10.1). In some areas, density differences create deep-ocean circulation, which is important for ocean mixing and nutrient recycling. In addition, the Moon and the Sun produce tides, which periodically raise and lower sea level. This chapter will examine these movements of ocean waters and their effects upon coastal regions.

Surface Circulation

Ocean currents are masses of ocean water that flow from one place to another. The amount of water can be large or small, currents can be at the surface or deep below, and the phenomena that create them can be simple or complex. In all cases, however, the currents that are generated involve water masses in motion.

Surface currents develop from friction between the ocean and the wind that blows across its surface. Some of these currents are short-lived and affect only small areas. Such water movements are responses to local or seasonal influences. Other surface currents are relatively permanent phenomena that extend over large portions of the oceans. These major horizontal movements of surface waters are closely related to the general circulation pattern of the atmosphere. This is clearly illustrated by comparing the pattern of global winds shown in Figure 13.16A (p. 361) and the position of Earth's principal surface ocean currents shown in Figure 10.2.

Ocean Circulation Patterns

Huge circular-moving current systems dominate the surfaces of the oceans. These large whirls of water within an ocean basin are called **gyres** (*gyros* = a circle). Figure 10.2 shows the world's five main gyres: the *North Pacific Gyre*, the *South Pacific Gyre*, the *North Atlantic Gyre*, the *South Atlantic Gyre*, and the *Indian Ocean Gyre* (which exists mostly within the Southern Hemisphere). The center of each gyre coincides with the subtropics at about 30 degrees north or south latitude, so they are often called *subtropical gyres*. Comparing Figure 13.16A to Figure 10.2 shows that there is a striking correspondence between the direction of surface-current flow and the major wind belts of the world.

As shown in Figure 10.2, subtropical gyres rotate clockwise in the Northern Hemisphere and counterclockwise in the Southern Hemisphere. Why do the gyres flow in different directions in the two hemispheres? Although wind is the force that generates surface currents, other factors also influence the movement of ocean waters. The most significant of these is the **Coriolis effect.** Because of Earth's rotation, currents are deflected to the *right* in the Northern Hemisphere and to the *left* in the Southern Hemisphere. (The Coriolis effect is more fully explained in Chapter 13.) As a consequence, gyres flow in opposite directions in the two different hemispheres.

Four main currents generally exist within each gyre (Figure 10.2). The North Pacific Gyre, for example, consists of the North Equatorial Current, the Kuroshio* Current, the North Pacific Current, and the California Current. The track-

*Kuroshio is pronounced "kuhr-ROH-shee-oh" and is sometimes known as the Japan Current. The term *Kuroshio* is Japanese for "black tide," in reference to its clear, lifeless waters.

Figure 10.1 Wind is not only responsible for creating waves, but it also provides the force that drives the ocean's surface circulation. (Photo by Bob Barbour/Minden Pictures)

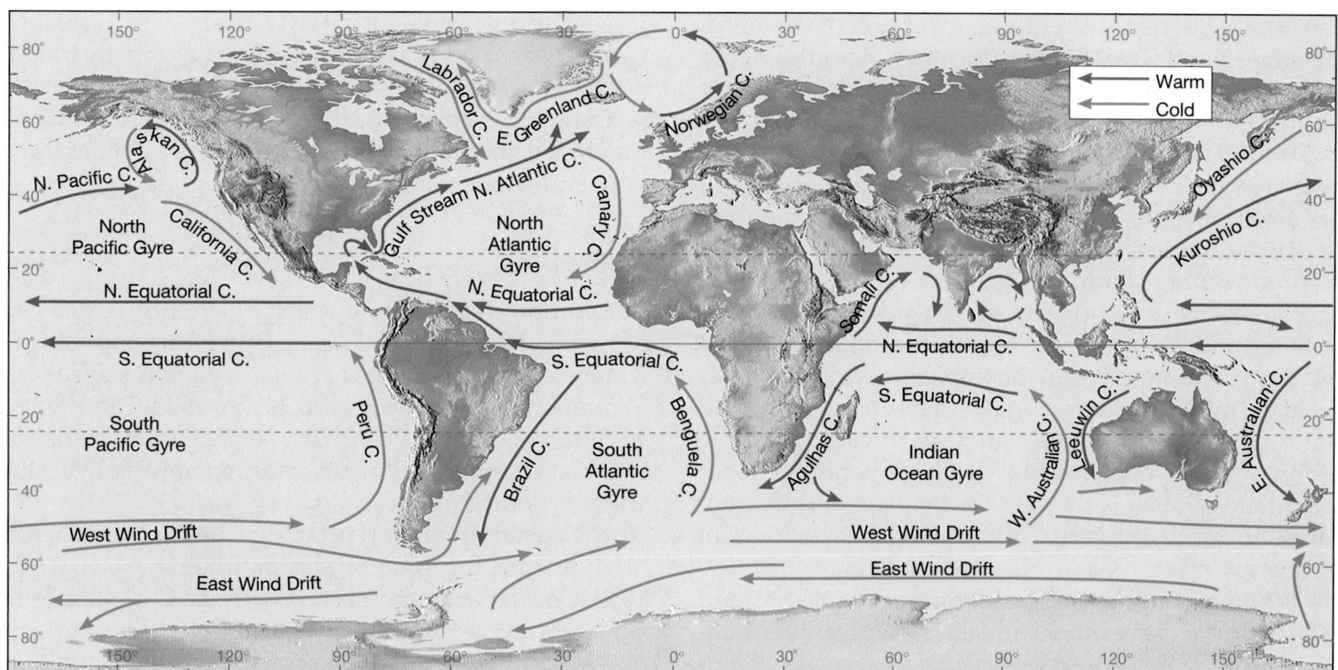

Figure 10.2 Average ocean surface currents in February–March. The ocean's circulation is organized into five major current gyres (large circular-moving loops of water), which exist in the North Pacific, South Pacific, North Atlantic, South Atlantic, and Indian oceans.

ing of floating objects that are released into the ocean intentionally or accidentally reveals that it takes about six years for the objects to go all the way around the loop.

In the North Atlantic, the North Equatorial Current is deflected northward through the Caribbean, where it becomes the Gulf Stream (Figure 10.2). As the Gulf Stream moves along the East Coast of the United States, it is strengthened by the prevailing westerly winds and is deflected to the east (to the right) between the Carolinas and New England. As it continues northeastward, it gradually widens and slows until it becomes a vast, slowly moving current known as the North Atlantic Current, which, because of its sluggish nature, is also known as the North Atlantic Drift.

As the North Atlantic Current approaches Western Europe, it splits, part of it moving northward past Great Britain, Norway, and Iceland, carrying heat to these otherwise chilly areas. The other part is deflected southward as the cool Canary Current. As the Canary Current moves southward, it eventually merges into the North Equatorial Current, completing the gyre. Because the North Atlantic Ocean basin is about half the size of the North Pacific, it takes floating objects about three years to go completely around this gyre.

The circular motion of gyres leaves a large central area that has no well-defined currents. In the North Atlantic, this zone of calmer waters is known as the Sargasso Sea, named for the large quantities of *Sargassum,* a type of floating seaweed encountered there.

The ocean basins in the Southern Hemisphere exhibit a similar pattern of flow as the Northern Hemisphere basins, with surface currents that are influenced by wind belts, the position of continents, and the Coriolis effect. In the South Atlantic and South Pacific, for example, surface ocean circulation is very much the same as in their Northern Hemisphere counterparts except that the direction of flow is counterclockwise (Figure 10.2).

The Indian Ocean exists mostly in the Southern Hemisphere, so it follows a surface circulation pattern similar to other Southern Hemisphere ocean basins (Figure 10.2). The small portion of the Indian Ocean in the Northern Hemisphere, however, is influenced by the seasonal wind shifts known as the summer and winter *monsoons* (*mausim* = season). When the winds change direction, the surface currents also reverse direction.

The West Wind Drift is the only current that completely encircles Earth (Figure 10.2). It flows around the ice-covered continent of Antarctica, where no large landmasses are in the way, so its cold surface waters circulate in a continuous loop. It moves in response to the Southern Hemisphere's prevailing westerly winds, and portions of it split off into the adjoining southern ocean basins.

Ocean Currents and Climate

Ocean currents have an important effect on climates. When currents from low-latitude regions move into higher latitudes, they transfer heat from warmer to cooler areas on Earth. In

Did You Know?

In 1768, as deputy postmaster of the colonies, Benjamin Franklin, together with a Nantucket ship captain, produced the first map of the Gulf Stream. His interest began when he realized that ships carrying the mail took two weeks longer going from England to America than in the other direction.

fact, the North Atlantic Current—an extension of the warm Gulf Stream—keeps Great Britain and much of northwestern Europe warmer during the winter than one would expect for their latitudes, which are similar to the latitudes of Alaska and Labrador. The prevailing westerly winds carry the moderating effects far inland. For example, Berlin, Germany (52 degrees north latitude), has an average January temperature similar to that experienced in New York City, which lies 12 degrees latitude farther south.

In contrast to warm ocean currents whose effects are felt mostly in the middle latitudes in winter, the influence of cold currents is most pronounced in the tropics or during summer months in the middle latitudes. Cold currents originate in cold high-latitude regions. As these currents travel toward the equator, they tend to moderate the warm temperatures of adjacent land areas. Such is the case for the Benguela Current along western Africa, the Peru Current along the west coast of South America, and the California Current (Figure 10.2).

In addition to influencing temperatures of adjacent land areas, cold currents have other climatic influences. For example, where tropical deserts exist along the west coasts of continents, cold ocean currents have a dramatic impact. The principal west coast deserts are the Atacama in Peru and Chile, and the Namib in southwestern Africa. The aridity along these coasts is intensified because the lower atmosphere is chilled by cold offshore waters. When this occurs, the air becomes very stable and resists the upward movement necessary to create precipitation-producing clouds. In addition, the presence of cold currents causes temperatures to approach and often reach the dew point, the temperature at which water vapor condenses. As a result, these areas are characterized by high relative humidities and much fog. Thus, not all tropical deserts are hot with low humidities and clear skies. Rather, the presence of cold currents transforms some tropical deserts into relatively cool, damp places that are often shrouded in fog.

Ocean currents also play a major role in maintaining Earth's heat balance. They accomplish this task by transferring heat from the tropics, where there is an excess of heat, to the polar regions, where a heat deficit exists (Figure 10.3). Ocean water movement accounts for about a quarter of this heat transport, and winds transport the remaining three-quarters.

Upwelling

In addition to producing surface currents, winds can also cause *vertical* water movements. **Upwelling,** the rising of cold water from deeper layers to replace warmer surface water, is a common wind-induced vertical movement. One type of upwelling, called *coastal upwelling,* is most characteristic along the west coasts of continents, most notably along California, western South America, and West Africa.

Coastal upwelling occurs in these areas when winds blow toward the equator and parallel to the coast (Figure 10.4). Coastal winds combined with the Coriolis effect cause surface water to move away from shore. As the surface layer moves away from the coast, it is replaced by water that "upwells" from below the surface. This slow upward movement of water from depths of 50 to 300 meters (165 to 1000 feet) brings water that is cooler than the original surface water and results in lower surface water temperatures near the shore.

For swimmers who are accustomed to the warm waters along the mid-Atlantic shore of the United States, a swim in the Pacific off the coast of central California can be a chilling surprise. In August, when temperatures in the Atlantic are 21°C (70°F) or higher, central California's surf is only about 15°C (60°F).

Upwelling brings greater concentrations of dissolved nutrients, such as nitrates and phosphates, to the ocean surface. These nutrient-enriched waters from below promote the growth of microscopic plankton, which in turn support ex-

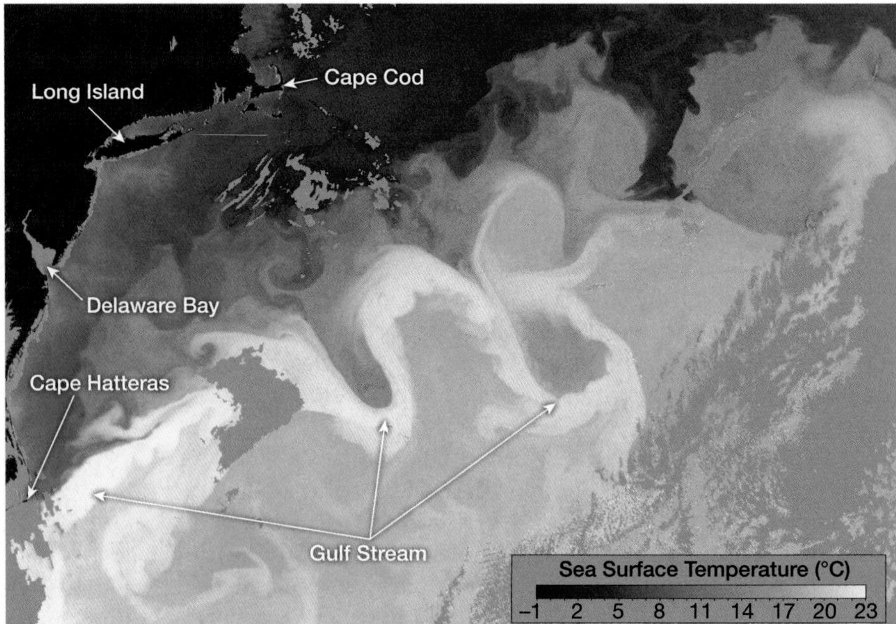

Figure 10.3 This satellite image shows sea-surface temperatures off the East Coast of the United States on April 18, 2005. The warm waters of the Gulf Stream (lighter shades) extend diagonally from bottom left to top right. The current transports heat from the tropics far into the North Atlantic. This image shows several deep bends in the path of the Gulf Stream. In fact, the northernmost of the two deep bends actually loops back on itself, creating a closed-off eddy. On the northern side of the current, cold waters (blue) dip southward into the Gulf Stream's warmth. Gray areas indicate clouds. (NASA)

Figure 10.4 Coastal upwelling occurs along the west coasts of continents where winds blow toward the equator and parallel to the coast. The Coriolis effect (deflection to the left in the Southern Hemisphere) causes surface water to move away from the shore, which brings cold, nutrient-rich water to the surface. This image from the SeaStar satellite shows chlorophyll concentration along the southwest coast of Africa (February 21, 2001). An instrument aboard the satellite detects changes in seawater color caused by changing concentrations of chlorophyll. High chlorophyll concentrations indicate high amounts of photosynthesis, which is linked to the upwelling nutrients. Red indicates high concentrations and blue low concentrations. (Provided by the SeaWiFS Project, NASA/Goddard Space Flight Center and ORBIMAGE)

tensive populations of fish and other marine organisms. Figure 10.4 is a satellite image that shows high productivity due to coastal upwelling off the southwest coast of Africa.

Deep-Ocean Circulation

In contrast to the largely horizontal movements of surface currents, deep-ocean circulation has a significant vertical component and accounts for the thorough mixing of deep-water masses. This component of ocean circulation is a response to density differences among water masses that cause denser water to sink and slowly spread out beneath the surface. Because the density variations that cause deep-ocean circulation are caused by differences in temperature and salinity, deep-ocean circulation is also referred to as **thermohaline** (*thermo* = heat, *haline* = salt) **circulation.**

An increase in seawater density can be caused by either a decrease in temperature or an increase in salinity. Density changes due to salinity variations are important in very high latitudes, where water temperature remains low and relatively constant.

Most water involved in deep-ocean currents (thermohaline circulation) begins in high latitudes at the surface. In these regions, surface water becomes cold and its salinity increases as sea ice forms (Figure 10.5). When this surface water becomes dense enough, it sinks, initiating deep-ocean currents. Once this water sinks, it is removed from the physical processes that increased its density in the first place, and so its temperature and salinity remain largely unchanged for the duration of the time it spends in the deep ocean.

Near Antarctica, surface conditions create the highest density water in the world. This cold saline brine slowly sinks to the seafloor, where it moves throughout the ocean basins in sluggish currents. After sinking from the surface of the ocean, deep waters will not reappear at the surface for an average of 500 to 2000 years.

A simplified model of ocean circulation is similar to a conveyor belt that travels from the Atlantic Ocean through the Indian and Pacific oceans and back again (Figure 10.6). In this model, warm water in the ocean's upper layers flows poleward, converts to dense water, and returns toward the equator as cold deep water that eventually upwells to complete the circuit. As this "conveyor belt" moves around the globe, it influences global climate by converting warm water to cold and liberating heat to the atmosphere.

Did You Know?

About half of the world's human population lives on or within 100 kilometers (60 miles) of a coast. The proportion of the U.S. population residing within 75 kilometers (45 miles) of a coast in 2010 is projected to be well in excess of 50 percent. The concentration of such large numbers of people so near the shore means that hurricanes and tsunamis place millions at risk.

The Shoreline: A Dynamic Interface

Shorelines are dynamic environments. Their topography, geologic makeup, and climate vary greatly from place to place. Continental and oceanic processes converge along coasts to create landscapes that frequently undergo rapid change. When it comes to the deposition of sediment, they are transition zones between marine and continental environments.

Figure 10.5 Sea ice in the Arctic Ocean. When seawater freezes, sea salts do not become part of the ice. Consequently, the salt content of the remaining seawater becomes more concentrated, which makes it denser and prone to sink. (Photo by Wayne Lynch/DRK Photo)

Nowhere is the restless nature of the ocean's water more noticeable than along the shore—the dynamic interface among air, land, and sea. An *interface* is a common boundary where different parts of a system interact. This is certainly an appropriate designation for the coastal zone, where we can see the rhythmic rise and fall of tides and observe waves rolling in and breaking. Sometimes, the waves are low and gentle. At other times, they pound the shore with awesome fury.

Although it may not be obvious, the shoreline is constantly being modified by waves. Crashing surf can erode the adjacent land. Wave activity also moves sediment toward and away from the shore, as well as along it. Such activity sometimes produces narrow sandbars that frequently change size and shape as storm waves come and go.

The nature of present-day shorelines is not just the result of the relentless attack of the land by the sea. Indeed, the shore has a complex character that results from multiple geologic processes. For example, practically all coastal areas were affected by the worldwide rise in sea level that accompanied the melting of glaciers at the close of the Pleistocene epoch. As the sea encroached landward, the shoreline retreated, becoming superimposed upon existing landscapes that had resulted from such diverse processes as stream erosion, glaciation, volcanic activity, and the forces of mountain building.

Today, the coastal zone is experiencing intensive human activity. Unfortunately, people often treat the shoreline as if it were a stable platform on which structures can safely be built. This attitude inevitably leads to conflicts between people and nature. As you will see, many coastal landforms, especially beaches and barrier islands, are relatively fragile, short-lived features that are inappropriate sites for development.

Waves

The Global Ocean
▼ Coastal Processes

Ocean waves are energy traveling along the interface between ocean and atmosphere, often transferring energy from a storm far out at sea over distances of several thousand kilometers. That's why even on calm days the ocean still has waves that

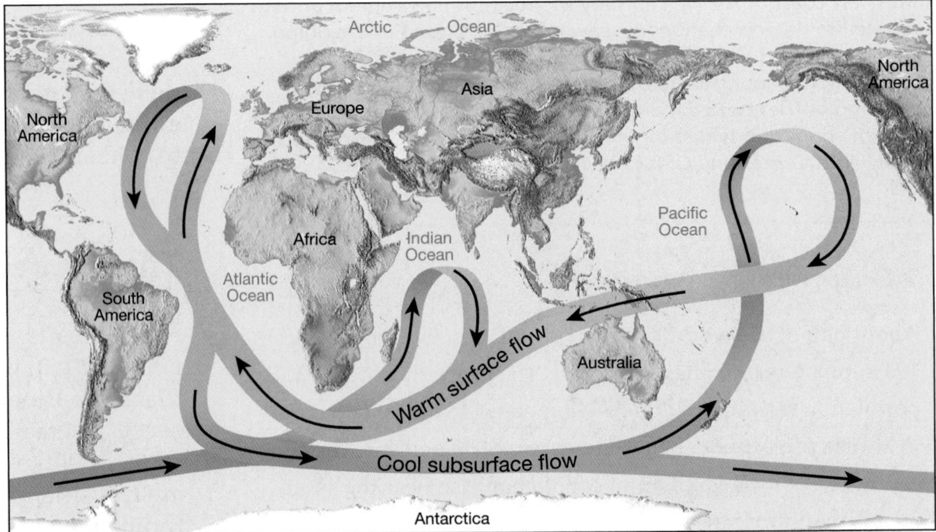

Figure 10.6 Idealized "conveyor belt" model of ocean circulation, which is initiated in the North Atlantic Ocean when warm water transfers its heat to the atmosphere, cools, and sinks below the surface. This water moves southward as a subsurface flow and joins water that encircles Antarctica. From here, this deep water spreads into the Indian and Pacific oceans, where it slowly rises and completes the conveyer as it travels along the surface into the North Atlantic Ocean.

travel across its surface. When observing waves, always remember that you are watching *energy* travel through a medium (water). If you make waves by tossing a pebble into a pond, or by splashing in a pool, or by blowing across the surface of a cup of coffee, you are imparting *energy* to the water, and the waves you see are just the visible evidence of the energy passing through.

Wind-generated waves provide most of the energy that shapes and modifies shorelines. Where the land and sea meet, waves that may have traveled unimpeded for hundreds or thousands of kilometers suddenly encounter a barrier that will not allow them to advance farther and must absorb their energy. Stated another way, the shore is the location where a practically irresistible force confronts an almost immovable object. The conflict that results is never ending and sometimes dramatic.

Wave Characteristics

Most ocean waves derive their energy and motion from the wind. When a breeze is less than 3 kilometers (2 miles) per hour, only small wavelets appear. At greater wind speeds, more stable waves gradually form and advance with the wind.

Characteristics of ocean waves are illustrated in Figure 10.7, which shows a simple, nonbreaking waveform. The tops of the waves are the *crests,* which are separated by *troughs.* Halfway between the crests and troughs is the *still water level,* which is the level that the water would occupy if there were no waves. The vertical distance between trough and crest is called the **wave height,** and the horizontal distance between successive crests (or troughs) is the **wavelength.** The time it takes one full wave—one wavelength—to pass a fixed position is the **wave period.**

The height, length, and period that are eventually achieved by a wave depend on three factors: (1) wind speed; (2) length of time the wind has blown; and (3) **fetch,** the distance that the wind has traveled across open water. As the quantity of energy transferred from the wind to the water increases, both the height and steepness of the waves increase. Eventually, a critical point is reached where waves grow so tall that they topple over, forming ocean breakers called *whitecaps.*

For a particular wind speed, there is a maximum fetch and duration of wind beyond which waves will no longer increase in size. When the maximum fetch and duration are reached for a given wind velocity, the waves are said to be "fully developed." The reason that waves can grow no further is that they are losing as much energy through the breaking of whitecaps as they are receiving from the wind.

When the wind stops or changes direction, or the waves leave the storm area where they were created, they continue on without relation to local winds. The waves also undergo a gradual change to *swells,* which are lower in height and longer in length and may carry a storm's energy to distant shores. Because many independent wave systems exist at the same time, the sea surface acquires a complex and irregular pattern, sometimes producing very large waves. The sea waves that are seen from shore are usually a mixture of swells from faraway storms and waves created by local winds.

Circular Orbital Motion

Waves can travel great distances across ocean basins. In one study, waves generated near Antarctica were tracked as they traveled through the Pacific Ocean basin. After more than 10,000 kilometers (over 6000 miles), the waves finally expended their energy a week later along the shoreline of the Aleutian Islands of Alaska. The water itself does not travel the entire distance, but the waveform does. As the wave travels, the water passes the energy along by moving in a circle. This movement is called *circular orbital motion.*

Observation of an object floating in waves reveals that it moves not only up and down but also slightly forward and backward with each successive wave. Figure 10.8 shows that a floating object moves up and backward as the crest approaches, up and forward as the crest passes, down and forward after the crest, down and backward as the trough

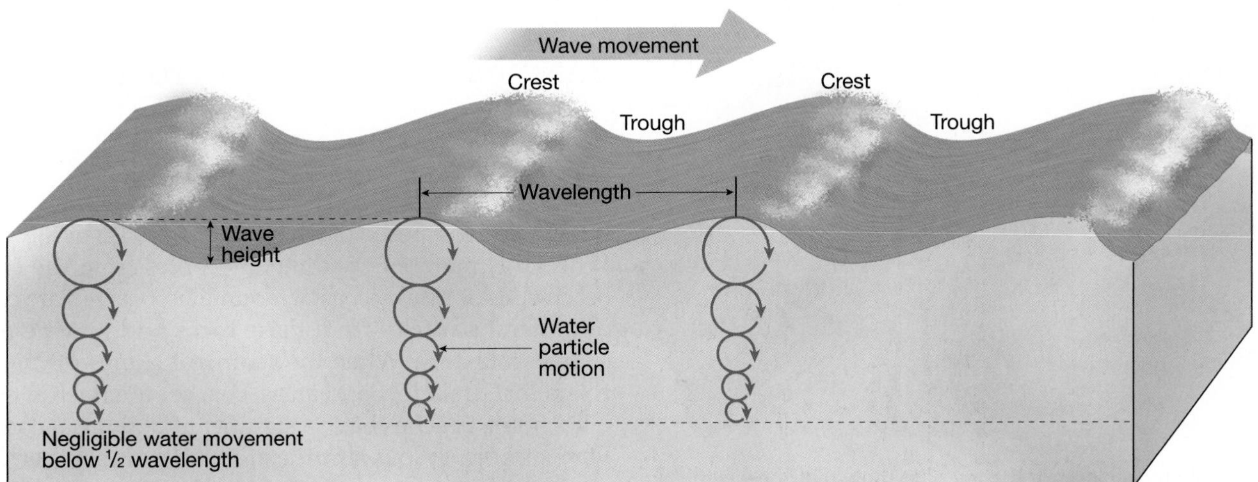

Figure 10.7 Diagrammatic view of an idealized nonbreaking ocean wave showing the basic parts of a wave as well as the movement of water particles at depth. Negligible water movement occurs below a depth equal to one-half the wavelength (*lower dashed line*).

approaches, and then it rises and moves backward again as the next crest advances. When the movement of the toy boat shown in Figure 10.8 is traced as a wave passes, it can be seen that the boat moves in a circle and it returns to essentially the same place. Circular orbital motion allows a waveform (the wave's shape) to move forward *through the water* while the individual water particles that transmit the wave move around in a circle. Wind moving across a field of wheat causes a similar phenomenon: The wheat itself does not travel across the field, but the waves do.

The energy contributed by the wind to the water is transmitted not only along the surface of the sea but also downward. However, beneath the surface the circular motion rapidly diminishes until, at a depth equal to one-half the wavelength measured from still water level, the movement of water particles becomes negligible. This depth is known as the *wave base.* The dramatic decrease of wave energy with depth is shown by the rapidly diminishing diameters of water-particle orbits in Figure 10.7.

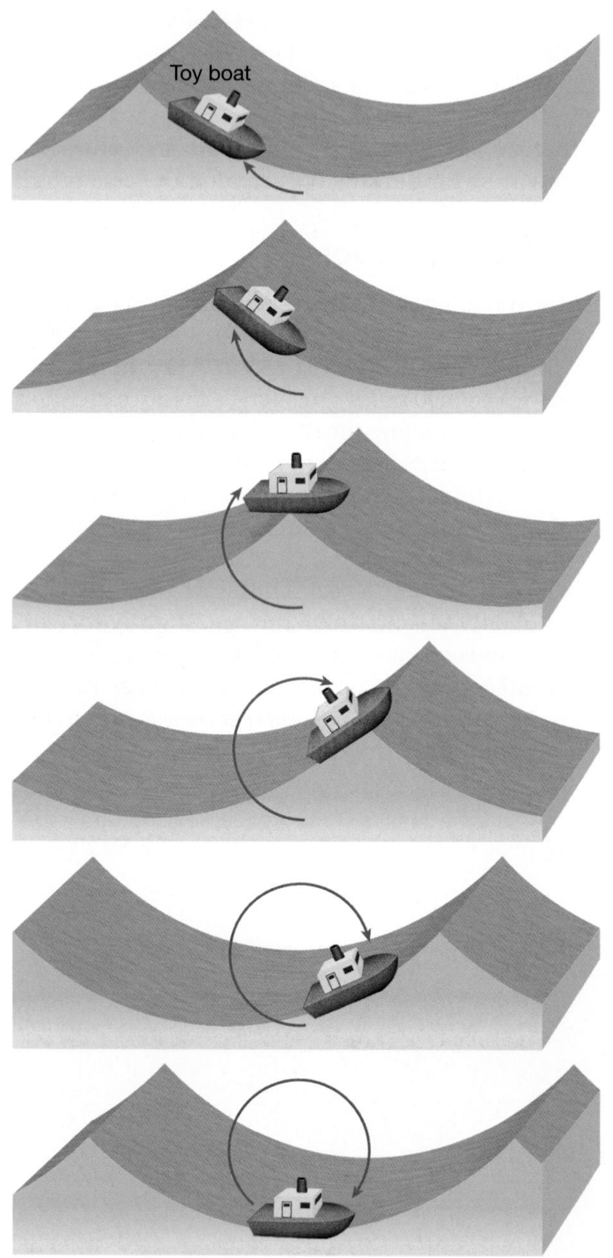

Figure 10.8 The movements of the toy boat show that the wave form advances, but the water does not advance appreciably from its original position. In this sequence, the wave moves from left to right as the boat (and the water in which it is floating) rotates in an imaginary circle.

Did You Know?

Rip currents are strong narrow surface or near-surface currents flowing seaward through the surf zone at nearly right angles to the shore. They represent the return of water that has been piled up on shore by incoming waves and can attain speeds of 7 to 8 kilometers (4 to 5 miles) per hour. Because they move faster than most people can swim, rip currents pose a hazard to swimmers.

Waves in the Surf Zone

As long as a wave is in deep water, it is unaffected by water depth (Figure 10.9, *left*). However, when a wave approaches the shore, the water becomes shallower and influences wave behavior. The wave begins to "feel bottom" at a water depth equal to its wave base. Such depths interfere with water movement at the base of the wave and slow its advance (Figure 10.9, *center*).

As a wave advances toward the shore, the slightly faster waves farther out to sea catch up, decreasing the wavelength. As the speed and length of the wave diminish, the wave steadily grows higher. Finally, a critical point is reached when the wave is too steep to support itself and the wave front collapses, or *breaks* (Figure 10.9, *right*), causing water to advance up the shore.

The turbulent water created by breaking waves is called **surf.** On the landward margin of the surf zone, the turbulent sheet of water from collapsing breakers, called *swash*, moves up the slope of the beach. When the energy of the swash has been expended, the water flows back down the beach toward the surf zone as *backwash.*

Beaches and Shoreline Processes

The Global Ocean
▼ Coastal Processes

Beaches are composed of whatever material is locally available. When this material—sediment—comes from the erosion of beach cliffs or nearby coastal mountains, beaches are composed of mineral particles from these rocks and may be relatively coarse in texture. When the sediment comes primarily from rivers that drain lowland areas, beaches are finer in texture.

Some beaches have a significant biological component. For example, in low-latitude areas such as southern Florida, where there are no mountains or other sources of rock-forming minerals nearby, most beaches are composed of shell fragments and the remains of organisms that live in coastal waters. Some beaches on volcanic islands in the open ocean

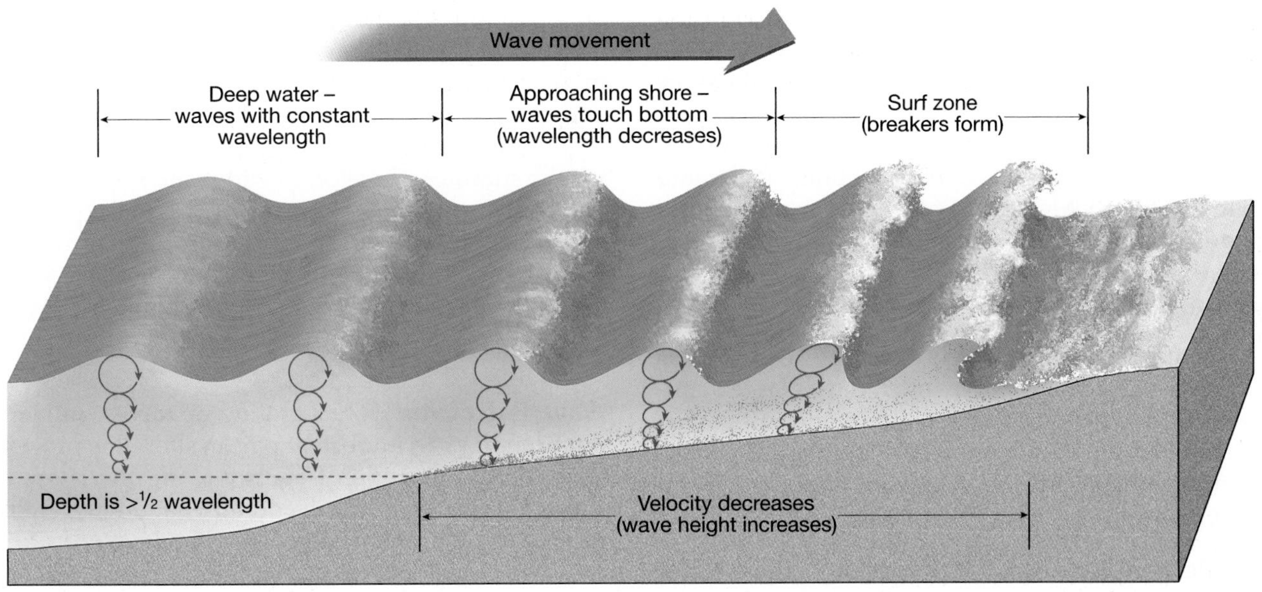

Figure 10.9 Changes that occur when a wave moves onto shore. The waves touch bottom as they encounter water depths less than half a wavelength. The wave speed decreases, and the waves stack up against the shore, causing the wavelength to decrease. This results in an increase in wave height to the point where the waves pitch forward and break in the surf zone.

are composed of black or green fragments of the basaltic lava that comprises the islands, or of coarse debris from coral reefs that develop around some islands in the low latitudes.

Regardless of the composition, the material that comprises the beach does not stay in one place. Instead, the waves that crash along the shoreline are constantly moving it. Thus, beaches can be thought of as material in transit along the shoreline.

Wave Erosion

During calm weather, wave action is minimal. During storms, however, waves can cause much erosion. The impact of large, high-energy waves against the shore can be awesome in its violence. Each breaking wave may hurl thousands of tons of water against the land, sometimes causing the ground literally to tremble. The pressures exerted by Atlantic waves in winter, for example, average nearly 10,000 kilograms per square meter (more than 2000 pounds per square foot). The force during storms is even greater.

It is no wonder that cracks and crevices are quickly opened in cliffs, coastal structures, and anything else that is subjected to these enormous shocks (Figure 10.10). Water is forced into every opening, causing air in the cracks to become highly compressed by the thrust of crashing waves. When the wave subsides, the air expands rapidly, dislodging rock fragments and enlarging and extending preexisting fractures.

In addition to the erosion caused by wave impact and pressure, **abrasion,** the sawing and grinding action of the water

Figure 10.10 The force of breaking waves can be powerful. Here, waves batter a seawall at Sea Bright, New Jersey. A seawall 5 to 6 meters (16 to 20 feet) high and 8 kilometers (5 miles) long was built to protect the town and the railroad that brought tourists to the beach. As you can see, after the wall was built, the beach narrowed dramatically. (Photo by Rafael Macia/Photo Researchers, Inc.)

armed with rock fragments, is also important. In fact, abrasion is probably more intense in the surf zone than in any other environment. Smooth rounded stones and pebbles along the shore are obvious reminders of the relentless grinding action of rock against rock in the surf zone (Figure 10.11A). Such fragments are also used as "tools" by the waves as they cut horizontally into the land (Figure 10.11B). Moreover, waves are very effective at breaking down rock material and supplying sand to beaches.

Sand Movement on the Beach

Beaches are sometimes called "rivers of sand." The reason is that the energy from breaking waves often causes large quantities of sand to move along the beach face and in the surf zone roughly parallel to the shoreline. Wave energy also causes sand to move perpendicular to (toward and away from) the shoreline.

Movement Perpendicular to the Shoreline If you stand ankle-deep in water at the beach, you will see that swash and backwash move sand toward and away from the shoreline. Whether there is a net loss or addition of sand depends on the level of wave activity. When wave activity is relatively light (less energetic waves), much of the swash soaks into the beach, which reduces the backwash. Consequently, the swash dominates and causes a net movement of sand up the beach face.

When high-energy waves prevail, the beach is saturated from previous waves, so much less of the swash soaks in. As a result, the beach erodes because backwash is strong and causes a net movement of sand toward open water.

Along many beaches, light wave activity is the rule during the summer. Therefore, a wide sandy beach gradually develops. During winter, when storms are frequent and more powerful, strong wave activity erodes and narrows the beach. A wide beach that may have taken months to build can be dramatically narrowed in just a few hours by the high-energy waves created by a strong winter storm.

Wave Refraction The bending of waves, called **wave refraction** plays an important part in shoreline processes. It affects the distribution of energy along the shore and thus strongly influences where and to what degree erosion, sediment transport, and deposition will take place.

Waves seldom approach the shore straight on. Rather, most waves move toward the shore at a slight angle. However, when they reach the shallow water of a smoothly sloping bottom, the wave crests are refracted (bent) and tend to line up nearly parallel to the shore. Such bending occurs because the part of the wave nearest the shore touches bottom and slows first, whereas the part of the wave that is still in deep water continues forward at its full speed. This causes wave

A.

B.

Figure 10.11 **A.** Abrasion can be intense in the surf zone. Smooth rounded stones along the shore are an obvious reminder of this fact. Garrapata State Park, California (Photo © by Carr Clifton) **B.** Sandstone cliff undercut by wave erosion at Gabriola Island, British Columbia, Canada. (Photo by Fletcher and Baylis/Photo Researchers, Inc.)

crests to become nearly parallel to the shore regardless of their original orientation.

Because of refraction, wave energy is concentrated against the sides and ends of headlands that project into the water, whereas wave attack is weakened in bays. This differential wave attack along irregular coastlines is illustrated in Figure 10.12. Because the waves reach the shallow water in front of the headland sooner than they do in adjacent bays, they are bent more nearly parallel to the protruding land and strike it from all three sides. By contrast, refraction in the bays causes waves to diverge and expend less energy. In these zones of weakened wave activity, sediments can accumulate and form sandy beaches. Over a long period, erosion of the headlands and deposition in the bays will straighten an irregular shoreline.

Longshore Transport Although waves are refracted, most still reach the shore at a slight angle. Consequently, the uprush of water from each breaking wave (the swash) is at an oblique angle to the shoreline. However, the backwash is straight down the slope of the beach. The effect of this pattern of water movement is to transport sediment in a zigzag pattern along the beach face (Figure 10.13). This movement is called **beach drift,** and it can transport sand and pebbles hundreds or even thousands of meters daily. However, a more typical rate is 5 to 10 meters (16 to 33 feet) per day.

Oblique waves also produce currents within the surf zone that flow parallel to the shore and move substantially more sediment than beach drift (Figure 10.13). Because the water here is turbulent, these **longshore currents** easily move the fine suspended sand and roll larger sand and gravel along the bottom. When the sediment transported by longshore currents is added to the quantity moved by beach drift, the total amount can be very large. At Sandy Hook, New Jersey, for example, the quantity of sand transported along the shore over a 48-year period averaged almost 680,000 metric tons (750,000 short tons) annually. For a 10-year period at Oxnard, California, more than 1.4 million metric tons (1.5 million short tons) of sediment moved along the shore each year.

Both rivers and coastal zones move water and sediment from one area (*upstream*) to another (*downstream*). As a result, the beach is often characterized as a "river of sand." Beach drift and longshore currents, however, move in a zigzag pattern, whereas rivers flow mostly in a turbulent, swirling fashion. Additionally, the direction of flow of longshore currents along a shoreline can change, whereas rivers flow in the same direction (downhill). Longshore currents change direction because waves approach the beach in different directions each season. Nevertheless, longshore currents generally flow southward along both the Atlantic and Pacific shores of the United States.

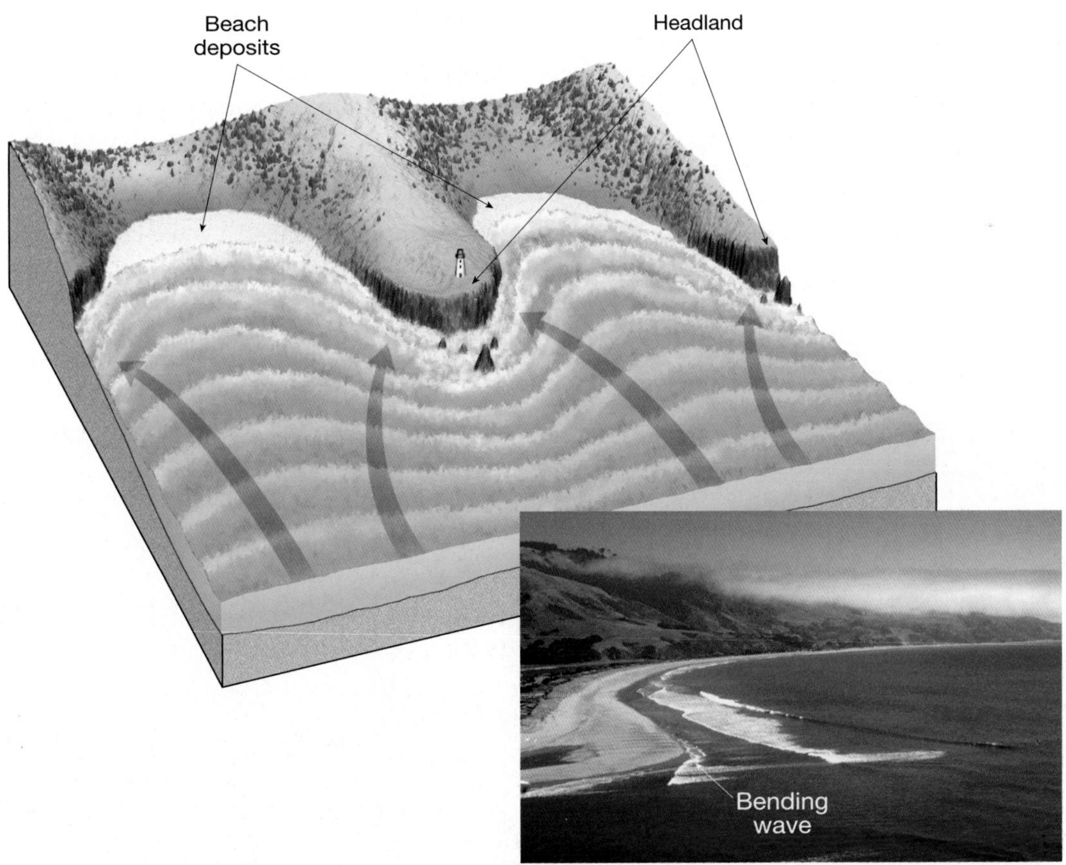

Figure 10.12 Wave refraction along an irregular coastline. As waves first touch bottom in the shallows off the headlands, they are slowed, causing the waves to refract and align nearly parallel to the shoreline. This causes wave energy to be concentrated at headlands (resulting in erosion) and dispersed in bays (resulting in deposition). (Photo by James E. Patterson)

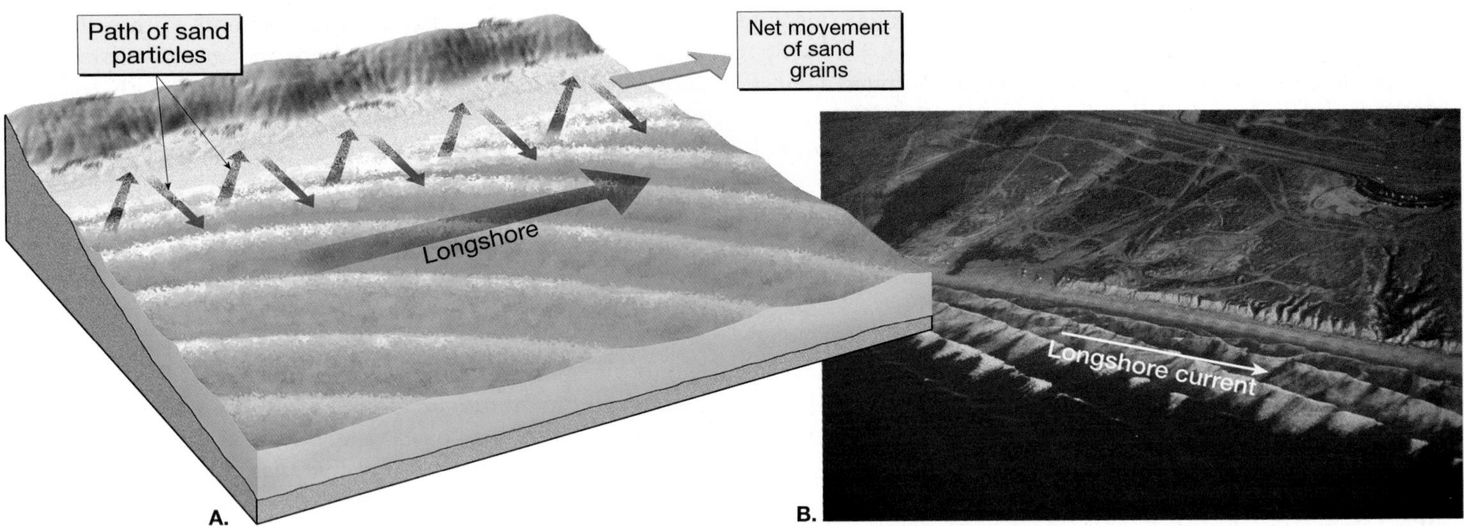

Figure 10.13 **A.** Beach drift and longshore currents are created by obliquely breaking waves. Beach drift occurs as incoming surf carries sand obliquely up the beach, while the water from spent waves carries it directly down the slope of the beach. Similar movements occur offshore in the surf zone to create the longshore current. These processes transport large quantities of material along the beach and in the surf zone. **B.** Waves approaching the beach at a slight angle near Oceanside, California, produce a longshore current moving from left to right. (Photo by John S. Shelton)

Shoreline Features

The Global Ocean
▼ Coastal Processes

A fascinating assortment of shoreline features can be observed along the world's coastal regions. These shoreline features vary depending on the type of rocks exposed along the shore, the intensity of waves, and the nature of coastal currents, as well as whether the coast is stable, sinking, or rising. Features that owe their origin primarily to the work of erosion are called *erosional features,* while deposits of sediment produce *depositional features.*

Erosional Features

Many coastal landforms owe their origin to erosional processes. Such erosional features are common along the rugged and irregular New England coast and along the steep shorelines of the West Coast of the United States.

Wave-Cut Cliffs, Wave-Cut Platforms, and Marine Terraces **Wave-cut cliffs,** as the name implies, originate by the cutting action of the surf against the base of coastal land. As erosion progresses, rocks overhanging the notch at the base of the cliff crumble into the surf and the cliff retreats. A relatively flat, benchlike surface, called a **wave-cut platform,** is left behind by the receding cliff (Figure 10.14, *left*). The platform broadens as

Figure 10.14 Wave-cut platform and marine terrace. A wave-cut platform is exposed at low tide along the California coast at Bolinas Point near San Francisco. To the right, uplift has produced an elevated wave-cut platform, called a marine terrace. (Photo by John S. Shelton)

Figure 10.15 Sea arch and sea stack at the tip of Mexico's Baja Peninsula. (Photo by Mark A. Johnson/CORBIS/The Stock Market)

wave attack continues. Some debris produced by the breaking waves remains along the water's edge as sediment on the beach, while the remainder is transported farther seaward. If a wave-cut platform is uplifted above sea level by tectonic forces, it becomes a **marine terrace** (Figure 10.14, *right*). Marine terraces are easily recognized by their gentle seaward-sloping shape and are often desirable sites for coastal development.

Sea Arches and Sea Stacks Headlands that extend into the sea are vigorously attacked by waves because of refraction. The surf erodes the rock selectively, wearing away the softer or more highly fractured rock at the fastest rate. At first, sea caves may form. When two caves on opposite sides of a headland unite, a **sea arch** results (Figure 10.15). Eventually, the arch falls in, leaving an isolated remnant, or **sea stack,** on the

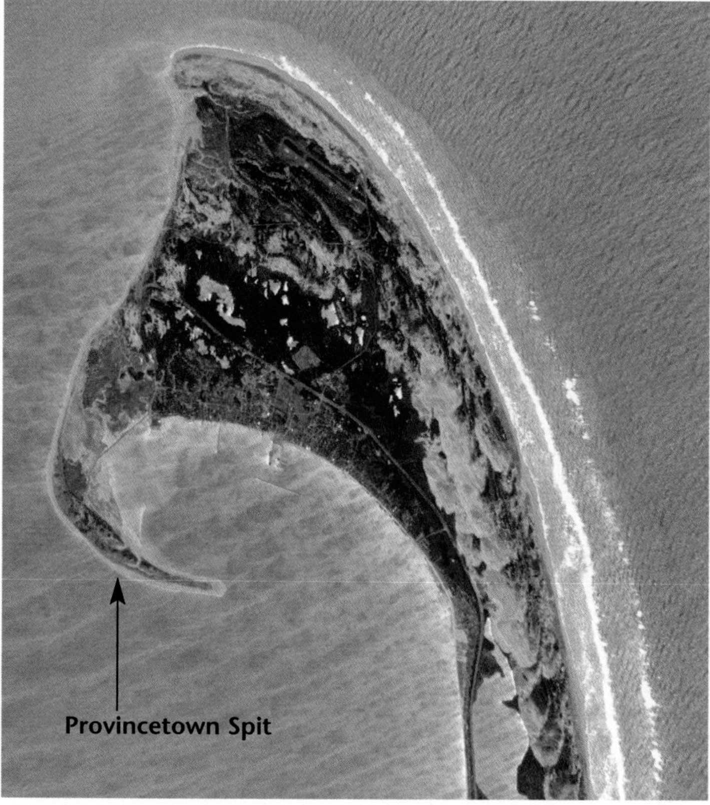

A.

B.

Figure 10.16 **A.** This photograph, taken from the International Space Station, shows Provincetown Spit located at the tip of Cape Cod. (NASA) **B.** High-altitude image of a well-developed spit and baymouth bar along the coast of Martha's Vineyard, Massachusetts. Also notice the tidal delta in the lagoon adjacent to the inlet through the baymouth bar. (Photo courtesy of USDA-ASCS)

Did You Know?

Along shorelines composed of unconsolidated material rather than hard rock, the rate of erosion by breaking waves can be extraordinary. In parts of Britain, where waves have the easy task of eroding glacial deposits of sand, gravel, and clay, the coast has been worn back 3 to 5 kilometers (2 to 3 miles) since Roman times (2000 years ago), sweeping away many villages and ancient landmarks.

wave-cut platform (Figure 10.15). In time, it, too, will be consumed by the action of the waves.

Depositional Features

Sediment eroded from the beach is transported along the shore and deposited in areas where wave energy is low. Such processes produce a variety of depositional features.

Spits, Bars, and Tombolos Where beach drift and longshore currents are active, several features related to the movement of sediment along the shore may develop. A **spit** is an elongated ridge of sand that projects from the land into the mouth of an adjacent bay. Often the end in the water hooks landward in response to the dominant direction of the longshore current (Figure 10.16A). The term **baymouth bar** is applied to a sandbar that completely crosses a bay, sealing it off from the open ocean (Figure 10.16B). Such a feature tends to form across bays where currents are weak, allowing a spit to extend to the other side. A **tombolo,** a ridge of sand that connects an island to the mainland or to another island, forms in much the same manner as a spit.

Barrier Islands The Atlantic and Gulf coastal plains are relatively flat and slope gently seaward. The shore zone is characterized by **barrier islands.** These low ridges of sand parallel the coast at distances from 3 to 30 kilometers (1.9 to 19 miles) offshore. From Cape Cod, Massachusetts, to Padre Island, Texas, nearly 300 barrier islands rim the coast (Figure 10.17).

Most barrier islands are from 1 to 5 kilometers (0.6 to 3 miles) wide and between 15 and 30 kilometers (9 to 19 miles) long. The tallest features are sand dunes, which usually reach heights of 5 to 10 meters (16 to 33 feet). The lagoons that separate these narrow islands from the shore are zones of relatively quiet water that allow small craft traveling between New York and northern Florida to avoid the rough waters of the North Atlantic.

Barrier islands probably formed in several ways. Some originated as spits that were subsequently severed from the mainland by wave erosion or by the general rise in sea level following the last episode of glaciation. Others were created when turbulent waters in the line of breakers heaped up sand that had been scoured from the bottom. Finally, some barrier islands may be former sand-dune ridges that originated

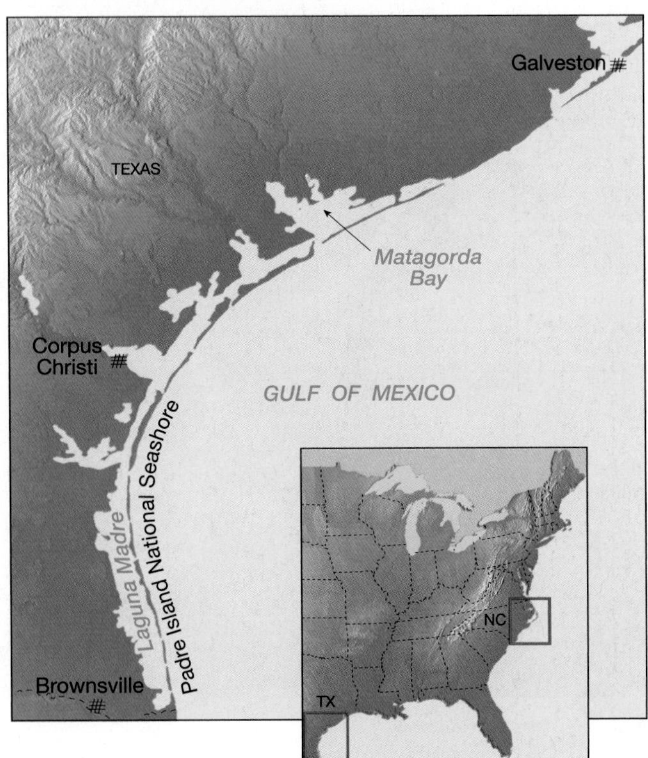

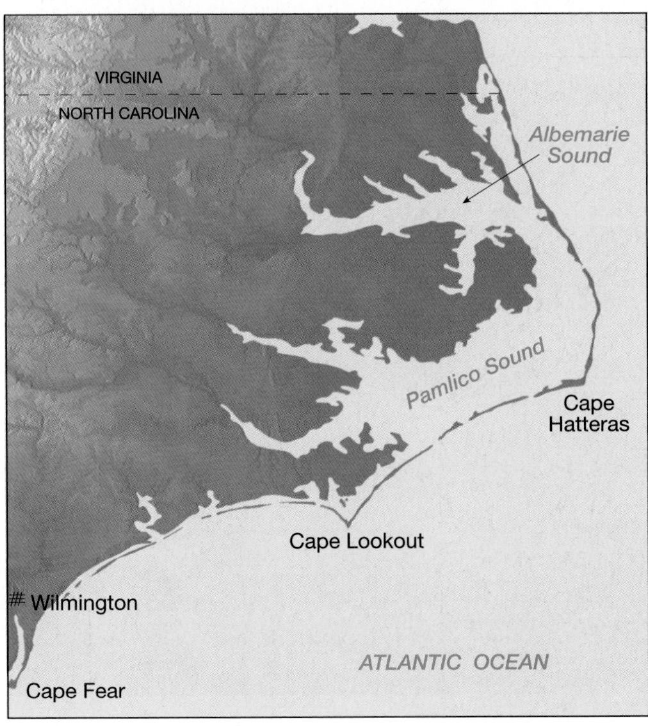

Figure 10.17 Nearly 300 barrier islands rim the Gulf and Atlantic coasts. The islands along the south Texas coast and along the coast of North Carolina are excellent examples.

along the shore during the last glacial period, when sea level was lower. As the ice sheets melted, sea level rose and flooded the area behind the beach–dune complex.

The Evolving Shore

A shoreline continually undergoes modification regardless of its initial configuration. At first, most coastlines are irregular, although the degree of and reason for the irregularity may differ considerably from place to place. Along a coastline that is characterized by varied geology, the pounding surf may initially increase its irregularity because the waves will erode the weaker rocks more easily than the stronger ones. However, if a shoreline remains stable, marine erosion and deposition will eventually produce a straighter, more regular coast.

Figure 10.18 illustrates the evolution of an initially irregular coast that remains relatively stable and shows many of the coastal features discussed in the previous section. As headlands are eroded and erosional features such as wave-cut cliffs and wave-cut platforms are created, sediment is produced that is carried along the shore by beach drift and longshore currents. Some material is deposited in the bays, while other debris is formed into depositional features such as spits and

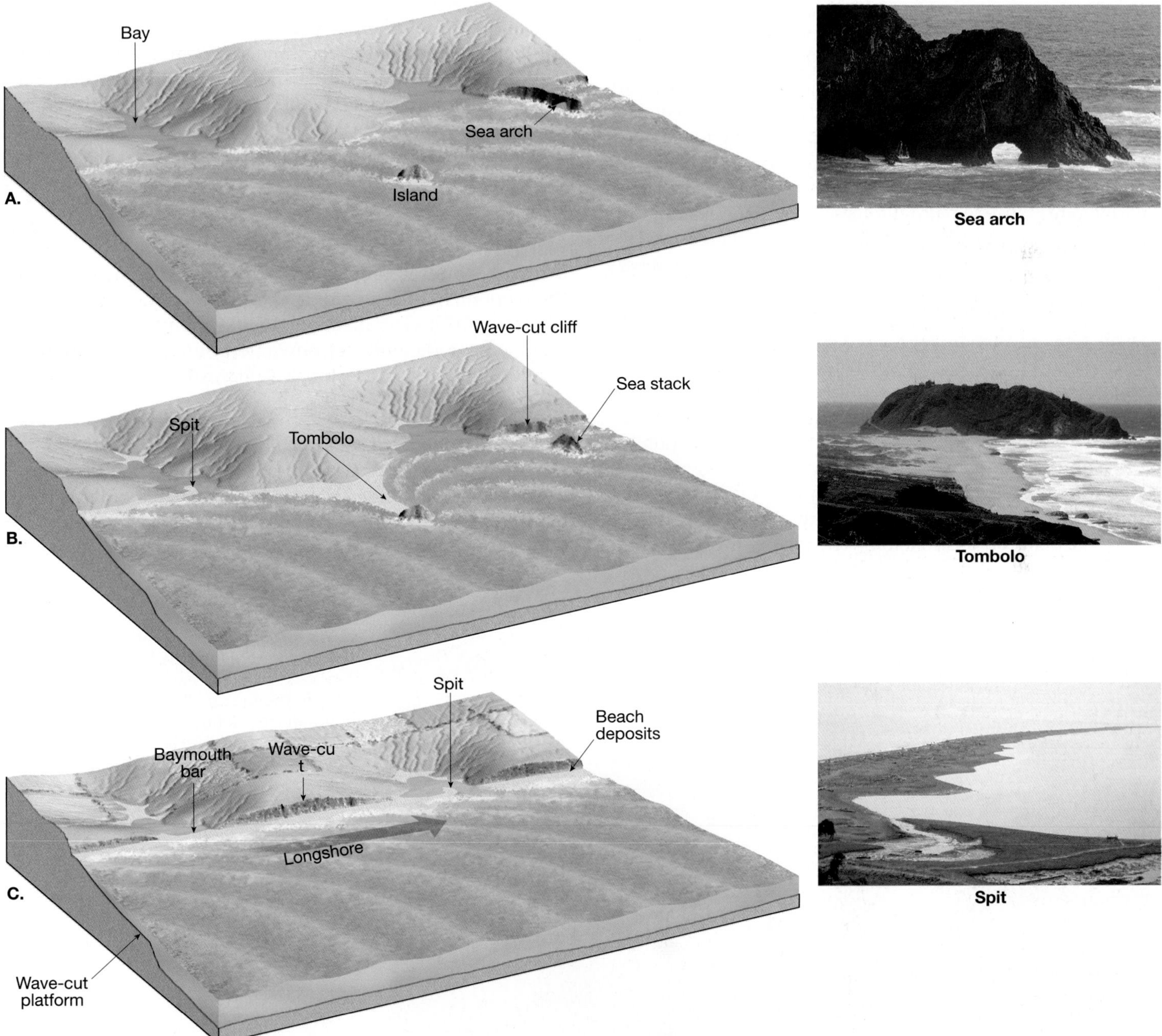

Figure 10.18 These diagrams illustrate the changes that can take place through time along an initially irregular coastline that remains relatively stable. The coastline shown in part **A** gradually evolves to **B** and then **C.** The diagrams also serve to illustrate many of the features described in the section on shoreline features. (Photos by E. J. Tarbuck)

baymouth bars. At the same time, rivers fill the bays with sediment. Ultimately, a smooth coast results.

Stabilizing the Shore

The Global Ocean
▼ Coastal Processes

The coastal zone today teems with human activity. Unfortunately, people often treat the shoreline as if it were a stable platform on which structures can be built safely. This approach jeopardizes both people and the shoreline because many coastal landforms are relatively fragile, short-lived features that are easily damaged by development. As anyone who has endured a strong coastal storm knows, the shoreline is not always a safe place to live (Figure 10.19).

Compared with other natural hazards, such as earthquakes, volcanic eruptions, and landslides, shoreline erosion appears to be a more continuous and predictable process that causes relatively modest damage to limited areas. In reality, the shoreline is one of Earth's most dynamic places that changes rapidly in response to natural forces. Storms, for example, are capable of eroding beaches and cliffs at rates that far exceed the long-term average. Such bursts of accelerated erosion not only have a significant impact on the natural evolution of a coast but can also have a profound impact on people who reside in the coastal zone. Erosion along the coast causes significant property damage. Huge sums are spent annually not only to repair damage but also to prevent or control erosion. Already a problem at many sites, shoreline erosion is certain to become increasingly serious as extensive coastal development continues.

Although the same processes cause change along every coast, not all coasts respond in the same way. Interactions among different processes and the relative importance of each process depend on local factors. The factors include (1) the proximity of a coast to sediment-laden rivers, (2) the degree of tectonic activity, (3) the topography and composition of the land, (4) prevailing winds and weather patterns, and (5) the configuration of the coastline and near-shore areas.

During the past 100 years, growing affluence and increasing demands for recreation have brought unprecedented development to many coastal areas. As both the number and the value of buildings have increased, so, too, have efforts to protect property from storm waves by stabilizing the shore.

Did You Know?

Ocean waves may someday be a significant source of electrical power. In November 2000, the world's first commercial wave-power station began operating off the Scottish coast. The 500-kilowatt station uses technology in which incoming waves push air up and down inside a partially submerged tube. The air rushing into and out of the top of the tube is used to drive a turbine.

Hard Stabilization

Structures built to protect a coast from erosion or to prevent the movement of sand along a beach are known as **hard stabilization.** Hard stabilization can take many forms and often results in predictable yet unwanted outcomes. Hard stabilization includes groins, breakwaters, and seawalls.

Groins To maintain or widen beaches that are losing sand, groins are sometimes constructed. A **groin** is a barrier built at a right angle to the beach to trap sand that is moving parallel to the shore. Groins are usually constructed of large rocks but may also be composed of wood. These structures often do their job so effectively that the longshore current beyond the

Figure 10.19 Erosion caused by strong storms forced abandonment of this highly developed shoreline area in Long Island, New York. Coastal areas are dynamic places that can change rapidly in response to natural forces. (Photo by Mark Wexler/Woodfin Camp & Associates)

Figure 10.20 A series of groins along the shoreline near Chichester, Sussex, England. Because groins trap sand on the upcurrent side, the movement of sand along this coast caused by the longshore current must be from lower right toward upper left. (Photo by Sandy Stockwell/London Aerial Photo Library/CORBIS)

groin becomes sand-starved. As a result, the current erodes sand from the beach on the downstream side of the groin.

To offset this effect, property owners downstream from the structure may erect a groin on their property. In this manner, the number of groins multiplies, resulting in a *groin field* (Figure 10.20). An example of such proliferation is the shoreline of New Jersey, where hundreds of these structures have been built. Because it has been shown that groins often do not provide a satisfactory solution, they are no longer the preferred method of keeping beach erosion in check.

Breakwaters and Seawalls Hard stabilization can also be built parallel to the shoreline. One such structure is a **breakwater,** which is designed to protect boats from the force of large breaking waves by creating a quiet water zone near the shore. However, when this is done, the reduced wave activity along the shore behind the structure may allow sand to accumulate. If this happens, the boat anchorage will eventually fill with sand while the downstream beach erodes and retreats. At Santa Monica, California, where the building of a breakwater created such a problem, the city uses a dredge to remove sand from the protected quiet water zone and deposit it farther downstream where longshore currents continue to move the sand down the coast (Figure 10.21).

Another type of hard stabilization built parallel to the shore is a **seawall,** which is designed to armor the coast and defend property from the force of breaking waves (see Figure 10.10, page 275). Waves expend much of their energy as they move across an open beach. Seawalls cut this process short by reflecting the force of unspent waves seaward. As a consequence, the beach to the seaward side of the seawall experiences significant erosion and may, in some instances, be eliminated entirely. Once the width of the beach is reduced, the seawall is subjected to even greater pounding by the waves. Eventually, this battering takes its toll on the seawall such that the seawall will fail and a larger, more expensive structure must be built to take its place.

The wisdom of building temporary protective structures along shorelines is increasingly questioned. The opinion of many coastal scientists and engineers is expressed in the following excerpt from a position paper that was developed as a result of a conference on America's eroding shoreline:

> It is now clear that halting the receding shoreline with protective structures benefits only a few and seriously degrades or destroys the natural beach and the value it holds for the majority. Protective structures divert the ocean's energy temporarily from private properties, but usually refocus the energy on the adjacent natural beaches. Many interrupt the natural sand flow in coastal currents, robbing many beaches of vital sand replacement.*

*"Strategy for Beach Preservation Proposed," *Geotimes* 3 (No. 12, December 1985): 15.

Figure 10.21 Aerial view of a breakwater at Santa Monica, California. The breakwater appears as a faint line in the water behind which many boats are anchored. The construction of the breakwater disrupted longshore transport and caused the seaward growth of the beach. (Photo by John S. Shelton)

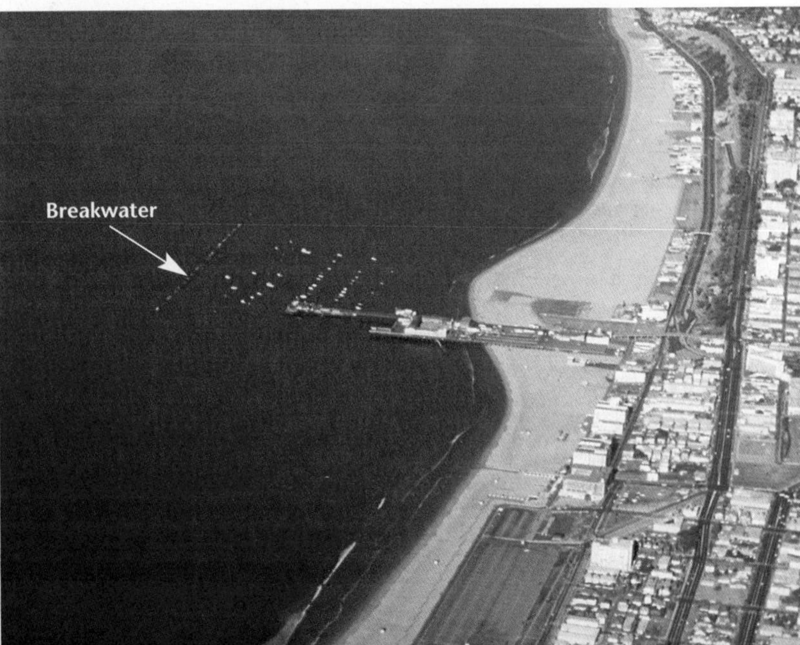

Alternatives to Hard Stabilization

Armoring the coast with hard stabilization has several potential drawbacks, including the cost of the structure and the loss of sand on the beach. Alternatives to hard stabilization include beach nourishment and relocation.

Beach Nourishment **Beach nourishment** represents an approach to stabilizing shoreline sands without hard stabilization. As the term implies, this practice simply involves the addition of large quantities of sand to the beach system (Figure 10.22). By building the beaches seaward, beach quality and storm protection are both improved. Beach nourishment, however, is not a permanent solution to the problem of shrinking beaches because the same processes that removed the sand in the first place will eventually remove the replacement sand as well. In addition, it can be very expensive because huge volumes of sand must be transported to the beach from offshore areas, nearby rivers, or other source areas. Orrin Pilkey, a respected coastal scientist, described the situation this way:

> Nourishment has been carried out on beaches on both sides of the continent, but by far the greatest effort in dollars and sand volume has been expended on the East Coast barrier islands between the southern shore of Long Island, N.Y., and southern Florida. Over this shoreline reach, communities have added a total of 500 million cubic yards of sand on 195 beaches in 680 separate instances. Some beaches . . . have been renourished more than 20 times each since 1965. Virginia Beach has been renourished more than 50 times. Nourished beaches typically cost between $2 million and $10 million per mile.*

In some instances, beach nourishment can lead to unwanted environmental effects. Beach replenishment at Waikiki Beach, Hawaii, involved replacing the natural coarse calcareous beach sand with softer, muddier sand. Destruction of the softer sand by breaking waves increased the water's turbidity and killed offshore coral reefs. Similar damage to local coral communities occurred after beach replenishment at Miami Beach.

Beach nourishment appears to be an economically viable long-range solution to the beach-preservation problem only in areas with dense development, large supplies of sand, relatively low wave energy, and reconcilable environmental issues. Unfortunately, few areas possess all these attributes.

Relocation Instead of building structures such as groins and seawalls to hold the shoreline in place or adding sand to replenish eroding beaches, another option is also available. Many coastal scientists and planners are calling for a policy shift from defending and rebuilding beaches and coastal property in high hazard areas to *relocating* storm-damaged or at-risk buildings and letting nature reclaim the beach. This approach is similar to the one adopted by the federal government for river floodplains following the devastating 1993 Mississippi River floods, in which vulnerable structures were abandoned and relocated on higher, safer ground.

Such proposals, of course, are controversial. People with significant shoreline investments shudder at the thought of not rebuilding and defending coastal developments from the erosional wrath of the sea. Others, however, argue that with sea level rising, the impact of coastal storms will only get worse in the decades to come. This group advocates that oft-damaged structures be relocated or abandoned to improve personal safety and to reduce costs. Such ideas will no doubt be the focus of much study and debate as states and communities evaluate and revise coastal land-use policies.

Erosion Problems Along U.S. Coasts

The shoreline along the Pacific Coast of the United States is strikingly different from that characterizing the Atlantic and Gulf Coast regions. Some of the differences are related to plate tectonics. The West Coast represents the leading edge of the North American plate; therefore, it experiences active uplift and deformation. By contrast, the East Coast is a tectonically

*"Beaches Awash with Politics," *Geotimes*, July 2005, pp. 38–39.

A.

B.

Figure 10.22 Miami Beach. **A.** Before beach nourishment. **B.** After beach nourishment. (Courtesy of the U.S. Army Corps of Engineers, Vicksburg District)

quiet region that is far from any active plate margin. Because of this basic geological difference, the nature of shoreline erosion problems along America's opposite coasts is different.

Atlantic and Gulf Coasts Much of the coastal development along the Atlantic and Gulf coasts has occurred on barrier islands. Typically, barrier islands consist of a wide beach that is backed by dunes and separated from the mainland by marshy lagoons. The broad expanses of sand and exposure to the ocean have made barrier islands exceedingly attractive sites for development. Unfortunately, development has taken place more rapidly than has our understanding of barrier island dynamics.

Because barrier islands face the open ocean, they receive the full force of major storms that strike the coast. When a storm occurs, the barriers absorb the energy of the waves primarily through the movement of sand. This process and the dilemma that results have been described as follows:

> Waves may move sand from the beach to offshore areas or, conversely, into the dunes; they may erode the dunes, depositing sand onto the beach or carrying it out to sea; or they may carry sand from the beach and the dunes into the marshes behind the barrier, a process known as *overwash.* The common factor is movement. Just as a flexible reed may survive a wind that destroys an oak tree, so the barriers survive hurricanes and nor'easters not through unyielding strength but by giving before the storm.
>
> This picture changes when a barrier is developed for homes or a resort. Storm waves that previously rushed harmlessly through gaps between the dunes now encounter buildings and roadways. Moreover, since the dynamic nature of the barriers is readily perceived only during storms, homeowners tend to attribute damage to a particular storm, rather than to the basic mobility of coastal barriers. With their homes or investments at stake, local residents are more likely to seek to hold the sand in place and the waves at bay than to admit that development was improperly placed to begin with.*

Pacific Coast In contrast to the broad, gently sloping coastal plains of the Atlantic and Gulf Coasts, much of the Pacific Coast is characterized by relatively narrow beaches that are backed by steep cliffs and mountain ranges. Recall that America's western margin is a more rugged and tectonically active region than the eastern margin. Because uplift continues, the rise in sea level in the West is not so readily apparent. Nevertheless, like the shoreline erosion problems facing the East's barrier islands, West Coast difficulties also stem largely from the alteration of natural systems by people.

A major problem facing the Pacific shoreline—particularly along southern California—is a significant narrowing of many beaches. The bulk of the sand on many of these beaches is supplied by rivers that transport it from the mountainous regions to the coast. Over the years, this natural flow of material to the coast has been interrupted by dams built for irrigation and flood control. The reservoirs effectively trap the sand that would otherwise nourish the beach environment. When the beaches were wider, they served to protect the cliffs behind them from the force of storm waves. Now, however, the waves move across the narrowed beaches without losing much energy and cause more rapid erosion of the sea cliffs.

Although the retreat of the cliffs provides material to replace some of the sand impounded behind dams, it also endangers homes and roads built on the bluffs. In addition, development atop the cliffs aggravates the problem. Urbanization increases runoff, which, if not carefully controlled, can result in serious bluff erosion. Watering lawns and gardens adds significant quantities of water to the slope. This water percolates downward toward the base of the cliff, where it may emerge in small seeps. This action reduces the slope's stability and facilitates mass wasting.

Shoreline erosion along the Pacific Coast varies considerably from one year to the next, largely because of the sporadic occurrence of storms. As a consequence, when the infrequent but serious episodes of erosion occur, the damage is often blamed on the unusual storms and not on coastal development or the sediment-trapping dams that may be great distances away. If, as predicted, sea level experiences a significant rise in the years to come because of global warming, increased shoreline erosion and sea-cliff retreat should be expected along many parts of the Pacific Coast.

*Frank Lowenstein, "Benches or Bedrooms—The Choice as Sea Level Rises," *Oceanus* 28 (No. 3, Fall 1985): 22.

Coastal Classification

The great variety of shorelines demonstrates their complexity. Indeed, to understand any particular coastal area, many factors must be considered, including rock types, size and direction of waves, frequency of storms, tidal range, and offshore topography. In addition, practically all coastal areas were affected by the worldwide rise in sea level that accompanied the melting of Ice Age glaciers at the close of the Pleistocene epoch. Finally, tectonic events that uplift or downdrop the land or change the volume of ocean basins must be taken into account. The large number of factors that influence coastal areas makes shoreline classification difficult.

Many geologists classify coasts based on changes that have occurred with respect to sea level. This commonly used classification system divides coasts into two very general categories: emergent and submergent. **Emergent coasts** develop either because an area experiences uplift or as a result of a drop in sea level. Conversely, **submergent coasts** are created when sea level rises or the land adjacent to the sea subsides.

Emergent Coasts

In some areas, the coast is clearly emergent because rising land or falling water levels expose wave-cut cliffs and marine terraces above sea level. Excellent examples include portions of coastal California where uplift has occurred in the recent geologic past. The elevated marine terrace in Figure 10.14 (p. 279) illustrates this situation. In the case of the Palos Verdes Hills, south of Los Angeles, California, seven different terrace levels exist, indicating at least seven episodes of uplift. The ever persistent sea is now cutting a new platform at the base of the cliff. If uplift follows, it, too, will become an elevated marine terrace.

Other examples of emergent shores include regions that were once buried beneath great ice sheets. When glaciers were present, their weight depressed the crust, and when the ice melted, the crust began gradually to spring back. Consequently, prehistoric shoreline features today are found high above sea level. The Hudson Bay region of Canada is one such area, portions of which are still rising at a rate of more than 1 centimeter (0.4 inch) annually.

Submergent Coasts

Other coastal areas show definite signs of submergence. Shorelines that have been submerged in the relatively recent past are often highly irregular because the sea typically floods the lower reaches of river valleys flowing into the ocean. The ridges separating the valleys, however, remain above sea level and project into the sea as headlands. These drowned river mouths, which are called **estuaries,** characterize many coasts today. Along the Atlantic coastline, Chesapeake and Delaware bays are examples of estuaries created by submergence (Figure 10.23). The picturesque coast of Maine, particularly in the vicinity of Acadia National Park, is another excellent example of an area that was flooded by the postglacial rise in sea level and transformed into a highly irregular submerged coastline.

Keep in mind, however, that most coasts have a complicated geologic history. With respect to sea level, many coasts have at various times emerged and then submerged again. Each time, they retain some of the features created during the previous event.

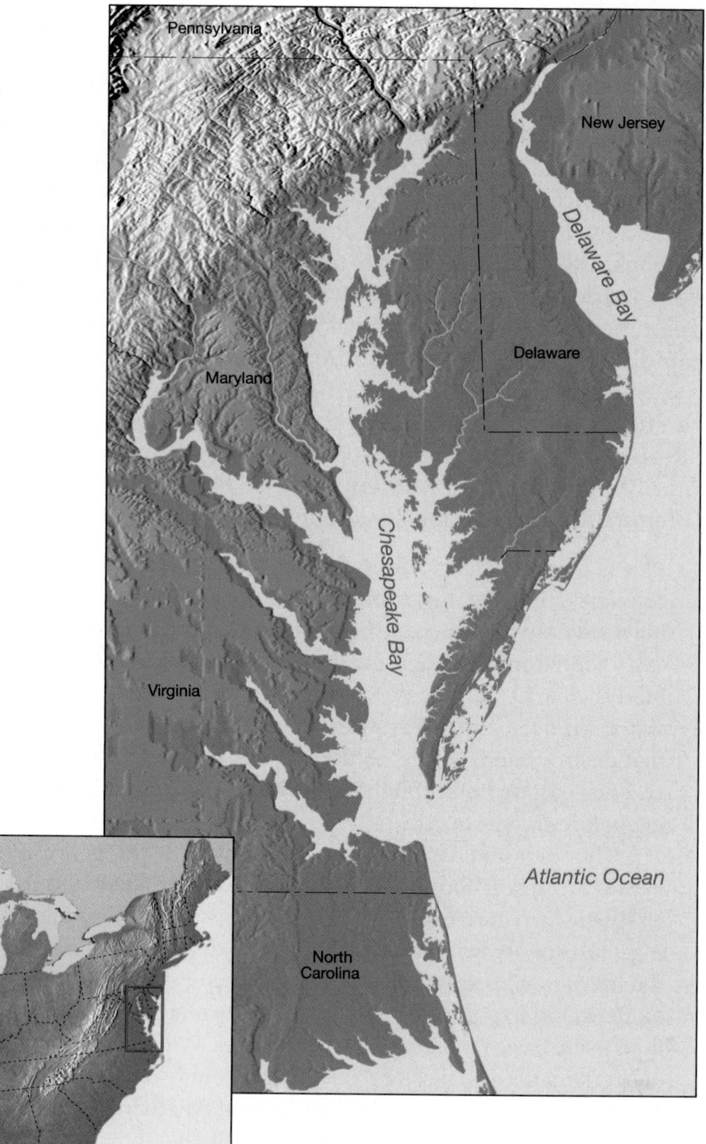

Figure 10.23 Major estuaries along the East Coast of the United States. The lower portions of many river valleys were flooded by the rise in sea level that followed the end of the last Ice Age, creating large estuaries such as Chesapeake and Delaware bays.

Tides

Tides are daily changes in the elevation of the ocean surface. Their rhythmic rise and fall along coastlines have been known since antiquity. Other than waves, they are the easiest ocean movements to observe (Figure 10.24).

Although known for centuries, tides were not explained satisfactorily until Sir Isaac Newton applied the law of gravitation to them. Newton showed that there is a mutual attractive force between two bodies, as between Earth and the Moon. Because both the atmosphere and the ocean are fluids and are free to move, both are deformed by this force. Hence, ocean tides result from the gravitational attraction exerted upon Earth by the Moon and, to a lesser extent, by the Sun.

Causes of Tides

To illustrate how tides are produced, consider Earth as a rotating sphere covered to a uniform depth with water (Figure 10.25). (Ignore the effect of the Sun for now.) It is easy to see how the Moon's gravitational force can cause the water to bulge on the side of Earth nearest the Moon. In addition, however, an equally large tidal bulge is produced on the *opposite* side of Earth.

Both tidal bulges are caused, as Newton discovered, by the pull of gravity. Gravity is inversely proportional to the square of the distance between two objects, meaning simply that it quickly weakens with distance. In this case, the two objects are the Moon and Earth. Because the force of gravity decreases with distance, the Moon's gravitational pull on Earth is slightly greater on the near side of Earth than on the far side. The result of this differential pulling is to stretch (elongate) the "solid" Earth very slightly. In contrast, the world ocean, which is mobile, is deformed quite dramatically by this effect to produce the two opposing tidal bulges.

Did You Know?

Tides can be used to generate electrical power by constructing a dam across the mouth of a bay or estuary in a coastal area having a large tidal range. The narrow opening between the bay and the open ocean magnifies the variations in water level as tides rise and fall and the strong in-and-out flow is used to drive turbines. The largest plant yet constructed is along the coast of France.

Figure 10.24 High tide and low tide on Nova Scotia's Minas Basin in the Bay of Fundy. The areas exposed during low tide and flooded during high tide are called *tidal flats.* Tidal flats here are extensive. (Courtesy Nova Scotia Department of Tourism and Culture)

Because the position of the Moon changes only moderately in a single day, the tidal bulges remain in place while Earth rotates "through" them. If you stand on the seashore for 24 hours, Earth will rotate you through alternating areas of higher and lower water. As you are carried into each tidal bulge, the tide rises, and as you are carried into the intervening troughs between the tidal bulges, the tide falls. Therefore, most places on Earth experience two high tides and two low tides each day.

In addition, the tidal bulges migrate as the Moon revolves around Earth about every $29\frac{1}{2}$ days. As a result, the tides—like the time of moonrise—shift about 50 minutes later each day. In essence, the tidal bulges exist in fixed positions relative to the Moon, which is slowly moving eastward.

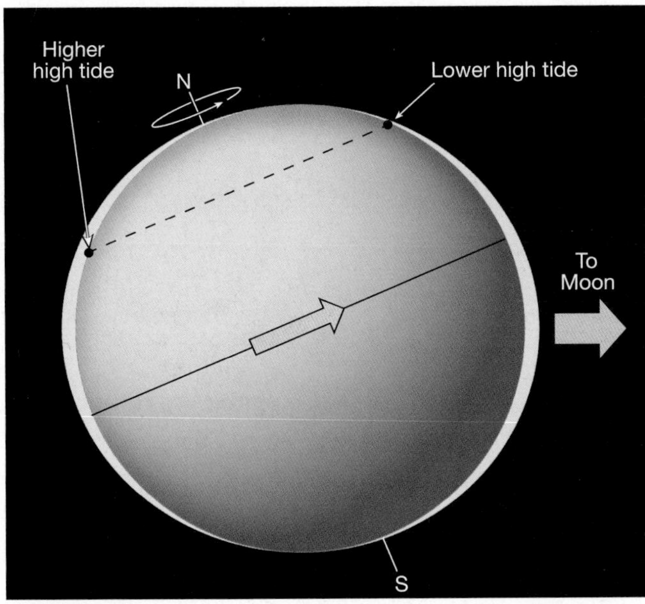

Figure 10.25 Idealized tidal bulges on Earth caused by the Moon. If Earth were covered to a uniform depth with water, there would be two tidal bulges: one on the side of Earth facing the Moon *(right)* and the other on the opposite side of Earth *(left).* Depending on the Moon's position, tidal bulges may be inclined relative to Earth's equator. In this situation, Earth's rotation causes two unequal high tides during a day.

Many locations may show an inequality between the high tides during a given day. Depending on the Moon's position, the tidal bulges may be inclined to the equator as in Figure 10.25. This figure illustrates that one high tide experienced by an observer in the Northern Hemisphere is considerably higher than the high tide half a day later. In contrast, a Southern Hemisphere observer would experience the opposite effect.

Monthly Tidal Cycle

The primary body that influences the tides is the Moon, which makes one complete revolution around Earth every $29\frac{1}{2}$ days. The Sun, however, also influences the tides. It is far larger than the Moon, but because it is much farther away, its effect is considerably less. In fact, the Sun's tide-generating effect is only about 46 percent that of the Moon's.

Near the times of new and full moons, the Sun and Moon are aligned and their forces are added together (Figure 10.26A). Accordingly, the combined gravity of these two tide-producing bodies causes larger tidal bulges (higher high tides) and larger tidal troughs (lower low tides), producing a large tidal range. These are called the **spring tides,** which have no connection with the spring season but occur twice a month during the times when the Earth–Moon–Sun system is aligned. Conversely, at about the time of the first and third quarters of the Moon, the gravitational forces of the Moon and Sun act on Earth at right angles, and each partially offsets the influence of the other (Figure 10.26B). As a result, the daily tidal range is less. These are called **neap tides,** which also occur twice each month. Each month, then, there are two spring tides and two neap tides, each about one week apart.

Tidal Patterns

So far, the basic causes and types of tides have been explained. Keep in mind, however, that these theoretical considerations cannot be used to predict either the height or the time of actual tides at a particular place. Many factors—including the shape of the coastline, the configuration of ocean basins, and water depth—greatly influence the tides. Consequently, tides at various locations respond differently

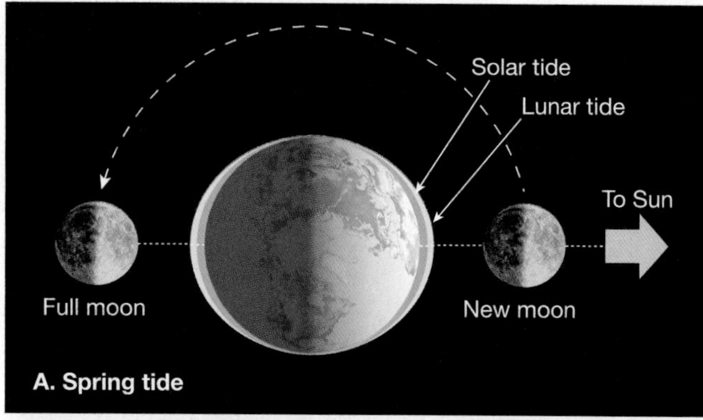

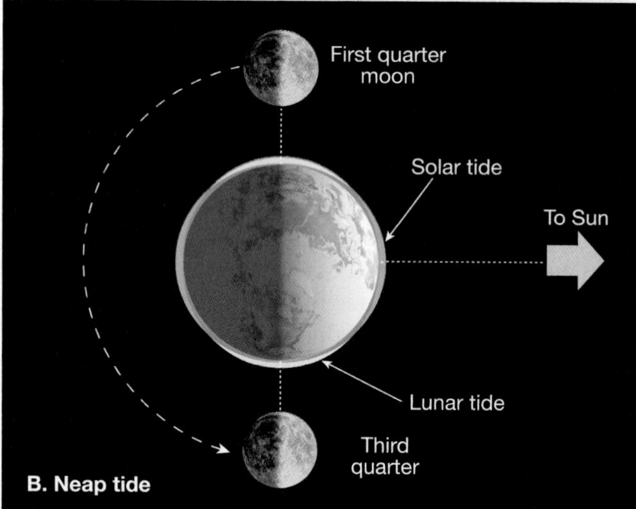

Figure 10.26 Earth-Moon-Sun positions and the tides. **A.** When the Moon is in the full or new position, the tidal bulges created by the Sun and Moon are aligned. There is a large tidal range on Earth, and spring tides are experienced. **B.** When the Moon is in the first- or third-quarter position, the tidal bulges produced by the Moon are at right angles to the bulges created by the Sun. Tidal ranges are smaller, and neap tides are experienced.

Did You Know?

The world's largest tidal range (difference between successive high and low tides) is found in the northern end of Nova Scotia's Bay of Fundy. Here, the maximum spring tidal range is about 17 meters (56 feet). This leaves boats "high and dry" during low tide (see Figure 10.24, p. 287).

to the tide-producing forces. Thus, the nature of the tide at any coastal location can be determined most accurately by actual observation. The predictions in tidal tables and tidal data on nautical charts are based on such observations.

Three main tidal patterns exist worldwide. A **diurnal** (*diurnal* = daily) **tidal pattern** is characterized by a single high tide and a single low tide each tidal day (Figure 10.27). Tides of this type occur along the northern shore of the Gulf of Mexico, among other locations. A **semidiurnal** (*semi* = twice, *diurnal* = daily) **tidal pattern** exhibits two high tides and two low tides each tidal day, with the two highs about the same height and the two lows about the same height (Figure 10.27). This type of tidal pattern is common along the Atlantic Coast of the United States. A **mixed tidal pattern** is similar to a semidiurnal pattern except that it is characterized by a large inequality in high-water heights, low-water heights, or both (Figure 10.27). In this case, there are usually two high and two low tides each day, with high tides of different heights and low tides of different heights. Such tides are prevalent along the Pacific Coast of the United States and in many other parts of the world.

Tidal Currents

Tidal current is the term used to describe the *horizontal* flow of water accompanying the rise and fall of the tides. These water movements induced by tidal forces can be important in some coastal areas. Tidal currents that advance into the coastal zone as the tide rises are called *flood currents*. As the tide falls, sea-

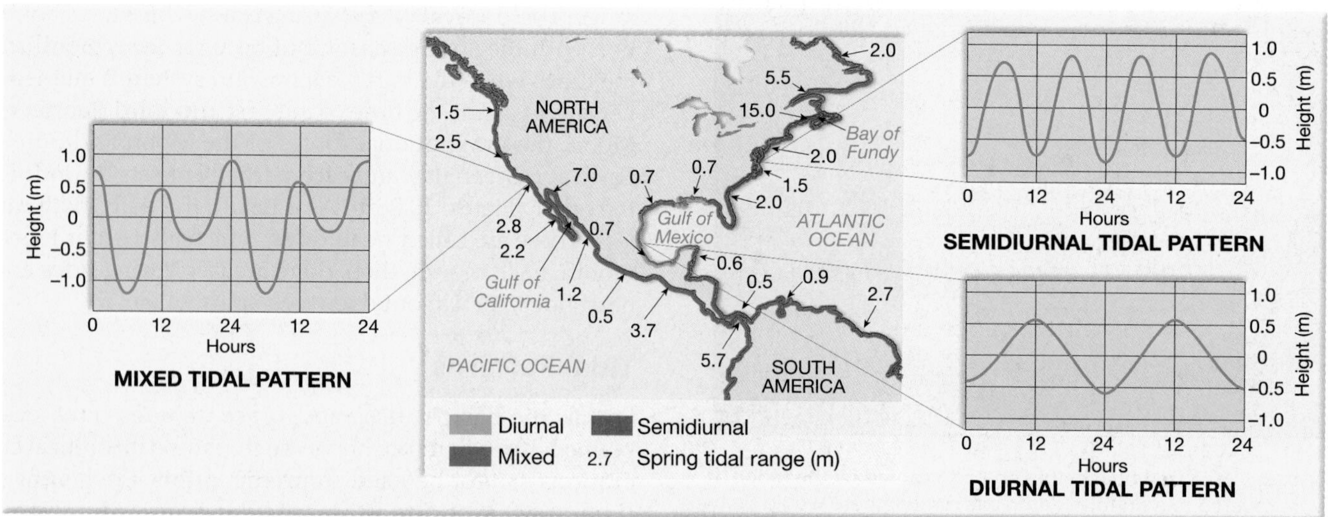

Figure 10.27 Tidal patterns and their occurrence along North and Central American coasts. A diurnal tidal pattern *(lower right)* shows one high and low tide each tidal day. A semidiurnal pattern *(upper right)* shows two highs and lows of approximately equal heights during each tidal day. A mixed tidal pattern *(left)* shows two highs and lows of unequal heights during each tidal day.

ward-moving water generates *ebb currents*. Periods of little or no current, called *slack water*, separate flood and ebb. The areas affected by these alternating tidal currents are called **tidal flats** (see Figure 10.24, p. 287). Depending on the nature of the coastal zone, tidal flats vary from narrow strips seaward of the beach to zones that may extend for several kilometers.

Although tidal currents are not important in the open sea, they can be rapid in bays, river estuaries, straits, and other narrow places. Off the coast of Brittany in France, for example, tidal currents that accompany a high tide of 12 meters (40 feet) may attain a speed of 20 kilometers (12 miles) per hour. Tidal currents are not generally considered to be major agents of erosion and sediment transport. Notable exceptions occur where tides move through narrow inlets, where they scour the narrow entrances to many harbors that would otherwise be blocked.

Sometimes deposits called **tidal deltas** are created by tidal currents (Figure 10.28). They may develop either as *flood deltas* landward of an inlet or as *ebb deltas* on the seaward side of an inlet. Because wave activity and longshore currents are reduced on the sheltered landward side, flood deltas are more common and actually more prominent (see Figure 10.16B, p. 280). They form after the tidal current moves rapidly through an inlet. As the current emerges into more open waters from the narrow passage, it slows and deposits its load of sediment.

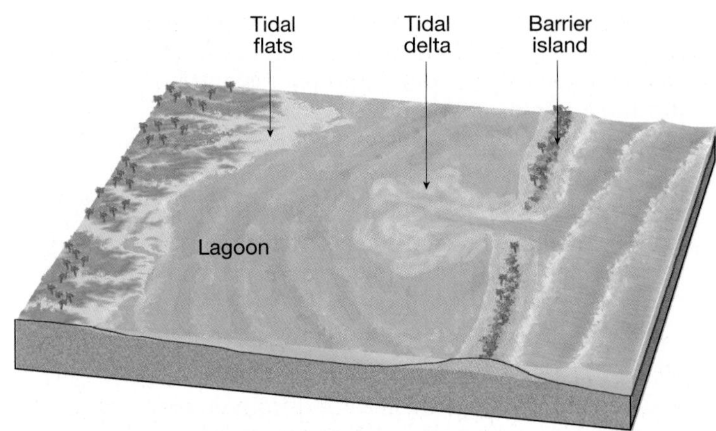

Figure 10.28 As a rapidly moving tidal current moves through a barrier island's inlet into the quiet waters of a lagoon, the current slows and deposits sediment, creating a tidal delta. Because this tidal delta has developed on the landward side of the inlet, it is called a *flood delta*.

The Chapter in Review

1. *The ocean's surface currents follow the general pattern of the world's major wind belts.* Surface currents are parts of huge, slowly moving loops of water called *gyres* that are centered in the subtropics of each ocean basin. The *positions of the continents* and the *Coriolis effect* also influence the movement of ocean water within gyres. Because of the Coriolis effect, subtropical gyres move *clockwise in the Northern Hemisphere* and *counterclockwise in the Southern Hemisphere*. Generally, *four main currents* comprise each subtropical gyre.

2. *Ocean currents* can have a significant *effect on climate*. Poleward-moving *warm ocean currents moderate winter temperatures* in the middle latitudes. Cold currents exert their greatest influence during summer in middle latitudes and year-round in the tropics. In addition to cooler temperatures, *cold currents are associated with greater fog frequency and drought*.

3. *Upwelling*, the rising of colder water from deeper layers, is a wind-induced movement that brings cold, nutrient-rich water to the surface. *Coastal upwelling* is most characteristic along the west coasts of continents.

4. In contrast to surface currents, *deep-ocean circulation* is governed by *gravity* and driven by *density differences*. The two factors that are most significant in creating a dense mass of water are *temperature* and *salinity*, so the movement of deep-ocean water is often termed *thermohaline circulation*. Most water involved in thermohaline circulation begins in high latitudes at the surface when the salinity of the cold water increases as a result of sea ice formation. This dense water sinks, initiating deep-ocean currents.

5. *Waves are moving energy* and *most ocean waves are initiated by the wind*. The three factors that influence the *height*, *wavelength*, and *period* of a wave are (1) *wind speed*; (2) *length of time the wind has blown*; and (3) *fetch*, the distance that the wind has traveled across open water. Once waves leave a storm area, they are termed *swells*, which are symmetrical, longer-wavelength waves.

6. As waves travel, *water particles transmit energy by circular orbital motion*, which extends to a depth equal to one-half the wavelength. When a wave travels into shallow water, it experiences physical changes that can cause the wave to collapse, or *break*, and form *surf*.

7. *Beaches are composed of any locally derived material* that is in transit along the shore. Wave erosion is caused by *wave impact pressure* and *abrasion* (the sawing and grinding action of water armed with rock fragments). The bending of waves is called *wave refraction*. Refraction causes wave impact to be concentrated against the sides and ends of headlands and dispersed in bays.

8. Most waves reach the shore at an angle. The uprush (swash) and backwash of water from each breaking wave move the sediment in a zigzag pattern along the beach. This movement is called *beach drift*. Oblique waves also produce *longshore currents* within the surf zone that flow parallel to the shore and transport more sediment than does beach drift.

9. *Erosional features* include *wave-cut cliffs* (which originate from the cutting action of the surf against the base of coastal land), *wave-cut platforms* (relatively flat, benchlike surfaces left behind by receding cliffs), and *marine terraces* (uplifted wave-cut platforms). Erosional features also include *sea arches* (formed when a headland is eroded and two sea caves from opposite sides unite), and *sea stacks* (formed when the roof of a sea arch collapses).

10. Some of the *depositional features* that form when sediment is moved by beach drift and longshore currents are *spits* (elongated ridges of sand that project from the land into the mouth of an adjacent bay), *baymouth bars* (sandbars that completely cross a bay), and *tombolos* (ridges of sand that connect an island to the mainland or to another island). Along the Atlantic and Gulf coastal plains, the coastal region is characterized by offshore *barrier islands*, which are low ridges of sand that parallel the coast.

11. Local *factors that influence shoreline erosion* are (1) the proximity of a coast to sediment-laden rivers, (2) the degree of tectonic activity, (3) the topography and composition of the land, (4) prevailing winds and weather patterns, and (5) the configuration of the coastline and near-shore areas.

12. *Hard stabilization* involves building hard, massive structures in an attempt to protect a coast from erosion or prevent the movement of sand along the beach. Hard stabilization includes *groins* (short walls constructed at a right angle to the shore to trap moving sand), *breakwaters* (structures built parallel to the shore to protect it from the force of large breaking waves), and *seawalls* (armoring the coast to prevent waves from reaching the area behind the wall). *Alternatives* to *hard stabilization* include *beach nourishment*, which involves the addition of sand to replenish eroding beaches, and *relocation* of damaged or threatened buildings.

13. Because of basic geologic differences, the *nature of shoreline erosion problems along America's Pacific and Atlantic/Gulf coasts is very different.* Much of the development along the Atlantic and Gulf coasts has occurred on barrier islands, which receive the full force of major storms. Much of the Pacific Coast is characterized by narrow beaches backed by steep cliffs and mountain ranges. A major problem facing the Pacific shoreline is the narrowing of beaches caused by dams that interrupt the natural flow of sand to the coast.

14. One commonly used classification of coasts is based upon changes that have occurred with respect to sea level. *Emergent coasts* often exhibit wave-cut cliffs and marine terraces and develop because an area experiences either uplift or a drop in sea level. Conversely, *submergent coasts* commonly display drowned river mouths called *estuaries* and are created when sea level rises or the land adjacent to the sea subsides.

15. *Tides*, the daily rise and fall in the elevation of the ocean surface at a specific location, are caused by the *gravitational attraction* of the Moon and, to a lesser extent, the Sun. The Moon and the Sun each produce a pair of *tidal bulges* on Earth. These tidal bulges remain in fixed positions relative to the generating bodies as Earth rotates through them, resulting in alternating high and low tides. *Spring tides* occur near the times of new and full moons when the Sun and Moon are aligned and their bulges are added together to produce especially high and low tides (a *large daily tidal range*). Conversely, *neap tides* occur at about the times of the first and third quarters of the Moon when the bulges of the Moon and Sun are at right angles, producing a *smaller daily tidal range.*

16. *Three main tidal patterns exist worldwide.* A *diurnal tidal pattern* exhibits one high and low tide daily; a *semidiurnal tidal pattern* exhibits two high and low tides daily of about the same height; and a *mixed tidal pattern* usually has two high and low tides of different heights daily.

17. *Tidal currents* are horizontal movements of water that accompany the rise and fall of the tides. *Tidal flats* are the areas that are affected by the advancing and retreating tidal currents. When tidal currents slow after emerging from narrow inlets, they deposit sediment that may eventually create *tidal deltas.*

Key Terms

abrasion (p. 275)
barrier island (p. 279)
baymouth bar (p. 278)
beach drift (p. 277)
beach nourishment (p. 284)
breakwater (p. 283)
Coriolis effect (p. 268)
diurnal tidal pattern (p. 288)
emergent coast (p. 285)
estuary (p. 286)
fetch (p. 273)
groin (p. 282)
gyre (p. 268)
hard stabilization (p. 282)
longshore current (p. 277)
marine terrace (p. 278)
mixed tidal pattern (p. 288)
neap tide (p. 287)
sea arch (p. 278)
sea stack (p. 278)
seawall (p. 283)
semidiurnal tidal pattern (p. 288)
spit (p. 278)
spring tide (p. 287)
submergent coast (p. 285)
surf (p. 274)
thermohaline circulation (p. 271)
tidal current (p. 288)
tidal delta (p. 289)
tidal flat (p. 289)
tide (p. 286)
tombolo (p. 279)
upwelling (p. 270)
wave-cut cliff (p. 278)
wave-cut platform (p. 278)
wave height (p. 273)
wavelength (p. 273)
wave period (p. 273)
wave refraction (p. 276)

Questions for Review

1. What is the primary driving force of surface ocean currents? How do the distribution of continents on Earth and the Coriolis effect influence these currents?
2. What is a *gyre?* How many currents exist within each gyre? Name the five subtropical gyres, and identify the primary surface currents that comprise each gyre.
3. How do ocean currents influence climate? Give at least one example.
4. Describe the process of coastal upwelling. Why is an abundance of marine life associated with these areas?
5. What is the driving force of deep-ocean circulation? Why is the movement of deep currents often termed *thermohaline circulation?*
6. List three factors that determine the height, length, and period of a wave.

7. Describe the motion of a floating object as a wave passes (see Figure 10.8, pp. 274).
8. Describe the physical changes that occur to a wave's speed, wavelength, and height as a wave moves into shallow water and breaks on the shore.
9. How do waves cause erosion?
10. What is wave refraction? What is the effect of this process along irregular coastlines?
11. What is beach drift, and how is it related to a longshore current? Why are beaches often called "rivers of sand"?
12. Describe the formation of the following shoreline features: wave-cut cliff, wave-cut platform, marine terrace, sea stack, spit, baymouth bar, and tombolo.
13. List three ways that barrier islands may form.
14. List the types of hard stabilization and describe what each is intended to do. What effect does each one have on the distribution of sand on the beach?
15. List two alternatives to hard stabilization, indicating potential problems with each one.
16. Relate the damming of rivers to the shrinking of beaches at some locations along the West Coast of the United States.
17. What observable features would lead you to classify a coastal area as emergent?
18. Are estuaries associated with submergent or emergent coasts? Explain.
19. Discuss the origin of ocean tides. Explain why the Sun's influence on Earth's tides is less than half that of the Moon's, even though the Sun is so much more massive than the Moon.
20. Explain how an observer can experience two unequal high tides during a day (see Figure 10.25, pp. 287).
21. How do diurnal, semidiurnal, and mixed tidal patterns differ?
22. Distinguish between a flood current and an ebb current. Of flood current, ebb current. Of flood current, ebb current, high slack water, and low slack water, when is the best time to navigate a boat in a shallow, rocky harbor?

Online Study Guide

The *Foundations of Earth Science* Website uses the resources and flexibility of the Internet to aid in your study of the topics in this chapter. Written and developed by Earth science instructors, this site will help improve your understanding of Earth science. Visit **http://www.prenhall.com/lutgens** and click on the cover of *Foundations of Earth Science* 5e to find:

- Online review quizzes.
- Critical thinking exercises.
- Links to chapter-specific Web resources.
- Internet-wide key term searches.

http://www.prenhall.com/lutgens

GEODe: Earth Science

GEODe: Earth Science makes studying more effective by reinforcing key concepts using animation, video, narration, interactive exercises, and practice quizzes. A copy is included with every copy of *Foundations of Earth Science 5e.*

933
Coastal Processes
Wave Erosion
• Main Menu
• Glossary
• QUIT
Reverse
Forward
Just as rivers do most of their erosional work during floods, so too do waves have their greatest impact during storms.

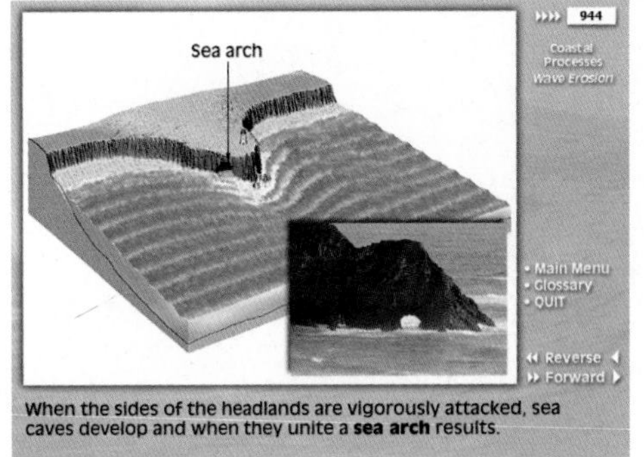
944
Sea arch
Coastal Processes
Wave Erosion
• Main Menu
• Glossary
• QUIT
Reverse
Forward
When the sides of the headlands are vigorously attacked, sea caves develop and when they unite a **sea arch** results.

UNIT VI EARTH'S DYNAMIC ATMOSPHERE

CHAPTER

11 Heating the Atmosphere

At 1886 meters (6288 feet), Mount Washington is the highest peak in the White Mountains and in fact the entire Northeast. It is a popular destination for hikers and is considered by some to be the "Home of the World's Worst Weather" due to its extreme cold, heavy snows, high winds, frequent icing, and dense fog. The observatory at its summit keeps detailed weather records. (Photo by Jim Salge)

FOCUS ON LEARNING

To assist you in learning the important concepts in this chapter, you will find it helpful to focus on the following questions:

1. What is weather? How is it different from climate?
2. What are the most important elements of weather and climate?
3. What are the major components of clean, dry air?
4. What is the extent and structure of the atmosphere?
5. What causes the seasons?
6. How is the atmosphere heated?
7. What causes temperature to vary from place to place?

Earth's atmosphere is unique. No other planet in our solar system has an atmosphere with the exact mixture of gases or the heat and moisture conditions necessary to sustain life as we know it. The gases that make up Earth's atmosphere and the controls to which they are subject are vital to our existence. In this chapter, we will begin our examination of the ocean of air in which we all must live. What is the composition of the atmosphere? Where does the atmosphere end and outer space begin? What causes the seasons? How is air heated? What factors control temperature variations over the globe?

Weather influences our everyday activities, our jobs, and our health and comfort (Figure 11.1). Many of us pay little attention to the weather unless we are inconvenienced by it or it adds to our enjoyment outdoors. Nevertheless, few other aspects of our physical environment affect our lives more than the phenomena we collectively call the weather.

The United States has the greatest variety of weather of any country in the world. Severe weather events such as tornadoes, flash floods, and intense thunderstorms, as well as hurricanes and blizzards, are collectively more frequent and damaging in the United States than in any other nation. Beyond its direct impact on the lives of individuals, the weather has a strong effect on the world economy by influencing agriculture, energy use, water resources, transportation, and industry.

Weather clearly influences our lives a great deal. Yet, it is also important to realize that people influence the atmosphere and its behavior as well. There are, and will continue to be, significant political and scientific decisions that must be made involving these impacts. Important examples are air pollution control and the effects of human activities on global climate and the atmosphere's protective ozone layer. We all need increased awareness and understanding of our atmosphere and its behavior.

Weather and Climate

Acted on by the combined effects of Earth's motions and energy from the Sun, our planet's formless and invisible envelope of air reacts by producing an infinite variety of weather, which in turn creates the basic patterns of global climates. Although not identical, weather and climate have much in common.

Weather is constantly changing, sometimes from hour to hour and at other times from day to day. It is a term that refers to the state of the atmosphere at a given time and place. Whereas changes in the weather are continuous and sometimes seemingly erratic, it is nevertheless possible to arrive at a generalization of these variations. Such a description of aggregate weather conditions is termed **climate.** It is based on observations that have been accumulated over many years. Climate is often defined simply as "average weather," but this is an inadequate definition. In order to more accurately portray the character of an area, variations and extremes must also be included, as well as the probabilities that such departures will take place. For example, it is not only necessary for farmers to know the average rainfall during the growing season, but it is also important to know the frequency of extremely wet and

Figure 11.1 There are few aspects of our physical environment that influence our daily lives more than the weather. **A.** Aerial view of Punta Gorda, Florida, after Hurricane Charley struck on August 13, 2004. This Category 4 storm, one of four to strike Florida in 2004, had sustained winds of 233 kilometers (145 miles) per hour. (Photo by Rhona Wise/EPA) **B.** Residents of Boston dig out after a blizzard dropped up to two feet of snow in the area on January 23, 2005. (AP Photo/Michael Dwyer) **C.** Exceptional winter rains in 2005 triggered many debris flows in Southern California, including this one at La Conchita that killed 10 people and destroyed 18 homes. (Photo by G. Delaurentis)

A.

B.

C.

extremely dry years. Thus, climate is the sum of all statistical weather information that helps describe a place or region.

Suppose you were planning a vacation trip to an unfamiliar place. You would probably want to know what kind of weather to expect. Such information would help as you selected clothes to pack and could influence decisions regarding activities you might engage in during your stay. Unfortunately, weather forecasts that go beyond a few days are not very dependable. Instead, you might ask someone who is familiar with the area about what kind of weather to expect. "Are thunderstorms common?" "Does it get cold at night?" "Are the afternoons sunny?" What you are seeking is information about the climate, the conditions that are typical for that place. Another useful source of information is the great variety of climate tables, maps, and graphs that are available. For example, the graph in Figure 11.2 shows average daily high and low temperatures for each month, as well as extremes for New York City.

Such information could no doubt help as you planned your trip. But it is important to realize that *climate data cannot predict the weather.* Although the place may usually (climatically) be warm, sunny, and dry during the time of your planned vacation, you may actually experience cool, overcast, and rainy weather. There is a well-known saying that summarizes this idea: "Climate is what you expect, but weather is what you get."

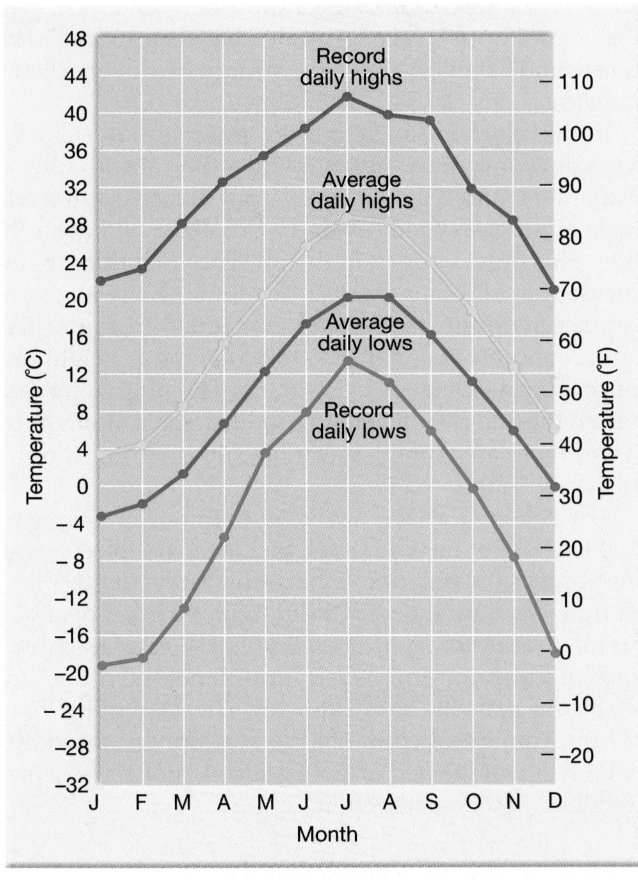

Figure 11.2 Graph showing daily temperature data for New York City. In addition to the average daily maximum and minimum temperatures for each month, extremes are also shown. As this graph shows, there can be significant departures from average.

Did You Know?

Because data from every part of the globe are needed to produce accurate weather forecasts, the World Meteorological Organization (WMO) was established by the United Nations to coordinate scientific activity related to weather and climate. It consists of 187 member states and territories. Its World Weather Watch provides up-to-the-minute standardized observations through member-operated observation systems. This global system involves 10 satellites, 10,000 land-observation and 7000 ship stations, as well as hundreds of automated data buoys and thousands of aircraft.

The nature of both weather and climate is expressed in terms of the same basic **elements**—those quantities or properties that are measured regularly. The most important are (1) air temperature, (2) humidity, (3) type and amount of cloudiness, (4) type and amount of precipitation, (5) air pressure, and (6) the speed and direction of the wind. These elements are the major variables by which weather patterns and climate types are depicted. Although you will study these elements separately at first, keep in mind that they are very much interrelated. A change in one of the elements often produces changes in the others.

Composition of the Atmosphere

Air is *not* a unique element or compound. Rather, **air** is a *mixture* of many discrete gases, each with its own physical properties, in which varying quantities of tiny solid and liquid particles are suspended.

Major Components

The composition of air is not constant; it varies from time to time and from place to place. If the water vapor, dust, and other variable components were removed from the atmosphere, we would find that its makeup is very stable worldwide, up to an altitude of about 80 kilometers (50 miles).

As you can see in Figure 11.3, clean, dry air is composed almost entirely of two gases—78 percent nitrogen and 21 percent oxygen. Although these gases are the most plentiful components of air and are of great significance to life on Earth, they are of little or no importance in affecting weather phenomena. The remaining 1 percent of dry air is mostly the inert gas argon (0.93 percent) plus tiny quantities of a number of other gases. Carbon dioxide, although present in only minute amounts (0.038 percent), is nevertheless an important constituent of air, because it has the ability to absorb heat energy radiated by Earth and thus influences the heating of the atmosphere.

Variable Components

Air includes many gases and particles that vary significantly from time to time and from place to place. Important examples include water vapor, dust particles, and ozone. Although

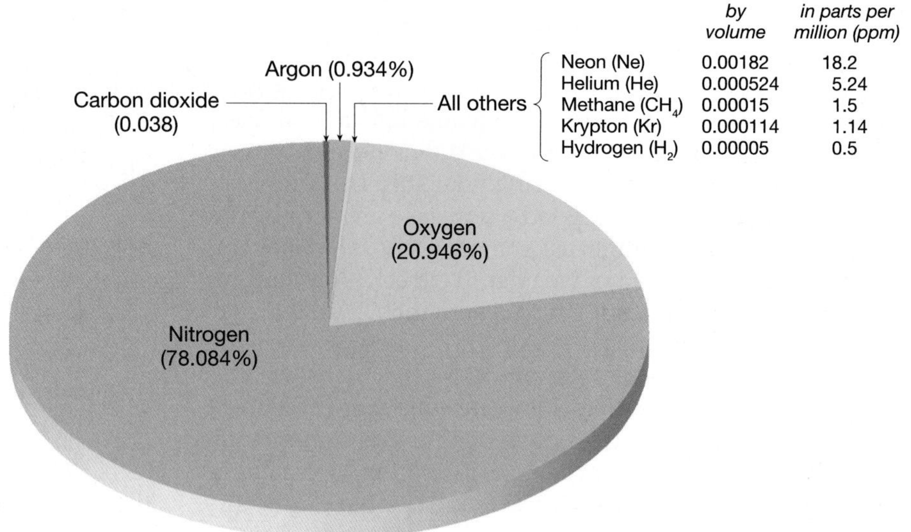

Figure 11.3 Proportional volume of gases comprising dry air. Nitrogen and oxygen clearly dominate.

usually present in small percentages, they can have significant effects on weather and climate.

Water Vapor The amount of water vapor in the air varies considerably, from practically none at all up to about 4 percent by volume. Why is such a small fraction of the atmosphere so significant? Certainly, the fact that water vapor is the source of all clouds and precipitation would be enough to explain its importance. However, water vapor has other roles. Like carbon dioxide, it has the ability to absorb heat energy given off by Earth as well as some solar energy. It is therefore important when we examine the heating of the atmosphere.

When water changes from one state to another (see Figure 12.2, p. 323), it absorbs or releases heat. This energy is termed *latent heat,* which means "hidden" heat. As we will see in later chapters, water vapor in the atmosphere transports this latent heat from one region to another, and it is the energy source that helps drive many storms.

Aerosols The movements of the atmosphere are sufficient to keep a large quantity of solid and liquid particles suspended within it. Although visible dust sometimes clouds the sky, these relatively large particles are too heavy to stay in the air for very long. Many other particles are microscopic and remain suspended for considerable periods of time. They may originate from many sources, both natural and human made, and include sea salts from breaking waves; fine soil blown into the air; smoke and soot from fires; pollen and microorganisms lifted by the wind, ash, and dust from volcanic eruptions; and more (Figure 11.4A). Collectively, these tiny solid and liquid particles are called **aerosols.**

From a meteorological standpoint, these tiny, often invisible particles can be significant. First, many act as surfaces on which water vapor can condense, an important function in the formation of clouds and fog. Second, aerosols can absorb or reflect incoming solar radiation. Thus, when an air pollution episode is occurring or when ash fills the sky following a volcanic eruption, the amount of sunlight reaching Earth's surface can be measurably reduced. Finally, aerosols contribute to an optical phenomenon we have all observed—the varied hues of red and orange at sunrise and sunset (Figure 11.4B).

Ozone Another important component of the atmosphere is *ozone.* It is a form of oxygen that combines three oxygen atoms into each molecule (O_3). Ozone is not the same as the oxygen we breathe, which has two atoms per molecule (O_2). There is very little ozone in the atmosphere, and its distribution is not uniform. It is concentrated between 10 and 50 kilometers (6 and 31 miles) above the surface in a layer called the *stratosphere.*

In this altitude range, oxygen molecules (O_2) are split into single atoms of oxygen (O) when they absorb ultraviolet radiation emitted by the Sun. Ozone is then created when a single atom of oxygen (O) and a molecule of oxygen (O_2) collide. This must happen in the presence of a third, neutral molecule that acts as a *catalyst* by allowing the reaction to take place without itself being consumed in the process. Ozone is concentrated in the 10- to 50-kilometer height range because a crucial balance exists there: The ultraviolet radiation from the Sun is sufficient to produce single atoms of oxygen, and there are enough gas molecules to bring about the required collisions.

The presence of the ozone layer in our atmosphere is crucial to those of us who dwell on Earth. The reason is that ozone absorbs the potentially harmful ultraviolet (UV) radiation from the Sun. If ozone did not filter a great deal of the ultraviolet radiation, and if the Sun's UV rays reached the surface of Earth undiminished, our planet would be uninhabitable for most life as we know it. Thus, anything that reduces the amount of ozone in the atmosphere could affect the well-being of life on Earth. Just such a problem exists and is described in the next section.

Ozone Depletion—A Global Issue

Although stratospheric ozone is 10 to 50 kilometers up, it is vulnerable to human activities. Chemicals we produce are breaking up ozone molecules in the stratosphere, weakening

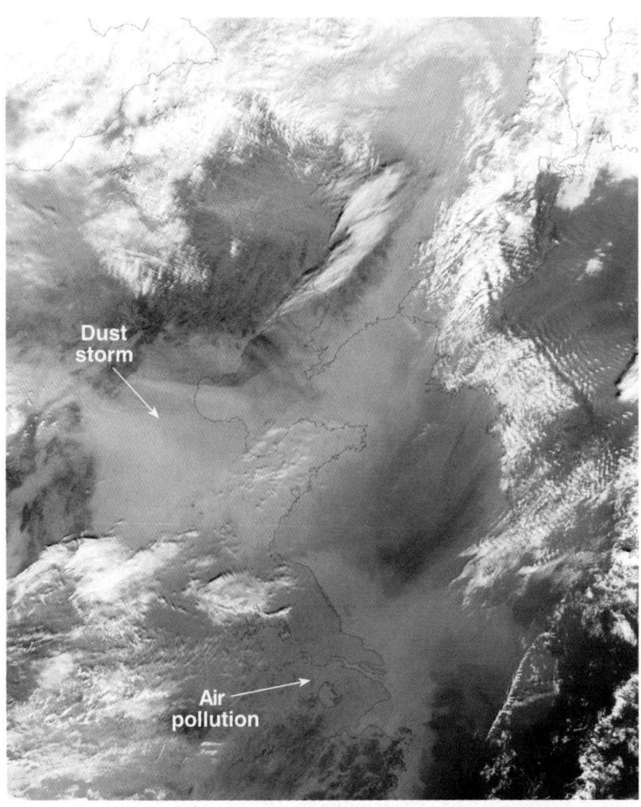

A.

B.

Figure 11.4 **A.** This satellite image from November 11, 2002, shows two examples of aerosols. First, a large dust storm is blowing across northeastern China toward the Korean Peninsula. Second, a dense haze toward the south (bottom center) is human-generated air pollution. (NASA) **B.** Dust in the air can cause sunsets to be especially colorful. (Photo by Steve Elmore/CORBIS/The Stock market)

our shield against UV rays. This loss of ozone is a serious global-scale environmental problem. Measurements over the past two decades confirm that ozone depletion is occurring worldwide and is especially pronounced above Earth's poles. You can see this effect over the South Pole in Figure 11.5.

Over the past half century, people have unintentionally placed the ozone layer in jeopardy by polluting the atmosphere. The offending chemicals are known as *chlorofluorocarbons* (*CFCs* for short). Over the decades, many uses were developed for CFCs: coolants for air-conditioning and refrigeration equipment, cleaning solvents for electronic components and computer chips, and propellants for aerosol sprays. CFCs are also used to produce certain plastic foams for products such as cups and home insulation.

Because CFCs are practically inert (that is, not chemically active) in the lower atmosphere, some of these gases gradually make their way to the ozone layer, where sunlight separates the chemicals into their constituent atoms. The chlorine atoms released this way break up some of the ozone molecules.

Because ozone filters out most of the UV radiation from the Sun, a decrease in its concentration permits more of these harmful wavelengths to reach Earth's surface. The most serious threat to human health is an increased risk of skin cancer. An increase in damaging UV radiation also can impair the human immune system as well as promote cataracts, a clouding of the eye lens that reduces vision and may cause blindness if not treated.

In response to this problem, an international agreement known as the *Montreal Protocol* was developed under the sponsorship of the United Nations to eliminate the production and use of CFCs.

Although relatively strong action was taken, CFC levels in the atmosphere will not drop rapidly. Once in the atmosphere, CFC molecules can take many years to reach the ozone layer, and once there, they can remain active for decades. This does not promise a near-term reprieve for the ozone layer. Even after CFC levels begin to decline, the ozone layer probably will not cleanse itself of most of the pollutants until sometime in the second half of the twenty-first century.

Did You Know?

Although naturally occurring ozone in the stratosphere is critical to life on Earth, it is regarded as a pollutant when produced at ground level because it can damage vegetation and be harmful to human health. Ozone is a major component in a noxious mixture of gases and particles called *photochemical smog.* It forms as a result of reactions triggered by sunlight that occur among pollutants emitted by motor vehicles and industrial sources.

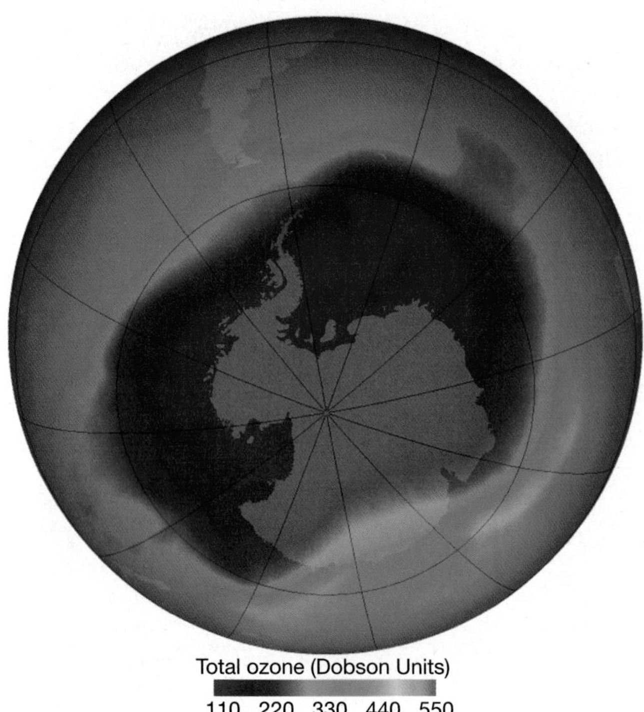

Figure 11.5 This satellite image shows ozone distribution in the Southern Hemisphere on September 24, 2006. The area of greatest depletion is called the "ozone hole." Dark blue colors correspond to the region with the sparsest ozone. Average values are about 300 Dobson Units. Any area below 220 Dobson Units is considered part of the hole. The ozone hole forms over Antarctica during the Southern Hemisphere spring. In 2006, it extended over about 29 million square kilometers (11.4 million square miles), an area slightly larger than all of North America. It tied the record for the largest ozone hole yet recorded. (NASA)

Height and Structure of the Atmosphere

Earth's Dynamic Atmosphere
▼ Heating The Atmosphere

To say that the atmosphere begins at Earth's surface and extends upward is obvious. However, where does the atmosphere end and outer space begin? There is no sharp boundary; the atmosphere rapidly thins as you travel away from Earth, until there are too few gas molecules to detect.

Pressure Changes

To understand the vertical extent of the atmosphere, let us examine the changes in the atmospheric pressure with height. Atmospheric pressure is simply the weight of the air above. At sea level, the average pressure is slightly more than 1000 millibars (mb). This corresponds to a weight of slightly more than 1 kilogram per square centimeter (14.7 pounds per square inch). Obviously, the pressure at higher altitudes is less (Figure 11.6).

One-half of the atmosphere lies below an altitude of 5.6 kilometers (3.5 miles). At about 16 kilometers (10 miles), 90 percent of the atmosphere has been transversed, and above 100 kilometers (62 miles), only 0.00003 percent of all the gases making up the atmosphere remain. Even so, traces of our atmosphere extend far beyond this altitude, gradually merging with the emptiness of space.

Temperature Changes

By the early twentieth century, much had been learned about the lower atmosphere. The upper atmosphere was partly known from indirect methods. Data from balloons and kites had revealed that the air temperature dropped with increasing height above Earth's surface. This phenomenon is felt by anyone who has climbed a high mountain and is obvious in pictures of snowcapped mountaintops rising above snow-free lowlands (Figure 11.7). We divide the atmosphere vertically into four layers on the basis of temperature (Figure 11.8).

Troposphere The bottom layer in which we live, where temperature decreases with an increase in altitude, is the **troposphere.** The term literally means the region where air "turns over," a reference to the turbulent weather in this lowermost zone. The troposphere is the chief focus of meteorologists, because it is in this layer that essentially all important weather phenomena occur.

The temperature decrease in the troposphere is called the **environmental lapse rate.** Although its average value is

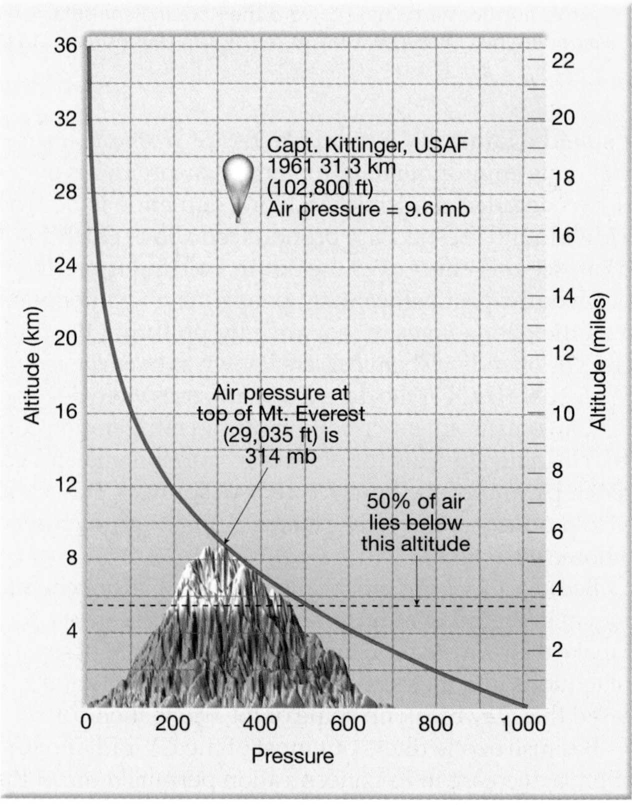

Figure 11.6 Atmospheric pressure variation with altitude. The rate of pressure decrease with an increase in altitude is not constant. Pressure decreases rapidly near Earth's surface and more gradually at greater heights. Put another way, the graph shows that the vast bulk of the gases making up the atmosphere is very near Earth's surface and that the gases gradually merge with the emptiness of space.

Figure 11.7 Temperatures drop with an increase in altitude in the troposphere. Therefore, it is possible to have snow on a mountaintop and warmer, snow-free lowlands below. Mount Kerkeslin, Jasper National Park, Alberta, Canada. (Photo by Carr Clifton/Minden Pictures)

6.5°C per kilometer (3.5°F per 1000 feet), a figure known as the *normal lapse rate,* its value is quite variable. Thus, to determine the actual environmental lapse rate for any particular time and place, as well as to gather information about vertical changes in pressure, wind, and humidity, radiosondes are used. The *radiosonde* is an instrument package that is attached to a balloon and transmits data by radio as it ascends through the atmosphere (Figure 11.9).

The thickness of the troposphere is not the same everywhere; it varies with latitude and the season. On the average, the temperature drop continues to a height of about 12 kilometers (7.4 miles). The outer boundary of the troposphere is the *tropopause.*

Stratosphere Beyond the tropopause is the **stratosphere.** In the stratosphere, the temperature remains constant to a height of about 20 kilometers (12 miles) and then begins a gradual increase that continues until the *stratopause,* at a height of about 50 kilometers (31 miles) above Earth's surface. Below the tropopause, atmospheric properties such as temperature and humidity are readily transferred by large-scale turbulence and mixing. Above the tropopause, in the stratosphere, they are not. The reason for the increased temperatures in the stratosphere is that the atmosphere's ozone is concentrated in this layer. Recall that ozone absorbs ultraviolet radiation from the Sun. As a consequence, the stratosphere is heated.

Mesosphere In the third layer, the **mesosphere,** temperatures again decrease with height until, at the *mesopause*—

Figure 11.8 Thermal structure of the atmosphere.

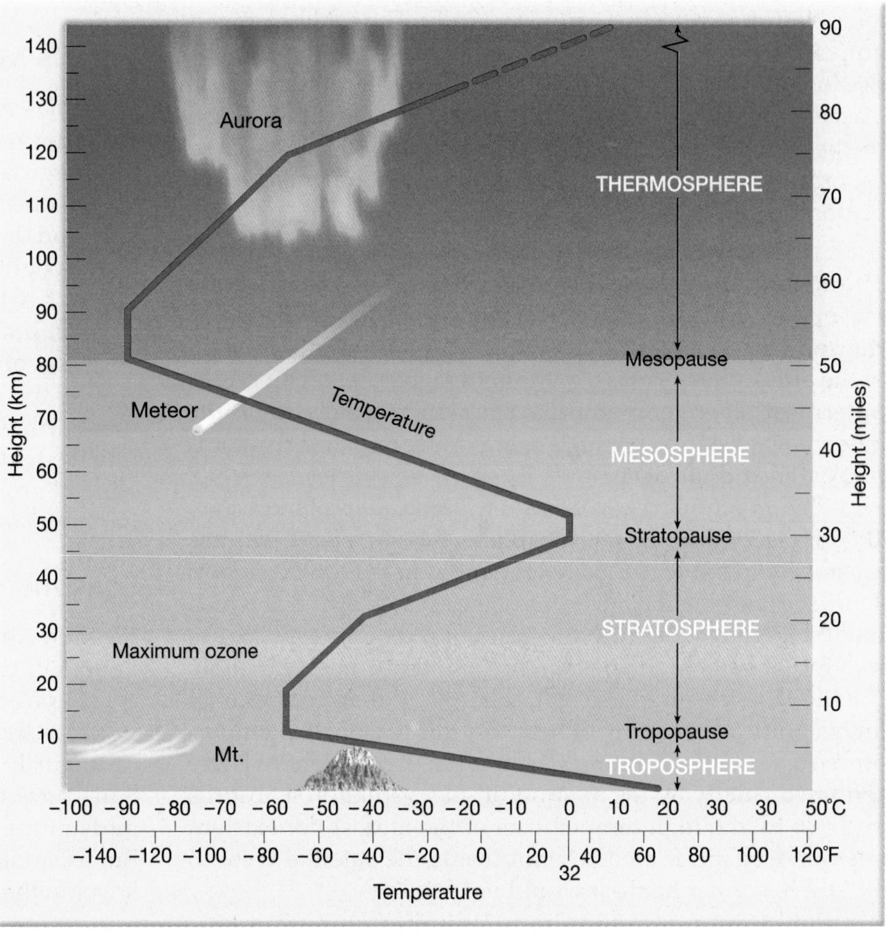

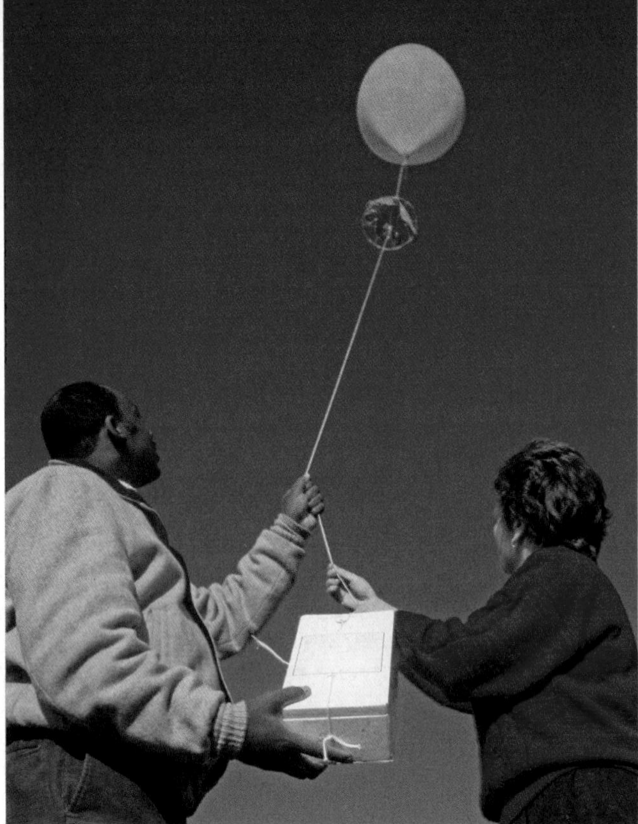

Figure 11.9 A lightweight package of instruments, the radiosonde, is carried aloft by a small weather balloon. Radiosondes supply data on vertical changes in temperature, pressure, and humidity. (Photo by Marc Burnett/Photo Researchers, Inc.)

approximately 80 kilometers (50 miles) above the surface—the temperature approaches −90°C (−130°F). The coldest temperatures anywhere in the atmosphere occur at the mesopause. Because accessibility is difficult, the mesosphere is one of the least explored regions of the atmosphere. It cannot be reached by the highest research balloons nor is it accessible to the lowest orbiting satellites. Recent technical developments are just beginning to fill this knowledge gap.

Thermosphere The fourth layer extends outward from the mesopause and has no well-defined upper limit. This is the **thermosphere,** a layer that contains only a minute fraction of the atmosphere's mass. In the extremely rarefied air of this outermost layer, temperatures again increase owing to the absorption of shortwave high-energy solar radiation by atoms of oxygen and nitrogen.

Temperatures rise to extremely high values of more than 1000°C (1832°F) in the thermosphere. But such temperatures are not comparable to those experienced near Earth's surface. Temperature is defined in terms of the average speed at which molecules move. Because the gases of the thermosphere are moving at very high speeds, the temperature is very high. But the gases are so sparse that, collectively, they possess only an insignificant quantity of heat. For this reason, the temperature of a satellite orbiting Earth in the thermosphere is determined chiefly by the amount of solar radiation it absorbs and not by the high temperature of the almost nonexistent surrounding air. If an astronaut inside the satellite were to expose his or her hand, it would not feel hot.

Earth–Sun Relationships

Always remember that nearly all of the energy that drives Earth's variable weather and climate comes from the Sun. Earth intercepts only a minute percentage of the energy given off by the Sun—less than 1 two-billionth. This may seem to be an insignificant amount until we realize that it is several hundred thousand times the electrical-generating capacity of the United States.

Solar energy is not distributed evenly over Earth's land-sea surface. The amount of energy received varies with latitude, time of day, and season of the year. Contrasting images of polar bears on ice rafts and palm trees along a remote tropical beach serve to illustrate the extremes. It is the unequal heating of Earth that creates winds and drives ocean currents. These movements, in turn, transport heat from the tropics toward the poles in an unending attempt to balance energy inequalities. The consequences of these processes are the phenomena we call weather.

If the Sun were to be "turned off," global winds would quickly subside. Yet, as long as the Sun shines, the winds will blow and the phenomena we know as weather will persist. So to understand how the atmosphere's dynamic weather machine works, we must first know why different latitudes receive varying quantities of solar energy and why the amount of solar energy changes to produce the seasons. As you will see, the variations in solar heating are caused by the motions of Earth relative to the Sun and by variations in Earth's land-sea surface.

Earth's Motions

Earth has two principal motions—rotation and revolution. **Rotation** is the spinning of Earth about its axis. The axis is an imaginary line running through the poles. Our planet rotates once every 24 hours, producing the daily cycle of daylight and darkness. At any moment, half of Earth is experiencing daylight, and the other half darkness. The line separating the dark half of Earth from the lighted half is called the **circle of illumination.**

Revolution refers to the movement of Earth in its orbit around the Sun. Hundreds of years ago, most people believed that Earth was stationary in space and that the Sun and stars revolved around our planet. We now know that Earth is traveling at more than 107,000 kilometers per hour (66,000 miles per hour) in its orbit about the Sun.

Seasons

We know that it is colder in winter than in summer. But why? Length of daylight certainly accounts for some of the difference. Long summer days expose us to more solar radiation, whereas short winter days expose us to less.

Furthermore, a gradual change in the angle of the noon Sun above the horizon is quite noticeable (Figure 11.10). At midsummer, the noon Sun is seen high above the horizon. But as summer gives way to autumn, the noon Sun appears lower in the sky and sunset occurs earlier each evening. What

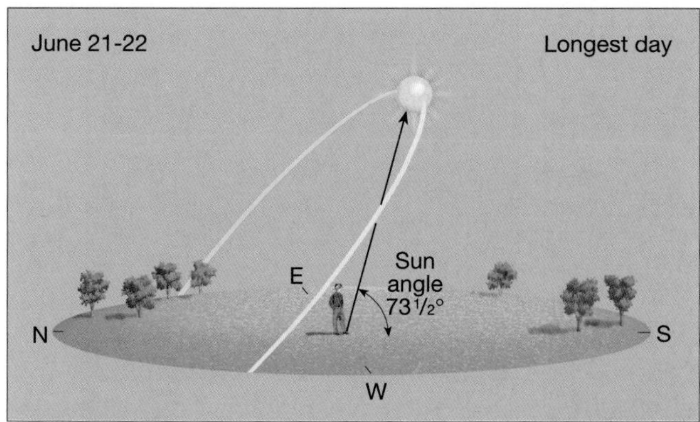

A. Summer solstice

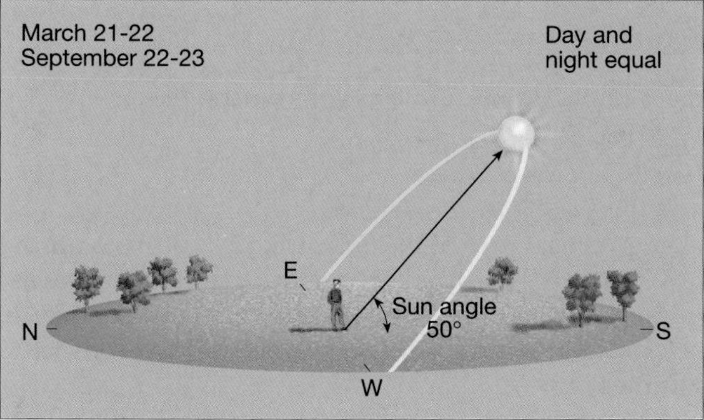

B. Spring or fall equinox

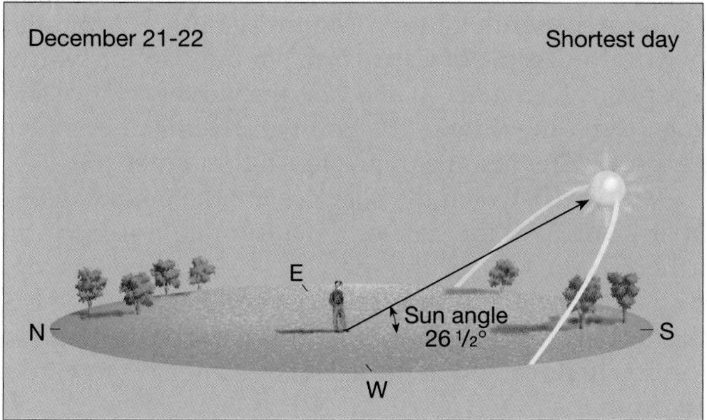

C. Winter solstice

Figure 11.10 Daily paths of the Sun for a place located at 40°N latitude for **A.** summer solstice, **B.** spring or fall equinox, and **C.** winter solstice. As we move from summer to winter, the angle of the noon Sun decreases from $73\frac{1}{2}$ to $26\frac{1}{2}$ degrees—a difference of 47 degrees. Notice also how the location of sunrise (east) and sunset (west) changes during a year.

we observe here is the annual shifting of the solar angle or *altitude* of the Sun.

The seasonal variation in the altitude of the Sun affects the amount of energy received at Earth's surface in two ways. First, when the Sun is high in the sky, the solar rays are most concentrated (you can see this in Figure 11.11A). The lower the angle, the more spread out and less intense is the solar radiation that reaches the surface (Figure 11.11B, C). To illustrate this principle, hold a flashlight at a right angle to a surface and then change the angle.

Second, and of lesser importance, the angle of the Sun determines the thickness of the atmosphere the rays must penetrate (Figure 11.12). When the Sun is directly overhead, the rays pass through a thickness of only 1 atmosphere. But rays entering at a 30-degree angle travel through twice this amount, and 5-degree rays travel through a thickness roughly equal to 11 atmospheres. The longer the path, the greater the chances that sunlight will be absorbed, reflected, or scattered by the atmosphere, all of which reduce the intensity at the surface. These same effects account for the fact that we cannot look directly at the midday Sun, but we can enjoy gazing at a sunset.

It is important to remember that Earth has a spherical shape. Hence, on any given day, only places located at a particular latitude receive vertical (90-degree) rays from the Sun. As we move either north or south of this location, the Sun's rays strike at an ever decreasing angle. Thus, the nearer a place is to the latitude receiving vertical rays of the Sun, the higher will be its noon Sun and the more intense will be the radiation it receives (Figure 11.12).

Earth's Orientation

What causes yearly fluctuations in Sun angle and length of daylight? Variations occur because Earth's orientation to the Sun continually changes as it travels along its orbit. Earth's axis (the imaginary line through the poles around which Earth rotates) is not perpendicular to the plane of its orbit around the Sun. Instead, it is tilted 23½ degrees from the perpendicular, as shown in Figure 11.12. This is termed the **inclination of the axis.** As you will see, if the axis were not inclined, we would have no seasonal changes. In addition, because the axis remains pointed in the same direction (toward the North Star) as Earth journeys around the Sun, the orientation of Earth's axis to the Sun's rays is constantly changing (Figure 11.13).

For example, on one day in June each year, Earth's position in orbit is such that the Northern Hemisphere is "leaning" 23½ degrees toward the Sun (Figure 11.13, left). Six months later, in December, when Earth has moved to the opposite side of its orbit, the Northern Hemisphere leans 23½ degrees away from the Sun (Figure 11.13, right). On days between these extremes, Earth's axis is leaning at amounts less than 23½ degrees to the rays of the Sun. This change in orientation causes the spot where the Sun's rays are vertical to make a yearly migration from 23½ degrees north of the equator to 23½ degrees south of the equator.

Did You Know?

The only state where the Sun is ever directly overhead is Hawaii because all of the other states are north of the Tropic of Cancer. Honolulu is about 21° north latitude and experiences a 90° Sun angle twice each year—once at noon on about May 27 and again at noon about July 20.

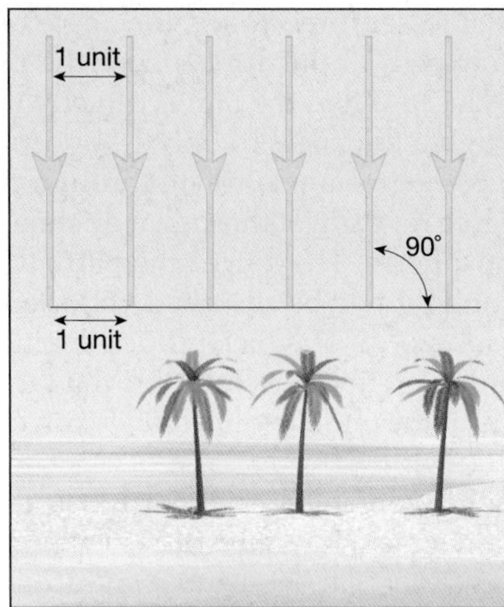

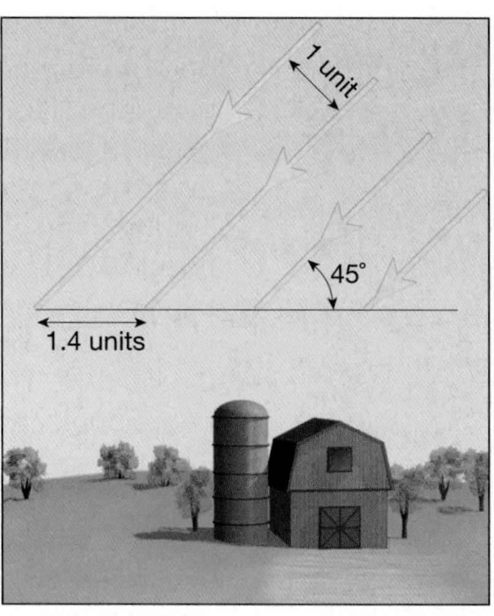

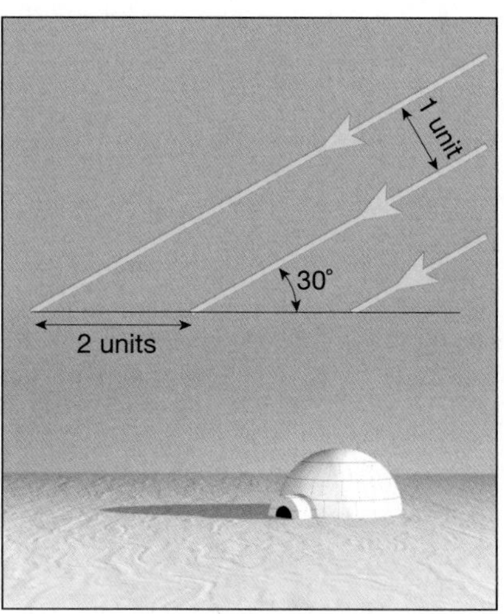

Figure 11.11 Changes in the Sun's angle cause variations in the amount of solar energy reaching Earth's surface. The higher the angle, the more intense the solar radiation.

In turn, this migration causes the angle of the noon Sun to vary by as much as 47 degrees (23½ + 23½) during the yearat places located poleward of latitude 23½ degrees. For example, a midlatitude city such as New York (about 40 degrees north latitude) has a maximum noon Sun angle of 73½ degrees when the Sun's vertical rays reach their farthest northward location in June and a minimum noon Sun angle of 26½ degrees six months later.

Solstices and Equinoxes

Historically, four days each year have been given special significance based on the annual migration of the direct rays of the Sun and its importance to the yearly weather cycle. On June 21 or 22, Earth is in a position such that the north end of its axis is tilted 23½ degrees *toward* the Sun (Figure 11.14A). At this time, the vertical rays of the Sun strike 23½ degrees north latitude (23½ degrees north of the equator), a latitude known as the **Tropic of Cancer.** For people in the Northern Hemisphere, June 21 or 22 is known as the **summer solstice,** the first "official" day of summer.

Six months later, on about December 21 or 22, Earth is in the opposite position, with the Sun's vertical rays striking at 23½ degrees south latitude (Figure 11.14B). This parallel is known as the **Tropic of Capricorn.** For those in the Northern Hemisphere, December 21 and 22 is the **winter solstice.** However, at the same time in the Southern Hemisphere, people are experiencing just the opposite—the summer solstice.

Midway between the solstices are the equinoxes. September 22 or 23 is the date of the **autumnal equinox** in the Northern Hemisphere, and March 21 or 22 is the date of the **spring equinox.** On these dates, the vertical rays of the Sun strike the equator (0 degrees latitude) because Earth is in such a position in its orbit that the axis is tilted neither toward nor away from the Sun (Figure 11.14C).

The length of daylight versus darkness is also determined by Earth's position in orbit. The length of daylight on June 21, the summer solstice in the Northern Hemisphere, is greater than the length of night. This fact can be established from Figure 11.14A by comparing the fraction of a given latitude that is on the "day" side of the circle of illumination with the fraction on the "night" side. The opposite is true for the winter solstice, when the nights are longer than the days. Again for comparison let us consider New York City, which has about 15 hours of daylight on June 21 and only about 9 hours on December 21 (you can see this in Figure 11.14 and Table 11.1). Also note from Table 11.1 that the farther north of the equator you are on June 21, the longer the period of daylight. When you reach the Arctic Circle (66½ degrees north latitude), the length of daylight is 24 hours. This is the land of the "midnight Sun," which does not set for about six months at the North Pole.

During an equinox (meaning "equal night"), the length of daylight is 12 hours *everywhere* on Earth, because the circle

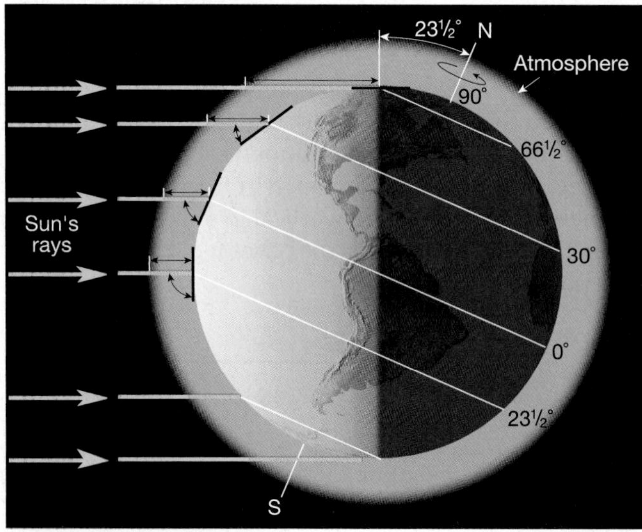

Figure 11.12 Rays striking Earth at a low angle (toward the poles) must travel through more of the atmosphere than rays striking at a high angle (around the equator) and thus are subject to greater depletion by reflection and absorption.

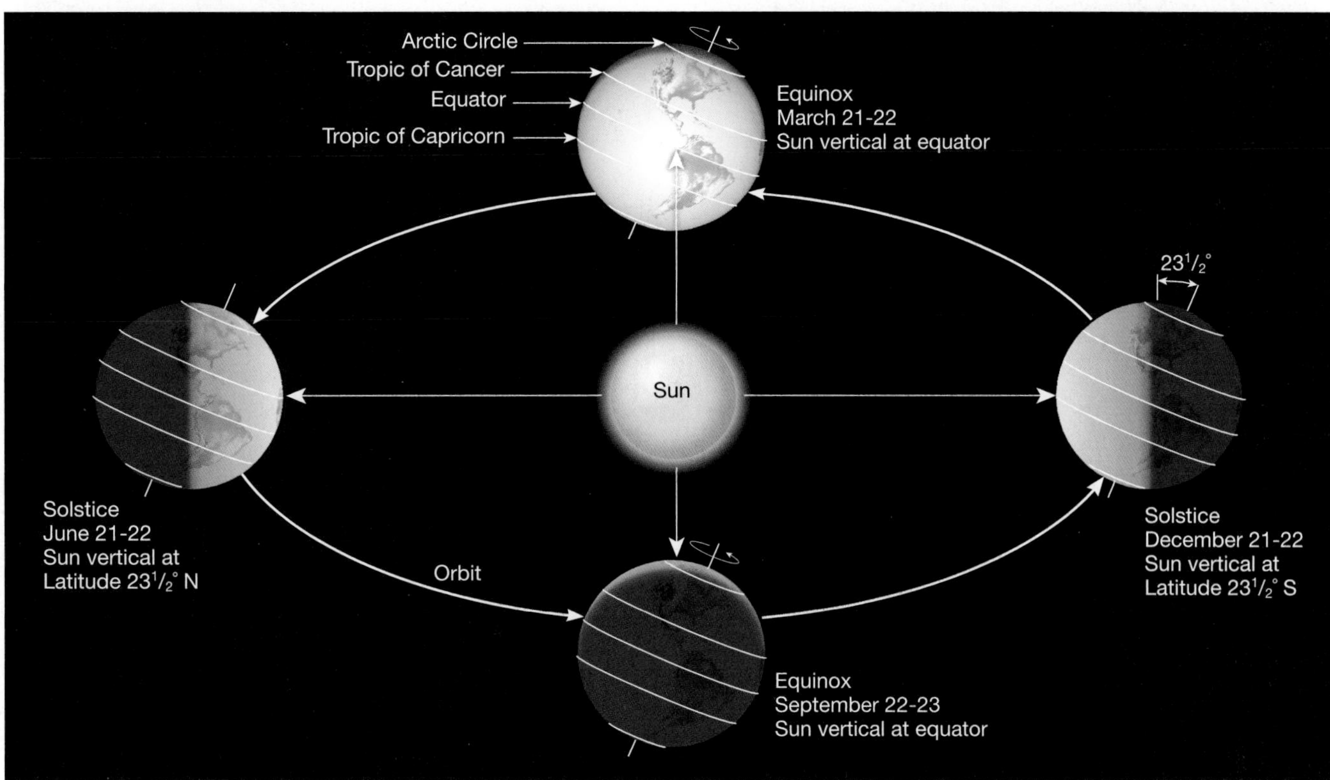

Figure 11.13 Earth-Sun relationships.

A. June Solstice (Northern Hemisphere summer)

B. December Solstice (Northern Hemisphere winter)

C. Spring/Autumnal Equinox

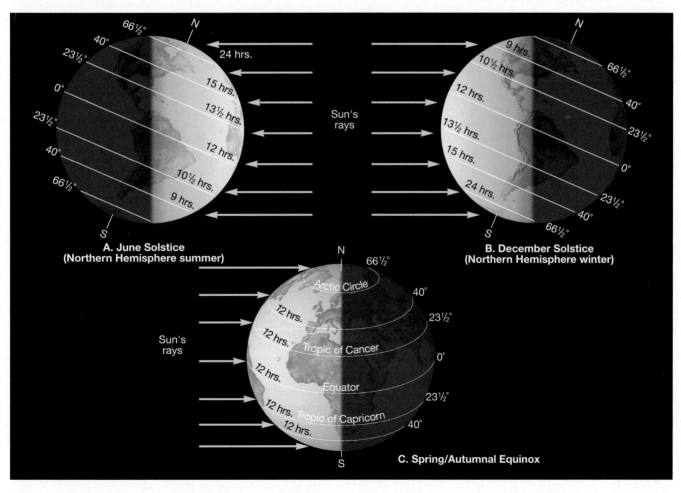

Figure 11.14 Characteristics of the solstices and equinoxes.

Table 11.1 Length of daylight

Latitude (degrees)	Summer Solstice	Winter Solstice	Equinoxes
0	12 h	12 h	12 h
10	12 h 35 min	11 h 25 min	12
20	13 h 12	10 h 48	12
30	13 h 56	10 h 04	12
40	14 h 52	9 h 08	12
50	16 h 18	7 h 42	12
60	18 h 27	5 h 33	12
70	24 h (for 2 mo)	0 00	12
80	24 h (for 4 mo)	0 00	12
90	24 h (for 6 mo)	0 00	12

of illumination passes directly through the poles, dividing the latitudes in half (Figure 11.14C).

As a review of the characteristics of the summer solstice for the Northern Hemisphere, examine Figure 11.14A and Table 11.1 and consider the following facts:

1. The solstice occurs on June 21 or 22.
2. The vertical rays of the Sun are striking the Tropic of Cancer (23½ degrees north latitude).
3. Locations in the Northern Hemisphere are experiencing their greatest length of daylight (opposite for the Southern Hemisphere).
4. Locations north of the Tropic of Cancer are experiencing their highest noon Sun angles (opposite for places south of the Tropic of Capricorn).
5. The farther you are north of the equator, the longer the period of daylight, until the Arctic Circle is reached, where daylight lasts for 24 hours (opposite for the Southern Hemisphere).

The facts about the winter solstice are just the opposite. It should now be apparent why a midlatitude location is warmest in the summer, for it is then that days are longest and Sun's altitude is highest.

In summary, seasonal variations in the amount of solar energy reaching places on Earth's surface are caused by the migrating vertical rays of the Sun and the resulting variations in Sun angle and length of daylight. These changes, in turn, cause the month-to-month variations in temperature observed at most locations outside the tropics. Figure 11.15 shows mean monthly temperatures for selected cities at different latitudes. Notice that the cities located at more poleward latitudes experience larger temperature differences from summer to winter than do cities located nearer the equator. Also notice that temperature minimums for Southern Hemisphere locations occur in July, whereas they occur in January for most places in the Northern Hemisphere.

All places at the same latitude have identical Sun angles and lengths of daylight. If the Earth-Sun relationships just described were the only controls of temperature, we

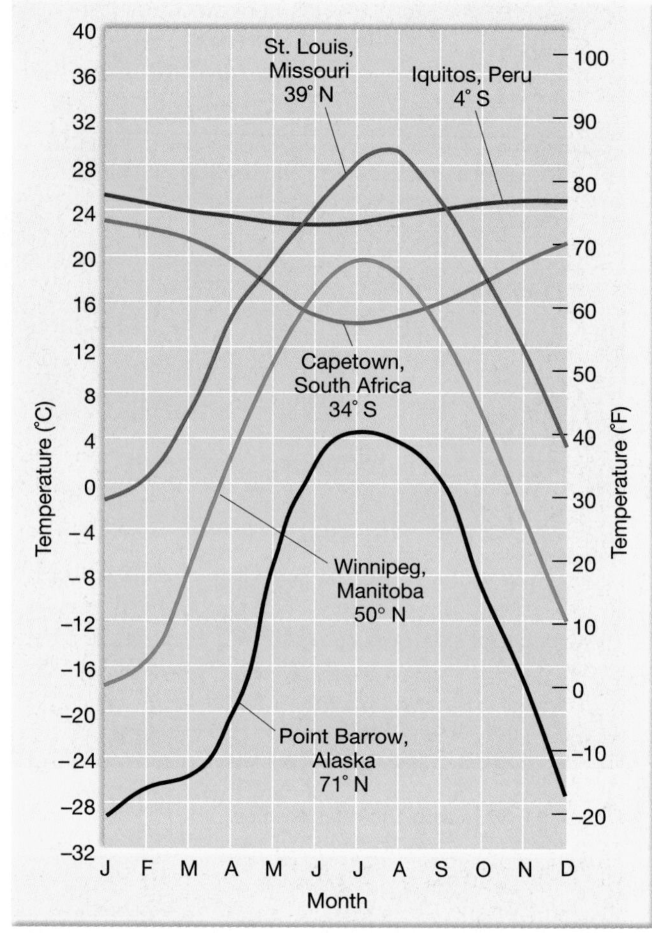

Figure 11.15 Mean monthly temperatures for five cities located at different latitudes. Note that Capetown, South Africa, experiences winter in June, July, and August.

would expect these places to have identical temperatures as well. Obviously, this is not the case. Although the altitude of the Sun is an important control of temperature, it is not the only control, as you will see.

Energy, Heat, and Temperature

The universe is made up of a combination of matter and energy. The concept of matter is easy to grasp because it is the "stuff" we can see, smell, and touch. Energy, on the other hand, is abstract and therefore more difficult to describe. For our purposes, we will define energy simply as *the capacity to do work.* We can think of work as being accomplished whenever matter is moved. You are likely familiar with some of the common forms of energy, such as thermal, chemical, nuclear, radiant (light), and gravitational energy. One type of energy is described as *kinetic energy,* which is energy of motion. Recall that matter is composed of atoms or molecules that are constantly in motion and therefore possesses kinetic energy.

Heat is a term that is commonly used synonymously with *thermal energy.* In this usage, heat is energy possessed by a material arising from the internal motions of its atoms or molecules. Whenever a substance is heated, its atoms move faster and faster, which leads to an increase in its heat content. **Temperature,** on the other hand, is related to the average kinetic energy of a material's atoms or molecules. Stated another way, the term *heat* generally refers to the quantity of energy present, whereas the word *temperature* refers to the intensity, that is, the degree of "hotness."

Heat and temperature are closely related concepts. Heat is the energy that flows because of temperature differences. In all situations, *heat is transferred from warmer to cooler objects.* Thus, if two objects of different temperature are in contact, the warmer object will become cooler and the cooler object will become warmer until they both reach the same temperature.

Mechanisms of Heat Transfer

Earth's Dynamic Atmosphere
▼ Heating The Atmosphere

The three mechanisms of heat transfer are conduction, convection, and radiation. Although we present them separately, all three processes go on simultaneously in the atmosphere. In addition, these mechanisms operate to transfer heat between Earth's surface (both land and water) and the atmosphere.

Conduction

Conduction is familiar to all of us. Anyone who has touched a metal spoon that was left in a hot pan has discovered that heat was conducted through the spoon. **Conduction** *is the transfer of heat through matter by molecular activity.* The energy of molecules is transferred through collisions between one molecule and another, with the heat flowing from the higher temperature to the lower temperature.

The ability of substances to conduct heat varies considerably. Metals are good conductors, as those of us who have touched hot metal have quickly learned (Figure 11.16). Air, on the other hand, is a very poor conductor of heat. Consequently, conduction is important only between Earth's surface and the air directly in contact with the surface. As a means of heat transfer for the atmosphere as a whole, conduction is the least significant.

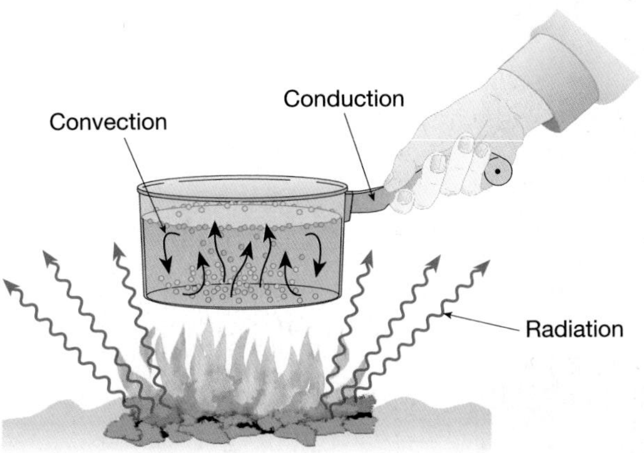

Figure 11.16 The three mechanisms of heat transfer: conduction, convection, and radiation.

Convection

Much of the heat transport that occurs in the atmosphere is carried on by convection. **Convection** *is the transfer of heat by mass movement or circulation within a substance.* It takes place in fluids (for example, liquids like the ocean and gases like air) where the atoms and molecules are free to move about.

The pan of water in Figure 11.16 illustrates the nature of simple convective circulation. Radiation from the fire warms the bottom of the pan, which conducts heat to the water near the bottom of the container. As the water is heated, it expands and becomes less dense than the water above. Because of this new buoyancy, the warmer water rises. At the same time, cooler, denser water near the top of the pan sinks to the bottom, where it becomes heated. As long as the water is heated unequally—that is, from the bottom up—the water will continue to "turn over," producing a *convective circulation.* In a similar manner, most of the heat acquired in the lowest layer of the atmosphere by way of radiation and conduction is transferred by convective flow.

On a global scale, convection in the atmosphere creates a huge, worldwide circulation of air. This is responsible for the redistribution of heat between hot equatorial regions and the frigid poles. This important process will be discussed in detail in Chapter 13.

Radiation

The third mechanism of heat transfer is **radiation.** As shown in Figure 11.16, radiation travels out in all directions from its source. Unlike conduction and convection, which need a medium to travel through, radiant energy readily travels through the vacuum of space. Thus, radiation is the heat-transfer mechanism by which solar energy reaches our planet.

From our everyday experience, we know that the Sun emits light and heat as well as the ultraviolet rays that cause suntan. Although these forms of energy comprise a major portion of the total energy that radiates from the Sun, they are only part of a large array of energy called radiation or **electromagnetic radiation.** This array, or spectrum, of electromagnetic energy is shown in Figure 11.17. All radiation, whether x-rays, microwaves, or radio waves, transmit energy through the vacuum of space at 300,000 kilometers (186,000 miles) per second and only slightly slower through our atmosphere.

Nineteenth-century physicists were so puzzled by the seemingly impossible phenomenon of energy traveling through the vacuum of space without a medium to transmit it, that they assumed that a material, which they named *ether,* existed between the Sun and Earth. This medium was thought to transmit radiant energy in much the same way that air transmits sound waves. Of course, this was incorrect. We now know that, like gravity, radiation requires no material to transmit it.

In some respects, the transmission of radiant energy parallels the motion of the gentle swells in the open ocean. Like ocean swells, electromagnetic waves come in various

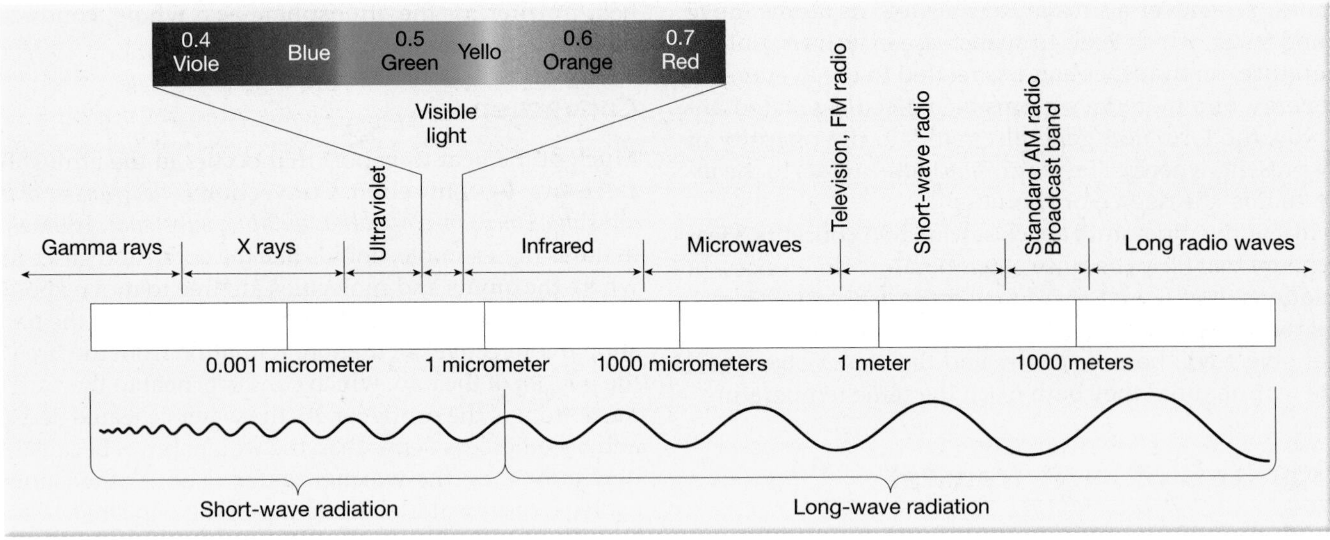

Figure 11.17 The electromagnetic spectrum, illustrating the wavelengths and names of various types of radiation.

sizes. For our purpose, the most important characteristic is their *wavelength,* or the distance from one crest to the next. Radio waves have the longest wavelengths, ranging to tens of kilometers, whereas gamma waves are the very shortest, being less than one-billionth of a centimeter long.

Visible light, as the name implies, is the only portion of the spectrum we can see. We often refer to visible light as "white" light since it appears "white" in color. However, it is easy to show that white light is really a mixture of colors, each corresponding to a particular wavelength (Figure 11.18). Using a prism, we can divide white light into a color array rainbow. Figure 11.17 shows that violet has the shortest wavelength—0.4 micrometer—and red has the longest—0.7 micrometer.

Located adjacent to red and having a longer wavelength is **infrared** radiation, which we cannot see but which we can detect as heat. The closest invisible waves to violet are called **ultraviolet (UV)** rays and are responsible for sunburn after an intense exposure to the Sun. Although we divide radiant energy into groups based on our ability to perceive them, all forms of radiation are basically the same. When any form of radiant energy is absorbed by an object, the result is an increase in molecular motion, which causes a corresponding increase in temperature.

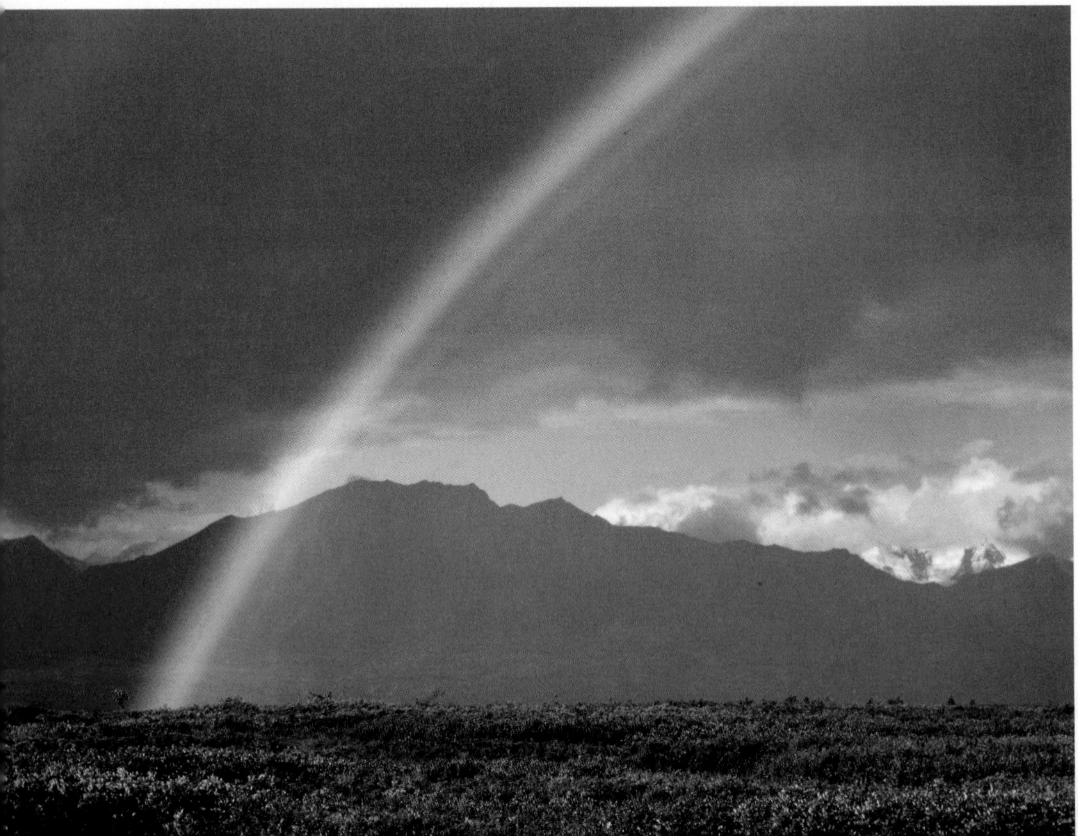

Figure 11.18 Visible light consists of an array of colors we commonly call the "colors of the rainbow." Rainbows are relatively common optical phenomena produced by the bending and reflection of light by drops of water. Denali National Park, Alaska. (Photo © by Carr Clifton Photography)

To better understand how the atmosphere is heated, it is useful to have a general understanding of the basic laws governing radiation:

1. *All objects, at whatever temperature, emit radiant energy.* Hence, not only hot objects such as the Sun but also Earth, including its polar ice caps, continually emit energy.
2. *Hotter objects radiate more total energy per unit area than do colder objects.*
3. *The hotter the radiating body, the shorter the wavelength of maximum radiation.* The Sun, with a surface temperature of about 5700°C, radiates maximum energy at 0.5 micrometer, which is in the visible range. The maximum radiation for Earth occurs at a wavelength of 10 micrometers, well within the infrared (heat) range. Because the maximum Earth radiation is roughly 20 times longer than the maximum solar radiation, Earth radiation is often called *long-wave radiation,* and solar radiation is called *short-wave radiation.*
4. *Objects that are good absorbers of radiation are good emitters as well.* Earth's surface and the Sun approach being perfect radiators because they absorb and radiate with nearly 100 percent efficiency for their respective temperatures. On the other hand, *gases are selective absorbers and radiators.* Thus, the atmosphere, which is nearly transparent (does not absorb) to certain wavelengths of radiation, is nearly opaque (a good absorber) to others. Our experience tells us that the atmosphere is transparent to visible light; hence, it readily reaches Earth's surface. This is not the case for the longer wavelength radiation emitted by Earth.

Figure 11.16 (p. 305) summarizes the various mechanisms of heat transfer. A portion of the radiant energy generated by the campfire is absorbed by the pan. This energy is readily transferred through the metal container by the process of conduction. Conduction also increases the temperature of the water at the bottom. Once warmed, this layer of water moves upward and is replaced by cool water descending from above. Thus, convection currents that redistribute the newly acquired energy throughout the pan are established. Meanwhile, the camper is warmed by radiation emitted by the fire and the pan. Furthermore, because metals are good conductors, the camper's hand is likely to be burned if he or she does not use a potholder. Like this example, the heating of Earth's atmosphere involves the processes of conduction, convection, and radiation, all of which occur simultaneously.

The Fate of Incoming Solar Radiation

Earth's Dynamic Atmosphere
▼ Heating The Atmosphere

When radiation strikes an object, three different results usually occur. First, some of the energy is *absorbed* by the object. Recall that when radiant energy is absorbed, it is converted to heat, which causes an increase in temperature. Second, substances such as water and air are transparent to certain wavelengths of radiation. Such materials simply *transmit* this energy. Radiation that is transmitted does not contribute energy to the object. Third, some radiation may "bounce off" the object without being absorbed or transmitted. *Reflection* and *scattering* are responsible for redirecting incoming solar radiation. In summary, *radiation may be absorbed, transmitted, or redirected (reflected or scattered).*

Figure 11.19 shows the fate of incoming solar radiation averaged for the entire globe. Notice that the atmosphere is quite transparent to incoming solar radiation. On average, about 50 percent of the solar energy reaching the top of the atmosphere is absorbed at Earth's surface. Another 30 percent is reflected back to space by the atmosphere, clouds, and reflective surfaces such as snow and water. The remaining 20 percent is absorbed by clouds and the atmosphere's gases. What determines whether solar radiation will be transmitted to the surface, scattered, reflected outward, or absorbed by the atmosphere? As you will see, it depends greatly on the wavelength of the energy being transmitted, as well as on the nature of the intervening material.

Reflection and Scattering

Reflection is the process whereby light bounces back from an object at the same angle at which it encounters a surface and with the same intensity (Figure 11.20A). By contrast, **scattering** produces a larger number of weaker rays that travel in different directions. Although scattering disperses light both forward and backward (*backscattering*), more energy is dispersed in the forward direction (Figure 11.20B).

Reflection and Earth's Albedo Energy is returned to space from Earth in two ways: reflection and emission of radiant

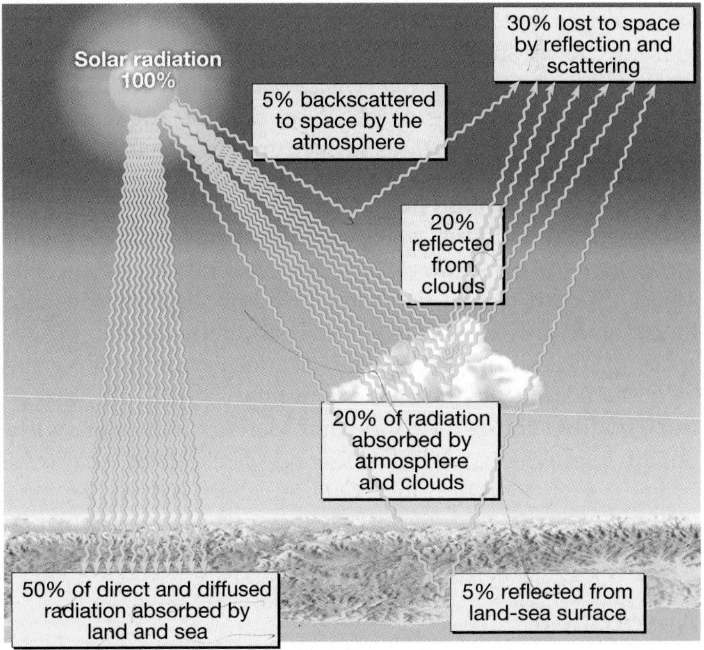

Figure 11.19 Average distribution of incoming solar radiation by percentage. More solar energy is absorbed by Earth's surface than by the atmosphere. Consequently, the air is not heated directly by the Sun but is heated indirectly from Earth's surface.

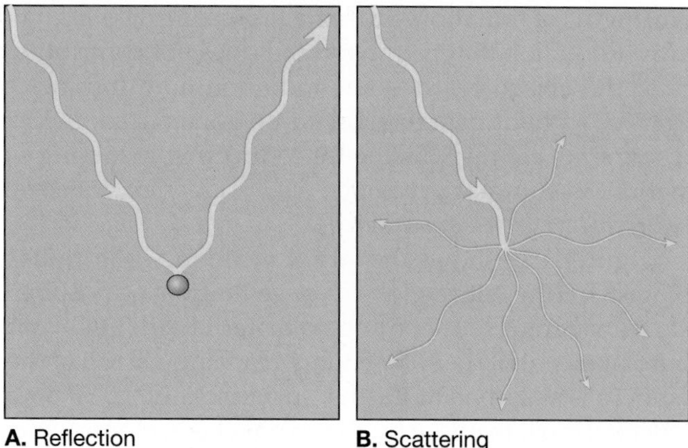

Figure 11.20 Reflection and scattering. **A.** Reflected light bounces back from a surface at the same angle at which it strikes that surface and with the same intensity. **B.** When a beam of light is scattered, it results in a larger number of weaker rays, traveling in all different directions. Usually more energy is scattered in the forward direction than is backscattered.

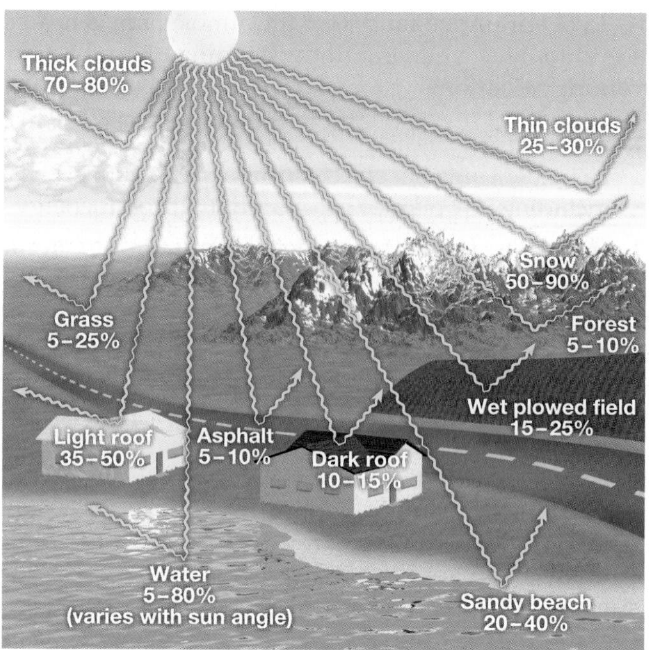

Figure 11.21 Albedo (reflectivity) of various surfaces. In general, light-colored surfaces tend to be more reflective than dark-colored surfaces and thus have higher albedos.

energy. The portion of solar energy that is reflected back to space leaves in the same short wavelengths in which it came to Earth. About 30 percent of the solar energy reaching the outer atmosphere is reflected back to space. Included in this figure is the amount sent skyward by backscattering. This energy is lost to Earth and does not play a role in heating the atmosphere.

The fraction of the total radiation that is reflected by a surface is called its **albedo.** Thus, the albedo for Earth as a whole (the *planetary albedo*) is 30 percent. However, the albedo from place to place as well as from time to time in the same locale varies considerably, depending on the amount of cloud cover and particulate matter in the air, as well as on the angle of the Sun's rays and the nature of the surface. A lower Sun angle means that more atmosphere must be penetrated, thus making the "obstacle course" longer and the loss of solar radiation greater (see Figure 11.12, p. 302). Figure 11.21 gives the albedo for various surfaces. Note that the angle at which the Sun's rays strike a water surface greatly affects its albedo.

Scattering Although incoming solar radiation travels in a straight line, small dust particles and gas molecules in the atmosphere scatter some of this energy in all directions. The result, called **diffused light,** explains how light reaches into the area beneath a shade tree, and how a room is lit in the absence of direct sunlight. Further, scattering accounts for the brightness and even the blue color of the daytime sky. In contrast, bodies such as the Moon and Mercury, which are without atmospheres, have dark skies and "pitch-black" shadows, even during daylight hours. Overall, about half of the solar radiation that is absorbed at Earth's surface arrives as diffused (scattered) light.

Absorption

As stated earlier, gases are selective absorbers, meaning that they absorb strongly in some wavelengths, moderately in others, and only slightly in still others. When a gas molecule absorbs light waves, the energy is transformed into internal molecular motion, which is detectable as a rise in temperature.

Nitrogen, the most abundant constituent in the atmosphere, is a poor absorber of all types of incoming solar radiation. Oxygen and ozone are efficient absorbers of ultraviolet radiation. Oxygen removes most of the shorter ultraviolet radiation high in the atmosphere, and ozone absorbs most of the remaining UV rays in the stratosphere. The absorption of UV radiation in the stratosphere accounts for the high temperatures experienced there. The only other significant absorber of incoming solar radiation is water vapor, which, along with oxygen and ozone, accounts for most of the solar radiation absorbed within the atmosphere.

For the atmosphere as a whole, none of the gases is an effective absorber of visible radiation. This explains why most visible radiation reaches Earth's surface and why we say that the atmosphere is transparent to incoming solar radiation. Thus, the atmosphere does not acquire the bulk of its energy directly from the Sun. Rather, it is heated chiefly by energy that is first absorbed by Earth's surface and then reradiated to the sky.

Heating the Atmosphere: The Greenhouse Effect

Earth's Dynamic Atmosphere
▼ Heating the Atmosphere

Approximately 50 percent of the solar energy that strikes the top of the atmosphere reaches Earth's surface and is absorbed. Most of this energy is then reradiated skyward. Because Earth has a much lower surface temperature than the Sun, the radiation that it emits has longer wavelengths than solar radiation.

The atmosphere as a whole is an efficient absorber of the longer wavelengths emitted by Earth (*terrestrial radiation*). Water vapor and carbon dioxide are the principal absorbing gases. Water vapor absorbs roughly five times more terrestrial radiation than do all of the other gases combined and accounts for the warm temperatures found in the lower troposphere, where it is most highly concentrated. Because the atmosphere is quite transparent to shorter-wavelength solar radiation and more readily absorbs longer-wavelength terrestrial radiation, the atmosphere is heated from the ground up rather than vice versa. This explains the general drop in temperature with increasing altitude experienced in the troposphere. The farther from the "radiator," the colder it becomes.

When the gases in the atmosphere absorb terrestrial radiation, they warm; but they eventually radiate this energy away. Some energy travels skyward, where it may be reabsorbed by other gas molecules, a possibility less likely with increasing height because the concentration of water vapor decreases with altitude. The remainder travels Earthward and is again absorbed by Earth. For this reason, Earth's surface is continually being supplied with heat from the atmosphere as well as from the Sun. Without these absorptive gases in our atmosphere, Earth would not be a suitable habitat for humans and numerous other life forms.

This very important phenomenon has been termed the **greenhouse effect** because it was once thought that greenhouses were heated in a similar manner (Figure 11.22). The gases of our atmosphere, especially water vapor and carbon dioxide, act very much like the glass in the greenhouse. They allow shorter-wavelength solar radiation to enter, where it is absorbed by the objects inside. These objects in turn radiate energy, but at longer wavelengths, to which glass is nearly opaque. The heat therefore is "trapped" in the greenhouse. However, a more important factor in keeping a greenhouse warm is the fact that the greenhouse itself prevents mixing of air inside with cooler air outside. Nevertheless, the term *greenhouse effect* is still used.

> *Did You Know?*
>
> If Earth's atmosphere had no greenhouse gases, its average surface temperature would be a frigid −18°C (−0.4°F) instead of the relatively comfortable 14.5°C (58°F) that it is today.

Global Warming

In the preceding section, you learned that carbon dioxide (CO_2) absorbs some of the radiation emitted by Earth and thus contributes to the greenhouse effect. Because CO_2 is an important heat absorber, it follows that a change in the atmosphere's CO_2 content could influence air temperature.

CO_2 Levels Are Rising

Earth's tremendous industrialization of the past two centuries has been fueled—and still is fueled—by burning fossil fuels: coal, natural gas, and petroleum (Figure 11.23). Combustion of these fuels has added great quantities of carbon dioxide to the atmosphere.

The use of coal and other fuels is the most prominent means by which humans add CO_2 to the atmosphere, but it is not the only way. The clearing of forests also contributes

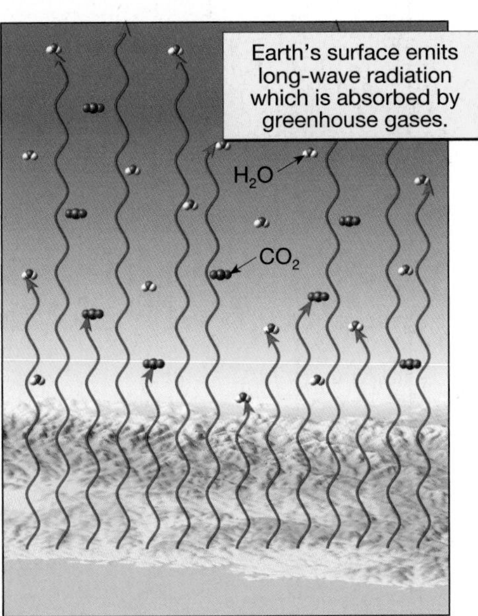

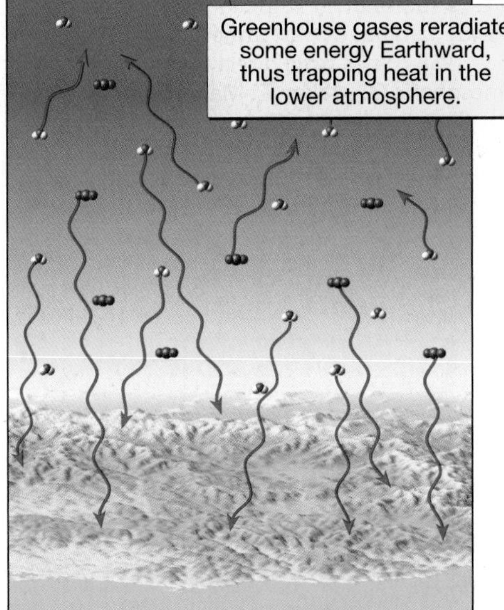

Figure 11.22 The heating of the atmosphere. Most of the short-wavelength radiation from the Sun passes through the atmosphere and is absorbed by Earth's land-sea surface. This energy is then emitted from the surface as longer-wavelength radiation, much of which is absorbed by certain gases in the atmosphere. Some of the energy absorbed by the atmosphere will be reradiated Earthward. This process, called the *greenhouse effect,* is responsible for keeping Earth's surface much warmer than it would be otherwise.

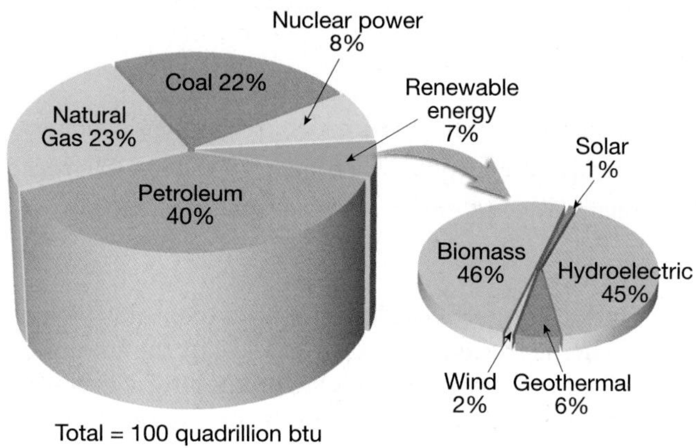

Figure 11.23 Paralleling the rapid growth of industrialization, beginning in the nineteenth century, has been the combustion of fossil fuels, which has added great quantities of carbon dioxide to the atmosphere. The graph shows energy consumption in the United States in 2004. The total was nearly 100 quadrillion Btu. A quadrillion is 10 raised to the 12th power, or a million million—a quadrillion Btu is a convenient unit for referring to U.S. energy use as a whole. (U.S. Department of Energy, Energy Information Administration)

Did You Know?

Carbon dioxide is not the only gas contributing to global warming. Scientists have come to realize that industrial and agricultural activities are causing a buildup of several trace gases including methane (CH_4) and nitrous oxide (N_2O) that may also play a significant role. These gases absorb wavelengths of outgoing Earth radiation that would otherwise escape into space. Taken together, the effects of these gases may be nearly as great as that of CO_2 in warming Earth.

substantially because CO_2 is released as vegetation is burned or decays. Deforestation is particularly pronounced in the tropics, where vast tracts are cleared for ranching and agriculture or subjected to inefficient commercial logging operations. According to UN estimates, the destruction of tropical forests exceeded 15 million hectares (38 million acres) per year during the 1990s.

Although some of the excess CO_2 is taken up by plants or is dissolved in the ocean, it is estimated that 45 to 50 percent remains in the atmosphere. Figure 11.24A shows CO_2 concentrations over the past thousand years based on ice-core records and (since 1958) measurements taken at Mauna Loa Observatory, Hawaii. The rapid increase in CO_2 concentration since the onset of industrialization is obvious and has closely followed the increase in CO_2 emissions from burning fossil fuels (Figure 11.24B).

The Atmosphere's Response

Given the increase in the atmosphere's carbon dioxide content, have global temperatures actually increased? The answer is yes. A report by the Intergovernmental Panel on Climate Change (IPCC)* indicates the following:

*Intergovernmental Panel on Climate Change, *Climate Change 2001: The Scientific Basis.* Cambridge, UK: Cambridge University Press, 2001, p. 2.

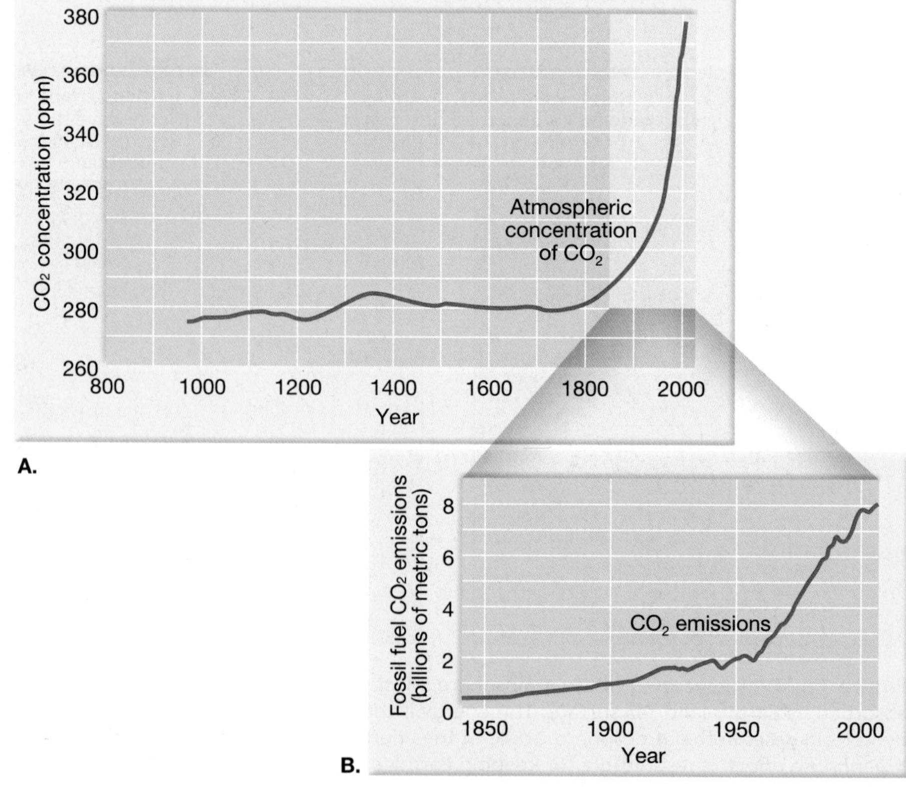

Figure 11.24 **A.** Carbon dioxide (CO_2) concentrations over the past 1000 years. Most of the record is based on data obtained from Antarctic ice cores. Bubbles of air trapped in the glacial ice provide samples of past atmospheres. The record since 1958 comes from direct measurements of atmospheric CO_2 taken at Mauna Loa Observatory, Hawaii. **B.** Fossil-fuel CO_2 emissions. The rapid increase in CO_2 concentrations since the onset of industrialization has followed closely the rise in CO_2 emissions from fossil fuels.

• During the twentieth century, the global average surface temperature increased by about 0.6°C (1°F).

• Globally, it is very likely that the 1990s was the warmest decade and 1998 and 2005 the warmest years since 1861. The next four warmest years all occurred after 2000 (Figure 11.25).

• New analyses of data for the Northern Hemisphere indicate that the increase in temperature in the twentieth century is likely to have been the largest in any century during the past 1000 years.

Are these temperature trends caused by human activities, or would they have occurred anyway? Scientists are cautious but seem convinced that human activities have played a significant role. An IPCC report in 1996 stated that "the balance of evidence suggests a discernible human influence on global climate."* Five years later, the IPCC stated that "there is new and stronger evidence that most of the warming observed over the last 50 years is attributable to human activities.** What about the future? By the year 2100, models project atmospheric CO_2 concentrations of 540 to 970 ppm. With such an increase, how will global temperatures change? Here is some of what the 2001 IPCC report has to say.†

• The globally averaged surface temperature is projected to increase by 1.4 to 5.8°C by the year 2100.

• The projected rate of warming is much larger than the observed changes during the twentieth century and is very likely to be without precedent during at least the last 10,000 years.

*Intergovernmental Panel on Climate Change, *Climate Change 1995: The Science of Climate Change.* New York: Cambridge University Press, 1996.
**IPCC, *Climate Change 2001, The Scientific Basis,* p. 10.
†IPCC, *Climate Change 2001: The Scientific Basis,* p. 13.

Did You Know?

The Intergovernmental Panel on Climate Change (IPCC) referred to in the discussion of global warming was established by the United Nations and the World Meteorological Organization to access the scientific, technical, and socioeconomic information that is relevant to understanding human-caused climate change. It is an authoritative group that provides periodic reports regarding the state of knowledge about causes and effects of climate change.

• It is very likely that nearly all land areas will warm more rapidly than the global average, particularly those at northern high latitudes in the cold season.

Some Possible Consequences

The effects of a rapid temperature change are a matter of great concern and much uncertainty. Because the climate system is so complex, predicting the distribution of particular regional changes is still very speculative. Nevertheless, plausible scenarios can be given for larger scales of space and time. One important impact of human-induced global warming is a probable rise in sea level. Potential weather changes include shifts in the paths of large-scale storms, which in turn would affect the distribution of precipitation and the occurrence of severe weather. Other possibilities include stronger tropical storms and increases in the frequency and intensity of heat waves and droughts (Table 11.2).

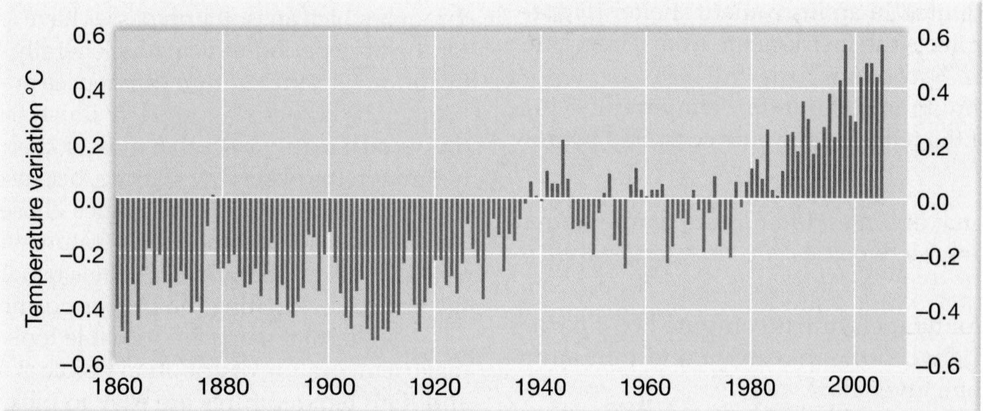

Figure 11.25 Annual average global temperature variations for the period 1860–2005. The basis for comparison is the average for the 1961–1990 period (the 0.0 line on the graph). Each narrow bar represents the departure of the global mean temperature from the 1961–1990 average for one year. For example, the global mean temperature for 1862 was more than 0.5°C (1°F) *below* the 1961–1990 average, whereas the global mean for 1998 was more than 0.5°C above. (Specifically, 1998 was 0.56°C warmer.) The bar graph clearly indicates that there can be *significant variations from year to year.* But the graph also shows a trend. Estimated global mean temperatures have been above the 1961–1990 average every year since 1978. Globally the 1990s was the warmest decade—and the years 1998, 2005, 2002, 2003, 2001, and 2004 the warmest years—since 1861. (Modified and updated after G. Bell et al. "Climate Assessment for 1998," *Bulletin of the American Meteorological Society,* Vol. 80, No. 5, May 1999, p. 54)

Table 11.2 Projected changes and effects of global warming in the twenty-first century (estimated probability)*

Higher maximum temperatures; more hot days and heat waves over nearly all land areas (*very likely*).
Higher minimum temperatures; fewer cold days, frost days, and cold waves over nearly all land areas (*very likely*).
More intense precipitation events (*very likely* over many areas).
Increased summer drying over most midlatitude continental interiors and associated risk of drought (*likely*).
Increase in tropical cyclone peak wind intensities, mean and peak precipitation intensities (*likely* over some areas).
Intensified droughts and floods associated with El Niño events in many different regions (*likely*).
Increased Asian summer monsoon precipitation variability (*likely*).
Increased intensity of midlatitude storms (*uncertain*).

**Very likely* indicates a probability of 90 to 99 percent. *Likely* indicates a probability of 67 to 90 percent.

Source: IPCC, *Climate Change* 2001.

Figure 11.26 This modern shelter contains an electrical thermometer called a *thermistor.* A shelter protects instruments from direct sunlight and allows for the free flow of air. (Photo by Bobbé Christopherson)

The changes that occur will probably take the form of gradual environmental shifts that will be imperceptible to most people from year to year. Although the changes may seem gradual, the effects will clearly have powerful economic, social, and political consequences.

For the Record: Air Temperature Data

Changes in air temperature are probably noticed by people more often than changes in any other element of weather. At a weather station, the temperature is read on a regular basis from instruments mounted in an instrument shelter (Figure 11.26). The shelter protects the instruments from direct sunlight and allows a free flow of air.

The daily maximum and minimum temperatures are the bases for much of the temperature data compiled by meteorologists:

1. By adding the maximum and minimum temperatures and then dividing by two, the *daily mean temperature* is calculated.
2. The *daily range* of temperature is computed by finding the difference between the maximum and minimum temperatures for a given day.
3. The *monthly mean* is calculated by adding together the daily means for each day of the month and dividing by the number of days in the month.
4. The *annual mean* is an average of the 12 monthly means.
5. The *annual temperature range* is computed by finding the difference between the highest and lowest monthly means.

Mean temperatures are particularly useful for making comparisons, whether on a daily, monthly, or annual basis. It is quite common to hear a weather reporter state, "Last month was the hottest July on record," or "Today, Chicago was 10 degrees warmer than Miami." Temperature ranges are also useful statistics, because they give an indication of extremes.

To examine the distribution of air temperatures over large areas, isotherms are commonly used. An **isotherm** is a line that connects points on a map that have the same temperature (*iso* = equal, *therm* = temperature). Therefore, all points through which an isotherm passes have identical temperatures for the time period indicated. Generally, isotherms representing 5° or 10° temperature differences are used, but any interval may be chosen. Figure 11.27 illustrates how isotherms are drawn on a map. Notice that most isotherms do not pass directly through the observing stations, because the station readings may not coincide with the values chosen for the isotherms. Only an occasional station temperature will be exactly the same as the value of the isotherm, so it is usually necessary to draw the lines by estimating the proper position between stations.

Isothermal maps are valuable tools because they clearly make temperature distribution visible at a glance. Areas of low and high temperatures are easy to pick out. In addition, the amount of temperature change per unit of distance, called the *temperature gradient,* is easy to visualize. Closely spaced isotherms indicate a rapid rate of temperature change, whereas more widely spaced lines indicate a more gradual rate of change. You can see this in Figure 11.27. The isotherms are closer in Colorado and Utah (steeper temperature gradient), whereas the isotherms are spread farther in Texas (gentler temperature gradient). Without isotherms, a map would be covered with numbers representing temperatures at dozens or hundreds of places, which would make patterns difficult to see.

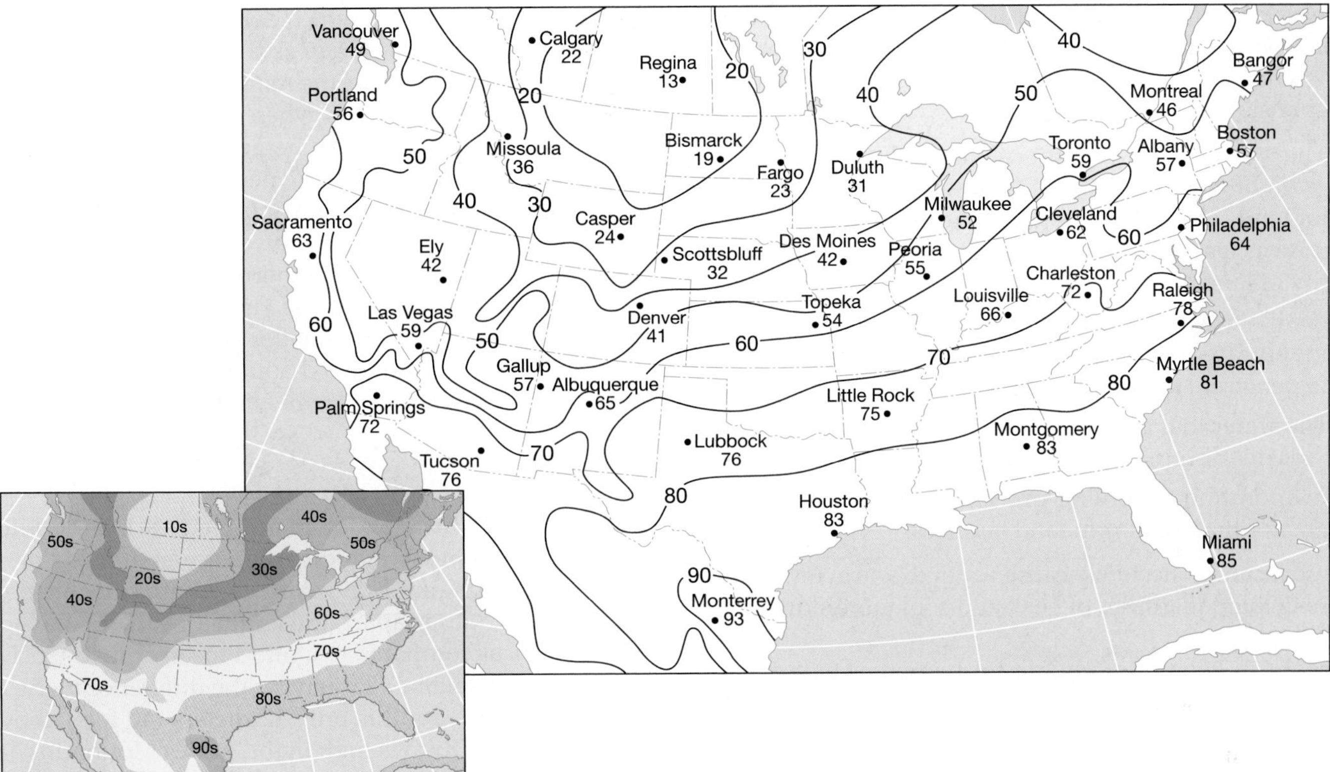

Figure 11.27 Temperature distribution using isotherms. Isotherms are lines that connect points of equal temperature. The temperatures on this map are in degrees Fahrenheit. Showing temperature distribution in this way makes patterns easier to see. On television, and in many newspapers, temperature maps are in color. Rather than labeling isotherms, the area *between* isotherms is labeled. For example, the zone between the 60° and 70° isotherms is labeled "60s."

Did You Know?

The highest accepted temperature record for the United States and the entire Western Hemisphere is 57°C (134°F). This long-standing record was set at Death Valley, California, on July 10, 1913. North America's coldest recorded temperature occurred at Snag in Canada's Yukon Territory. This remote outpost experienced a temperature or −63°C (−81°F) on February 3, 1947.

Why Temperatures Vary: The Controls of Temperature

A *temperature control* is any factor that causes temperature to vary from place to place and from time to time. Earlier in this chapter, we examined the single greatest cause for temperature variations—differences in the receipt of solar radiation. Because variations in Sun angle and length of daylight are a function of latitude, they are responsible for warm temperatures in the tropics and colder temperatures at more poleward locations. Of course, seasonal temperature changes at a given latitude occur as the Sun's vertical rays migrate toward and away from a place during the year.

But latitude is not the only control of temperature; if it were, we would expect that all places along the same parallel of latitude would have identical temperatures. This is clearly not the case. For example, Eureka, California, and New York City are both coastal cities at about the same latitude, and both have an average mean temperature of 11°C (52°F). However, New York City is 9°C (16°F) warmer than Eureka in July and 10°C cooler in January. In another example, two cities in Ecuador—Quito and Guayaquil—are relatively close to one another, but the mean annual temperatures at these two cities differ by 12°C (21°F). To explain these situations and countless others, we must understand that factors other than latitude exert a strong influence on temperature. Among the most important are the differential heating of land and water, altitude, geographic position, and ocean currents.*

Land and Water

The heating of Earth's surface directly influences the heating of the air above. Therefore, to understand variations in air temperature, we must understand the variations in heating properties of the different surfaces that Earth presents to the Sun—soil, water, trees, ice, and so on. Different land surfaces absorb varying amounts of incoming solar energy, which in turn cause variations in the temperature of the air above. The greatest contrast, however, is not between different land surfaces, but between land and water. *Land heats more rapidly and to higher temperatures than water and cools more rapidly and to*

*For a discussion of the effects of ocean currents on temperatures, see Chapter 10.

lower temperatures than water. Temperature variations, therefore, are considerably greater over land than over water. Why do land and water heat and cool differently? Several factors are responsible:

1. The *specific heat* (amount of energy needed to raise 1 gram of a substance 1°C) is far greater for water than for land. Thus, water requires a great deal more heat to raise its temperature the same amount as an equal quantity of land.
2. Land surfaces are opaque, so heat is absorbed only at the surface. Water, being more transparent, allows heat to penetrate to a depth of many meters.
3. The water that is heated often mixes with water below, thus distributing the heat through an even larger mass.
4. Evaporation (a cooling process) from water bodies is greater than that from land surfaces.

All these factors collectively cause water to warm more slowly, store greater quantities of heat, and cool more slowly than land.

Monthly temperature data for two cities will demonstrate the moderating influence of a large water body and the extremes associated with land (Figure 11.28). Vancouver, British Columbia, is located along a windward coast, whereas Winnipeg, Manitoba, is in a continental position far from the influence of water. Both cities are at about the same latitude and thus experience similar Sun angles and lengths of daylight. Winnipeg, however, has a mean January temperature that is 20°C (36°F) lower than Vancouver's. Winnipeg's July mean is 2°C (3.6°F) higher than Vancouver's. Although their latitudes are nearly the same, Winnipeg, which has no water influence, experiences much greater temperature extremes than Vancouver, which does. The key to Vancouver's moderate year-round climate is the Pacific Ocean.

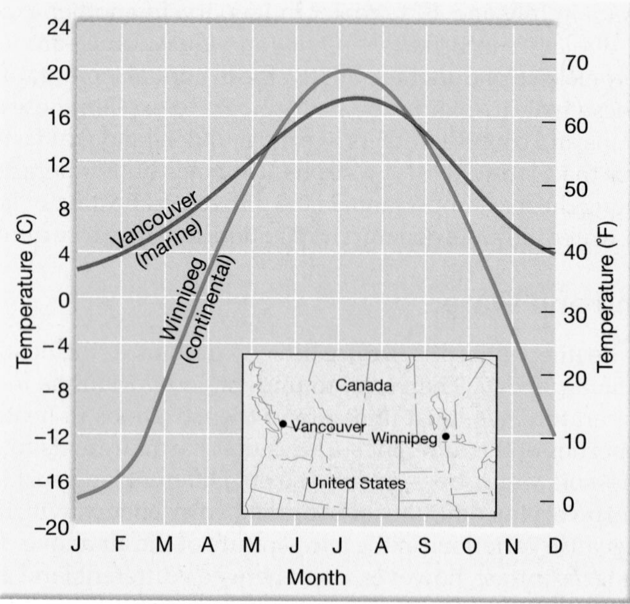

Figure 11.28 Mean monthly temperatures for Vancouver, British Columbia, and Winnipeg, Manitoba. Vancouver has a much smaller annual temperature range owing to the strong marine influence of the Pacific Ocean. The curve for Winnepeg illustrates the greater extremes associated with an interior location.

On a different scale, the moderating influence of water may also be demonstrated when temperature variations in the Northern and Southern hemispheres are compared. In the Northern Hemisphere, 61 percent is covered by water, and land accounts for the remaining 39 percent. However, in the Southern Hemisphere, 81 percent is covered by water and 19 percent by land. The Southern Hemisphere is correctly called the *water hemisphere* (see Figure 9.1, p. 246). Table 11.3 portrays the considerably smaller annual temperature variations in the water-dominated Southern Hemisphere as compared with the Northern Hemisphere.

Altitude

The two cities in Ecuador mentioned earlier—Quito and Guayaquil—demonstrate the influence of altitude on mean temperatures (Figure 11.29). Although both cities are near the equator and not far apart, the annual mean at Guayaquil is 25°C (77°F), as compared to Quito's mean of 13°C (55°F). The difference is explained largely by the difference in the cities' elevations: Guayaquil is only 12 meters (40 feet) above sea level, whereas Quito is high in the Andes Mountains at 2800 meters (9200 feet).

Recall that temperatures drop an average of 6.5°C per kilometer (3.5°F per 1000 feet) in the troposphere; thus, cooler temperatures are to be expected at greater heights (see Figure 11.8, p. 299). Yet, the magnitude of the difference is not totally explained by the normal lapse rate. If this figure were used, we would expect Quito to be about 18°C (nearly 33°F) cooler than Guayaquil; the difference, however, is only 12°C. The fact that high-altitude places, such as Quito, are warmer than the value calculated using the normal lapse rate results from the absorption and reradiation of solar energy by the ground surface.

Geographic Position

The geographic setting can greatly influence the temperatures experienced at a specific location. A coastal location where prevailing winds blow from the ocean onto the shore (a *windward* coast) experiences considerably different temperatures than does a coastal location where prevailing winds blow from the land toward the ocean (a *leeward* coast). The

Table 11.3 Variation in mean annual temperature range (°C) with latitude

Latitude	Northern Hemisphere	Southern Hemisphere
0	0	0
15	3	4
30	13	7
45	23	6
60	30	11
75	32	26
90	40	31

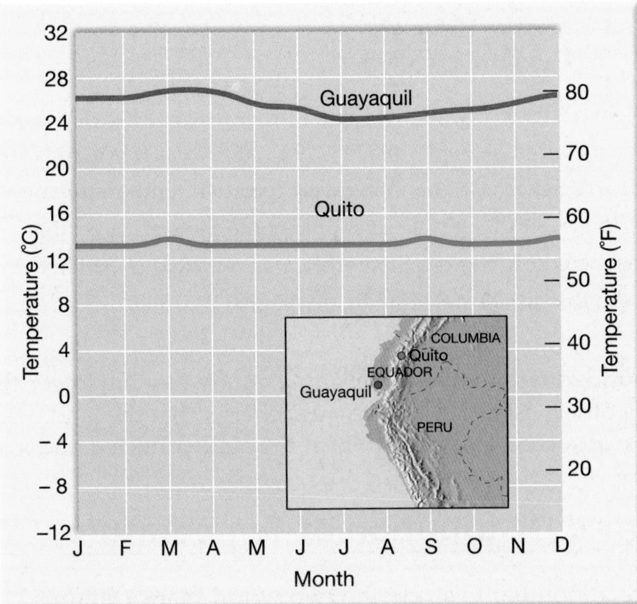

Figure 11.29 Because Quito is high in the Andes Mountains, it experiences much cooler temperatures than Guayaquil, which is near sea level. Because both cities are near the equator, they have a negligible annual temperature range.

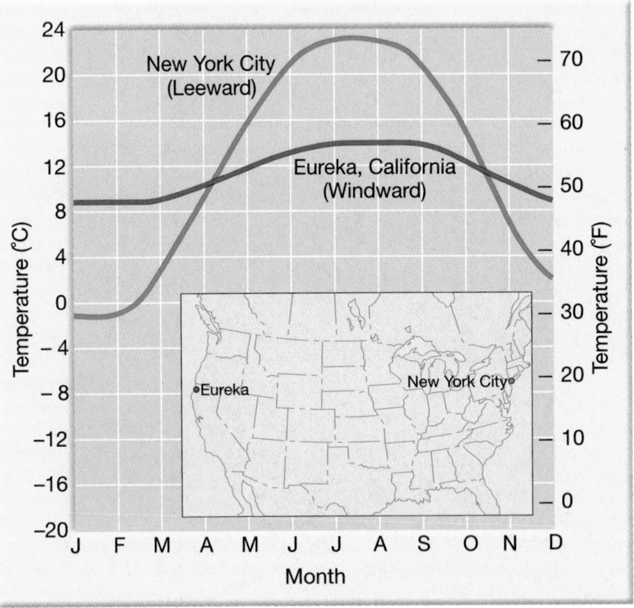

Figure 11.30 Monthly mean temperatures for Eureka, California, and New York City. Both cities are coastal and located at about the same latitude. Because Eureka is strongly influenced by prevailing winds from the ocean and New York City is not, the annual temperatures range at Eureka is much smaller.

windward coast will experience the full moderating influence of the ocean—cool summers and mild winters—compared to an inland location at the same latitude.

A leeward coast, however, will have a more continental temperature regime because the winds do not carry the ocean's influence onshore. Eureka, California, and New York City, the two cities mentioned earlier, illustrate this aspect of geographic position. The annual temperature range at New York City is 19°C (34°F) greater than Eureka's (Figure 11.30).

Seattle and Spokane, both in Washington State, illustrate a second aspect of geographic position—mountains that act as barriers. Although Spokane is only about 360 kilometers (220 miles) east of Seattle, the towering Cascade Range separates the cities. Consequently, Seattle's temperatures show a marked marine influence, but Spokane's are more typically continental (Figure 11.31). Spokane is 7°C (13°F) cooler than Seattle in January and 4°C (7°F) warmer than Seattle in July. The annual range at Spokane is 11°C (20°F) greater than at Seattle. The Cascade Range effectively cuts Spokane off from the moderating influence of the Pacific Ocean.

Cloud Cover and Albedo

You may have noticed that clear days are often warmer than cloudy ones and that clear nights are usually cooler than cloudy ones. This demonstrates that cloud cover is another factor that influences temperature in the lower atmosphere. Studies using satellite images show that at any particular time, about half of our planet is covered by clouds. Cloud cover is important because many clouds have a high albedo; therefore, clouds reflect a significant portion of the sunlight that strikes them back into space (see Figure 11.21, p. 308). By reducing the amount of incoming solar radiation, daytime temperatures will be lower than if the clouds were not present and the sky were clear.

At night, clouds have the opposite effect as during daylight. They act as a blanket by absorbing radiation emitted by Earth's surface and reradiating a portion of it back to the surface. Consequently, some of the heat that otherwise would have been lost remains near the ground. Thus, nighttime air temperatures do not drop as low as they would on a clear

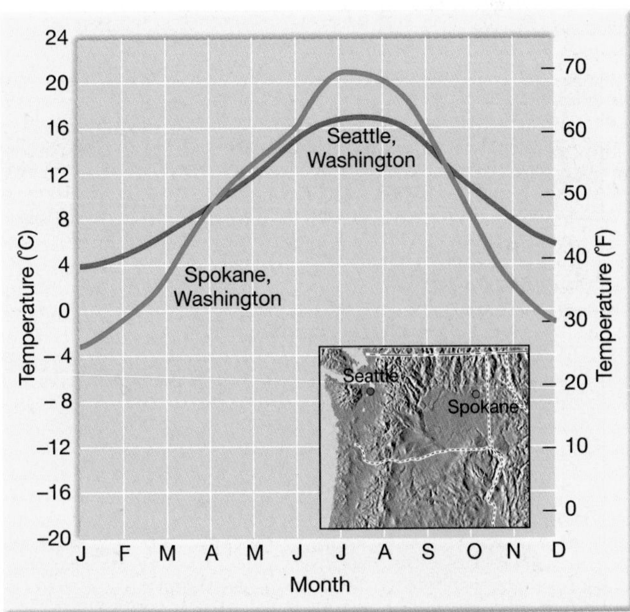

Figure 11.31 Monthly mean temperatures for Seattle and Spokane, Washington. Because the Cascade Mountains cut off Spokane from the moderating influence of the Pacific Ocean, its annual temperature range is greater than Seattle's.

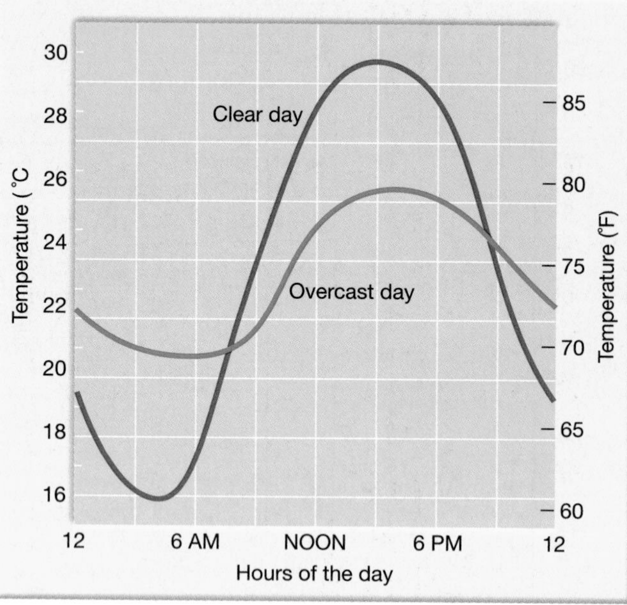

Figure 11.32 The daily cycle of temperature at Peoria, Illinois, for two July days. Clouds reduce daily temperature range. During daylight hours, clouds reflect solar radiation back to space. Therefore, the maximum temperature is lower than if the sky were clear. At night, the minimum temperature will not fall as low because clouds retard the loss of heat.

night. The effect of cloud cover is to reduce the daily temperature range by lowering the daytime maximum and raising the nighttime minimum (Figure 11.32).

Clouds are not the only phenomenon that increases albedo and thereby reduces air temperatures. We also recognize that snow- and ice-covered surfaces have high albedos. This is one reason why mountain glaciers do not melt away in the summer and why snow may still be present on a mild spring day. In addition, during the winter, when snow covers the ground, daytime maximums on a sunny day are lower than they otherwise would be because energy that the land would have absorbed and used to heat the air is reflected and lost.

Did You Know?

The world's record-high temperature is nearly 59°C (138°F). It was recorded on September 13, 1922, at Azizia, Libya, in North Africa's Sahara Desert. The lowest recorded temperature on Earth is −89°C (−129°F). It should come as no surprise that this incredibly frigid temperature was recorded in Antarctica, at the Russian Vostok Station, on August 24, 1960.

World Distribution of Temperature

Take a moment to study the two world maps (Figures 11.33 and 11.34). From hot colors near the equator to cool colors toward the poles, these maps portray sea-level temperatures in the seasonally extreme months of January and July. Temperature distribution is shown by using isotherms. On these maps, you can study global temperature patterns and the effects of the controlling factors of temperature, especially latitude, the distribution of land and water, and ocean currents. Like most isothermal maps of large regions, all temperatures on these world maps have been reduced to sea level to eliminate the complications caused by differences in altitude.

On both maps, the isotherms generally trend east-west and show a decrease in temperatures poleward from the tropics. They illustrate one of the most fundamental aspects of world temperature distribution: that the effectiveness of incoming solar radiation in heating Earth's surface and the atmosphere above is largely a function of latitude.

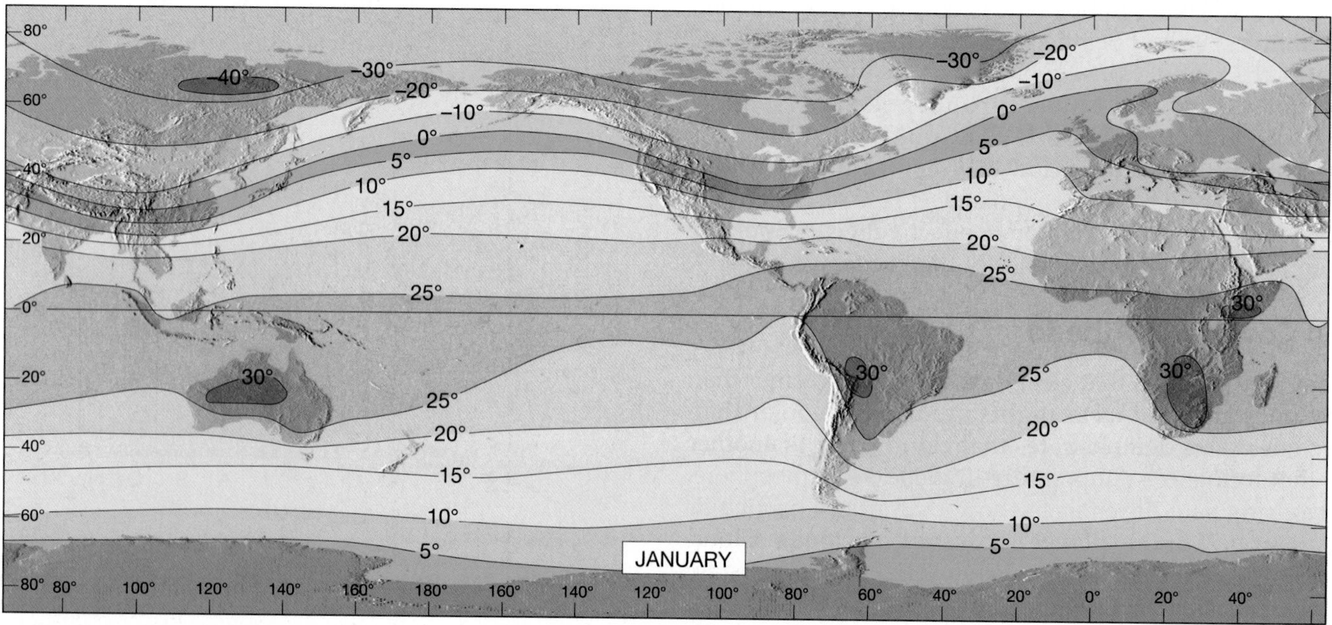

Figure 11.33 World mean sea-level temperatures in January in degrees Celsius.

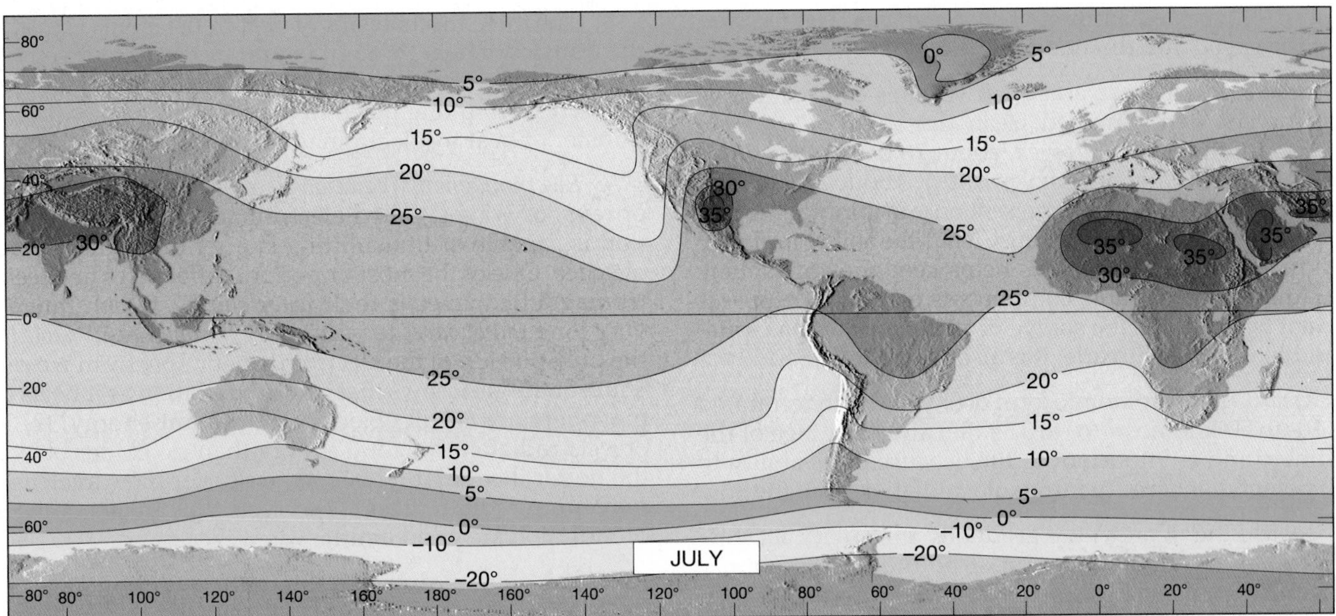

Figure 11.34 World mean sea-level temperatures in July in degrees Celsius.

Moreover, there is a latitudinal shifting of temperatures caused by the seasonal migration of the Sun's vertical rays. To see this, compare the color bands by latitude on the two maps. On the January map, the "hot spots" of 30°C are *south* of the equator, but in July they have shifted *north* of the equator.

If latitude were the only control of temperature distribution, our analysis could end here, but this is not the case. The added effect of the differential heating of land and water is also reflected on the January and July temperature maps. The warmest and coldest temperatures are found over land. Because temperatures do not fluctuate as much over water as over land, the seasonal north-south migration of isotherms is greater over the continents than over the oceans.

In addition, it is clear that the isotherms in the Southern Hemisphere, where there is little land and the oceans predominate, are much straighter and more stable than in the Northern Hemisphere, where they bend sharply northward in July and southward in January over the continents.

Isotherms also reveal the presence of ocean currents. Warm currents cause isotherms to be deflected poleward, whereas cold currents cause an equatorward bending. The horizontal transport of water poleward warms the overlying air and results in air temperatures that are higher than otherwise would be expected for the latitude. Conversely, currents moving toward the equator produce air temperatures cooler than expected.

Because Figures 11.33 and 11.34 show the seasonal extremes of temperature, they can be used to evaluate variations in the annual range of temperature from place to place. A comparison of the two maps shows that a station near the equator will record a very small annual temperature range. This is true because such a station experiences little variation in the length of daylight and always has a relatively high Sun angle. By contrast, a site in the middle latitudes experiences much wider variations in Sun angle and length of daylight and thus larger temperature variations. Therefore, we can state that the annual temperature range increases with an increase in latitude.

Land and water also affect seasonal temperature variations, especially outside the tropics. A continental location must endure hotter summers and colder winters than a coastal location. Consequently, the annual range will increase with an increase in continentality.

Did You Know?

A classic example of the effect of high latitude and continentality on annual temperature range is Yakutsk, a city in Siberia, approximately 60° north latitude and far from the influence of water. As a result, Yakutsk has an average annual temperature range of 62.2°C (112°F), one of the greatest in the world.

The Chapter in Review

1. *Weather* is the state of the atmosphere at a particular place for a short period of time. *Climate*, on the other hand, is a generalization of the weather conditions of a place over a long period of time.

2. The most important *elements*—those quantities or properties that are measured regularly—of weather and climate are (1) air *temperature*, (2) *humidity*, (3) type and amount of *cloudiness,* (4) type and amount of *precipitation,* (5) air *pressure,* and (6) the speed and direction of the *wind.*

3. If water vapor, dust, and other variable components of the atmosphere were removed, clean, dry air would be composed almost entirely of *nitrogen,* about 78 percent of the atmosphere by volume, and *oxygen* (O_2), about 21 percent. *Carbon dioxide* (CO_2), although present only in minute amounts (0.036 percent), is important because it has the ability to absorb heat radiated by Earth and thus helps keep the atmosphere warm. Among the variable components of air, *water vapor* is important because it is the source of all clouds and precipitation, and like carbon dioxide, it is also a heat absorber.

4. *Ozone* (O_3), the triatomic form of oxygen, is concentrated in the 10- to 50-kilometer (6- to 31-mile) altitude range of the atmosphere and is important to life because of its ability to absorb potentially harmful ultraviolet radiation from the Sun.

5. Because the atmosphere gradually thins with increasing altitude, it has no sharp upper boundary but simply blends into outer space. Based on temperature, the atmosphere is divided vertically into four layers. The *troposphere* is the lowermost layer. In the troposphere, temperature usually decreases with increasing altitude. This *environmental lapse rate* is variable but averages about 6.5°C per kilometer (3.5°F per 1000 feet). Essentially, all important weather phenomena occur in the troposphere. Beyond the troposphere is the *stratosphere,* which exhibits warming because of absorption of UV radiation by ozone. In the *mesophere,* temperatures again decrease. Upward from the mesosphere is the *thermosphere,* a layer with only a tiny fraction of the atmosphere's mass and no well-defined upper limit.

6. The two principal motions of Earth are (1) *rotation,* the spinning about its axis that produces the daily cycle of daylight and darkness; and (2) *revolution,* the movement in its orbit around the Sun.

7. Several factors acting together cause the seasons. Earth's axis is inclined $23\frac{1}{2}°$ from the perpendicular to the plane of its orbit around the Sun and remains pointed in the same direction (toward the North Star) as Earth journeys around the Sun. As a consequence, Earth's orientation to the Sun continually changes. Fluctuations in the Sun angle and length of daylight brought about by Earth's changing orientation to the Sun cause the seasons.

8. The three mechanisms of heat transfer are (1) *conduction,* the transfer of heat through matter by molecular activity; (2) *convection,* the transfer of heat by the movement of a mass or substance from one place to another; and (3) *radiation,* the transfer of heat by electromagnetic waves.

9. *Electromagnetic radiation* is energy emitted in the form of rays, or waves, called electromagnetic waves. All radiation is capable of transmitting energy through the vacuum of space. One of the most important differences between electromagnetic waves is their *wavelengths,* which range from very long *radio waves* to very short *gamma rays. Visible light* is the only portion of the electromagnetic spectrum we can see. Some of the basic laws that govern radiation as it heats the atmosphere are (1) all objects emit radiant energy; (2) hotter objects radiate more total energy than do colder objects; (3) the hotter the radiating body, the shorter the wavelengths of maximum radiation; and (4) objects that are good absorbers of radiation are good emitters as well.

10. The general drop in temperature with increasing altitude in the troposphere supports the fact that *the atmosphere is heated from the ground up.* Approximately 50 percent of the solar energy that strikes the top of the atmosphere reaches Earth's surface and is absorbed. Earth releases the absorbed radiation in the form of long-wave radiation. The atmospheric absorption of this long-wave *terrestrial radiation,* primarily by water vapor and carbon dioxide, is responsible for heating the atmosphere.

11. Carbon dioxide, an important heat absorber in the atmosphere, is one of several gases that influence *global warming.* Some consequences of global warming could be (1) shifts in temperature and rainfall patterns, (2) a gradual rise in sea level, (3) changing storm tracks and a higher frequency and greater intensity of hurricanes, and (4) an increase in the frequency and intensity of heat waves and droughts.

12. The factors that cause temperature to vary from place to place, also called the *controls of temperature,* are (1) differences in the *receipt of solar radiation* due to latitude, (2) the unequal heating and cooling of *land and water,* (3) *altitude,* (4) *geographic position,* and (5) *ocean currents.*

13. Temperature distribution is shown on a map by using *isotherms,* which are lines that connect equal temperatures.

Key Terms

aerosols (p. 296)
air (p. 295)
albedo (p. 308)
circle of illumination (p. 300)
climate (p. 294)
conduction (p. 305)
convection (p. 305)
diffused light (p. 308)
element (of weather and climate) (p. 295)
environmental lapse rate (p. 298)
equinox (spring or autumnal) (p. 302)
greenhouse effect (p. 309)
heat (p. 304)
inclination of the axis (p. 301)
infrared (p. 306)
isotherm (p. 312)
mesosphere (p. 299)
radiation or electromagnetic radiation (p. 305)
reflection (p. 307)
revolution (p. 300)
rotation (p. 300)
scattering (p. 307)
solstice (summer or winter) (p. 302)
stratosphere (p. 299)
temperature (p. 305)
thermosphere (p. 300)
Tropic of Cancer (p. 302)
Tropic of Capricorn (p. 302)
troposphere (p. 298)
ultraviolet (UV) (p. 306)
visible light (p. 306)
weather (p. 294)

Questions for Review

1. Distinguish between weather and climate.
2. List the basic elements of weather and climate.
3. What are the two major components of clean, dry air (Figure 11.3, p. 296)? Are they important meteorologically?
4. Why are water vapor and aerosols important constituents of our atmosphere?
5. a. Why is ozone important to life on Earth?
   b. What are CFCs, and what is their connection to the ozone problem?
   c. What is the most serious threat to human health of a decrease in the stratosphere's ozone?
6. The atmosphere is divided vertically into four layers on the basis of temperature. List the names of these layers in order (from lowest to highest) and describe how temperatures change in each layer.
7. If the temperature at sea level were 23°C what would the air temperature be at a height of 2 kilometers *under average conditions?*
8. Why do temperatures rise in the stratosphere?
9. Briefly explain the primary cause of the seasons.
10. After examining Table 11.1 (p. 304), write a general statement that relates the season, the latitude, and the length of daylight.
11. Describe the relationship between the temperature of a radiating body and the wavelengths it emits.
12. Distinguish among the three basic mechanisms of heat transfer.
13. Figure 11.19 (p. 307) illustrates what happens to incoming solar radiation. The percentages shown, however, are only global averages. In particular, the amount of solar radiation reflected (albedo) may vary considerably. What factors might cause variations in albedo?
14. Describe the basic process by which Earth's atmosphere is heated.
15. Why has the atmosphere's carbon dioxide level been rising for more than 150 years?
16. How are temperatures in the lower atmosphere likely to change as carbon dioxide levels continue to increase? Why?
17. List three possible consequences of global warming.
18. Quito, Ecuador, is located on the equator and is not a coastal city. It has an average annual temperature of only 13°C (55°F). What is the likely cause for this low average temperature?
19. In what ways can geographic position be considered a control of temperature?

Online Study Guide

The *Foundations of Earth Science* Web site uses the resources and flexibility of the Internet to aid in your study of the topics in this chapter. Written and developed by Earth science instructors, this site will help improve your understanding of Earth science. Visit **http://www.prenhall.com/lutgens** and click on the cover of *Foundations of Earth Science 5e* to find:

- Online review quizzes.
- Critical thinking exercises.
- Links to chapter-specific Web resources.
- Internet-wide key-term searches.

http://www.prenhall.com/lutgens

GEODe: Earth Science

GEODe: Earth Science makes studying more effective by reinforcing key concepts using animation, video, narration, interactive exercises, and practice quizzes. A copy is included with every copy of *Foundations of Earth Science 5e.*

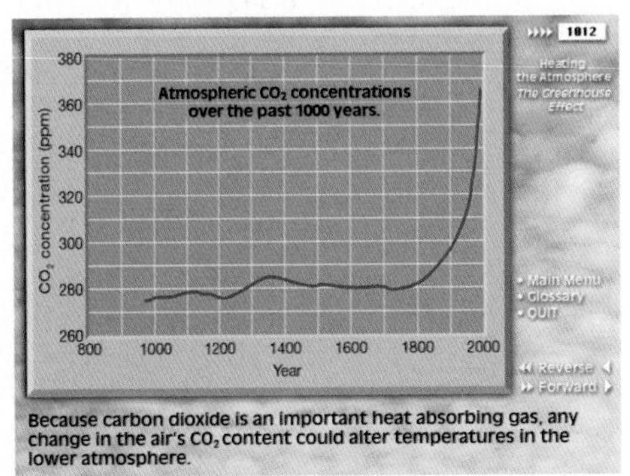

CHAPTER

12

Moisture, Clouds, and Precipitation

Lightning display associated with a thunderstorm (cumulonimbus clouds) near Colorado Springs, Colorado. (Photo by Sean Cayton/The Image Works)

FOCUS ON LEARNING

To assist you in learning the important concepts in this chapter, you will find it helpful to focus on the following questions:

1. Which processes cause water to change from one state of matter to another?
2. What is humidity? What is the most common method used to express humidity?
3. What is the basic cloud-forming process?
4. What controls the stability of the atmosphere?
5. What are three mechanisms that initiate the vertical movement of air?
6. What are the necessary conditions for condensation?
7. Which two criteria are used for cloud classification?
8. What is fog? How do fogs form?
9. How is precipitation produced in a cloud? What are the different forms of precipitation?

Water vapor is an odorless, colorless gas that mixes freely with the other gases of the atmosphere. Unlike oxygen and nitrogen—the two most abundant components of the atmosphere—water can change from one state of matter to another (solid, liquid, or gas) at the temperatures and presures experienced on Earth. (By contrast, nitrogen will not condense to a liquid unless its temperature is lowered to −196°C [−371°F]) Because of this unique property, water freely leaves the oceans as a gas and returns again as a liquid (Figure 12.1)

As you observe day-to-day weather changes, you might ask: Why is it generally more humid in the summer than in the winter? Why do clouds form on some occasions but not on others? Why do some clouds look thin and harmless whereas others form gray and ominous towers? Answers to these questions involve the role of water vapor in the atmosphere, the central theme of this chapter.

Water's Changes of State

Earth's Dynamic Atmosphere
▼ Moisture and Cloud Formation

Water is the only substance that exists in the atmosphere as a solid (ice), liquid, and gas (water vapor). It is made of hydrogen and oxygen atoms that are bonded together to form water molecules (H_2O). In all three states of matter (even ice), these molecules are in constant motion—the higher the temperature, the more vigorous the movement. The chief difference among liquid water, ice, and water vapor is the arrangement of the water molecules.

Ice, Liquid Water, and Water Vapor

Ice is composed of water molecules that are held together by mutual molecular attractions. The molecules form a tight, orderly network, as shown in Figure 12.2. As a consequence, the water molecules in ice are not free to move relative to one another but rather vibrate about fixed sites. When ice is heated, the molecules oscillate more rapidly. When the rate of molecular movement increases sufficiently, the bonds between some of the water molecules are broken, resulting in melting.

In the liquid state, water molecules are still tightly packed but are moving fast enough that they are able to slide past one another. As a result, liquid water is fluid and will take the shape of its container.

As liquid water gains heat from its environment, some of the molecules will acquire enough energy to break the remaining molecular attractions and escape from the surface, becoming water vapor. Water-vapor molecules are widely spaced compared to liquid water and exhibit very energetic random motion. What distinguishes a gas from a liquid is its compressibility (and expandability). For example, you can easily put more and more air into a tire and increase its volume only slightly. However, do not try to put 10 gallons of gasoline into a five-gallon can.

To summarize, when water changes state, it does not turn into a different substance; only the distances and interactions among the water molecules change.

Latent Heat

Whenever water changes state, heat is exchanged between water and its surroundings. When water evaporates, heat is absorbed (Figure 12.2). Meteorologists often measure heat en-

Figure 12.1 People caught in a downpour. (Photo by Mary Fulton/Getty Images; Inc.—Liaison)

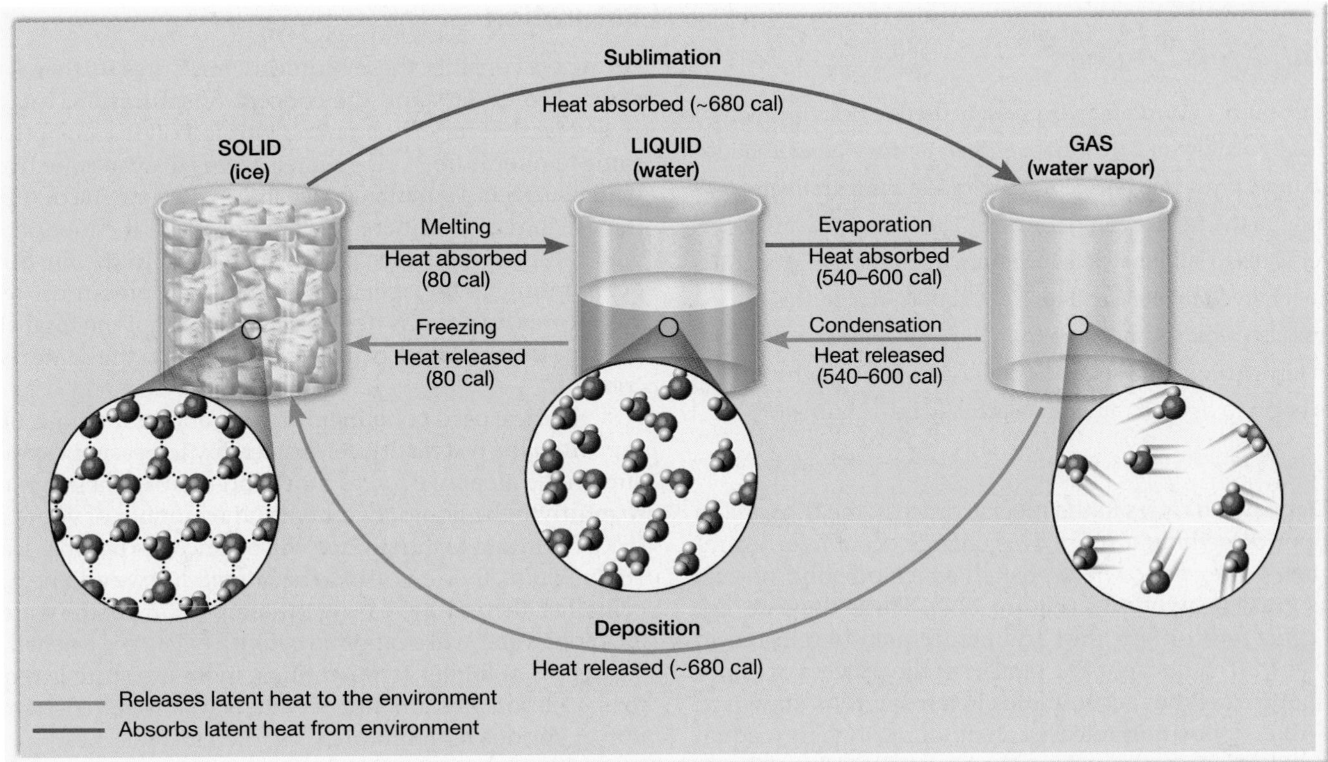

Figure 12.2 Changes of state always involve an exchange of heat. The numbers shown here are the approximate amounts for the change of 1 gram of water from one state of matter to another.

ergy in calories. One **calorie,** is the amount of heat required to raise the temperature of 1 gram of water 1°C (1.8°F). Thus, when 10 calories of heat are absorbed by 1 gram of water, the molecules vibrate faster and a 10°C (18°F) temperature rise occurs.

Under certain conditions, heat may be added to a substance without an accompanying rise in temperature. For example, when a glass of ice water is warmed, the temperature of the ice-water mixture remains a constant 0°C (32°F) until all the ice has melted. If adding heat does not raise the temperature, where does this energy go? In this case, the added energy went to break the molecular attractions between the water molecules in the ice cubes.

Because the heat used to melt ice does not produce a temperature change, it is referred to as **latent heat.** (*Latent* means "hidden," like the latent fingerprints hidden at a crime scene.) This energy can be thought of as being stored in liquid water, and it is not released to its surroundings as heat until the liquid returns to the solid state.

It requires 80 calories to melt one gram of ice, an amount referred to as *latent heat of melting. Freezing,* the reverse process, releases these 80 calories per gram to the environment as *latent heat of fusion.*

Evaporation and Condensation We saw that heat is absorbed when ice is converted to liquid water. Heat is also absorbed during **evaporation,** the process of converting a liquid to a gas (vapor). The energy absorbed by water molecules during evaporation is used to give them the motion needed to escape the surface of the liquid and become a gas. This energy is referred to as the *latent heat of vaporization.* During the process of evaporation, it is the higher-temperature (faster-moving) molecules that escape the surface. As a result, the average molecular motion (temperature) of the remaining water is reduced—hence, the common expression "evaporation is a cooling process." You have undoubtedly experienced this cooling effect on stepping dripping wet from a swimming pool or bathtub. The energy used to evaporate water comes from the skin—and cools the body.

Condensation, the reverse process, occurs when water vapor changes to the liquid state. During condensation, water vapor molecules release energy (*latent heat of condensation*) in an amount equivalent to what was absorbed during evaporation. When condensation occurs in the atmosphere, it results in the formation of such phenomena as fog and clouds.

As you will see, latent heat plays an important role in many atmospheric processes. In particular, when water vapor condenses to form cloud droplets, latent heat of condensation is released, warming the surrounding air and giving it buoyancy. When the moisture content of air is high, this process can spur the growth of towering storm clouds.

Sublimation and Deposition You are probably least familiar with the last two processes illustrated in Figure 12.2—sublimation and deposition. **Sublimation** is the conversion of a solid directly to a gas without passing through the liquid state. Examples you may have observed include the gradual shrinking of unused ice cubes in the freezer and the rapid conversion of dry ice (frozen carbon dioxide) to wispy clouds that quickly disappear.

Did You Know?

"Freezer burn" is a common expression used to describe foods that have been left in a frost-free refrigerator for extended periods of time. Frost-free refrigerators circulate comparatively dry air through the freezer compartments. This causes ice on the freezer walls to sublimate (change from a solid to a gas) so it can be removed by the circulating air. Unfortunately, this process also removes moisture from those frozen foods that are not in airtight containers. Over a period of a few months these foods begin to dry out rather than actually being burned.

Deposition refers to the reverse process, the conversion of a vapor directly to a solid. This change occurs, for example, when water vapor is deposited as ice on solid objects such as grass or windows (Figure 12.3). These deposits are called *white frost* or *hoar frost* and are frequently referred to simply as *frost.* A household example of the process of deposition is the frost that accumulates in a freezer. As shown in Figure 12.2, deposition releases an amount of energy equal to the total amount released by condensation and freezing.

Humidity: Water Vapor in the Air

Earth's Dynamic Atmosphere
▼ Moisture and Cloud Formation

Water vapor constitutes only a small fraction of the atmosphere, varying from as little as one-tenth of 1 percent up to about 4 percent by volume. But the importance of water in the air is far greater than these small percentages would indicate. Indeed, scientists agree that *water vapor* is the most important gas in the atmosphere when it comes to understanding atmospheric processes.

Humidity is the general term for the amount of water vapor in air. Meteorologists employ several methods to express the water-vapor content of the air; we will examine three: mixing ratio, relative humidity, and dew-point temperature.

Figure 12.3 Frost on a window pane is an example of deposition. (Photo by D. Cavagnaro/DRK Photo)

Saturation

Before we consider these humidity measures further, it is important to understand the concept of **saturation.** Imagine a closed jar half full of water and half full of dry air, both at the same temperature. As the water begins to evaporate from the water surface, a small increase in pressure can be detected in the air above. This increase is the result of the motion of the water-vapor molecules that were added to the air through evaporation. In the open atmosphere, this pressure is termed **vapor pressure** and is defined as that part of the total atmospheric pressure that can be attributed to the water-vapor content.

In the closed container, as more and more molecules escape from the water surface, the steadily increasing vapor pressure in the air above forces more and more of these molecules to return to the liquid. Eventually, the number of vapor molecules returning to the surface will balance the number leaving. At that point, the air is said to be *saturated.* If we add heat to the container, thereby increasing the temperature of the water and air, more water will evaporate before a balance is reached. Consequently, at higher temperatures, more moisture is required for saturation. The amount of water vapor required for saturation at various temperatures is shown in Table 12.1.

Mixing Ratio

Not all air is saturated, of course. Thus, we need ways to express how humid a parcel of air is. One method specifies the amount of water vapor contained in a unit of air. The **mixing ratio** is the mass of water vapor in a unit of air compared to the remaining mass of dry air.

$$\text{mixing ratio} = \frac{\text{mass of water vapor (grams)}}{\text{mass of dry air (kilograms)}}$$

Table 12.1 Amount of water vapor (grams) required to saturate a kilogram of air at various temperatures

Temperature		
(°C)	(°F)	Grams of Water Vapor Per Kilogram of Air
−40	−40	0.1
−30	−22	0.3
−20	−4	0.75
−10	14	2
0	32	3.5
5	41	5
10	50	7
15	59	10
20	68	14
25	77	20
30	86	26.5
35	95	35
40	104	47

Table 12.1 shows the mixing ratios of saturated air at various temperatures. For example, at 25°C (77°F), a saturated parcel of air (one kilogram) would contain 20 grams of water vapor.

Because the mixing ratio is expressed in units of mass (usually in grams per kilogram), it is not affected by changes in pressure or temperature. However, the mixing ratio is time consuming to measure by direct sampling. Other methods are employed to express the moisture content of the air. These include relative humidity and dew-point temperature.

Relative Humidity

The most familiar and, unfortunately, the most misunderstood term used to describe the moisture content of air is relative humidity. **Relative humidity** *is a ratio of the air's actual water-vapor content compared with the amount of water vapor required for saturation at that temperature (and pressure).* Thus, relative humidity indicates how near the air is to saturation, rather than the actual quantity of water vapor in the air.

To illustrate, we see from Table 12.1 that at 25°C (77°F), air is saturated when it contains 20 grams of water vapor per kilogram of air. Thus, if the air contains only 10 grams per kilogram on a 25°C day, the relative humidity is expressed as 10/20 or 50 percent. If air with a temperature of 25°C had awater-vapor content of 20 grams per kilogram, the relative humidity would be expressed at 20/20 or 100 percent. When the relative humidity reaches 100 percent, the air is saturated.

Because relative humidity is based on the air's water-vapor content, as well as the amount of moisture required for saturation, it can be changed in either of two ways. First, relative humidity can be changed by the addition or removal of water vapor. Second, because the amount of moisture required for saturation is a function of air temperature, relative humidity varies with temperature. (Recall that the amount of water vapor required for saturation is temperature dependent, such that at higher temperatures it takes more water vapor to saturate air than at lower temperatures.)

Adding or Subtracting Moisture Notice in Figure 12.4 that when water vapor is added to a parcel of air, its relative humidity increases until saturation occurs (100 percent relative humidity). What if even more moisture is added to this parcel of saturated air? Does the relative humidity exceed 100 percent? Normally, this situation does not occur. Instead, the excess water vapor condenses to form liquid water.

In nature, moisture is added to the air mainly via evaporation from the oceans. Plants, soil, and smaller bodies of water also make substantial contributions.

Changes with Temperature Examine Figure 12.5 carefully. Note in Part A that when air at 20°C contains 7 grams of water vapor per kilogram of air, it has a relative humidity of 50 percent. When the air is cooled from 20°C to 10°C as shown in Part B, it still contains 7 grams of water vapor, but the relative humidity increases from 50 percent to 100 percent. We can conclude from this that when the water-vapor content remains constant, *a decrease in temperature results in an increase in relative humidity.*

What happens when the air is cooled further, below the temperature at which saturation occurs? Part C of Figure 12.5 illustrates this situation. Notice from Table 12.1 that when the flask is cooled to 0°C, the air is saturated at 3.5 grams of water vapor per kilogram of air. Because this flask originally contained 7 grams of water vapor, 3.5 grams of water vapor

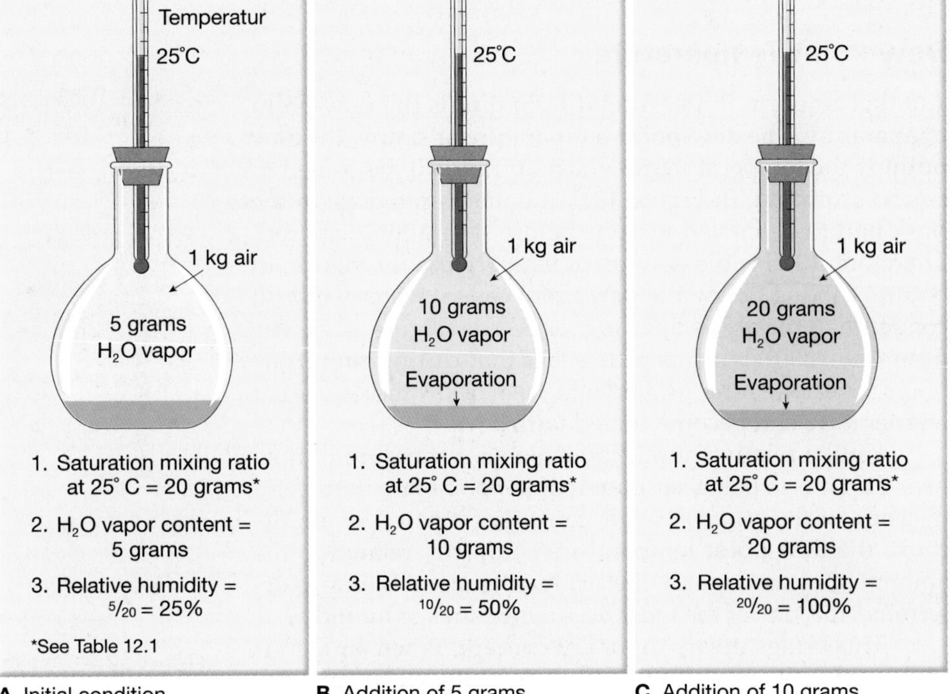

Figure 12.4 Relative humidity. At a constant temperature, the relative humidity will increase as water vapor is added to the air. The saturation mixing ratio remains constant at 20 grams per kilogram and the relative humidity rises from 25 to 100 percent as the water vapor content increases.

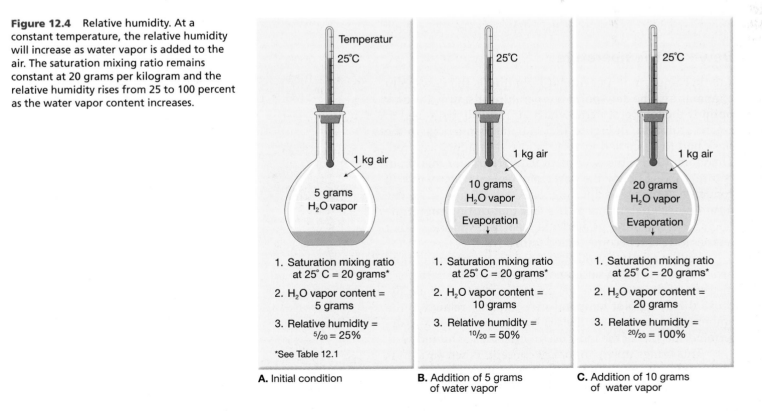

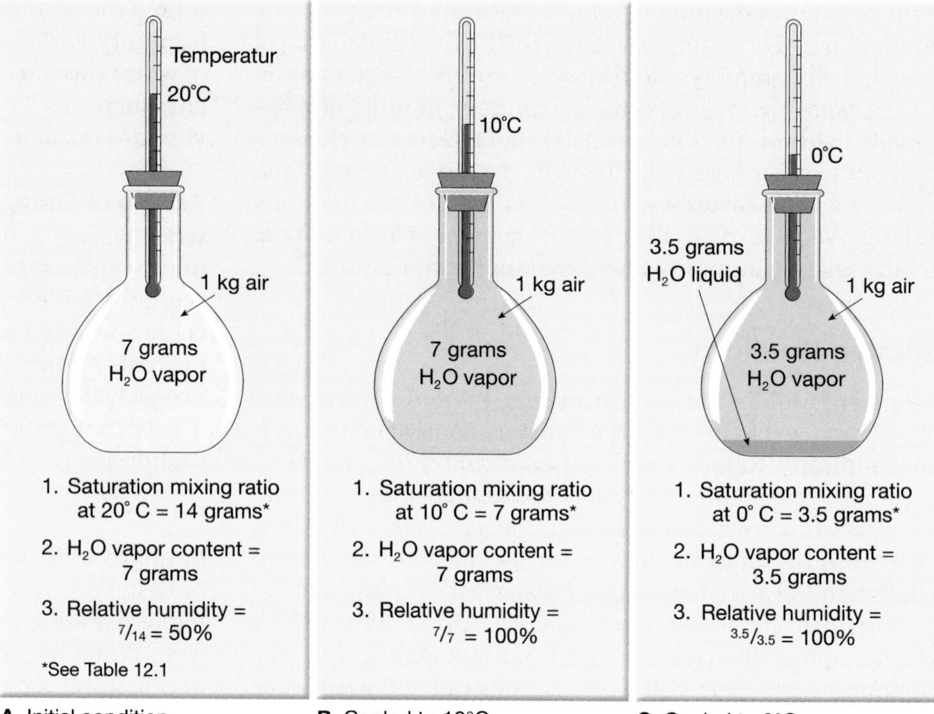

Figure 12.5 Relative humidity varies with temperature. When the water-vapor content (mixing ratio) remains constant, the relative humidity can be changed by increasing or decreasing the air temperature. In this example, when the temperature of the air in the flask was lowered from 20°C to 10°C, the relative humidity increased from 50 percent to 100 percent. Thus, 10°C is the dew point. Further cooling (from 10°C to 0°C) causes one-half of the water vapor to condense. In nature, cooling of air below its dew point generally causes condensation in the form of clouds, dew, or fog.

have to go somewhere, because there is no room for them in the cooler air. They will condense to form liquid droplets that collect on the walls of the container. The relative humidity of the air inside remains at 100 percent.

We can summarize the effects of temperature on relative humidity as follows: When the water-vapor content of air remains at a constant level, a decrease in air temperature results in an increase in relative humidity, and an increase in temperature causes a decrease in relative humidity. In Figure 12.6 the variations in temperature and relative humidity during a typical day demonstrate the relationship just described.

Dew-Point Temperature

Another important measure of humidity is the dew-point temperature. The **dew-point temperature** or simply the **dew point** is the temperature to which air needs to be cooled to reach saturation. In Figure 12.5, the unsaturated air in the flask had to be cooled to 10°C before saturation occurred. Therefore, 10°C is the dew-point temperature for this air. In nature, cooling below the dew point causes water vapor to condense, typically as dew, fog, or clouds (Figure 12.7). The term *dew point* stems from the fact that during nighttime hours, objects near the ground often cool below the dew-point temperature and become coated with dew.

Unlike relative humidity, which is a measure of how near the air is to being saturated, dew-point temperature is a measure of the *actual moisture* content of a parcel of air. Because the dew-point temperature is directly related to the amount of water vapor in the air, and because it is easy to determine, it is one of the most useful measures of humidity.

This brings up an important concept. When air aloft is cooled below its dew point, some of the water vapor condenses to form clouds. The moisture in clouds is made up of *liquid*

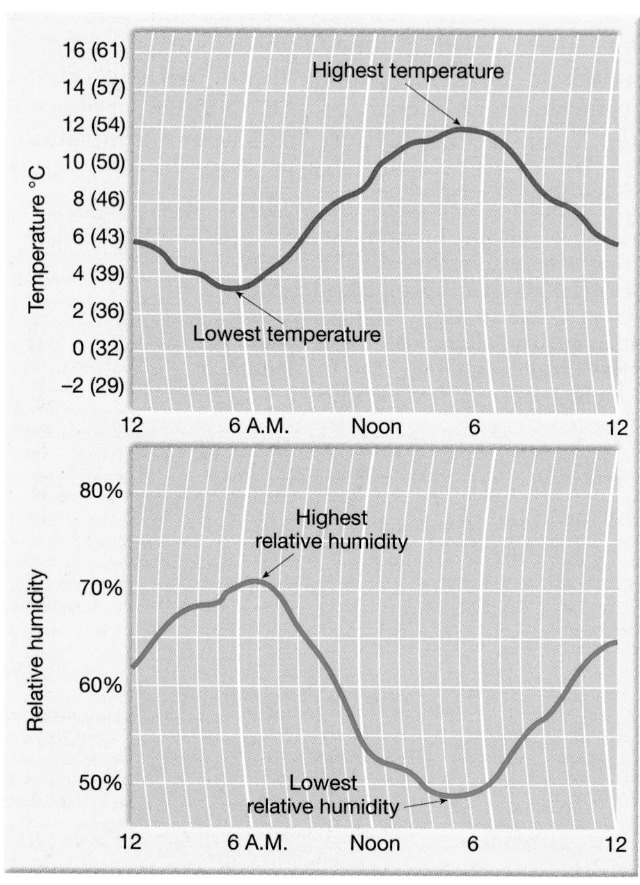

Figure 12.6 Typical daily variations in temperature and relative humidity during a spring day at Washington, D.C. When temperature increases, relative humidity drops (see midafternoon) and vice versa.

Figure 12.7 Condensation, or "dew," occurs when a cold drinking glass chills the surrounding layer of air below the dew-point temperature. (Photo © by Dorling Kindersley)

droplets, (or ice crystals) and is no longer part of the *water-vapor* content of the air. (Clouds are not water vapor; they are liquid water droplets or ice crystals too tiny to fall to Earth.)

Because dew point is the temperature at which saturation occurs, we can conclude that *high dew-point temperatures indicate moist air,* and *low dew-point temperatures indicate dry air.* In fact, for every 10°C (18°F) increase in dew-point temperature, the air's water-vapor content approximately doubles. This fact can be derived by examining Table 12.1 and noting that warm saturated air having a temperature of 20°C (68°F) contains twice the water vapor as cooler saturated air having a temperature of 10°C (50°F).

In summary, dew point is the temperature to which air would have to be cooled to reach saturation, and relative humidity indicates how near the air is to being saturated.

Did You Know?

Contrary to popular belief, frost is not frozen dew. Rather, white frost (*hoar frost*) forms on occasions when saturation occurs at temperature of 0°C (32°F) or below (a temperature called the *frost point*). Thus, frost forms when water vapor changes directly from a gas into a solid (ice) without entering the liquid state. This process, called *deposition,* produces delicate patterns of ice crystals that often decorate windows during winter.

Measuring Humidity

Relative humidity is commonly measured using a **hygrometer.** One type of hygrometer, called a *sling psychrometer,* consists of two identical thermometers mounted side by side (Figure 12.8). One thermometer, the *dry-bulb,* gives the current air temperature. The other, called the *wet-bulb* thermometer, has a thin muslin wick tied around the end (see ends of thermometers in Figure 12.8).

To use the psychrometer, the cloth sleeve is saturated with water and a continuous current of air is passed over the wick. This is done either by spinning or swinging the instrument freely in the air or by fanning air past it. As a consequence, water evaporates from the wick, and the heat absorbed by the evaporating water makes the temperature of the wet bulb drop. The loss of heat that was required to evaporate water from the wet bulb lowers the thermometer reading.

The amount of cooling that takes place is directly proportional to the dryness of the air. The drier the air, the more moisture evaporates. The more heat the evaporating water absorbs, the greater the cooling. Therefore, the larger the difference that is observed between the thermometer readings, the lower the relative humidity; the smaller the difference,

Figure 12.8 Sling psychrometer. This instrument is used to determine both relative humidity and dew point. The dry-bulb thermometer gives the current air temperature. The wet-bulb thermometer is covered with a cloth wick that is dipped in water. The thermometers are spun until the temperature of the wet-bulb thermometer stops declining. Then the thermometers are read and the data used in conjunction with the tables in Appendix C (pp. 000–000). (Photo by E. J. Tarbuck)

Did You Know?

As you might expect, the most humid cities in the United States are located near the ocean in regions that experience frequent onshore breezes. The record belongs to Quillayute, Washington, with an average relative humidity of 83 percent. However, many coastal cities in Oregon, Texas, Louisiana, and Florida also have average relative humidities that exceed 75 percent. Coastal cities in the Northeast tend to be somewhat less humid because they often experience air masses that originate over the drier, continental interior.

the higher the relative humidity. If the air is saturated, no evaporation will occur; and the two thermometers will have identical readings.

To determine the precise relative humidity from the thermometer readings, a standard table is used (Appendix C, pp. 455–456). With the same information, but using a different table, the dew-point temperature may also be calculated.

A different type of hygrometer is used in remote-sensing instrument packages such as radiosondes that transmit upper-air observations back to ground stations. The *electric hygrometer* contains an electrical conductor coated with a moisture-absorbing chemical. It works on the principle that the passage of current varies as the relative humidity varies.

The Basis of Cloud Formation: Adiabatic Cooling

Earth's Dynamic Atmosphere
▼ Moisture and Cloud Formation

Up to this point, we have considered basic properties of water vapor and how its variability is measured. This section will examine some of the important roles that water vapor plays in weather, especially in the formation of clouds.

Fog and Dew versus Cloud Formation

Recall that condensation occurs when water vapor changes to a liquid. This condensation may form dew, fog, or clouds. Although these three forms are different, all require saturated air to develop. As indicated earlier, saturation occurs either when sufficient water vapor is added to the air or, more commonly, when the air is cooled to its dew point.

Near Earth's surface, heat is readily exchanged between the ground and the air above. During evening hours, the surface radiates heat away, and the surface and adjacent air cool rapidly. This "radiation cooling" accounts for the formation of dew and some types of fog. Thus, surface cooling that occurs after sunset accounts for some condensation. However, cloud formation often takes place during the warmest part of the day. Some other mechanism must operate aloft that cools air sufficiently to generate clouds.

Adiabatic Temperature Changes

The process that is responsible for most cloud formation is easily demonstrated if you have ever pumped up a bicycle tire and noticed that the pump barrel became quite warm. The heat you felt was the consequence of the work you did on the air to compress it. When energy is used to compress air, the motion of the gas molecules increases and, therefore, the temperature of the air rises. Conversely, air that is allowed to escape from a bicycle tire *expands and cools* because the expanding air pushes (does work on) the surrounding air and must cool by an amount equivalent to the energy expended.

You have probably experienced the cooling effect of a propellant gas expanding as you applied hair spray or spray deodorant. As the compressed gas in the aerosol can is released, it quickly expands and cools. This drop in temperature occurs *even though heat is neither added nor subtracted.* Such variations are known as **adiabatic temperature changes** and result when air is compressed or allowed to expand. In summary, *when air is allowed to expand, it cools, and when it is compressed, it warms.*

Adiabatic Cooling and Condensation

To simplify the following discussion, it helps to imagine a volume of air enclosed in a thin elastic cover. Meteorologists call this imaginary volume of air a **parcel.** We typically consider a parcel to be a few hundred cubic meters in volume and assume that it acts independently of the surrounding air. It is also assumed that no heat is transferred into or out of the parcel. Although highly idealized, over short time spans, a parcel of air behaves in a manner much like an actual volume of air moving vertically in the atmosphere.

Dry Adiabatic Rate As you travel from Earth's surface upward through the atmosphere, atmospheric pressure rapidly diminishes because there are fewer and fewer gas molecules. Thus, any time a parcel of air moves upward, it passes through regions of successively lower pressure, and the ascending air expands. As it expands, it cools adiabatically. Unsaturated air cools at the constant rate of 10°C for every 1000 meters of ascent (5.5°F per 1000 feet).

Conversely, descending air comes under increasingly higher pressures, compresses, and is heated 10°C for every 1000 meters of descent. This rate of cooling or heating applies only to unsaturated air and is known as the **dry adiabatic rate.**

Wet Adiabatic Rate If a parcel of air rises high enough, it will eventually cool sufficiently to reach the dew point, where the process of condensation begins. From this point on along its ascent, *latent heat of condensation* stored in the water vapor will be liberated. Although the air will continue to cool after condensation begins, the released latent heat works against the adiabatic process, thereby reducing the rate at which the air cools. This slower rate of cooling caused by the addition of latent heat is called the **wet adiabatic rate** of cooling. Because the amount of latent heat released depends on the quantity of moisture present in the air, the wet adiabatic rate varies from 5°C per 1000 meters for air with a high moisture content to 9°C per 1000 meters for dry air.

Figure 12.9 illustrates the role of adiabatic cooling in the formation of clouds. Note that from the surface up to the condensation level, the air cools at the dry adiabatic rate. The wet adiabatic rate begins at the condensation level.

Processes that Lift Air

Earth's Dynamic Atmosphere
▼ Moisture and Cloud Formation

To review, when air rises, it expands and cools adiabatically. If air is lifted sufficiently, it will eventually cool to its dew-point temperature, saturation will occur, and clouds will develop. But why does air rise on some occasions and not on others?

It turns out that, in general, air tends to resist vertical movement. Therefore, air located near the surface tends to stay near the surface, and air aloft tends to remain aloft. Exceptions to this rule, as we will see, include conditions in the atmosphere that give air sufficient buoyancy to rise without the aid of outside forces. In many situations, however, when you see clouds forming, there is some mechanical phenomenon at work that forces the air to rise (at least initially).

We will be looking at four mechanisms that cause air to rise:

1. *Orographic lifting*—air is forced to rise over a mountainous barrier.
2. *Frontal wedging*—warmer, less dense air is forced over cooler, denser air.
3. *Convergence*—a pileup of horizontal air flow results in upward movement.
4. *Localized convective lifting*—unequal surface heating causes localized pockets of air to rise because of their buoyancy.

Orographic Lifting

Orographic lifting occurs when elevated terrains, such as mountains, act as barriers to the flow of air (Figure 12.10). As air ascends a mountain slope, adiabatic cooling often generates clouds and copious precipitation. In fact, many of the rainiest places in the world are located on windward mountain slopes.

By the time air reaches the leeward side of a mountain, much of its moisture has been lost. If the air descends, it warms adiabatically, making condensation and precipitation even less likely. As shown in Figure 12.10, the result can be a **rainshadow desert.** The Great Basin Desert of the western United States lies only a few hundred kilometers from the Pacific Ocean, but it is effectively cut off from the ocean's moisture by the imposing Sierra Nevada (Figure 12.11). The Gobi Desert of Mongolia, the Takla Makan of China, and the Patagonia Desert of Argentina are other examples of deserts that exist because they are on the leeward sides of mountains.

Did You Know?

Many of the rainiest places in the world are located on windward mountain slopes. A station at Mount Waialeale, Hawaii, records the highest average annual rainfall, some 1234 centimeters (486 inches). The greatest recorded rainfall for a single 12-month period occurred at Cherrapunji, India, where an astounding 2647 centimeters (1042 inches, over 86 feet) fell. Much of this rainfall occurred in the month of July—a record 930 centimeters (366 inches, more than 30 feet). This is 10 times more rain than Chicago receives in an average year.

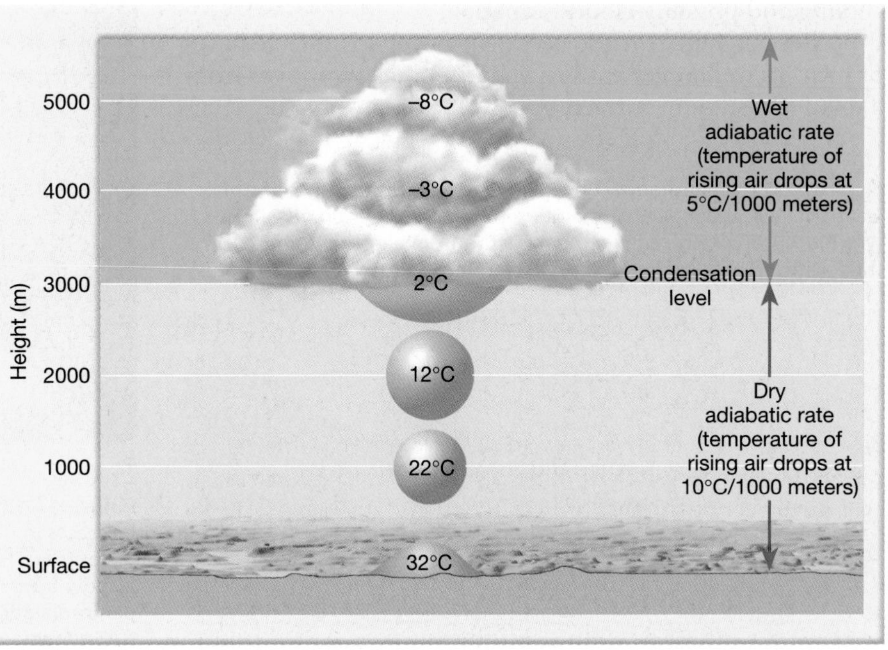

Figure 12.9 Rising air cools at the dry adiabatic rate of 10°C per 1000 meters until the air reaches the dew point and condensation (cloud formation) begins. As air continues to rise, the latent heat released by condensation reduces the rate of cooling. The wet adiabatic rate is therefore always less than the dry adiabatic rate.

Figure 12.10 Orographic lifting occurs where air is forced over a topographic barrier.

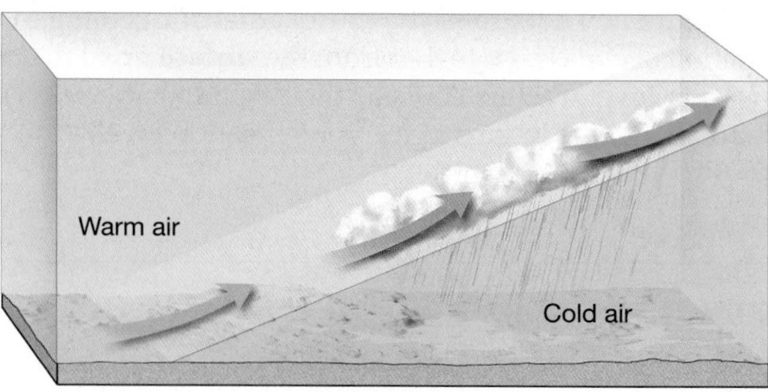

Figure 12.12 Frontal wedging. Colder, denser air acts as a barrier over which warmer, less dense air rises.

Frontal Wedging

If orographic lifting were the only mechanism that forced air aloft, the relatively flat central portion of North America would be an expansive desert instead of the nation's breadbasket. Fortunately, this is not the case.

In central North America, masses of warm and cold air collide, producing a **front.** The cooler, denser air acts as a barrier over which the warmer, less dense air rises. This process, called **frontal wedging,** is illustrated in Figure 12.12.

It should be noted that weather-producing fronts are associated with storm systems called *middle-latitude cyclones.* Because these storms are responsible for producing a high percentage of the precipitation in the middle latitudes, we will examine them closely in Chapter 14.

Convergence

We saw that the collision of contrasting air masses forces air to rise. In a more general sense, whenever air in the lower atmosphere flows together, lifting results. This phenomenon is called **convergence.** When air flows in from more than one direction, it must go somewhere. Because it cannot go down, it goes up (Figure 12.13A) This, of course, leads to adiabatic cooling and possibly cloud formation.

Convergence can also occur whenever an obstacle slows or restricts horizontal air flow (wind). When air moves from a relatively smooth surface, such as the ocean, onto an irregular landscape, its speed is reduced. The result is a pileup of air (convergence). This is similar to what happens when people leave a well-attended sporting event and pileup results at the exits. When air converges, the air molecules do not simply squeeze closer together (like people); rather, there is a net upward flow.

The Florida peninsula provides an excellent example of the role that convergence can play in initiating cloud devel-

Figure 12.11 Rainshadow desert. The arid conditions in California's Death Valley can be partially attributed to the adjacent mountains, which orographically remove the moisture from air originating over the Pacific. (Photo by James E. Patterson/James Patterson Collection)

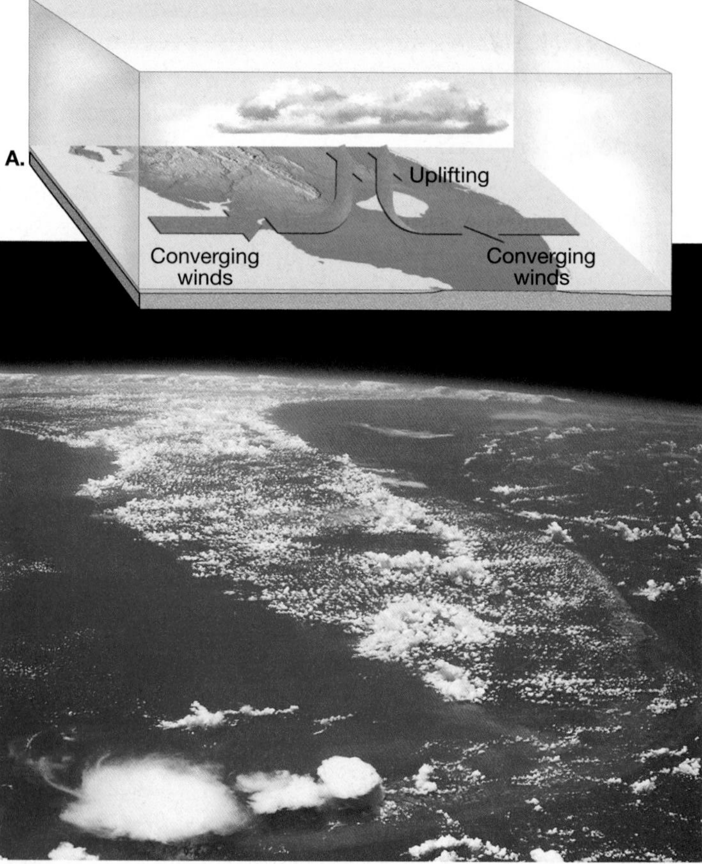

Figure 12.13 Convergence. **A.** When surface air converges, it increases in height to allow for the decreased area it occupies. **B.** Southern Florida viewed from the space shuttle. On warm days, airflow from the Atlantic Ocean and Gulf of Mexico onto the Florida peninsula generates many mid-afternoon thunderstorms. (Photo by NASA/Media Services)

opment and precipitation. On warm days, the air flow is from the ocean to the land along both coasts of Florida. This leads to a pileup of air along the coasts and general convergence over the peninsula (Figure 12.13B). This pattern of air movement and the lifting that results are aided by intense solar heating of the land. This is why the peninsula of Florida experiences the greatest frequency of mid-afternoon thunderstorms in the United States.

More importantly, convergence as a mechanism of forceful lifting is a major contributor to the weather associated with middle-latitude cyclones and hurricanes. The low-level horizontal air flow associated with these systems is inward and upward around their centers. These important weather producers will be covered in more detail later, but for now remember that convergence near the surface results in a general upward flow.

Localized Convective Lifting

On warm summer days, unequal heating of Earth's surface may cause pockets of air to be warmed more than the surrounding air. For instance, air above a paved parking lot will be warmed more than the air above an adjacent wooded park. Consequently, the parcel of air above the parking lot, which is warmer (less dense) than the surrounding air, will be buoyed upward (Figure 12.14). These rising parcels of warmer air are called *thermals.* Birds such as hawks and eagles use these thermals to carry them to great heights where they can gaze down on unsuspecting prey. People have learned to employ these rising parcels using hang gliders as a way to "fly."

The phenomenon that produces rising thermals is called **localized convective lifting.** When these warm parcels of air rise above the condensation level, clouds form, which can produce midafternoon rain showers. The accompanying rains, although occasionally heavy, are of short duration and widely scattered.

Although localized convective lifting by itself is not a major producer of precipitation, the added buoyancy that results from surface heating contributes significantly to the lifting initiated by the other mechanisms. It should also be remembered that even though the other mechanisms force air to rise, convective lifting occurs because the air is warmer (less dense) than the surrounding air and rises for the same reasons as does a hot-air balloon.

The Weathermaker: Atmospheric Stability

Earth's Dynamic Atmosphere
▼ Moisture and Cloud Formation

When air rises, it cools and eventually produces clouds. Why do clouds vary so much in size, and why does the resulting precipitation vary so much? The answers are closely related to the *stability* of the air.

Recall that a parcel of air can be thought of as having a thin flexible cover that allows it to expand but prevents it from mixing with the surrounding air. If this parcel were forced to rise, its temperature would *decrease because of expansion.* By comparing the parcel's temperature to that of the surrounding air, we can determine the stability of the bubble. If the parcel's temperature is lower than that of the surrounding environment, it will be denser, and if allowed to move freely, it would sink to its original position. Air of this type, called **stable air,** resists vertical displacement.

If, however, our imaginary rising parcel were *warmer* and hence less dense than the surrounding air, it would continue to rise until it reached an altitude where its temperature equalled that of its surroundings. This is exactly how a hot-air balloon works, rising as long as it is warmer and less dense than the surrounding air (Figure 12.15). This type of air is classified as **unstable air.** In summary, stability is a property of air that describes its tendency to remain in its original position (stable) or to rise (unstable).

Types of Stability

The stability of air is determined by examining the temperature of the atmosphere at various heights. Recall from Chapter 11 that this measure is termed the *environmental lapse rate.* The environmental lapse rate is the temperature of the atmosphere as determined from observations made by radiosondes and

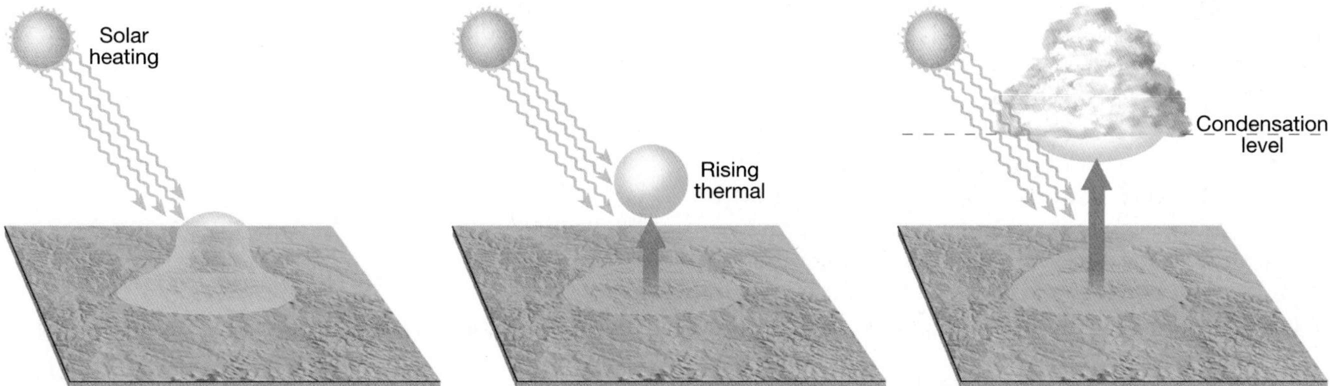

Figure 12.14 Localized convective lifting. Unequal heating of Earth's surface causes pockets of air to be warmed more than the surrounding air. These buoyant parcels of hot air rise, producing thermals, and if they reach the condensation level, clouds form.

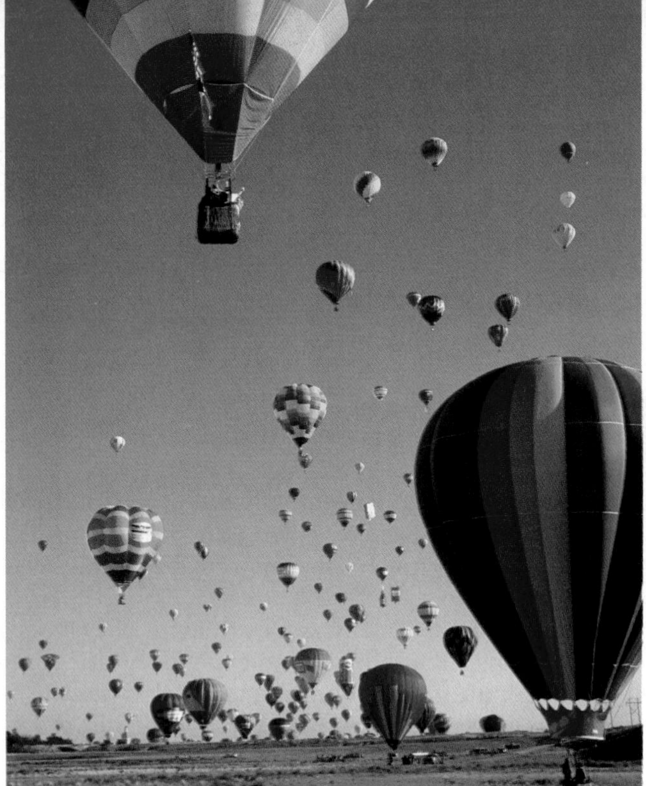

Figure 12.15 As long as air is warmer than its surroundings, it will rise. Hot-air balloons rise up through the atmosphere for this reason. (Photo by Barbara Cushman Rowell/Mountain Light Photography, Inc.)

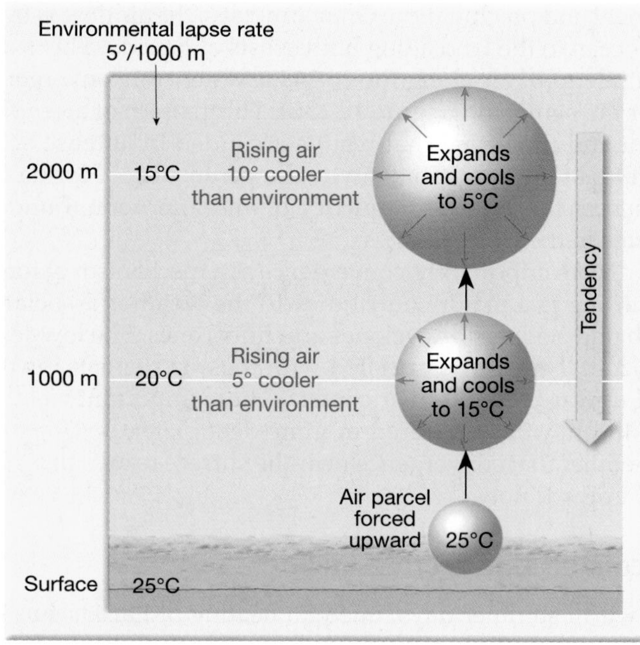

Figure 12.16 In a stable atmosphere, as an unsaturated parcel of air is lifted, it expands and cools at the dry adiabatic rate of 10°C per 1000 meters. Because the temperature of the rising parcel of air is lower than that of the surrounding environment, it will be heavier and, if allowed to do so, will sink to its original position.

Did You Know?

According to National Weather Service Records, Rochester, New York, which averages nearly 239 centimeters (94 inches) of snow annually, is the snowiest city in the United States. However, Buffalo, New York, is a close runner-up.

aircraft. It is important not to confuse this with *adiabatic temperature changes,* which are changes in temperature due to expansion or compression as a parcel of air rises or descends.

To illustrate, we will examine a situation in which the environmental lapse rate is 5°C per 1000 meters (Figure 12.16). When air at the surface has a temperature of 25°C, the air at 1000 meters will be 5 degrees cooler, or 20°C, the air at 2000 meters will have a temperature of 15°C, and so forth. At first glance, it appears that the air at the surface is less dense than the air at 1000 meters, because it is 5 degrees warmer. However, if the air near the surface is unsaturated and were to rise to 1000 meters, it would expand and cool at the dry adiabatic rate of 10°C per 1000 meters. Therefore, upon reaching 1000 meters, its temperature would have dropped 10°C. Being 5 degrees cooler than its environment, it would be denser and tend to sink to its original position. Hence, we say that the air near the surface is potentially cooler than the air aloft and, therefore, will not rise on its own. The air just described is *stable* and resists vertical movement.

Absolute Stability Stated quantitatively, **absolute stability** prevails when the environmental lapse rate is less than the wet adiabatic rate. Figure 12.17 depicts this situation using an environmental lapse rate of 5°C per 1000 meters and a wet adiabatic rate of 6°C per 1000 meters. Note that at 1000 meters, the temperature of the surrounding air is 15°C, while the rising parcel of air has cooled to 10°C and is, therefore, the denser air. Even if this stable air were to be forced above the condensation level, it would remain cooler and denser than its environment, and it would tend to return to the surface.

Absolute Instability At the other extreme, air is said to exhibit **absolute instability** when the environmental lapse rate is greater than the dry adiabatic rate. As shown in Figure 12.18, the ascending parcel of air is always warmer than its environment and will continue to rise because of its own buoyancy. However, absolute instability is generally found near Earth's surface. On hot, sunny days, the air above some surfaces, such as shopping center parking lots, is heated more than the air over adjacent surfaces. These invisible pockets of more intensely heated air, being less dense than the air aloft, will rise like a hot-air balloon. This phenomenon produces the small fluffy clouds we associate with fair weather. Occasionally, when the surface air is considerably warmer than the air aloft, clouds with considerable vertical development can form.

Did You Know?

The record for the lowest average annual precipitation belongs to Arica, Chile, which receives 0.08 centimeter (0.03 inch) of rainfall per year. Over a span of 59 years, this region in South America received a total of less than 5 centimeters (2 inches) of precipitation.

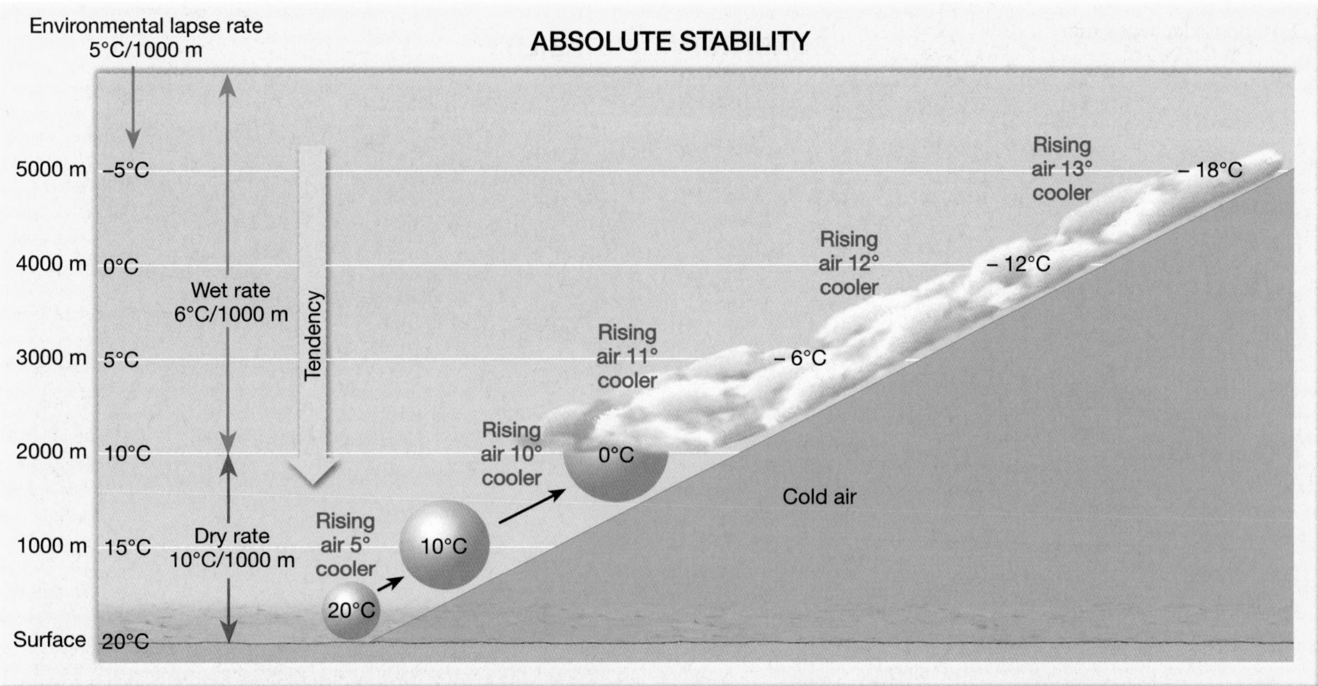

Figure 12.17 Absolute stability prevails when the environmental lapse rate is less than the wet adiabatic rate. The rising parcel of air is always cooler and denser than the surrounding air. When stable air is forced to rise, it produces flat, layered clouds.

Conditional Instability A more common type of atmospheric instability is called **conditional instability.** This occurs when moist air has an environmental lapse rate between the dry and wet adiabatic rates (between 5 and 10° C per 1000 meters). Simply, the atmosphere is said to be conditionally unstable when it is *stable* for an *unsaturated* parcel of air, but *unstable* for a *saturated* parcel of air. Notice in Figure 12.19 that the rising parcel of air is cooler than the surrounding air for nearly 3000 meters. With the addition of latent heat above the lifting condensation level, the parcel becomes warmer than the surrounding air. From this point along its ascent, the parcel will continue to rise because of its own buoyancy, without

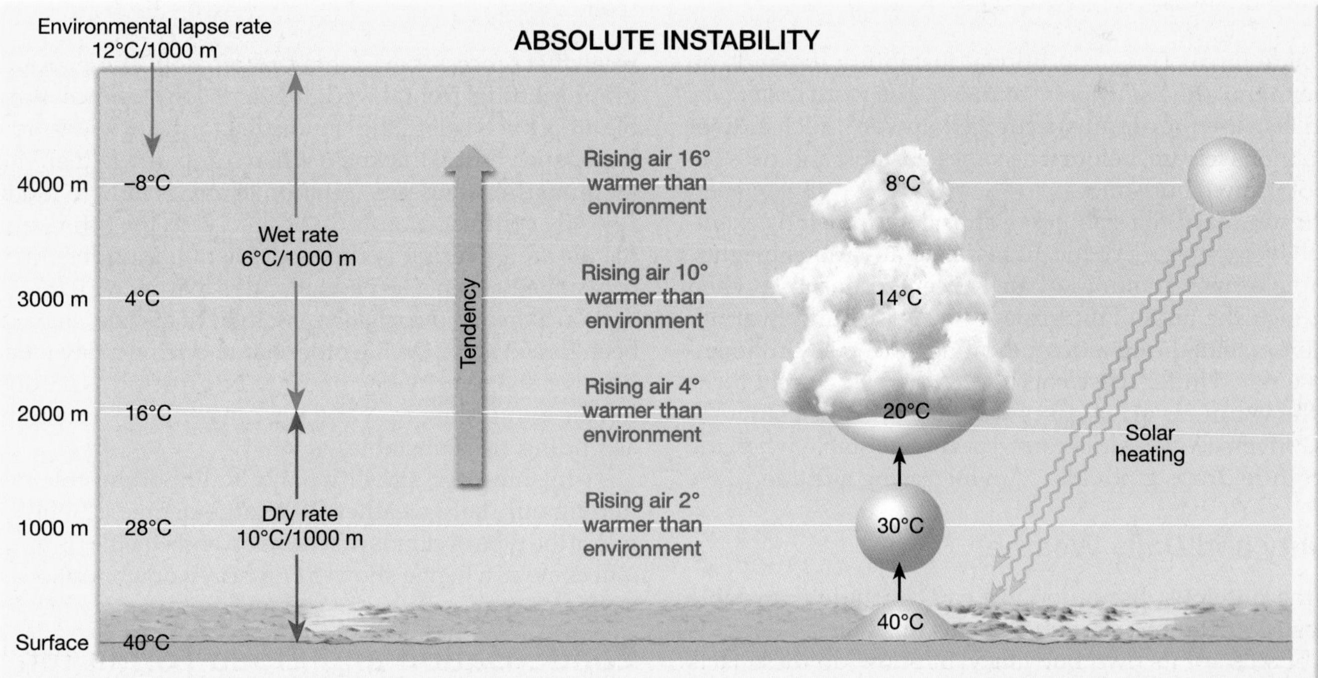

Figure 12.18 Illustration of *absolute instability* that develops when solar heating causes the lowermost layer of the atmosphere to be warmed to a higher temperature than the air aloft. The result is a steep environmental lapse rate that renders the atmosphere unstable.

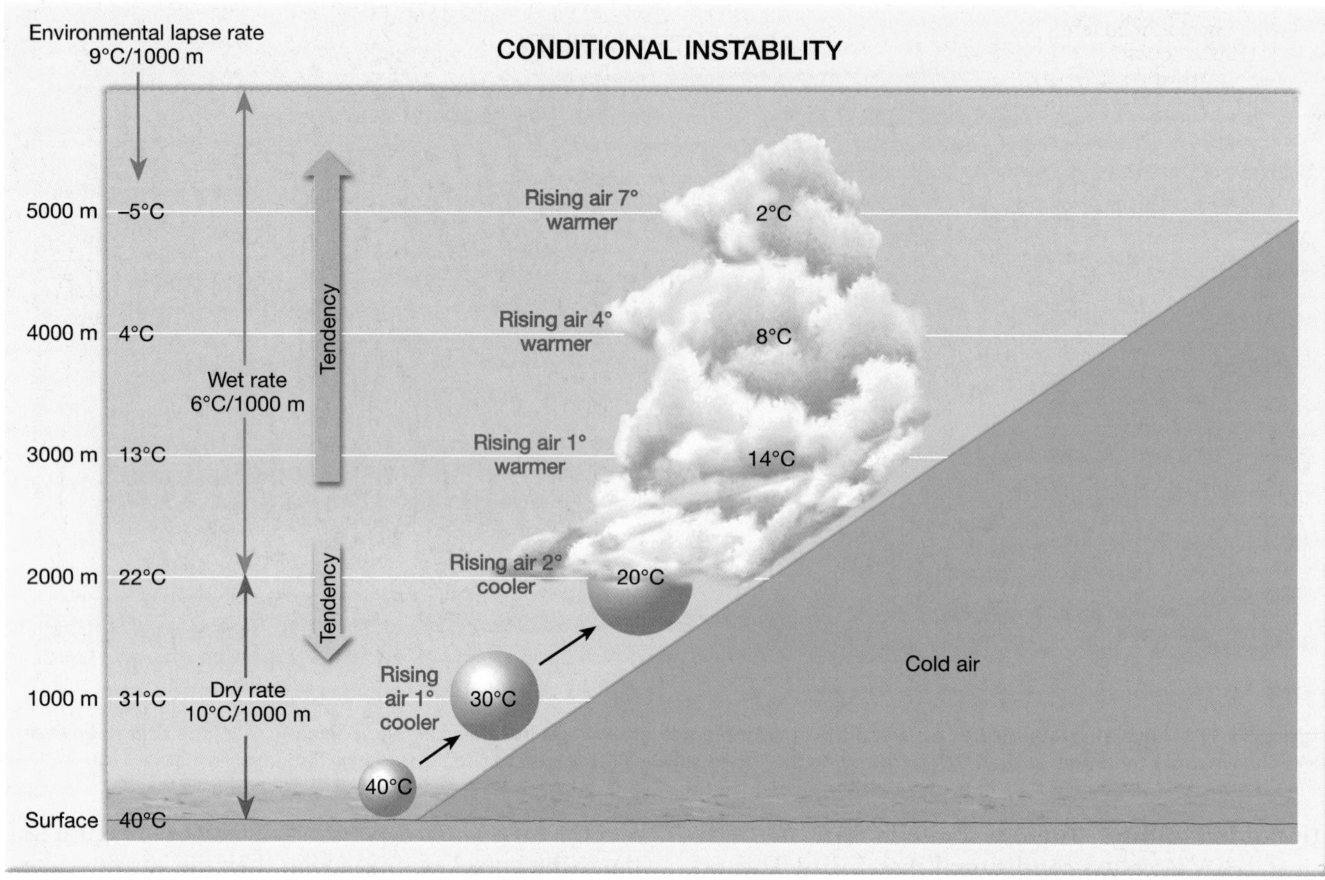

Figure 12.19 Illustration of *conditional instability* where warm air is forced to rise along a frontal boundary. Note that the environmental lapse rate of 9°C per 1000 meters lies between the dry and wet adiabatic rates. The parcel of air is cooler than the surrounding air up to nearly 3000 meters, where its tendency is to sink toward the surface (stable). Above this level, however, the parcel is warmer than its environment and will rise because of its own buoyancy (unstable). Thus, when conditionally unstable air is forced to rise, the result can be towering cumulus clouds.

an outside force. Thus, conditional instability depends on whether or not the rising air is saturated. The word *conditional* is used because the air must be forced upward, such as over mountainous terrain, before it becomes unstable and rises because of its own buoyancy.

In summary, the stability of air is determined by examining the temperature of the atmosphere at various heights. In simple terms, a column of air is deemed unstable when the air near the bottom of this layer is significantly warmer (less dense) than the air aloft, indicating a steep environmental lapse rate. Under these conditions, the air actually turns over; the warm air below rises and displaces the colder air aloft. Conversely, the air is considered to be stable when the temperature drops gradually with increasing altitude.

Stability and Daily Weather

From the previous discussion, we can conclude that stable air resists vertical movement, whereas unstable air ascends freely because of its own buoyancy. But how do these facts manifest themselves in our daily weather?

Because stable air resists upward movement, we might conclude that clouds will not form when stable conditions prevail in the atmosphere. Although this seems reasonable, recall that processes exist that *force* air aloft. These include orographic lifting, frontal wedging, and convergence. When stable air is forced aloft, the clouds that form are widespread and have little vertical thickness when compared to their horizontal dimension, and precipitation, if any, is light to moderate.

By contrast, clouds associated with the lifting of unstable air are towering and often generate thunderstorms and occasionally even a tornado. For this reason, we can conclude that on a dreary, overcast day with light drizzle, stable air has been forced aloft. On the other hand, during a day when cauliflower-shaped clouds appear to be growing as if bubbles of hot air are surging upward, we can be fairly certain that the ascending air is unstable.

In summary, stability plays an important role in determining our daily weather. To a large degree, stability determines the type of clouds that develop and whether precipitation will come as a gentle shower or a heavy downpour.

Condensation and Cloud Formation

To review briefly, condensation occurs when water vapor in the air changes to a liquid. The result of this process may be dew, fog, or clouds. For any of these forms of condensation to occur, the air must be saturated. Saturation occurs most commonly when air is cooled to its dew point, or less often when water vapor is added to the air.

Generally, there must be a surface on which the water vapor can condense. When dew occurs, objects at or near the ground, such as grass and car windows, serve this purpose. But when condensation occurs in the air above the ground, tiny bits of particulate matter, known as **condensation nuclei,** serve as surfaces for water-vapor condensation. These nuclei are important, for in their absence, a relative humidity well in excess of 100 percent is needed to produce clouds.

Condensation nuclei such as microscopic dust, smoke, and salt particles (from the ocean) are profuse in the lower atmosphere. Because of this abundance of particles, relative humidity rarely exceeds 101 percent. Some particles, such as ocean salt, are particularly good nuclei because they absorb water. These particles are termed **hygroscopic** ("water-seeking") **nuclei.**

When condensation takes place, the initial growth rate of cloud droplets is rapid. It diminishes quickly because the excess water vapor is quickly absorbed by the numerous competing particles. This results in the formation of a cloud consisting of millions upon millions of tiny water droplets, all so fine that they remain suspended in air. When cloud formation occurs at below-freezing temperatures, tiny ice crystals form. Thus, a cloud can consist of water droplets, ice crystals, or both.

The slow growth of cloud droplets by additional condensation and the immense size difference between cloud droplets and raindrops suggest that condensation alone is not responsible for the formation of drops large enough to fall as rain. We will first examine clouds and then return to the questions of how precipitation forms.

Types of Clouds

Clouds are among the most conspicuous and observable aspects of the atmosphere and its weather. **Clouds** are a form of condensation best described as *visible aggregates of minute droplets of water or tiny crystals of ice.* In addition to being prominent and sometimes spectacular features in the sky, clouds are of continual interest to meteorologists, because they provide a visible indication of what is going on in the atmosphere. Anyone who observes clouds with the hope of recognizing different types often finds that there are a bewildering variety of these familiar white and gray masses streaming across the sky. Still, once one comes to know the basic classification scheme for clouds, most of the confusion vanishes.

Clouds are classified on the basis of their *form* and *height* (Figure 12.20). Three basic forms are recognized: cirrus, cumulus, and stratus.

- **Cirrus** clouds are high, white, and thin. They can occur as patches composed of small cells or as delicate veil-like sheets or extended wispy fibers that often have a feathery appearance.
- **Cumulus** clouds consist of globular individual cloud masses. They normally exhibit a flat base and have the appearance of rising domes or towers. Such clouds are frequently described as having a cauliflower structure.
- **Stratus** clouds are best described as sheets or layers that cover much or all of the sky. While there may be minor breaks, there are no distinct individual cloud units.

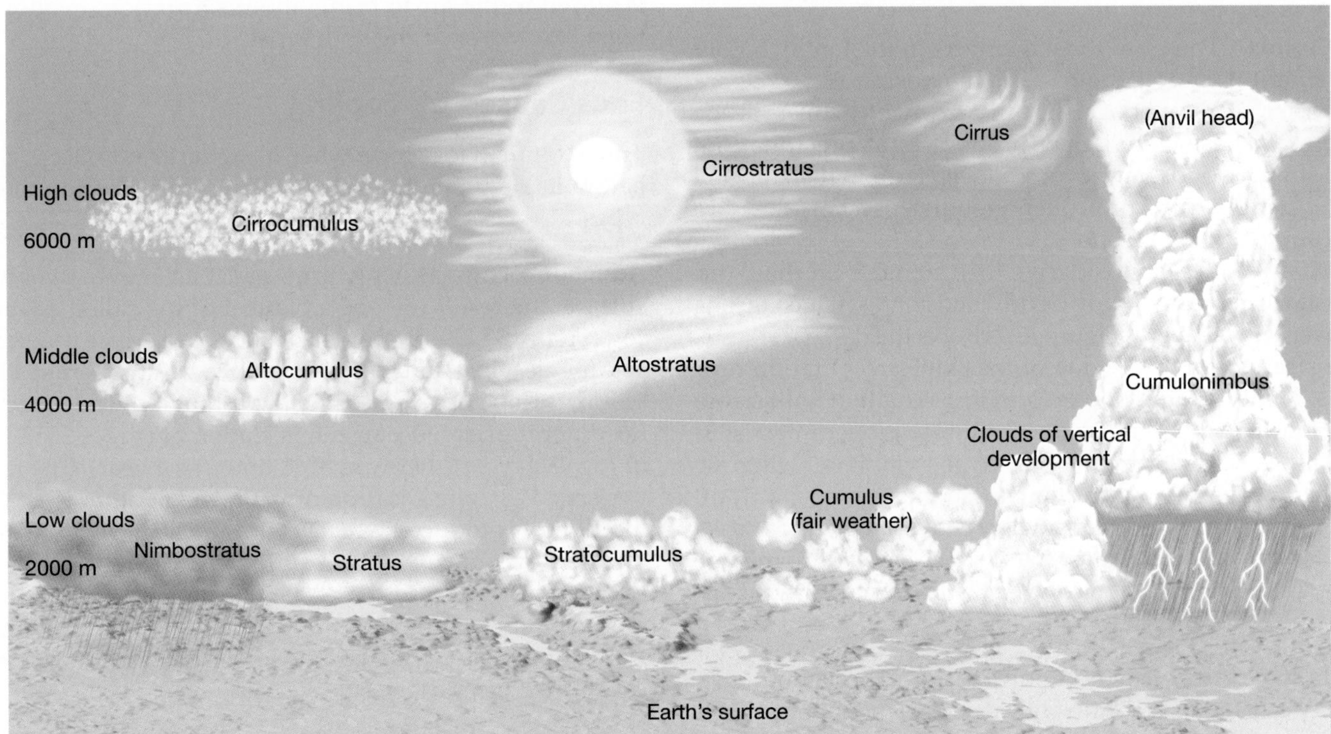

Figure 12.20 Classification of clouds according to height and form. (After Ward's Natural Science Establishment, Inc., Rochester, NY)

All other clouds reflect one of these three basic forms or are combinations or modifications of them.

Three levels of cloud heights are recognized: high, middle, and low (Figure 12.20). **High clouds** normally have bases above 6000 meters (20,000 feet), **middle clouds** generally occupy heights from 2000 to 6000 meters (6500 to 20,000 feet), and **low clouds** form below 2000 meters (6500 feet). The altitudes listed for each height category are not hard and fast. There is some seasonal as well as latitudinal variation. For example, at high latitudes or during cold winter months in the midlatitudes, high clouds are often found at lower altitudes.

High Clouds Three cloud types make up the family of high clouds (above 6000 meters [20,000 feet]): cirrus, cirrostratus, and cirrocumulus. *Cirrus* clouds are thin and delicate and sometimes appear as hooked filaments called "mares' tails" (Figure 12.21A). As the names suggest, *cirrocumulus* clouds consist of fluffy masses (Figure 12.21B), whereas *cirrostratus* clouds are flat layers (Figure 12.21C). Because of the low temperatures and small quantities of water vapor present at high altitudes, all high clouds are thin and white and are made up of ice crystals. Further, these clouds are not considered precipitation makers. However, when cirrus clouds are followed by cirrocumulus clouds and increased sky coverage, they may warn of impending stormy weather.

Middle Clouds Clouds that appear in the middle range (2000 to 6000 meters [6500 to 20,000 feet]) have the prefix *alto-* as part of their name. *Altocumulus* clouds are composed of globular masses that differ from cirrocumulus clouds in that they are larger and denser (Figure 12.21D). *Altostratus* clouds create a uniform white to grayish sheet covering the sky with the Sun or Moon visible as a bright spot (Figure 12.21E). Infrequent light snow or drizzle may accompany these clouds.

Low Clouds There are three members in the family of low clouds: stratus, stratocumulus, and nimbostratus. *Stratus* are a uniform foglike layer of clouds that frequently covers much of the sky. These clouds may produce light precipitation. When stratus clouds develop a scalloped bottom that appears as long parallel rolls or broken globular patches, they are called *stratocumulus* clouds.

Nimbostratus clouds derive their name from the Latin *nimbus*, which means "rainy cloud," and *stratus*, which means "to cover with a layer" (Figure 12.21F). As the name suggests, nimbostratus clouds are one of the chief precipitation producers. Nimbostratus clouds form in association with stable conditions. We might not expect clouds to grow or persist in stable air, yet cloud growth of this type is common when air is forced to rise, as occurs along a mountain range, a front, or near the center of a cyclone where converging winds cause air to ascend. Such forced ascent of stable air leads to the formation of a stratified cloud layer that is large horizontally compared to its depth.

Clouds of Vertical Development Some clouds do not fit into any one of the three height categories mentioned. Such clouds have their bases in the low height range but often extend upward into the middle or high altitudes. Clouds in this category are called **clouds of vertical development.** They are all related to one another and are associated with unstable air. Although *cumulus* clouds are often connected with "fair" weather (Figure 12.26G), they may grow dramatically under the proper circumstances. Once upward movement is triggered, acceleration is powerful and clouds with great vertical extent form. The end result is often a towering cloud, called a *cumulonimbus,* that may produce rain showers or a thunderstorm (Figure 12.26H).

Definite weather patterns can often be associated with particular clouds or certain combinations of cloud types, so it is important to become familiar with cloud descriptions and characteristics.

Fog

Fog is generally considered to be an atmospheric hazard. When is it light, visibility is reduced to 2 or 3 kilometers (1 or 2 miles). However, when it is dense, visibility may be cut to a few dozen meters or less, making travel by any mode not only difficult but often dangerous. Officially, visibility must be reduced to 1 kilometer (0.6 mile) or less before fog is reported. While this figure is arbitrary, it does provide an objective criterion for comparing fog frequencies at different locations.

Fog is defined as a *cloud with its base at or very near the ground.* There is no physical difference between a fog and a cloud; their appearance and structure are the same. The essential difference is the method and place of formation. Whereas clouds result when air rises and cools adiabatically, most fogs are the consequence of radiation cooling or the movement of air over a cold surface. (The exception is upslope fog.) In other circumstances, fogs form when enough water vapor is added to the air to bring about saturation (evaporation fogs). We will examine both types.

Fogs Caused by Cooling

Three common fogs form when air at Earth's surface is chilled below its dew point. A fog is often a combination of these types.

Advection Fog When warm, moist air moves over a cool surface, the result may be a blanket of fog called **advection fog** (Figure 12.22). Examples of such fogs are very common. The foggiest location in the United States, and perhaps in the world, is Cape Disappointment, Washington. The name is indeed appropriate, because this station averages 2552 hours of fog each year (there are 8760 hours in a year). The fog experienced at Cape Disappointment, and at other West Coast locations, is produced when warm, moist air from the Pacific Ocean moves over the cold California Current and is then carried onto shore by the prevailing winds. Advection fogs are also quite common in the winter season when warm air from the Gulf of Mexico blows across cold, often snow-covered surfaces of the Midwest and East.

Radiation Fog As its name implies, **radiation fog** forms on cool, clear, calm nights, when Earth's surface cools rapidly

A. Cirrus

B. Cirrocumulus

C. Cirrostratus

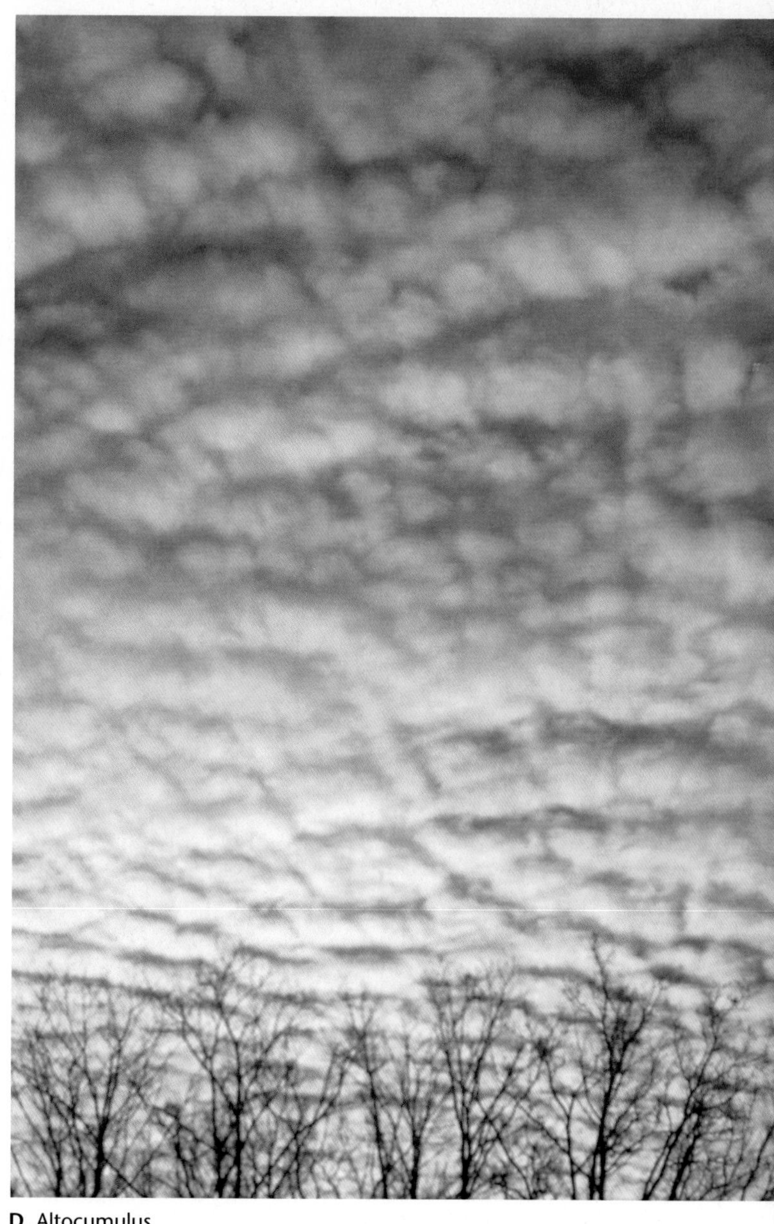

D. Altocumulus

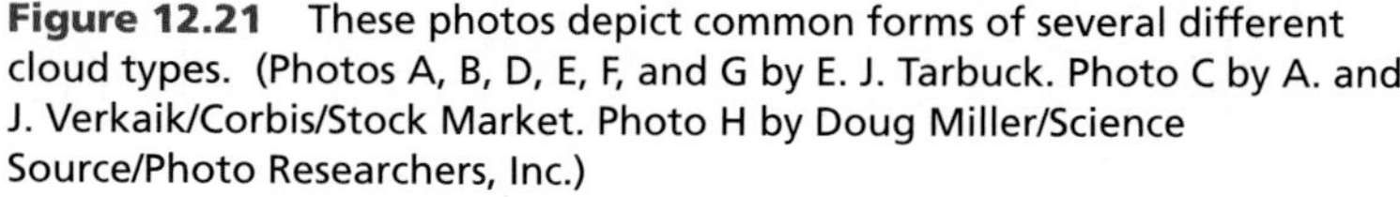

Figure 12.21 These photos depict common forms of several different cloud types. (Photos A, B, D, E, F, and G by E. J. Tarbuck. Photo C by A. and J. Verkaik/Corbis/Stock Market. Photo H by Doug Miller/Science Source/Photo Researchers, Inc.)

E. Altostratus

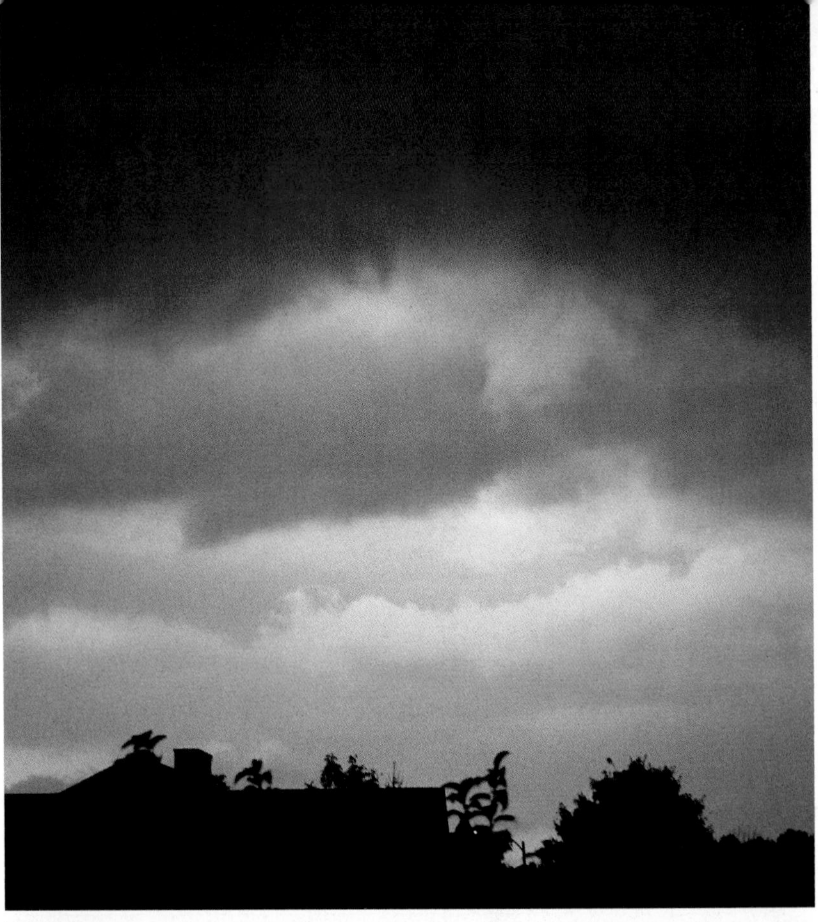

F. Nimbostratus

G. Cumulus

H. Cumulonimbus

Figure 12.21 (continued)

Figure 12.22 Advection fog rolling into San Francisco Bay. (Photo by Ed Pritchard/Getty Images Inc.—Stone Allstock)

by radiation. As the night progresses, a thin layer of air in contact with the ground is cooled below its dew point. As the air cools and becomes denser, it drains into low areas, resulting in "pockets" of fog. The largest pockets are often river valleys, where thick accumulations may occur (Figure 12.23).

Upslope Fog **Upslope fog** is created when relatively humid air moves up a gradually sloping plain or up the steep slopes of a mountain. Because of this upward movement, air expands and cools adiabatically. If the dew point is reached, an extensive layer of fog may form. In the United States, the Great Plains offers an excellent example. When humid easterly or southeasterly winds blow westward from the Mississippi River upslope toward the Rocky Mountains, the air gradually rises, resulting in an adiabatic temperature decrease of about 12°C (22°F). When the difference between the air temperature and dew point of westward-moving air is less than 12°C (22°F), an extensive fog can result in the western plains.

Evaporation Fogs

When the saturation of air occurs primarily because of the addition of water vapor, the resulting fogs may be called *evaporation fogs.* Two types of evaporation fogs are recognized: steam fog and frontal, or precipitation, fog.

Steam Fog When cool air moves over warm water, enough moisture may evaporate from the water surface to produce saturation. As the rising water vapor meets the cold air, it immediately recondenses and rises with the air that is being warmed from below. Because the water has a steaming appearance, the phenomenon is called **steam fog.** Steam fog is fairly common over lakes and rivers in the fall and early winter, when the water may still be relatively warm and the air is rather crisp (Figure 12.24). Steam fog is often shallow because as the steam rises, it reevaporates in the unsaturated air above.

Did You Know?

On cold days when you "see your breath," you are actually creating steam fog. The moist air that you exhale saturates a small volume of cold air, causing tiny droplets to form. Like most steam fogs, the droplets quickly evaporate as the "fog" mixes with the unsaturated air around it.

Frontal or Precipitation Fog When frontal wedging occurs, warm air is lifted over colder air. If the resulting clouds yield rain, and the cold air below is near the dew point, enough rain will evaporate to produce fog. A fog formed in this manner is called **frontal fog,** or **precipitation fog.** The result is a more or less continuous zone of condensed water droplets reaching from the ground up through the clouds.

In summary, both steam fog and frontal fog result from the addition of moisture to a layer of air. The air is usually cool or cold and already near saturation. Because air's capacity to hold water vapor at low temperatures is small, only a

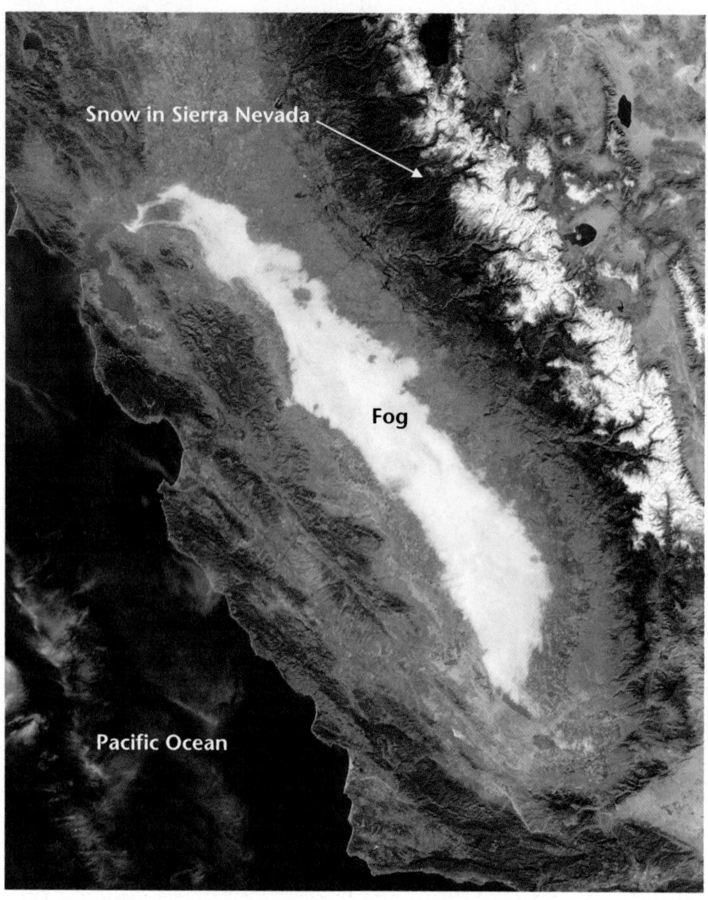

Figure 12.23 Satellite image of dense fog in California's San Joaquin Valley on November 20, 2002. This early morning radiation fog was responsible for several car accidents in the region, including a 14-car pileup. The white areas to the east of the fog are the snow-capped Sierra Nevadas. (NASA)

relatively modest amount of evaporation is necessary to produce saturated conditions and fog.

Precipitation

All clouds contain water. But why do some produce precipitation while others drift placidly overhead? This simple question perplexed meteorologists for many years. First, cloud droplets are very small, averaging less than 10 micrometers in diameter (for comparison, a human hair is about 75 micrometers in diameter). Because of their small size, cloud droplets fall incredibly slowly. In addition, clouds are made up of many billions of these droplets, all competing for the available water vapor; thus, their continued growth via condensation is extremely slow. So, what causes precipitation?

How Precipitation Forms

A raindrop large enough to reach the ground without completely evaporating contains roughly 1 million times more water than a single cloud droplet. Therefore, for precipitation to form, millions of cloud droplets must somehow coalesce (join together) into drops large enough to sustain themselves during their descent to the surface. Two mechanisms have been proposed to explain this phenomenon: the ice crystal process and the collision-coalescence process.

Ice Crystal Process Meteorologists discovered that in the mid to high latitudes, precipitation often forms in clouds where the temperature is below 0°C (32°F). Although we would expect all cloud droplets to freeze in subzero temperatures, only those droplets that make contact with solid particles that have a particular structure (called *ice nuclei*) actually freeze. The remaining liquid droplets are said to be *supercooled* (below 0°C [32°F], but still liquid).

When ice crystals and supercooled water droplets coexist in a cloud, the stage is set to generate precipitation. It turns

Figure 12.24 Steam fog rising from upper St. Regis Lake, Adirondack Mountains, New York. (Photo by Jim Brown/CORBIS/Stock Market)

out that ice crystals collect the available water vapor at a much faster rate than does liquid water. Thus, ice crystals grow larger at the expense of the water droplets. Eventually, this process generates ice crystals large enough to fall as snowflakes. During their descent, these ice crystals get bigger as they intercept supercooled cloud droplets that freeze on them. When the surface temperature is about 4°C (39°F) or higher, snowflakes usually melt before they reach the ground and continue their descent as rain.

Collision-Coalescence Process Precipitation can form in warm clouds that contain large hygroscopic ("water-seeking") condensation nuclei, such as salt particles. Relatively large droplets form on these large nuclei. Because the bigger droplets fall faster, they collide and join with smaller water droplets. After many collisions, the droplets are large enough to fall to the ground as rain.

Forms of Precipitation

Because atmospheric conditions vary greatly from place to place as well as seasonally, several different forms of precipitation are possible (Table 12.2). Rain and snow are the most common and familiar, but other forms of precipitation are important as well. The occurrence of sleet, glaze, or hail is often associated with important weather events. Although limited in occurrence and sporadic in both time and space, these forms, especially glaze and hail, can cause considerable damage.

Rain and Drizzle In meteorology, the term **rain** is restricted to drops of water that fall from a cloud and that have a diameter of at least 0.5 millimeter (0.002 inch). Most rain originates either in nimbostratus clouds or in towering cumulonimbus clouds that are capable of producing unusually heavy rainfalls known as *cloudbursts.* Raindrops rarely exceed about 5 millimeters (0.2 inch). Larger drops do not survive, because surface tension, which holds the drops together, is exceeded by the frictional drag of the air. Consequently, large raindrops regularly break apart into smaller ones.

Fine, uniform drops of water having a diameter less than 0.5 millimeter (0.02 inch) are called *drizzle.* Drizzle can be so fine that the tiny drops appear to float, and their impact is almost imperceptible. Drizzle and small raindrops generally are produced in stratus or nimbostratus clouds where precipitation may be continuous for several hours or, on rare occasions, for days.

Snow **Snow** is precipitation in the form of ice crystals (snowflakes) or, more often, aggregates of crystals. The size, shape, and concentration of snowflakes depend to a great extent on the temperature at which they form.

Recall that at very low temperatures, the moisture content of air is small. The result is the generation of very light and fluffy snow made up of individual six-sided ice crystals. This is the "powder" that downhill skiers talk so much about. By contrast, at temperatures warmer than about −5°C (23°F), the ice crystals join together into larger clumps consisting of tangled aggregates of crystals. Snowfalls consisting of these composite snowflakes are generally heavy and have a high moisture content, which makes them ideal for making snowballs.

Table 12.2 Forms of precipitation.

Type	Appropriate Size	State of Matter	Description
Mist	0.005 to 0.05 mm	Liquid	Droplets large enough to be felt on the face when air is moving 1 meter/second. Associated with stratus clouds.
Drizzle	Less than 0.5 mm	Liquid	Small uniform drops that fall from stratus clouds, generally for several hours.
Rain	0.5 to 5 mm	Liquid	Generally produced by nimbostratus or cumulonimbus clouds. When heavy, it can show high variability from one place to another.
Sleet	0.5 to 5 mm	Solid	Small, spherical to lumpy ice particles that form when raindrops freeze while falling through a layer of subfreezing air. Because the ice particles are small, damage, if any, is generally minor. Sleet can make travel hazardous.
Glaze	Layers 1 mm to 2 cm thick	Solid	Produced when supercooled raindrops freeze on contact with solid objects. Glaze can form a thick coating of ice having sufficient weight to seriously damage trees and power lines
Rime	Variable	Solid	Deposits usually consisting of ice feathers that point into the wind. These delicate, frostlike accumulations form as supercooled cloud or fog droplets encounter objects and freeze on contact.
Snow	1 mm to 2 cm	Solid	The crystalline nature of snow allows it to assume many shapes including six-sided crystals, plates, and needles. Produced in supercooled clouds where water vapor is deposited as ice crystals that remain frozen during their descent.
Hail	5 mm to 10 cm	Solid	Precipitation in the form of hard, rounded pellets or irregular lumps of ice. Produced in large convective, cumulonimbus clouds, where frozen ice particles and supercooled water coexist.
Graupel	2 to 5 mm	Solid	Sometimes called soft hail, graupel forms when rime collects on snow crystals to produce irregular masses of "soft" ice. Because these particles are softer than hailstones, they normally flatten out upon impact.

Sleet and Glaze **Sleet** is a wintertime phenomenon and refers to the fall of small particles of ice that are clear to translucent. For sleet to be produced, a layer of air with temperatures above freezing must overlie a subfreezing layer near the ground. When raindrops, which are often melted snow, leave the warmer air and encounter the colder air below, they freeze and reach the ground as small pellets of ice the size of the raindrops from which they formed.

On some occasions, when the vertical distribution of temperatures is similar to that associated with the formation of sleet, **freezing rain** or **glaze** results instead. In such situations, the subfreezing air near the ground is not thick enough to allow the raindrops to freeze. The raindrops, however, do become supercooled as they fall through the cold air and turn to ice upon colliding with solid objects. The result can be a thick coating of ice having sufficient weight to break tree limbs, down power lines, and make walking or driving extremely hazardous (Figure 12.25).

Hail **Hail** is precipitation in the form of hard, rounded pellets or irregular lumps of ice. Moreover, large hailstones often consist of a series of nearly concentric shells of differing densities and degrees of opaqueness (Figure 12.26). Most hailstones have diameters between 1 centimeter (0.4 inch, pea size) and 5 centimeters (2 inches, golf ball size), although some can be as big as an orange or larger. Occasionally, hailstones weighing a pound or more have been reported. Many of these were probably composites of several stones frozen together.

Figure 12.25 Glaze forms when supercooled raindrops freeze on contact with objects. In January 1998, an ice storm of historic proportions caused enormous damage in New England and southeastern Canada. Nearly five days of freezing rain (glaze) left millions without electricity—some for as long as a month. (Photo by Syracuse Newspapers/The Image Works)

The largest hailstone ever recorded in the United States fell during violent thunderstorms that pounded southeastern Nebraska on June 22, 2003. In the town of Aurora, a 17.8 centimeter (7 inch) wide chunk of ice, almost as large as a volleyball, was recovered. However, this hailstone was not the heaviest hailstone ever measured. Apparently, a chunk of this stone was broken off when it hit the gutter of a house.

The heaviest hailstone in North America fell on Coffeyville, Kansas, in 1970. Having a 14 centimeter (5.5 inch) diameter, this hailstone weighed 766 grams (1.69 pounds). Even heavier hailstones have reportedly been recorded in Bangladesh, where a hailstorm in 1987 killed more than 90 people. It is estimated that large hailstones hit the ground at speeds exceeding 160 kilometers (100 miles) per hour.

The destructive effects of large hailstones are well known, especially to farmers whose crops have been devastated in a few minutes and to people whose windows and roofs have been damaged (Figure 12.27). In the United States, hail damage each year can run into the hundreds of millions of dollars.

Hail is produced only in large cumulonimbus clouds where updrafts can sometimes reach speeds approaching 160 kilometers (100 miles) per hour, and where there is an abundant supply of supercooled water. Hailstones begin as small embryonic ice pellets that grow by collecting supercooled water droplets as they fall through the cloud. If they encounter a strong updraft, they may be carried upward again

Figure 12.26 A cross section of the Coffeyville hailstone. This huge hailstone fell over Kansas in 1970 and weighed 0.75 kilogram (1.65 pounds). (National Center for Atmospheric Research/University Corporation for Atmospheric Research/National Science Foundation)

Figure 12.27 Hail damage to a used-car lot in Fort Worth, Texas, May 6, 1995. This storm, which packed high winds and hail the size of baseballs, killed at least nine people and injured more than 100 as it swept through the northern part of the state. (Photo by Ron Heflin/AP Wide World Photos)

and begin the downward journey anew. Each trip through the supercooled portion of the cloud may be represented by an additional layer of ice. Hailstones can also form from a single descent through an updraft. Either way, the process continues until the hailstone encounters a downdraft or grows too heavy to remain suspended by the thunderstorm's updraft.

Rime **Rime** is a deposit of ice crystals formed by the freezing of supercooled fog or cloud droplets on objects whose surface temperature is below freezing. When rime forms on trees, it adorns them with its characteristic ice feathers, which can be spectacular to behold (Figure 12.28). Objects such as pine needles act as freezing nuclei, causing the supercooled droplets to freeze on contact. When the wind is blowing, only the windward surfaces of objects will accumulate the layer of rime.

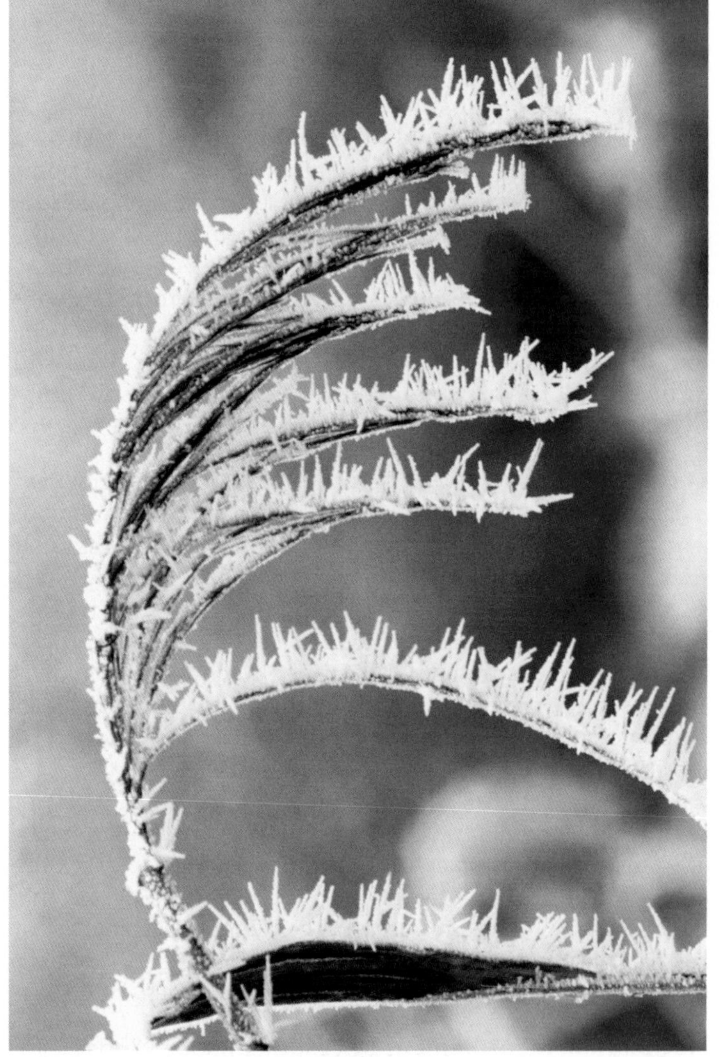

Figure 12.28 Rime consists of delicate ice crystals that form when supercooled fog or cloud droplets freeze on contact with objects. (Photo by John Cancalosi/Stock Boston)

Measuring Precipitation

The most common form of precipitation, rain, is the easiest to measure. Any open container having a consistent cross section throughout can be a rain gauge. In general practice, however, more sophisticated devices are used so that small amounts of rainfall can be measured accurately and losses from evaporation can be reduced. The *standard rain gauge* (Figure 12.29) has a diameter of about 20 centimeters (8 inches) at the top. Once the water is caught, a funnel conducts the rain into a cylindrical measuring tube that has a cross-sectional area only one-tenth as large as the receiver. Consequently, rainfall depth is magnified 10 times, which allows for accurate measurements to the nearest 0.025 centimeter (0.01 inch). The narrow opening also minimizes evaporation. When the amount of rain is less than 0.025 centimeter, it is reported as a *trace* of precipitation.

Measurement Errors In addition to the standard rain gauge, several types of recording gauges are routinely used. These

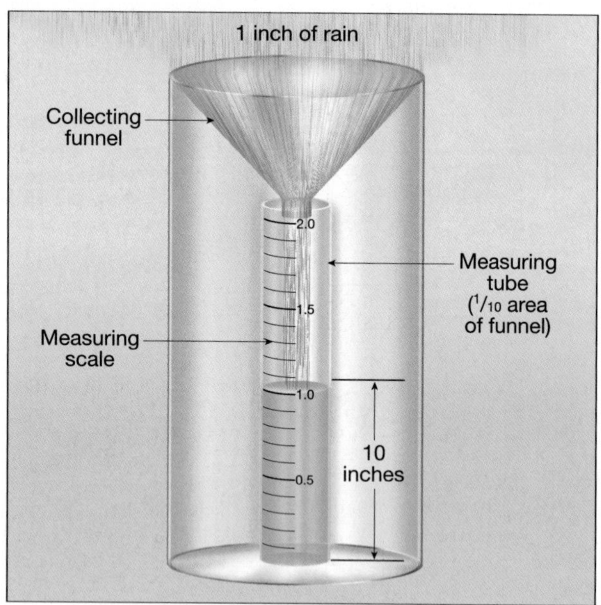

Figure 12.29 Precipitation measurement. The standard rain gauge allows for accurate rainfall measurement to the nearest 0.025 centimeter (0.01 inch). Because the cross-sectional area of the measuring tube is only one-tenth as large as the collector, rainfall is magnified 10 times.

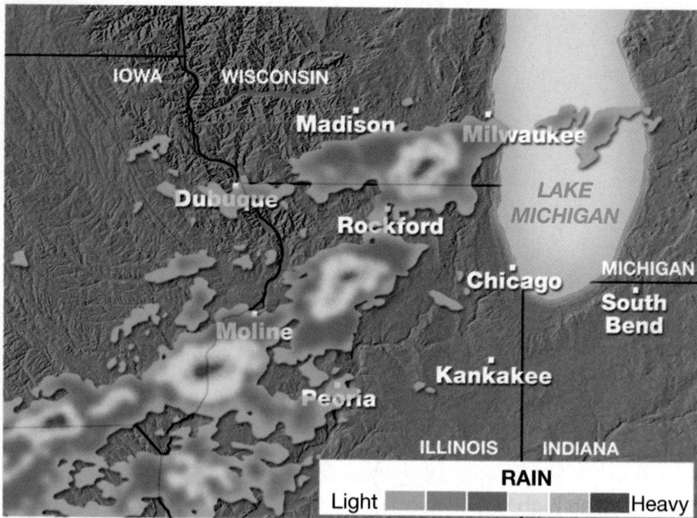

Figure 12.30 Weather radar display commonly seen on TV weathercasts. Colors indicate different intensities of precipitation. Note the band of heavy precipitation extending from northeastern Iowa to Milwaukee, Wisconsin.

Did You Know?

Despite the impressive lake-effect snowstorms recorded in cities such as Buffalo and Rochester, New York, the greatest accumulations of snow generally occur in the mountainous regions of the western United States. The record for a single season goes to Mount Baker ski area north of Seattle, Washington, where 2896 centimeters (1140 inches) of snow fell during the winter of 1998–1999.

instruments record not only the amount of rain but also its time of occurrence and intensity (amount per unit of time).

No matter which rain gauge is used, proper exposure is critical. Errors arise when the gauge is shielded from obliquely falling rain by buildings, trees, or other high objects. Hence, the instrument should be at least as far away from such obstructions as the objects are high. Another cause of error is the wind. It has been shown that with increasing wind and turbulence, it becomes more difficult to collect a representative quantity of rain.

Measuring Snowfall When snow records are kept, two measurements are normally taken: depth and water equivalent. Usually the depth of snow is measured with a calibrated stick. The actual measurement is simple, but choosing a *representative* spot often poses a dilemma. Even when winds are light or moderate, snow drifts freely. As a rule, it is best to take several measurements in an open place away from trees and obstructions and then average them. To obtain the water equivalent, samples may be melted and then weighed or measured as rain.

The quantity of water in a given volume of snow is not constant. A general ratio of 10 units of snow to 1 unit of water is often used when exact information is not available, but the actual water content of snow may deviate widely from this figure. It may take as much as 30 centimeters (12 inches) of light and fluffy dry snow or as little as 4 centimeters (1.6 inches) of wet snow to produce 1 centimeter (0.4 inch) of water.

Precipitation Measurement by Weather Radar Today's TV weathercasts show helpful maps like the one in Figure 12.30 to depict precipitation patterns. The instrument that produces these images is the *weather radar.*

The development of radar has given meteorologists an important tool to probe storm systems that may be up to a few hundred kilometers away. All radar units have a transmitter that sends out short pulses of radio waves. The specific wavelengths that are used depend on the objects the user wants to detect. When radar is used to monitor precipitation, wavelengths between 3 and 10 centimeters (1 and 4 inches) are employed.

These wavelengths can penetrate small cloud droplets but are reflected by larger raindrops, ice crystals, or hailstones. The reflected signal, called an *echo,* is received and displayed on a TV monitor. Because the echo is "brighter" when the precipitation is more intense, modern radar is able to depict not only the regional extent of the precipitation but also the rate of rainfall. Figure 12.30 is a typical radar display in which colors show precipitation intensity. Weather radar is also an important tool for determining the rate and direction of storm movement.

The Chapter in Review

1. *Water vapor,* an odorless, colorless gas, changes from one state of matter (solid, liquid, or gas) to another at the temperatures and pressures experienced near Earth's surface. The processes involved in *changing the state of matter* of water are *evaporation, condensation, melting, freezing, sublimation,* and *deposition.* During each change, *latent* (hidden) *heat* is either absorbed or released.

2. *Humidity* is the general term to describe the amount of water vapor in the air. The methods used to express humidity quantitatively include (1) *mixing ratio,* the mass of water vapor in a unit of air compared to the remaining mass of dry air; (2) *vapor pressure,* that part of the total atmospheric pressure attributable to its water-vapor content; (3) *relative humidity,* the ratio of the air's actual water-vapor content compared to the amount of water vapor required for saturation at that temperature; and (4) *dew point,* that temperature to which a parcel of air would need to be cooled to reach saturation. When air is saturated, the pressure exerted by the water vapor, called the *saturation vapor pressure,* produces a balance between the number of water molecules leaving the surface of the water and the number returning. Because the saturation vapor pressure is temperature dependent, at higher temperatures more water vapor is required for saturation to occur.

3. *Relative humidity can be changed in two ways.* One is by *adding or subtracting water vapor.* The second is by *changing air temperature.* When air is cooled, its relative humidity increases.

4. The basic cloud-forming process is the cooling of air as it rises. The cooling results from expansion because the pressure becomes progressively lower with height. Air temperature changes brought about by compressing or expanding the air are called *adiabatic temperature changes.* Unsaturated air warms by compression and cools by expansion at the rather constant rate of 10°C per 1000 meters of altitude change, a figure called the *dry adiabatic rate.* If air rises high enough, it will cool sufficiently to cause condensation and form a cloud. From this point on, air that continues to rise will cool at the *wet adiabatic rate,* which varies from 5°C to 9°C per 1000 meters of ascent. The difference in the two rates results because the condensing water vapor releases *latent heat,* thereby slowing the rate at which the rising air cools.

5. Four mechanisms that can initiate the vertical movement of air are (1) *orographic lifting,* which occurs when flowing air encounters elevated terrains, such as mountains, and the air rises to go over the barrier; (2) *frontal wedging,* when cool air acts as a barrier over which warmer, less dense air rises; (3) *convergence,* which happens when volumes of air flow together and a general upward movement of air occurs; and (4) *localized convective lifting,* which occurs when unequal surface heating causes localized pockets of air to rise because of their buoyancy.

6. The *stability of air* is determined by examining the temperature of the atmosphere at various altitudes. Air is said to be *unstable* when the *environmental lapse rate* (the rate of temperature decrease with increasing altitude in the troposphere) is greater than the *dry adiabatic rate.* Stated differently, a column of air is unstable when the air near the bottom is significantly warmer (less dense) than the air aloft. When stable air is forced aloft, precipitation, if any, is light, whereas unstable air generates towering clouds and stormy conditions.

7. For condensation to occur, air must be saturated. Saturation takes place either when air is cooled to its dew point, which most commonly happens, or when water vapor is added to the air. There must also be a surface on which the water vapor can condense. In cloud and fog formation, tiny particles called *condensation nuclei* serve this purpose.

8. *Clouds* are classified on the basis of their *appearance* and *height.* The three basic forms are *cirrus* (high, white, thin, wispy fibers), *cumulus* (globular, individual cloud masses), and *stratus* (sheets or layers that cover much or all of the sky). The four categories based on height are *high clouds* (bases normally above 6000 meters [20,000 feet]), *middle clouds* (from 2000 to 6000 meters [6500 to 20,000 feet]), *low clouds* (below 2000 meters [6500 feet]), and *clouds of vertical development.*

9. *Fog* is a cloud with its base at or very near the ground. Fogs form when air is cooled below its dew point or when enough water vapor is added to the air to bring about saturation. Various types of fog include *advection fog, radiation fog, upslope fog, steam fog,* and *frontal fog* (or *precipitation fog*).

10. For *precipitation* to form, millions of cloud droplets must somehow join together into large drops. Two mechanisms for the formation of precipitation have been proposed: First, in clouds where the temperatures are below freezing, ice crystals form and fall as snowflakes. At lower altitudes the snowflakes melt and become raindrops before they reach the ground. Second, large droplets form in warm clouds that contain large hygroscopic ("water-seeking") nuclei, such as salt particles. As these big droplets descend, they collide and join with smaller water droplets. After many collisions, the droplets are large enough to fall to the ground as rain.

11. The forms of precipitation include *rain, snow, sleet, freezing rain (glaze), hail,* and *rime.*

12. Rain, the most common form of precipitation, is probably easiest to measure. A common instrument is the *standard rain gauge,* which is read directly. Several types of *recording gauges,* which record the intensity and amount of rain are also in use. The two most common measurements of snow are depth and water equivalent. Although the quantity of water in a given volume of snow is not constant, a general ratio of 10 units of snow to 1 unit of water is often used when exact information is not available.

Key Terms

absolute instability (p. 332)
absolute stability (p. 332)
adiabatic temperature change (p. 328)
advection fog (p. 336)
calorie (p. 323)
cirrus (p. 335)
cloud (p. 335)
clouds of vertical development (p. 336)
condensation (p. 323)
condensation nuclei (p. 335)
conditional instability (p. 333)
convergence (p. 330)
cumulus (p. 335)
deposition (p. 324)
dew-point temperature (p. 326)
dry adiabatic rate (p. 328)
evaporation (p. 323)
fog (p. 336)
freezing rain (p. 342)
front (p. 330)
frontal fog (p. 339)
frontal wedging (p. 330)
glaze (p. 342)
hail (p. 342)
high clouds (p. 336)
humidity (p. 324)
hygrometer (p. 327)
hygroscopic nuclei (p. 335)
latent heat (p. 323)
localized convective lifting (p. 331)
low clouds (p. 336)
middle clouds (p. 336)
mixing ratio (p. 324)
orographic lifting (p. 329)
parcel (p. 328)
precipitation fog (p. 339)
radiation fog (p. 336)
rain (p. 341)
rainshadow desert (p. 329)
relative humidity (p. 325)
rime (p. 343)
saturation (p. 324)
sleet (p. 342)
snow (p. 341)
stable air (p. 331)
steam fog (p. 339)
stratus (p. 335)
sublimation (p. 323)
unstable air (p. 331)
upslope fog (p. 339)
vapor pressure (p. 324)
wet adiabatic rate (p. 328)

Questions for Review

1. Summarize the processes by which water changes from one state to another. Indicate whether heat energy is absorbed or liberated.
2. After studying Table 12.1 (p. 324), write a generalization relating temperature and the capacity of air to hold water vapor.
3. How do relative humidity and mixing ratio differ?
4. Referring to Figure 12.6 (p. 326), answer the following questions:
   a. During a typical day, when is the relative humidity highest? Lowest?
   b. At what time of day would dew most likely form?
5. If the air temperature remains unchanged and the mixing ratio decreases, how will relative humidity change?
6. On a cold winter day when the temperature is −10°C (14°F) and the relative humidity is 50 percent, what is the mixing ratio (refer to Table 12.1, p. 324)? What is the mixing ratio for a day when the temperature is 20°C (68°F) and the relative humidity is 50 percent?
7. Explain the principle of the sling psychrometer.
8. On a warm summer day when the relative humidity is high, it may seem even warmer than the thermometer indicates. Why do we feel so uncomfortable on a "muggy" day?
9. Why does air cool when it rises through the atmosphere?
10. Explain the difference between environmental lapse rate and adiabatic cooling.
11. If unsaturated air at 23°C were to rise, what would its temperature be at 500 meters? If the dew-point temperature at the condensation level were 13°C, at what altitude would clouds begin to form?
12. Why does the adiabatic rate of cooling change when condensation begins? Why is the wet adiabatic rate not a constant figure?
13. How do orographic lifting and frontal wedging act to force air to rise?
14. Explain why the Great Basin area of the western United States is so dry. What term is applied to such a situation?
15. How does stable air differ from unstable air? Describe the general nature of the clouds and precipitation expected with each.
16. What is the function of condensation nuclei in cloud formation? What is the function of the dew point?
17. As you drink an ice-cold beverage on a warm day, the outside of the glass or bottle becomes wet. Explain.
18. What is the basis for the classification of clouds?
19. Why are high clouds always thin?
20. List five types of fog and discuss the details of their formation.
21. What is the difference between precipitation and condensation?
22. List the forms of precipitation and the circumstances of their formation.
23. Sometimes, when rainfall is light, the amount is reported as a trace. When this occurs, how much (or little) rain has fallen?

Online Study Guide

The *Foundations of Earth Science* Web site uses the resources and flexibility of the Internet to aid in your study of the topics in this chapter. Written and developed by Earth science instructors, this site will help improve your understanding of Earth science. Visit **http://www.prenhall.com/lutgens** and click on the cover of *Foundations of Earth Science 5e* to find:

- Online review quizzes.
- Critical thinking exercises.
- Links to chapter-specific Web resources.
- Internet-wide key-term searches.

http://www.prenhall.com/lutgens

GEODe: Earth Science

GEODe: Earth Science makes studying more effective by reinforcing key concepts using animation, video, narration, interactive exercises, and practice quizzes. A copy is included with every copy of *Foundations of Earth Science 5e.*

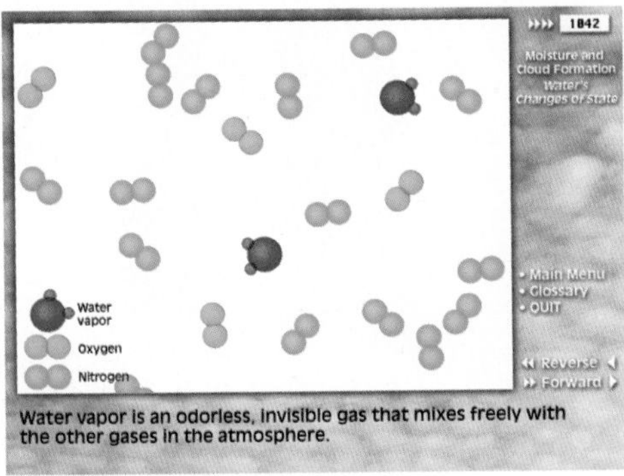

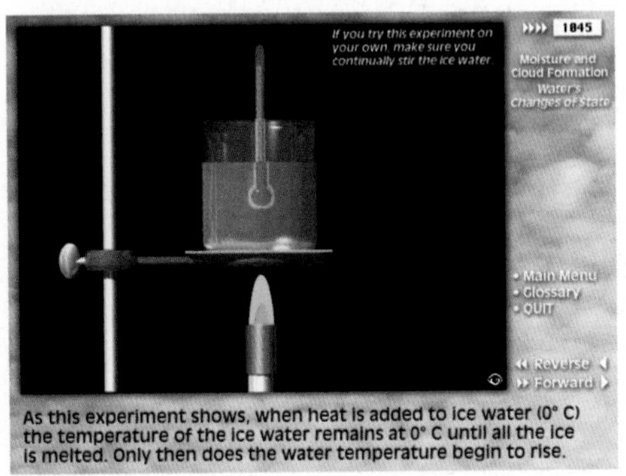

CHAPTER

13 The Atmosphere in Motion

Sailboats in a Norwegian fiord. (Photo by The Image Bank/Getty Images, Inc.)

FOCUS ON LEARNING

To assist you in learning the important concepts in this chapter, you will find it helpful to focus on the following questions:

1. What is air pressure, and how does it change with altitude?
2. What force creates wind, and what other factors influence it?
3. What are the two types of pressure centers? What is the air movement and general weather associated with each type?
4. What is the idealized global circulation?
5. What is the general atmospheric circulation in the midlatitudes?
6. What are the names and causes of some local winds?

Of the various elements of weather and climate, changes in air pressure are the least noticeable. When listening to a weather report, we are generally interested in moisture conditions (humidity and precipitation), temperature, and perhaps wind. It is the rare individual who wonders about air pressure. Although the hour-to-hour and day-to-day variations in air pressure are generally not noticed by people, they are very important in producing changes in our weather. Variations in air pressure from place to place cause the movement of air we call wind and are a significant factor in weather forecasting (Figure 13.1). As we will see, air pressure is closely tied to the other elements of weather in a cause-and-effect relationship.

Understanding Air Pressure

Earth's Dynamic Atmosphere
▼ Air Pressure and Wind

In Chapter 11 we noted that **air pressure** is simply the pressure exerted by the weight of air above. Average air pressure at sea level is about 1 kilogram per square centimeter, or 14.7 pounds per square inch. This is roughly the same pressure that is produced by a column of water 10 meters (33 feet) in height. With some simple arithmetic, you can calculate that the air pressure exerted on the top of a small (50 centimeter by 100 centimeter [20 inch by 40 inch]) school desk exceeds 5000 kilograms (11,000 pounds), or about the weight of a 50-passenger school bus. Why doesn't the desk collapse under the weight of the ocean of air above? Simply, air pressure is exerted in all directions—down, up, and sideways. Thus, the air pressure pushing down on the desk exactly balances the air pressure pushing up on the desk.

To visualize this phenomenon better, imagine a tall aquarium that has the same dimensions as the desktop. When this aquarium is filled to a height of 10 meters (33 feet), the water pressure at the bottom equals 1 atmosphere (14.7 pounds per square inch). Now, imagine what will happen if this aquarium is placed on top of our student desk so that all the force is directed downward. Compare this to what results when the desk is placed inside the aquarium and allowed to sink to the bottom. In the latter example, the desk survives because the water pressure is exerted in all directions, not just downward as in our earlier example. The desk, like your body, is "built" to withstand the pressure of 1 atmosphere. It is important to note that although we do not generally notice the pressure exerted by the ocean of air around us, except when ascending or descending in an elevator or airplane, it is nonetheless substantial. The pressurized suits used by

Figure 13.1 Strong winds blowing snow during a blizzard in Minnesota. (Photo by Jim Brandendurg/Minden Pictures)

astronauts on space walks are designed to duplicate the atmospheric pressure experienced at Earth's surface. Without these protective suits to keep body fluids from boiling away, astronauts would perish in minutes.

The concept of air pressure can also be understood if we examine the behavior of gas molecules. Gas molecules, unlike those of the liquid and solid phases, are not bound to one another but are freely moving about, filling all the space available to them. When two gas molecules collide, which happens frequently under normal conditions, they bounce off each other like elastic balls. If a gas is confined to a container, this motion is restricted by its sides, much like the walls of a handball court redirect the motion of the handball. The continuous bombardment of gas molecules against the sides of the container exerts an outward push that we call air pressure. Although the atmosphere is without walls, it is confined from below by Earth's surface and effectively from above because the force of gravity prevents its escape. Here, we can define *air pressure* as the force exerted against a surface by the continuous collision of gas molecules.

Measuring Air Pressure

Earth's Dynamic Atmosphere
▼ Air Pressure and Wind

When meteorologists measure atmospheric pressure, they employ a unit called the *millibar.* Standard sea-level pressure is 1013.2 millibars. Although the millibar has been the unit of measure on all U.S. weather maps since January 1940, the media use "inches of mercury" to describe atmospheric pressure. In the United States, the National Weather Service converts millibar values to inches of mercury for public and aviation use.

Inches of mercury is easy to understand. The use of mercury for measuring air pressure dates from 1643, when Torricelli, a student of the famous Italian scientist Galileo, invented the **mercury barometer.** Torricelli correctly described the atmosphere as a vast ocean of air that exerts pressure on us and all objects about us. To measure this force, he filled a glass tube, which was closed at one end, with mercury. The tube was then inverted into a dish of mercury (Figure 13.2). Torricelli found that the mercury flowed out of the tube until the weight of the column was balanced by the pressure that the atmosphere exerted on the surface of the mercury in the dish. In other words, the weight of mercury in the column equaled the weight of the same diameter column of air that extended from the ground to the top of the atmosphere.

When air pressure increases, the mercury in the tube rises. Conversely, when air pressure decreases, so does the height of the mercury column. With some refinements, the mercurial barometer invented by Torricelli is still the standard pressure-measuring instrument used today. Standard atmospheric pressure at sea level equals 29.92 inches of mercury.

The need for a smaller and more portable instrument for measuring air pressure led to the development of the

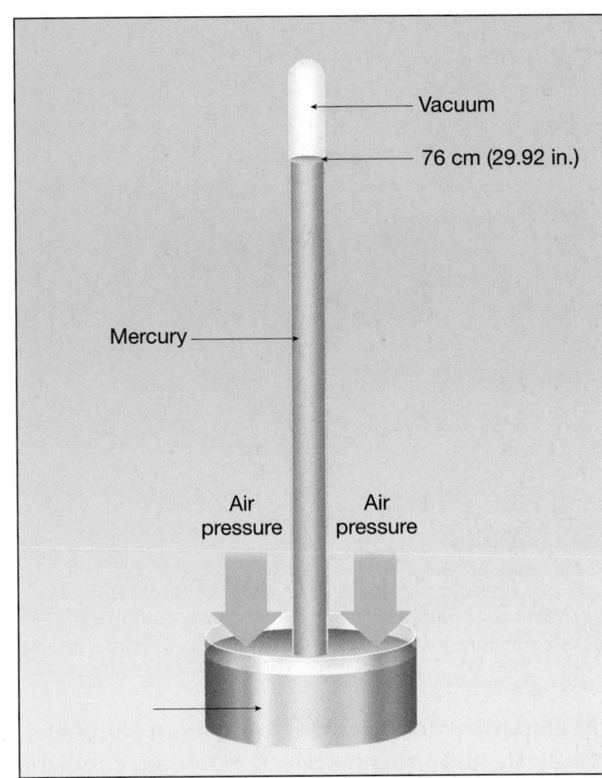

Figure 13.2 Simple mercury barometer. The weight of the column of mercury is balanced by the pressure exerted on the dish of mercury by the air above. If the pressure decreases, the column of mercury falls; if the pressure increases, the column rises.

aneroid barometer (*aneroid* means "without liquid"). Instead of having a mercury column held up by air pressure, the aneroid barometer uses a partially evacuated metal chamber (Figure 13.3). The chamber is extremely sensitive to variations in air pressure and changes shape, compressing as the pressure increases and expanding as the pressure decreases. A series of levers transmits the movements of the chamber to a pointer on a dial that is calibrated to read in inches of mercury and/or millibars.

As shown in Figure 13.3, the face of an aneroid barometer intended for home use is inscribed with words such as *fair, change, rain,* and *stormy.* Notice that "fair" corresponds with high-pressure readings, whereas "rain" is associated with low pressures. Barometric readings, however, may not always indicate the weather. The dial may point to "fair" on a rainy day or to "rain" on a fair day. To "predict" the local weather, the change in air pressure over the past few hours is more important than the current pressure reading. Falling pressure is often associated with increasing cloudiness and the possibility of precipitation, whereas rising air pressure generally indicates clearing conditions. It is useful to remember, however, that particular barometer readings or trends do not always correspond to specific types of weather.

One advantage of the aneroid barometer is that it can easily be connected to a recording mechanism. The resulting instrument is a **barograph,** which provides a continuous record of pressure changes with the passage of time (Figure 13.4).

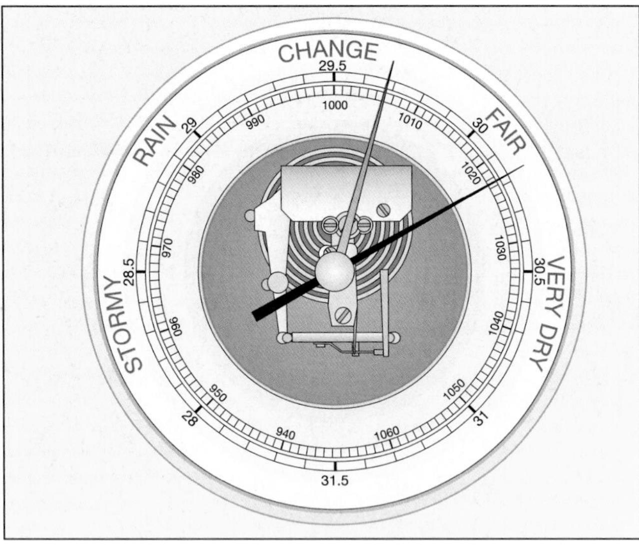

Figure 13.3 Aneroid barometer. The aneroid barometer has a partially evacuated chamber that changes shape, compressing as atmospheric pressure increases and expanding as pressure decreases.

Another important adaptation of the aneroid barometer is its use to indicate altitude for aircraft, mountain climbers, and mapmakers.

Did You Know?

A barometer could be constructed using any number of different liquids, including water. The problem with a water-filled barometer is size. Because water is 13.6 times less dense than mercury, the height of a water column at standard sea-level pressure would be 13.6 times taller than that of a mercurial barometer. Consequently, a water-filled barometer would need to be nearly 34 feet tall.

Figure 13.4 An aneroid barograph makes a continuous record of pressure changes. (Photo courtesy of Qualimetrics, Inc., Sacramento, California)

Factors Affecting Wind

Earth's Dynamic Atmosphere
▼ Air Pressure and Wind

In Chapter 12, we examined the upward movement of air and its role in cloud formation. As important as vertical motion is, far more air moves horizontally, the phenomenon we call **wind.** What causes wind?

Simply stated, wind is the result of horizontal differences in air pressure. *Air flows from areas of higher pressure to areas of lower pressure.* You may have experienced this when opening a vacuum-packed can of coffee. The noise you hear is caused by air rushing from the higher pressure outside the can to the lower pressure inside. Wind is nature's attempt to balance such inequalities in air pressure. Because unequal heating of Earth's surface generates these pressure differences, *solar radiation is the ultimate energy source for most wind.*

If Earth did not rotate, and if there were no friction between moving air and Earth's surface, air would flow in a straight line from areas of higher pressure to areas of lower pressure. But because Earth does rotate and friction does exist, wind is controlled by the following combination of forces: (1) pressure gradient force, (2) Coriolis effect, and (3) friction. We will now examine each of these factors.

Pressure-Gradient Force

Pressure differences create wind, and the greater these differences, the greater the wind speed. Over Earth's surface, variations in air pressure are determined from barometric readings taken at hundreds of weather stations. These pressure data are shown on a weather map using **isobars,** lines that connect places of equal air pressure (Figure 13.5). The spacing of isobars indicates the amount of pressure change occurring over a given distance and is expressed as the **pressure gradient.**

To visualize a pressure gradient, think of it as the slope of a hill. A steep pressure gradient, like a steep hill, causes greater acceleration of an air parcel than does a weak pressure gradient. Thus, the relationship between wind speed and the pressure gradient is straightforward: *Closely spaced isobars indicate a steep pressure gradient and high winds, whereas widely spaced isobars indicate a weak pressure gradient and light winds.* Figure 13.5 illustrates the relationship between the spacing of isobars and wind speed. Notice that wind speeds are greater in Ohio, Kentucky, Michigan, and Illinois, where isobars are more closely spaced, than in the western states, where isobars are more widely spaced.

The pressure gradient is the driving force of wind, and it has both magnitude and direction. Its magnitude is determined from the spacing of isobars. The direction of force is always from areas of higher pressure to areas of lower pressure and at right angles to the isobars. Once the air starts to move, the Coriolis effect and friction come into play, but then only to modify the movement, not to produce it.

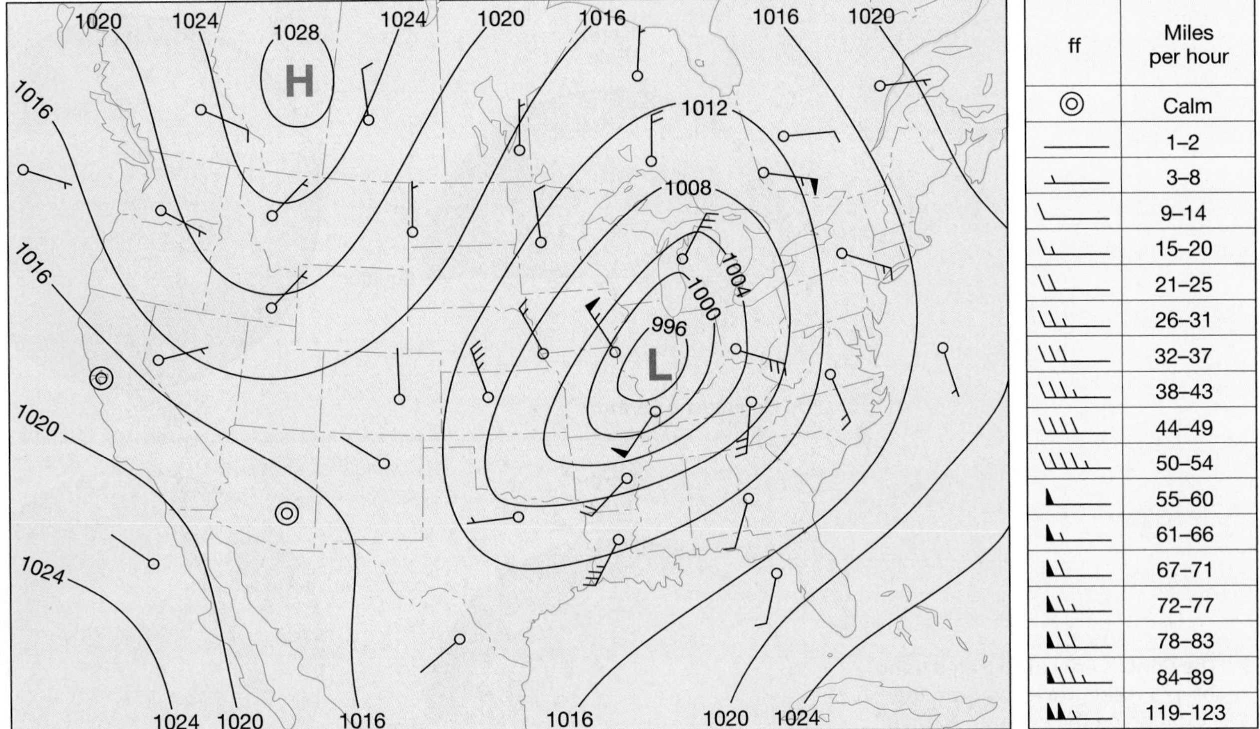

Figure 13.5 The black lines are isobars that connect places of equal barometric pressure. They show the distribution of pressure on weather maps. The lines usually curve and often close around cells of high and low pressure. Flags indicate the expected airflow surrounding cells of high and low pressure and are plotted like flags flying with the wind. Wind speed is indicated by flags and "feathers," as shown along the right-hand side of this drawing.

Did You Know?

Approximately 0.25 percent (one-quarter of one percent) of the solar energy that reaches the lower atmosphere is transformed into wind. Although it is just a minuscule percentage, the absolute amount of energy is enormous. According to one estimate, North Dakota alone is theoretically capable of providing enough wind-generated power to meet more than one-third of the electricity demand of the United States.

Coriolis Effect

Figure 13.5 shows the typical air movements associated with high- and low-pressure systems. As expected, the air moves out of the regions of higher pressure and into the regions of lower pressure. However, the wind does not cross the isobars at right angles as the pressure gradient force directs it to do. The direction deviates as a result of Earth's rotation. This has been named the **Coriolis effect** after the French scientist who first thoroughly described it.

All free-moving objects or fluids, including the wind, are deflected to the *right* of their path of motion in the Northern Hemisphere and to the *left* in the Southern Hemisphere. The reason for this deflection can be illustrated by imagining the path of a rocket launched from the North Pole toward a target located on the equator (Figure 13.6). If the rocket took an hour to reach its target, during its flight Earth would have rotated 15 degrees to the east. To someone standing on Earth, it would look as if the rocket veered off its path and hit Earth 15 degrees west of its target. The true path of the rocket is straight and would appear so to someone out in space looking at Earth. It was Earth turning under the rocket that gave it its *apparent* deflection.

Note that the rocket was deflected to the right of its path of motion because of the counterclockwise rotation of the Northern Hemisphere. In the Southern Hemisphere, the effect is reversed. Clockwise rotation produces a similar deflection, but to the *left* of the path of motion. The same deflection is experienced by wind regardless of the direction it is moving.

We attribute the apparent shift in wind direction to the Coriolis effect. This deflection (1) is always directed at right angles to the direction of airflow; (2) affects only wind direction, not wind speed; (3) is affected by wind speed (the stronger the wind, the greater the deflection); and (4) is strongest at the poles and weakens equatorward, becoming nonexistent at the equator.

It is of interest to point out that any free-moving object will experience a deflection caused by the Coriolis effect. This fact was dramatically discovered by the United States Navy in World War II. During target practice, long-range guns on battleships continually missed their targets by as much as several hundred yards until ballistic corrections were made for the changing position of a seemingly stationary target. Over a short distance, however, the Coriolis effect is relatively small.

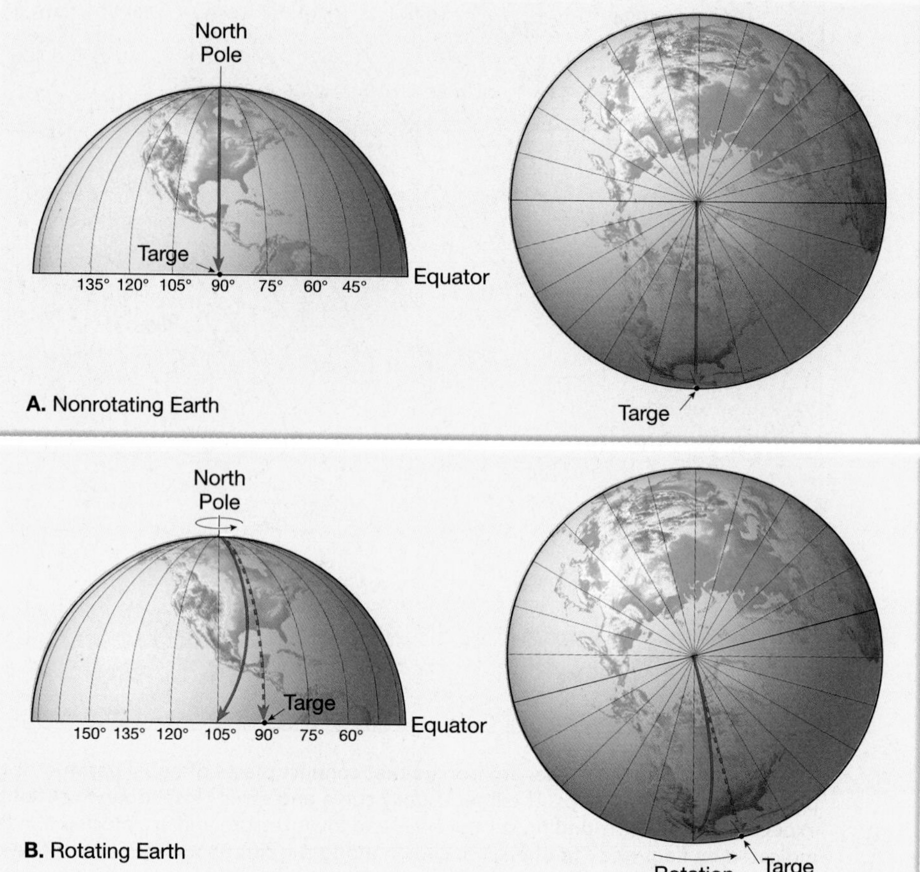

Figure 13.6 The Coriolis effect illustrated using a one-hour flight of a rocket traveling from the North Pole to a location on the equator. **A.** On a nonrotating Earth, the rocket would travel straight to its target. **B.** However, Earth rotates 15 degrees each hour. Thus, although the rocket travels in a straight line, when we plot the path of the rocket on Earth's surface, it follows a curved path that veers to the right of the target.

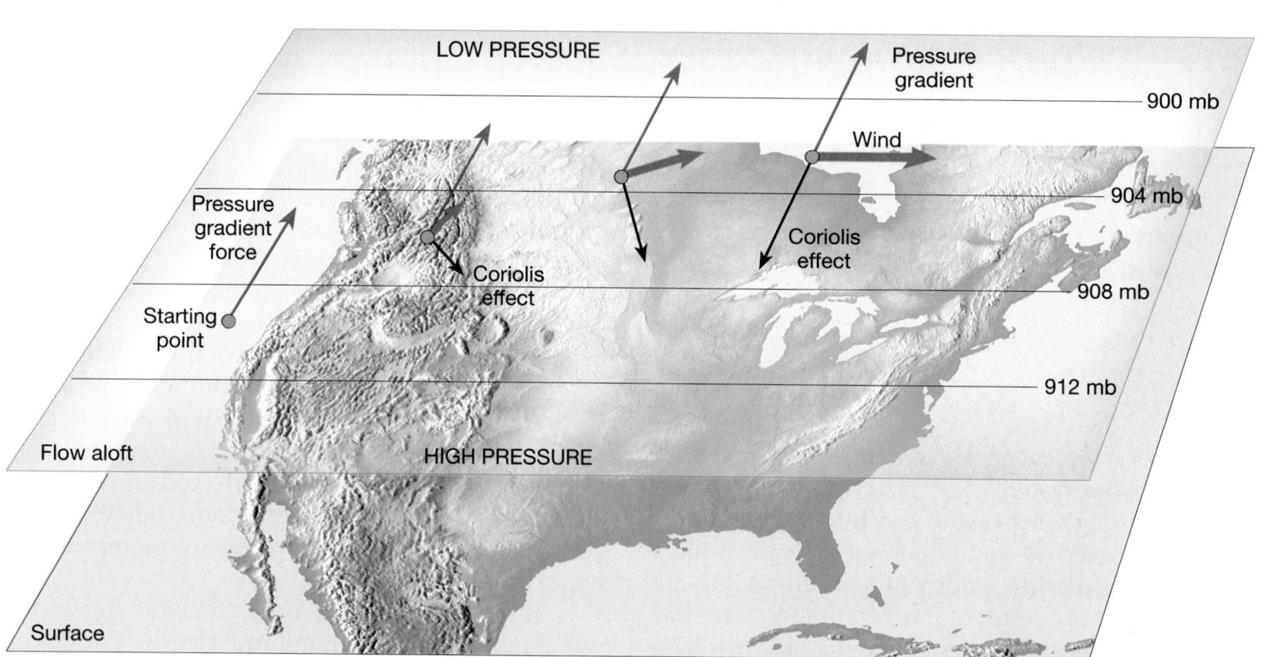

Figure 13.7 The geostrophic wind. The only force acting on a stationary parcel of air is the pressure-gradient force. Once the air begins to accelerate, the Coriolis effect deflects it to the right in the Northern Hemisphere. Greater wind speeds result in a stronger Coriolis effect (deflection) until the flow is parallel to the isobars. At this point, the pressure-gradient force and Coriolis effect are in balance and the flow is called a *geostrophic wind.* It is important to note that in the "real" atmosphere, airflow is continually adjusting for variations in the pressure field. As a result, the adjustment to geostrophic equilibrium is much more irregular than shown.

Friction with Earth's Surface

The effect of friction on wind is important only within a few kilometers of Earth's surface. Friction acts to slow air movement and, as a consequence, alters wind direction. To illustrate friction's effect on wind direction, let us look at a situation in which it has no role. Above the friction layer, the pressure gradient force and Coriolis effect work together to direct the flow of air. Under these conditions, the pressure gradient force causes air to start moving across the isobars. As soon as the air starts to move, the Coriolis effect acts at right angles to this motion. The faster the wind speed, the greater the deflection.

Eventually, the Coriolis effect will balance the pressure gradient force and the wind will blow parallel to the isobars (Figure 13.7). Upper-air winds generally take this path and are called **geostrophic winds.** The lack of friction with Earth's surface allows geostrophic winds to travel at higher speeds than do surface winds. This can be observed in Figure 13.8 by noting the wind flags, many of which indicate winds of 50 to 100 miles per hour.

The most prominent features of upper-level flow are the **jet streams.** First encountered by high-flying bombers during

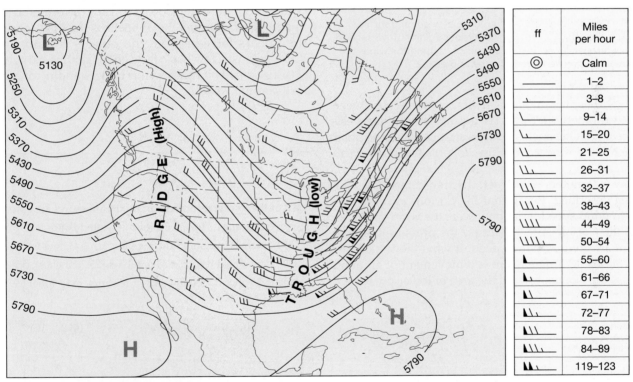

Did You Know?

The Coriolis effect could potentially affect the outcome of a baseball game. A ball hit a horizontal distance of 100 meters (330 feet) in four seconds down the right field line will be deflected 1.5 centimeters (more than one-half inch) to the right. This could be just enough to turn a potential home run into a foul ball!

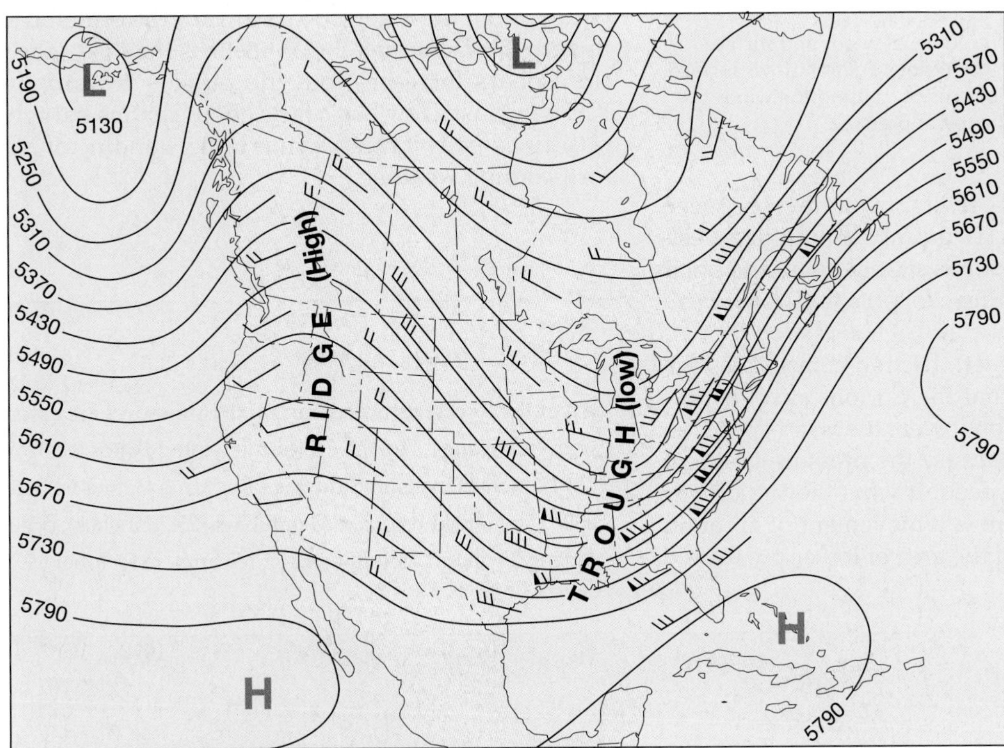

ff	Miles per hour
◎	Calm
	1–2
	3–8
	9–14
	15–20
	21–25
	26–31
	32–37
	38–43
	44–49
	50–54
	55–60
	61–66
	67–71
	72–77
	78–83
	84–89
	119–123

A. Upper-level weather chart

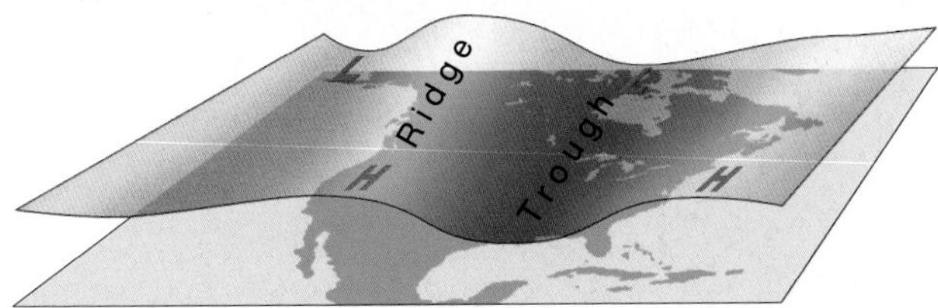

B. Representation of upper-level chart

Figure 13.8 Upper-air weather chart. **A.** This simplified weather chart shows the direction and speed of the upper-air winds. Note from the flags that the airflow is almost parallel to the contours. Like most upper-air charts, this one shows variations in the height (in meters) at which a selected pressure (500 millibars) is found, instead of showing variations in pressure at a fixed height like surface maps. Do not let this confuse you, because there is a simple relationship between height contours and pressure. Places experiencing 500-millibar pressure at higher altitudes (toward the south) are experiencing higher pressures than places where the height contours indicate lower altitudes. Thus, *higher-elevation* contours indicate *higher* pressures, and *lower-elevation* contours indicate *lower* pressures. **B.** Representation of the 500-millibar surface shown in the upper-air weather chart above.

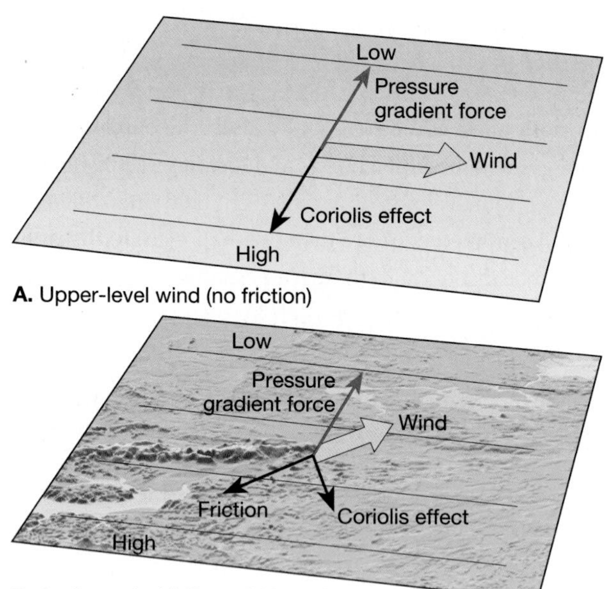

Figure 13.9 Comparison between upper-level winds and surface winds showing the effects of friction on airflow. Friction slows surface wind speed, which weakens the Coriolis effect, causing the winds to cross the isobars and move toward the lower pressure.

World War II, these fast-moving "rivers" of air travel between 120 and 240 kilometers (75 and 150 miles) per hour in a west-to-east direction. One such stream is situated over the polar front, which is the zone separating cool polar air from warm subtropical air.

Below 600 meters (2000 feet), friction complicates the airflow just described. Recall that the Coriolis effect is proportional to wind speed. Friction lowers the wind speed, so it reduces the Coriolis effect. Because the pressure gradient force is not affected by wind speed, it wins the tug of war shown in Figure 13.9. The result is a movement of air at an angle across the isobars toward the area of lower pressure.

The roughness of the terrain determines the angle of airflow across the isobars. Over the smooth ocean surface, friction is low and the angle is small. Over rugged terrain, where friction is higher, the angle that air makes as it flows across the isobars can be as great as 45 degrees.

In summary, upper airflow is nearly parallel to the isobars, while the effect of friction causes the surface winds to move more slowly and cross the isobars at an angle.

Highs and Lows

Earth's Dynamic Atmosphere
▼ Air Pressure and Wind

Among the most common features on any weather map are areas designated as pressure centers. **Cyclones,** or **lows,** are centers of low pressure, and **anticyclones,** or **highs,** are high-pressure centers. As Figure 13.10 illustrates, the pressure decreases from the outer isobars toward the center in a cyclone. In an anticyclone, just the opposite is the case—the values of the isobars increase from the outside toward the center. Knowing just a few basic facts about centers of high and low pressure greatly increases your understanding of current and forthcoming weather.

Did You Know?

All of the lowest recorded barometric pressures are associated with hurricanes. The record for the United States is 882 millibars (26.12 inches) measured during Hurricane Wilma in October 2005. The world record, 870 millibars (25.70 inches), occurred during Typhoon Tip (a Pacific hurricane), in October 1979.

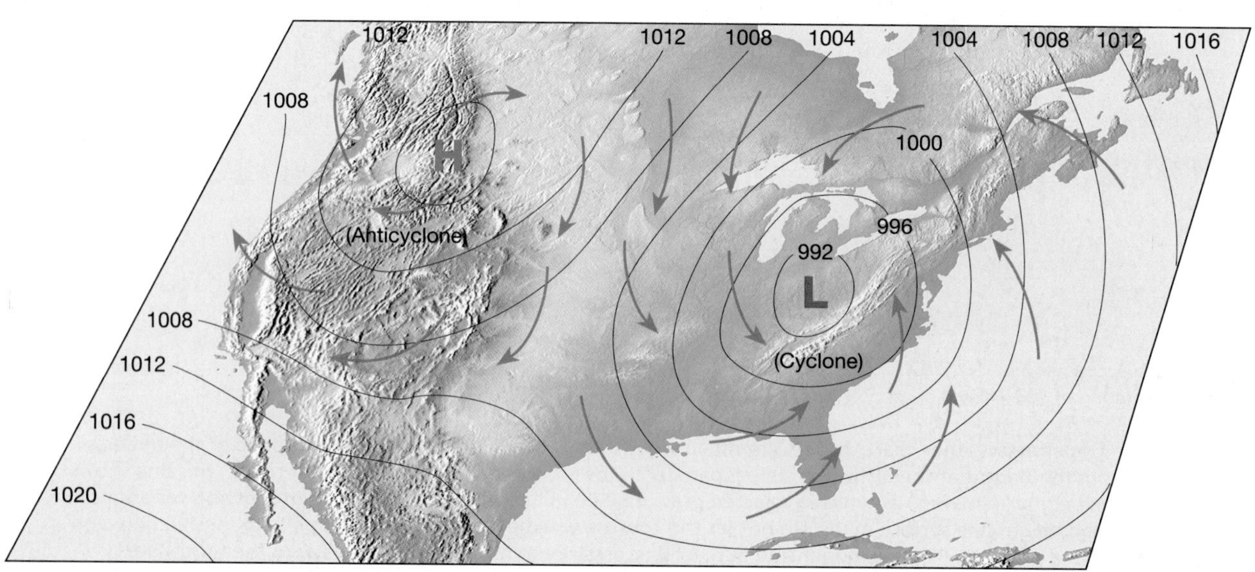

Figure 13.10 Cyclonic and anticyclonic winds in the Northern Hemisphere. Arrows show that winds blow into and counterclockwise around a low. By contrast, around a high, winds blow outward and clockwise.

Cyclonic and Anticyclonic Winds

From the preceding section, you learned that the two most significant factors that affect wind are the pressure-gradient force and the Coriolis effect. Winds move from higher pressure to lower pressure and are deflected to the right or left by Earth's rotation. When these controls of airflow are applied to pressure centers in the Northern Hemisphere, the result is that winds blow inward and counterclockwise around a low (Figure 13.11A). Around a high, they blow outward and clockwise (Figure 13.10).

In the Southern Hemisphere, the Coriolis effect deflects the winds to the left; therefore, winds around a low blow clockwise, and winds around a high move counterclockwise (Figure 13.11B). In either hemisphere, friction causes a net inflow (**convergence**) around a cyclone and a net outflow (**divergence**) around an anticyclone.

Weather Generalizations About Highs and Lows

Rising air is associated with cloud formation and precipitation, whereas subsidence produces clear skies. This section will discuss how the movement of air can itself create pressure change and generate winds. After doing so, we will examine the relation between horizontal and vertical flow, and their effects on the weather.

Let us first consider a surface low-pressure system where the air is spiraling inward. The net inward transport of air causes a shrinking of the area occupied by the air mass, a process that is termed *horizontal convergence.* Whenever air converges horizontally, it must pile up—that is, increase in height to allow for the decreased area it now occupies. This generates a "taller" and therefore heavier air column. Yet, a surface low can exist only as long as the column of air exerts less pressure than that occurring in surrounding regions. We seem to have encountered a paradox—a low-pressure center causes a net accumulation of air, which increases its pressure. Consequently, a surface cyclone should quickly eradicate itself in a manner not unlike what happens when a vacuum-packed can is opened.

For a surface low to exist for very long, compensation must occur at some layer aloft. For example, surface convergence could be maintained if divergence (spreading out) aloft occurred at a rate equal to the inflow below. Figure 13.12 shows the relationship between surface convergence and divergence aloft that is needed to maintain a low-pressure center.

Divergence aloft may even exceed surface convergence, thereby resulting in intensified surface inflow and accelerated vertical motion. Thus, divergence aloft can intensify storm centers as well as maintain them. On the other hand, inadequate divergence aloft permits surface flow to "fill" and weaken the accompanying cyclone.

Note that surface convergence about a cyclone causes a net upward movement. The rate of this vertical movement is slow, generally less than 1 kilometer (0.6 mile) per day. Nevertheless, because rising air often results in cloud formation

A.

B.

Figure 13.11 Cyclonic circulation in the Northern and Southern hemispheres. The cloud patterns in these images allow us to "see" the circulation pattern in the lower atmosphere. **A.** This satellite image shows a large low-pressure center in the Gulf of Alaska on August 17, 2004. The cloud pattern clearly shows an inward and *counterclockwise* spiral. **B.** This image from March 26, 2004, shows a strong cyclonic storm in the South Atlantic near the coast of Brazil. The cloud pattern reveals an inward and *clockwise* circulation. (NASA)

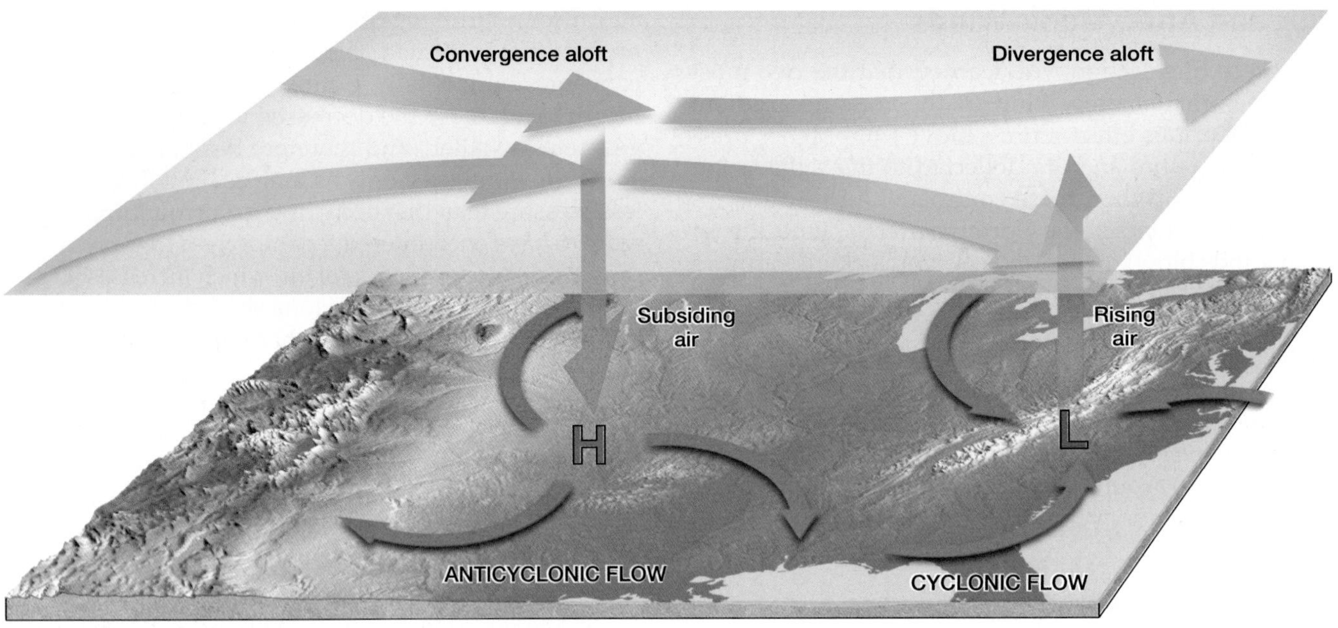

Figure 13.12 Airflow associated with surface cyclones and anticyclones. A low, or cyclone, has converging surface winds and rising air, causing cloudy conditions. A high, or anticyclone, has diverging surface winds and descending air, which leads to clear skies and fair weather.

and precipitation, a low-pressure center is generally related to unstable conditions and stormy weather (Figure 13.13A).

As often as not, it is divergence aloft that creates a surface low. Spreading out aloft initiates upflow in the atmosphere directly below, eventually working its way to the surface, where inflow is encouraged.

Like their cyclonic counterparts, anticyclones must be maintained from above. Outflow near the surface is accompanied by convergence aloft and general subsidence of the air column (Figure 13.12). Because descending air is compressed and warmed, cloud formation and precipitation are unlikely in an anticyclone. Thus, fair weather can usually

A.

B.

Figure 13.13 These two photographs illustrate the basic weather generalizations associated with pressure centers. **A.** A sea of umbrellas on a rainy day in Shanghai, China. Centers of low pressure are frequently associated with cloudy conditions and precipitation. (Photo by Getty Images, Inc.— Stone Allstock) **B.** By contrast, clear skies and "fair" weather may be expected when an area is under the influence of high pressure. Sunbathers on a beach at Cape Henlopen, Delaware. (Photo by Mark Gibson/DRK Photo)

be expected with the approach of a high-pressure center (Figure 13.13B).

For reasons that should now be obvious, it has been common practice to print on household barometers the words "stormy" at the low-pressure end and "fair" on the high-pressure end. By noting whether the pressure is rising, falling, or steady, we have a good indication of what the forthcoming weather will be. Such a determination, called the **pressure,** or **barometric, tendency,** is a useful aid in short-range weather prediction.

You should now be better able to understand why television weather reporters emphasize the positions and projected paths of cyclones and anticyclones. The "villain" on these weather programs is always the low-pressure center, which produces "bad" weather in any season. Lows move in roughly a west-to-east direction across the United States and require a few days to more than a week for the journey. Because their paths can be somewhat erratic, accurate prediction of their migration is difficult, although essential, for short-range forecasting.

Meteorologists must also determine if the flow aloft will intensify an embryo storm or act to suppress its development. Because of the close tie between conditions at the surface and those aloft, a great deal of emphasis has been placed on the importance and understanding of the total atmospheric circulation, particularly in the midlatitudes. We will now examine the workings of Earth's general atmospheric circulation and then again consider the structure of the cyclone in light of this knowledge.

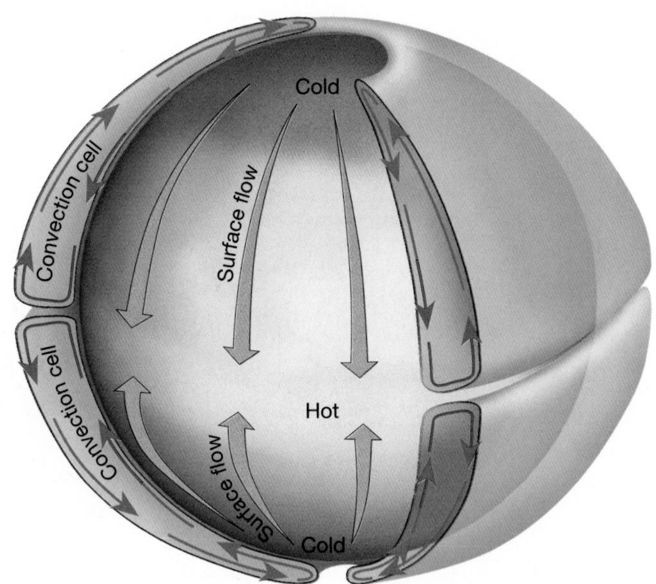

Figure 13.14 Global circulation on a nonrotating Earth. A simple convection system is produced by unequal heating of the atmosphere on a nonrotating Earth.

General Circulation of the Atmosphere

As noted, the underlying cause of wind is unequal heating of Earth's surface. In tropical regions, more solar radiation is received than is radiated back to space. In polar regions, the opposite is true—less solar energy is received than is lost. Attempting to balance these differences, the atmosphere acts as a giant heat-transfer system, moving warm air poleward and cool air equatorward. On a smaller scale, but for the same reason, ocean currents also contribute to this global heat transfer. The general circulation is complex, and a great deal has yet to be explained. We can, however, develop a general understanding by first considering the circulation that would occur on a nonrotating Earth having a uniform surface. We will then modify this system to fit observed patterns.

Circulation on a Nonrotating Earth

On a hypothetical nonrotating planet with a smooth surface of either all land or all water, two large thermally produced cells would form (Figure 13.14). The heated equatorial air would rise until it reached the tropopause, which acts like a lid and deflects the air poleward. Eventually, this upper-level airflow would reach the poles, sink, spread out in all directions at the surface, and move back toward the equator. Once there, it would be reheated and start its journey over again. This hypothetical circulation system has upper-level air flowing poleward and surface air flowing equatorward.

If we add the effect of rotation, this simple convection system will break down into smaller cells. Figure 13.15 illustrates the three pairs of cells proposed to carry on the task of heat redistribution on a rotating planet. The polar and tropical cells retain the characteristics of the thermally generated convection described earlier. The nature of the midlatitude circulation is more complex and will be discussed in more detail in a later section.

Idealized Global Circulation

Near the equator, the rising air is associated with the pressure zone known as the **equatorial low**—a region marked by abundant precipitation. As the upper-level flow from the equatorial low reaches 20 to 30 degrees latitude, north or south, it sinks back toward the surface. This subsidence and associated adiabatic heating produce hot, arid conditions. The center of this zone of subsiding dry air is the **subtropical high,** which encircles the globe near 30 degrees latitude, north and south (Figure 13.15). The great deserts of Australia, Arabia, and Africa exist because of the stable, dry condition caused by the subtropical highs.

At the surface, airflow is outward from the center of the subtropical high. Some of the air travels equatorward and is deflected by the Coriolis effect, producing the reliable **trade winds.** The remainder travels poleward and is also deflected, generating the prevailing **westerlies** of the midlatitudes. As the westerlies move poleward, they encounter the cool **polar easterlies** in the region of the **subpolar low.** The interaction of these warm and cool winds produces the stormy belt known as the **polar front.** The source region for the variable polar easterlies is the **polar high.** Here, cold polar air is subsiding and spreading equatorward.

In summary, this simplified global circulation is dominated by four pressure zones. The subtropical and polar highs

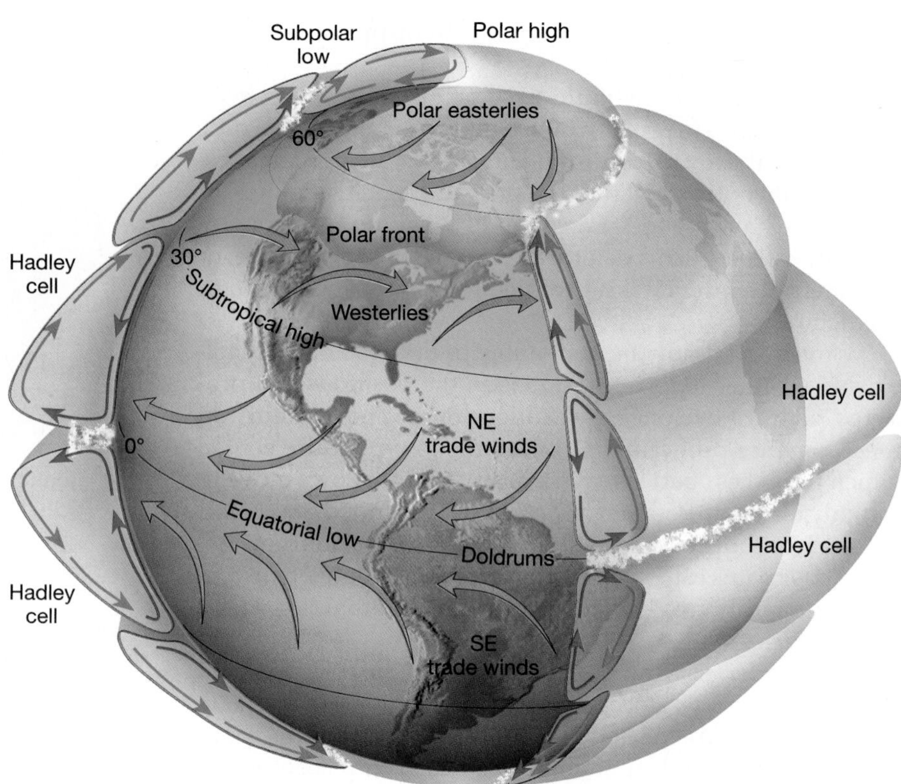

Figure 13.15 Idealized global circulation.

Did You Know?

According to the American Wind Energy Association, by late 2006 U.S. wind energy installations exceeded 10,000 megawatts in capacity. That is enough electricity to supply 2.5 million average American households.

are areas of dry subsiding air that flows outward at the surface, producing the prevailing winds. The low-pressure zones of the equatorial and subpolar regions are associated with inward and upward airflow accompanied by clouds and precipitation.

Influence of Continents

Up to this point, we have described the surface pressure and associated winds as continuous belts around Earth. However, the only truly continuous pressure belt is the subpolar low in the Southern Hemisphere, where the ocean is uninterrupted by landmasses. At other latitudes, particularly in the Northern Hemisphere, where landmasses break up the ocean surface, large seasonal temperature differences disrupt the pattern. Figure 13.16 shows the resulting pressure and wind patterns for January and July. The circulation over the oceans is dominated by semipermanent cells of high pressure in the subtropics and cells of low pressure over the subpolar regions. The subtropical highs are responsible for the trade winds and westerlies, as mentioned earlier.

The large landmasses, on the other hand, particularly Asia, become cold in the winter and develop a seasonal high-pressure system from which surface flow is directed off the land (Figure 13.16). In the summer, the opposite occurs; the landmasses are heated and develop a low-pressure cell, which permits air to flow onto the land. These seasonal changes in wind direction are known as the **monsoons.** During warm months, areas such as India experience a flow of warm, water-laden air from the Indian Ocean, which produces the rainy summer monsoon. The winter monsoon is dominated by dry continental air. A similar situation exists, but to a lesser extent, over North America.

In summary, the general circulation is produced by semipermanent cells of high and low pressure over the oceans and is complicated by seasonal pressure changes over land.

The Westerlies

The circulation in the midlatitudes, the zone of the westerlies, is complex and does not fit the convection system proposed for the tropics. Between 30 and 60 degrees latitude, the general west-to-east flow is interrupted by the migration of cyclones and anticyclones. In the Northern Hemisphere, these cells move from west to east around the globe, creating an anticyclonic (clockwise) flow or a cyclonic (counterclockwise) flow in their area of influence. A close correlation exists between the paths taken by these surface pressure systems and the position of the upper-level airflow, indicating that the upper air steers the movement of cyclonic and anticyclonic systems.

Among the most obvious features of the flow aloft are the seasonal changes. The steep temperature gradient across the middle latitudes in the winter months corresponds to a

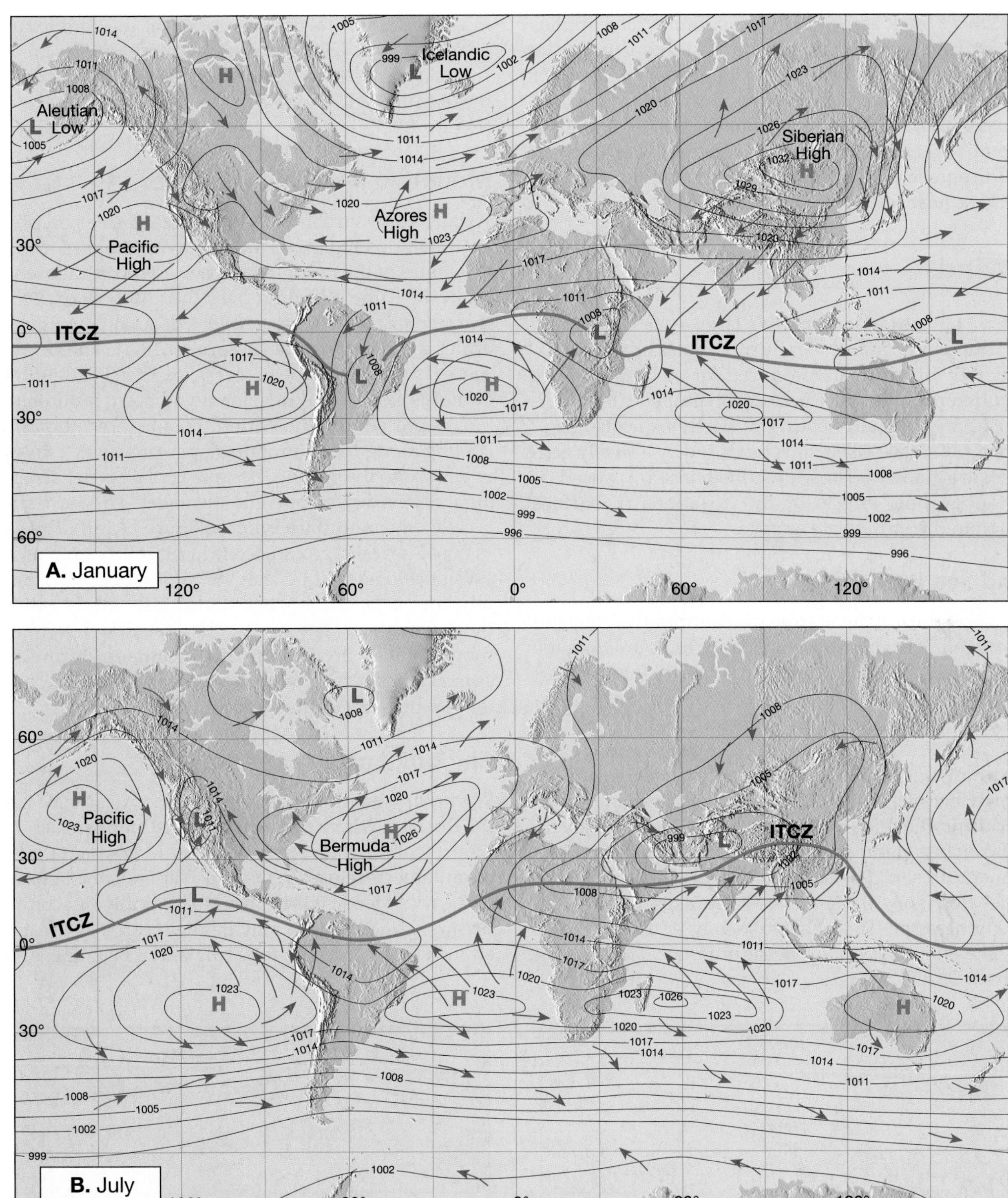

Figure 13.16 Average surface barometric pressure in millibars for **A.** January and **B.** July, with associated winds. The pressure zones clearly migrate north-south following the Sun's vertical rays. The line labeled "ITCZ" refers to the *intertropical convergence zone* and corresponds to the equatorial low. This region of ascending moist hot air is associated with abundant precipitation. It gets its name because it is the region where the trade winds converge.

stronger flow aloft. In addition, the polar jet stream fluctuates seasonally such that its average position migrates southward with the approach of winter, and northward as summer nears. By midwinter, the jet core may penetrate as far south as central Florida.

Because the paths of low-pressure centers are guided by the flow aloft, we can expect the southern tier of states to experience more of their stormy weather in the winter season. During the hot summer months, the storm track is across the northern states, and some cyclones never leave Canada. The northerly storm track associated with summer also applies to Pacific storms, which move toward Alaska during the warm months, thus producing an extended dry season for much of the West Coast. The number of cyclones generated is seasonal as well, with the largest number occurring in the cooler months when the temperature gradients are greatest. This fact is in agreement with the role of cyclonic storms in the distribution of heat across the midlatitudes.

Local Winds

Now that we have examined Earth's large-scale circulation, let us turn briefly to winds that influence much smaller areas. Remember that all winds are produced for the same reason: pressure differences that arise because of temperature differences caused by unequal heating of Earth's surface. *Local winds* are simply small-scale winds produced by a locally generated pressure gradient. Those described here are caused either by topographic effects or by variations in surface composition in the immediate area.

Land and Sea Breezes

In coastal areas during the warm summer months, the land is heated more intensely during the daylight hours than the adjacent body of water. As a result, the air above the land surface heats, expands, and rises, creating an area of lower pressure. A **sea breeze** then develops, because cooler air over the water (higher pressure) moves toward the warmer land (lower pressure) (Figure 13.17A). The sea breeze begins to develop shortly before noon and generally reaches its greatest intensity during the mid- to late afternoon. These relatively cool winds can be a significant moderating influence on afternoon temperatures in coastal areas. Small-scale sea breezes can also develop along the shores of large lakes. People who live in a city near the Great Lakes, such as Chicago, recognize this lake effect, especially in the summer. They are reminded daily by weather reports of the cool temperatures near the lake as compared to warmer outlying areas.

At night, the reverse may take place. The land cools more rapidly than the sea, and the **land breeze** develops (Figure 13.17B).

Mountain and Valley Breezes

A daily wind similar to land and sea breezes occurs in many mountainous regions. During daylight hours, the air along the slopes of the mountains is heated more intensely than the air at the same elevation over the valley floor. Because this warmer air is less dense, it glides up along the slope and generates a **valley breeze** (Figure 13.18A). The occurrence of these daytime upslope breezes can often be identified by the cumulus clouds that develop on adjacent mountain peaks.

After sunset, the pattern may reverse. Rapid radiation cooling along the mountain slopes produces a layer of cooler air next to the ground. Because cool air is denser than warm air, it drains downslope into the valley. This movement of air is called the **mountain breeze** (Figure 13.18B). The same type of cool air drainage can occur in places that have very modest slopes. The result is that the coldest pockets of air are usually found in the lowest spots. Like many other winds, mountain and valley breezes have seasonal preferences. Although valley breezes are most common during the warm season when solar heating is most intense, mountain breezes tend to be more dominant in the cold season.

Chinook and Santa Ana Winds

Warm, dry winds sometimes move down the east slopes of the Rockies, where they are called **chinooks.** Such winds are often created when a strong pressure gradient develops in a mountainous region. As the air descends the leeward slopes of the mountains, it is heated adiabatically (by compression). Because condensation may have occurred as the air ascended the windward side, releasing latent heat, the air descend-

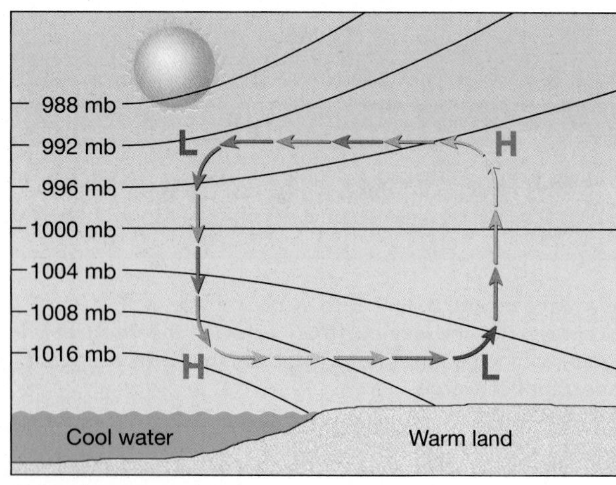

A. Sea breeze

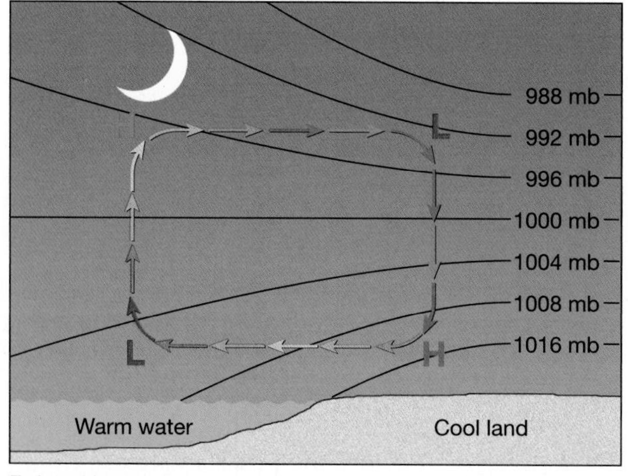

B. Land breeze

Figure 13.17 A sea breeze and a land breeze. **A.** During the daylight hours, the air above the land heats and expands, creating an area of lower pressure. Cooler and denser air over the water moves onto the land, generating a *sea breeze.* **B.** At night, the land cools more rapidly than the water, generating an offshore flow called a *land breeze.*

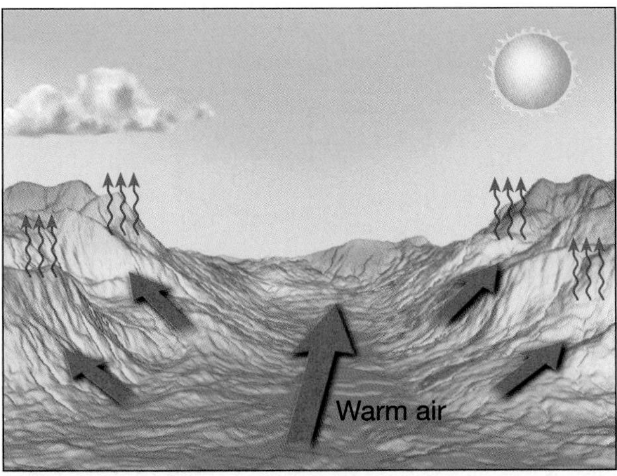

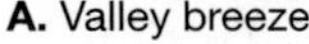

A. Valley breeze

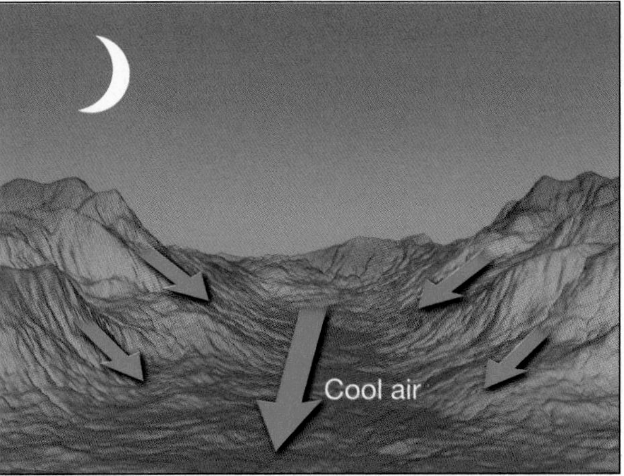

B. Mountain breeze

Figure 13.18 Valley and mountain breezes. **A.** Heating during the daylight hours warms the air along the mountain slopes. This warm air rises, generating a valley breeze. **B.** After sunset, cooling of the air near the mountain can result in cool air drainage into the valley, producing the mountain breeze.

Did You Know?

"Snow eaters" is a local term for chinooks, the warm, dry winds that descend the eastern slopes of the Rockies. These winds have been known to melt more than a foot of snow in a single day. A chinook that moved through Granville, North Dakota, on February 21, 1918, caused temperatures to rise from −33°F to 50°F, an increase of 83°F!

ing the leeward slope will be warmer and drier than it was at a similar elevation on the windward side. Although the temperature of these winds is generally less than 10°C (50°F), which is not particularly warm, they occur mostly in the winter and spring when the affected areas may be experiencing below-freezing temperatures. Thus, by comparison, these dry, warm winds often bring a drastic change. When the ground has a snow cover, these winds are known to melt it in short order. The word *chinook* literally means "snow eater."

A chinooklike wind that occurs in southern California is the **Santa Ana.** These hot, desiccating winds greatly increase the threat of fire in this already dry area (Figure 13.19).

How Wind Is Measured

Two basic wind measurements, direction and speed, are particularly significant to the weather observer. Regarding direction, *winds are always labeled by the direction from which they blow.* A north wind blows *from* the north *toward* the south, an east wind *from* the east *toward* the west. The instrument most commonly used to determine wind direction is the **wind vane** (Figure 13.20, right). This instrument, a common sight on many buildings, always points *into* the wind. The wind direction is often shown on a dial that is connected to the wind vane. The dial indicates wind direction, either by points of the compass (N, NE, E, SE, etc.) or by a scale of 0° to 360°. On the latter scale, 0° or 360° are both north, 90° is east, 180° is south, and 270° is west.

Wind speed is commonly measured using a **cup anemometer** (Figure 13.20, upper left). The wind speed is read from a dial much like the speedometer of an automobile. Places where winds are steady and speeds are relatively high are potential sites for tapping wind energy.

When the wind consistently blows more often from one direction than from any other, it is called a **prevailing wind.** You may be familiar with the prevailing westerlies that dominate the circulation in the midlatitudes. In the United States, for example, these winds consistently move the "weather" from west to east across the continent. Embedded within this general eastward flow are cells of high and low pressure with

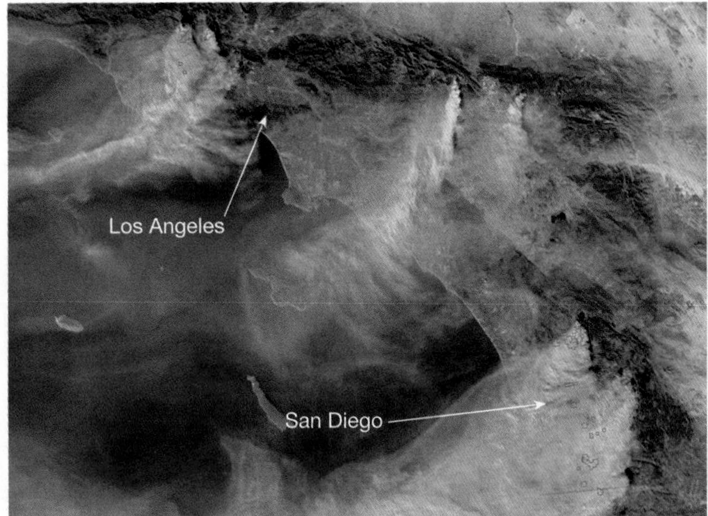

Figure 13.19 This satellite image shows strong Santa Ana winds fanning the flames of several large wildfires in Southern California on October 27, 2003. These fires scorched more than 740,000 acres and destroyed more than 3000 homes. (NASA)

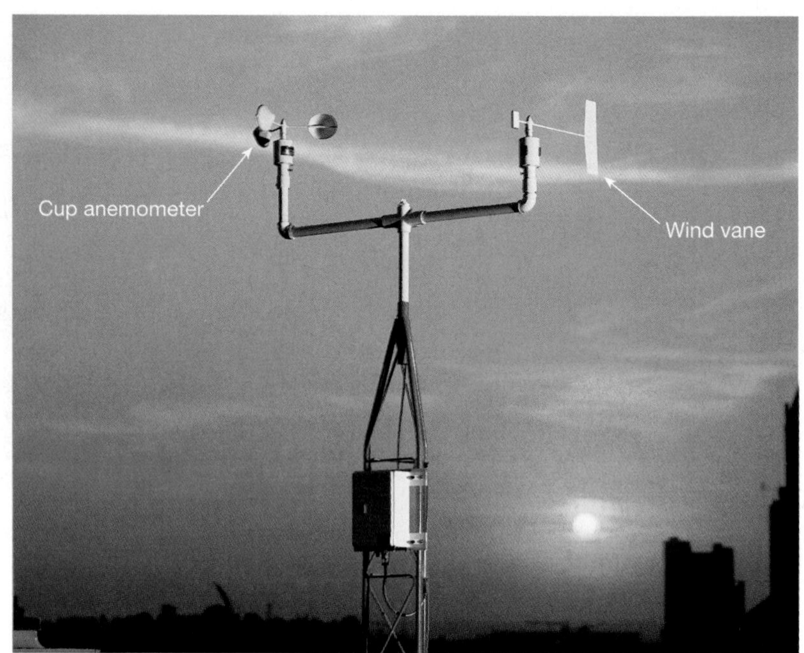

Figure 13.20 Wind vane (right) and cup anemometer (left). The wind vane shows wind direction and the anemometer measures wind speed. (Photo by Belfort Instrument Company)

their characteristic clockwise and counterclockwise flow. As a result, the winds associated with the westerlies, as measured at the surface, often vary considerably from day to day and from place to place. By contrast, the direction of airflow associated with the belt of trade winds is much more consistent, as can be seen in Figure 13.21.

By knowing the locations of cyclones and anticyclones in relation to where you are, you can predict the changes in wind direction that will occur as a pressure center moves past. Because changes in wind direction often bring changes in temperature and moisture conditions, the ability to predict the winds can be very useful. In the Midwest, for example, a north wind may bring cool, dry air from Canada, whereas a south wind may bring warm, humid air from the Gulf of Mexico. Sir Francis Bacon summed it up nicely when he wrote, "Every wind has its weather."

Did You Know?

The highest wind speed recorded at a surface station is 372 kilometers (231 miles) per hour, measured April 12, 1934, at Mount Washington, New Hampshire (see Chapter 11 opening photo p. 292). Located at an elevation of 1886 meters (6288 feet), the observatory atop Mount Washington has an average wind speed of 56 kilometers (35 miles) per hour. Faster wind speeds have undoubtedly occurred, but no instruments were in place to record them.

Figure 13.21 Wind roses showing the percentage of time airflow is from various directions. **A.** Wind frequency for the winter in the eastern United States. **B.** Wind frequency for the winter in northern Australia. Note the reliability of the southeast trades in Australia as compared to the westerlies in the eastern United States. (Data from G. T. Trewartha)

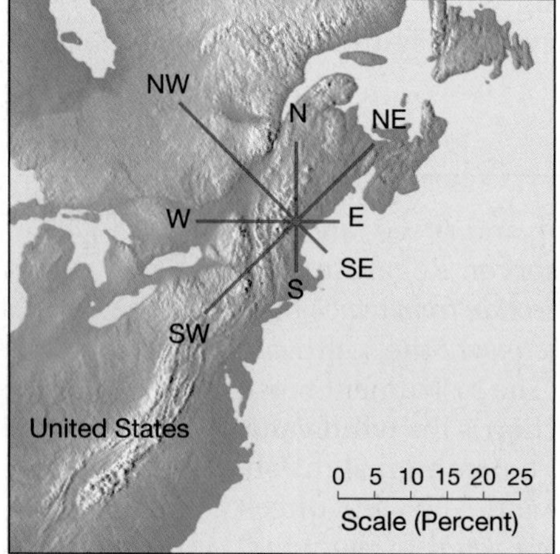

A. Westerlies (winter)

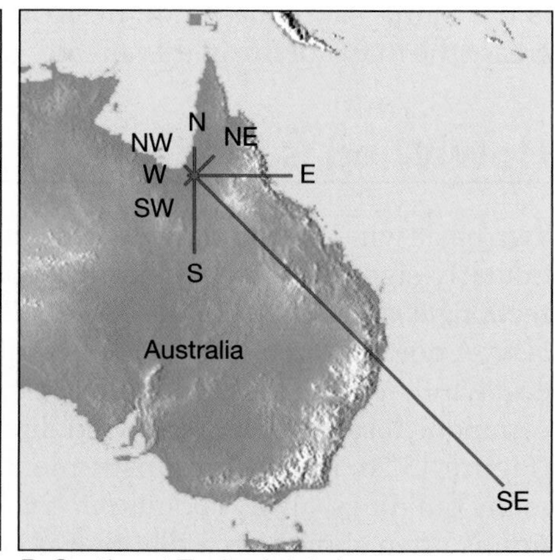

B. Southeast Trades (winter)

The Chapter in Review

1. Air has weight: At sea level, it exerts a pressure of 1 kilogram per square centimeter (14.7 pounds per square inch). *Air pressure* is the force exerted by the weight of air above. With increasing altitude, there is less air above to exert a force, and thus air pressure decreases with altitude, rapidly at first, then much more slowly. The unit used by meteorologists to measure atmospheric pressure is the *millibar. Standard sea-level pressure* is expressed as 1013.2 millibars. *Isobars* are lines on a weather map that connect places of equal air pressures.

2. A *mercury barometer* measures air pressure using a column of mercury in a glass tube sealed at one end and inverted in a dish of mercury. It measures atmospheric pressure in *inches of mercury,* the height of the column of mercury in the barometer. Standard atmospheric pressure at sea level equals 29.92 inches of mercury. As air pressure increases, the mercury in the tube rises, and when air pressure decreases, so does the height of the column of mercury. *Aneroid* ("without liquid") *barometers* consist of partially evacuated metal chambers that compress as air pressure increases and expand as pressure decreases.

3. *Wind* is the horizontal flow of air from areas of higher pressure toward areas of lower pressure. Winds are controlled by the following combination of forces: (1) the *pressure-gradient force* (amount of pressure change over a given distance), (2) *Coriolis effect* (deflective effect of Earth's rotation, to the right in the Northern Hemisphere and to the left in the Southern Hemisphere), and (3) *friction* with Earth's surface (slows the movement of air and alters wind direction).

4. The two types of pressure centers are (1) *cyclones,* or *lows* (centers of low pressure), and (2) *anticyclones,* or *highs* (high-pressure centers). In the Northern Hemisphere, winds around a low (cyclone) are counterclockwise and inward. Around a high (anticyclone), they are clockwise and outward. In the Southern Hemisphere, the Coriolis effect causes winds to be clockwise around a low and counterclockwise around a high. Because air rises and cools adiabatically in a low-pressure center, cloudy conditions and precipitation are often associated with their passage. In a high-pressure center, descending air is compressed and warmed; therefore, cloud formation and precipitation are unlikely in an anticyclone, and "fair" weather is usually expected.

5. Earth's *global pressure zones* include the *equatorial low, subtropical high, subpolar low,* and *polar high.* The *global surface winds* associated with these pressure zones are the *trade winds, westerlies,* and *polar easterlies.*

6. Particularly in the Northern Hemisphere, large seasonal temperature differences over continents disrupt the idealized, or zonal, global patterns of pressure and wind. In winter, large, cold landmasses develop a seasonal high-pressure system from which surface airflow is directed off the land. In summer, landmasses are heated and a low-pressure system develops over them, which permits air to flow onto the land. These seasonal changes in wind direction are known as - *monsoons.*

7. In the middle latitudes, between 30 and 60 degrees latitude, the general west-to-east flow of the westerlies is interrupted by the migration of cyclones and anticyclones. The paths taken by these cyclonic and anticyclonic systems is closely correlated to upper-level airflow and the polar jet stream. The average position of the polar *jet stream,* and hence the paths followed by cyclones, migrates southward with the approach of winter and northward as summer nears.

8. *Local winds* are small-scale winds produced by a locally generated pressure gradient. Local winds include *sea and land breezes* (formed along a coast because of daily pressure differences over land and water), *valley and mountain breezes* (in mountainous areas where the air along slopes heats differently than does the air at the same elevation over the valley floor), and *chinook and Santa Ana winds* (warm, dry winds created when air descends the leeward side of a mountain and warms by compression).

9. The two basic wind measurements are *direction* and *speed.* Winds are always labeled by the direction *from* which they blow. Wind direction is measured with a *wind vane,* and wind speed is measured using a *cup anemometer.*

Key Terms

air pressure (p. 350)
aneroid barometer (p. 351)
anticyclone (high) (p. 356)
barograph (p. 351)
barometric tendency (p. 359)
chinook (p. 362)
convergence (p. 357)
Coriolis effect (p. 353)
cup anemometer (p. 363)
cyclone (low) (p. 356)
divergence (p. 357)
equatorial low (p. 359)
geostrophic wind (p. 355)
isobar (p. 352)
jet stream (p. 355)
land breeze (p. 362)
mercury barometer (p. 351)
monsoon (p. 360)
mountain breeze (p. 362)
polar easterlies (p. 359)
polar front (p. 359)
polar high (p. 359)
pressure gradient (p. 352)
pressure tendency (p. 359)
prevailing wind (p. 363)
Santa Ana (p. 363)
sea breeze (p. 362)
subpolar low (p. 359)
subtropical high (p. 359)
trade winds (p. 359)
valley breeze (p. 362)
westerlies (p. 359)
wind (p. 352)
wind vane (p. 363)

Questions for Review

1. What is standard sea-level pressure in millibars? In inches of mercury? In pounds per square inch?
2. Mercury is 13 times heavier than water. If you built a barometer using water rather than mercury, how tall would it have to be to record standard sea-level pressure (in inches of water)?
3. Describe the principle of the aneroid barometer.
4. What force is responsible for generating wind?
5. Write a generalization relating the spacing of isobars to the speed of wind.
6. How does the Coriolis effect modify air movement?
7. Contrast surface winds and upper-air winds in terms of speed and direction.
8. Describe the weather that usually accompanies a drop in barometric pressure and a rise in barometric pressure.
9. Sketch a diagram (isobars and wind arrows) showing the winds associated with surface cyclones and anticyclones in both the Northern and Southern hemispheres.
10. If you live in the Northern Hemisphere and are directly west of a cyclone, what will the wind direction most probably be? What will the wind direction be if you are west of an anticyclone?
11. The following questions relate to the global pattern of air pressure and winds.
    a. From which pressure zone do the trade winds diverge?
    b. Which prevailing wind belts converge in the stormy region known as the polar front?
    c. Which pressure belt is associated with the equator?
12. What influence does upper-level airflow seem to have on surface pressure systems?
13. Describe the monsoon circulation of India.
14. What is a local wind? List three examples.
15. A northeast wind is blowing *from* the ___ (direction) *toward* the ___ (direction).

Online Study Guide

The *Foundations of Earth Science* Web site uses the resources and flexibility of the Internet to aid in your study of the topics in this chapter. Written and developed by Earth science instructors, this site will help improve your understanding of Earth science. Visit **http://www.prenhall.com/lutgens** and click on the cover of *Foundations of Earth Science 5e* to find:

- Online review quizzes.
- Critical thinking exercises.
- Links to chapter-specific Web resources.
- Internet-wide key-term searches.

http://www.prenhall.com/lutgens

GEODe: Earth Science

GEODe: Earth Science makes studying more effective by reinforcing key concepts using animation, video, narration, interactive exercises, and practice quizzes. A copy is included with every copy of *Foundations of Earth Science 5e.*

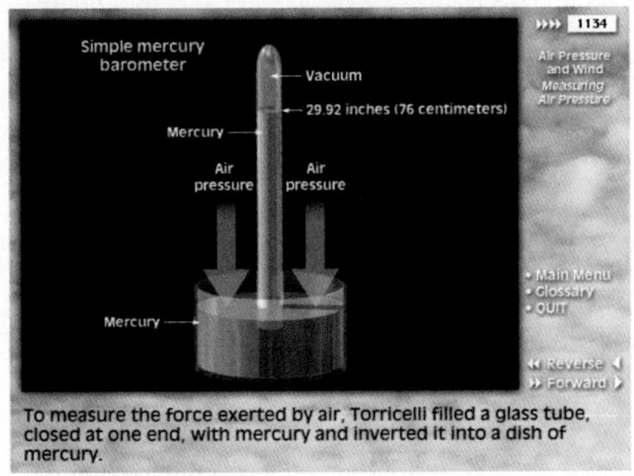

To measure the force exerted by air, Torricelli filled a glass tube, closed at one end, with mercury and inverted it into a dish of mercury.

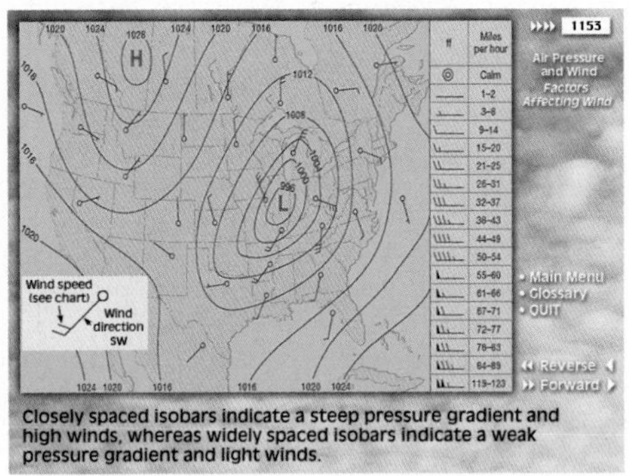

Closely spaced isobars indicate a steep pressure gradient and high winds, whereas widely spaced isobars indicate a weak pressure gradient and light winds.

CHAPTER 14

Weather Patterns and Severe Weather

Hurricane Rita became a Category 5 hurricane late on September 21, 2005, with sustained winds of 275 kilometers (170 miles) per hour and a central pressure of 897 millibars, making it the third most powerful Atlantic basin storm ever measured. When this image was taken about mid-day on September 22, the storm was slightly weaker. (NASA)

FOCUS ON LEARNING

To assist you in learning the important concepts in this chapter, you will find it helpful to focus on the following questions:

1. What is an air mass?
2. How are air masses classified? What is the general weather associated with each air-mass type?
3. What are fronts? How do warm fronts and cold fronts differ?
4. What are the primary weather producers in the middle latitudes? What are the weather patterns associated with these systems?
5. What atmospheric conditions produce thunderstorms, tornadoes, and hurricanes?

A.

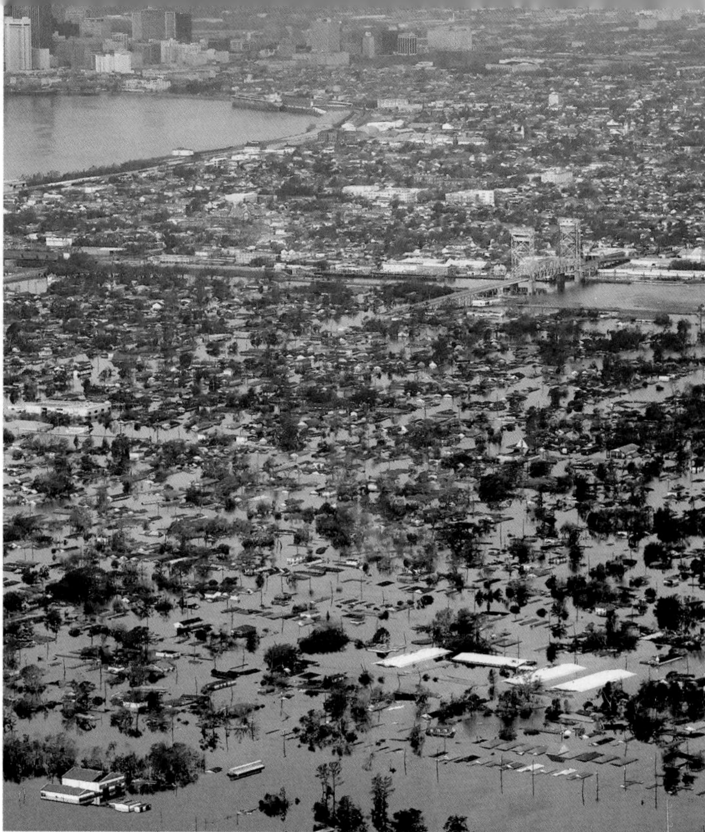

B.

Figure 14.1 A. Tornadoes are intense and destructive local storms of short duration that cause many deaths each year. (Photo by Alan R. Moller/Getty Images, Inc.— Stone Allstock) **B.** An estimated 80 percent of New Orleans was flooded after several levees failed in the wake of Hurricane Katrina in August 2005. (David J. Phillip/AP Photo)

Tornadoes and hurricanes rank among nature's most destructive forces (Figure 14.1). Each spring, newspapers report the death and destruction left in the wake of a "band" of tornadoes. During late summer and fall, we hear occasional news reports about hurricanes. Storms with names such as Katrina, Rita, Charley, and Ivan make front-page headlines. Thunderstorms, although less intense and far more common than tornadoes and hurricanes, will also be part of our discussion of severe weather in this chapter. Before looking at violent weather, however, we will study those atmospheric phenomena that most often affect our day-to-day weather: air masses, fronts, and traveling middle-latitude cyclones. Here, we will see the interplay of the elements of weather discussed in Chapters 11, 12, and 13.

Air Masses

Earth's Dynamic Atmosphere
▼ Basic Weather Patterns

For many people who live in the middle latitudes, which include much of the United States, summer heat waves and winter cold spells are familiar experiences. In the first instance, several days of high temperatures and oppressive humidities may finally end when a series of thunderstorms pass through the area, followed by a few days of relatively cool relief. By contrast, the clear skies that often accompany a span of frigid subzero days may be replaced by thick gray clouds and a period of snow as temperatures rise to levels that seem mild when compared to those that existed just a day earlier. In both examples, what was experienced was a period of generally constant weather conditions followed by a relatively short period of change, and then the reestablishment of a new set of weather conditions that remained for perhaps several days before changing again.

What Is an Air Mass?

The weather patterns just described are the result of movements of large bodies of air, called air masses. An **air mass,** as the term implies, is an immense body of air, typically 1600 kilometers (1000 miles) or more across and perhaps several kilometers thick, that is characterized by a similarity of temperature and moisture at any given altitude. When this air moves out of its region of origin, it will carry these temperatures and moisture conditions elsewhere, eventually affecting a large portion of a continent.

An excellent example of the influence of an air mass is illustrated in Figure 14.2, which shows a cold, dry mass from northern Canada moving southward. With a beginning temperature of −46°C (−51°F), the air mass warms to −33°C (−27°F) by the time it reaches Winnipeg. It continues to warm as it moves southward through the Great Plains and into Mexico. Throughout its southward journey, the air mass becomes warmer. But it also brings some of the coldest weather of the winter to the places in its path. Thus, the air mass is modified, but it also modifies the weather in the areas over which it moves.

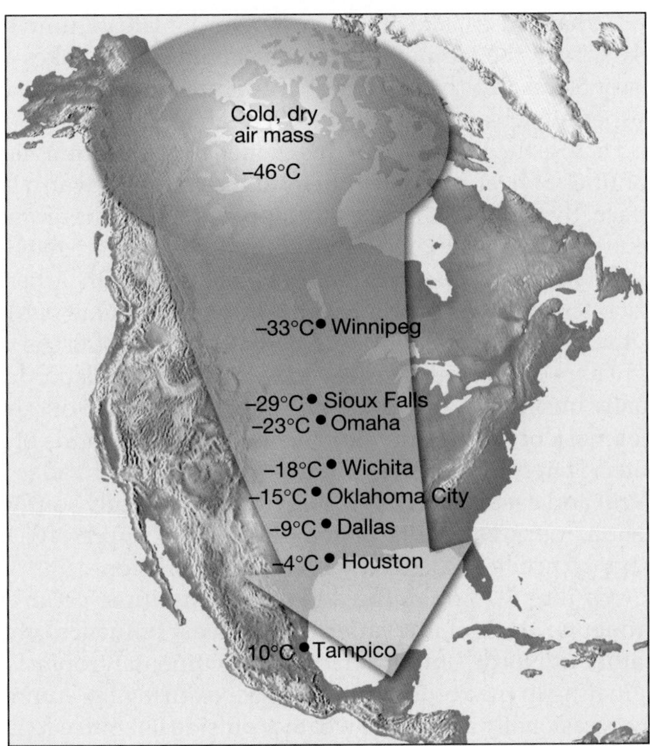

Figure 14.2 As this frigid Canadian air mass moved southward, it brought some of the coldest weather of the winter. (After Tom L. McKnight, *Physical Geography,* 5th ed. Upper Saddle River, NJ: Prentice Hall, 1996, p. 174)

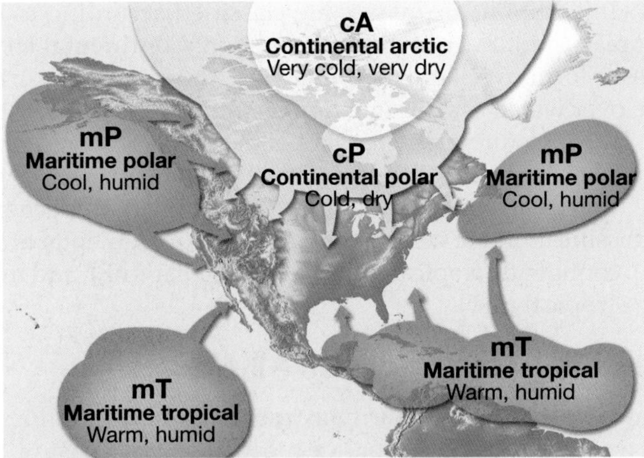

A. Winter pattern

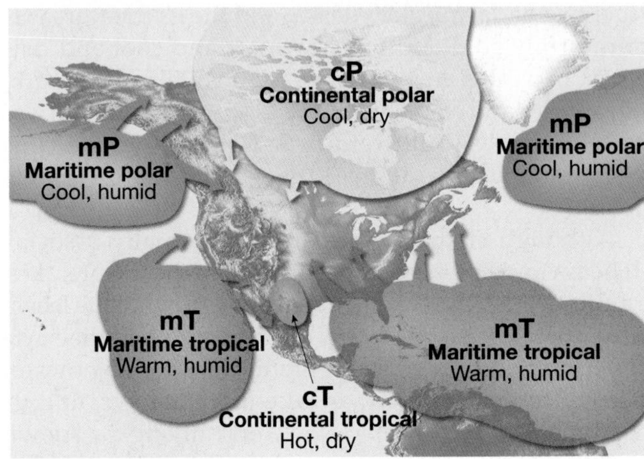

B. Summer pattern

Figure 14.3 Air-mass source regions for North America. Source regions are largely confined to subtropical and subpolar locations. The fact that the middle latitudes are the site where cold and warm air masses clash, often because the converging winds of a traveling cyclone draw them together, means that this zone lacks the conditions necessary to be a source region. The differences between polar and arctic are relatively small and serve to indicate the degree of coldness of the respective air masses. By comparing the summer and winter maps, it is clear that the extent and temperature characteristics fluctuate.

The horizontal uniformity of an air mass is not perfect, of course. Because air masses extend over large areas, small differences in temperature and humidity from place to place occur. Still, the differences observed within an air mass are small compared to the rapid changes experienced across air-mass boundaries.

Since it may take several days for an air mass to move across an area, the region under its influence will probably experience fairly constant weather, a situation called **air-mass weather.** Certainly, there may be some day-to-day variations, but the events will be very unlike those in an adjacent air mass.

The air-mass concept is an important one because it is closely related to the study of atmospheric disturbances. Most disturbances in the middle latitudes originate along the boundary zones that separate different air masses.

Source Regions

When a portion of the lower atmosphere moves slowly or stagnates over a relatively uniform surface, the air will assume the distinguishing features of that area, particularly with regard to temperature and moisture conditions.

The area where an air mass acquires its characteristic properties of temperature and moisture is called its **source region.** The source regions that produce air masses that influence North America are shown in Figure 14.3.

Air masses are classified according to their source region. **Polar (P)** and **arctic (A) air masses** originate in high latitudes toward Earth's poles, whereas those that form in low latitudes are called **tropical (T) air masses.** The designation *polar, arctic,* or *tropical* gives an indication of the temperature characteristics of an air mass. *Polar* and *arctic* indicate cold, and *tropical* indicates warm.

Did You Know?

When a fast-moving frigid air mass advances from the Canadian arctic into the northern Great Plains, temperatures have been known to plunge 20° to 30°C (40° to 50°F) in a matter of just a few hours. One notable example is a drop of 55.5°C (100°F), from 6.7°C to −48.8°C (44° to −56°F), in 24 hours at Browning, Montana, on January 23–24, 1916.

In addition, air masses are classified according to the nature of the surface in the source region. **Continental (c) air masses** form over land, and **maritime (m) air masses** originate over water. The designation *continental* or *maritime* suggests the moisture characteristics of the air mass. Continental air is likely to be dry, and maritime air humid.

The basic types of air masses according to this scheme of classification are continental polar (cP), continental arctic (cA), continental tropical (cT), maritime polar (mP), and maritime tropical (mT).

Weather Associated with Air Masses

Continental polar and maritime tropical air masses influence the weather of North America most, especially east of the Rocky Mountains. Continental polar air masses originate in northern Canada, interior Alaska, and the Arctic—areas that are uniformly cold and dry in winter and cool and dry in summer. In winter, an invasion of continental polar air brings the clear skies and cold temperatures we associate with a cold wave as it moves southward from Canada into the United States. In summer, this air mass may bring a few days of cooling relief.

Although cP air masses are not normally associated with heavy precipitation, those that cross the Great Lakes in late autumn and winter sometimes bring snow to the leeward shores. These localized storms often form when the surface weather map indicates no apparent cause for a snowstorm to occur. These are known as **lake-effect snows,** and they make Buffalo and Rochester, New York, among the snowiest cities in the United States (Figure 14.4).

What causes lake-effect snow? During late autumn and early winter, the temperature contrast between the lakes and adjacent land areas can be large.* The temperature contrast can be especially great when a very cold cP air mass pushes southward across the lakes. When this occurs, the air acquires large quantities of heat and moisture from the relatively warm lake surface. By the time it reaches the opposite shore, the air mass is humid and unstable, and heavy snow showers are likely.

Maritime tropical air masses affecting North America most often originate over the warm waters of the Gulf of Mexico, the Caribbean Sea, or the adjacent Atlantic Ocean. As you might expect, these air masses are warm, moisture-laden, and usually unstable. Maritime tropical air is the source of much, if not most, of the precipitation in the eastern two-thirds of the United States. In summer, when an mT air mass invades the central and eastern United States, and occasionally southern Canada, it brings the high temperatures and oppressive humidity typically associated with its source region.

Of the two remaining air masses, maritime polar and continental tropical, the latter has the least influence on the weather of North America. Hot, dry continental tropical air, originating in the Southwest and Mexico during the summer, only occasionally affects the weather outside its source region.

During winter, maritime polar air masses coming from the North Pacific often originate as continental polar air masses in Siberia. The cold, dry cP air is transformed into relative-

*Recall that land cools more rapidly and to lower temperatures than water. See the discussion of "Land and Water" in the section on "Why Temperatures Vary: The Controls of Temperature" in Chapter 11.

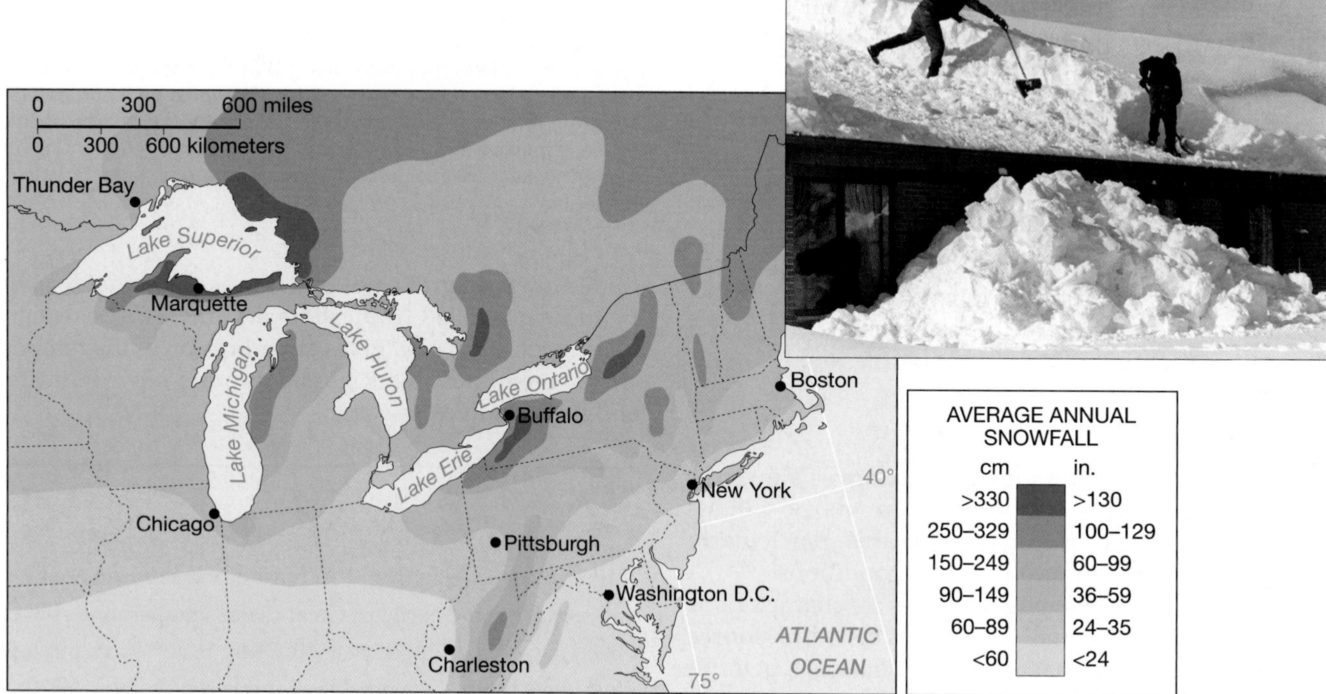

Figure 14.4 The snowbelts of the Great Lakes are easy to pick out on this snowfall map. (Data from NOAA) The photo was taken following a six-day lake-effect snowstorm in November 1996 that dropped 175 centimeters (nearly 69 inches) of snow on Chardon, Ohio, setting a new state record. (Photo by Tony Dejak/AP/Wide World Photos)

Did You Know?

Buffalo, New York, located along the eastern shore of Lake Erie (see Figure 14.4) is famous for its lake-effect snows. One of the most memorable events took place between December 24, 2001, and January 1, 2002. This storm, which set a record as the longest-lasting lake-effect event, buried Buffalo with 207.3 centimeters (81.6 inches) of snow. Prior to this storm, the record for the *entire month* of December had been 173.7 centimeters (68.4 inches)!

ly mild, humid, unstable mP air during its long journey across the North Pacific. As this mP air arrives at the western shore of North America, it is often accompanied by low clouds and shower activity. When this air advances inland against the western mountains, orographic uplift produces heavy rain or snow on the windward slopes of the mountains. Maritime polar air also originates in the North Atlantic off the coast of eastern Canada and occasionally influences the weather of the northeastern United States. In winter, when New England is on the northern or northwestern side of a passing low, the counterclockwise cyclonic winds draw in maritime polar air. The result is a storm characterized by snow and cold temperatures, known locally as a *nor'easter.*

Fronts

Earth's Dynamic Atmosphere
▼ Basic Weather Patterns

Fronts are boundaries that separate different air masses, one warmer than the other and often higher in moisture content. Fronts can form between any two contrasting air masses. Considering the vast size of the air masses involved, fronts are relatively narrow, being 15- to 200-kilometer-wide (9- to 120-mile) bands of discontinuity. On the scale of a weather map, they are generally narrow enough to be represented by a broad line (as in Figure 14.8, p. 376).

Above Earth's surface, the front slopes at a low angle so that warmer air overlies cooler air (Figure 14.5). In the ideal case, the air masses on both sides of the front move in the same direction and at the same speed. Under this condition, the front acts simply as a barrier, moving along between the two contrasting air masses. Generally, however, an air mass on one side of a front is moving faster relative to the frontal boundary than the air mass on the other side. Thus, one air mass actively advances into another and "clashes" with it. In fact, these boundaries were tagged *fronts* during World War I by Norwegian meteorologists, who visualized them as analogous to battle lines between two armies. It is along these "battlegrounds" that cyclonic circulation (centers of low pressure) develop and generate much of the precipitation and severe weather in the middle latitudes.

As one air mass advances into another, limited mixing occurs along the frontal surface, but for the most part, the air masses retain their distinct identities as one is displaced upward over the other. No matter which air mass is advancing, *it is always the warmer (less dense) air that is forced aloft,* while *the cooler (denser) air acts as the wedge upon which lifting takes place.* The term **overrunning** is generally applied to warm air gliding up along a cold air mass. We will now take a look at different types of fronts.

Warm Fronts

When the surface position of a front moves so that warm air occupies territory formerly covered by cooler air, it is called a **warm front** (Figure 14.5). On a weather map, the surface position of a warm front is shown by a red line with red semicircles protruding into the cooler air.

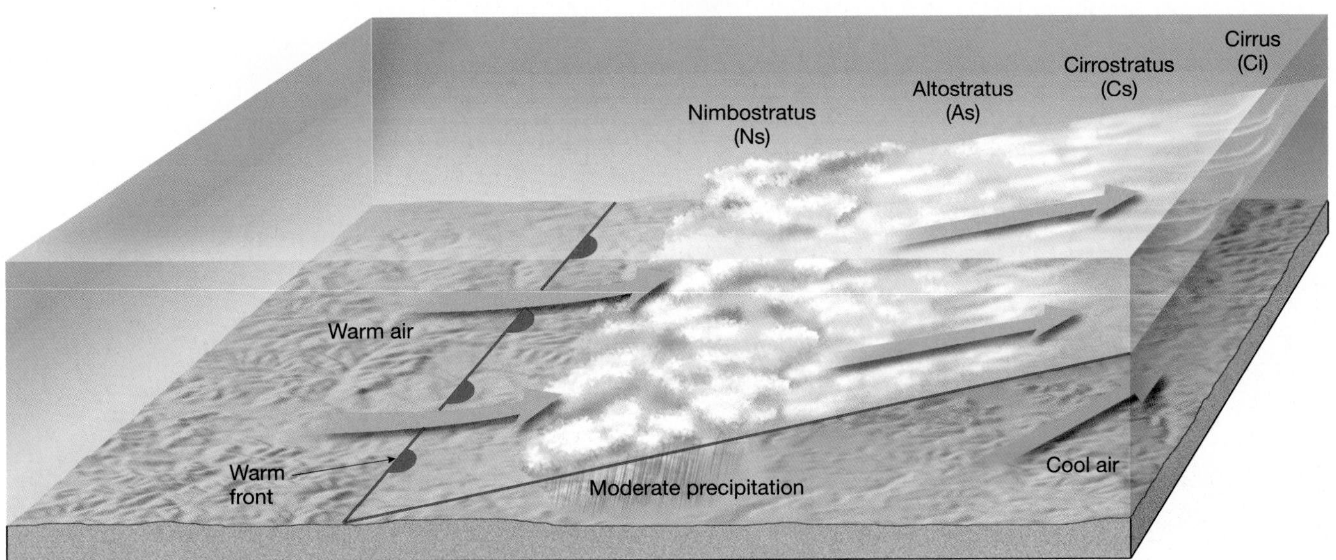

Figure 14.5 Warm front produced as warm air glides up over a cold air mass. Precipitation is moderate and occurs within a few hundred kilometers of the surface front.

East of the Rockies, warm tropical air often enters the United States from the Gulf of Mexico and overruns receding cool air. As the cold air retreats, friction with the ground slows the advance of the surface position of the front more so than its position aloft. Stated another way, less dense, warm air has a hard time displacing denser, cold air. For this reason, the boundary separating these air masses acquires a very gradual slope. The average slope of a warm front is about 1:200, which means that if you are 200 kilometers (120 miles) ahead of the surface location of a warm front, you will find the frontal surface at a height of 1 kilometer (0.6 mile).

As warm air ascends the retreating wedge of cold air, it expands and cools adiabatically to produce clouds and, frequently, precipitation. The sequence of clouds shown in Figure 14.5 typically precedes a warm front. The first sign of the approach of a warm front is the appearance of cirrus clouds overhead. These high clouds form 1000 kilometers (600 miles) or more ahead of the surface front, where the overrunning warm air has ascended high up the wedge of cold air.

As the front nears, cirrus clouds grade into cirrostratus, which blend into denser sheets of altostratus. About 300 kilometers (180 miles) ahead of the front, thicker stratus and nimbostratus clouds appear and rain or snow begins. Because of their slow rate of advance and very low slope, warm fronts usually produce light to moderate precipitation over a large area for an extended period. Warm fronts, however, are occasionally associated with cumulonimbus clouds and thunderstorms. This occurs when the overrunning air is unstable, and the temperatures on opposite sides of the front contrast sharply. At the other extreme, a warm front associated with a dry air mass could pass unnoticed at the surface.

A gradual increase in temperature occurs with the passage of a warm front. The increase is most noticeable when there is a large temperature difference between the adjacent air masses. The moisture content and stability of the encroaching warm air mass largely determine when clear skies will return. During the summer, cumulus, and occasionally cumulonimbus, clouds are embedded in the warm unstable air mass that follows the front. Precipitation from these clouds can be heavy but is usually scattered and of short duration.

Cold Fronts

When dense cold air is actively advancing into a region occupied by warmer air, the boundary is called a **cold front** (Figure 14.6). As with warm fronts, friction tends to slow the surface position of a cold front more so than its position aloft. However, because of the relative positions of the adjacent air masses, the cold front steepens as it moves. On the average, cold fronts are about twice as steep as warm fronts, having a slope of perhaps 1:100. In addition, cold fronts advance at speeds around 35 to 50 kilometers (20 to 35 miles) per hour compared to 25 to 35 kilometers (15 to 20 miles) per hour for warm fronts. These two differences—rate of movement and steepness of slope—largely account for the more violent nature of cold-front weather compared to the weather generally accompanying a warm front.

As a cold front approaches, generally from the west or northwest, towering clouds often can be seen in the distance. Near the front, a dark band of ominous clouds foretells the ensuing weather. The forceful lifting of air along a cold front is often so rapid that the latent heat released when water vapor condenses will increase the air's buoyancy appreciably. The heavy downpours and vigorous wind gusts associated with mature cumulonimbus clouds frequently result. Because a cold front produces roughly the same amount of lifting as does a warm front, but over a shorter distance, the intensity of precipitation is greater, but the duration is shorter. In addition, a marked temperature drop and a wind shift from the south to west or northwest accompany the passage of the front. The sometimes violent weather and sharp temperature contrast along the cold front are symbolized on a weather map by a blue line with blue triangle-shaped points that extend into the warmer air mass (Figure 14.6).

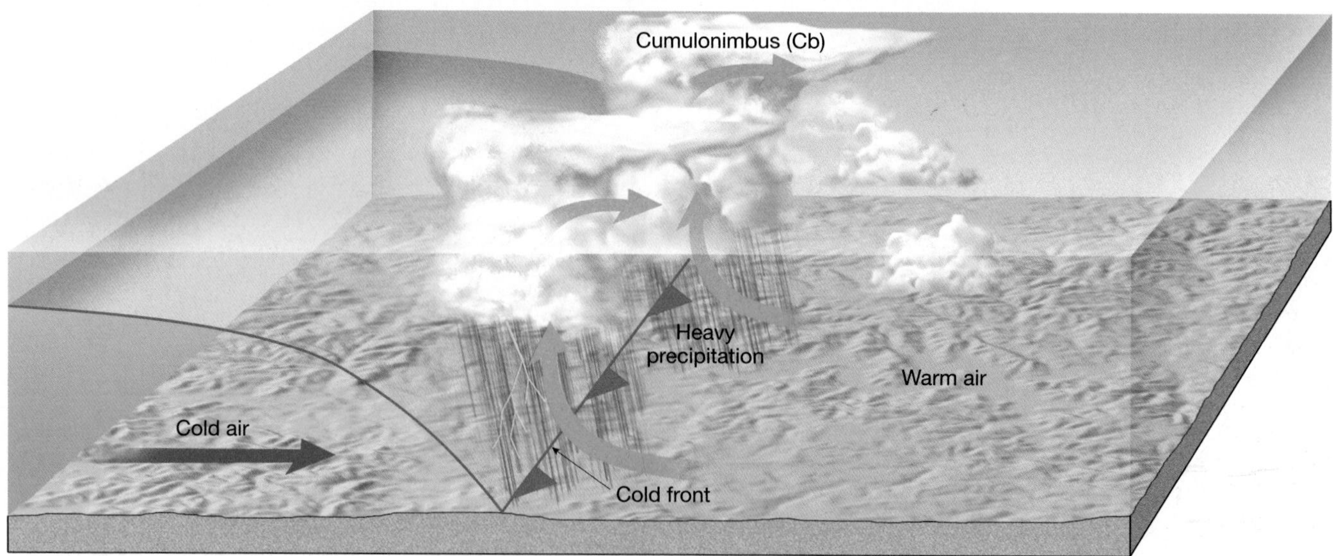

Figure 14.6 Fast-moving cold front and cumulonimbus clouds. Thunderstorms may occur if the warm air is unstable.

The weather behind a cold front is dominated by a subsiding and relatively cold air mass. Thus, clearing usually begins soon after the front passes. Although the compression of air due to subsidence causes some adiabatic heating, the effect on surface temperatures is minor. In winter, the long, cloudless nights that often follow the passage of a cold front allow for abundant radiation cooling that reduces surface temperatures. When a cold front moves over a relatively warm area, surface heating can produce shallow convection. This, in turn, may generate low cumulus or stratocumulus clouds behind the front.

Stationary Fronts and Occluded Fronts

Occasionally, the flow on both sides of a front is neither toward the cold air mass nor toward the warm air mass, but almost parallel to the line of the front. Thus, the surface position of the front does not move. This condition is called a **stationary front.** On a weather map, stationary fronts are shown with blue triangular points on one side of the front and red semicircles on the other. At times, some overrunning occurs along a stationary front, most likely causing gentle to moderate precipitation.

The fourth type of front is the **occluded front:** An active cold front overtakes a warm front, as shown in Figure 14.7. As the advancing cold air wedges the warm front upward, a new front emerges between the advancing cold air and the air over which the warm front is gliding. The weather of an occluded front is generally complex. Most precipitation is associated with the warm air being forced aloft. When conditions are suitable, however, the newly formed front is capable of initiating precipitation of its own.

A word of caution is in order concerning the weather associated with various fronts. Although the preceding discussion will help you recognize the weather patterns associated with fronts, remember that these descriptions are generalizations. The weather generated along any individual front may or may not conform fully to this idealized picture. Fronts, like all aspects of nature, never lend themselves to classification as easily as we would like.

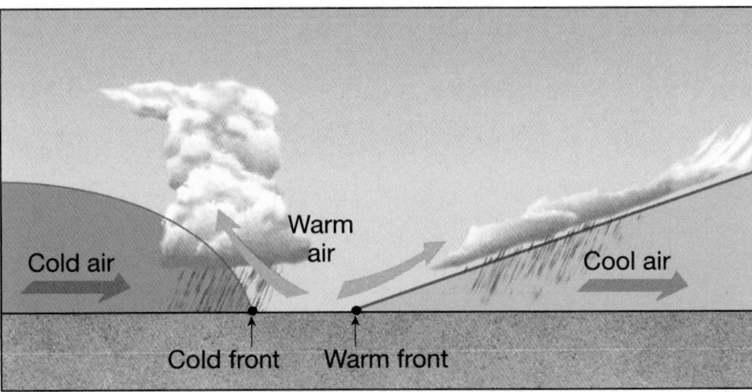

A.

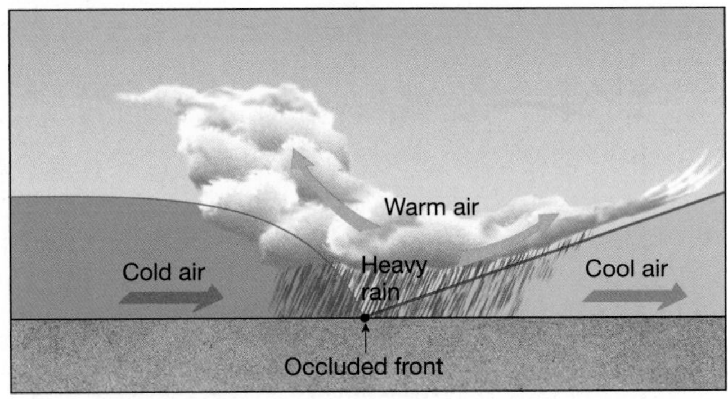

B.

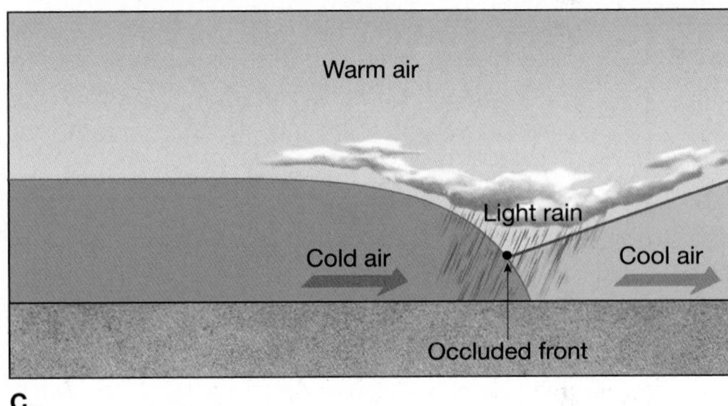

C.

Figure 14.7 Stages in the formation of an occluded front.

The Middle-Latitude Cyclone

Earth's Dynamic Atmosphere
▼ Basic Weather Patterns

So far, we have examined the basic elements of weather as well as the dynamics of atmospheric motions. We are now ready to apply our knowledge of these diverse phenomena to an understanding of day-to-day weather patterns in the middle latitudes. For our purposes, *middle latitudes* refers to the region between southern Florida and Alaska. The primary weather producers here are **middle-latitude,** or **midlatitude, cyclones.** On weather maps, such as those used on the Weather Channel, they are shown by an **L,** meaning *low-pressure system.*

Middle-latitude cyclones are large centers of low pressure that generally travel from west to east (Figure 14.8). Lasting from a few days to more than a week, these weather systems have a counterclockwise circulation with an airflow inward toward their centers. Most middle-latitude cyclones also have a cold front extending from the central area of low pressure, and frequently a warm front as well. Convergence and forceful lifting initiate cloud development and frequently cause abundant precipitation.

As early as the 1800s, it was known that middle-latitude cyclones were the bearers of precipitation and severe weather. But it was not until the early part of the 1900s that a model was developed to explain how cyclones form. A group of Norwegian scientists formulated and published this model in 1918. The model was created primarily from near-surface observations.

Years later, as data from the middle and upper troposphere and from satellite images became available, some modifications were necessary. Yet, this model is still a useful

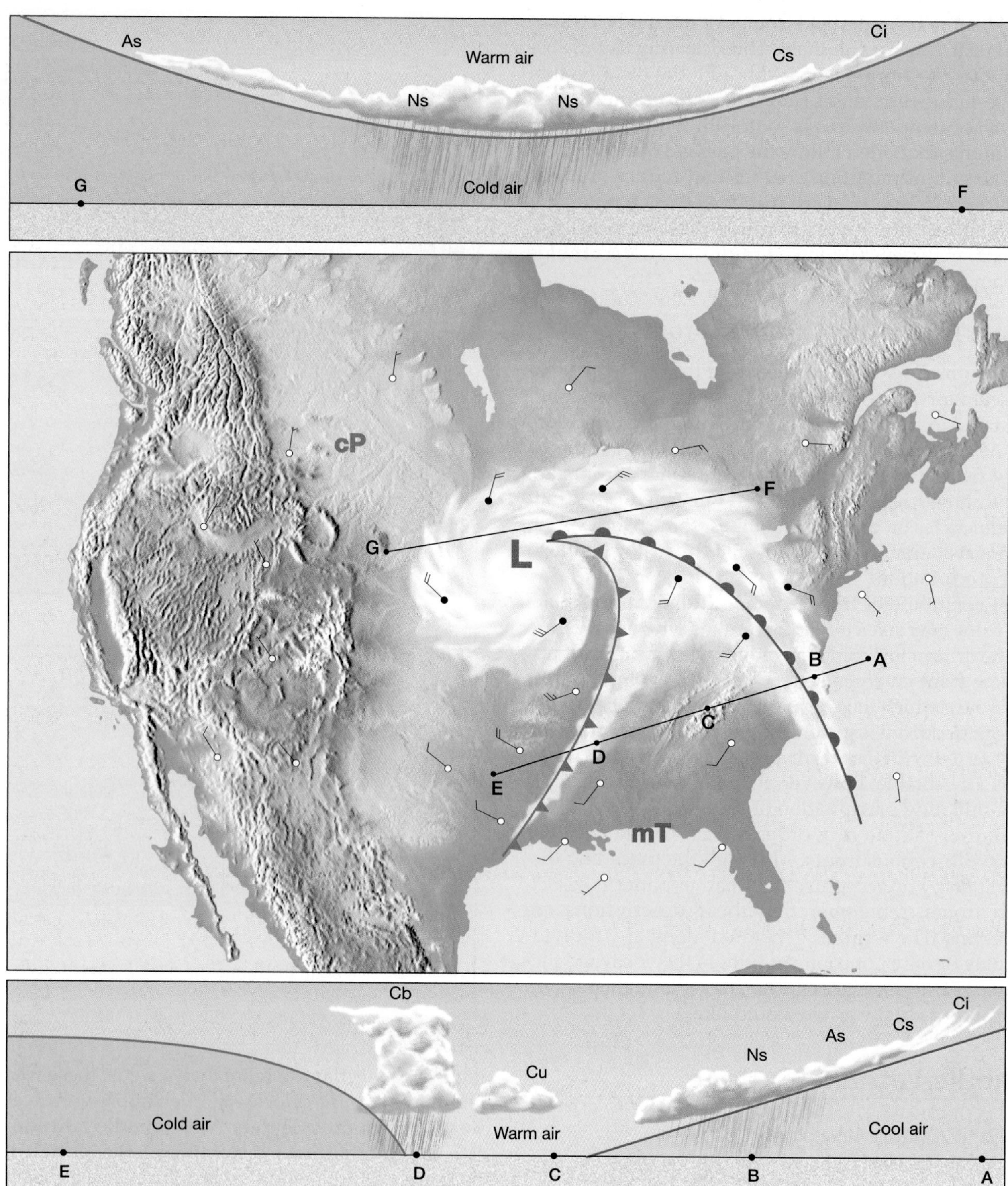

Figure 14.8 Cloud patterns typically associated with a mature middle-latitude cyclone. The middle section is a map view. Note the cross-section lines (F–G, A–E). Above the map is a vertical cross section along line F–G. Below the map is a section along A–E. For cloud abbreviations, refer to Figures 14.4 and 14.5.

working tool for interpreting the weather. If you keep this model in mind when you observe changes in the weather, the changes will no longer come as a surprise. You should begin to see some order in what had appeared to be disorder, and you might even occasionally "predict" the impending weather.

Idealized Weather of a Middle-Latitude Cyclone

The middle-latitude cyclone model provides a useful tool for examining the weather patterns of the middle latitudes. Figure 14.8 illustrates the distribution of clouds and thus the regions of possible precipitation associated with a mature system. Compare this drawing to the satellite image shown in Figure 14.9. It is easy to see why we often refer to the cloud pattern of a midlatitude cyclone as having a "comma" shape.

Guided by the westerlies aloft, cyclones generally move eastward across the United States, so we can expect the first signs of their arrival in the west. However, often in the region of the Mississippi valley, cyclones begin a more northeasterly path and occasionally move directly northward. A midlatitude cyclone typically requires two to four days to move completely across a given region. During that brief period, abrupt changes in atmospheric conditions may be experienced. This is particularly true in the winter and spring, when the largest temperature contrasts occur across the middle latitudes.

Using Figure 14.8 as a guide, we will now consider these weather producers and what we should expect from them as they move over an area. To facilitate our discussion, Figure 14.8 includes two profiles along lines A–E and F–G.

- Imagine the change in weather as you move along profile A–E. At point A, the sighting of high cirrus clouds would be the first sign of the approaching cyclone. These high clouds can precede the surface front by 1000 kilometers (600 miles) or more, and they generally will be accompanied by falling pressure. As the warm front advances, a lowering and thickening of the cloud deck is noticed.
- Usually within 12 to 24 hours after the first sighting of cirrus clouds, light precipitation begins (point B). As the front nears, the rate of precipitation increases, a rise in temperature is noticed, and winds begin to change from east or southeast to south or southwest.
- With the passage of the warm front, an area is under the influence of a maritime tropical air mass (point C). Generally, the region affected by this sector of the cyclone experiences warm temperatures, south or southwest winds, and generally clear skies, although fair-weather cumulus or altocumulus are not uncommon.
- The relatively warm, humid weather of the warm sector passes quickly and is replaced by gusty winds and precipitation generated along the cold front. The approach of a rapidly advancing cold front is marked by a wall of dark clouds (point D). Severe weather accompanied by heavy precipitation, hail, and an occasional tornado is a definite possibility at this time of year. The passage of the cold front is easily detected by a wind shift; the southwest winds are replaced by winds from the west to northwest and by a pronounced drop in temperature. Also, rising pressure hints of the subsiding cool, dry air behind the front.
- Once the front passes, skies clear as cooler air invades the region (point E). Often a day or two of almost cloudless deep blue skies occurs unless another cyclone is edging into the region.

A very different set of weather conditions will prevail in those regions north of the storm's center along profile F–G of Figure 14.8. Often the storm reaches its greatest intensity in this zone, and the area along profile F–G receives the brunt of the storm's fury. Temperatures remain cold during the passage of the system, and heavy snow, sleet, and/or freezing rain may develop during the winter months.

The Role of Airflow Aloft

When the earliest studies of midlatitude cyclones were made, little was known about the nature of the airflow in the middle and upper troposphere. Since then, a close relationship has been established between surface disturbances and the flow aloft. Airflow aloft plays an important role in maintaining cyclonic and anticyclonic circulation. In fact, more often than not, these rotating surface wind systems are actually generated by upper-level flow.

Recall that the airflow around a cyclone (low-pressure system) is inward, a fact that leads to mass convergence, or coming together (Figure 14.10). The resulting accumulation of

Figure 14.9 Satellite view of a mature cyclone over the eastern United States. It is easy to see why we often refer to the cloud pattern of a cyclone as having a "comma" shape. (Photo courtesy of John Jensenius/National Weather Service)

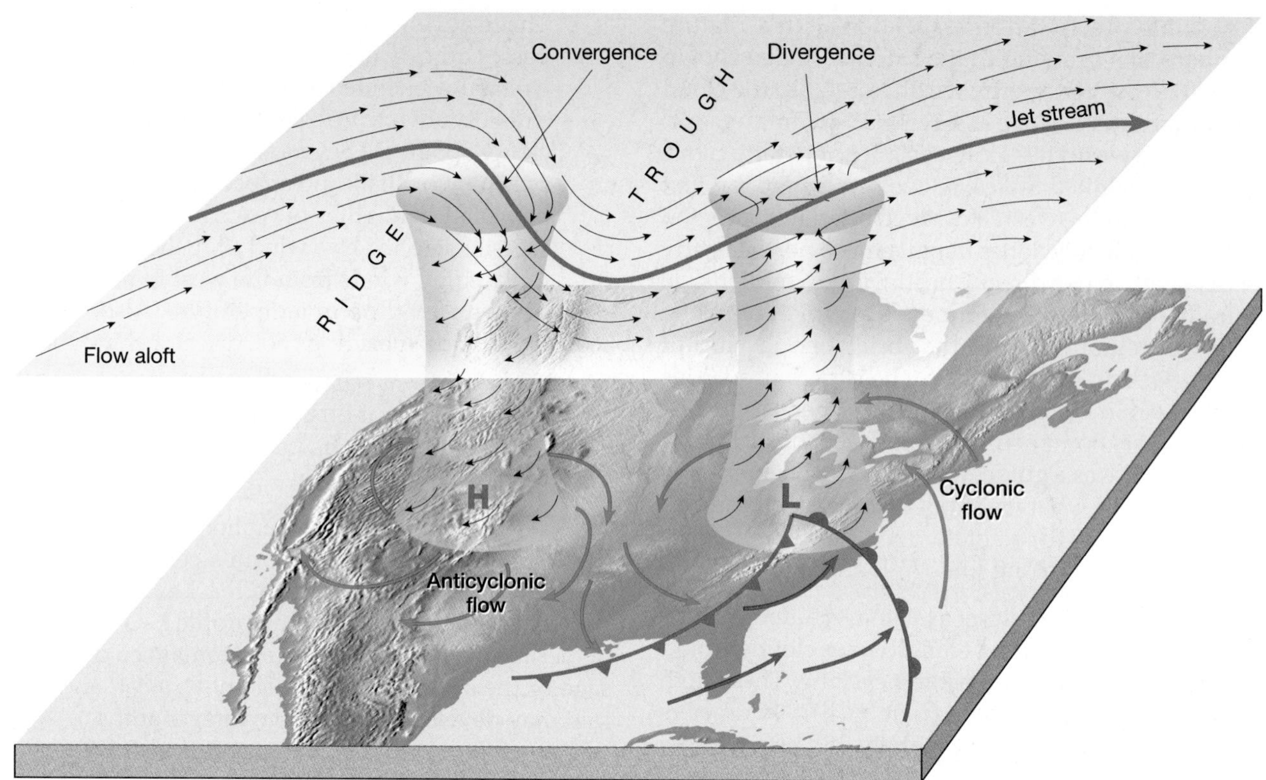

Figure 14.10 Idealized diagram depicting the support that divergence and convergence aloft provide to cyclonic and anticyclonic circulation at the surface. Divergence aloft initiates upward air movement, reduced surface pressure, and cyclonic flow. On the other hand, convergence along the jet stream results in general subsidence of the air column, increased surface pressure, and anticyclonic surface winds.

air must be accompanied by a corresponding increase in surface pressure. Consequently, we might expect a low pressure system to "fill" rapidly and be eliminated, just as the vacuum in a coffee can is quickly dissipated when we open it. However, this does not occur. On the contrary, cyclones often exist for a week or longer. For this to happen, surface convergence must be offset by a mass outflow at some level aloft (Figure 14.10). As long as divergence (spreading out) aloft is equal to or greater than the surface inflow, the low pressure and its accompanying convergence can be sustained.

Because cyclones are bearers of stormy weather, they have received far more attention than anticyclones. Nevertheless, a close relation exists, which makes it difficult to separate any discussion of these two types of pressure systems. The surface air that feeds a cyclone, for example, generally originates as air flowing out of an anticyclone. Consequently, cyclones and anticyclones typically are found adjacent to each other. Like the cyclone, an anticyclone depends on the flow far above to maintain its circulation. Divergence at the surface is balanced by convergence aloft and general subsidence of the air column (Figure 14.10).

What's in a Name?

Up to now we have examined the middle-latitude cyclones, which play such an important role in causing day-to-day weather changes. Yet the use of the term *cyclone* is often confusing. To many people, the term implies only an intense storm, such as a tornado or a hurricane. When a hurricane unleashes its fury on India or Bangladesh, for example, it is usually reported in the media as a cyclone (the term denoting a hurricane in that part of the world).

Similarly, tornadoes are referred to as cyclones in some places. This custom is particularly common in portions of the Great Plains of the United States. Recall that in *The Wizard of Oz,* Dorothy's house was carried from her Kansas farm to the land of Oz by a cyclone. Indeed, the nickname for the athletic teams at Iowa State University is the *Cyclones.* Although hurricanes and tornadoes are, in fact, cyclones, the vast majority of cyclones are *not* hurricanes or tornadoes. The term *cyclone* simply refers to the circulation around any low-pressure center, no matter how large or intense it is.

Tornadoes and hurricanes are both smaller and more violent than middle-latitude cyclones. Middle-latitude cyclones can have a diameter of 1600 kilometers (1000 miles) or more. By contrast, hurricanes average only 600 kilometers (375 miles) across, and tornadoes, with a diameter of just 0.25 kilometer (0.16 mile) are much too small to show up on a weather map.

The thunderstorm, a much more familiar weather event, hardly needs to be distinguished from tornadoes, hurricanes, and midlatitude cyclones. Unlike the flow of air about these latter storms, the circulation associated with thunderstorms is characterized by strong up-and-down movements. Winds in the vicinity of a thunderstorm do not follow the inward spiral of a cyclone, but they are typically variable and gusty.

Although thunderstorms form "on their own" away from cyclonic storms, they also form in conjunction with cyclones. For instance, thunderstorms are frequently spawned along the cold front of a midlatitude cyclone, where on rare occasions a tornado may descend from the thunderstorm's cumulonimbus tower. Hurricanes also generate widespread thunderstorm activity. Thus, thunderstorms are related in some manner to all three types of cyclones mentioned here.

Thunderstorms

Thunderstorms are the first of three severe weather types we will examine in this chapter. Sections on tornadoes and hurricanes follow.

Severe weather has a fascination that everyday weather phenomena cannot provide. The lightning display and booming thunder generated by a severe thunderstorm can be a spectacular event that elicits both awe and fear (Figure 14.11). Of course, hurricanes and tornadoes also attract a great deal of much deserved attention. A single tornado outbreak or hurricane can cause many deaths as well as billions of dollars in property damage. In a typical year, the United States experiences thousands of violent thunderstorms, hundreds of floods and tornadoes, and several hurricanes. According to the National Weather Service, during the 30-year span from 1975 through 2004, tornadoes took an average of 65 lives each year, whereas hurricanes were responsible for about one-quarter as many deaths. Surprisingly (to many), lightning and flash floods were deadlier. During the 1975–2004 period, lightning annually claimed 66 lives, and floods were responsible for an average of 107 fatalities each year.

Thunderstorm Occurrence

Almost everyone has observed various small-scale phenomena that result from the vertical movements of relatively warm, unstable air. Perhaps you have seen a dust devil over an open field on a hot day whirling its dusty load to great heights. Or maybe you have noticed a bird glide effortlessly skyward on an invisible thermal of hot air. These examples illustrate the dynamic thermal instability that occurs during the development of a thunderstorm.

A **thunderstorm** is simply a storm that generates lightning and thunder. It frequently produces gusty winds, heavy rain, and hail. A thunderstorm may be produced by a single cumulonimbus cloud and influence only a small area, or it may be associated with clusters of cumulonimbus clouds covering a large area.

Thunderstorms form when warm, humid air rises in an unstable environment. Various mechanisms can trigger the upward air movement needed to create thunderstorm-producing cumulonimbus clouds. One mechanism, the unequal heating of Earth's surface, significantly contributes to the formation of *air-mass thunderstorms.* These storms are associated with the scattered puffy cumulonimbus clouds that commonly form *within* maritime tropical air masses and produce scattered thunderstorms on summer days. Such storms are usually short-lived and seldom produce strong winds or hail.

Did You Know?

Beginning in the late 1980s, it has been possible to detect cloud-to-ground lightning flashes in real time across the country. Since 1989, the National Lightning Detection Network has recorded an average of about 20 million cloud-to-ground flashes per year in the contiguous 48 states. In addition, about half of all flashes have more than one ground-strike point, so at least 30 million points on the ground are struck in an average year. Besides cloud-to-ground flashes, there are roughly 5 to 10 times as many flashes within clouds.

Figure 14.11 This lightning display occurred near Colorado Springs, Colorado. Cumulonimbus clouds can produce lightning, thunderstorms, and other forms of severe weather. (Photo by Sean Cayton/The Image Works)

In contrast, another type of thunderstorm not only benefits from uneven surface heating but is associated with the lifting of warm air, as occurs along a front or a mountain slope. Moreover, diverging winds aloft frequently contribute to the formation of these storms because they tend to draw air from lower levels upward beneath them. Some of the thunderstorms of this type may produce high winds, damaging hail, flash floods, and tornadoes. Such storms are described as *severe*.

At any given time, an estimated 2000 thunderstorms are in progress on Earth (Figure 14.12). As we would expect, the greatest number occur in the tropics, where warmth, plentiful moisture and instability are always present. About 45,000 thunderstorms take place each day, and more than 16 million occur annually around the world. The lightning from these storms strikes Earth 100 times each second (Figure 14.12A). Annually, the United States experiences about 100,000 thunderstorms and millions of lightning strikes. A glance at Figure 14.12B shows that thunderstorms are most frequent in Florida and the eastern Gulf Coast region, where such activity is recorded between 70 and 100 days each year. The region on the east side of the Rockies in Colorado and New Mexico is next, with thunderstorms occurring on 60 to 70 days each year. Most of the rest of the nation experiences thunderstorms on 30 to 50 days annually. Clearly, the western margin of the United States has little thunderstorm activity. The same is true for the northern tier of states and for Canada, where warm, moist, unstable mT air seldom penetrates.

Did You Know?

According to the National Severe Storms Laboratory (NSSL) in Norman, Oklahoma, lightning has been known to strike more than 16 kilometers (10 miles) from a storm in an area of clear skies.

Stages of Thunderstorm Development

All thunderstorms require warm, moist air, which, when lifted, will release sufficient latent heat to provide the buoyancy necessary to maintain its upward flight. This instability and associated buoyancy are triggered by a number of different processes, yet most thunderstorms have a similar life history.

Because instability and buoyancy are enhanced by high surface temperatures, thunderstorms are most common in the afternoon and early evening (Figure 14.13A). However, sur-

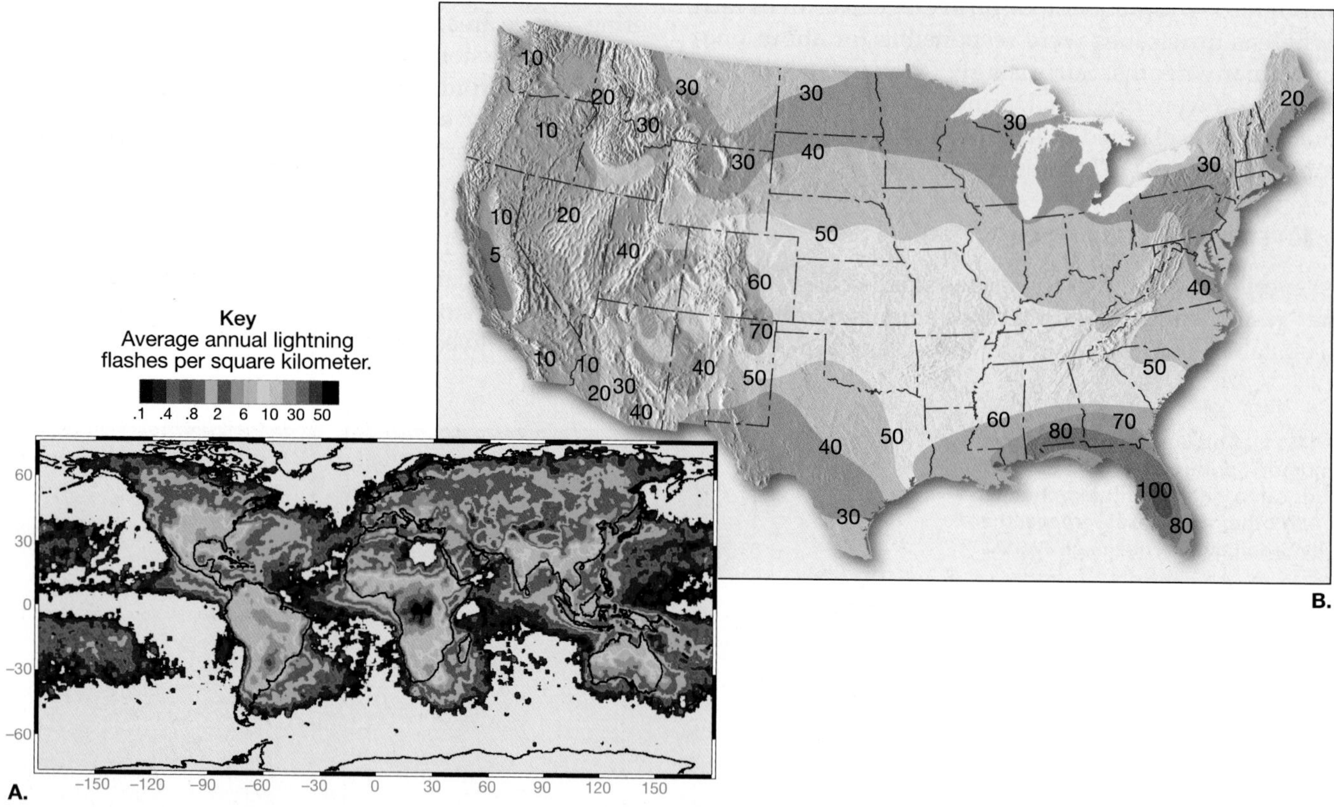

Figure 14.12 **A.** Data from space-based optical sensors show the worldwide distribution of lightning, with color variations indicating the average annual number of lightning flashes per square kilometer. The map includes data obtained from April 1995 to March 2000 from NASA's Optical Transient Detector, and from December 1997 to November 2000 from NASA's Lightning Imaging Sensor. Both are satellite-based sensors that use high-speed cameras capable of detecting brief lightning flashes even under daytime conditions. (NASA) **B.** Average number of days each year with thunderstorms. The humid subtropical climate that dominates the southeastern United States receives much of its precipitation in the form of thunderstorms. Most of the Southeast averages 50 or more days each year with thunderstorms. (Environmental Data Service, NOAA)

A.

B.

Figure 14.13 **A.** Buoyant thermals often produce fair-weather cumulus clouds that soon evaporate into the surrounding air, making it more humid. As this process of cumulus development and evaporation continues, the air eventually becomes sufficiently humid, so that newly forming clouds do not evaporate but continue to grow. (Photo by Henry Lansford/Photo Researchers, Inc.) **B.** This developing cumulonimbus cloud became a towering August thunderstorm over central Illinois. (Photo by E. J. Tarbuck)

face heating alone is not sufficient for the growth of towering cumulonimbus clouds. A solitary cell of rising hot air produced by surface heating could, at best, produce a small cumulus cloud, which would evaporate within 10 to 15 minutes.

The development of 12,000-meter (40,000-foot) (or on rare occasions 18,000-meter [60,000-foot]) cumulonimbus towers requires a continual supply of moist air (Figure 14.13B). Each new surge of warm air rises higher than the last, adding to the height of the cloud (Figure 14.14). These updrafts must occasionally reach speeds over 100 kilometers (60 miles) per hour to accommodate the size of hailstones they are capable of carrying upward. Usually within an hour, the amount and size of precipitation that has accumulated is too much for the updrafts to support, and consequently downdrafts develop in one part of the cloud, releasing heavy precipitation. This represents the most active stage of the thunderstorm. Gusty winds, lightning, heavy precipitation, and sometimes hail are experienced.

Eventually the warm, moist air supplied by updrafts ceases as downdrafts dominate throughout the cloud. The cooling effect of falling precipitation, coupled with the influx of colder air aloft, marks the end of the thunderstorm activity. The life span of a typical cumulonimbus cell within a thunderstorm complex is only about an hour, but as the storm moves, fresh supplies of warm, water-laden air generate new cells to replace those that are dissipating.

Tornadoes

Tornadoes are local storms of short duration that must be ranked high among nature's most destructive forces (see Figure 14.1A, p. 370). Their sporadic occurrence and violent winds cause many deaths each year. Tornadoes are violent windstorms that take the form of a rotating column of air or *vortex* that extends downward from a cumulonimbus cloud.

Pressures within some tornadoes have been estimated to be as much as 10 percent lower than immediately outside the storm. Drawn by the much lower pressure in the center of the vortex, air near the ground rushes into the tornado from all directions. As the air streams inward, it is spiraled upward around the core until it eventually merges with the

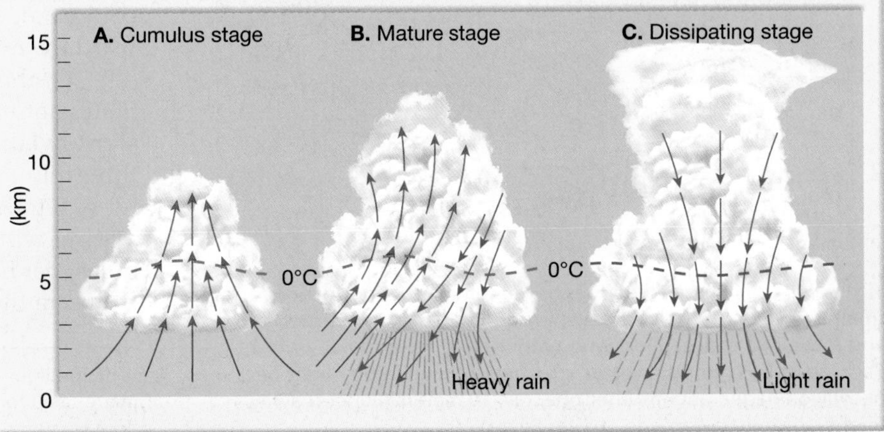

Figure 14.14 Stages in the development of a thunderstorm. During the cumulus stage, strong updrafts act to build the storm. The mature stage is marked by heavy precipitation and cool downdrafts in part of the storm. When the warm updrafts disappear completely, precipitation becomes light, and the cloud begins to evaporate.

airflow of the parent thunderstorm deep in the cumulonimbus tower. Because of the tremendous pressure gradient associated with a strong tornado, maximum winds can sometimes approach 480 kilometers (300 miles) per hour.

Some tornadoes consist of a single vortex, but within many stronger tornadoes are smaller whirls called *suction vortices* that rotate within the main vortex (Figure 14.15). Suction vortices have diameters of only about 10 meters (33 feet) and rotate very rapidly. This structure accounts for occasional observations of virtually total destruction of one building while another one, just 10 meters (33 feet) away, suffers little damage.

Tornado Occurrence and Development

Tornadoes form in association with severe thunderstorms that produce high winds, heavy (sometimes torrential) rainfall, and often damaging hail. Fortunately, less than 1 percent of all thunderstorms produce tornadoes. Nevertheless, a much higher number must be monitored as potential tornado producers. Although meteorologists are still not sure what triggers tornado formation, it is apparent that they are the product of the interaction between strong updrafts in the thunderstorm and the winds in the troposphere.

Tornadoes can form in any situation that produces severe weather, including cold fronts and tropical cyclones (hurricanes). The most intense tornadoes are usually those that form in association with huge thunderstorms called *supercells.* An important precondition linked to tornado formation in severe thunderstorms is the development of *a mesocyclone*—a vertical cylinder of rotating air, typically about 3 to 10 kilometers (2 to 6 miles) across, that develops in the updraft of a severe thunderstorm (Figure 14.16). The formation of this large vortex often precedes tornado formation by 30 minutes or so.

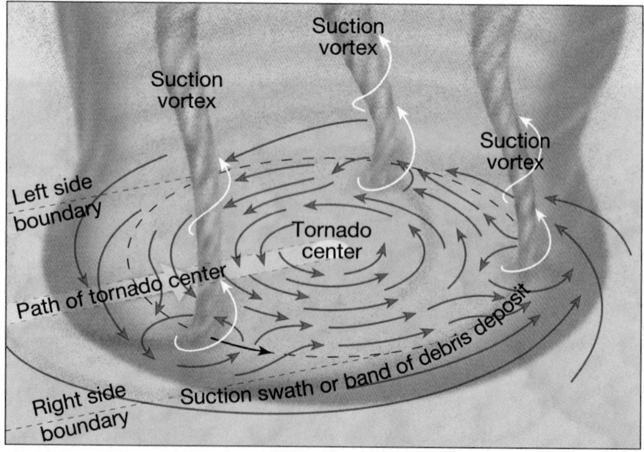

Figure 14.15 Some tornadoes have multiple suction vortices. These small and very intense vortices are roughly 10 meters (30 feet) across and move in a counterclockwise path around the tornado center. Because of this multiple vortex structure, one building might be heavily damaged and another one, just 10 meters (30 feet) away, might suffer little damage. (After Fujita)

The formation of a mesocyclone does not necessarily mean that tornado formation will follow. Only about half of all mesocyclones produce tornadoes. Forecasters cannot determine in advance which mesocyclones will spawn tornadoes.

General Atmospheric Conditions Severe thunderstorms—and, hence, tornadoes—are most often spawned along the cold front of a middle-latitude cyclone or in association with a supercell thunderstorm such as pictured in Figure 14.16D, E. Throughout spring, air masses associated with middle-latitude cyclones are most likely to have greatly contrasting conditions. Continental polar air from Canada may still be very cold and dry, whereas maritime tropical air from the Gulf of Mexico is warm, humid, and unstable. The greater the contrast when these air masses meet, the more intense the storm. These two contrasting air masses are most likely to meet in the central United States, because there is no significant natural barrier separating the center of the country from the Arctic or the Gulf of Mexico. Consequently, this region generates more tornadoes than any other area of the country or, in fact, the world. The map in Figure 14.17, which depicts tornado incidence in the United States for a 27-year period, readily substantiates this fact.

Tornado Climatology An average of about 1200 tornadoes were reported annually in the United States between 1990 and 2004. Still, the actual number that occurs from one year to the next varies greatly. During this 15-year span, for example, yearly totals ranged from a low of 941 in 2002 to a high of 1819 in 2004.

Tornadoes occur during every month of the year. April through June is the period of greatest tornado frequency in the United States; the number is lowest during December and January (Figure 14.17 graph inset). Of the 40,522 confirmed tornadoes reported over the contiguous 48 states during the 50-year period from 1950 through 1999, an average of almost 6 per day occurred during May. At the other extreme, a tornado was reported only about every other day in December and January.

Profile of a Tornado An average tornado has a diameter of between 150 and 600 meters (500 and 2000 feet), travels across the landscape at approximately 45 kilometers (30 miles) per hour, and cuts a path about 10 kilometers (6 miles) long.* Because tornadoes usually occur slightly ahead of a cold front, in the zone of southwest winds, most move toward the northeast.

Of the hundreds of tornadoes reported in the United States annually, more than half are comparatively weak and short-lived. Most of these small tornadoes have lifetimes of three minutes or less and paths that seldom exceed 1 kilometer (0.6 mile) in length and 100 meters (330 feet) in width. Typical wind speeds are on the order of 150 kilometers (90 miles) per hour or less. On the other end of the tornado spectrum are the infrequent and often long-lived violent torna-

*The 10-kilometer (6-mile) figure applies to documented tornadoes. Because many small tornadoes go undocumented, the real average path of all tornadoes is unknown but is shorter than 10 kilometers.

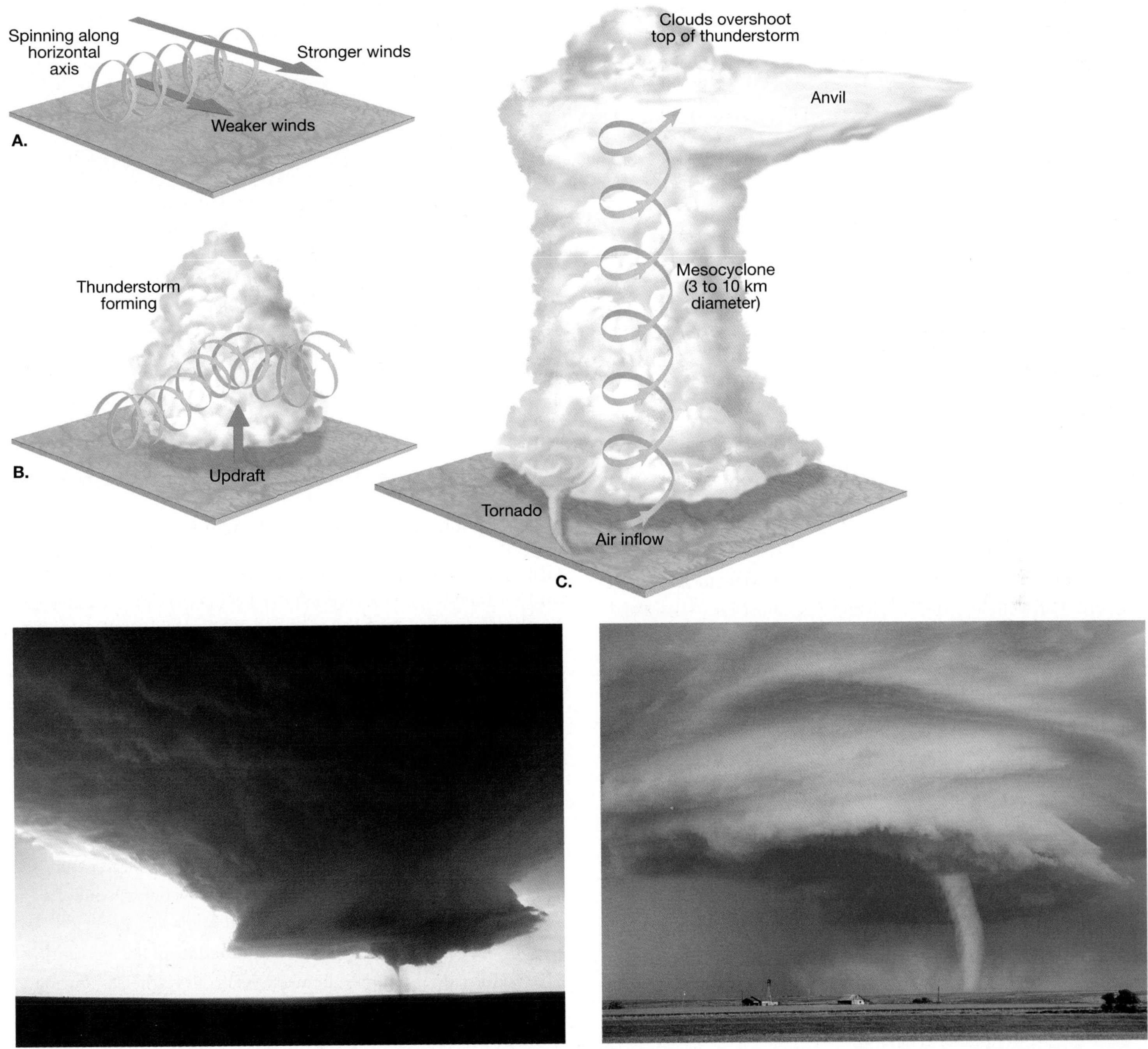

Figure 14.16 The formation of a mesocyclone often precedes tornado formation. **A.** Winds are stronger aloft than at the surface (called speed wind shear), producing a rolling motion about a horizontal axis. **B.** Strong thunderstorm updrafts tilt the horizontally rotating air to a nearly vertical alignment. **C.** The mesocyclone, a vertical cylinder of rotating air, is established. **D.** If a tornado develops, it will descend from a slowly rotating wall cloud in the lower portion of the mesocyclone. (Photo © Howard B. Bluestein, professor of meteorology) **E.** A tornado is visible when it contains condensation or when it contains dust and debris. The appearance is often the result of both. When the column of air is aloft and does not produce damage, the visible portion is properly called a *funnel cloud.* This tornado in New Mexico near the Texas Panhandle touched down out of a rotating supercell thunderstorm. (Photo by A. T. Willett/Alamy)

does. Although large tornadoes constitute only a small percentage of the total reported, their effects are often devastating. Such tornadoes may exist for periods in excess of three hours and produce an essentially continuous damage path more than 150 kilometers (90 miles) long and perhaps a kilometer or more wide. Maximum winds range beyond 500 kilometers (310 miles) per hour.

Tornado Destruction

The potential for tornado destruction depends largely upon the strength of the winds generated by the storm. Because tornadoes generate the strongest winds in nature, they have accomplished many seemingly impossible tasks, such as driving a piece of straw through a thick wooden plank and uprooting

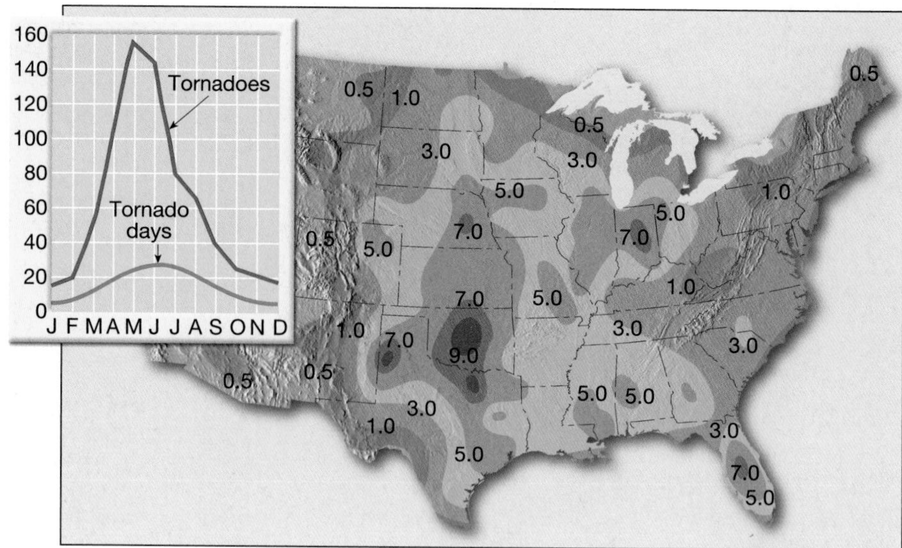

Figure 14.17 The map shows average annual tornado incidence per 26,000 square kilometers (10,000 square miles) for a 27-year period. The graph shows the average number of tornadoes and tornado days each month in the United States for the same period.

Did You Know?

One tornado easily ranks above all others as the single most dangerous and destructive. Known as the Tri-state tornado, it occurred on March 18, 1925. Starting in southeastern Missouri, the tornado remained on the ground for 352 kilometers (219 miles), finally ending in Indiana. The losses included 695 dead and 2027 injured. Property losses were also great, with several small towns almost totally destroyed.

huge trees. Although it may seem impossible for winds to cause some of the extensive damage attributed to tornadoes, tests in engineering facilities have repeatedly demonstrated that winds in excess of 320 kilometers (200 miles) per hour are capable of incredible feats (Figure 14.18).

One commonly used guide to tornado intensity is the *Fujita intensity scale*, or simply the *F-scale* (Table 14.1). Because tornado winds cannot be measured directly, a rating on the F-scale is determined by assessing the worst damage produced by a storm. Although widely used, the F-scale is not perfect. Estimating tornado intensity based on damage alone does not take into account the structural integrity of the objects hit by a tornado. A well-constructed building can withstand very high winds, whereas a poorly built structure can suffer devastating damage from the same or even weaker winds.

Although the greatest part of tornado damage is caused by violent winds, most tornado injuries and deaths result from flying debris. In the United States, the average annual death toll from tornadoes is about 60 to 70 people, but the actual number of deaths each year can depart significantly from the average. On April 3–4, 1974, for example, an outbreak of 148 tornadoes brought death and destruction to a 13-state region east of the Mississippi River. More than 300 people died and nearly 5500 people were injured in this worst disaster in half a century. Most tornadoes, however, do not result in a loss of life. In one statistical study that examined a 29-year period, 689 tornadoes resulted in deaths. This figure represents slightly less than 4 percent of the total 19,312 reported storms.

Although the percentage of tornadoes resulting in death is small, each tornado is potentially lethal. When tornado fatalities and storm intensities are compared, the results are quite interesting: The majority (63 percent) of tornadoes are weak (F0 and F1), and the number of storms decreases as tornado intensity increases. The distribution of tornado fatalities, however, is just the opposite. Although only 2 percent of tornadoes are classified as violent (F4 and F5), they account for nearly 70 percent of the deaths.

Did You Know?

The strongest tornadoes (F5 on the Fujita scale) are rare. During the 30-year span, from 1970 through 1999, only 26 out of a total of 28,913 tornadoes were classified in the F5 category. That is just 0.09 percent (nine one-hundredths of one percent)! No F5 tornadoes reported in many years. Nevertheless, during a single 16-hour span, April 3–4, 1974, there were seven F5 storms.

Tornado Forecasting

Because severe thunderstorms and tornadoes are small and relatively short-lived phenomena, they are among the most difficult weather features to forecast precisely. Nevertheless, the prediction, detection, and monitoring of such storms are among the most important services provided by professional meteorologists. Both the timely issuance and dissemination of watches and warnings are critical to the protection of life and property.

The Storm Prediction Center (SPC) located in Norman, Oklahoma, is part of the National Weather Service (NWS) and the National Centers for Environmental Prediction (NCEP). Its mission is to provide timely and accurate forecasts and watches for severe thunderstorms and tornadoes.

A.

B.

Figure 14.18 **A.** The force of the wind during a tornado near Wichita, Kansas, in April 1991 was enough to drive this piece of metal into a utility pole. (Photo by John Sokich/NOAA) **B.** Damage from a tornado that struck Xenia, Ohio, on September 20, 2000. (Photo by Craig Holman/*Columbus Dispatch*/Reuters NewMedia, Inc./CORBIS/ Bettmann)

Severe thunderstorm outlooks are issued several times daily. *Day 1* outlooks identify those areas likely to be affected by severe thunderstorms during the next 6 to 30 hours, and *day 2* outlooks extend the forecast through the following day. Both outlooks describe the type, coverage, and intensity of the severe weather expected. Many local NWS field offices also issue severe weather outlooks that provide a more local description of the severe weather potential for the next 12 to 24 hours.

Table 14.1 Fujita Intensity Scale

Scale	Wind Speed Km/Hr	Wind Speed Mi/Hr	Damage
F0	<116	<72	Light damage
F1	116–180	72–112	Moderate damage
F2	181–253	113–157	Considerable damage
F3	254–332	158–206	Severe damage
F4	333–419	207–260	Devastating damage
F5	>419	>260	Incredible damage

Tornado Watches and Warnings **Tornado watches** alert the public to the possibility of tornadoes over a specified area for a particular time interval. Watches serve to fine-tune forecast areas already identified in severe weather outlooks. A typical watch covers an area of about 65,000 square kilometers (25,000 square miles) for a four- to six-hour period. A tornado watch is an important part of the tornado alert system because it sets in motion the procedures necessary to deal adequately with detection, tracking, warning, and response. Watches are generally reserved for organized severe weather events where the tornado threat will affect at least 26,000 square kilometers (10,000 square miles) and/or persist for at least three hours. Watches typically are not issued when the threat is thought to be isolated and/or short-lived.

Whereas a tornado watch is designed to alert people to the possibility of tornadoes, a **tornado warning** is issued by local offices of the National Weather Service when a tornado has actually been sighted in an area or is indicated by weather radar. It warns of a high probability of imminent danger. Warnings are issued for much smaller areas than for watches, usually covering portions of a county or counties. In addition, they are in effect for much shorter periods, typically 30 to 60 minutes. Because a tornado warning may be based on an actual sighting, warnings are occasionally issued after a tornado has already developed. However, most warnings are issued prior to tornado formation, sometimes by several tens of minutes, based on Doppler radar data and/or spotter reports of funnel clouds.

If the direction and the approximate speed of the storm are known, an estimate of its most probable path can be made. Because tornadoes often move erratically, the warning area is fan-shaped downwind from the point where the tornado has been spotted. Improved forecasts and advances in technology have contributed to a significant decline in tornado deaths over the past 50 years.

Doppler Radar Many of the difficulties that once limited the accuracy of tornado warnings have been reduced or eliminated by an advancement in radar technology called **Doppler radar.** Doppler radar not only performs the same tasks as conventional radar but also has the ability to detect motion directly (Figure 14.19). Doppler radar can detect the initial formation and subsequent development of a *mesocyclone,* the intense rotating wind system in the lower part of a thunderstorm that frequently precedes tornado development. Almost all mesocyclones produce damaging hail, severe winds, or tornadoes. Those that produce tornadoes (about 50 percent) can sometimes be distinguished by their stronger wind speeds and their sharper gradients of wind speeds.

It should also be pointed out that not all tornado-bearing storms have clear-cut radar signatures and that other storms can give false signatures. Detection, therefore, is sometimes a subjective process and a given display could be interpreted in several ways. Consequently, trained observers will continue to form an important part of the warning system in the foreseeable future.

Although some operational problems exist, the benefits of Doppler radar are many. As a research tool, it is not only providing data on the formation of tornadoes but is also helping meteorologists gain new insights into thunderstorm development, the structure and dynamics of hurricanes, and air-turbulence hazards that plague aircraft. As a practical tool for tornado detection, Doppler radar has significantly improved our ability to track thunderstorms and issue warnings.

Hurricanes

Most of us view the weather in the tropics with favor. Places such as Hawaii and the islands of the Caribbean are known for their lack of significant day-to-day variations. Warm breezes, steady temperatures, and rains that come as heavy but brief tropical showers are expected. It is ironic that these relatively tranquil regions sometimes produce the most violent storms on Earth.

The whirling tropical cyclones that on occasion have wind speeds attaining 300 kilometers (185 miles) per hour are known in the United States as **hurricanes**—the greatest storms on Earth. Out at sea, they can generate 15-meter (50-foot) waves capable of inflicting destruction hundreds of kilometers from their source. Should a hurricane smash onto land, strong winds coupled with extensive flooding can impose billions of dollars in damage and great loss of life (Figure 14.20 and Figure 14.1B, p. 370).

The vast majority of hurricane-related deaths and damage are caused by relatively infrequent, yet powerful, storms. A storm that pounded an unsuspecting Galveston, Texas, in 1900 was responsible for at least 8000 deaths. It was not just the deadliest U.S. hurricane ever, but the deadliest natural disaster of *any kind* to affect the United States. Of course, the deadliest and most costly storm in recent memory occurred in August 2005, when Hurricane Katrina devastated the Gulf Coast of Louisiana, Mississippi, and Alabama and took an estimated 1300 lives. Although hundreds of thousands fled before the storm made landfall, thousands of others were caught by the storm. In addition to the human suffering and tragic loss of life that were left in the wake of Hurricane Katrina, the financial losses caused by the storm are practically incalculable. Some suggest that the final accounting could reach or exceed $300 billion.

The devastating hurricane season of 2005 came on the heels of an extraordinary 2004 season. Storms claimed more than 3100 lives (mostly in Haiti), and property losses in the United States topped $44 billion. The most notable storms were hurricanes Charley, Frances, Ivan, and Jeanne, all of which struck the state of Florida in August and September. Three of the storm tracks (Charley, Frances, and Jeanne) intersected in central Florida, whereas Hurricane Ivan struck the state's panhandle as it made landfall along the Gulf Coast (Figure 14.21). The Bahamas and many Caribbean islands were also hit hard. Each hurricane season, storms like those just described remind us of our vulnerability to these great forces of nature.

Profile of a Hurricane

Most hurricanes form between the latitudes of 5 and 20 degrees over all the tropical oceans except the South Atlantic and the eastern South Pacific. The North Pacific has the greatest number of storms, averaging 20 each year. Fortunately for those living in the coastal regions of the southern and eastern

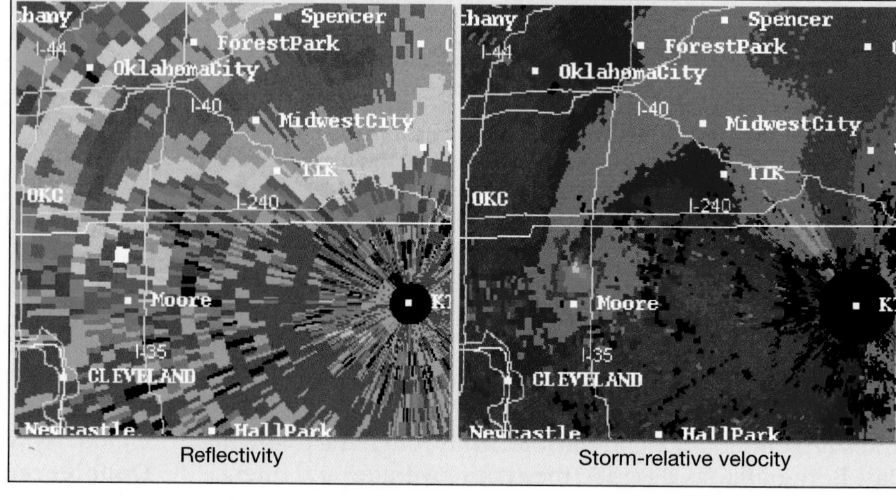

Figure 14.19 This is a dual Doppler radar image of an F5 tornado near Moore, Oklahoma, on May 3, 1999. The left image (reflectivity) shows precipitation in the supercell thunderstorm. The right image shows motion of the precipitation along the radar beam, that is, how fast rain or hail is moving toward or away from the radar. In this example, the radar was unusually close to the tornado—close enough to make out the signature of the tornado itself (most of the time only the weaker and larger mesocyclone is detected). (After NOAA)

United States, fewer than five hurricanes, on the average, develop annually in the warm sector of the North Atlantic.

These intense tropical storms are known in various parts of the world by different names. In the western Pacific, they are called *typhoons,* and in the Indian Ocean, including the Bay of Bengal and Arabian Sea, they are simply called *cyclones.* In the following discussion, these storms will be referred to as hurricanes. The term *hurricane* is derived from Huracan, a Carib god of evil.

Although many tropical disturbances develop each year, only a few reach hurricane status. By international agreement, a hurricane has wind speeds in excess of 119 kilometers (74 miles) per hour and a rotary circulation. Mature hurricanes average 600 kilometers (375 miles) across, although they can range in diameter from 100 kilometers (60 miles) up to about 1500 kilometers (930 miles). From the outer edge to the center, the barometric pressure has on occasion dropped 60 millibars, from 1010 millibars to 950 millibars. The lowest pressures ever recorded in the Western Hemisphere are associated with these storms.

A steep pressure gradient generates the rapid, inward-spiraling winds of a hurricane. As the air rushes toward the center of the storm, its velocity increases. This occurs for the same reason that skaters with their arms extended spin faster as they pull their arms in close to their bodies.

As the inward rush of warm, moist surface air approaches the core of the storm, it turns upward and ascends in a ring of cumulonimbus towers (Figure 14.22). This doughnut-shaped wall of intense convective activity surrounding the center of the storm is called the **eye wall.** It is here that the greatest wind speeds and heaviest rainfall occur. Surrounding the eye wall are curved bands of clouds that trail away in a spiral fashion. Near the top of the hurricane, the airflow is outward, carrying the rising air away from the storm center, thereby providing room for more inward flow at the surface.

At the very center of the storm is the **eye** of the hurricane (Figure 14.22). This well-known feature is a zone where precipitation ceases and winds subside. It offers a brief but deceptive break from the extreme weather in the enormous curving wall clouds that surround it. The air within the eye gradually descends and heats by compression, making it the warmest part of the storm. Although many people believe that the eye is characterized by clear blue skies, this is usually not the case, because the subsidence in the eye is seldom

A.

B.

Figure 14.20 Storm surge damage along the Mississippi coast caused by Hurricane Katrina. At some locations, the surge exceeded 7.5 meters (25 feet). **A.** This concrete slab is all that remains of an apartment house in Biloxi. (Photo by Barry Williams/Getty Images) **B.** Houses and vehicles litter the railway near Pass Christian. (Phil Coale/AP Photo)

Did You Know?

Hurricane season is different in different parts of the world. The Atlantic hurricane season officially extends from June through November. More than 97 percent of tropical activity in that region occurs during this six-month span. The "heart" of the season is August through October. During these three months, 87 percent of the minor hurricane (categories 1 and 2) days and 96 of the major hurricane (categories 3, 4, and 5) days occur. Peak activity is in early to mid-September.

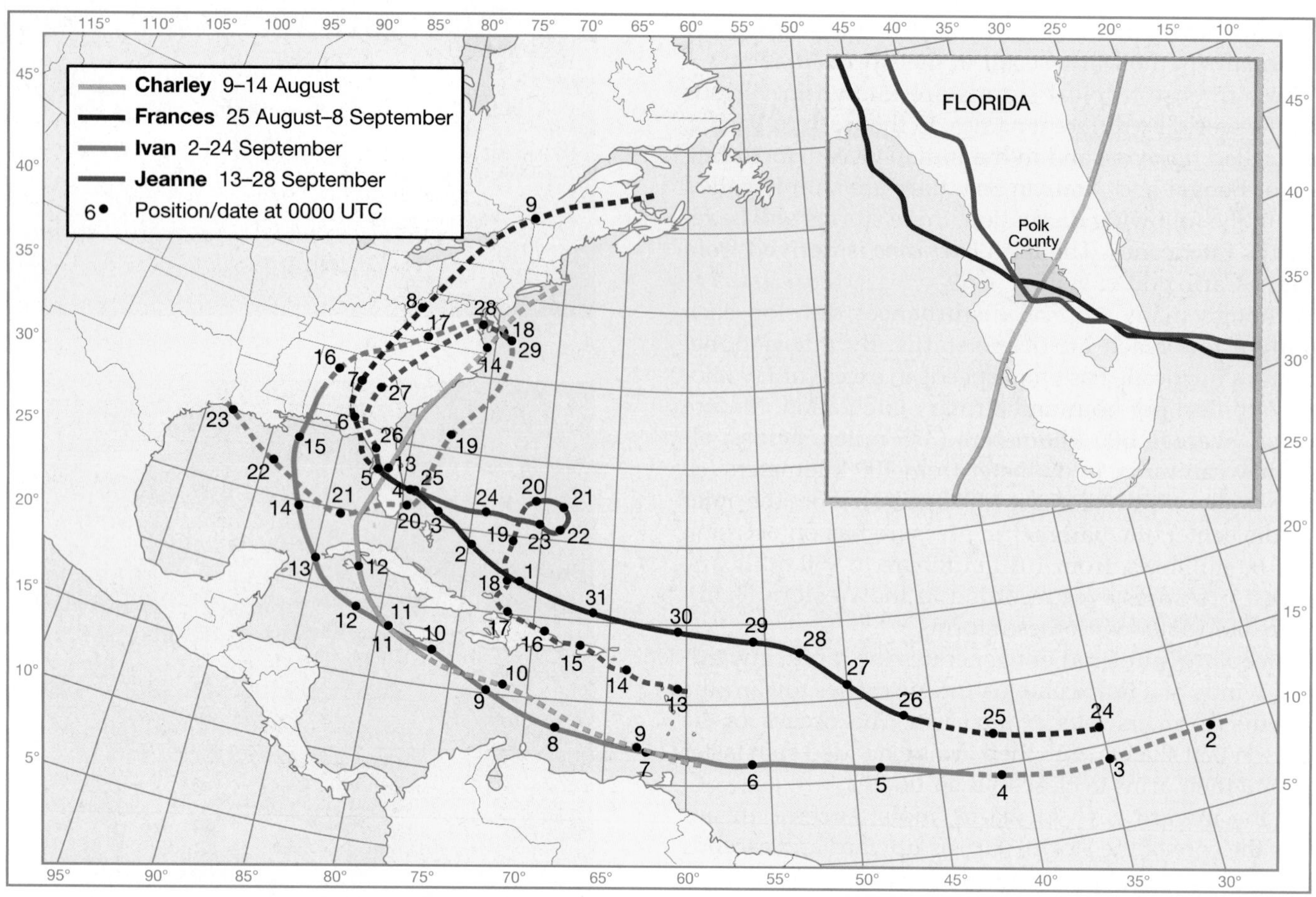

Figure 14.21 Tracks of the four major hurricanes to hit Florida in 2004. Hurricanes Charley, Frances, and Jeanne intersected over Polk County, Florida (see inset). Ivan made landfall near Mobile, Alabama, but greatly affected Florida's western panhandle. Each solid line shows the path of a storm when it had hurricane status. The dashed portion of each line traces the storm's path before and after hurricane stage. (After National Hurricane Center)

strong enough to produce cloudless conditions. Although the sky appears much brighter in this region, scattered clouds at various levels are common.

Hurricane Formation and Decay

A hurricane is a heat engine that is fueled by the latent heat liberated when huge quantities of water vapor condense. The amount of energy produced by a typical hurricane in just a single day is truly immense. The release of latent heat warms the air and provides buoyancy for its upward flight. The result is to reduce the pressure near the surface, which encourages a more rapid inward flow of air. To get this engine started, a large quantity of warm, moisture-laden air is required, and a continual supply is needed to keep it going.

Hurricanes develop most often in the late summer when ocean waters have reached temperatures of 27°C (80°F) or higher and thus are able to provide the necessary heat and moisture to the air. This ocean-water temperature requirement accounts for the fact that hurricanes do not form over the relatively cool waters of the South Atlantic and the eastern South Pacific. For the same reason, few hurricanes form poleward of 20 degrees of latitude. Although water temperatures are sufficiently high, hurricanes do not form within 5 degrees of the equator, because the Coriolis effect is too weak to initiate the necessary rotary motion.

Many tropical storms begin as disorganized arrays of clouds and thunderstorms that develop weak pressure gradients but exhibit little or no rotation. Such areas of low-level convergence and lifting are called *tropical disturbances*. Most of the time, these zones of convective activity die out. However, tropical disturbances occasionally grow larger and develop a strong cyclonic rotation.

What happens on those occasions when conditions favor hurricane development? As latent heat is released from the clusters of thunderstorms that make up the tropical disturbance, areas within the disturbance get warmer. As a consequence, air density lowers and surface pressure drops, creating a region of weak low pressure and cyclonic circulation. As pressure drops at the storm center, the pressure gradient steepens. If you were watching an animated weather map of the storm, you would see the isobars get closer together. In response, surface wind speeds increase and bring additional supplies of moisture to nurture storm growth. The water vapor condenses, releasing latent heat, and the heated air rises. Adiabatic cooling of rising air triggers more conden-

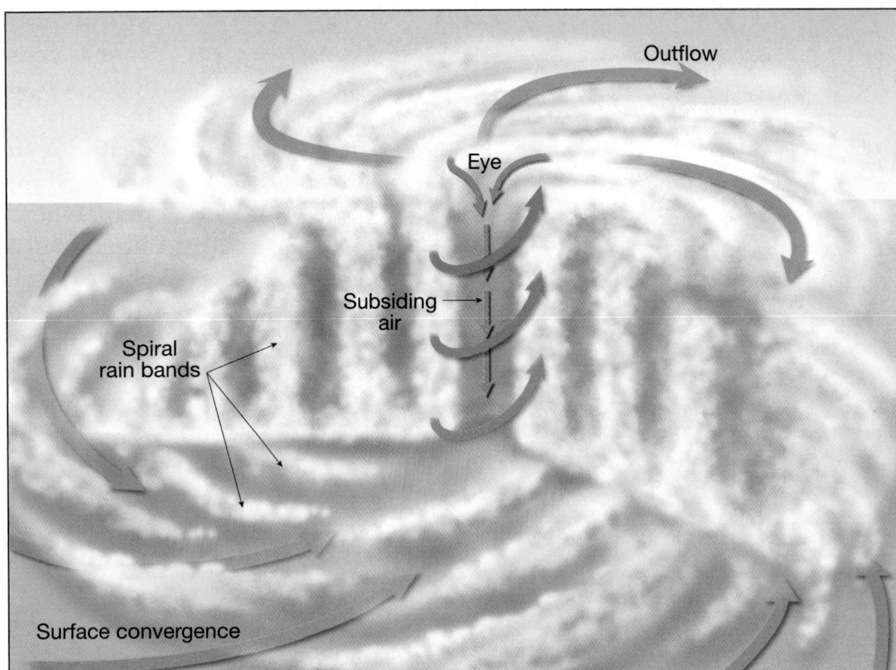

Figure 14.22 Cross section of a hurricane. Note that the vertical dimension is greatly exaggerated. The eye, the zone of relative calm at the center of the storm, is a distinctive hurricane feature. Sinking air in the eye warms by compression. Surrounding the eye is the eye wall, the zone where winds and rain are most intense. Tropical moisture spiraling inward creates rain bands that pinwheel around the storm center. Outflow of air at the top of the hurricane is important because it prevents the convergent flow at lower levels from filling in the storm. (After NOAA)

sation and the release of more latent heat, which causes a further increase in buoyancy. And so it goes.

Meanwhile, at the top of the storm, air is diverging. Without this outward flow up top, the inflow at lower levels would soon raise surface pressures (that is, fill in the low) and thwart storm development.

Many tropical disturbances occur each year, but only a few develop into full-fledged hurricanes. By international agreement, lesser tropical cyclones are placed in different categories, based on wind strength. When a cyclone's strongest winds do not exceed 61 kilometers (38 miles) per hour, it is called a **tropical depression.** When winds are between 61 and 119 kilometers (38 and 74 miles) per hour, the cyclone is termed a **tropical storm.** It is during this phase that a name is given (Andrew, Floyd, Opal, etc.). Should the tropical storm become a hurricane, the name remains the same. Each year, between 80 and 100 tropical storms develop around the world. Of these, usually half or more eventually become hurricanes.

Hurricanes diminish in intensity whenever they (1) move over ocean waters that cannot supply warm, moist tropical air; (2) move onto land; or (3) reach a location where the large-scale flow aloft is unfavorable. Whenever a hurricane moves onto land, it loses its punch rapidly. The most important reason for this rapid demise is the fact that the storm's source of warm, moist air is cut off. When an adequate supply of water vapor does not exist, condensation and the release of latent heat must diminish. In addition, friction from the increased roughness of the land surface rapidly slows surface wind speeds. This factor causes the winds to move more directly into the center of the low, thus helping to eliminate the large pressure differences.

Hurricane Destruction

A location only a few hundred kilometers from a hurricane—just one day's striking distance away—may experience clear skies and virtually no wind. Prior to the age of weather satellites, such a situation made the task of warning people of impending storms very difficult (Figure 14.23).

Although the amount of damage caused by a hurricane depends on several factors, including the size and population density of the area affected and the shape of the ocean bottom near the shore, the most significant factor is the strength of the storm itself.

Based on the study of past storms, the **Saffir-Simpson scale** was established to rank the relative intensities of hurricanes (Table 14.2). Predictions of hurricane severity and damage are usually expressed in terms of this scale. When a tropical storm becomes a hurricane, the National Weather Service assigns it a scale (category) number. Category as-

Did You Know?

If the entire alphabetical list of names for hurricanes and tropical storms for a particular year is exhausted, the naming system moves on to using letters of the Greek alphabet (alpha, beta, gamma, etc.). This situation never arose until the record-breaking 2005 hurricane season when Tropical Storm Alpha, Hurricane Beta, tropical storms Gamma and Delta, Hurricane Epsilon, and Tropical Storm Zeta occurred after Hurricane Wilma.

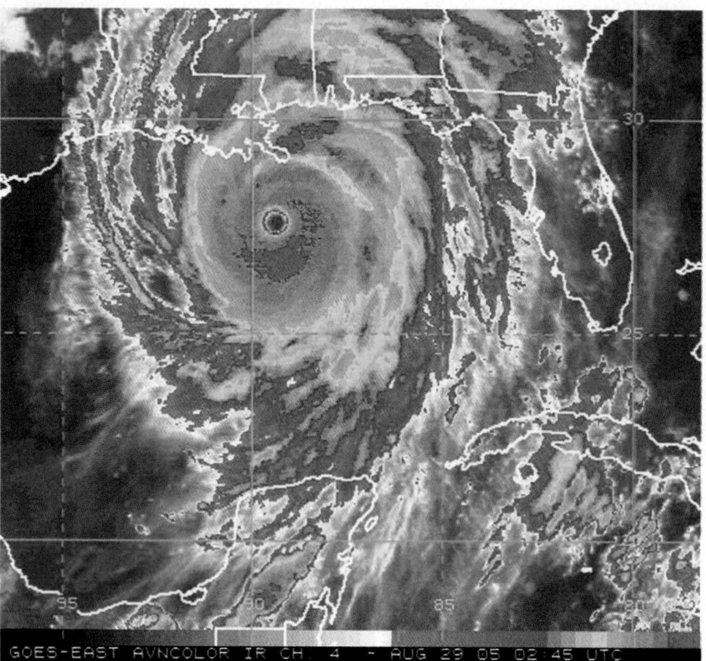

Figure 14.23 Satellites are invaluable tools for tracking storms and gathering atmospheric data. This color-enhanced infrared image from the *GOES-East* satellite shows Hurricane Katrina several hours before making landfall on August 29, 2005. The most intense activity is associated with the colors red and orange. (NOAA)

signments are based on observed conditions at a particular stage in the life of hurricane and are viewed as estimates of the amount of damage a storm would cause if it were to make landfall without changing size or strength. As conditions change, the category of a storm is reevaluated, so that public-safety officials can be kept informed. By using the Saffir-Simpson scale, the disaster potential of a hurricane can be monitored and appropriate precautions can be planned and implemented.

A rating of 5 on the scale represents the worst storm possible, and a 1 is least severe. Storms that fall into category 5 are rare. Only three storms this powerful are known to have hit the continental United States: Andrew struck Florida in 1992, Camille pounded Mississippi in 1969, and a Labor Day hurricane struck the Florida Keys in 1935. Damage caused by hurricanes can be divided into three categories: (1) wind damage, (2) storm surge, and (3) inland freshwater flooding.

Table 14.2 Saffir-Simpson Hurricane Scale

Scale Number (category)	Central Pressure (millibars)	Winds (km/hr)	Storm Surge (meters)	Damage
1	≥980	119–153	1.2–1.5	Minimal
2	965–979	154–177	1.6–2.4	Moderate
3	945–964	178–209	2.5–3.6	Extensive
4	920–944	210–250	3.7–5.4	Extreme
5	<920	>250	>5.4	Catastrophic

Storm Surge The most devastating damage in the coastal zone is caused by the storm surge (see Figure 14.20, p. 387). It not only accounts for a large share of coastal property losses but is also responsible for 90 percent of all hurricane-caused deaths. A **storm surge** is a dome of water 65 to 80 kilometers (40 to 50 miles) wide that sweeps across the coast near the point where the eye makes landfall. If all wave activity were smoothed out, the storm surge is the height of the water above normal tide level. Thus, a storm surge commonly adds 2 to 3 meters (6 to 10 feet) to normal tide heights—to say nothing of the tremendous wave activity superimposed atop the surge.

In the delta region of Bangladesh, for example, the land is mostly less than 2 meters (6 feet) above sea level. When a storm surge superimposed upon normal high tide inundated that area on November 13, 1970, the official death toll was 200,000; unofficial estimates ran to 500,000. This was one of the worst disasters of modern times. In May 1991, a similar event again struck Bangladesh. This time, the storm took the lives of at least 135,000 people and devastated coastal villages in its path.

The most important factor responsible for the development of a storm surge is the piling up of ocean water by strong onshore winds The hurricane's winds gradually push water toward the shore, causing sea level to elevate while also churning up violent wave activity.

As a hurricane advances toward the coast in the Northern Hemisphere, storm surge is always most intense on the right side of the eye where winds are blowing *toward* the shore. In addition, on this side of the storm, the forward movement of the hurricane also contributes to the storm surge. In Figure 14.24, assume a hurricane with peak winds of 175 kilometers (109 miles) per hour is moving toward the shore at 50 kilometers (31 miles) per hour. The net wind speed on the right side of the advancing storm is 225 kilometers (140 miles) per hour. On the left side, the hurricane's winds are blowing opposite the direction of storm movement, so the net winds are blowing *away* from the coast at 125 kilometers (78 miles) per hour. Along the shore facing the left side of the oncoming hurricane, the water level may actually decrease as the storm makes landfall.

Wind Damage Destruction caused by wind is perhaps the most obvious of the classes of hurricane damage. For some structures, the force of the wind is sufficient to cause total ruin. Mobile homes are particularly vulnerable. In addition, the strong winds can create a dangerous barrage of flying debris. In regions with good building codes, wind damage is usually not as catastrophic as storm-surge damage. However, hurricane-force winds affect a much larger area than does storm surge and can cause huge economic losses. For example, winds associated with Hurricane Andrew in 1992 produced more than $25 billion of damage in southern Florida and Louisiana.

Hurricanes sometimes produce tornadoes that contribute to the storm's destructive power. Studies have shown that more than half of the hurricanes that make landfall produce at least one tornado. In 2004, the number of tornadoes

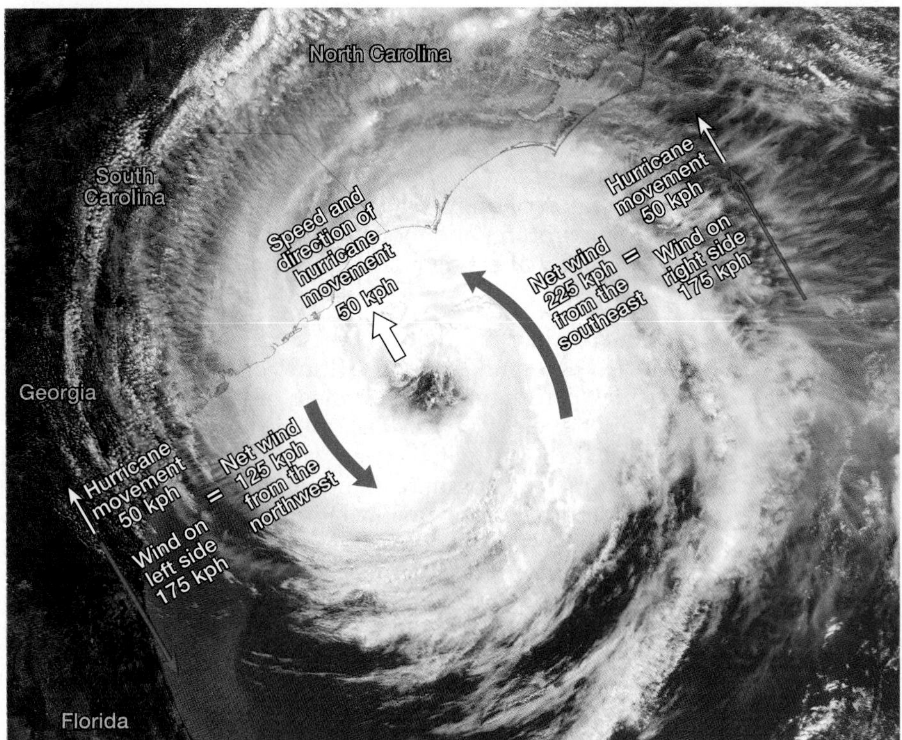

Figure 14.24 Winds associated with a Northern Hemisphere hurricane that is advancing toward the coast. This hypothetical storm, with peak winds of 175 kilometers (109 miles) per hour, is moving toward the coast at 50 kilometers (31 miles) per hour. On the right side of the advancing storm, the 175-kilometer-per-hour winds are in the same direction as the movement of the storm (50 kilometers per hour). Therefore, the *net* wind speed on the right side of the storm is 225 kilometers (140 miles) per hour. On the left side, the hurricane's winds are blowing opposite the direction of storm movement, so the *net* winds of 125 kilometers (78 miles) per hour are away from the coast. Storm surge will be greatest along that part of the coast hit by the right side of the advancing hurricane.

associated with tropical storms and hurricanes was extraordinary. Tropical Storm Bonnie and five landfalling hurricanes—Charley, Frances, Gaston, Ivan, and Jeanne—produced nearly 300 tornadoes that affected the southeast and mid-Atlantic states. Hurricane Frances produced 117 tornadoes—the most ever reported from one hurricane. The large number of hurricane-generated tornadoes in 2004 helped make this a record-breaking year—surpassing the previous record by more than 300.*

Inland Flooding The torrential rains that accompany most hurricanes represent a third significant threat—flooding. While the effects of storm surge and strong winds are concentrated in coastal areas, heavy rains may affect places hundreds of kilometers from the coast for several days after the storm has lost its hurricane-force winds.

Hurricanes weaken rapidly as they move inland, yet the remnants of the storm can still yield 15 to 30 centimeters (6 to 12 inches) or more of rain as they move inland. A good example of such destruction was Hurricane Floyd. In September 1999, this storm dumped more than 48 centimeters (19 inches) of rain on Wilmington, North Carolina—33.98 centimeters (13.38 inches) of it in a single 24-hour span. In August and September 2004, when the remnants of hurricanes Charley, Frances, Ivan, and Jeanne moved northward from Florida and the Gulf Coast, they brought huge rains and floods from Alabama and Georgia through the Carolinas and beyond.

To summarize, extensive damage and loss of life in the coastal zone can result from storm surge, torrential rains, and strong winds. When loss of life occurs, it is commonly caused by the storm surge, which can devastate entire barrier islands and low-lying land along the coast. Although wind damage is usually not as catastrophic as that caused by the storm surge, it affects a much larger area. Where building codes are inadequate, economic losses can be especially severe. Because hurricanes weaken as they move inland, most wind damage occurs within 200 kilometers (125 miles) of the coast. Far from the coast, a weakening storm can produce extensive flooding long after the winds have diminished below hurricane levels. Sometimes the damage from inland flooding exceeds storm-surge destruction.

*K. L. Gleason, et al. "U.S. Tornado Records" in *Bulletin of the American Meteorological Society,* Vol. 86, No. 6, June 2005, p. 551.

The Chapter in Review

1. An *air mass* is a large body of air, usually 1600 kilometers (1000 miles) or more across, that is characterized by a *sameness of temperature and moisture at any given altitude.* When this air moves out of its region of origin, called the *source region,* it will carry these temperatures and moisture conditions elsewhere, perhaps eventually affecting a large portion of a continent.

2. Air masses are classified according to (1) the nature of the surface in the source region and (2) the latitude of the source region. *Continental (c)* designates an air mass of land origin, with the air likely to be dry; *a maritime (m)* air mass originates over water and, therefore, will be relatively humid. *Polar (P)* and *arctic (A)* air masses originate in high latitudes and are cold. *Tropical (T)* air masses form in low latitudes and are warm. According to this classification scheme, the *four basic types of air masses are continental polar (cP), continental tropical (cT), maritime polar (mP), and maritime tropical (mT).* Continental polar (cP) and maritime tropical (mT) air masses influence the weather of North America most, especially east of the Rocky Mountains. Maritime tropical air is the source of much, if not most, of the precipitation received in the eastern two-thirds of the United States.

3. *Fronts* are boundaries that separate air masses of different densities, one warmer and often higher in moisture content than the other. A *warm front* occurs when the surface position of the front moves so that warm air occupies territory formerly covered by cooler air. Along a warm front, a warm air mass overrides a retreating mass of cooler air. As the warm air ascends, it cools adiabatically to produce clouds and frequently light-to-moderate precipitation over a large area. A *cold front* forms where cold air is actively advancing into a region occupied by warmer air. Cold fronts are about twice as steep and move more rapidly than do warm fronts. Because of these two differences, precipitation along a cold front is generally more intense and of shorter duration than precipitation associated with a warm front.

4. The primary weather producers in the middle latitudes are *large centers of low pressure* that generally travel from *west to east,* called *middle-latitude cyclones.* These *bearers of stormy weather,* which last from a few days to a week, have a *counterclockwise circulation* pattern in the Northern Hemisphere, with an *inward flow of air* toward their centers. Most middle-latitude cyclones have a *cold front and frequently a warm front* extending from the central area of low pressure. *Convergence* and *forceful lifting along the fronts* initiate cloud development and frequently cause precipitation. The particular weather experienced by an area depends on the path of the cyclone.

5. *Thunderstorms* are caused by the upward movement of warm, moist, unstable air. They are associated with cumulonimbus clouds that generate heavy rainfall, lightning, thunder, and occasionally hail and tornadoes.

6. *Tornadoes,* destructive local storms of short duration, are violent windstorms associated with severe thunderstorms that take the form of a rotating column of air that extends downward from a cumulonimbus cloud. Tornadoes are most often spawned along the cold front of a middle-latitude cyclone, most frequently during the spring months.

7. *Hurricanes,* the greatest storms on Earth, are tropical cyclones with wind speeds in excess of 119 kilometers (74 miles) per hour. These complex tropical disturbances develop over tropical ocean waters and are fueled by the latent heat liberated when huge quantities of water vapor condense. Hurricanes form most often in late summer when ocean-surface temperatures reach 27°C (80°F) or higher and thus are able to provide the necessary heat and moisture to air. Hurricanes diminish in intensity whenever they (1) move over cool ocean water that cannot supply adequate heat and moisture, (2) move onto land, or (3) reach a location where large-scale flow aloft is unfavorable. Hurricane damage is of three main types: (1) storm surge, (2) wind damage, and (3) inland flooding.

Key Terms

air mass (p. 370)
air-mass weather (p. 371)
arctic (A) air mass (p. 371)
cold front (p. 374)
continental (c) air mass (p. 372)
Doppler radar (p. 385)
eye (p. 387)
eye wall (p. 387)
front (p. 373)
hurricane (p. 386)
lake-effect snow (p. 372)
maritime (m) air mass (p. 372)
middle-latitude or midlatitude cyclone (p. 375)
occluded front (p. 375)
overrunning (p. 373)
polar (P) air mass (p. 371)
Saffir–Simpson scale (p. 389)
source region (p. 371)
stationary front (p. 375)
storm surge (p. 390)
thunderstorm (p. 379)
tornado (p. 381)
tornado warning (p. 385)
tornado watch (p. 385)
tropical (T) air mass (p. 371)
tropical depression (p. 389)
tropical storm (p. 389)
warm front (p. 373)

Questions for Review

1. Describe the weather associated with a continental polar air mass in the winter and in the summer. When would this air mass be most welcome in the United States?
2. What are the characteristics of a maritime tropical air mass? Where are the source regions for the maritime tropical air masses that affect North America? Where are the source regions for the maritime polar air masses?
3. Describe the weather along a cold front where very warm, moist air is being displaced.
4. For each weather element listed, describe the changes that an observer experiences when a middle-latitude cyclone passes with its center north of the observer: wind direction, pressure tendency, cloud type, cloud cover, precipitation, temperature.
5. Describe the weather conditions an observer would experience if the center of a middle-latitude cyclone passed to the south.
6. Briefly explain how the flow aloft aids the formation of cyclones at the surface.
7. What is the primary requirement for the formation of thunderstorms?
8. Based on your answer to Question 7, where would you expect thunderstorms to be most common on Earth? Where specifically in the United States?
9. Why do tornadoes have such high wind speeds?
10. What general atmospheric conditions are most conducive to the formation of tornadoes?
11. When is "tornado season"? That is, during what months is tornado activity most pronounced?
12. Distinguish between a *tornado watch* and a *tornado warning.*
13. Which has stronger winds, a tropical storm or a tropical depression?
14. Why does the intensity of a hurricane diminish rapidly when it moves onto land?
15. Name the three broad categories of hurricane damage. Which category is responsible for the highest percentage of hurricane-related deaths?
16. A hurricane has slower wind speeds than does a tornado, yet it inflicts more total damage. How might this be explained?

Online Study Guide

The *Foundations of Earth Science* Web site uses the resources and flexibility of the Internet to aid in your study of the topics in this chapter. Written and developed by Earth science instructors, this site will help improve your understanding of Earth science. Visit http://www.prenhall.com/lutgens and click on the cover of *Foundations of Earth Science 5e* to find:

- Online review quizzes.
- Critical thinking exercises.
- Links to chapter-specific Web resources.
- Internet-wide key-term searches.

http://www.prenhall.com/lutgens

GEODe: Earth Science

GEODe: Earth Science makes studying more effective by reinforcing key concepts using animation, video, narration, interactive exercises, and practice quizzes. A copy is included with every copy of *Foundations of Earth Science 5e.*

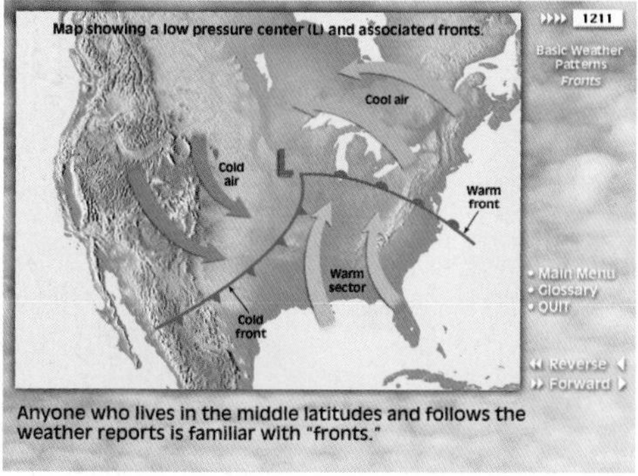

UNIT VII EARTH'S PLACE IN THE UNIVERSE

CHAPTER 15

The Nature of the Solar System

FOCUS ON LEARNING

To assist you in learning the important concepts in this chapter, you will find it helpful to focus on the following questions:

1. What was the geocentric theory of the universe held by many early Greeks?
2. What were the contributions to modern astronomy of Nicolaus Copernicus, Tycho Brahe, Johannes Kepler, Galileo Galilei, and Isaac Newton?
3. What are the two groups of planets in the solar system? What are the general characteristics of the planets in each group?
4. How is the solar system thought to have formed?
5. What are the major features of the lunar surface?
6. What are some distinguishing features of each planet in the solar system?
7. What are the minor members of the solar system?

Time-lapse photograph of the night sky over the Keck Telescope on Hawaii's Mauna Kea volcano. (Photo by Roger Ressmeyer/CORBIS)

Earth is one of eight planets and numerous smaller bodies that orbit the Sun. The Sun is part of a much larger family of perhaps 100 billion stars that comprise the Milky Way, which in turn is only one of billions of galaxies in an incomprehensibly large universe. This view of Earth's position in space is considerably different from that held only a few hundred years ago, when our planet was thought to occupy a privileged position as the center of the universe. This chapter will trace the fascinating events that led to modern astronomy. In addition, it will examine the formation and structure of the solar system.

Ancient Astronomy

Long before recorded history, which began about 5000 years ago, people were aware of the close relationship between events on Earth and the positions of heavenly bodies, the Sun in particular. People noted that changes in the seasons and floods of great rivers such as the Nile in Egypt occurred when the celestial bodies, including the Sun, Moon, planets, and stars, reached a particular place in the heavens. Early agrarian cultures, which were dependent on the weather, believed that if the heavenly objects could control the seasons, they must also strongly influence all Earthly events. This belief undoubtedly was the reason why early civilizations began keeping records of the positions of celestial objects (Figure 15.1). The Chinese, Egyptians, and Babylonians in particular are noted for this.

A study of Chinese archives shows that the Chinese recorded every appearance of the famous Halley's comet for at least 10 centuries. However, because this comet appears only once in a lifetime—once every 76 years—they were unable to link these appearances to establish that what they saw was the same object each time. Thus, like most ancients, the Chinese considered comets to be mystical. Comets were seen as bad omens and were blamed for a variety of disasters, from wars to plagues (Figure 15.2).

Early Greeks

The "Golden Age" of early astronomy (600 B.C.–A.D. 150) was centered in Greece. The early Greeks have been criticized, and rightly so, for using philosophical arguments to explain natural phenomena. However, they did rely on observational data as well. The basics of geometry and trigonometry, which they had developed, were used to measure the sizes and distances of the largest-appearing bodies in the heavens—the Sun and the Moon.

Many astronomical discoveries have been credited to the Greeks. They held the **geocentric** ("Earth-centered") view, believing that Earth was a sphere that stayed motionless at the center of the universe. Orbiting Earth were the Moon, the Sun, and the known planets—Mercury, Venus, Mars, Jupiter, and Saturn. Beyond the planets was a transparent, hollow **celestial sphere,** on which the stars traveled daily around Earth (this is how it looks, but, of course, the effect is actually caused by Earth's rotation about its axis). Some early Greeks realized that the motion of the stars could be explained just as easily by a rotating Earth, but they rejected that idea, because Earth exhibited no sense of motion and seemed too large to be movable. In fact, proof of Earth's rotation was not demonstrated until 1851.

To the Greeks, all of the heavenly bodies, except seven, appeared to remain in the same relative position to one another. These seven "wanderers" (*planetai* in Greek) included the Sun, the Moon, Mercury, Venus, Mars, Jupiter, and Saturn. Each was thought to have a circular orbit around Earth. Although this system was incorrect, the Greeks refined it to the point that it explained the apparent movements of all celestial bodies.

Although many of the Greek discoveries were lost during the Middle Ages, the Earth-centered view that the Greeks proposed became established in Europe. Presented in its finest form by Claudius Ptolemy, this geocentric outlook became known as the **Ptolemaic system.**

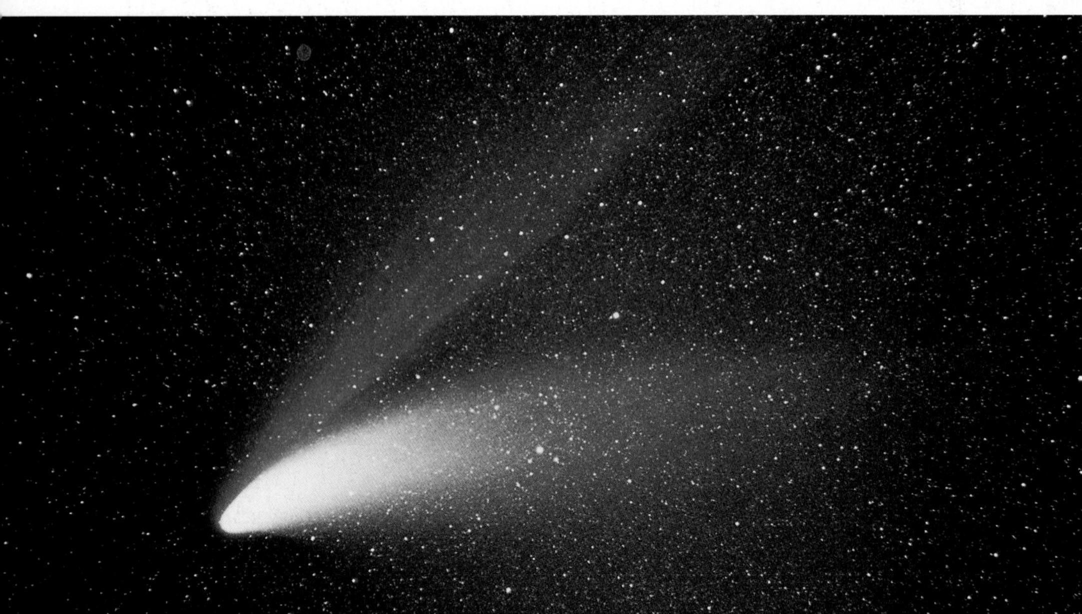

Figure 15.1 Comet Hale-Bopp. The two tails are about 16 to 24 million kilometers (10 to 15 million miles) long. (Peoria Astronomical Society, Inc., photograph by Eric Clifton and Greg Neaveill)

Figure 15.2 The Bayeux Tapestry that hangs in Bayeux, France, shows the apprehension caused by Halley's comet in A.D. 1066. This event preceded the defeat of King Harold by William the Conqueror. ("Sighting of a comet." Detail from Bayeux Tapestry. Musée de la Tapisserie, Bayeux. "With special authorization of the City of Bayeux." Bridgeman-Giraudon/Art Resources, NY)

Ptolemy's Model

Much of our knowledge of Greek astronomy comes from a 13-volume treatise, *Almagest* ("the great work"), which was compiled by Ptolemy in A.D. 141 and which survived thanks to the work of Arab scholars. In this work, Ptolemy is credited with developing a model of the universe that accounted for the observable motions of the planets (Figure 15.3). The precision with which his model was able to predict planetary motion is attested to by the fact that it went virtually unchallenged, in principle if not in detail, for nearly 13 centuries.

Figure 15.3 View of the universe according to Ptolemy, second-century A.D. **A.** Ptolemy believed that the star-studded celestial sphere made a daily trip around a motionless Earth. In addition, he proposed that the Sun, Moon, and planets made trips of various lengths along individual orbits. **B.** Retrograde motion as explained by Ptolemy.

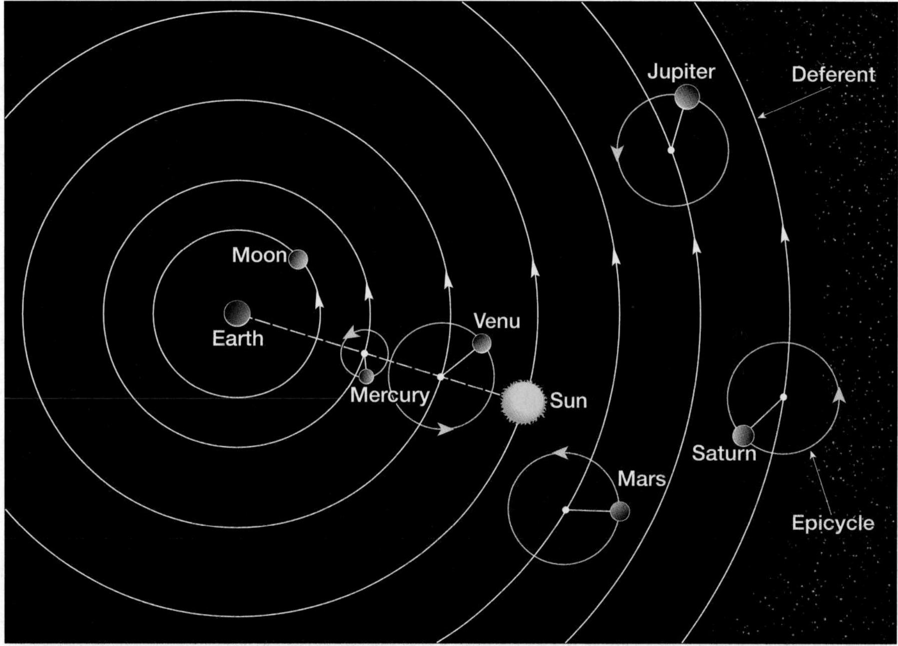

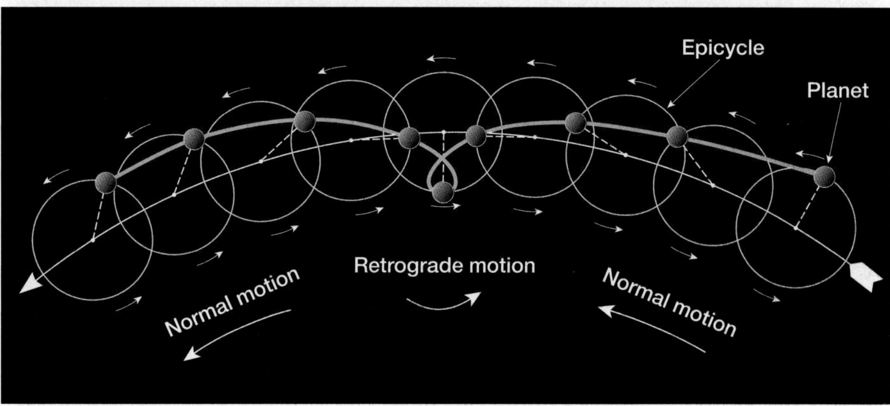

In the Greek tradition, the Ptolemaic model had the planets moving in circular orbits around a motionless Earth. (The Greeks considered the circle to be the pure and perfect shape.) However, the motion of the planets, as seen against the background of stars, is not so simple. Each planet, if watched night after night, moves slightly eastward among the stars. Periodically, each planet appears to stop, reverse direction for a period of time, and then resume an eastward motion. The apparent westward drift is called **retrograde motion.** This rather odd apparent motion results from the combination of the motion of Earth and the planet's own motion around the Sun.

The retrograde motion of Mars is shown in Figure 15.4. Earth has a faster orbital speed than does Mars, so it overtakes its neighbor. While doing so, Mars *appears* to be moving backward, in retrograde motion. This is analogous to what a race-car driver sees out the side window when passing a slower car. The slower planet, like the slower car, appears to be going backward, although its actual motion is in the same direction as the faster-moving body.

It is much more difficult to represent retrograde motion accurately using the incorrect Earth-centered model, but Ptolemy was able to do so (Figure 15.3B). Rather than using a simple circle for each planet's orbit, he showed that the planets orbited on small circles (*epicycles*), revolving along large circles (*deferents*). By trial and error, he found the right combinations of circles to produce the amount of retrograde motion observed for each planet. (An interesting note is that almost any closed curve can be produced by the combination of two circular motions, a fact that can be verified by anyone who has used the Spirograph © design-drawing toy.)

It is a tribute to Ptolemy's genius that he was able to account for the planets' motions as well as he did, considering that he used an incorrect model. Some suggest that he did not mean his model to represent reality, but only to be used for calculating the positions of the heavenly bodies. We will probably never know his intentions. However, the Roman Catholic Church, which dominated European thought for centuries, accepted Ptolemy's theory as the correct representation of the heavens, and this created problems for those who found fault with it.

Did You Know?

According to the Ptolemaic (Earth-centered) model of the universe, all of the heavenly bodies revolved around Earth. When Galileo, using a crude telescope, saw four moons revolving around Jupiter, he knew that Earth was not the center of all motion. Consequently, at least one of the basic tenets of the Ptolemaic model had to be incorrect. Astronomers soon demonstrated that the other tenets of the Earth-centered model were also inconsistent with observations.

The Birth of Modern Astronomy

Modern astronomy was not born overnight. Its development involved a break from deeply entrenched philosophical and religious views that was brought about by the discovery of a new and greater universe governed by discernible laws. Let us now look at the work of five noted scientists involved in this transition: Nicolaus Copernicus, Tycho Brahe, Johannes Kepler, Galileo Galilei, and Sir Isaac Newton.

Nicolaus Copernicus

For almost 13 centuries after the time of Ptolemy, very few astronomical advances were made in Europe. The first great astronomer to emerge after the Middle Ages was Nicolaus Copernicus (1473–1543) from Poland (Figure 15.5). Copernicus became convinced that Earth is a planet, just like the other five then-known planets. The daily motions of the heavens, he reasoned, could be better explained by a rotating Earth.

Having concluded that Earth is a planet, Copernicus reconstructed the solar system with the Sun at the center and

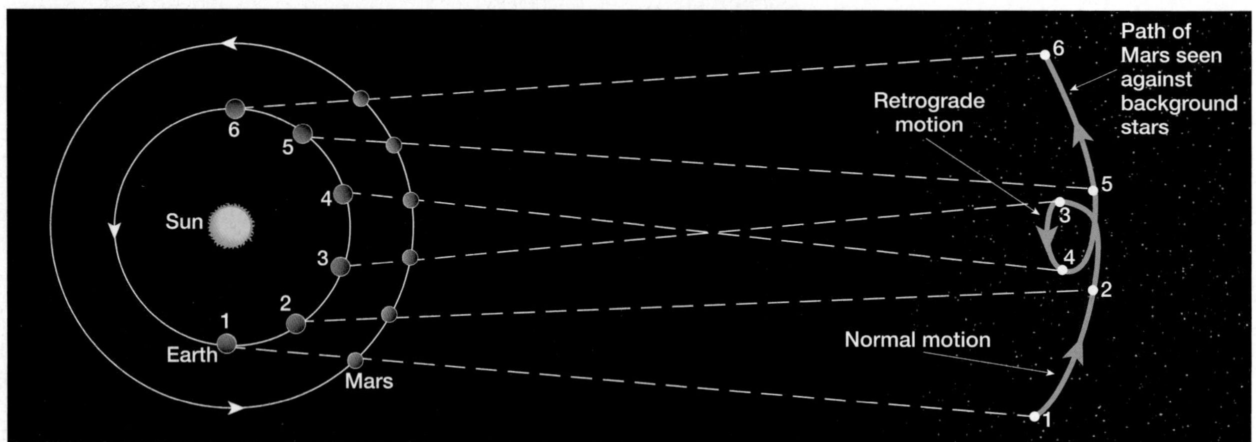

Figure 15.4 Retrograde (backward) motion of Mars as seen against the background of distant stars. When viewed from Earth, Mars moves eastward among the stars each day and then periodically appears to stop and reverse direction. This apparent westward drift happens because Earth has a faster orbital speed than does Mars and overtakes it. As this occurs, Mars appears to be moving backward; that is, it exhibits retrograde motion.

Figure 15.5 Polish astronomer Nicolaus Copernicus (1473–1543) believed that Earth was just another planet. (Yerkes Observatory Photograph/University of Chicago)

the planets Mercury, Venus, Earth, Mars, Jupiter, and Saturn orbiting around it. This was a major break from the ancient idea that a motionless Earth lies at the center of all movement. However, Copernicus retained a link to the past and used circles, which were considered to be the perfect geometric shape, to represent the orbits of the planets. Although these circular orbits were close to reality, they did not quite match what people saw. Unable to get satisfactory agreement between predicted locations of the planets and their observed positions, Copernicus found it necessary to add epicycles like those used by Ptolemy. The discovery that the planets have *elliptical* orbits would wait another century for the insights of Johannes Kepler.

Also like his predecessors, Copernicus used philosophical justifications to support his point of view. Copernicus's monumental work, *De Revolutionibus, Orbium Coelestium (On the Revolution of the Heavenly Spheres),* which set forth his controversial ideas, was published as he lay on his deathbed. Hence, he never suffered the criticism that befell many of his followers.

Tycho Brahe

Tycho Brahe (1546–1601) was born of Danish nobility three years after the death of Copernicus. Reportedly, Tycho became interested in astronomy while viewing a solar eclipse that had been predicted by astronomers. He persuaded King Fredrich II to establish an observatory near Copenhagen, which Tycho headed. There, he designed and built pointers (the telescope would not be invented for a few more decades), which he used for 20 years to systematically measure the locations of the heavenly bodies. These observations, particularly of Mars, were far more precise than any made previously and are his legacy to astronomy (Figure 15.6).

Tycho did not believe in the Copernican (Sun-centered) system, because he was unable to observe an apparent shift in the position of stars that would be caused by Earth's motion. His argument went like this: If Earth does revolve along an orbit around the Sun, the position of a nearby star, when observed from extreme points in Earth's orbit six months apart, should shift with respect to the more distant stars. His *idea* was correct, and this apparent shift of the stars is called *stellar parallax* (see Figure 16.2, p. 433).

Figure 15.6 Tycho Brahe (1546–1601) in his observatory, Uraniborg, on the Danish island of Hveen. Tycho (central figure) and the background are painted on the wall of the observatory within the arc of the sighting instrument called a *quadrant.* In the far right, Tycho can be seen "sighting" a celestial object through the "hole" in the wall. Tycho's accurate measurements of Mars enabled Johannes Kepler to formulate his three laws of planetary motion. (Courtesy of Thomas Clarke, McLaughlin Planetarium/Royal Ontario Museum, copyright ROM)

The principle of parallax is easy to visualize: Close one eye, and with your index finger vertical, use your eye to line up your finger with some distant object. Now, without moving your finger, view the object with your other eye and notice that the object's position appears to shift. The farther away you hold your finger, the less the object's position seems to shift. Herein lay the flaw in Tycho's argument. He was right about parallax, but because the distance to even the nearest stars is enormous compared to the width of Earth's orbit, the shift that occurs is too small to be noticed by using the first primitive telescopes, let alone the unaided eye.

With the death of his patron, the King of Denmark, Tycho was forced to leave his observatory. It was probably his arrogant and extravagant nature that caused a conflict with the next ruler, so Tycho moved to Prague in what is now the Czech Republic. Here, in the last year of his life, he acquired an able assistant, Johannes Kepler. Kepler retained most of the observations made by Tycho and put them to exceptional use. Ironically, the data Tycho collected to refute the Copernican view would later be used by Kepler to support it.

Johannes Kepler

If Copernicus ushered out the old astronomy, Johannes Kepler (1571–1630) ushered in the new (Figure 15.7). Armed with Tycho's data, a good mathematical mind, and, of greater importance, a strong faith in the accuracy of Tycho's work, Kepler derived three basic laws of planetary motion. The first two laws resulted from his inability to fit Tycho's observations of Mars to a circular orbit. Unwilling to concede that the discrepancies were due to observational error, he searched for another solution. This endeavor led him to discover that the orbit of Mars is not a perfect circle but is elliptical (Figure 15.6). About the same time, he realized that the orbital speed of Mars varies in a predictable way. As it approaches the Sun, it speeds up, and as it moves away from the Sun, it slows down.

In 1609, after almost a decade of work, Kepler proposed his first two laws of planetary motion:

1. The path of each planet around the Sun is an ellipse, with the Sun at one focus (Figure 15.8). The other focus is symmetrically located at the opposite end of the ellipse.
2. Each planet revolves so that an imaginary line connecting it to the Sun sweeps over equal areas in equal intervals of time (Figure 15.9). This law of equal areas expresses geometrically the variations in orbital speeds of the planets.

Figure 15.9 illustrates the second law. Note that in order for a planet to sweep equal areas in the same amount of time, it must travel more rapidly when it is nearer the Sun and more slowly when it is farther from the Sun.

Kepler was religious and believed that the Creator made an orderly universe. The uniformity he tried to find eluded him for nearly a decade. Then, in 1619, he published his third law in *The Harmony of the Worlds.*

Figure 15.7 German astronomer Johannes Kepler (1571–1630) helped establish the era of modern astronomy by deriving three laws of planetary motion. (National Museum of American History/Smithsonian Institution)

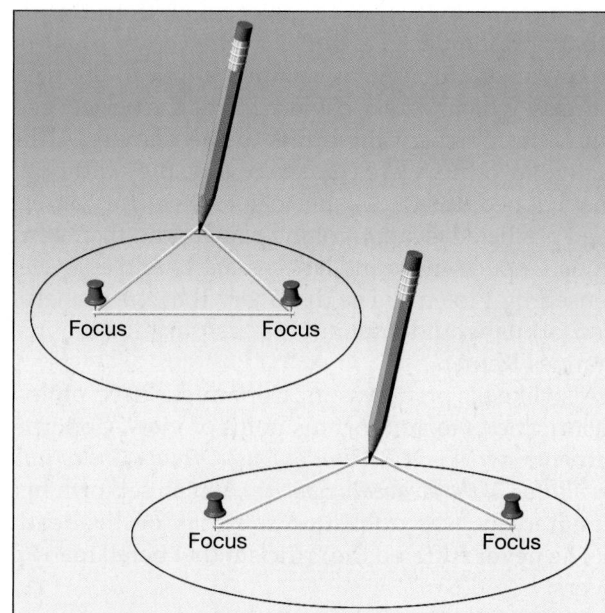

Figure 15.8 Ellipses with various eccentricities. Using two straight pins for foci and a loop of string, trace out a curve while keeping the string taut, and you will have drawn an ellipse. The farther the pins (the foci) are moved apart, the more flattened (more eccentric) is the resulting ellipse.

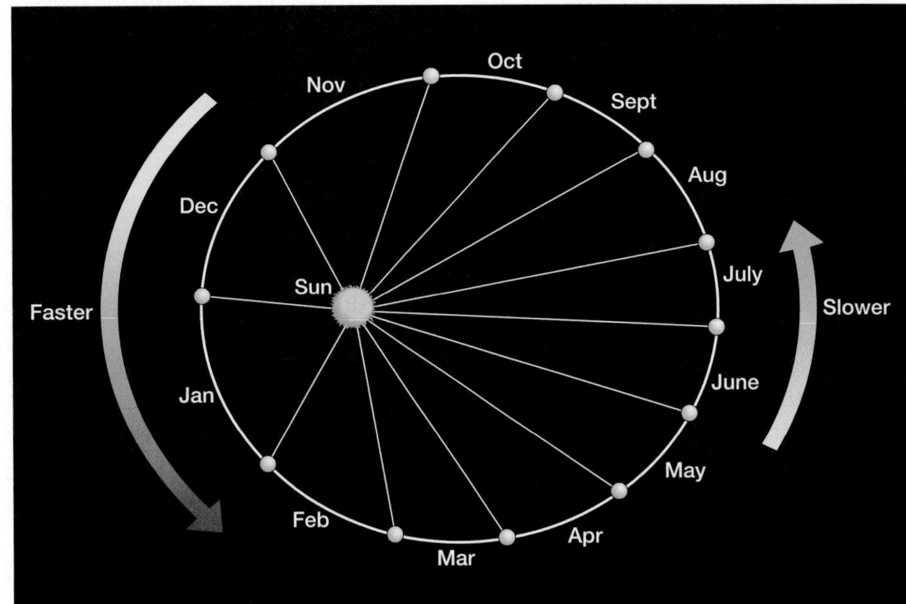

Figure 15.9 Kepler's law of equal areas. A line connecting a planet (Earth) to the Sun sweeps out an area in such a manner that equal areas are swept out in equal times. Thus, Earth revolves (moves in orbit) slower when it is farther from the Sun (aphelion) and faster when it is closest (perihelion). The eccentricity of Earth's orbit is greatly exaggerated in this diagram.

3. The orbital periods of the planets and their distances to the Sun are proportional. In its simplest form, the orbital period of revolution is measured in Earth years, and the planet's distance to the Sun is expressed in terms of Earth's mean distance to the Sun. The latter "yardstick" is called the **astronomical unit (AU)** and averages about 150 million kilometers (93 million miles). Using these units, Kepler's third law states that the planet's orbital period squared is equal to its mean solar distance cubed ($p^2 = a^3$). Consequently, the solar distances of the planets can be calculated when their periods of revolution are known. For example, Mars has a period of 1.88 years, which squared equals 3.54. The cube root of 3.54 is 1.52, and that is the distance to Mars in astronomical units (Table 15.1).

Kepler's laws assert that the planets revolve around the Sun and, therefore, support the Copernican view. Kepler, however, did fall short of determining the *forces* that act to produce the planetary motion he had so ably described. That task would remain for Galileo Galilei and Sir Isaac Newton.

Table 15.1 Period of revolution and solar distances of planets

Planet	Solar Distance (AU)*	Period (years)
Mercury	0.39	0.24
Venus	0.72	0.62
Earth	1.00	1.00
Mars	1.52	1.88
Jupiter	5.20	11.86
Saturn	9.54	29.46
Uranus	19.18	84.01
Neptune	30.06	164.80

*AU = astronomical unit.

Galileo Galilei

Galileo Galilei (1564–1642) was the greatest Italian scientist of the Renaissance (Figure 15.10). He was a contemporary of Kepler and, like Kepler, strongly supported the Copernican theory of a Sun-centered solar system. Galileo's greatest

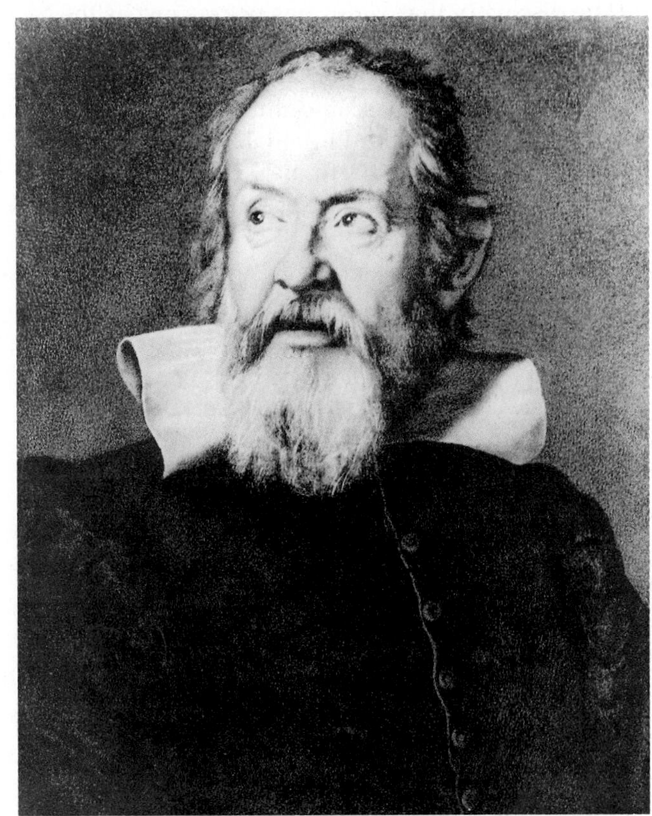

Figure 15.10 Italian scientist Galileo Galilei (1564–1642) used a new invention, the telescope, to observe the Sun, Moon, and planets in more detail than ever before. (Yerkes Observatory Photograph/ University of Chicago)

contributions to science were his descriptions of the behavior of moving objects, which he derived from experimentation. The method of using experiments to determine natural laws had essentially been lost since the time of the early Greeks.

All astronomical discoveries before Galileo's time were made without the aid of a telescope. In 1609, Galileo heard that a Dutch lens maker had devised a system of lenses that magnified objects. Apparently without ever seeing a telescope, Galileo constructed his own, which magnified distant objects to three times the size seen by the unaided eye. He immediately made others, the best having a magnification of about 30 times.

With the telescope, Galileo was able to view the universe in a new way. He made many important discoveries that supported the Copernican view of the universe, including the following:

1. The discovery of four satellites, or moons, orbiting Jupiter. Galileo accurately determined their periods of revolution, which range from 2 to 17 days (Figure 15.11). This finding dispelled the old idea that Earth was the only center of motion in the universe; for here, plainly visible, was another center of motion—Jupiter.
2. The discovery, through observation, that the planets are circular disks rather than just points of light, as was previously thought. This indicated that the planets might be Earthlike.
3. The discovery that Venus has phases just as the Moon does, demonstrating that Venus orbits its source of light—the Sun. He saw that Venus appears smallest when it is in full phase and thus is farthest from Earth (Figure 15.12). In the Ptolemaic system, the orbit of Venus lies between Earth and the Sun, which means that only the crescent phase of Venus could be seen from Earth.
4. The discovery that the Moon's surface is not a smooth glass sphere, as the ancients had proclaimed. Rather, Galileo saw mountains, craters, and plains. He thought the plains might be bodies of water, and this idea was strongly promoted by others, as we can tell from the names given to these features (Sea of Tranquility, Sea of Storms, etc.).
5. The discovery that the Sun (the viewing of which may have caused the eye damage that later blinded him) had sunspots (dark regions caused by slightly lower temperatures). He tracked the movement of these spots and

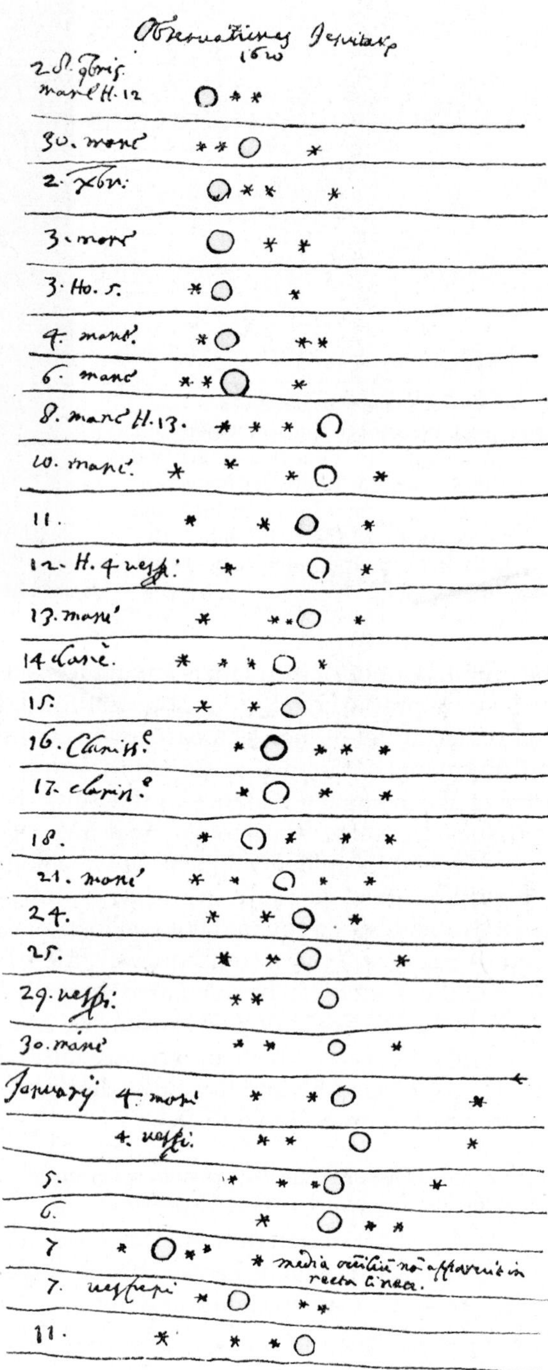

Figure 15.11 Sketch by Galileo of how he saw Jupiter and its four largest satellites through his telescope. The positions of Jupiter's four largest moons (drawn as stars) change nightly. You can observe these same changes with binoculars. (Yerkes Observatory Photograph/University of Chicago)

Did You Know?

Through experimentation, Galileo discovered that the acceleration of falling objects does not depend on their weight. According to some accounts, Galileo made this discovery by dropping balls of iron and wood from the Leaning Tower of Pisa to show that they would fall together and hit the ground at the same time. Despite the popularity of this legend, Galileo probably did not attempt this experiment. In fact, it would have been inconclusive because of the effect of air resistance. However, nearly four centuries later, this experiment was performed most dramatically on the airless Moon when David Scott, an *Apollo 15* astronaut, demonstrated that a feather and a hammer do, indeed, fall at the same rate.

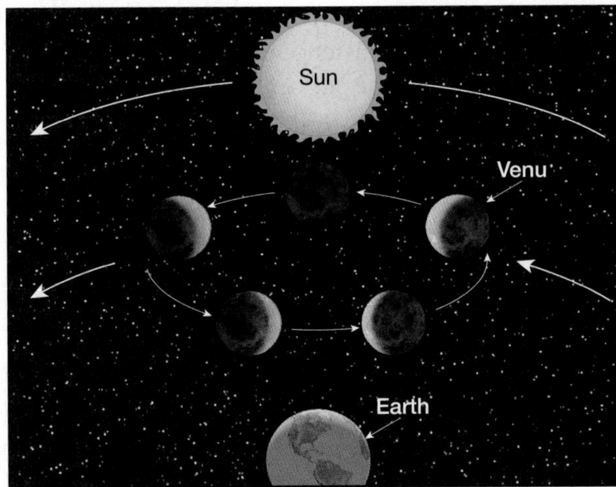

A. Phases of Venus as seen from Earth in the Earth-centered model.

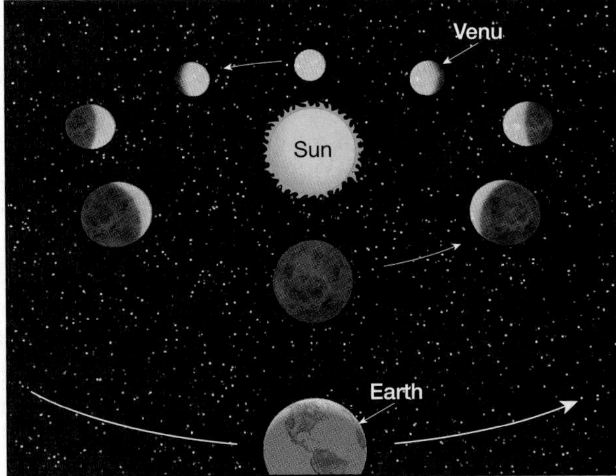

B. Phases of Venus as seen from Earth in the Sun-centered model.

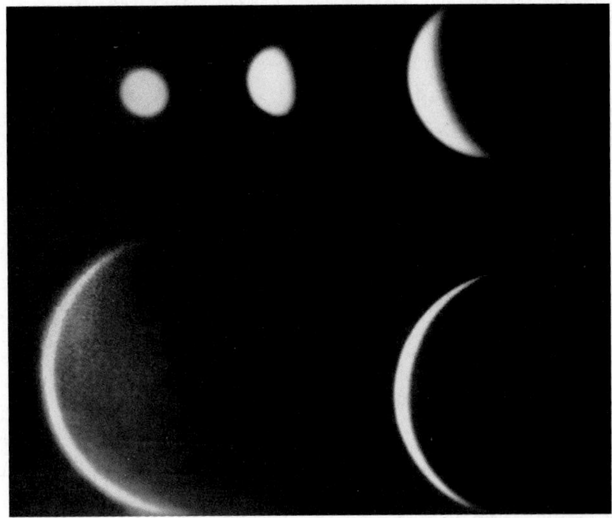

C.

Figure 15.12 Using a telescope, Galileo discovered that Venus has phases just as the Moon does. **A.** In the Ptolemaic (Earth-centered) system, the orbit of Venus lies between those of the Sun and Earth. Thus, in an Earth-centered solar system, only the crescent phase of Venus would be visible from Earth. **B.** In the Copernican (Sun-centered) system, Venus orbits the Sun, and all of the phases of Venus should be visible from Earth. **C.** As Galileo observed, Venus goes through a series of Moonlike phases. Further, Venus appears smallest during the full phase because it is farthest from Earth, and largest in the crescent phase because it is closest to Earth. This verified Galileo's belief that the Sun was the center of the solar system. (Photo courtesy of Lowell Observatory)

estimated the rotational period of the Sun as just under a month. Hence, another heavenly body was found to have both "blemishes" and rotational motion.

In 1616, the Catholic Church condemned the Copernican theory as contrary to Scripture, and Galileo was told to abandon it. Unwilling to accept this verdict, Galileo began writing his most famous work, *Dialogue of the Great World Systems.* Despite poor health, he completed the project and in 1630 went to Rome, seeking permission from Pope Urban VIII to publish. Because the book was a dialogue that expounded both the Ptolemaic and Copernican systems, publication was allowed. However, Galileo's enemies were quick to realize that he was promoting the Copernican view at the expense of the Ptolemaic system. Sale of the book was quickly halted, and Galileo was called before the Inquisition. Tried and convicted of proclaiming doctrines contrary to religious doctrine, he was sentenced to permanent house arrest, under which he remained for the last 10 years of his life. Later, as scientific evidence that supported the Copernican system was discovered, the Catholic Church allowed Galileo's works to be published.

Sir Isaac Newton

Sir Isaac Newton (1642–1727) was born in the year of Galileo's death (Figure 15.13). His many accomplishments in mathematics and physics led a successor to say that "Newton was the greatest genius that ever existed."

Although Kepler and those who followed attempted to explain the forces involved in planetary motion, their explanations were less than satisfactory. Kepler believed that some force pushed the planets along in their orbits. Galileo, however,

Figure 15.13 English scientist Sir Isaac Newton (1642–1727) explained gravity as the force that holds planets in orbit around the Sun. (Yerkes Observatory Photograph/University of Chicago)

correctly reasoned that no force is required to keep an object in motion. Galileo proposed that the natural tendency for a moving object (that is unaffected by an outside force) is to continue moving at a uniform speed and in a straight line. This concept, *inertia*, was later formalized by Newton as his first law of motion.

The problem, then, was not to explain the force that keeps the planets moving but rather to determine the force that *keeps them from going in a straight line out into space.* It was to this end that Newton conceptualized the force of *gravity.* At the early age of 23, he envisioned a force that extends from Earth into space and holds the Moon in orbit around Earth. Although others had theorized the existence of such a force, he was the first to formulate and test the *law of universal gravitation.* It states:

> Every body in the universe attracts every other body with a force that is directly proportional to their masses and inversely proportional to the square of the distance between them.

Thus, the gravitational force decreases with distance, so that two objects 3 kilometers apart have 3^2, or 9, times less gravitational attraction than if the same objects were 1 kilometer apart.

The law of gravitation also states that the greater the mass of the object, the greater its gravitational force. For example, the large mass of the Moon has a gravitational force strong enough to cause ocean tides on Earth, whereas the tiny mass of a communications satellite has no measurable effect on Earth.

With his laws of motion, Newton proved that the force of gravity, combined with the tendency of a planet to remain in straight-line motion, results in the elliptical orbits discovered by Kepler. Earth, for example, moves forward in its orbit about 30 kilometers (18.5 miles) each second, and during the same second, the force of gravity pulls it toward the Sun 0.5 centimeter (1/8 inch). Therefore, as Newton concluded, it is the combination of Earth's forward motion and its "falling" motion that defines its orbit (Figure 15.14). If gravity were somehow eliminated, Earth would move in a straight line out into space. On the other hand, if Earth's forward motion suddenly stopped, gravity would pull it directly toward the Sun.

With the acceptance of Newton's explanation of planetary motion, it became clear that the planets were more similar to Earth than to the stars. This realization stimulated a great deal of interest in observational astronomy. Interest in the planets was further enhanced by the possibility of discovering evidence of intelligent life elsewhere in the solar system.

The Planets: An Overview

The Sun is the hub of a huge rotating system of eight classical planets, their satellites, and numerous smaller asteroids, comets, meteoroids, and dwarf planets. An estimated 99.85

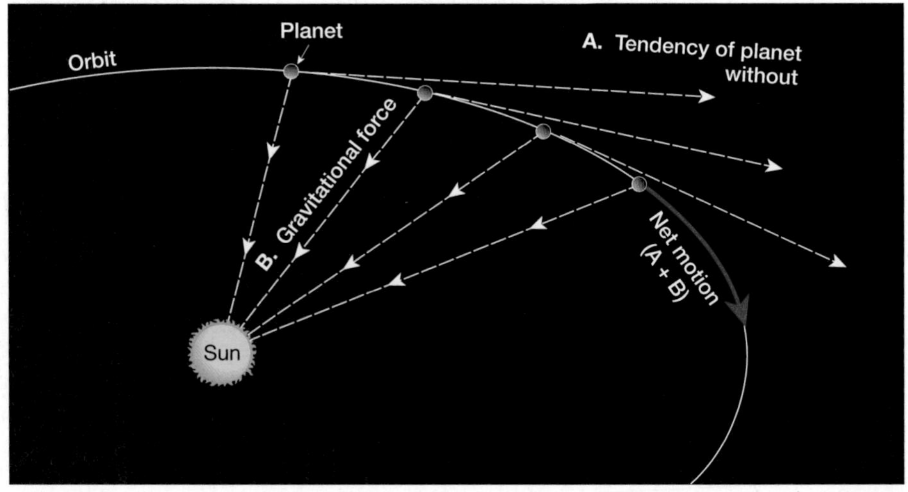

Figure 15.14 Orbital motion of Earth and other planets.

percent of the mass of our solar system is contained within the Sun. The planets collectively make up most of the remaining 0.15 percent. The planets, traveling outward from the Sun, are Mercury, Venus, Earth, Mars, Jupiter, Saturn, Uranus, and Neptune (Figure 15.15). Pluto was recently reclassified as a dwarf planet.

Formation of the Planets

The following scenario describes the most widely accepted views of the origin of our solar system. Although this model is presented as fact, keep in mind that like all scientific hypotheses, this one is subject to revision and even outright rejection. Nevertheless, it remains the most consistent set of ideas to explain what we observe today.

Our scenario begins about 14 billion years ago with the *Big Bang,* an incomprehensibly large explosion that sent all matter of the universe flying outward at incredible speeds. In time, the debris from this explosion, which was almost entirely hydrogen and helium, began to cool and condense into the first stars and galaxies. It was in one of these galaxies, the Milky Way, that our solar system and planet Earth took form.

The orderly nature of our solar system leads most researchers to conclude that Earth and the other planets formed at essentially the same time and from the same primordial material as the Sun. The **nebular hypothesis** proposes that the bodies of our solar system evolved from an enormous rotating cloud called the **solar nebula** (Figure 15.16). Besides the hydrogen and helium atoms generated during the Big Bang, the solar nebula consisted of microscopic dust grains and the ejected matter of long-dead stars. (Nuclear fusion in stars converts hydrogen and helium into the other elements found in the universe.)

Nearly 5 billion years ago, this huge cloud of gases and minute grains of heavier elements began to slowly contract as a result of the gravitational interactions among its particles. Some external influence, such as a shock wave traveling from a catastrophic explosion (*supernova*), may have triggered the collapse. As this slowly spiraling nebula contracted, it rotated faster and faster for the same reason ice skaters do when they draw their arms toward their bodies. Eventually the inward pull of gravity came into balance with the outward force caused by the rotational motion of the nebula (Figure 15.16). By this time the once vast cloud had assumed a flat disk shape with a large concentration of material at its center called the *protosun* (pre-Sun). (Astronomers are fairly confident that the nebular cloud formed a disk because similar structures have been detected around other stars.)

Within this spinning disk, matter gradually formed clumps of material that collided, stuck together, and grew into asteroid-sized objects called **planetesimals.** The composition of these planetesimals depended largely on their location with respect to the protosun. (Temperatures were greater near the protosun and much lower in the outer reaches of the

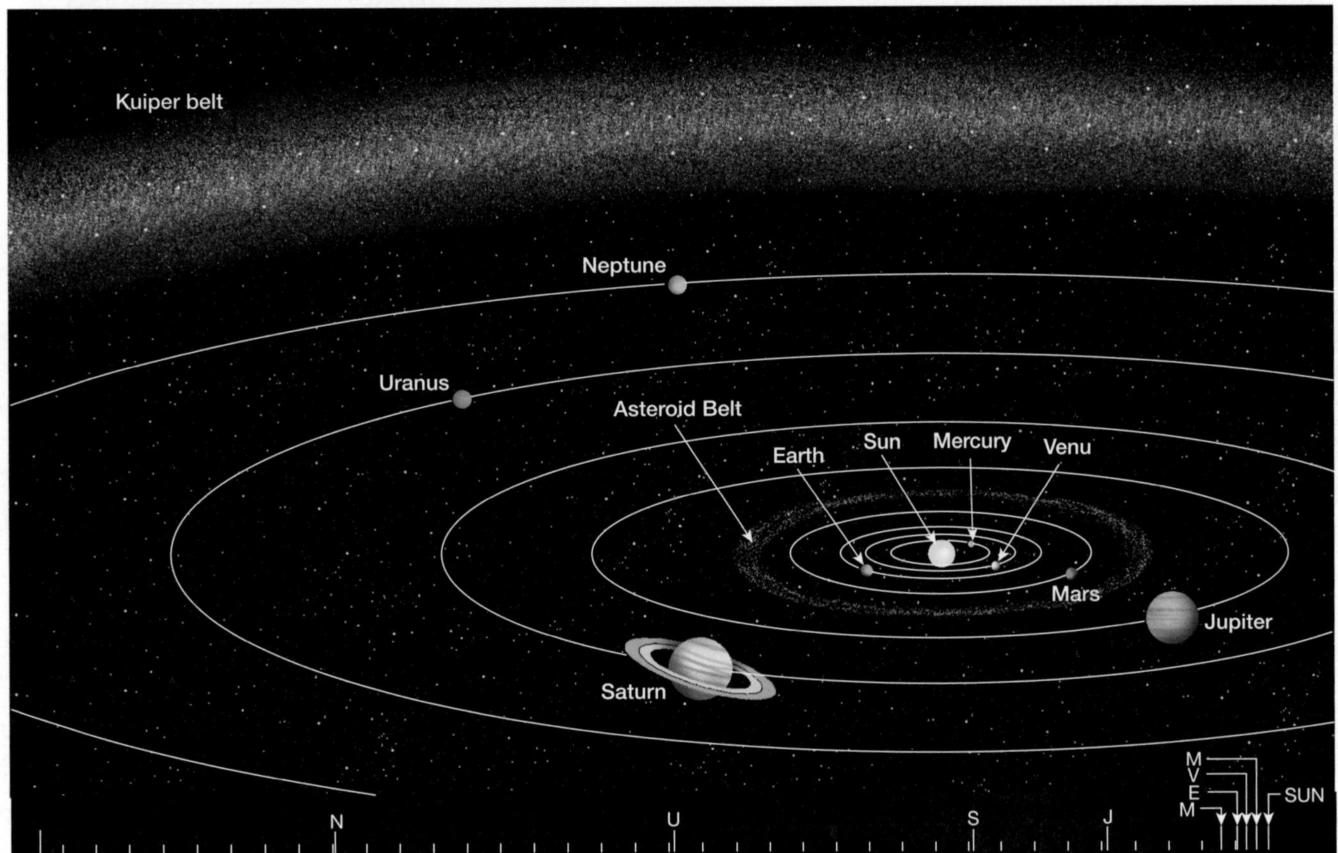

Figure 15.15 Orbits of the planets (not to scale). Orbits are shown to scale along bottom of diagram.

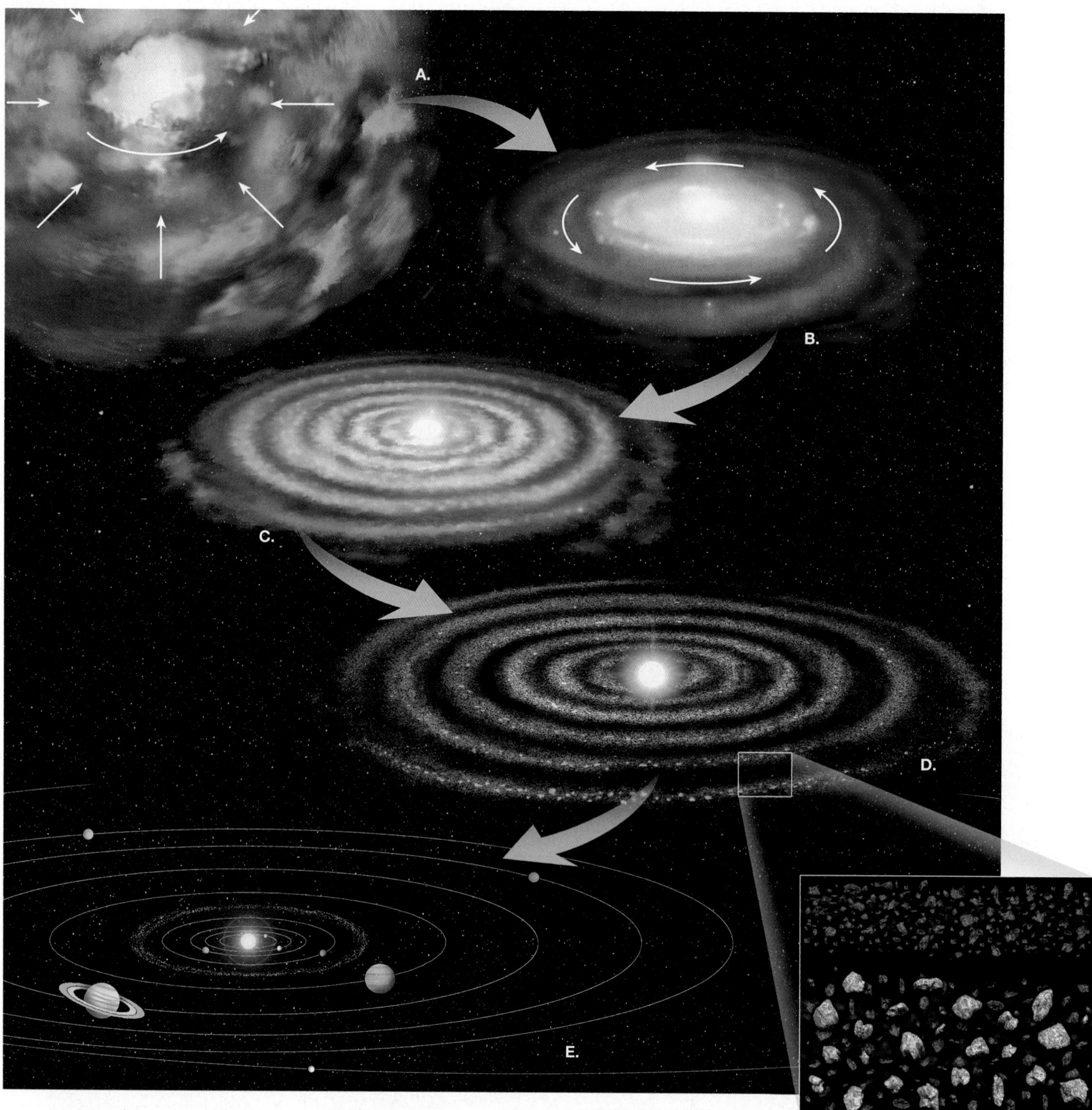

Figure 15.16 Formation of the solar system according to the nebular hypothesis. **A.** The birth of our solar system began as dust and gases (nebula) started to gravitationally collapse. **B.** The nebula contracted into a rotating disk that was heated by the conversion of gravitational energy into thermal energy. **C.** Cooling of the nebular cloud caused rocky and metallic material to condense into tiny solid particles. **D.** Repeated collisions caused the dust-size particles to gradually coalesce into asteroid-size bodies. **E.** Within a few million years, these bodies accreted into the planets.

disk.) This was critical since only those materials that could condense—form solid or liquid clumps—in a particular location would be available to form planetesimals.

Near the present orbit of Mercury, only metallic grains condensed—it was simply too hot for anything else to exist. Farther out, near Earth's orbit, metallic as well as rocky substances condensed, and beyond Mars, ices of water, carbon dioxide, ammonia, and methane formed. It was from these clumps of matter that the planetesimals formed and through repeated collisions and accretion (sticking together) grew into eight **protoplanets** and their moons. It took roughly a billion years after the protoplanets formed to gravitationally sweep the solar system clear of interplanetary debris. This was a period of intense bombardment that is clearly visible on the Moon and elsewhere in the solar system. Only a small amount of the interplanetary matter escaped capture by a planet or moon and became the asteroids, comets, and meteoroids.

Terrestrial and Jovian Planets

Careful examination of Table 15.2 shows that the planets fall quite nicely into two groups: the **terrestrial** (Earthlike) **planets** (Mercury, Venus, Earth, and Mars), and the **Jovian** (Jupiter-like) **planets** (Jupiter, Saturn, Uranus, and Neptune). Pluto was recently demoted to a *dwarf planet*—a new class of solar system objects that have an orbit around the Sun but share their space with other celestial bodies. We will consider the nature of dwarf planets later in the chapter.

The most obvious difference between the terrestrial and the Jovian planets is their size (Figure 15.17). The largest terrestrial planets (Earth and Venus) have diameters only one-quarter as great as the diameter of the smallest Jovian planet (Neptune). Also, their masses are only 1/17 as great as Neptune's. Hence, the Jovian planets are often called *giants.* Because of their relative locations, the four Jovian planets are also referred to as the **outer planets,** whereas the terrestrial planets are called the **inner planets.** As we will see, there appears to be a correlation between the location of these planets within the solar system and their sizes.

Other dimensions in which the terrestrial and the Jovian planets differ include density, chemical makeup, and rate of rotation. The densities of the terrestrial planets average about five times the density of water, whereas the Jovian planets have densities that average only 1.5 times that of water. One of the outer planets, Saturn, has a density only 0.7 times that of water, which means that Saturn would float if placed in a large enough water tank! Differences in the chemical

Did You Know?

Although it was long suspected, it was not until recently that the presence of extrasolar planets has been verified. Astronomers have found these bodies by measuring the telltale wobbles of nearby stars. The first apparent planet outside the solar system was discovered in 1995, orbiting the star 51 Pegasi, 42 light-years from Earth. Since that time, numerous Jupiter-size bodies have been identified, most of them surprisingly close to the stars they orbit.

Table 15.2 Planetary Data

Planet	Symbol	Mean Distance from Sun: AU*	Mean Distance from Sun: Millions of Miles	Mean Distance from Sun: Millions of Kilometers	Period of Revolution	Inclination of Orbit	Orbital Velocity: mi/s	Orbital Velocity: km/s
Mercury	☿	0.39	36	58	88^{d}	7°00′	29.5	47.5
Venus	♀	0.72	67	108	225^{d}	3°24′	21.8	35.0
Earth	⊕	1.00	93	150	365.25^{d}	0°00′	18.5	29.8
Mars	♂	1.52	142	228	687^{d}	1°51′	14.9	24.1
Jupiter	♃	5.20	483	778	12^{yr}	1°18′	8.1	13.1
Saturn	♄	9.54	886	1427	29.5^{yr}	2°29′	6.0	9.6
Uranus	♅	19.18	1783	2870	84^{yr}	0°46′	4.2	6.8
Neptune	♆	30.06	2794	4497	165^{yr}	1°46′	3.3	5.3

Planet	Period of Rotation	Diameter: Miles	Diameter: Kilometers	Relative Mass (Earth = 1)	Average Density (g/cm³)	Polar Flattening (%)	Eccentricity†	Number of Known Satellites**
Mercury	59^{d}	3015	4878	0.06	5.4	0.0	0.206	0
Venus	244^{d}	7526	12,104	0.82	5.2	0.0	0.007	0
Earth	$23^{h}56^{m}04^{s}$	7920	12,756	1.00	5.5	0.3	0.017	1
Mars	$24^{h}37^{m}23^{s}$	4216	6794	0.11	3.9	0.5	0.093	2
Jupiter	$9^{h}50^{m}$	88,700	143,884	317.87	1.3	6.7	0.048	63
Saturn	$10^{h}14^{m}$	75,000	120,536	95.14	0.7	10.4	0.056	56
Uranus	$17^{h}14^{m}$	29,000	51,118	14.56	1.2	2.3	0.047	27
Neptune	$16^{h}03^{m}$	28,900	50,530	17.21	1.7	1.8	0.009	13

* AU = astronomical unit, Earth's mean distance from the Sun.

** Includes all satellites discovered as of August 2006.

†Eccentricity is a measure of the amount an orbit deviates from a circular shape. The larger the number is, the less circular the orbit.

Figure 15.17 The planets drawn to scale.

compositions of the planets are largely responsible for these density differences.

The Compositions of the Planets

The substances that make up the planets are divided into three compositional groups based on their melting points: *gases*, *rocks*, and *ices*.

1. **Gases,** hydrogen and helium, are those with melting points near absolute zero (0 Kelvin). These two gases were the most abundant constituents of the solar nebula.
2. **Rocks** are principally silicate minerals and metallic iron, which have melting points that exceed 700°C (1290°F).
3. **Ices** include ammonia, methane, carbon dioxide, and water. They have intermediate melting points (for example, water has a melting point of 0°C [32°F]).

The terrestrial planets are dense, consisting mostly of rocky and metallic substances, with minor amounts of ices. The Jovian planets, on the other hand, contain large amounts of gases (hydrogen and helium) and ices (mostly water, ammonia, and methane). This accounts for their low densities. The Jovian planets also contain substantial amounts of rocky and metallic materials, which are concentrated in their central cores.

The Atmospheres of the Planets

The Jovian planets have very thick atmospheres of hydrogen, helium, methane, and ammonia. By contrast, the terrestrial planets, including Earth, have meager atmospheres at best. There are two reasons for this difference.

The first reason is the location of each planet within the solar nebula during its formation. The outer planets formed where the temperature was low enough to allow water vapor, ammonia, and methane to condense into ices. Hence, the Jovian planets contain large amounts of these volatiles. However, in the inner regions of the developing solar system, the environment was too hot for ices to survive. Consequently, one of the long-standing questions for the nebular hypothesis was "How did Earth acquire water and other volatile gases?" The answer seems to be that during its protoplanet stage, Earth was bombarded with icy fragments (planetesimals) that originated beyond the orbit of Mars.

But why do Mercury and our Moon lack an atmosphere? Like the inner planets, they surely would have been bombarded by icy bodies. That brings us to the second reason that the inner planets (and the Moon) lack substantial atmospheres. A planet's ability to retain an atmosphere depends on its mass and its temperature. Simply stated, more massive planets have a better chance of retaining their atmospheres because atoms and molecules need a higher speed to escape. On the Moon, the **escape velocity** is only 2.4 km/s compared with more than 11 km/s for Earth.

Because of their strong gravitational fields, the Jovian planets have escape velocities that are much higher than that of Earth, which is the largest terrestrial planet. Consequently, it is much more difficult for gases to escape from the outer planets. Also, because the molecular motion of a gas is temperature-dependent, at the low temperatures of the Jovian planets, even the lightest gases (hydrogen and helium) are unlikely to acquire the speed needed to escape.

By contrast, a comparatively warm body with a small surface gravity, such as Mercury and our Moon, is unable to hold even heavy gases such as carbon dioxide and radon. (Mercury does have trace amounts of gas present.) The slightly larger terrestrial planets of Earth, Venus, and Mars retain some heavy gases such as water vapor, nitrogen, and carbon dioxide, but even their atmospheres make up only a very small portion of their total mass.

Earth's Moon

Earth now has hundreds of satellites, but only one, the Moon, is natural. Although other planets have moons, our planet-satellite system is unique in the solar system because Earth's Moon is unusually large compared to its parent planet. The diameter of the Moon is 3475 kilometers (2160 miles), about one-fourth of Earth's 12,756 kilometers (7926 miles).

From a calculation of the Moon's mass, its density is 3.3 times that of water. This density is comparable to that of *mantle* rocks on Earth but is considerably less than Earth's average density, which is 5.5 times that of water. Geologists have suggested that this difference can be accounted for if the Moon's iron core is small.

The gravitational attraction at the lunar surface is one-sixth of that experienced on Earth's surface (a 150-pound person on Earth weighs only 25 pounds on the Moon although they still have the same mass). This difference allows an astronaut to carry a heavy life-support system with relative ease. If not burdened with such a load, an astronaut could jump six times higher than on Earth.

The Lunar Surface

When Galileo first pointed his telescope toward the Moon, he saw two different types of terrains–dark lowlands and brighter, highly cratered highlands (Figure 15.18). Because the dark regions resembled seas on Earth, they were called **maria** (*mar* = sea, singular **mare**). Today, we know that the maria are not oceans but, instead, are flat plains that resulted from immense outpourings of fluid basaltic lavas. By contrast, the light-colored areas resemble Earth's continents, so the first observers dubbed them **terrae** (Latin for "land"). These areas are now generally referred to as **lunar highlands,** because they are elevated several kilometers above the maria. Together, the arrangement of terrae and maria result in the well-known "face of the moon."

Impact Craters The most obvious features of the lunar surface are craters. They are so profuse that craters-within-craters are the rule! The larger ones in the lower portion of Figure 15.18 are about 250 kilometers (155 miles) in diameter, roughly

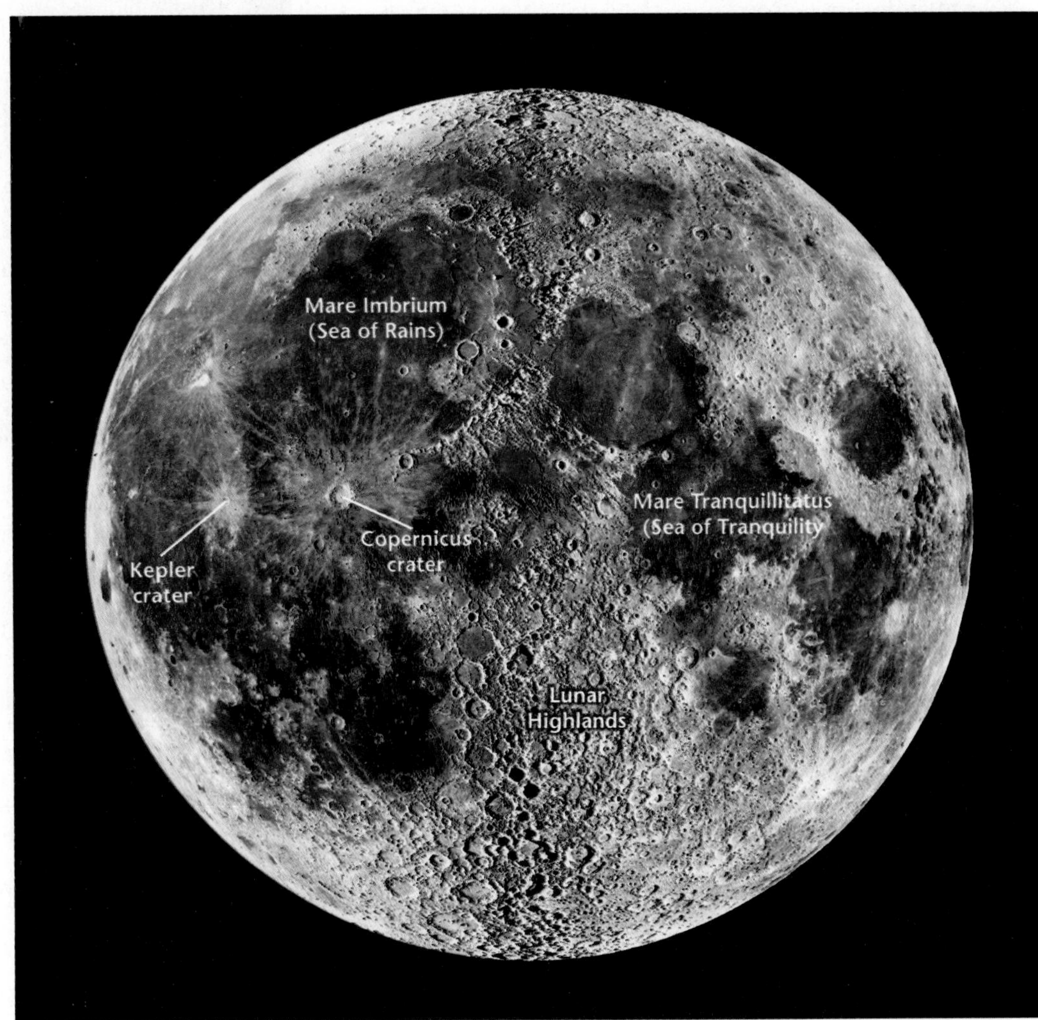

Figure 15.18 Telescopic view from Earth of the lunar surface. (Photo © UC Regents/Lick Observatory)

the width of Indiana. **Impact craters** are produced by the impact of rapidly moving debris (meteoroids, asteroids, and comets), a phenomenon that was considerably more common in the early history of the solar system than it is today.

By contrast, Earth has only about a dozen easily recognized impact craters. This difference can be attributed to Earth's atmosphere, erosion, and tectonic processes. Friction with the air burns up small debris and makes large meteoroids smaller before they reach the ground. In addition, evidence for most of the craters that formed in Earth's history has been obliterated by erosional or tectonic processes.

The formation of an impact crater is illustrated in Figure 15.19. Upon impact, the high-speed meteoroid compresses the material it strikes; then, almost instantaneously, the compressed rock rebounds, ejecting material from the crater. This process is analogous to the splash that occurs when a rock is dropped into water. Craters excavated by objects several kilometers across often exhibit a central peak, as seen in the large crater in Figure 15.20. Most of the ejected material (*ejecta*) lands in the crater or nearby, where it builds a rim around it. Depending on the size of the meteoroid, the heat generated by the impact may be sufficient to melt some of the impacted rock. Astronauts have brought back samples of glass beads produced in this manner, as well as rock formed when broken fragments and dust were welded together by the heat of an impact.

A meteoroid only 3 meters (10 feet) in diameter can blast out a 150-meter-wide (500-foot-wide) crater. A few of the large craters, such as Kepler and Copernicus, shown in Figure 15.18, formed from the bombardment of bodies 1 kilometer (0.6 mile) or more in diameter. These two large craters are thought to be relatively young because of the bright *rays* (splash marks) that radiate outward for hundreds of kilometers.

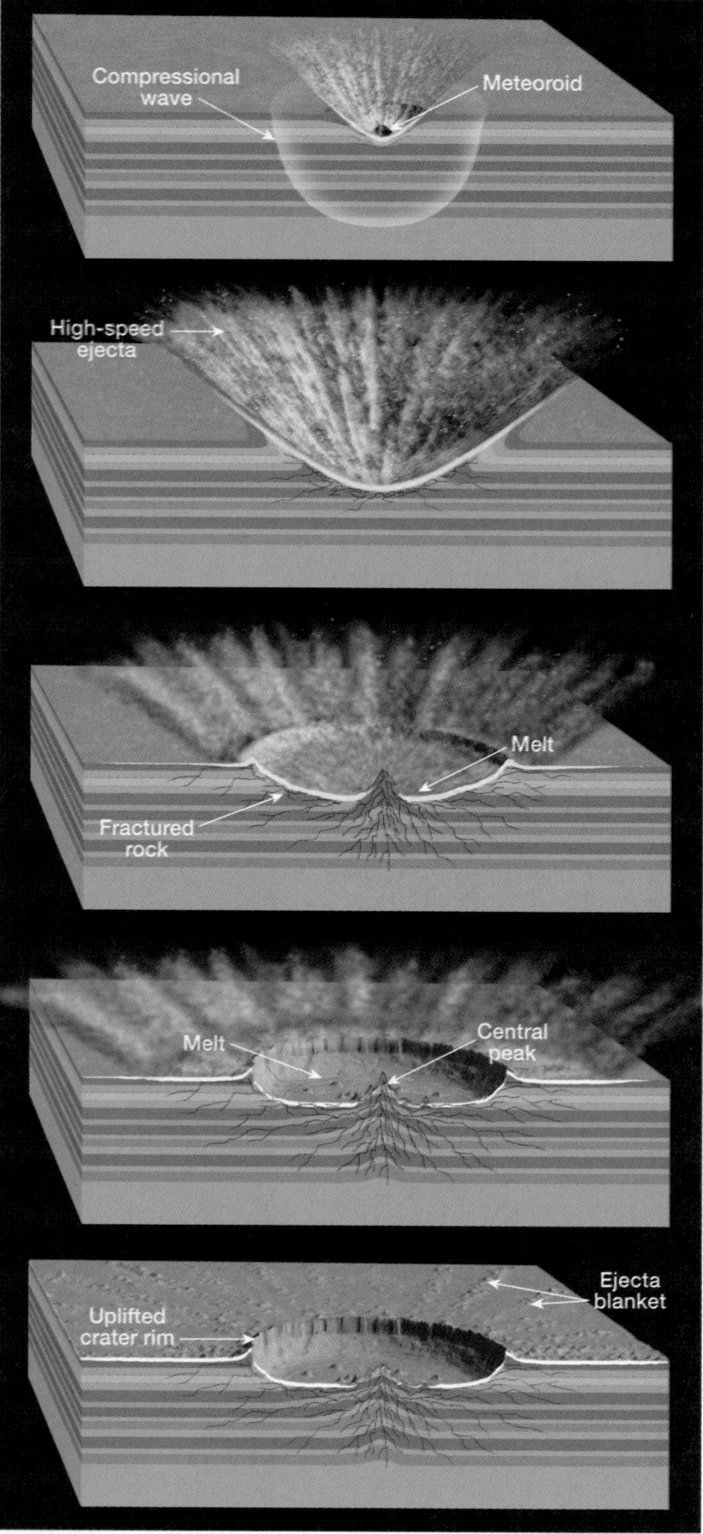

Figure 15.19 Formation of an impact crater. The energy of the rapidly moving meteoroid is transformed into heat energy and compressional waves. The rebound of the compressed rock causes debris to be ejected from the crater, and the heat melts some material, producing glass beads. Small, secondary craters are formed by the material "splashed" from the impact crater. (After E. M. Shoemaker)

Highlands and "Seas" Densely pockmarked highland areas make up most of the lunar surface (Figure 15.21). In fact, most of the back side of the Moon is characterized by such topography. (Only astronauts have directly observed the farside, because the Moon rotates on its axis once with each revolution around Earth, always keeping the same side facing Earth.) As seen in Figure 15.18, p. 409 highlands consist of an apparently endless sequence of overlapping craters. Indeed, the great number of impact craters is evidence of the Moon's violent early history. This activity crushed and repeatedly mixed at least the upper few kilometers of the lunar crust. As a result, the highlands are very rugged.

The highlands are made of plutonic rocks that contain more than 90 percent plagioclase feldspar that rose like "scum" from a magma ocean early in lunar history. Maria, on the other hand, are composed of volcanic rocks with a mafic composition and, hence, are similar to flood basalts on Earth.

The dark, flat maria make up only about 16 percent of the Moon's landscape and are concentrated on the side of the Moon facing Earth. (Figure 15.18). More than 4 billion years ago, asteroids having diameters as large as Rhode Island excavated several huge craters on the lunar surface. Because the crust was sufficiently fractured, magma began to bleed out. Apparently, the craters were flooded with layer upon layer of very fluid lavas resembling those of the Columbia Plateau in

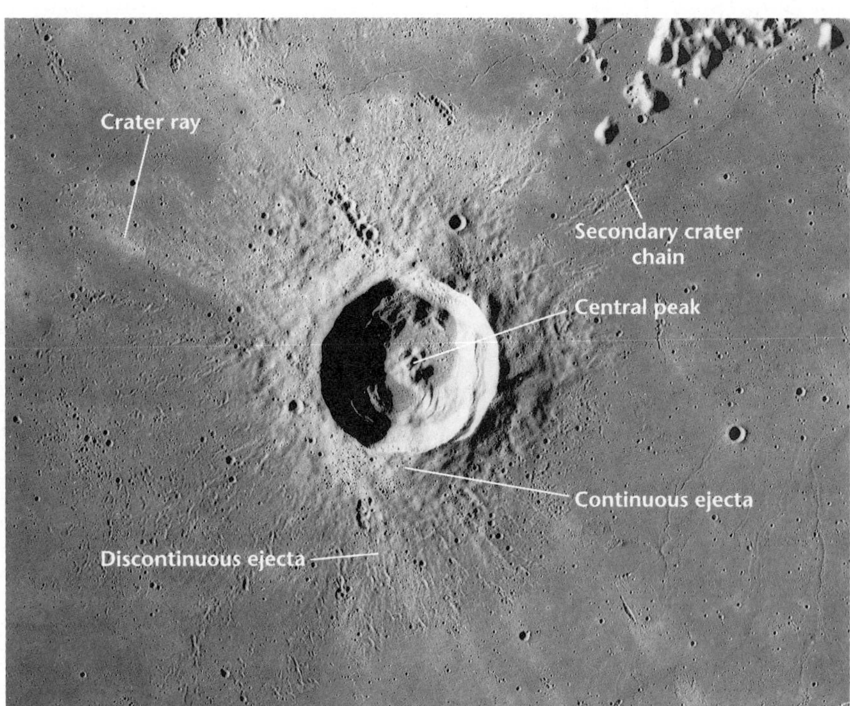

Figure 15.20 The 20-kilometer-wide (12-mile-wide) lunar crater Euler in the southwestern part of Mare Imbrium. Clearly visible are the rays, central peak, secondary craters, and the large accumulation of ejecta near the crater rim. (NASA)

the Pacific Northwest. In most cases, these lavas welled up long after the craters formed. It seems that volcanism occurred because the crust was thinner and more fractured in these regions rather than because of heat generated by the impact itself.

Weathering and Erosion The Moon has no atmosphere or flowing water. Therefore, the processes of weathering and erosion that continually modify Earth's surface are virtually lacking on the Moon. In addition, tectonic forces are no longer active on the Moon, so earthquakes and volcanic eruptions do not occur. However, because the Moon is unprotected by an atmosphere, a different kind of erosion occurs: Tiny particles from space (micrometeorites) continually bombard its surface and ever so gradually smooth the landscape.

Both the maria and terrae are mantled with a layer of gray, unconsolidated debris derived from a few billion years of meteoric bombardment (Figure 15.22). This soil-like layer, properly called **lunar regolith** (*rhegos* = blanket, *lithos* = stone) is composed of igneous rocks, breccia, glass beads, and fine *lunar dust*. In the maria that have been explored by *Apollo* astronauts, the lunar regolith is apparently slightly more than 3 meters (10 feet) thick.

Impact Hypothesis for Moon's Origin

Although the Moon is our nearest planetary neighbor and astronauts have sampled its surface, much is still unknown about its origin. The most widely favored scenario is that during the

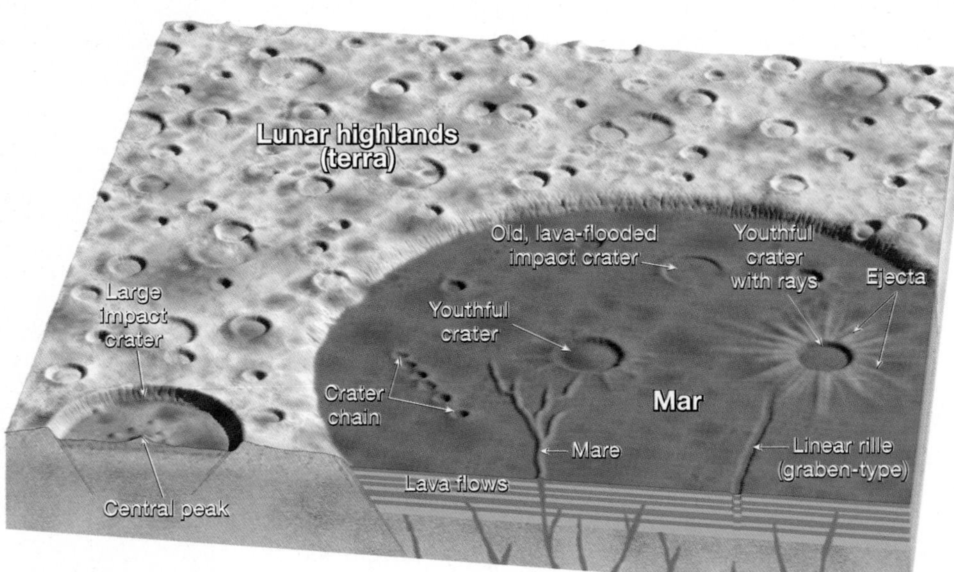

Figure 15.21 Block diagram illustrating major topographic features on the lunar surface.

Figure 15.22 Astronaut Harrison Schmitt sampling the lunar surface. Notice the footprints (inset) in the lunar "soil." (NASA)

formative period of the solar system, a glancing collision occurred between a Mars-sized body and a youthful semimolten Earth. (Collisions of this type were probably frequent events in the early solar system.) During such a catastrophic collision, most of the ejected debris would have collected in an orbit around Earth. Gradually, this debris coalesced to form the Moon. Computer simulations show that most of the ejected material would have come from the mantles of both the impacting object and Earth. (Recall that the mantle makes up 82 percent of Earth's volume.) Further, if the impacting object had an iron core, this dense material would eventually have been incorporated into Earth's core. The *impact theory* is consistent with the Moon's internal structure—the Moon has a large mantle and a small core.

The Planets: A Brief Tour

Mercury: The Innermost Planet

Mercury, the innermost and smallest planet, is hardly larger than Earth's Moon and is smaller than three other moons in the solar system. Mercury revolves quickly but rotates slowly. One full day-night cycle on Earth takes 24 hours, but on Mercury it requires 179 Earth-days. Thus, a night on Mercury lasts for about three months and is followed by three months of daylight. Nighttime temperatures drop as low as −173°C (−280°F) and noontime temperatures exceed 427°C (800°F), hot enough to melt tin and lead. Mercury has the greatest temperature extremes of any planet. The odds of life as we know it existing on Mercury are nil.

Like our own Moon, Mercury absorbs most of the sunlight that strikes it, reflecting only 6 percent into space (Figure 15.23). In contrast, the Earth reflects about 30 percent of the light that strikes it, much of it from clouds. The low reflectivity of sunlight from Mercury is characteristic of terrestrial bodies that have virtually no atmosphere. The minuscule amounts of gas present on Mercury may have originated as ionized gas emitted from the Sun, from the ices of a recent comet impact, from surface rocks subject to solar winds, and/or from outgassing of the planet's interior.

The probe *Messenger,* scheduled to arrive in 2011, will investigate the composition of Mercury's core and the nature of its magnetic field. Mercury is very dense (5.4 grams per cubic centimeter) which implies that it contains a large iron core

Figure 15.23 Photomosaic of Mercury. This view of Mercury is remarkably similar to the "far side" of Earth's Moon. (NASA)

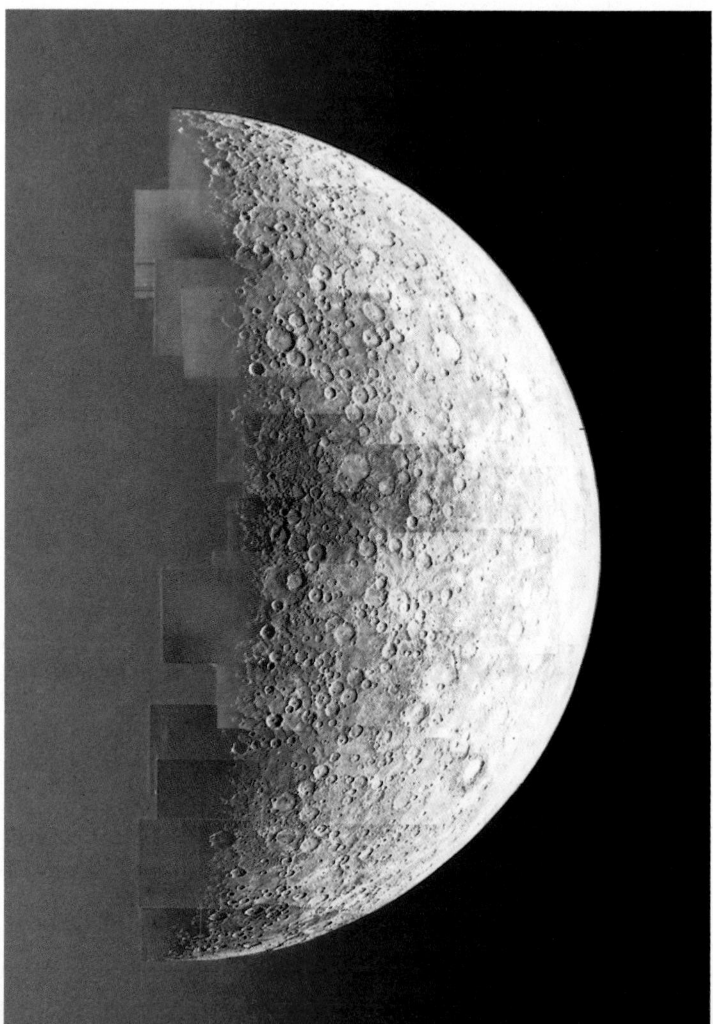

Did You Know?

Planets come and planets go! The ancient Greeks recognized seven celestial bodies that moved against the background pattern of stars: the Sun, the Moon, Mercury, Venus, Mars, Jupiter, and Saturn. They called them *planets* or "wanderers." After Nicolas Copernicus persuaded astronomers that the Sun, rather than Earth, lies at the center of the solar system, Earth was added to the list of planets and the Sun and Moon were deleted. During the early 1800s, 15 of the largest asteroids along with Uranus and Neptune were added to the list of known planets. In 1852, astronomers deleted asteroids from the list of planets. If we still counted asteroids as planets, our solar system would consist of more than 135,000 known planets. In 2006, Pluto was also demoted from planetary status after a larger Kuiper belt object was discovered.

for its size. Mercury has cratered highlands, much like the Moon, and vast smooth terrains that resemble maria. Mercury also has long scarps that cut across the plains and numerous craters. These scarps are thought to be the result of crustal shortening as the planet cooled and shrank early in its history. The shrinking caused the crust to fracture.

Venus: The Veiled Planet

Venus, second only to the Moon in brilliance in the night sky, is named for the goddess of love and beauty. It orbits the Sun in a nearly perfect circle once every 225 Earth-days. Venus is similar to Earth in size, density, mass, and location in the solar system. Thus, it has been referred to as "Earth's twin." Because of these similarities, it is hoped that a detailed study of Venus will provide geologists with a better understanding of Earth's evolutionary history.

Venus is shrouded in thick clouds impenetrable to visible light. The Venusian atmosphere contains an opaque cloud layer about 25 kilometers (15 miles) thick and has an atmospheric pressure that is 90 times that at Earth's surface. Before the advent of space vehicles, Venus was considered to be a potentially hospitable site for living organisms. However, evidence from space probes indicates otherwise. The surface of Venus reaches temperatures as great as 480°C (900°F), and the Venusian atmosphere is 97 percent carbon dioxide. Only scant water vapor and nitrogen have been detected. This hostile environment makes it unlikely that life as we know it exists on Venus.

The composition of the Venusian core is likened to that of Earth. However, since Venus only has a small induced magnetic field, the internal dynamics must be very different. Mantle convection still operates on Venus, but the processes of plate tectonics, which recycle rigid lithosphere, do not appear to have contributed to the present Venusian topography. Tectonic activity on Venus seems to be restricted to upwelling (mantle plumes) and downwelling of material in the planet's interior, rather than lateral movement and subduction of lithospheric plates.

Radar mapping by the unmanned *Magellan* spacecraft and by instruments on Earth has revealed a varied topography with features somewhat between those of Earth and Mars (Figure 15.24). Radar pulses in the microwave range are sent toward the Venusian surface, and the heights of plateaus and mountains are measured by timing the return of the radar echo. These data have confirmed that basaltic volcanism and tectonic deformation are the dominant processes operating on Venus. Further, based on the low density of impact craters, volcanism and tectonic deformation must have been very active during the recent geologic past.

About 80 percent of the Venusian surface consists of subdued plains that are covered by volcanic flows. Some lava channels extend hundreds of kilometers; one meanders 6800 kilometers (4225 miles) across the planet. Thousands of volcanic structures have been identified, mostly small shield volcanoes, although more than 1500 volcanoes greater than 20 kilometers (12 miles) across have been mapped (Figure 15.25). One is Sapas Mons, 400 kilometers (250 miles) across and 1.5 kilometers (0.9 mile) high. Many flows from this volcano erupted from its flanks rather than the summit, in the manner of Hawaiian shield volcanoes. Only 8 percent of the Venusian surface consists of highlands that may be likened to continental areas on Earth.

Figure 15.24 This global view of the surface of Venus is computer generated from two years of Magellan Project radar mapping. The twisting bright features that cross the globe are highly fractured mountains and canyons of the eastern Aphrodite highland. (Courtesy of NASA/JPL)

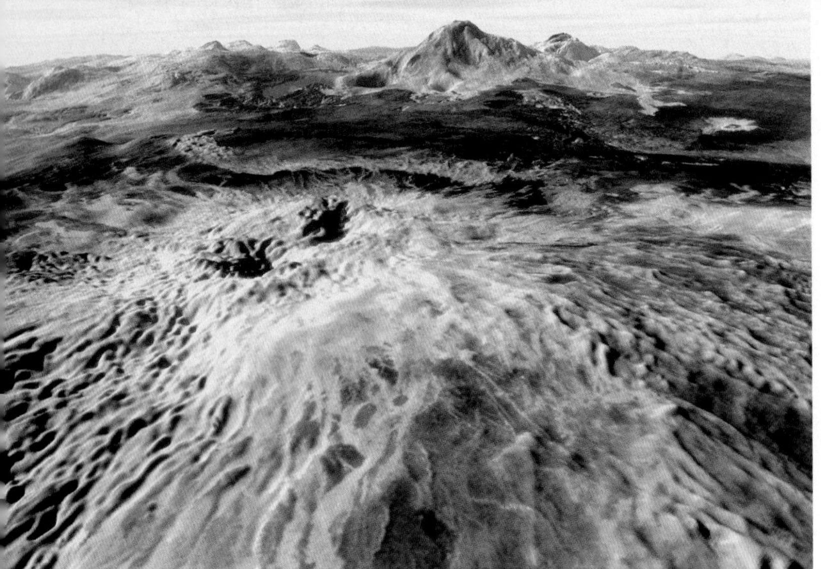

Figure 15.25 Computer-generated image of Venus. Near the horizon is Maat Mons, a large volcano. The bright feature below is the summit of Sapas Mons. (Courtesy David P. Anderson/SMU/NASA/Science Photo Library/Photo Researchers, Inc.)

Mars: The Red Planet

Mars has evoked greater interest than any other planet, both for scientists and for nonscientists. When one imagines intelligent life on other worlds, little green Martians may come to mind. Interest in Mars stems mainly from this planet's accessibility to observation. All other planets within telescopic range have their surfaces hidden by clouds, except for Mercury, whose nearness to the Sun makes viewing difficult. Mars is approximately half the size of the Earth and revolves around the Sun in 687 Earth-days. Through the telescope, Mars appears as a reddish ball interrupted by some permanent dark regions that change intensity during the Martian year. The most prominent telescopic features of Mars are its brilliant white polar caps, resembling Earth's.

The Martian Atmosphere The Martian atmosphere has only 1 percent the density of Earth's, and it is primarily carbon dioxide with tiny amounts of water vapor. Data from Mars probes confirm that the polar caps of Mars are made of water ice, covered by a thin layer of frozen carbon dioxide. As winter nears in either hemisphere, we see the equatorward growth of that hemisphere's ice cap as temperatures drop to −125°C (−193°F), and additional carbon dioxide is deposited.

Although the atmosphere of Mars is very thin, extensive dust storms occur and may cause the color changes observed from Earth-based telescopes. Hurricane-force winds up to 270 kilometers (170 miles) per hour can persist for weeks. Images from *Viking 1* and *Viking 2* revealed a Martian landscape remarkably similar to a rocky desert on Earth, with abundant dunes and impact craters partially filled with dust.

Mars's History Although the Martian core is believed to be solid because of the lack of a global magnetic field, evidence suggests that some of the planet's oldest rocks formed in the presence of a strong magnetic field. In the past, Mars may have had a hot interior with a molten core.

Most Martian surface features are old by Earth standards. Evidence suggests that weathering accounts for almost all surface changes during the last 3.5 billion years. Denser cratering in the highlands indicates that these areas are very old, whereas lighter cratering in the lowlands suggest volcanic eruptions 3.7 to 3.8 billion years ago. The highly cratered southern hemisphere is probably similar in age to the lunar highlands (nearly 4.5 billion years old). Even the relatively fresh-appearing volcanic features of the northern hemisphere are most likely older than several hundred million years. These facts and the absence of Marsquake recordings by *Viking* seismographs point to a tectonically dead planet.

Mars's Dramatic Geologic Past *Mariner 9,* the first spacecraft to orbit another planet, reached Mars in 1971 amid a raging dust storm. When the dust cleared, images of Mars's northern hemisphere revealed numerous large volcanoes. The biggest, Olympus Mons, is the size of Ohio and 23 kilometers (75,000 feet) tall, nearly three times higher than Mount Everest. This gigantic volcano was last active about 100 million years ago and resembles Hawaiian shield volcanoes on Earth (Figure 15.26).

Why are the volcanoes on Mars many times larger than even the biggest volcanic structures on Earth? The largest volcanoes on terrestrial planets tend to form where plumes of hot rock rise from deep within its interior. Earth is tectonically active, with moving plates that keep the crust in constant motion. The Hawaiian Islands, for example, consist of a chain of shield volcanoes that formed as the Pacific plate moved over a relatively stationary mantle plume. On Mars, volca-

Figure 15.26 Image of Mons Olympus, an inactive shield volcano on Mars that covers an area about the size of the state of Ohio. (Courtesy of the U.S. Geological Survey, U.S. Department of the Interior)

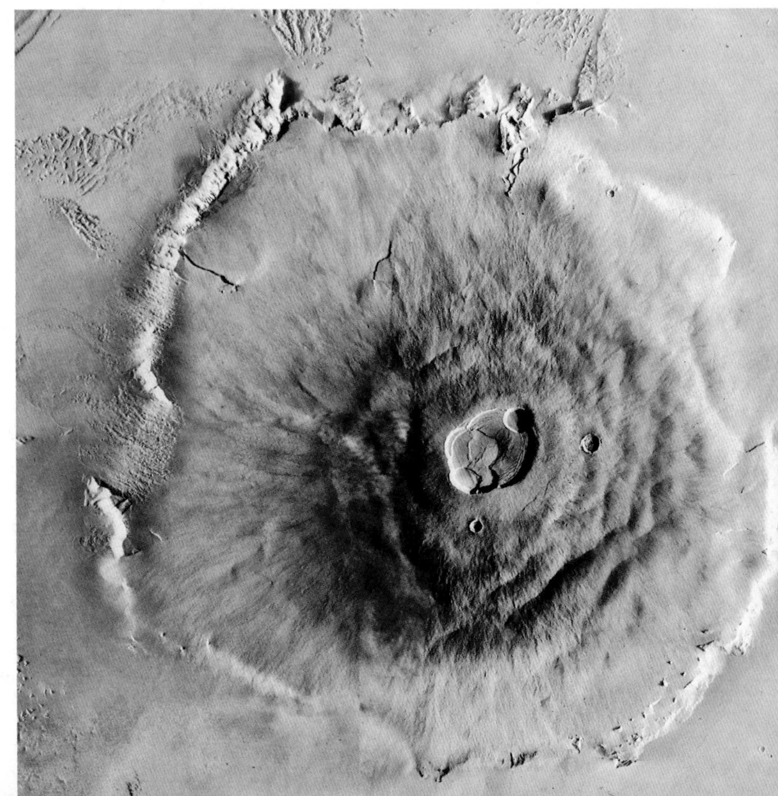

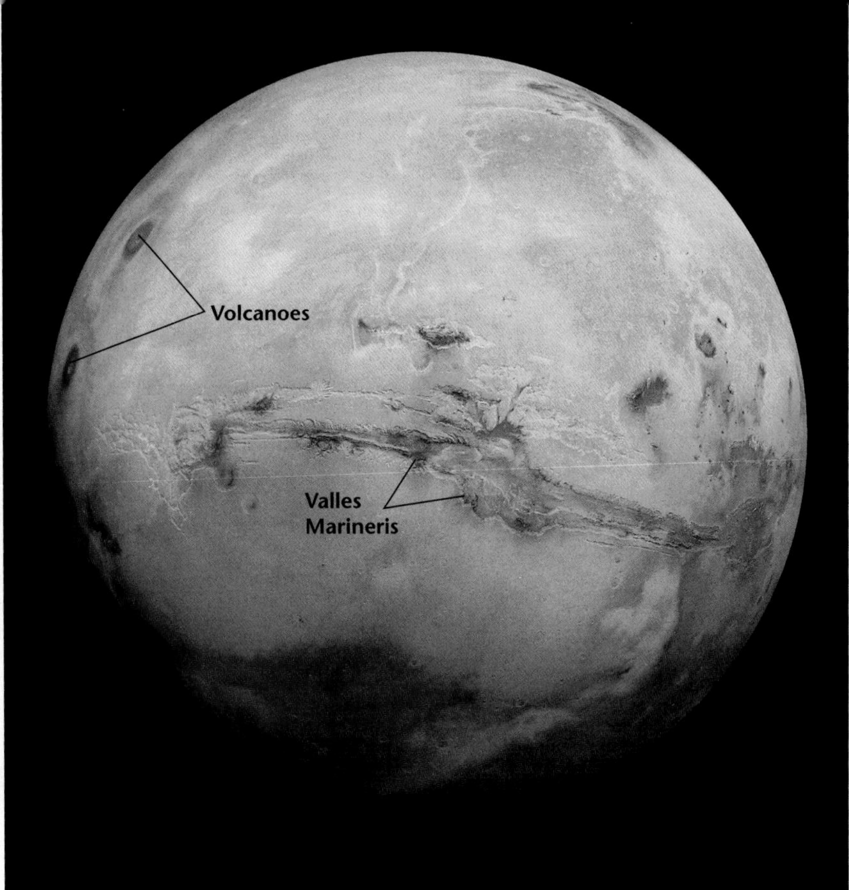

Figure 15.27 This image shows the entire Valles Marineris canyon system, more than 2000 kilometers (1240 miles) long and up to 8 kilometers (5 miles) deep. The dark red spots on the left edge of the image are huge volcanoes, each about 25 kilometers (15.5 miles) high. (Courtesy of U.S. Geological Survey, U.S. Department of the Interior)

noes such as Olympus Mons have grown to great size because the crust there remains stationary. Successive eruptions occur from a mantle plume at the same location and add to the bulk of a single volcano rather than producing several smaller structures, as occurs on Earth.

Another surprising find made by *Mariner 9* was the existence of several canyons that dwarf even Earth's Grand Canyon of the Colorado River. One of the largest, Valles Marineris, is thought to have formed by slippage of material along huge faults in the outer crustal layer (Figure 15.27). In this respect, it would be comparable to the rift valleys of East Africa.

Water on Mars? Liquid water does not appear to exist anywhere on the Martian surface. However, poleward of about 30 degrees latitude, ice can be found within a meter of the surface, and in the polar regions, it forms small permanent ice caps. In addition, considerable evidence indicates that in the first billion years of the planet's history, liquid water flowed on the surface, creating valleys and related features. In particular, some areas of Mars exhibit treelike drainage patterns similar to those created by streams on Earth. When these streamlike channels were first discovered, some observers speculated that a thick, water-rich atmosphere capable of generating torrential downpours once existed on Mars. Other researchers, however, were skeptical.

Today, most planetary geologists agree that running water was involved in carving at least some of the valleys on Mars. Some examples can be seen on an image from the *Mars Global Surveyor* in Figure 15.28A. Researchers have proposed that melting of subsurface ice caused springlike seeps to emerge along this valley wall, slowly creating the gullies shown.

Other channels have streamlike banks and contain numerous teardrop-shaped islands (15.28B). These valleys appear to have been cut by catastrophic floods that had discharge rates that were more than 1000 times greater than those of the Missippippi River. Most of these large flood channels emerge from areas of chaotic topography that appear to have formed when the surface collapsed. The most likely source of water for these flood valleys is the melting of subsurface ice. If the meltwater was trapped beneath a thick layer of permafrost, pressure could mount until a catastrophic release of groundwater occurred. As a result, the overlying surface layer would collapse, creating the chaotic terrain.

Not all Martian valleys appear to be the result of water released by the melting of subsurface ice. Some that exhibit branching treelike patterns that closely resemble the drainage networks on Earth were likely part of an active hydrologic cycle.

Recent evidence for surface water comes from the *Spirit* and *Opportunity* rovers. *Opportunity* investigated geologic structures similar to formations created by water on Earth—layered sedimentary rock, playas, and lake beds. Minerals that form only in the presence of water such as hydrated sulfates and micas were detected. Small spheres of hematite, called "blueberries," that probably precipitated from water to form lake sediments were also found. Although these findings are new, water does not appear to have significantly altered the topography of Mars for billions of years with the exception of the polar regions. Near the poles, the *Mars Express Orbiter* and *Mars Odyssey* detected geologically recent glacial and snow deposits. Polygonal patterned ground similar to features associated with permafrost landscapes on Earth are also commonly seen on Mars (Figure 15.29).

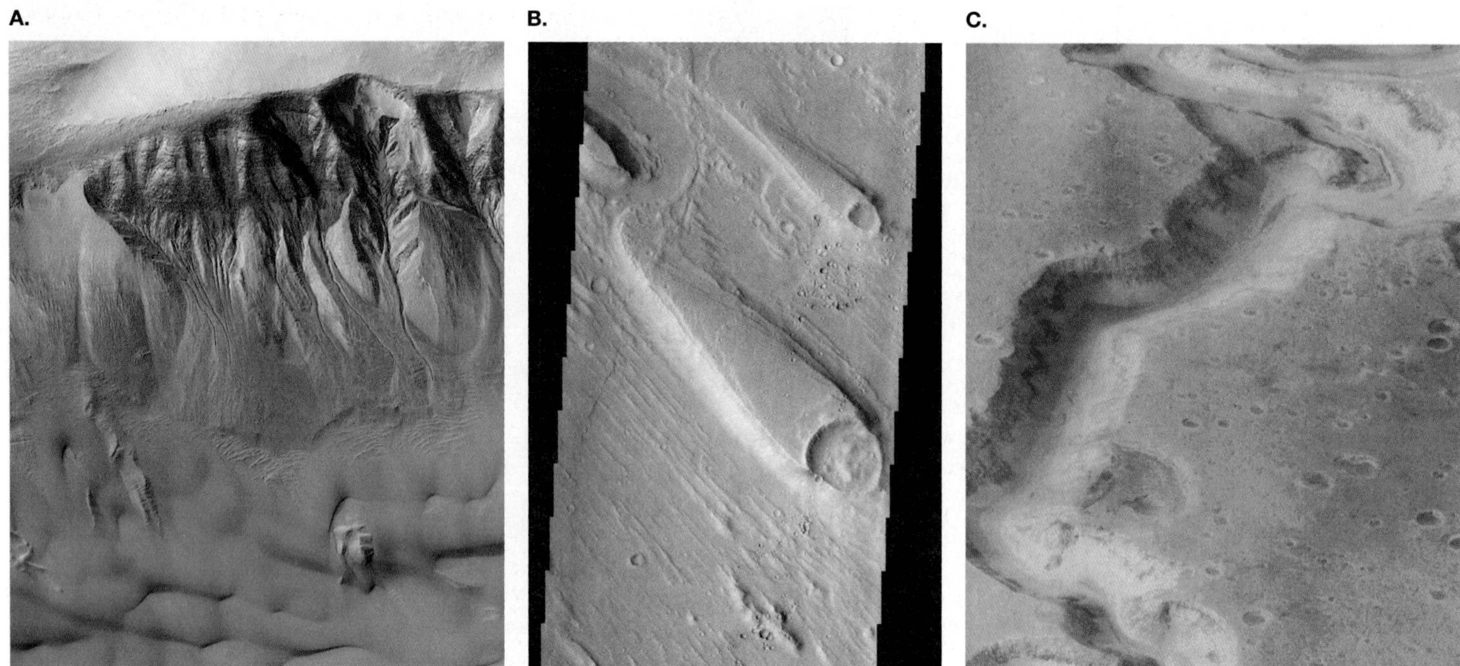

Figure 15.28 Evidence for erosion by running water. **A.** Image from *Mars Global Surveyor* showing crater wall with large gullies that may have been cut by liquid water, mixed with soil, rocks, and ice. (NASA/JPL photo courtesy of National Geographic) **B.** Streamlined islands in Ares Valles formed where running water encountered obstacles along its path. (NASA) **C.** Terraces and a small central channel suggest that Nanedi Ballis may have been carved by running water. (NASA)

Jupiter: Lord of the Heavens

Jupiter, truly a giant among planets, has a mass two and a half times greater than the combined mass of all the remaining planets, satellites, and asteroids. In fact, had Jupiter been about 10 times larger, it would have evolved into a small star. Despite its great size, however, it is only 1/800 as massive as the Sun.

Jupiter revolves around the Sun once every 12 Earth-years, and it rotates more rapidly than any other planet, completing one rotation in slightly less than 10 hours. The effect of this fast spin is to make the equatorial region bulge and to make the polar dimension contract (see the Polar Flattening column in Table 15.2, p. 407).

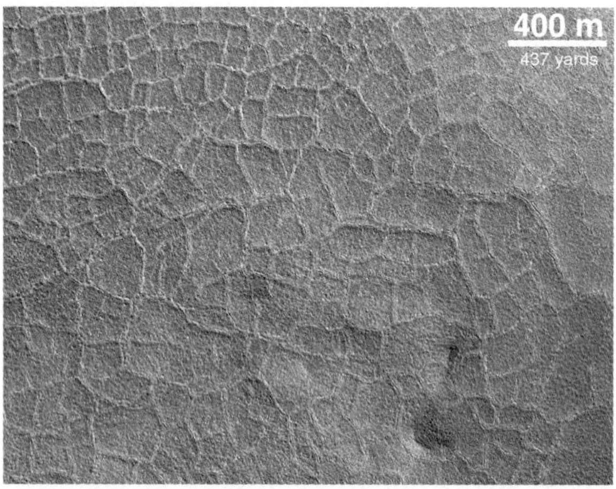

Figure 15.29 Polygonal patterned ground, shown here on Mars, is also a common feature in areas of permafrost in polar latitudes on Earth. (Courtesy of JPL/NASA)

Atmosphere, Structure, and Composition When viewed through a telescope or binoculars, Jupiter appears to be covered with alternating bands of multicolored clouds aligned parallel to its equator (Figure 15.30). The most striking feature is the *Great Red Spot* in the southern hemisphere (Figure 15.30). The Great Red Spot has been a prominent feature since it was first seen more than three centuries ago. When *Voyager 2* swept by Jupiter in 1979, the Great Red Spot was the size of two Earth-size circles placed side by side. On occasion, it has grown even larger. Images from *Pioneer 11* as it moved near Jupiter's cloud tops in 1974 indicated that the Great Red Spot is a counterclockwise-rotating storm caught between two jetstream-like bands flowing in opposite directions. This huge, hurricane-like storm makes a complete rotation about once every 12 days.

As on other planets, the winds on Jupiter are the product of differential heating, which generates vertical convective motions in the atmosphere. Jupiter's convective flow produces alternating dark-colored *belts* and light-colored *zones* as shown in Figure 15.31. The light clouds (zones) are regions where warm material is ascending and cooling, whereas cool material is sinking in the dark belts. This convective circulation, along with Jupiter's rapid rotation, generates the high-speed, east-west flow observed between the belts and zones. Unlike winds on Earth, which are driven by solar energy, Jupiter gives off nearly twice as much heat as it receives from the Sun. Thus, it is the heat emanating from Jupiter's interior that produces the huge convection currents observed in its atmosphere.

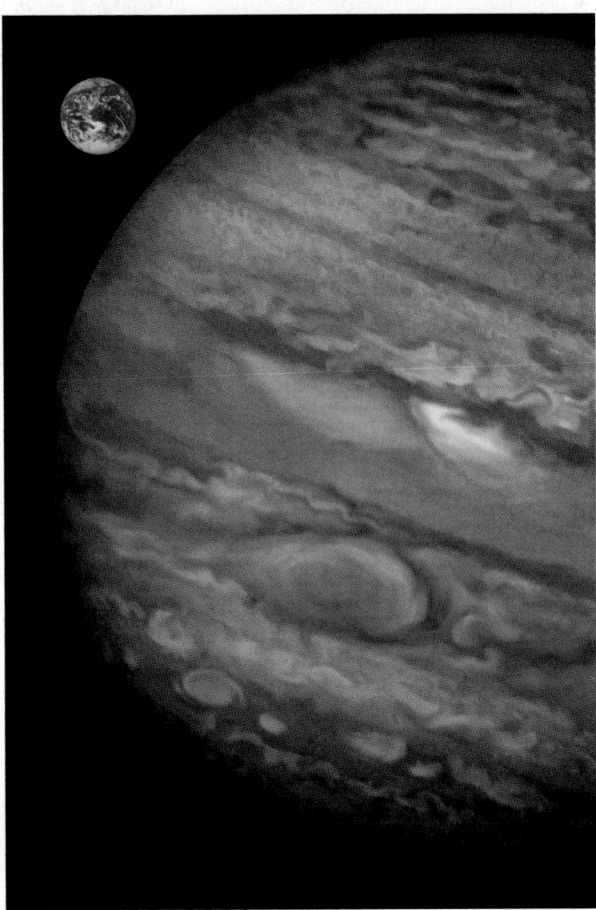

Figure 15.30 Artist's view of Jupiter with Great Red Spot visible in its southern hemisphere. Earth for scale.

Jupiter's atmosphere is composed mainly of hydrogen and helium but also contains lesser amounts of methane, ammonia, and water, which form clouds composed of liquid droplets or ice crystals. Atmospheric pressure at the top of the clouds is equal to sea-level pressure on Earth. Because of Jupiter's immense gravity, the pressure increases rapidly toward its surface. At 1000 kilometers (620 miles) below the clouds, the pressure is great enough to compress hydrogen gas into a liquid. Consequently, Jupiter's surface is thought to be a gigantic ocean of liquid hydrogen. Less than halfway into Jupiter's interior, extreme pressures cause the liquid hydrogen to turn into *liquid metallic* hydrogen. The fast rotation and liquid metallic core are a possible explanation for the intense magnetic field surrounding Jupiter. Jupiter is also believed to contain as much rocky and metallic material as is found in the terrestrial planets, probably located in a central core.

Jupiter's Moons Jupiter's satellite system, consisting of the 63 moons discovered so far, resembles a miniature solar system. The four largest satellites, discovered by Galileo, travel in nearly circular orbits around the planet, with periods of from 2 to 17 Earth-days (Figure 15.32). The two largest Galilean satellites, Callisto and Ganymede, surpass Mercury in size, whereas the two smaller ones, Europa and Io, are about the size of Earth's Moon. The Galilean moons can be observed with binoculars or a small telescope and are interesting in their own right.

Images from *Voyagers 1* and *2* revealed, to the surprise of almost everyone, that each of the four Galilean satellites is a unique geologic world (Figure 15.32). A further surprise from the *Galileo* mission was that the composition of each satellite is strikingly different, which implies a different evolution for each one. For example, Ganaymede's has a dynamic core that generates a strong magnetic field not observed on the other satellites.

The innermost of the Galilean moons, Io, is perhaps the most volcanically active body in our solar system. In all, more than 80 active sulfurous volcanic centers have been discovered. Umbrella-shaped plumes have been observed rising from the surface of Io to heights approaching 200 kilometers (120 miles) (Figure 15.33A). The heat source for volcanic activity is tidal energy generated by a relentless "tug of war" between Io and Jupiter and the other Galilean satellites. The gravitational field of Jupiter and the other nearby satellites

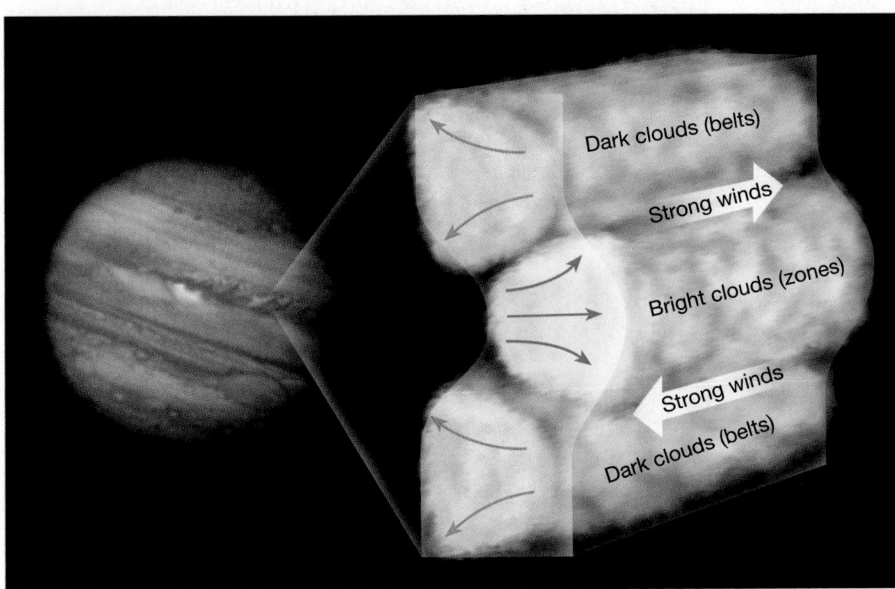

Figure 15.31 The structure of Jupiter's atmosphere. The areas of light clouds (zones) are regions where gases are ascending and cooling. Sinking dominates the flow in the darker cloud layers (belts). This convective circulation, along with the rapid rotation of the planet, generates the high-speed winds observed between the belts and zones.

A. Io B. Europa C. Ganymede D. Callisto

Figure 15.32 Jupiter's four largest moons (from left to right) called the Galilean moons because they were discovered by Galileo. **A.** The innermost moon, Io, is one of only three volcanically active bodies in the solar system. **B.** Europa, smallest of the Galilean moons, has an icy surface that is criss-crossed by many linear features. **C.** Ganymede, the largest Jovian satellite, contains cratered areas, smooth regions, and areas covered by numerous parallel grooves. **D.** Callisto, the outermost of the Galilean satellites, is densely cratered, much like Earth's moon. (Courtesy of NASA/NGS Image Collection)

pulls and pushes on Io's tidal bulge as its slightly eccentric orbit takes it alternately closer to and farther from its parent planet. This gravitational flexing of Io is transformed into heat energy (similar to the back-and-forth bending of a paper clip) and results in Io's spectacular sulfurous volcanic eruptions. Moreover, lava thought to be mostly composed of silicate minerals regularly erupts on its surface (Figure 15.33B).

In addition, Jupiter has numerous satellites that are very small (about 20 kilometers [12 miles] in diameter), revolve in a direction that is opposite (*retrograde motion*) to that of the largest moons, and have orbits that are steeply inclined to the Jovian equator. These satellites appear to be asteroids that passed near enough to be captured gravitationally by Jupiter.

Jupiter's Rings One of the interesting aspects of the *Voyager 1* mission was a study of Jupiter's ring system. By analyzing how these rings scatter light, researchers determined that the rings are composed of fine, dark particles, similar in size to smoke particles. Further, the faint nature of the rings indicates that these minute particles are widely dispersed. The main ring is composed of particles believed to be fragments blasted by meteorite impacts from the surfaces of Metis and Adrastea, two small moons of Jupiter. Impacts on Jupiter's moon's Amalthea and Thebe are believed to be the sources of the outer Gossamer ring.

Saturn: The Elegant Planet

Requiring 29.46 Earth-years to make one revolution, Saturn is almost twice as far from the Sun as Jupiter, yet its atmosphere, composition, and internal structure are believed to be remarkably similar to Jupiter's. The most prominent feature of Saturn is its system of rings (Figure 15.34), first seen by Galileo in 1610. With his primitive telescope, the rings appeared as two small bodies adjacent to the planet. Their ring nature was determined 50 years later by the Dutch astronomer Christian Huygens.

Figure 15.33 A volcanic eruption on Jupiter's moon Io. **A.** This plume of volcanic gases and debris is rising more than 100 kilometers (60 miles) above Io's surface. **B.** The bright area on the left side of the image is newly erupted hot lava. (NASA)

A.

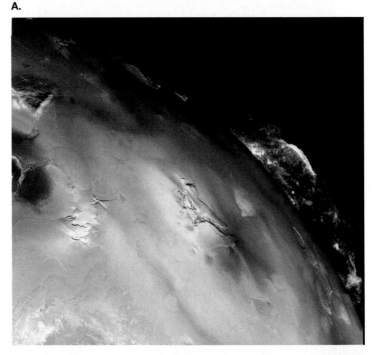

B.

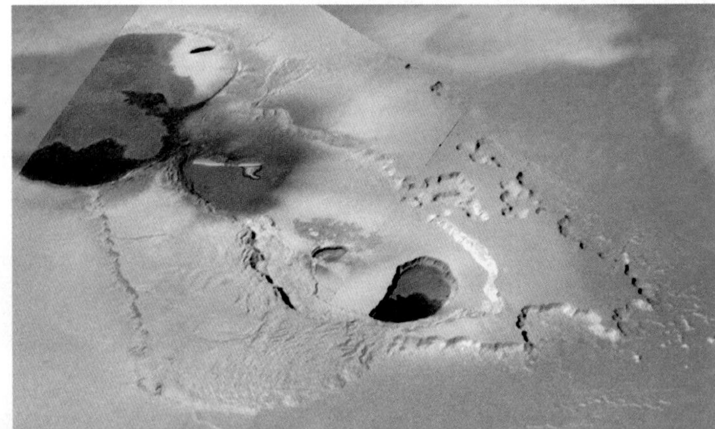

Structure of Saturn Saturn's atmosphere is quite dynamic, with winds roaring at up to 1500 kilometers (930 miles) per hour. Cyclonic "storms" similar to Jupiter's Great Red Spot occur in Saturn's atmosphere, as does intense lightning. Although the atmosphere is nearly 75 percent hydrogen and 25 percent helium, the clouds are composed of ammonia, ammonia hydrosulfide, and water, each segregated by temperature. The central core is composed of rock and ice layered with liquid metallic hydrogen, and then liquid hydrogen. Like Earth, Saturn's magnetic field is believed to be created within the core. In this process, helium condenses in the liquid hydrogen layers, releasing the heat necessary for convection.

In 1980 and 1981, the nuclear-powered *Voyagers 1* and *2* space probes came within 100,000 kilometers (62,000 miles) of Saturn. More information was gained in a few days than had been acquired in all the time since Galileo first viewed this elegant planet telescopically. More recently, observations from ground-based telescopes and the *Hubble Space Telescope* have added to our knowledge of Saturn's ring system. In 1995 and 1996, when the positions of Earth and Saturn allowed the rings to be viewed edge-on, reducing the glare from the main rings, Saturn's faintest rings and satellites became visible.

Planetary Ring Systems Until the recent discovery that Jupiter, Uranus, and Neptune also have ring systems, this phenomenon was thought to be unique to Saturn. Although the four known ring systems differ in detail, they share many attributes. They all consist of multiple concentric rings separated by gaps of various widths. In addition, each ring is composed of individual particles—"moonlets" of ice and rock—that circle the planet while regularly impacting one another.

Most rings fall into one of two categories based on particle density. Saturn's main rings (designated A and B in Figure 15.34) and the bright rings of Uranus are tightly packed and contain moonlets that range in size from a few centimeters (pebble size) to several meters (house size). These particles are thought to collide frequently as they orbit the parent planet. Despite the fact that Saturn's dense rings stretch across several hundred kilometers, they are very thin, perhaps less than 100 meters (330 feet) from top to bottom.

At the other extreme, the faintest rings, such as Jupiter's ring system and Saturn's outermost ring (designated E in Figure 15.34), are composed of very fine (smoke-size) particles that are widely dispersed. In addition to having very low particle densities, these rings tend to be thicker than Saturn's bright rings.

Recent studies have shown that the moons that coexist with the rings play a major role in determining their structure. In particular, the gravitational influence of these moons tends to shepherd the ring particles by altering their orbits. The narrow rings appear to be the work of satellites located on either side that confine the ring by pulling back particles that try to escape.

More importantly, the ring particles are believed to be debris ejected from these moons, including the volcanic "ash" that spews from Jupiter's moon Io. Recent evidence from the *Cassini* mission shows the moon Enceladus has a hot spot with geysers spewing water ice that contributes to Saturn's E ring. It is possible that material is continually being recycled between the rings and the ring moons. The moons gradually sweep up particles, which are subsequently ejected by collisions with large chunks of ring material, or perhaps by energetic collisions with other moons. It seems, then, that

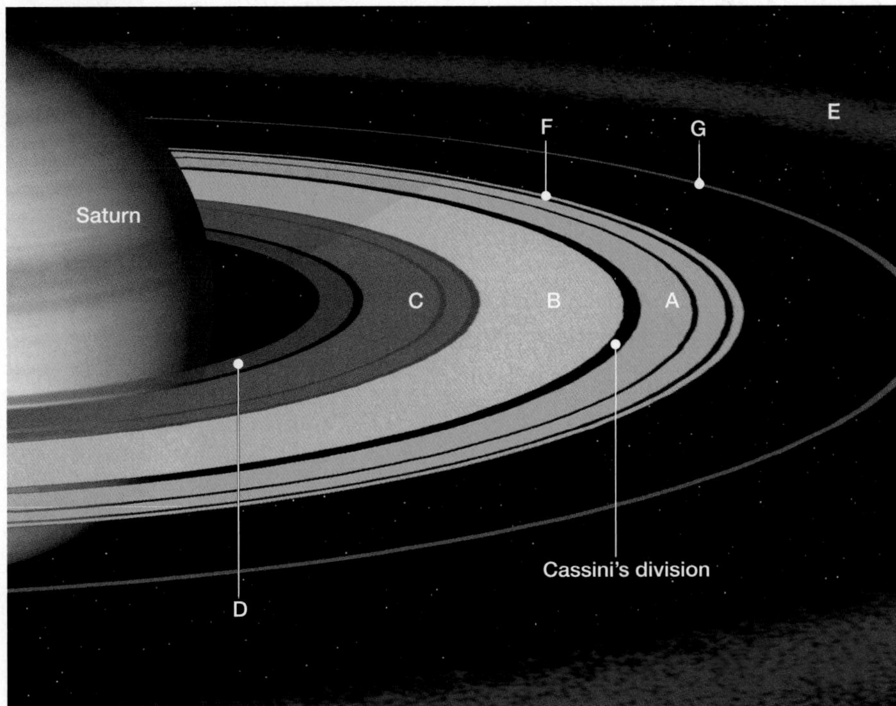

Figure 15.34 A view of the dramatic ring system of Saturn.

planetary rings are not the timeless features that we once thought; rather, they continually reinvent themselves.

The origin of planetary ring systems is still being debated. Perhaps the rings formed out of a flattened cloud of dust and gases that encircled the parent planet. In this scenario, the rings formed at the same time and from the same material as the planets and moons. Perhaps the rings formed later, when a moon or large asteroid was gravitationally pulled apart after straying too close to a planet. Yet another hypothesis suggests that a foreign body blasted apart one of the planet's moons; the fragments of which would tend to jostle one another and form a flat, thin ring. Researchers expect more light to be shed on the origin of planetary rings as the *Cassini* spacecraft continues its four-year tour of Saturn.

Uranus and Neptune: The Twins

While Earth and Venus have many similar traits, Uranus and Neptune are nearly twins, having similar structures and compositions. They are less than 1 percent different in diameter (about four times the size of Earth), and they are both bluish in appearance, which is attributable to the methane in their atmospheres (Figures 15.35 and 15.36). Uranus and Neptune take 84 and 165 Earth-years, respectively, to complete one revolution around the Sun. The core composition is similar to that of the other gas giants with a rocky silicate and iron core, but with less liquid metallic hydrogen and more ice than Jupiter and Saturn. Neptune, however, is colder, because it is half again as distant from the Sun as is Uranus.

Figure 15.35 This image of Uranus was sent back to Earth by *Voyager 2* as it passed by this planet on January 24, 1986. Taken from a distance of nearly 1 million kilometers (620,000 miles) little detail of its atmosphere is visible, except for a few streaks (clouds) in the northern hemisphere. (NASA)

Uranus: The Sideways Planet A unique feature of Uranus is that it rotates "on its side." Its axis of rotation, instead of being generally perpendicular to the plane of its orbit, like that of the other planets, lies nearly parallel to the plane of its orbit. Its rotational motion, therefore, has the appearance of a rolling ball, rather than the toplike spinning of other planets. This is likely due to a huge impact early in the planet's evolution that literally knocked Uranus on its side. Because

Did You Know?

Uranus is often referred to as the "sideways planet." The most likely explanation for the unusual sideways spin of Uranus is that it started out spinning the same way as the other planets, but then its spin was altered by a giant impact—which was probably a common occurrence when the planets were first formed. However, a giant impact would be difficult to verify because it would not have left any crater on Uranus, which has no solid surface. Like many events that occurred early on in the formation of our solar system, the reason for Uranus's sideways spin may never be known for certain.

Figure 15.36 This image of Neptune shows the Great Dark Spot (left center). Also visible are bright cirruslike clouds that travel at high speed around the planet. A second oval spot is at 54° south latitude on the lower right portion of the image. (Photo courtesy of the Jet Propulsion Laboratory/NASA Headquarters)

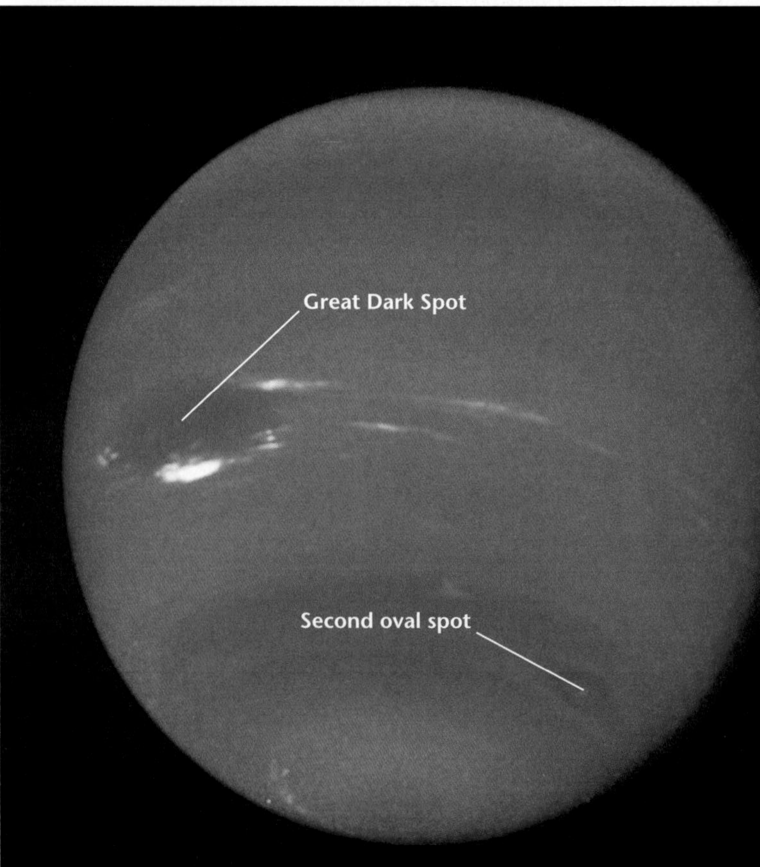

the axis of Uranus is inclined more than 90 degrees, the Sun is nearly overhead at one of its poles once each revolution, and then half a revolution later it is overhead at the other pole.

A surprise discovery in 1977 revealed that Uranus has a ring system. This find occurred as Uranus passed in front of a distant star and blocked its view, a process called *occultation* (*occult* = hidden). Observers saw the star "wink" briefly five times (meaning five rings) before the primary occultation and again five times afterward. Later studies indicate that Uranus has *at least* nine distinct belts of debris orbiting its equatorial region.

Spectacular views from *Voyager 2* of the five largest moons of Uranus show quite varied terrains. Some have long, deep canyons and linear scars, whereas others possess large, smooth areas on otherwise crater-riddled surfaces. The Jet Propulsion Laboratory described Miranda, the innermost of the five largest moons, as having a greater variety of landforms than any body yet examined in the solar system. These features indicate that Miranda was recently geologically active.

Neptune: The Windy Planet Even when the most powerful telescope is focused on Neptune, it appears as a bluish fuzzy disk. Until the 1989 *Voyager 2* encounter, astronomers knew very little about this planet. However, the 12-year, nearly 3-billion-mile journey of *Voyager 2* provided investigators with so much new information about Neptune and its satellites that years will still be needed to analyze it all.

Neptune has a dynamic atmosphere, much like those of Jupiter and Saturn (Figure 15.36). Winds exceeding 1000 kilometers (620 miles) per hour encircle the planet, making it one of the windiest places in the solar system. It also has an Earth-size blemish called the *Great Dark Spot* that is reminiscent of Jupiter's Great Red Spot and is assumed to be a large rotating storm. About five years after the *Voyager 2* encounter, when the *Hubble Space Telescope* viewed Neptune, the spot had vanished and was replaced by another dark spot in the planet's northern hemisphere.

Perhaps most surprising are white, cirruslike clouds that form a layer about 50 kilometers (30 miles) above the main cloud deck, probably frozen methane. Six new satellites were discovered in the *Voyager* images, bringing Neptune's family to 8, and more recent observations bring the total to 13. All of the newly discovered moons orbit the planet in a direction opposite that of the two larger satellites. *Voyager* images also revealed a ring system around Neptune.

Triton, Neptune's largest moon, is a most interesting object. It is the only large moon in the solar system that exhibits retrograde motion. This indicates that Triton formed independently of Neptune and was gravitationally captured.

Triton, along with other moons of the Jovian planets, exhibits one of the most amazing manifestations of volcanism, the eruption of ices. This type of volcanism is termed **cryovolcanism** (from the Greek *Kryos*, meaning "frost") and refers to the eruption of magmas derived from the partial melting of ice rather than silicate rocks. Triton's icy magma originates from a mixture of water-ice, methane, and probably ammonia. When partially melted, this mixture behaves just as molten rock does on Earth. In fact, upon reaching the surface, these magmas can generate quiet outpourings of ice lavas, or even explosive eruptions. When volatiles are released instantaneously, an explosive eruptive column will generate the ice equivalent of volcanic ash. In 1989, *Voyager 2* detected active plumes on Triton that rose 8 kilometers (5 miles) above the surface and were blown downwind for more than 100 kilometers (62 miles). In other environments, ice lavas develop that can flow great distances from their source—not unlike the fluid basaltic flows on Hawaii.

Minor Members of the Solar System: Asteroids, Comets, Meteoroids, and Dwarf Planets

In the vast spaces separating the eight planets as well as the expanses that extend to the outer reaches of the solar system are countless small chunks of debris, ranging from several hundred kilometers in diameter down to minute grains of dust. The objects found in this vast interplanetary space include *asteroids, comets, meteoroids,* and *dwarf planets.* Asteroids and meteoroids are fragments of rocky and metallic material with compositions somewhat like the terrestrial planets. They are distinguished from each other on the basis of size—those larger than 100 meters (330 feet) are asteroids, whereas anything smaller is a meteoroid. By contrast, comets are predominantly ices, with only small amounts of rocky material. The newest class of solar-system objects—dwarf planets—includes Ceres, the largest known asteroid, and Pluto, which is thought to have a composition similar to Neptune's icy moon, Triton.

Asteroids: Planetesimals

Asteroids are small fragments (planetesimals) about 4.5 billion years old left over from the formation of the solar system. Most asteroids have a lower density than was first thought. This suggests that asteroids are porous bodies and not solid rocky or metallic objects, but more like a "rubble pile" of fragments bound together by gravity. The largest asteroid, Ceres, is 940 kilometers (580 miles) in diameter, but most of the 100,000 known asteroids are much smaller.

Most asteroids lie roughly midway between the orbits of Mars and Jupiter in the region known as the **asteroid belt** (Figure 15.37). Some travel along eccentric orbits that take them very near the Sun, and a few larger ones regularly pass close to Earth and the Moon. Many of the recent large impact craters on the Moon and Earth were probably the result of collisions with asteroids. About 2000 Earth-crossing asteroids are known, one-third of which are more than one kilometer (0.6 mile) in diameter. Inevitably, future Earth-asteroid collisions will occur.

Because most asteroids have irregular shapes, planetary geologists first speculated that they might be fragments of a broken planet that once orbited between Mars and Jupiter

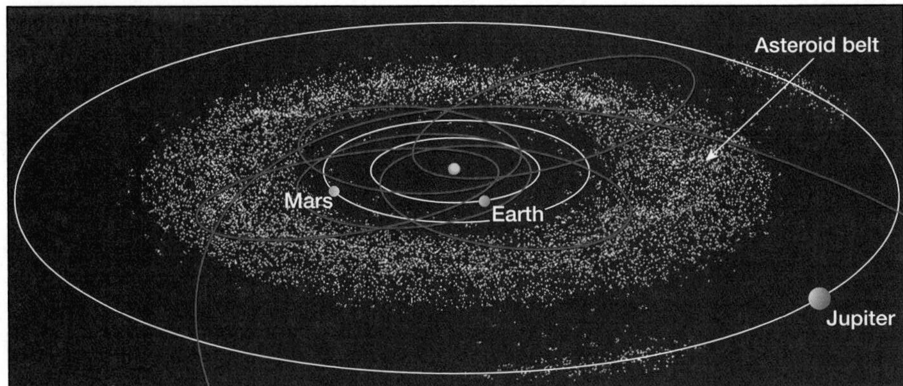

Figure 15.37 The orbits of most asteroids lie between Mars and Jupiter. Also shown are the orbits of a few known near-Earth asteroids. Perhaps a thousand or more asteroids have near-Earth orbits. Luckily, only a few dozen are thought to be larger than 1 kilometer (0.6 mile) in diameter.

(Figure 15.37). However, the total mass of the asteroids is estimated to be only $\frac{1}{1000}$ that of Earth, which itself is not a large planet. Today, most researchers agree that asteroids are the leftover debris from the solar nebula. Because of their location near Jupiter, with its huge gravitational field that continuously disrupted their motion, these planetesimals never accreted into a planet.

In February 2001, an American spacecraft became the first visitor to an asteroid. Although it was not designed for landing, *NEAR Shoemaker* landed successfully and generated information that has planetary geologists intrigued and perplexed. Images obtained as the spacecraft drifted toward the surface of Eros revealed a barren, rocky surface composed of particles ranging in size from fine dust to boulders up to 10 meters (33 feet) across (Figure 15.38). Researchers unexpectedly discovered that fine debris tends to concentrate in the low areas where it forms flat deposits that resemble ponds. Surrounding the low areas, the landscape is marked by an abundance of large boulders.

One of several hypotheses to explain the boulder-strewn topography is seismic shaking, which causes the boulders to move upward as the finer material sinks. This is analogous to what happens when a can of mixed nuts is shaken—the larger nuts rise to the top while the smaller pieces settle to the bottom.

Indirect evidence from meteorites suggests asteroids might retain much of the heat generated from impact events. Some may even have completely melted, which would cause them to differentiate into a dense iron and nickel core and a rocky mantle. In November 2005, a Japanese probe, *Hayabusa*, landed on a small near-Earth asteroid named 25143 Itokawa and is scheduled to return samples to Earth by June 2010.

Comets: Dirty Snowballs

Comets, like asteroids, are also left over from the formation of the solar system. Unlike asteroids, comets are composed of ices (water, ammonia, methane, carbon dioxide, and carbon monoxide) that hold together small pieces of rocky and metallic materials, thus the nickname "dirty snowballs." Comets are among the most interesting and unpredictable bodies in the solar system. Many comets travel in very elongated orbits that carry them far beyond Pluto. These comets take hundreds of thousands of years to complete a single orbit around the Sun. However, a few *short-period comets* (those having orbital periods of less than 200 years), such as Halley's comet, make regular encounters with the inner solar system.

When first observed, a comet appears very small, but as it approaches the Sun, solar energy begins to vaporize the ices, producing a glowing head called the **coma** (Figure 15.39). The size of the coma varies greatly from one comet to another. Extremely rare ones exceed the size of the Sun, but most approximate the size of Jupiter. Within the coma, a small glowing nucleus with a diameter of only a few kilometers can sometimes be detected. As comets approach the Sun, some, but not all, develop a tail that extends for millions of kilometers. Despite the enormous size of their tails and

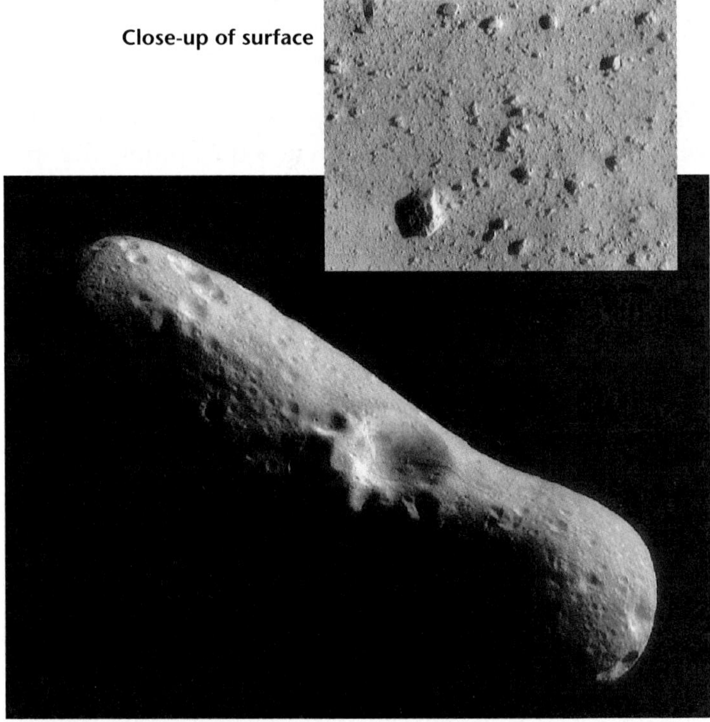

Figure 15.38 Image of asteroid Eros obtained by the *NEAR Shoemaker* probe. Inset shows a close-up of Eros's barren rocky surface. (NASA)

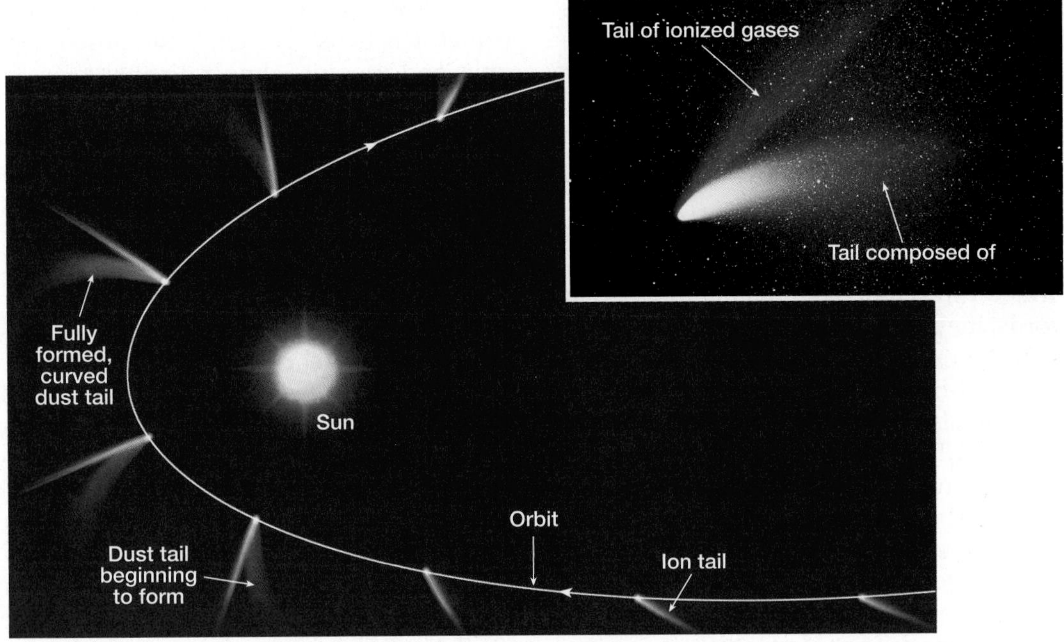

Figure 15.39 Orientation of a comet's tail as it orbits the Sun. Inset photo of comet Hale-Bopp. The two tails seen in the photograph are between 16 and 24 million kilometers (10 million and 15 million miles) long. (Peoria Astronomical Society photograph by Eric Clifton and Craig Neaveill)

comas, comets are relatively small members of the solar system.

The fact that the tail of a comet points away from the Sun in a slightly curved manner (Figure 15.39) led early astronomers to propose that the Sun has a repulsive force that pushes the particles of the coma away, thus forming the tail. Today, two solar forces are known to contribute to this formation. One, *radiation pressure,* pushes dust particles away from the coma. The second, known as *solar wind,* is responsible for moving the ionized gases, particularly carbon monoxide. Sometimes a single tail composed of both dust and ionized gases is produced, but often two tails are observed (Figure 15.39).

As a comet moves away from the Sun, the gases forming the coma recondense, the tail disappears, and the comet returns to cold storage. Material that was blown from the coma to form the tail is lost from the comet forever. Consequently, it is believed that most comets cannot survive more than a few hundred close orbits of the Sun. Once all the gases are expelled, the remaining material—a swarm of tiny metallic and stony particles—continues the orbit without a coma or a tail.

Most comets are found in two regions of the outer solar system. The short-period comets are thought to orbit beyond Neptune in a region called the **Kuiper belt,** in honor of astronomer Gerald Kuiper, who had predicted their existence (Figure 15.40). (During the past decade, more than a hundred of these icy bodies have been discovered.) Like the asteroids in the inner solar system, most Kuiper belt comets move in nearly circular orbits that lie roughly in the same plane as the planets. A chance collision between two Kuiper belt comets, or the gravitational influence of one of the Jovian planets, may occasionally alter the orbit of a comet enough to send it to the inner solar system, and into our view.

Unlike Kuiper belt comets, long-period comets have orbits that are *not* confined to the plane of the solar system. These comets appear to be distributed in all directions from the Sun, forming a spherical shell around the solar system, called the **Oort cloud,** after the Dutch astronomer Jan Oort.

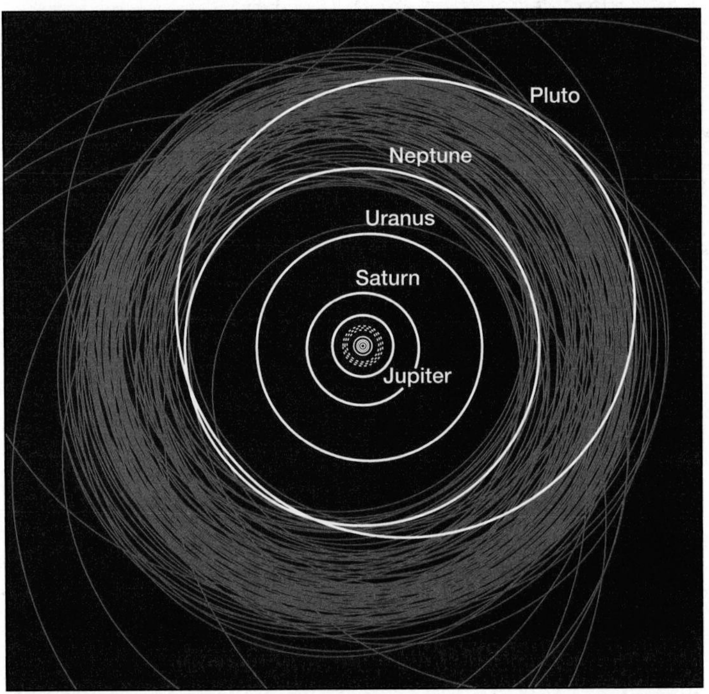

Figure 15.40 The orbits of some known Kuiper belt objects.

Millions of comets are believed to orbit the Sun at distances greater than 10,000 times the Earth-Sun distance. The gravitational effect of a distant passing star is believed to send an occasional Oort cloud comet into a highly eccentric orbit that carries it toward the Sun. However, only a tiny fraction of Oort cloud comets have orbits that bring them into the inner solar system.

Why are comets found in both the Kuiper belt and the Oort cloud? It appears that comets, like asteroids, are planetesimal bodies formed by accretion of material in the solar nebula. In other words, they are leftovers from the planetary formation process. In contrast to asteroids that formed in the inner solar system of mostly metallic and rocky materials, comets are icy planetesimals that formed in the region of the giant planets and beyond. The ones that formed beyond the orbit of Neptune and did not accrete into a large planetary body make up the Kuiper belt objects.

The icy planetesimals of the Oort cloud, on the other hand, are thought to have formed in the region of the Jovian planets. While the vast majority of icy objects would have accreted into protoplanets, some did not. Under the strong gravitational fields of the giants, their orbits would have been greatly altered. Some cometary objects would have been hurled toward the inner solar system, where they collided with the terrestrial planets. The remainder were "thrown out" into space to form the Oort cloud located at the outermost reaches of the solar system.

The most famous short-period comet is Halley's Comet. Its orbital period averages 76 years, and every one of its 29 appearances since 240 B.C. has been recorded by Chinese astronomers. This record is a testimonial to their dedication as astronomical observers and to the endurance of their culture. When seen in 1910, Halley's Comet had developed a tail nearly 1.6 million kilometers (1 million miles) long and was visible during the daytime.

In 1986, the unspectacular showing of Halley's Comet was a disappointment to many people in the Northern Hemisphere. Yet, it was during this most recent visit to the inner solar system that a great deal of new information was learned about this most famous of comets. The new data were gathered by space probes sent to rendezvous with the comet. Most notably, the European probe *Giotto* approached to within 600 kilometers (370 miles) of the comet's nucleus and obtained the first close-up images of comet structure.

We now know that the nucleus of Halley's Comet is potato-shaped and 16 by 8 kilometers (10 by 5 miles) in size. The surface is irregular and full of craterlike pits. Gases and dust that escape from the nucleus to form the coma and tail appear to gush from its surface as bright jets or streams. Only about 10 percent of the comet's total surface was emitting these jets at the time of the rendezvous. The remaining surface area of the comet appeared to be covered with a dark layer that may be organic material.

In 1997, the comet Hale-Bopp made for spectacular viewing around the globe. As comets go, the nucleus of Hale-Bopp was unusually large, about 40 kilometers (25 miles) in diameter. As shown in Figure 15.39, two tails nearly 24 million kilometers (15 million miles) long extended from this comet. The bluish gas-tail is composed of positively charged ions, and it points almost directly away from the Sun. The brighter tail is composed of dust and other rocky debris. Because the rocky material is more massive than the ionized gases, it is less affected by the solar wind and follows a different trajectory away from the comet.

Meteoroids: Visitors to Earth

Nearly everyone has seen a **meteor,** popularly (but inaccurately) called a "shooting star." This streak of light lasts from an eyeblink to a few seconds and occurs when a small solid particle, a **meteoroid,** enters Earth's atmosphere from interplanetary space. Friction between the meteoroid and the air heats both and produces the light we see. Most meteoroids originate from any one of the following three sources: (1) interplanetary debris that was not gravitationally swept up by the planets during the formation of the solar system, (2) material that is continually being lost from the asteroid belt, or (3) the solid remains of comets that once passed through Earth's orbit. A few meteoroids are believed to be fragments of the Moon, or possibly Mars, that were ejected when an asteroid impacted these bodies.

Meteoroids less than about a meter in diameter generally vaporize before reaching Earth's surface. Some, called *micrometeorites,* are so tiny that their rate of fall becomes too slow to cause them to burn up, so they drift down as space dust. Each day, the number of meteoroids that enter Earth's atmosphere must reach into the thousands. After sunset on a clear night, a half dozen or more are bright enough to be seen with the naked eye each hour from anywhere on Earth.

Occasionally, meteor sightings increase dramatically to 60 or more per hour. These displays, called **meteor showers,** result when Earth encounters a swarm of meteoroids traveling in the same direction and at nearly the same speed as Earth. The close association of these swarms to the orbits of some short-term comets strongly suggests that they represent material lost by these comets (Table 15.3). Some swarms not associated with the orbits of known comets are probably the remains of the nucleus of a long-defunct comet. The notable Perseid meteor shower that occurs each year around August 12 is believed to be the remains of the Comet 1862 III, which has a period of 110 years.

Meteoroids that are the remains of comets tend to be small and only occasionally reach the ground. Most meteoroids large enough to survive the heated fall are thought to originate among the asteroids, where chance collisions modify their orbits and send them toward Earth. Earth's gravity does the rest.

The remains of meteoroids, when found on Earth, are referred to as **meteorites** (Figure 15.41). A few very large meteoroids have blasted out craters on Earth's surface that strongly resemble those on the lunar surface. The most famous is Meteor Crater in Arizona (Figure 15.42). This huge cavity is about 1.2 kilometers (0.75 mile) across and 170 meters (560 feet) deep and has an upturned rim that rises 50 meters (165 feet) above the surrounding countryside. More than

Table 15.3 Major meteor showers

Shower	Approximate Dates	Associated Comet
Quadrantids	January 4–6	—
Lyrids	April 20–23	Comet 1861 I
Eta Aquarids	May 3–5	Halley's comet
Delta Aquarids	July 30	—
Perseids	August 12	Comet 1862 III
Draconids	October 7–10	Comet Giacobini-Zinner
Orionids	October 20	Halley's comet
Taurids	November 3–13	Comet Encke
Andromedids	November 14	Comet Biela
Leonids	November 18	Comet 1866 I
Geminids	December 4–16	—

30 tons of iron fragments have been found in the immediate area, but attempts to locate the main body have been unsuccessful. Based on the amount of erosion, the impact likely occurred within the last 50,000 years.

Prior to the Moon rocks brought back by lunar explorers, meteorites were the only extraterrestrial materials that could be directly examined. Meteorites are classified by their composition: (1) *irons,* mostly iron with 5 to 20 percent nickel; (2) *stony* silicate minerals with inclusions of other minerals; and (3) *stony-irons* which are mixtures. Although stony meteorites are much more common, people tend to find irons. This is understandable because metallic meteorites withstand the impact better, weather more slowly, and are much easier for a layperson to distinguish from terrestrial rocks. Iron meteorites are probably fragments of once-molten cores of large asteroids or small planets.

Figure 15.41 Iron meteorite found near Meteor Crater, Arizona. (Courtesy of Meteor Crater Enterprises, Inc., northern Arizona, USA)

Did You Know?

In the last few decades, it has become increasingly clear that comets and asteroids have collided with Earth far more frequently than was previously known. The dangers of living with these small but deadly objects from space came to public attention in 1989 when an asteroid nearly 1 kilometer (0.6 mile) across shot past Earth. It was a near miss as it crossed our orbit just six hours ahead of Earth. Traveling at 71,000 kilometers per hour (44,000 miles per hour), it could have produced a crater 8 kilometers (5 miles) in diameter and perhaps 2 kilometers (1 mile) deep. As an observer noted, "Sooner or later it will be back." Statistics show that collisions of this tremendous magnitude should take place every few hundred million years and could have drastic consequences for life on Earth.

One type of meteorite, called a *carbonaceous chondrite,* contains simple amino acids and other organic compounds, which are the basic building blocks of life. This discovery confirms similar findings in observational astronomy, which indicate that numerous organic compounds exist in the frigid realm of outer space.

If meteorites represent the makeup of Earthlike planets, as some planetary geologists suggest, then Earth must contain a much larger percentage of iron than is indicated by surface rocks. This is one reason that geologists suggest that Earth's core must be mostly iron and nickel. In addition, radiometric dating of meteorites indicates that our solar system's age certainly exceeds 4.5 billion years. This "old age" has been confirmed by data obtained from lunar samples.

Dwarf Planets

Since Pluto's discovery in 1930, it has been a mystery on the edge of the solar system. At first, Pluto was thought to be about as large as Earth, but as better images were obtained, Pluto's diameter was estimated to be a little less than one-half that of Earth. Then, in 1978, astronomers discovered that Pluto has a satellite (Charon), whose brightness combined with its parent made Pluto appear much larger than it really was (Figure 15.43). Recent images obtained by the *Hubble Space Telescope* established the diameter of Pluto at only 2300 kilometers (1430 miles). This is about one-fifth that of Earth and less than half that of Mercury, long considered the runt of the solar system. In fact, seven moons, including Earth's moon, are larger than Pluto.

Even more attention was given to Pluto's status as a planet when in 1992 astronomers discovered another icy body in orbit beyond Neptune. Soon hundreds of these Kuiper belt objects were discovered forming a band of small objects, similar to the asteroid belt between Mars and Jupiter. However, these orbiting bodies are made up of dust and ices, like

Figure 15.42 Meteor Crater, near Winslow, Arizona. This impact feature is about 1.2 kilometers (0.75 mile) across, and 170 meters (560 feet) deep. The solar system is cluttered with meteoroids and other objects that can strike Earth with explosive force. (Photo by Michael Collier)

comets, rather than metallic and rocky substances, like asteroids. Many other planetary objects, some larger than Pluto, are thought to exist in this belt of icy worlds found beyond the orbit of Neptune.

The International Astronomical Union, a group that has the power to determine whether or not Pluto is a planet, voted August 24, 2006, to add a new class of planets called **dwarf planets.** These include celestial bodies that orbit around the Sun, are essentially round due to their self-gravity, but are not the only objects to occupy their area of space. By this definition, Pluto is recognized as a dwarf planet and the prototype of a new category of planetary objects. Other dwarf planets include Eris (formerly known as 2003 UB313), a Kuiper belt object, and Ceres, the largest known asteroid.

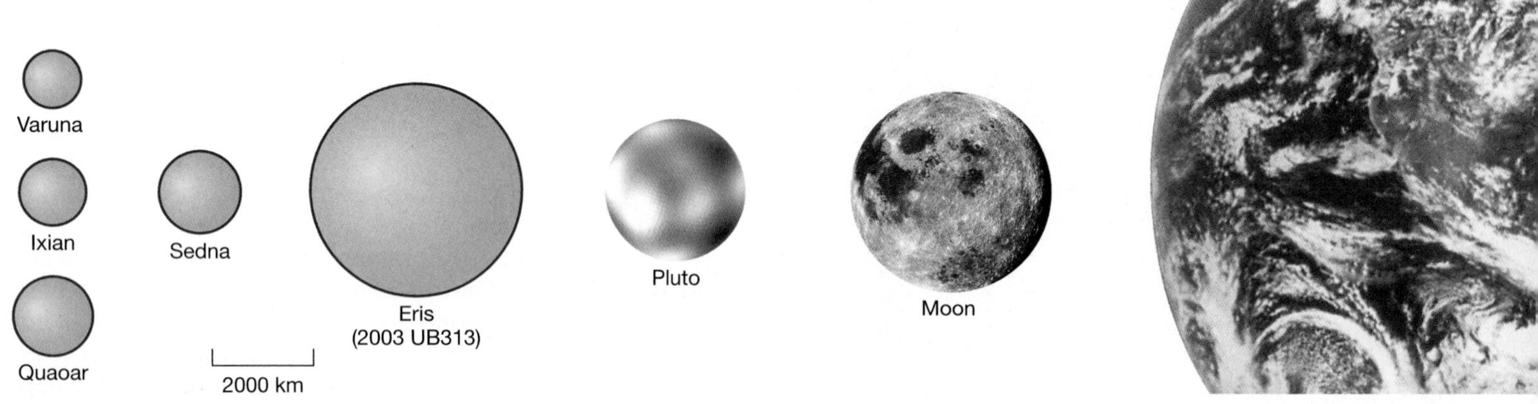

Figure 15.43 Pluto, a dwarf planet, compared to some large Kuiper belt objects. The Earth-Moon system added for scale. (NASA)

This is not the first time a planet has been demoted. In the mid-1800s, astronomy textbooks listed nearly two dozen planets in our solar system, including the asteroids Vesta, Juno, Ceres, and Pallas. Soon astronomers discovered hundreds of other such "planets," which made it clear that these small bodies represent another class of objects separate from the planets. Consequently, the number of major planets fell to eight. (Neptune had just been discovered, but Pluto's discovery did not occur until 1930.)

It is now clear that Pluto was unique among the classical planets, being very different from the four rocky innermost planets, and unlike the four gaseous giants. With the new classification, astronomers believe hundreds of additional dwarf planets might be found. *New Horizons,* the first spacecraft to explore the outer solar system, was launched in January 2006. It is scheduled to fly by Pluto in July 2015 and continue on to explore the Kuiper belt.

The Chapter in Review

1. Early Greeks held the *geocentric* ("Earth-centered") view of the universe, believing that Earth was a sphere that stayed motionless at the center of the universe. Orbiting Earth were the Moon, the Sun, and the known planets—Mercury, Venus, Mars, Jupiter, and Saturn. To the early Greeks, the stars traveled daily around Earth on a transparent, hollow *celestial sphere.* In A.D. 141, *Claudius Ptolemy* documented this geocentric view called the Ptolemaic system.

2. Modern astronomy evolved through the work of many dedicated individuals during the 1500s and 1600s. *Nicolaus Copernicus* (1473–1543) reconstructed the solar system with the Sun at the center and the planets orbiting around it, but he erroneously continued to use circles to represent the orbits of planets. *Tycho Brahe*'s (1546–1601) observations were far more precise than any made previously and are his legacy to astronomy. *Johannes Kepler* (1571–1630) ushered in the new astronomy with his three laws of planetary motion. After constructing his own telescope, *Galileo Galilei* (1564–1642) made many important discoveries that supported the Copernican view of a Sun-centered solar system. *Sir Isaac Newton* (1642–1727) developed laws of motion and proved that the force of gravity, combined with the tendency of an object to move in a straight line, causes elliptical orbits for planets.

3. The planets can be arranged into two groups: the *terrestrial* (Earthlike) *planets* (Mercury, Venus, Earth, and Mars) and the *Jovian* (Jupiter-like) *planets* (Jupiter, Saturn, Uranus, and Neptune). When compared to the Jovian planets, the terrestrial planets are smaller, more dense, contain proportionally more rocky material, and have slower rates of rotation.

4. The *nebular hypothesis* describes the formation of the solar system. The planets and the Sun began forming about 5 billion years ago from a large cloud of dust and gases called a *nebula.* As the nebular cloud contracted, it began to rotate and flatten into a disk. Material that was gravitationally pulled toward the center became the *protosun.* Within the rotating disk, small centers formed, called *protoplanets,* sweeping up more and more of the nebular debris. Because of their high temperatures and weak gravitational fields, the inner planets (Mercury, Venus, Earth, and Mars) were unable to accumulate much of the lighter components (hydrogen, ammonia, methane, and water) of the nebula. However, because of the very cold temperatures existing far from the Sun, the fragments from which the Jovian planets formed contained a high percentage of ices—water, carbon dioxide, ammonia, and methane.

5. The lunar surface exhibits several types of features. Most *craters* were produced by the impact of rapidly moving debris (meteoroids). Bright, densely cratered highlands make up most of the lunar surface. Dark, fairly smooth lowlands are called *maria.* Maria basins are enormous impact craters that have been flooded with layer upon layer of fluid basaltic lava. All lunar terrains are mantled with a soil-like layer of gray, unconsolidated debris, called *lunar regolith,* which has been derived from a few billion years of meteoric bombardment.

6. Closest to the Sun, *Mercury* is small and dense, has no atmosphere, and exhibits the greatest temperature extremes of any planet. *Venus,* the brightest planet in the sky, has a thick, heavy, carbon dioxide atmosphere; a surface of relatively subdued plains and inactive volcanoes; a surface atmospheric pressure 90 times that of Earth's; and surface temperatures of 475°C (890°F). *Mars,* the red planet, has a carbon dioxide atmosphere only 1 percent as dense as Earth's, extensive winds and dust storms, numerous inactive volcanoes, many large canyons, and several valleys of debatable origin resembling drainage patterns similar to stream valleys on Earth. *Jupiter,* the largest planet, rotates rapidly, has a banded appearance, a Great Red Spot that varies in size, a ring system, and at least 16 moons (one of which, Io, is volcanically active). *Saturn,* best known for its rings, also has a dynamic atmosphere, with winds up to 1500 kilometers per hour (930 miles per hour) and "storms" similar to Jupiter's Great Red Spot. *Uranus* and *Neptune* are often called the twins, because of their similar structures and compositions. Uranus is unique in rotating "on its side." Neptune has white, cirruslike clouds above its main cloud deck and an Earth-size Great Dark Spot, assumed to be a large rotating storm similar to Jupiter's Great Red Spot.

7. The minor members of the solar system include *asteroids, comets, meteoroids,* and *dwarf planets.* Most asteroids lie between the orbits of Mars and Jupiter. Asteroids are leftover rocky and metallic debris from the solar nebula that never accreted into a planet. Comets are made of ices (water, ammonia, methane, carbon dioxide, and carbon monoxide) with small pieces of rocky and metallic material. Many travel in elongated orbits that carry them beyond Pluto. Meteoroids, small solid particles that travel through interplanetary space, become *meteors* when they enter Earth's atmosphere and vaporize with a flash of light. *Meteor showers* occur when Earth encounters a swarm of meteoroids, probably material lost by a comet. *Meteorites* are the remains of meteoroids found on Earth. Recently, Pluto was placed into a new class of solar system objects called *dwarf planets.*

Key Terms

asteroid (p. 421)
asteroid belt (p. 422)
astronomical unit (AU) (p. 401)
celestial sphere (p. 396)
coma (p. 423)
comet (p. 422)
cryovolcanism (p. 421)
dwarf planet (p. 426)
escape velocity (p. 408)
geocentric (p. 396)
impact crater (p. 410)
inner planet (p. 407)
Jovian planet (p. 407)
Kuiper belt (p. 423)
lunar highlands (p. 409)
lunar regolith (p. 411)
maria (mare) (p. 409)
meteor (p. 424)
meteorite (p. 425)
meteoroid (p. 424)
meteor shower (p. 424)
nebular hypothesis (p. 405)
Oort cloud (p. 424)
outer planet (p. 407)
planetesimals (p. 405)
protoplanets (p. 406)
Ptolemaic system (p. 396)
retrograde motion (p. 398)
solar nebula (p. 405)
terrae (p. 409)
terrestrial planet (p. 407)

Questions for Review

1. Why did the ancients think that celestial objects had some control over their lives?
2. Describe what produces the retrograde motion of Mars. What geometric arrangement did Ptolemy use to explain this motion?
3. What major change did Copernicus make in the Ptolemaic system?
4. What was Tycho Brahe's contribution to science?
5. Did Galileo invent the telescope?
6. Explain how Galileo's discovery of a rotating Sun supported the Copernican view of a Sun-centered universe.
7. Newton learned that the orbits of the planets are the result of two actions. Describe these actions.
8. Compare the rotational periods of the terrestrial and Jovian planets.
9. How fast (in kilometers per hour) does a location on the equator of Jupiter rotate? Note that Jupiter's circumference equals approximately 452,000 kilometers.
10. By what criteria are the planets placed into either the Jovian or terrestrial group?
11. What are the three types of materials thought to make up the planets? How are they different? How does their distribution account for the density differences between the terrestrial and Jovian planetary groups?
12. Briefly describe the events that are thought to have led to the formation of the solar system.
13. How are the maria of the Moon thought to be similar to the Columbia Plateau?
14. What surface features does Mars have that are also common on Earth?
15. Why are astrobiologists excited about evidence that groundwater has seeped over the Martian surface?
16. What is the nature of Jupiter's Great Red Spot?
17. What is distinctive about Jupiter's satellite Io?
18. Why are the four outer satellites of Jupiter thought to have been captured?
19. What three bodies in the solar system exhibit volcano-like activity?
20. What do you think would happen if Earth passed through the tail of a comet?
21. Where are most comets thought to reside? What eventually becomes of comets that orbit close to the Sun?
22. Where are most asteroids found?
23. It has been estimated that Halley's comet has a mass of 100 billion tons. Further, this comet is estimated to lose 100 million tons of material during the few months that its orbit brings it close to the Sun. With an orbital period of 76 years, what is the maximum remaining life span of Halley's comet?
24. What are the three main sources of meteoroids?
25. Compare a meteoroid, meteor, and meteorite.
26. Why are meteorite craters more common on the Moon than on Earth, even though the Moon is much smaller?

Online Study Guide

The *Foundations of Earth Science* Web site uses the resources and flexibility of the Internet to aid in your study of the topics in this chapter. Written and developed by Earth science instructors, this site will help improve your understanding of Earth science. Visit http://www.prenhall.com/lutgens and click on the cover of *Foundations of Earth Science 5e* to find:

- Online review quizzes.
- Critical thinking exercises.
- Links to chapter-specific Web resources.
- Internet-wide key-term searches.

http://www.prenhall.com/lutgens

GEODe: Earth Science

GEODe: Earth Science makes studying more effective by reinforcing key concepts using animation, video, narration, interactive exercises, and practice quizzes. A copy is included with every copy of *Foundations of Earth Science 5e.*

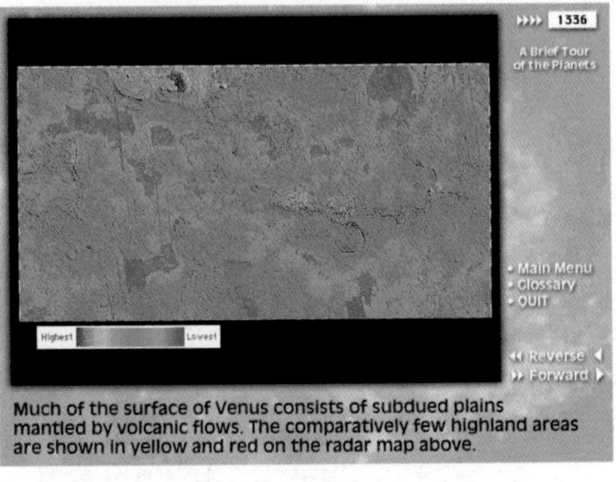

Much of the surface of Venus consists of subdued plains mantled by volcanic flows. The comparatively few highland areas are shown in yellow and red on the radar map above.

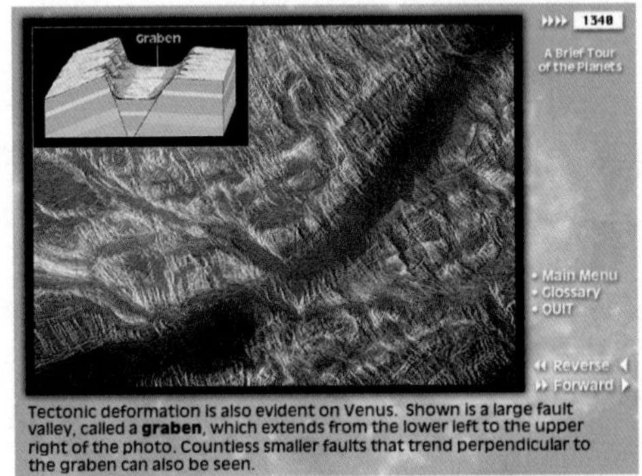

Tectonic deformation is also evident on Venus. Shown is a large fault valley, called a **graben**, which extends from the lower left to the upper right of the photo. Countless smaller faults that trend perpendicular to the graben can also be seen.

CHAPTER 16

Beyond the Solar System

Stars embedded in clouds of dust and gases produce colorful emission nebulae. (Royal Observatory, Edinburgh/Anglo-Australian Telescope Board/Science Photo Library/Photo Researchers, Inc.)

FOCUS ON LEARNING

To assist you in learning the important concepts in this chapter, you will find it helpful to focus on the following questions:

1. How is the principle of parallax used to measure the distance to a star?
2. What are the major intrinsic properties of stars?
3. What are the size categories of stars?
4. What are the different types of nebulae?
5. What is the most plausible model for stellar evolution? What are the stages in the life cycle of a star?
6. What are the possible final states that a star may assume after it consumes its nuclear fuel and collapses?
7. What is our own Milky Way Galaxy like?
8. What are the different types of galaxies?
9. What is the Doppler shift?
10. What is the Big Bang Theory for the origin of the universe?

The star Proxima Centauri is about 4.3 light-years away, roughly 100 million times farther than the Moon. Yet, other than our own Sun, it is the closest star to Earth. This fact suggests how incomprehensibly large the universe is. What is the nature of this vast cosmos beyond our solar system? Are the stars distributed randomly, or are they organized into distinct clusters? Do stars move, or are they permanently fixed features, like lights strung out against the black cloak of outer space? Does the universe extend infinitely in all directions, or is it bounded? To consider these questions, this chapter will examine the universe by taking a census of the stars—the most numerous objects in the night sky.

Properties of Stars

The Sun is the only star whose surface we can observe. Yet, a great deal is known about the universe beyond our solar system. This knowledge hinges on the fact that stars, and even gases in the "empty" space between stars, radiate energy in all directions into space (Figure 16.1). The key to understanding the universe is to collect this radiation and unravel the secrets it holds. Astronomers have devised many ingenious methods to do just that. We will begin by examining stellar distances and some intrinsic properties of stars, including *color, brightness, mass, temperature,* and *size.*

Measuring Distances to the Stars

Measuring the distance to a star is very difficult. Obviously, we cannot journey to the star. Nevertheless, astronomers have developed some *indirect* methods of measuring stellar distances. The most basic of these measurements is called *stellar parallax.*

Stellar parallax is the extremely slight back-and-forth shifting in the apparent position of a nearby star due to the orbital motion of Earth. The principle of parallax is easy to visualize. Close one eye, and with your index finger in a vertical position, use your eye to line up your finger with some distant object. Now, without moving your finger, view the object with your other eye and notice that its position appears to have changed. The farther away you hold your finger, the less its position seems to shift.

In practice, parallax is determined by photographing a nearby star against the background of distant stars. Then, six months later, when Earth has moved halfway around its orbit, a second photograph is taken. When these photographs are compared, the position of the nearby star appears to have shifted with respect to the background stars. Figure 16.2 illustrates this shift and the parallax angle determined from it. The nearest stars have the largest parallax angles, whereas those of distant stars are too slight to measure. Recall that the medieval astronomer Tycho Brahe was unable to detect stellar parallax, leading him to reject the idea that Earth orbits the Sun.

It should be emphasized that parallax angles are *very* small. The parallax angle to the nearest star, Proxima Centauri, is less than 1 second of arc, which equals $\frac{1}{3600}$ of a degree. To put this in perspective, fully extend your arm and raise your little finger. Your finger is roughly 1 degree wide. Try doing this on a moonlit night, covering the Moon with your finger. The moon is only about $\frac{1}{2}$ degree wide. Now, imagine detecting a movement that is only $\frac{1}{3600}$ as wide as your finger. It should be apparent why Tycho Brahe, without the aid of a telescope, was unable to observe stellar parallax.

The distances to stars are so large that the conventional units, such as kilometers or astronomical units, are often too

Figure 16.1 Lagoon Nebula. It is in glowing clouds like these that gases and dust particles become concentrated into stars. (Courtesy of National Optical Astronomy Observatories)

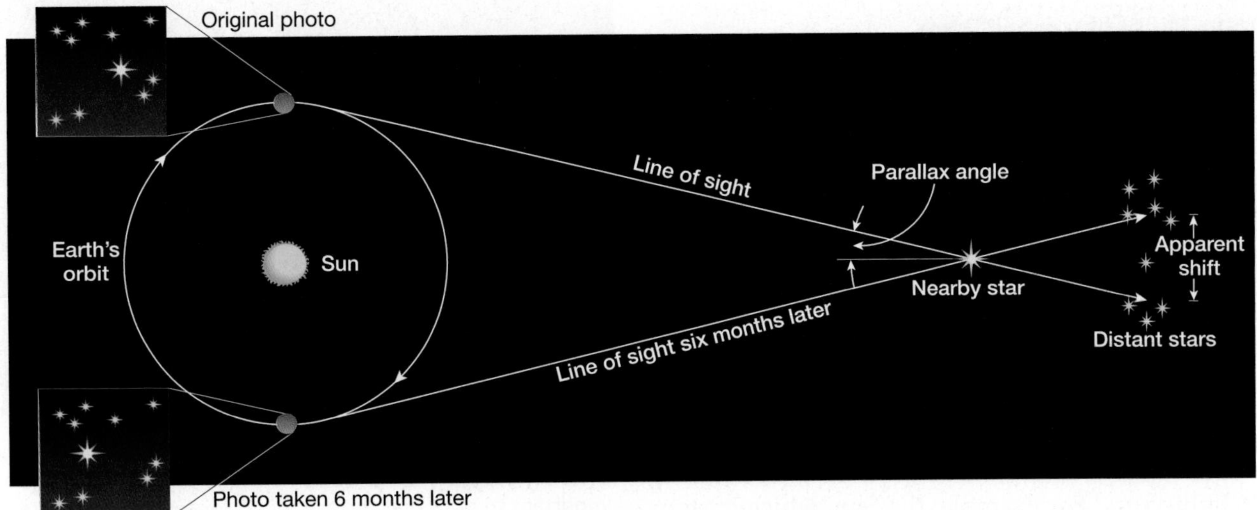

Figure 16.2 Geometry of stellar parallax. The parallax angle shown here is enormously exaggerated to illustrate the principle. Because the distance to even the nearest stars is thousands of times greater than the Earth-Sun distance, the triangles that astronomers work with are very long and narrow, making the angles that are measured extremely small.

cumbersome to use. A better unit to express stellar distance is the **light-year,** which is the distance light travels in one Earth year—about 9.5 trillion kilometers (5.8 trillion miles).

In principle, the method used to measure stellar distances is elementary and was known to the ancient Greeks. But in practice, measurements are greatly complicated because of the tiny angles involved and because the Sun, and the star being measured, also move through space. The first accurate stellar parallax was not determined until 1838. Even today, parallax angles for only a few thousand of the nearest stars are known with certainty.

Most stars have such small parallax shifts that accurate measurement is not possible. Fortunately, a few other methods have been derived for estimating distances to these stars. Also, the Hubble Space Telescope, which is not hindered by Earth's turbulent, light-distorting atmosphere, is expected to obtain accurate parallax distances for many additional remote stars.

Stellar Brightness

Three factors control the apparent brightness of a star as seen from Earth: *how big* it is, *how hot* it is, and *how far away* it is. The stars in the night sky are a grand assortment of sizes, temperatures, and distances, so their brightnesses vary widely.

Apparent Magnitude Stars have been classified according to their brightness ever since the second century B.C., when Hipparchus placed about 1000 of them into six categories. The measure of a star's brightness is called its **magnitude.** Some stars may appear dimmer than others only because they are farther away. Therefore, a star's brightness, *as it appears when viewed from Earth,* has been termed its **apparent magnitude.**

When numbers are used to designate relative brightness, the larger the magnitude number, the dimmer the star. *Stars that appear the brightest are of the first magnitude, while the faintest stars visible to the unaided eye are of the sixth magnitude.* With the invention of the telescope, many stars fainter than the sixth magnitude were discovered.

In the mid-1800s, a method was developed for comparing the brilliance of stars using magnitude. Just as we can compare the brightness of a 50-watt lightbulb to that of a 100-watt bulb, we can compare the brightness of stars having different magnitudes. It was determined that a first-magnitude star was about 100 times brighter than a sixth-magnitude star. Therefore, on the scale that was devised, two stars that differ by five magnitudes have a ratio in brightness of 100 to 1. Hence, a seventh-magnitude star is 100 times brighter than a twelfth-magnitude star.

It follows, then, that the brightness ratio of two stars differing by only one magnitude is about 2.5.* Thus, a star of the first magnitude is about 2.5 times brighter than a star of the second magnitude. Table 16.1 shows some magnitude differences and the corresponding brightness ratios.

Because some stars are brighter than first-magnitude stars, zero and negative magnitudes were introduced. The brightest star in the night sky, Sirius, has an apparent magnitude of −1.4, about 10 times brighter than a first-magnitude star. On this scale, the Sun has an apparent magnitude of −26.7. At its brightest, Venus has a magnitude of −4.3. At the other end of the scale, the 5-meter (200-inch) Hale Telescope can view stars with an apparent magnitude of 23, approximately 100 million times dimmer than stars that are visible to the unaided eye.

Absolute Magnitude Astronomers are also interested in the "true" brightness of stars, called their **absolute magnitude.** Stars of the same luminosity, or brightness, usually do not have the same *apparent* magnitude, because their distances from us are not equal. To compare their true, or intrinsic, brightness, astronomers determine what magnitude

*Calculations: $2.512 \times 2.512 \times 2.512 \times 2.512 \times 2.512$, or 2.512 raised to the fifth power, equals 100.

Table 16.1 Ratios of star brightness

Difference in Magnitude	Brightness Ratio
0.5	1.6:1
1	2.5:1
2	6.3:1
3	16:1
4	40:1
5	100:1
10	10,000:1
20	100,000,000:1

the stars would have if they were at a standard distance of about 32.6 light-years. For example, the Sun, which has an apparent magnitude of −26.7, would have an absolute magnitude of about 5 if located at a distance of 32.6 light-years. Thus, stars with absolute magnitudes greater than 5 (smaller numerical value) are intrinsically brighter than the Sun, but because of their distance, they appear much dimmer.

Table 16.2 lists the absolute and apparent magnitudes of some stars as well as their distances from Earth. Most stars have an absolute magnitude between −5 and 15, which puts the Sun near the middle of this range.

Stellar Color and Temperature

The next time you are outdoors on a clear night, take a good look at the stars and note their color (Figure 16.3). Some that are quite colorful can be found in the constellation Orion. Of the two brightest stars in Orion, Betelgeuse (α Orionis) is definitely red, whereas Rigel (β Orionis) appears blue.

Very hot stars with surface temperatures above 30,000 K (kelvin) emit most of their energy in the form of short-wavelength light and, therefore, appear blue. Red stars, on the other hand, are much cooler, generally less than 3000 K, and most of their energy is emitted as longer-wavelength red light. Stars with temperatures between 5000 and 6000 K appear yellow, like the Sun. Because color is primarily a manifestation of a star's temperature, this characteristic provides the astronomer with useful information about a star.

Table 16.2 Distance, apparent magnitude, and absolute magnitude of some stars

Name	Distance (Light-years)	Apparent Magnitude*	Absolute Magnitude*
Sun	NA	−26.7	5.0
Alpha Centauri	4.27	0.0	4.4
Sirius	8.70	−1.4	1.5
Arcturus	36	−0.1	−0.3
Betelgeuse	520	0.8	−5.5
Deneb	1600	1.3	−6.9

*The more negative, the brighter; the more positive, the dimmer.

Figure 16.3 Time-lapse photograph of stars in the constellation Orion. These star trails show some of the various star colors. It is important to note that the eye sees color somewhat differently than does photographic film. (Courtesy of National Optical Astronomy Observatories)

Binary Stars and Stellar Mass

One of the night sky's best-known constellations, the Big Dipper, appears at first glance to consist of seven stars. But individuals with good eyesight can resolve the second star in the handle of the Big Dipper (Mizar) as two stars. During the 1700s, astronomers used their new tool, the telescope, to discover numerous such star pairs. One of the stars in the pair is usually fainter than the other, and for this reason, it was considered to be farther away. In other words, the stars were not considered true pairs but were thought only to lie in the same line of sight.

In the early 1800s, careful examination of numerous star pairs by William Herschel revealed that many actually orbit each other. The two stars are in fact united by their mutual gravitation. These pairs of stars, in which the members are far enough apart to be resolved telescopically, are called *visual binaries*. The idea of one star orbiting another may seem unusual, but evidence indicates that more than 50 percent of the stars in the universe occur in pairs or multiples.

Binary stars are used to determine the star property most difficult to calculate—its mass. The mass of a body can be established if it is gravitationally attached to a partner, which is the case for any binary star system. Binary stars orbit each other around a common point called the *center of mass* (Figure 16.4). For stars of equal mass, the center of mass lies exactly halfway between them. If one star is more massive than its partner, their common center will be located closer to the more massive one. Thus, if the sizes of their orbits can be observed, a determination of their individual masses can be made. You can experience this relationship on a seesaw by trying to balance a person who has a much greater mass.

For illustration, when one star has an orbit half the size (radius) of its companion, it is twice as massive. If their combined masses are equal to three times the mass of the Sun, then the larger will be twice as massive as the Sun, and the smaller will have a mass equal to that of the Sun. Most stars have masses that range between 1/10 and 50 times the mass of our Sun.

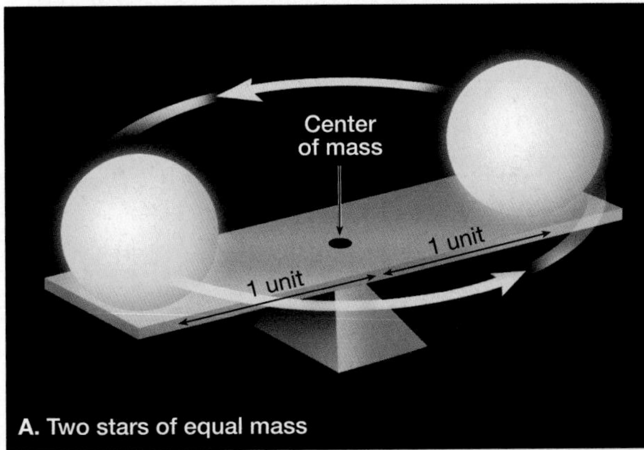

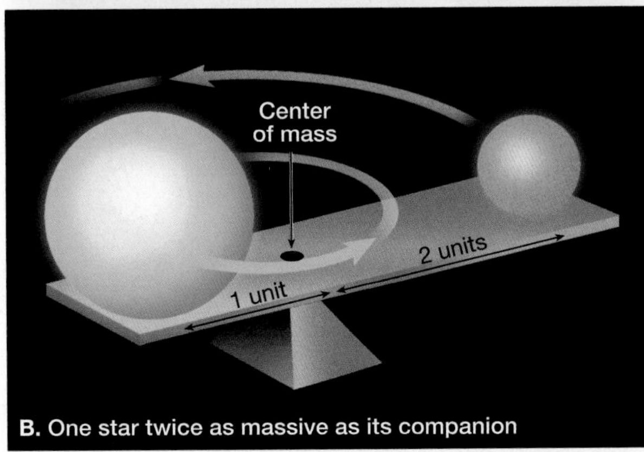

Figure 16.4 Binary stars orbit each other around their common center of mass. **A.** For stars of equal mass, the center of mass lies exactly halfway between them. **B.** If one star is twice as massive as its companion, it is twice as close to their common center of mass. Therefore, more massive stars have proportionately smaller orbits than do their less massive companions.

Hertzsprung-Russell Diagram

Early in this century, Einar Hertzsprung and Henry Russell independently studied the relation between the true brightness (absolute magnitude) and the temperature of stars. From this research, each developed a graph, now called a **Hertzsprung-Russell (H-R)** diagram, that exhibits these intrinsic stellar properties. By studying H–R diagrams, we learn a great deal about the sizes, colors, and temperatures of stars.

To produce an H-R diagram, astronomers survey a portion of the sky and plot each star according to its luminosity (brightness) and temperature (Figure 16.5). Notice that the stars in Figure 16.5 are not uniformly distributed. Rather, about 90 percent of all stars fall along a band that runs from the upper-left corner to the lower-right corner of an H-R diagram. These "ordinary" stars are called **main-sequence stars.** As shown in Figure 16.5, *the hottest main-sequence stars are intrinsically the brightest, and vice versa.*

The luminosity of the main-sequence stars is also related to their mass. The hottest (blue) stars are about 50 times more massive than our Sun, whereas the coolest (red) stars are only $\frac{1}{10}$ as massive. Therefore, on the H-R diagram, the main-sequence stars appear in a decreasing order, from *hotter, more massive* blue stars to *cooler, less massive* red stars.

Note the location of our Sun in Figure 16.5. The Sun is a yellow, main-sequence star with an absolute magnitude of about 5. Because the magnitude of a vast majority of main-sequence stars lies between -5 and 15, and because the Sun falls midway in this range, the Sun is often considered an average star. However, more main-sequence stars are cooler and less massive than our Sun.

Just as all humans do not fall into the normal size range, some stars do not fit in with the main-sequence stars. Above and to the right of the main sequence in the H-R diagram in Figure 16.5 lies a group of very luminous stars called *giants,* or, on the basis of their color, **red giants.** The size of these giants can be estimated by comparing them with stars of known size that have the same surface temperature. We know that objects having equal surface temperatures radiate the same amount of energy per unit area. Therefore, any difference in the brightness of two stars having the same surface temperature is attributable to their relative sizes.

As an example, let us compare the Sun, which has been assigned a luminosity of 1, with another yellow star that has a luminosity of 100. Because both stars have the same surface temperature, both radiate the same amount of energy per unit area. Therefore, for the more luminous star to be 100 times brighter than the Sun, it must have 100 times more surface area. It should be clear why stars whose plots fall in the upper-right position of an H–R diagram are called *giants.*

Some stars are so large that they are called **supergiants.** Betelgeuse, a bright-red supergiant in the constellation Orion, has a radius about 800 times that of the Sun. If this star were at the center of the solar system, it would extend beyond the orbit of Mars, and Earth would find itself inside the star! Other red giants that are easy to locate in our sky are Arcturus in the constellation Bootes and Antares in Scorpius.

In the lower-central portion of the H-R diagram, the opposite situation occurs. These stars are much fainter than main-sequence stars of the same temperature, and by using the same reasoning as before, they must be much smaller. Some probably approximate Earth in size. This group has come to be called **white dwarfs,** although not all are white.

Soon after the first H-R diagrams were developed, astronomers realized their importance in interpreting stellar evolution. Just like living things, a star is born, ages, and dies. Since almost 90 percent of the stars lie on the main sequence, we can be relatively certain that stars spend most of their active years as main-sequence stars. Only a few percent are giants, and perhaps 10 percent are white dwarfs.

Next, we will examine variable stars and the nature of interstellar matter and then return to the topic of stellar evolution.

Variable Stars

Stars that fluctuate in brightness are known as variables. Some, called **pulsating variables,** fluctuate regularly in brightness by expanding and contracting in size. The importance of one

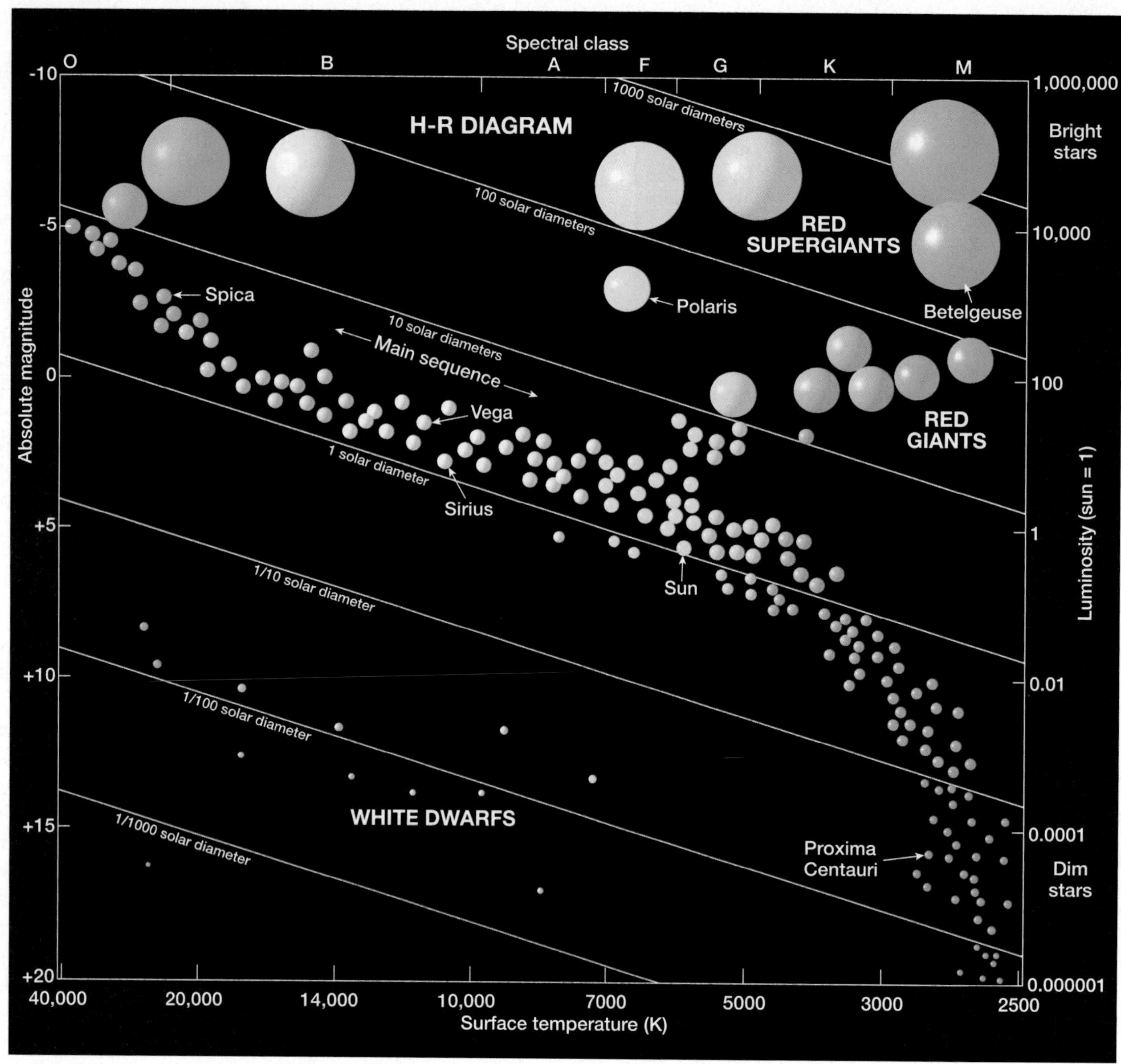

Figure 16.5 Idealized Hertzsprung-Russell (H-R) diagram on which stars are plotted according to temperature and absolute magnitude.

member of this group (cepheid variables) is in determining stellar distances.

The most spectacular variables belong to a group known as **eruptive variables.** When one of the explosive events associated with these stars occurs, it appears as a sudden brightening of a star, called a **nova** (Figure 16.6). The term *nova*, meaning "new," was used by the ancients because these stars were unknown to them before their abrupt increase in luminosity.

During the outburst, the outer layer of the star is ejected at high speed. The "cloud" of ejected material can occasionally be captured photographically (Figure 16.7). A nova generally reaches maximum brightness in a couple of days, remains bright for only a few weeks, then slowly returns in a year or so to its original brightness. Because the star returns to its prenova brightness, we can assume that only a small amount of its mass is lost during the flareup. Some stars have experienced more than one such event. In fact, the process probably occurs repeatedly.

The modern explanation of novae proposes that they occur in binary systems consisting of an expanding red giant and a hot white dwarf that are in close proximity. Hydrogen-rich gas from the oversized giant encroaches near enough to the white dwarf to be gravitationally transferred. Eventually, enough of the hydrogen-rich gas is added to the dwarf to cause it to ignite explosively. Such a thermonuclear reaction rapidly heats and expands the outer gaseous envelope of the hot dwarf to produce a nova event. In a relatively short time, the white dwarf returns to its prenova state, where it remains inactive until the next buildup occurs.

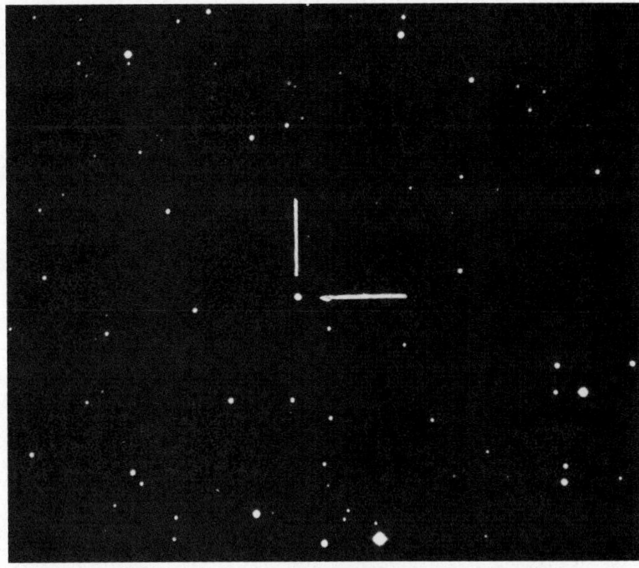

Figure 16.6 Photographs of Nova Herculis (a nova in the constellation Hercules), taken about 2 months apart, showing the decrease in brightness. (Courtesy of Lick Observatory)

Interstellar Matter

Between existing stars is "the vacuum of space." However, it is far from being a pure vacuum, for it is populated with accumulations of dust and gases. The name applied to these concentrations of interstellar matter is **nebula** (Latin for "cloud"). If this interstellar matter is close to a very hot (blue) star, it will glow and is called a **bright nebula.** The two main types of bright nebulae are known as *emission nebulae* and *reflection nebulae.*

Emission nebulae are gaseous masses that consist largely of hydrogen. They absorb *ultraviolet radiation* emitted by an embedded or nearby hot star. Because these gases are under very low pressure, they reradiate, or emit, this energy as visible light. This conversion of ultraviolet light to visible light is known as *fluorescence,* an effect you observe daily in fluorescent lights. A well-known emission nebula easily seen with binoculars is the sword of the hunter in the constellation Orion (Figure 16.8).

Reflection nebulae, as the name implies, merely reflect the light of nearby stars (Figure 16.9). Reflection nebulae are thought to be composed of rather dense clouds of large particles called **interstellar dust.** This view is supported by the fact that atomic gases with low densities could not reflect light sufficiently to produce the glow observed.

When a dense cloud of interstellar material is not close enough to a bright star to be illuminated, it is referred to as a **dark nebula.** Exemplified by the Horsehead Nebula in Orion, dark nebulae appear as opaque objects silhouetted against a bright background (Figure 16.10). Dark nebulae can also be seen as starless regions—"holes in the heavens"—when viewing the Milky Way.

Although nebulae appear dense, they actually consist of thinly scattered matter. Because of their enormous size, however, the total mass of rarefied particles and molecules may be many times that of the Sun. Interstellar matter is of great interest to astronomers because it is from this material that stars and planets are formed.

Figure 16.7 Expanding a "cloud" around Nova Persei. (Courtesy of Hale Observatories. Copyright by the California Institute of Technology and Carnegie Institution of Washington)

Stellar Evolution

The idea of describing how a star is born, ages, and then dies may seem a bit presumptuous, for many of these objects have life spans that surely exceed billions of years. However, by studying stars of different ages, astronomers have been able to piece together a plausible model for stellar evolution.

The method that was used to create this model is analogous to what an alien being, upon reaching Earth, might do to determine the developmental stages of human life. By examining a large number of humans, this stranger would be able to observe the birth of human life, the activities of children and adults, and the death of the elderly. From this

Figure 16.8 The Orion Nebula is a well-known emission nebula. Bright enough to be seen by the naked eye, the Orion Nebula is located in the sword of the hunter in the constellation of the same name. (Courtesy of National Optical Astronomy Observatories)

information, the alien would attempt to put the stages of human development into their proper sequence. Based on the relative abundance of humans in each stage of development, it would even be possible to conclude that humans spend more of their lives as adults than as toddlers. In a similar fashion, astronomers have pieced together the story of the stars.

Figure 16.9 A faint blue reflection nebula, in the Pleiades star cluster, is caused by the reflection of starlight from dust in the nebula. The Pleiades star cluster, just visible to the naked eye in the constellation Taurus, is spectacular when viewed through binoculars or a small telescope. (Palomar Observatories/California Institute of Technology [Caltech])

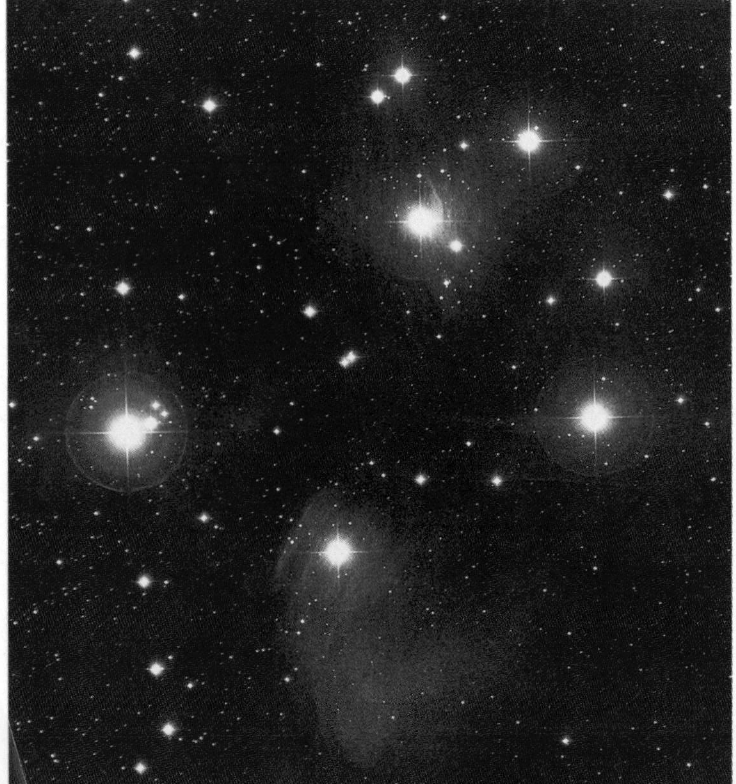

Did You Know?

In a few billion years, the Sun will exhaust the remaining hydrogen fuel in its core, an event that will trigger hydrogen fusion in the surrounding shell. As a result, the Sun's outer envelope will expand, producing a red giant that is hundreds of times larger and more luminous. The intense solar radiation will cause Earth's oceans to boil, and the solar winds will drive away Earth's atmosphere. In another billion years, the Sun will expel its outermost layer, producing a spectacular planetary nebula, while its interior will collapse to form a dense, small (planet-size), white dwarf. Because of its small size, the Sun's energy output will be less than 1 percent of its current level. Gradually, the Sun will emit its remaining thermal energy, eventually becoming a cold, nonluminous body.

Simply, stars exist because of gravity. The mutual gravitational attraction of particles in a thin, gaseous cloud causes the cloud to collapse. As the cloud is squeezed to unimaginable pressures, its temperature rises, igniting its nuclear furnace, and a star is born. A star is a ball of very hot gases, caught between the opposing forces of gravity trying to contract it and thermal nuclear energy trying to expand it. Eventually, all of a star's nuclear fuel will be exhausted and gravity takes over, collapsing the stellar remnant into a small, dense body.

Star Birth

The birthplaces of stars are dark, cool, interstellar clouds, which are comparatively rich in dust and gases (Figure 16.11). In the neighborhood of the Milky Way, these gaseous clouds

Figure 16.10 The Horsehead Nebula, a dark nebula in a region of glowing nebulosity in Orion. (© Anglo-Australian Observatory/Royal Observatory Edinburgh. Photograph from UK Schmidt plates by David Malin)

consist of 92 percent hydrogen, 7 percent helium, and less than 1 percent of the remaining heavier elements. By some mechanism not yet fully understood, these thin gaseous clouds become concentrated enough to begin to gravitationally contract. One proposal to explain the triggering of stellar formation is a shock wave traveling from a catastrophic explosion (supernova) of a nearby star. But, regardless of the force that initiates the concentration of interstellar matter, once it is accomplished, mutual gravitational attraction of the particles squeeze the cloud, pulling every particle toward the center. As the cloud shrinks, gravitational energy (potential energy) is converted into energy of motion, or heat energy.

The initial contraction spans a million or so years. With the passage of time, the temperature of this gaseous body slowly rises, eventually reaching a temperature sufficiently high to cause it to radiate energy from its surface in the form of long-wavelength red light. Because this large red object is not hot enough to engage in nuclear fusion, it is not yet a star. The name **protostar** is applied to these bodies.

Protostar Stage

During the protostar phase, gravitational contraction continues, slowly at first, then much more rapidly (Figure 16.11). This collapse causes the core of the developing star to heat much more intensely than the outer envelope. When the core has reached at least 10 million K, the pressure within is so great that groups of four hydrogen nuclei are fused together into single helium nuclei. Astronomers refer to this nuclear reaction as **hydrogen burning** because an enormous amount of energy is released. However, keep in mind that thermonuclear "burning" is not burning in the usual chemical sense.

Heat from hydrogen fusion causes the stellar gases to increase their motion. This, in turn, results in an increase in the outward gas pressure. At some point, this outward pressure exactly balances the inward force of gravity. When this balance is reached, the star becomes a stable *main-sequence star* (Figure 16.11). Stated another way, a stable main-sequence star is balanced between two forces: *gravity,* which is trying to squeeze it into a smaller sphere, and *gas pressure,* which is trying to expand it.

Main-Sequence Stage

From this point in the evolution of a main-sequence star until its death, the internal gas pressure struggles to offset the unrelenting force of gravity. Typically, hydrogen burning continues for a few billion years and provides the outward pressure required to support the star from gravitational collapse.

Different stars age at different rates. Hot, massive blue stars radiate energy at such an enormous rate that they substantially deplete their hydrogen fuel in only a few million years. By contrast, the very smallest (red) main-sequence stars may remain stable for hundreds of billions of years. A yellow star, such as the Sun, remains a main-sequence star for about 10 billion years. Because the solar system is about 5 billion

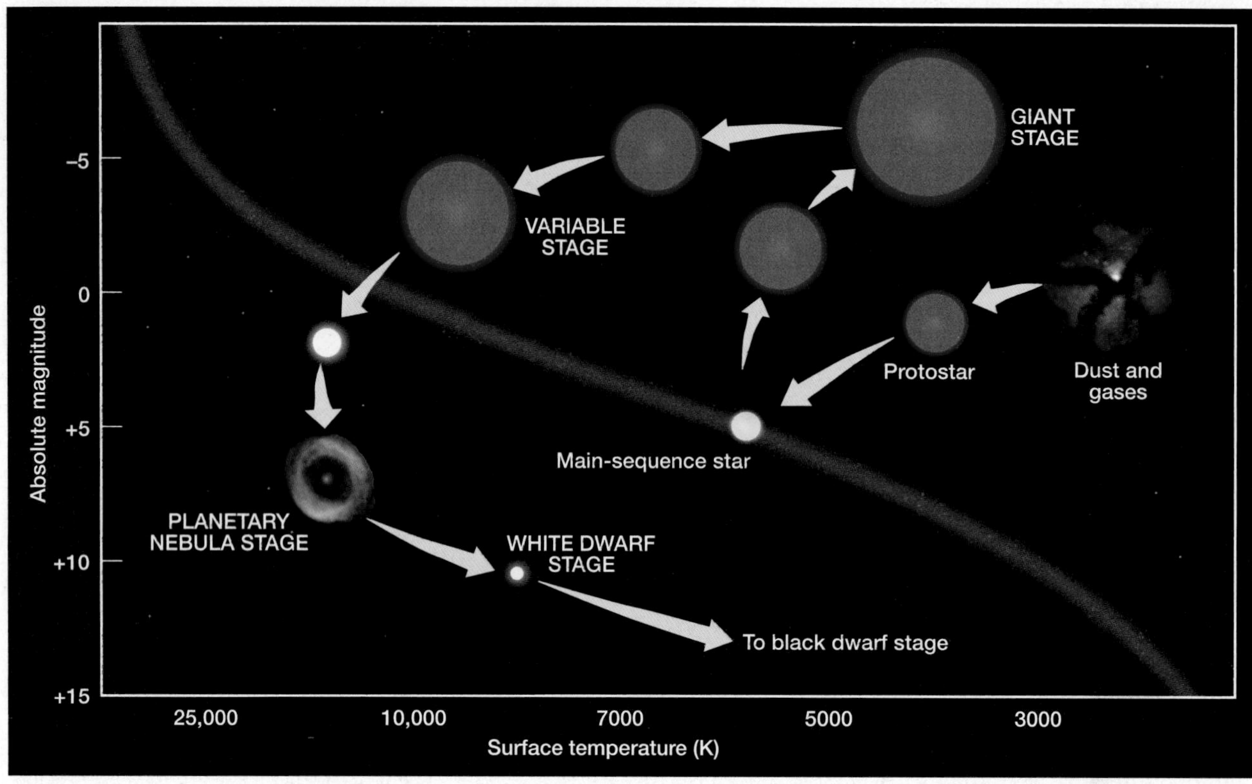

Figure 16.11 Diagram of stellar evolution on H-R diagram for a star about as massive as the Sun.

years old, it is comforting to know that our Sun will remain stable for another 5 billion years.

An average star spends 90 percent of its life as a hydrogen-burning, main-sequence star. Once the hydrogen fuel in the star's core is depleted, it evolves rapidly and dies. However, with the exception of the least-massive (red) stars, a star can delay its death by burning other fuels and becoming a giant.

Red Giant Stage

The evolution to the red giant stage results because the zone of hydrogen burning continually migrates outward, leaving behind an inert helium core. Eventually, all of the hydrogen in the star's core is consumed. While hydrogen fusion is still progressing in the star's outer shell, no fusion is taking place in the core. Without a source of energy, the core no longer has enough pressure to support itself against the inward force of gravity. As a result, the core begins to contract.

Although the core cannot generate nuclear energy, it does grow hotter by converting gravitational energy into heat energy. Some of this energy is radiated outward, initiating a more vigorous level of hydrogen fusion in the star's outer shell. This energy in turn heats and enormously expands the star's outer envelope, producing a giant body hundreds to thousands of times its main-sequence size (Figure 16.11).

As the star expands, its surface cools, which explains the star's reddish appearance. Eventually, the star's gravitational force will stop this outward expansion. Once again, the two opposing forces, gravity and gas pressure, will be in balance, and this gaseous mass will be a stable but much larger star. Some red giants overshoot the equilibrium point and rebound like an overextended spring. Such stars continue to oscillate in size, becoming variable stars.

While the envelope of a red giant expands, the core continues to collapse and heat until it reaches 100 million K. At this incredible temperature, it is hot enough to initiate a nuclear reaction in which helium is converted to carbon. Thus, a red giant consumes both hydrogen and helium to produce energy. In stars more massive than our Sun, still other thermonuclear reactions occur that generate all the elements on the periodic table up to number 26, iron. Nuclear "burning" of elements heavier than iron requires an additional source of energy to keep the reaction progressing. Hence, these elements are not produced in ordinary stars.

Eventually, all the usable nuclear fuel in these giants will be consumed. Our Sun, for example, will spend less than a billion years as a giant, and the more massive stars will pass through this stage even more rapidly. The force of gravity will again control the star's destiny as it squeezes the star into the smallest, most dense piece of matter possible.

Burnout and Death

Most of the events of stellar evolution discussed thus far are well documented. What happens to a star after the red-giant phase is more speculative. We do know that a star, regardless of its size, must eventually exhaust all of its usable nuclear fuel and collapse in response to its immense gravitational force. With this in mind, we will now consider the final stage of stars in three different mass categories.

Death of Low-Mass Stars Stars less than one-half the mass of the Sun (0.5 solar mass) consume their fuel at a comparatively slow rate (Figure 16.12A). Consequently, these small, *cool red stars* may remain on the main sequence for up to 100 billion years. Because the interior of a low-mass star never attains sufficiently high temperatures and pressures to fuse helium, its only energy source is hydrogen fusion. Thus, low-mass stars never evolve to become bloated red giants. Rather, they remain as stable main-sequence stars until they consume their hydrogen fuel and collapse into a hot, dense *white dwarf.* As we will see, white dwarfs are small, compact objects unable to support nuclear burning.

Death of Medium-mass (Sun-like) Stars Main-sequence stars with masses ranging between half that of the Sun and three times that of the Sun evolve in essentially the same way (Figure 16.12B). During their giant phase, Sun-like stars fuse hydrogen and helium fuel at an accelerated rate. Once this fuel is exhausted, these stars (like low-mass stars) collapse into an Earth-size body of great density—a white dwarf. The density of a white dwarf is as great as physics will allow short of destroying protons and electrons.

The gravitational energy supplied to a collapsing white dwarf is reflected in its high surface temperature. However, without a source of nuclear energy, a white dwarf becomes cooler and dimmer as it continually radiates its remaining thermal energy into space.

During their collapse from red giants to white dwarfs, medium-mass stars are thought to cast off their bloated outer atmosphere, creating an expanding spherical cloud of gas. The remaining hot, central white dwarf heats the gas cloud, causing it to glow. These often beautiful, gleaming spherical clouds are called **planetary nebulae.** A good example of a planetary nebula is the Helix nebula in the constellation Aquarius (Figure 16.13). This nebula appears as a ring because our line of sight through the center traverses less gaseous material than at the nebula's edge. It is, nevertheless, spherical in shape.

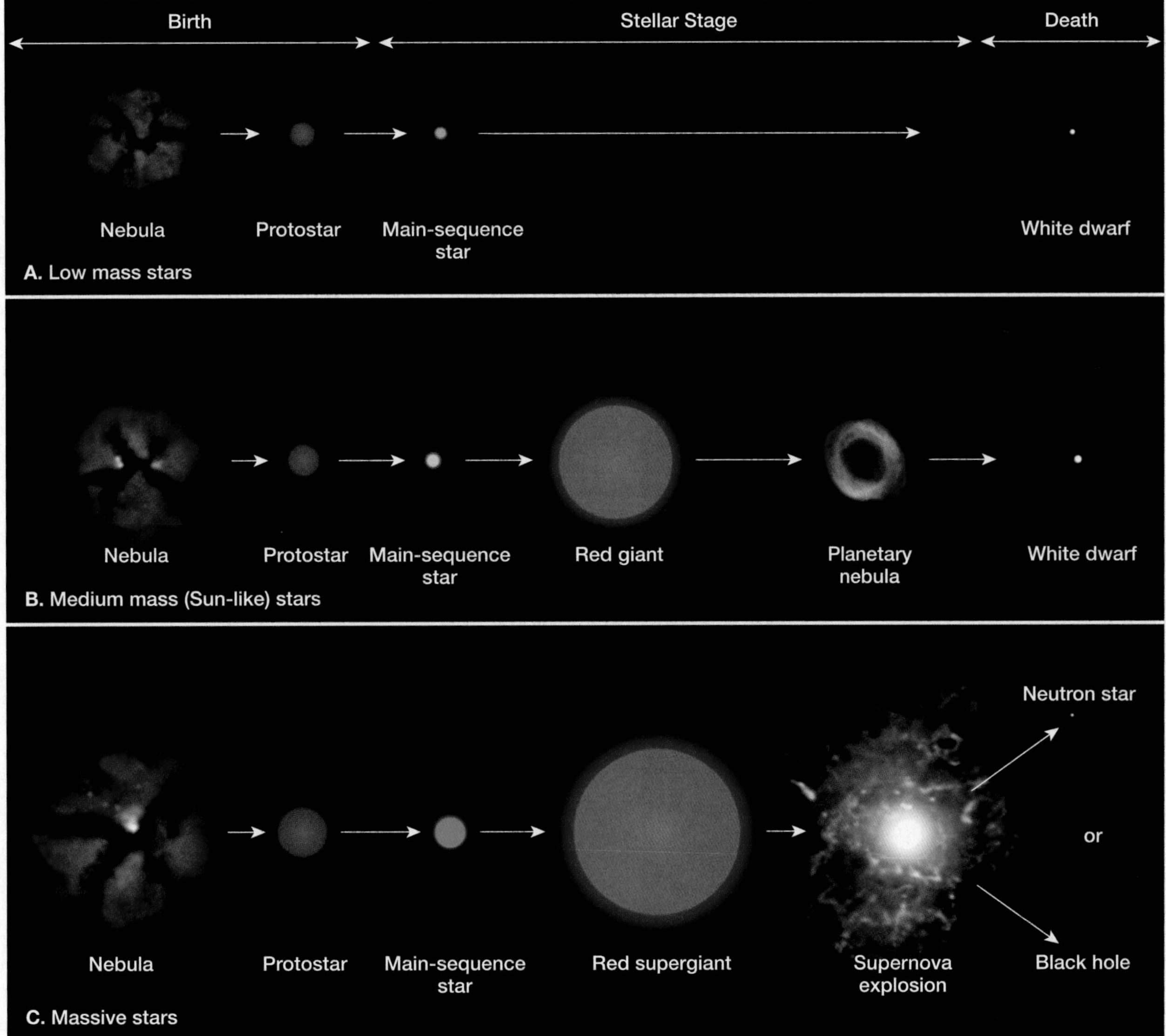

Figure 16.12 The evolutionary stages of stars having various masses.

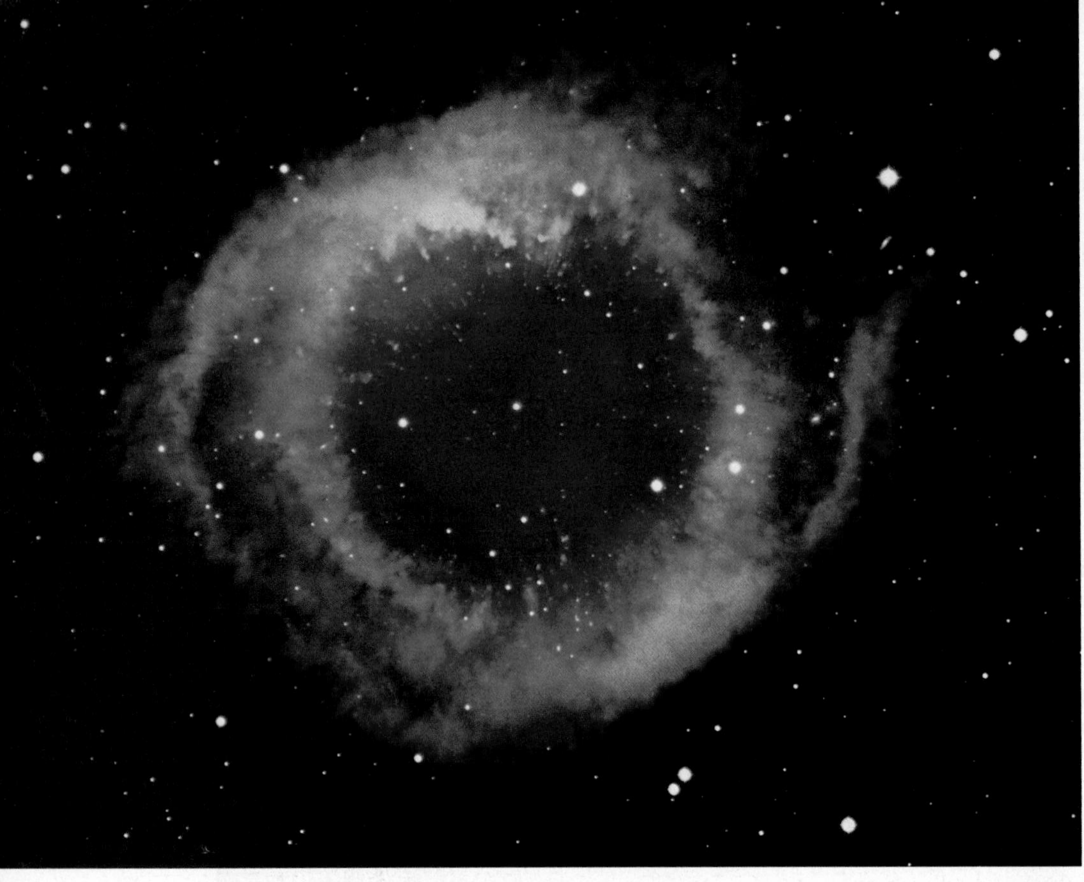

Figure 16.13 The Helix Nebula, the nearest planetary nebula to our solar system. A planetary nebula is the ejected outer envelope of a Sun-like star that formed during the star's collapse from a red giant to a white dwarf. (© Anglo-Australian Observatory, photography by David Malin)

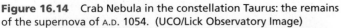

Figure 16.14 Crab Nebula in the constellation Taurus: the remains of the supernova of A.D. 1054. (UCO/Lick Observatory Image)

Death of Massive Stars In contrast to Sun-like stars, which expire gracefully, stars exceeding three solar masses have relatively short life spans and terminate in a brilliant explosion called a **supernova** (Figure 16.12C). During a supernova event, a star becomes millions of times brighter than its prenova stage. If one of the nearest stars to Earth produced such an outburst, its brilliance would surpass that of the Sun. Supernovae are rare; none have been observed in our galaxy since the advent of the telescope, although Tycho Brahe and Galileo each recorded one about 30 years apart. An even larger supernova was recorded in A.D. 1054 by the Chinese. Today, the remnant of this great outburst is the Crab Nebula, shown in Figure 16.14.

A supernova event is thought to be triggered when a massive star consumes most of its nuclear fuel. Without a heat engine to generate the gas pressure required to balance

---

Did You Know?

The first naked-eye supernova in 383 years was discovered in the southern sky in February 1987. This stellar explosion was officially named SN 1987A (SN stands for supernova, and 1987A indicates that it was the first supernova observed in 1987). Naked-eye supernovae are rare. Only a few have been recorded in historic times. Arab observers saw one in 1006, and the Chinese recorded one in 1054 at the present location of the Crab Nebula. In addition, the astronomer Tycho Brahe observed a supernova in 1572, and Kepler saw one shortly thereafter in 1604.

its immense gravitational field, it collapses. This implosion is of cataclysmic proportion, resulting in a shock wave that moves out from the star's interior. This energetic shock wave destroys the star and blasts the outer shell into space, generating the supernova event.

Theoretical work predicts that during a supernova, the star's interior condenses into a very hot object, possibly no larger than 20 kilometers (12 miles) in diameter. These incomprehensibly dense bodies have been named *neutron stars.* Some supernovae events are thought to produce even smaller, and most intriguing, objects called *black holes.* We will consider the nature of neutron stars and black holes in the "Stellar Remnants" section.

H-R Diagrams and Stellar Evolution

The Hertzsprung-Russell (H-R) diagrams have been helpful in formulating and testing models of stellar evolution. They are also useful for illustrating the changes that take place in an individual star during its life span. Figure 16.11 (p. 439) shows on an H-R diagram the evolution of a star about the size of our Sun. Keep in mind that the star does not physically move along this path, but rather that its position on the H-R diagram represents the color (temperature) and absolute magnitude (brightness) of the star at various stages in its evolution.

For example, on the H-R diagram, a protostar would be located to the right and above the main sequence (Figure 16.11). It is formed to the right because of its relatively cool surface temperature (red color), and above because it would be more luminous than a main-sequence star of the same color, a fact attributable to its large size. Careful examination of Figure 16.11 should help you visualize the evolutionary changes experienced by a star the size of our Sun. In addition, Table 16.3 provides a summary of the evolutionary history of stars having various masses.

Stellar Remnants

Eventually, all stars consume their nuclear fuel and collapse into one of three final states—white dwarf, neutron star, or black hole. Although different in some ways, these small, compact objects are all composed of incomprehensibly dense material and all have extreme surface gravity.

White Dwarfs

White dwarfs are extremely small stars with densities greater than any known terrestrial material. It is believed that white dwarfs were once low-mass or medium-mass stars whose internal heat was able to keep these gaseous bodies from collapsing under their own gravitational force.

Although some white dwarfs are no larger than Earth, the mass of such a dwarf can equal 1.4 times that of the Sun. Thus, their densities may be a million times greater than that of water. A spoonful of such matter would weigh several tons. Densities this great are possible only when electrons are displaced inward from their regular orbits, around an atom's nucleus, allowing the atoms to take up less than the "normal" amount of space. Material in this state is called **degenerate matter.**

In degenerate matter, the atoms have been squeezed together so tightly that the electrons are displaced much nearer to the nucleus. Degenerate matter uses electrical repulsion rather than molecular motion to support itself from total collapse. Although atomic particles in degenerate matter are much closer together than in normal Earth matter, they still are not packed as tightly as possible. Stars made of matter that has an even greater density are thought to exist.

As a star contracts into a white dwarf, its surface becomes very hot, sometimes exceeding 25,000 K. Even so, without a source of energy, it can only become cooler and dimmer. Although none have been observed, the terminal stage of a white dwarf must be a small, cold, nonluminous body called a *black dwarf.*

Neutron Stars

A study of white dwarfs produced what might at first appear to be a surprising conclusion. The smallest white dwarfs are the most massive, and the largest are the least massive. The explanation for this is that a more massive star, because of its greater gravitational force, is able to squeeze itself into a smaller, more densely packed object than can a less massive star. Thus, the smallest white dwarfs were produced from the collapse of larger, more massive stars than were the larger white dwarfs.

This conclusion led to the prediction that stars smaller and more massive than white dwarfs must exist. Named **neutron stars,** these objects are thought to be the remnants

Table 16.3 Summary of evolution for stars of various masses

Initial Mass of Interstellar Cloud (Sun = 1)	Main-Sequence Stage	Giant Phase	Evolution After Giant Phase	Terminal State (Final Mass*)
0.001	None (Planet)	No	None	Planet (0.001)
0.1	Red	No	None	White dwarf (0.1)
1–3	Yellow	Yes	Planetary nebula	White dwarf (<1.4)
6	White	Yes	Supernova	Neutron star (1.4–3)
20	Blue	Yes/Supergiant	Supernova	Black hole (>3.0)

*These mass numbers are estimates.

of supernova events. In a white dwarf, the electrons are pushed close to the nucleus, whereas in a neutron star, the electrons are forced to combine with protons to produce neutrons (hence the name). If Earth were to collapse to the density of a neutron star, it would have a diameter equivalent to the length of a football field. A pea-size sample of this matter would weigh 100 million tons. This is approximately the density of an atomic nucleus; thus, neutron stars can be thought of as large atomic nuclei.

During a supernova implosion, the envelope of the star is ejected (Figure 16.15), while the core collapses into a very hot neutron star about 20 kilometers (12 miles) in diameter. Although neutron stars have high surface temperatures, their small size would greatly limit their luminosity. Consequently, locating one visually would be extremely difficult.

However, theory predicts that a neutron star would have a strong magnetic field. Further, as a star collapses, it will rotate faster, for the same reason ice skaters rotate faster as they pull in their arms. If the Sun were to collapse to the size of a neutron star, it would increase its rate of rotation from once every 25 days to nearly 1000 times per second. Radio waves generated by this rotating star would be concentrated into two narrow zones that would align with the star's magnetic poles. Consequently, the star would resemble a rapidly rotating beacon emitting strong radio waves. If Earth happened to be in the path of these beacons, the star would appear to blink on and off, or pulsate, as the waves swept past.

In the early 1970s, a source that radiates short pulses of radio energy, called a **pulsar** (pulsating radio source), was discovered in the Crab Nebula. Visual inspection of this radio source revealed it to be a small star centered in the nebula. The pulsar found in the Crab Nebula is undoubtedly the remains of the supernova of A.D. 1054 (see Figure 16.14). Thus, the first neutron star was discovered.

Figure 16.15 Veil Nebula in the constellation Cygnus is the remnant of an ancient supernova implosion. (Palomar Observatories/California Institute of Technology [Caltech])

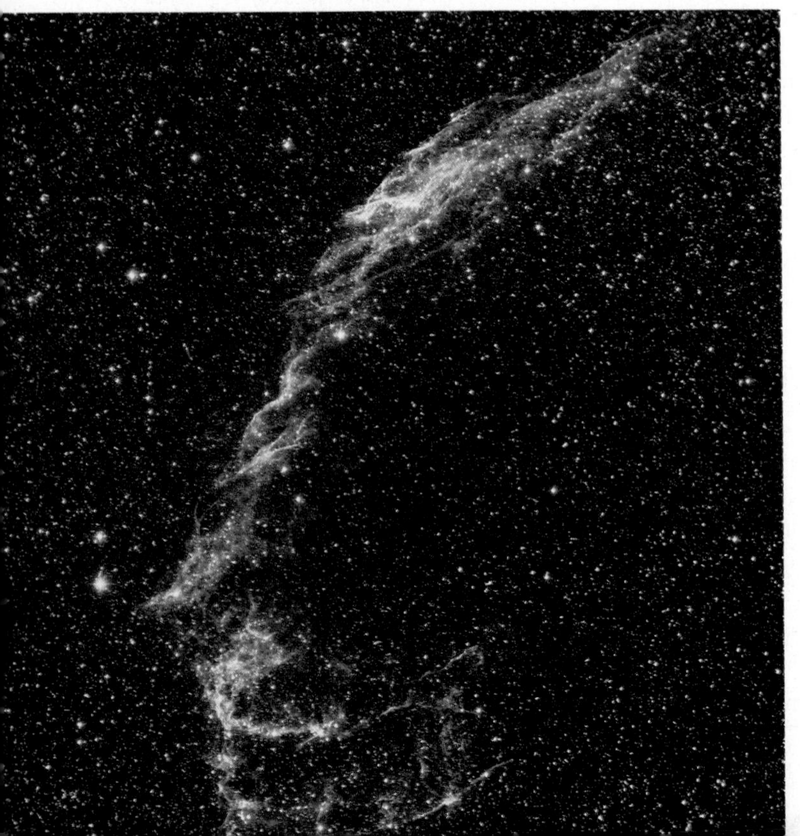

Black Holes

Are neutron stars made of the most dense materials possible? No. During a supernova event, remnants of stars greater than three solar masses apparently collapse into objects even smaller and denser than neutron stars. Even though these objects would be very hot, their surface gravity would be so immense that even light could not escape the surface. Consequently, they would literally disappear from sight. These incredible bodies have appropriately been named **black holes.** Anything that moved too near a black hole would be swept in by its irresistible gravity and devoured forever.

How can astronomers find an object whose gravitational field prevents the escape of all matter and energy? One strategy is to seek evidence of matter being rapidly swept into a region of apparent nothingness. Theory predicts that as matter is pulled into a black hole, it should become very hot and emit a flood of x-rays before being engulfed. Because isolated black holes would not have a source of matter to engulf, astronomers first looked at binary star systems.

A likely candidate for a black hole is Cygnus X-1, a strong x-ray source in the constellation Cygnus. In this case, the x-ray source can be observed orbiting a supergiant companion with a period of 5.6 days. It appears that gases are pulled from this companion and spiral into the disk-shaped structure around the black hole (Figure 16.16). The result is a stream of x-rays. Because x-rays cannot penetrate our atmosphere efficiently, the existence of black holes was not confirmed until the late twentieth century. The first x-ray sources were discovered in 1971 by detectors on satellites. Cygnus X-1 is such a source.

The Milky Way Galaxy

On a clear, moonless night away from city lights, you can see a truly marvelous sight—our own Milky Way Galaxy (Figure 16.17). With his telescope, Galileo discovered that this band of light was produced by countless individual stars that the unaided eye is unable to resolve. We now realize that the Sun is actually a part of this vast system of stars, which number about 100 billion. The "milky" appearance of our galaxy results because the solar system is located within the flat *galactic disk.* Thus, when it is viewed from the "inside," a higher concentration of stars appears in the direction of the galactic plane than in any other direction. You can see this in the edge-on view in Figure 16.18B.

When astronomers began telescopically surveying the stars located along the plane of the Milky Way, they found that equal numbers lay in every direction. Could Earth actually be at the center of the galaxy? A better explanation was advanced. Imagine that the trees in an enormous forest represent the stars in the galaxy. After hiking into this forest a short distance, you look around. What you see is an equal

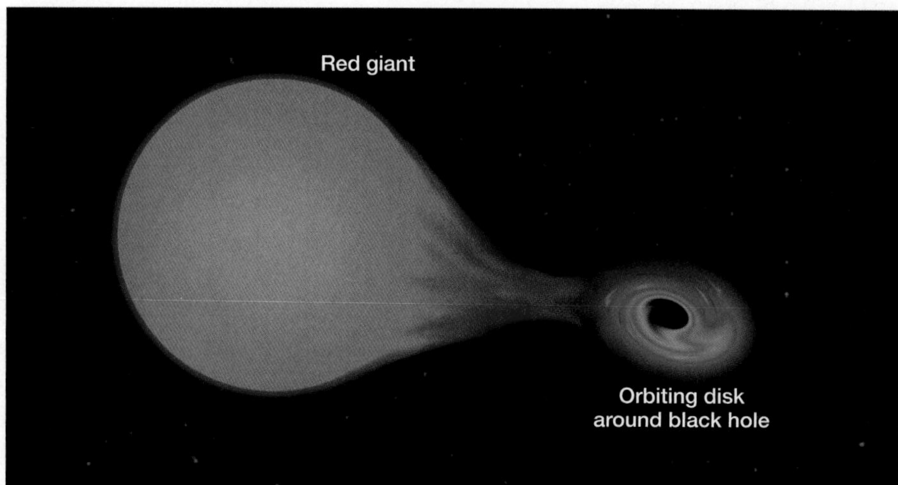

Figure 16.16 This illustration shows how astronomers believe a binary pair (red giant/black hole) might function.

number of trees in every direction. Are you really in the center of the forest? Not necessarily; anywhere in the forest, except at the very edge, you will seem to be in the middle.

Structure of the Milky Way Galaxy

Attempts to inspect the Milky Way visually are hindered by the large quantities of interstellar matter that lie in our line of sight. Nevertheless, with the aid of radio telescopes, the gross structure of our galaxy has been determined. The Milky Way is a rather large spiral galaxy whose disk is about 100,000 light-years wide and about 10,000 light-years thick at the nucleus (Figure 16.18A). As viewed from Earth, the center of the galaxy lies beyond the constellation Sagittarius.

Radio telescopes reveal the existence of at least three distinct *spiral arms,* with some showing splintering (Figure 16.19). The Sun is positioned in one of these arms about two-thirds of the way from the center, at a distance from the hub of about 30,000 light-years. The stars in the arms of the Milky Way rotate around the *galactic nucleus,* with the most outward ones moving the slowest, such that the ends of the arms appear to trail. The Sun and the arm it is in require about 200 million years for each orbit around the nucleus.

Surrounding the galactic disk is a nearly spherical *halo* made of very tenuous gas and numerous globular clusters. These star clusters do not participate in the rotating motion of the arms but rather have their own orbits that carry them through the disk. Although some clusters are very dense, they pass among the stars of the arms with plenty of room to spare.

Galaxies

In the mid-1700s, German philosopher Immanuel Kant proposed that the telescopically visible fuzzy patches of light scattered among the stars were actually distant galaxies like the Milky Way. Kant described them as "island universes." Each galaxy, he believed, contained billions of stars and, as such, was a universe in itself. The weight of opinion, however, favored the hypothesis that they were dust and gas clouds (nebulae) within our galaxy.

This matter was not resolved until the 1920s, when American astronomer Edwin Hubble was able to locate within one of these fuzzy patches some unique stars that are known to be intrinsically very bright. Because these stars appeared only very faintly in the telescope, Hubble believed they must lie outside the Milky Way.

This fuzzy patch, which lies over a million light-years away, was named the Great Galaxy in Andromeda (Figure 16.20). Hubble had extended the universe far beyond the limits of our imagination, to include hundreds of billions of galaxies, each containing hundreds of billions of stars. It has been said that a million galaxies are found in that portion of the sky bounded by the cup of the Big Dipper. There are more stars in the heavens than grains of sand in all the beaches on Earth.

Types of Galaxies

From the hundreds of billions of galaxies, three basic types have been identified: spiral, elliptical, and irregular.

Figure 16.17 Panorama of our galaxy, the Milky Way. Notice the dark bands caused by the presence of interstellar dark nebulae. (Courtesy of Axel Mellinger)

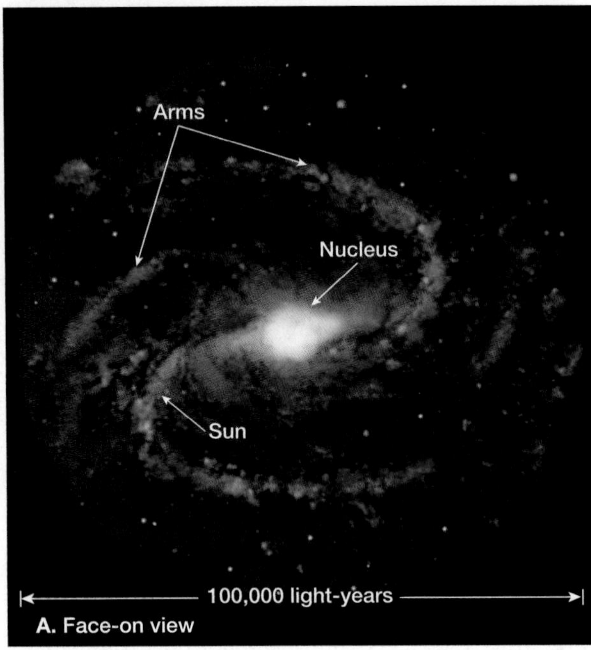

Figure 16.18 Structure of the visible portion of the Milky Way Galaxy.

Spiral Galaxies The Milky Way and the Great Galaxy in Andromeda are examples of fairly large **spiral galaxies** (Figure 16.21). Andromeda can be seen with the unaided eye as a fuzzy fifth-magnitude object. Typically, spiral galaxies are disk-shaped, with a somewhat greater concentration of stars near their centers, but there are numerous variations. Viewed broadside, arms are often seen extending from the central nucleus and sweeping gracefully away. The outermost stars of these arms rotate most slowly, giving the galaxy the appearance of a fireworks pinwheel.

One type of spiral galaxy, however, has the stars arranged in the shape of a bar, which rotates as a rigid system. This requires that the outer stars move faster than the inner ones, a fact not easy for astronomers to reconcile with the laws of motion. Attached to each end of these bars are curved spiral arms. These have become known as **barred spiral galaxies** (Figure 16.22). Spiral galaxies are generally quite large, ranging from 20,000 to about 125,000 light-years in diameter. About 10 percent of all galaxies are thought to be barred spirals, and another 20 percent are regular spiral galaxies like the Milky Way.

Elliptical Galaxies The most abundant group, making up 60 percent of the total, is the **elliptical galaxies.** These are gen-

Figure 16.19 If the Milky Way were photographed from a distance, it might appear like the image of spiral galaxy NGC 2997 shown here. (© Anglo-Australian Observatory, photography by David Malin)

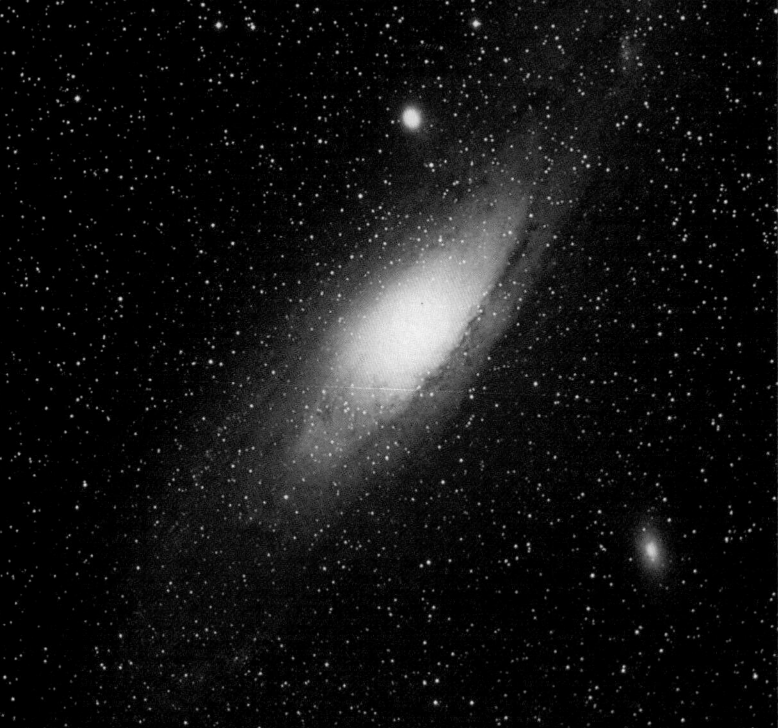

Figure 16.20 Great Galaxy, a spiral galaxy, in the constellation Andromeda. The two bright spots above and below are dwarf elliptical galaxies. (Palomar Observatories/California Institute of Technology [Caltech])

erally smaller than spiral galaxies. Some are so much smaller, in fact, that the term *dwarf* has been applied to them. Because these dwarf galaxies are not visible at great distances, a survey of the sky reveals more of the conspicuous large spiral galaxies. Although most elliptical galaxies are small, the very largest known galaxies (200,000 light-years in diameter) are also elliptical. As their name implies, elliptical galaxies have an ellipsoidal shape that ranges to nearly spherical, and they lack spiral arms. The two dwarf companions of Andromeda shown in Figure 16.20 are elliptical galaxies.

Irregular Galaxies Only 10 percent of the known galaxies lack symmetry and are classified as **irregular galaxies.** The best-known irregular galaxies, the Large and Small Magellanic Clouds in the Southern Hemisphere, are easily visible with the unaided eye. Named after the explorer Ferdinand Magellan, who observed them when he circumnavigated Earth in 1520, they are our nearest galactic neighbors—only 150,000 light-years away.

One of the major differences among the galactic types is the age of the stars they contain. Irregular galaxies consist mostly of young stars, whereas elliptical galaxies contain old stars. The Milky Way and other spiral galaxies consist of both young and old stars, with the youngest located in the arms.

Galactic Clusters

Once astronomers discovered that stars were associated in groups, they set out to determine whether galaxies were also grouped or just randomly distributed throughout the universe. They found that, like stars, galaxies are grouped in **galactic clusters** (Figure 16.23). Some abundant clusters contain thousands of galaxies. Our own, called the **Local Group,** contains at least 28 galaxies. Of these, 3 are spirals, 11 are irregulars, and 14 are ellipticals. Galactic clusters also reside in huge swarms called *superclusters.* From visual observations, it appears that superclusters may be the largest entities in the universe.

A.

B.

Figure 16.21 Two views illustrating the idealized structure of spiral galaxies. (Left photo courtesy of Smithsonian Astrophysical Observatory; right photo courtesy of Hansen Planetarium/U.S. Naval Observatory)

Figure 16.22 Barred spiral galaxy. (Palomar Observatories/California Institute of Technology [Caltech])

Red Shifts

You probably have noticed the change in pitch of a car horn or ambulance siren as it passes by. When it is approaching, the sound seems to have a higher-than-normal pitch, and when it is moving away, the pitch sounds lower than normal. This effect, which occurs for all wave motion, including sound and light waves, was first explained by Christian Doppler in 1842 and is called the **Doppler effect.** The reason for the difference in pitch is that it takes time for the wave to be emitted. If the source of the wave is moving away, the beginning of the wave is emitted nearer to you than the end of the wave, effectively "stretching" the wave. This gives it a longer wavelength (Figure 16.24). The opposite is true for an approaching source.

In the case of light, when a source is moving away, its light appears redder than it actually is, because its waves appear lengthened. Objects approaching have their light waves shifted toward the blue (shorter wavelength). Therefore, the Doppler effect reveals whether Earth and another celestial body are approaching or leaving each other. In addition, the amount of shift allows us to calculate the *rate* at which the relative movement is occurring. Large Doppler shifts indicate higher velocities; smaller Doppler shifts indicate slower velocities.

Figure 16.23 A cluster of galaxies located about 1 million light-years from Earth. (NASA)

Expanding Universe

One of the most important discoveries of modern astronomy was made in 1929 by Edwin Hubble. Observations completed several years earlier revealed that most galaxies have Doppler shifts toward the red end of the spectrum. Recall that red shift occurs because the light waves are "stretched," indicating that Earth and the source are moving away from each other. Hubble set out to explain the predominance of red shift.

Hubble realized that dimmer galaxies were probably farther away than brighter galaxies. Thus, he tried to determine if there is a relation between the distances to galaxies and their red shifts. Using estimated distances based on relative brightness and the observed Doppler red shifts, Hubble discovered that galaxies that exhibit the greatest red shifts are the most distant.

A consequence of the universal red shift is that it predicts that most galaxies (except for a few nearby) are receding from us. Recall that the amount of Doppler red shift is dependent on the velocity at which the object is moving away. Greater red shifts indicate faster recessional velocities. Because more distant galaxies have greater red shifts, Hubble concluded that they must be retreating from us at greater velocities. This idea is currently called **Hubble's law** and states that galaxies are receding from us at a speed that is proportional to their distance.

Hubble was surprised at this discovery because it implied that the most distant galaxies are moving away from us many times faster than those nearby. What type of cosmological theory can explain this fact? It was soon realized that an *expanding universe* can adequately account for the observed red shifts.

To help visualize the nature of this expanding universe, consider a commonly used analogy. Imagine a loaf of raisin bread dough that has been set out to rise for a few hours (Figure 16.25). As the dough doubles in size, so does the distance between all of the raisins. However, the raisins that were originally farther apart traveled a greater distance in the same time span than those located closer together. We conclude, therefore, that in an expanding universe, as in our analogy, those objects located farther apart recede at greater velocities.

Another feature of the expanding universe can be demonstrated using the raisin bread analogy. No matter which raisin you select, it will move away from all the other raisins. Likewise, no matter where one is located in the universe, every other galaxy (except those in the same cluster) will be receding. Edwin Hubble had indeed advanced our understanding of the universe. The Hubble Space Telescope is named in his honor.

The Big Bang

Did the universe have a beginning? Will it have an end? Cosmologists are trying to answer these questions, and that makes them a rare breed.

First and foremost, any viable theory regarding the origin of the universe must account for the fact that all galaxies (except for the very nearest) are moving away from us. Be-

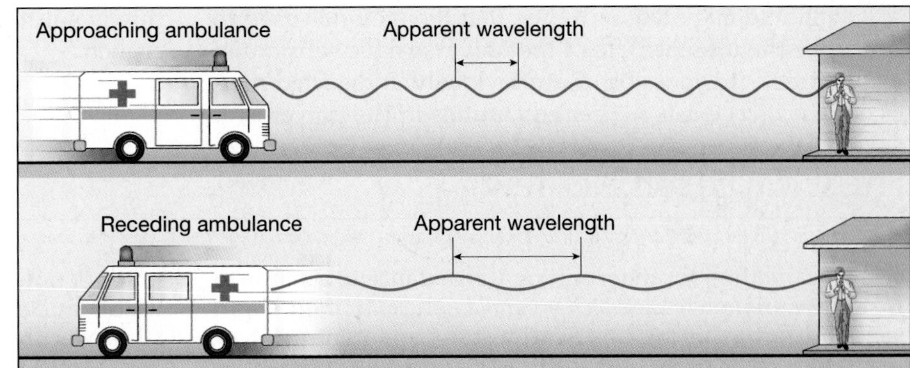

Figure 16.24 The Doppler effect, illustrating the apparent lengthening and shortening of wavelengths caused by the relative motion between a source and an observer.

cause all galaxies appear to be moving away from Earth, is our planet in the center of the universe? Probably not, because if we are not even in the center of our own solar system, and our solar system is not even in the center of our galaxy, it seems unlikely that we could be in the center of the universe.

A more probable explanation exists: Imagine a balloon with paper-punch "dots" glued to its surface. When the balloon is inflated, each dot spreads apart from every other dot. Similarly, if the universe is expanding, every galaxy would be moving away from every other galaxy.

This concept of an expanding universe led to the widely accepted **Big Bang** Theory. According to this theory, the entire universe was at one time confined to a dense, hot, supermassive ball. Then, nearly 14 billion years ago, a cataclysmic explosion occurred, hurling this material in all directions. The Big Bang marks the inception of the universe; all matter and space were created at this instant. The ejected masses of gas cooled and condensed, forming the stars that compose the galactic systems we now observe fleeing from their birthplaces.

If the universe began with a big bang, how will it end? One view is that the universe will last forever. In this scenario, the stars will slowly burn out, being replaced by invisible degenerate matter and black holes that will travel outward through an endless, dark cold universe. The other possibility is that the outward flight of the galaxies will slow and eventually stop. Gravitational contraction would follow, causing the galaxies to collide and coalesce into the high-energy, high-density mass from which the universe began. This fiery death of the universe, the big bang operating in reverse, has been called the "big crunch."

Whether or not the universe will expand forever, or eventually collapse upon itself, depends on its average density. If the average density of the universe is more than an amount known as its *critical density* (about one atom for every cubic meter), the gravitational field is sufficient to stop the outward expansion and cause the universe to contract. On the other hand, if the density of the universe is less than the critical value, it will expand forever. Current estimates of the density of the universe place it below the critical density, which predicts an ever-expanding, or *open, universe.* Additional support for an open universe comes from studies that indicate the universe is expanding faster now than in the past. Hence, the view currently favored by most cosmologists is an expanding universe with no ending point.

Did You Know?

The name Big Bang was originally coined by cosmologist Fred Hoyle as a sarcastic comment on the believability of the theory. The Big Bang Theory proposes that our universe began as a violent explosion, from which the universe continues to expand, evolve, and cool. Through decades of experimentation and observation, scientists have gathered substantial evidence that supports this theory. Despite this fact, the Big Bang Theory, like all other scientific theories, can never be proved. It is always possible that a future observation will disprove a previously accepted theory. Nevertheless, the Big Bang has replaced all alternative theories and remains the only widely accepted scientific model for the origin of the universe.

Figure 16.25 Illustration of the raisin bread analogy of an expanding universe. As the dough rises, raisins originally farther apart travel a greater distance in the same time span than those located closer together. Thus, raisins (like galaxies in a uniform expanding universe) that are located farther apart move away from each other more rapidly than those located nearer to one another.

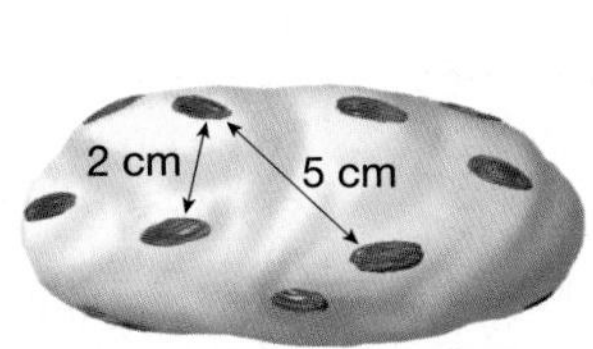

A. Raisin bread dough before it rises.

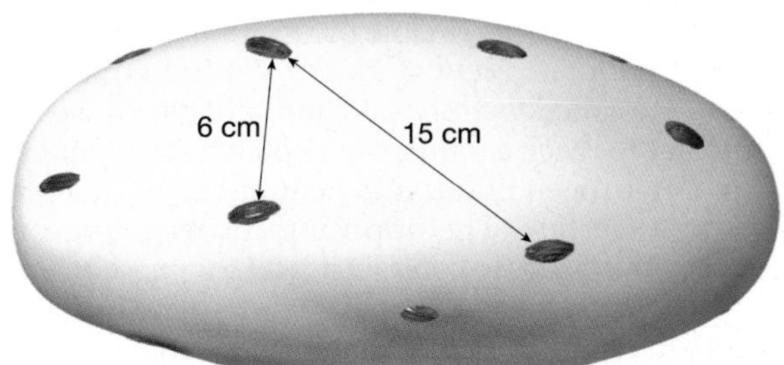

B. Raisin bread dough a few hours later.

It should be noted, however, that the methods used to determine the ultimate fate of the universe have substantial uncertainties. It is possible that previously undetected matter (*dark matter*) exists in great quantities in the universe. If this is so, the galaxies could, in fact, collapse in the "big crunch."

Absence of evidence is not evidence of absence.

—Anonymous

The Chapter in Review

1. One method for determining the distance to a star is to use a measurement called *stellar parallax,* the extremely slight back-and-forth shifting in a nearby star's apparent position due to the orbital motion of Earth. *The farther away a star is, the less its parallax.* A unit used to express stellar distance is the *light-year,* which is the distance light travels in a year—about 9.5 trillion kilometers (5.8 trillion miles).

2. The intrinsic properties of stars include *brightness, color, temperature, mass,* and *size.* Three factors control the brightness of a star as seen from Earth: how big it is, how hot it is, and how far away it is. *Apparent magnitude* is how bright a star appears from Earth. *Absolute magnitude* is the "true" brightness of a star if it were at a standard distance of about 32.6 light-years. Color reveals a star's temperature. Very hot stars (surface temperatures above 30,000 K) appear blue; red stars are much cooler (surface temperatures generally less than 3000 K). Stars with surface temperatures between 5000 and 6000 K appear yellow, like the Sun. *Binary stars* are two stars kept together by gravitational attraction and revolving around a common center of mass. This center is used to determine the mass of the individual stars.

3. A *Hertzsprung-Russell (H-R) diagram* is constructed by plotting the absolute magnitudes and temperatures of stars on a graph. A great deal about the sizes of stars can be learned from H-R diagrams. Stars that fall in the upper-right position of an H-R diagram are called *giants,* luminous stars of large radii. *Supergiants* are very large. Very small *white dwarf* stars fall in the lower-central portion of an H-R diagram. Ninety percent of all stars, called *main-sequence stars,* fall in a band that runs from the upper-left corner to the lower-right corner of an H-R diagram.

4. New stars are born out of enormous accumulations of dust and gases, called *nebulae,* that are scattered between existing stars. A *bright nebula* glows because the matter is close to a very hot (blue) star. The main types of bright nebulae are *emission nebulae* (which derive their visible light from the fluorescence of ultraviolet light from a star in or near the nebula) and *reflection nebulae* (relatively dense dust clouds in interstellar space that reflect the light of nearby stars). When a nebula is not close enough to a bright star to be illuminated, it is a *dark nebula.*

5. Stars are born when their nuclear furnaces begin operating under the unimaginable pressures and temperatures in collapsing nebulae. New stars not yet hot enough for nuclear fusion are called *protostars.* When collapse causes the core of a protostar to reach a temperature of at least 10 million K, the fusion of hydrogen nuclei into helium begins in a process called *hydrogen burning.* The opposing forces acting on a star are *gravity,* which tries to contract it, and *gas pressure (thermal nuclear energy),* which tries to expand it. When the two forces are balanced, the star becomes a stable main-sequence star. When the hydrogen in a star's core is consumed, its outer envelope expands enormously to form a *red giant* star, hundreds to thousands of times larger than its main-sequence size. When all the usable nuclear fuel in these giants is exhausted and gravity takes over, the stellar remnant collapses into a small dense body.

6. The *final fate of a star is determined by its mass.* Stars with less than one-half the mass of the Sun collapse into hot, dense white dwarf stars. Medium-mass stars (between 0.5 and 3.0 times the mass of the Sun) become red giants, collapse, and end up as white dwarf stars, often surrounded by expanding spherical clouds of glowing gas called *planetary nebulae.* Stars more than three times the mass of the Sun terminate in a brilliant explosion called a *supernova.* Supernovae events can produce small, extremely dense *neutron stars,* composed entirely of subatomic particles called neutrons; or even smaller and more dense *black holes,* objects that have such immense gravity that light cannot escape their surface.

7. The *Milky Way Galaxy* is a large, disk-shaped, *spiral galaxy* about 100,000 light-years wide and about 10,000 light-years thick at the center. It has three distinct *spiral arms* of stars, with some showing splintering. The Sun is positioned in one of these arms about two-thirds of the way from the galactic center, a distance of about 30,000 light-years. Surrounding the galactic disk is a nearly spherical halo made of very tenuous gas and numerous *globular clusters* (nearly spherical groups of densely packed stars).

8. The various types of galaxies include (1) *spiral galaxies,* which are typically disk-shaped, with a somewhat greater concentration of stars near their centers and often having arms of stars extending from their central nucleus; (2) *elliptical galaxies,* the most abundant type, which have an ellipsoidal shape that ranges to nearly spherical and lack spiral arms; and (3) *irregular galaxies,* which lack symmetry and account for only 10 percent of the known galaxies.

9. The *Doppler effect* is the apparent change in wavelength of radiation caused by the motions of the source and the observer. By applying Doppler effect to the light of galaxies, galactic motion can be determined. Most galaxies have Doppler shifts toward the red end of the spectrum, indicating increasing distance. The amount of Doppler shift is dependent on the velocity at which the object is moving. Because the most distant galaxies have the greatest red shifts, Edwin Hubble concluded in the early 1900s that they were retreating from us faster than were closer galaxies. It was soon realized that an *expanding universe* can adequately account for the observed red shifts.

10. The belief in the expanding universe led to the widely accepted *Big Bang Theory.* According to this theory, the entire universe was at one time confined in a dense, hot, supermassive concentration. Almost 14 billion years ago, a cataclysmic explosion hurled this material in all directions, creating all matter and space. Eventually, the ejected masses of gas cooled and condensed, forming the stellar systems we now observe fleeing from their place of origin.

Key Terms

absolute magnitude (p. 433)
apparent magnitude (p. 433)
barred spiral galaxy (p. 446)
Big Bang (p. 449)
black hole (p. 444)
bright nebula (p. 437)
dark nebula (p. 437)
degenerate matter (p. 443)
Doppler effect (p. 448)
elliptical galaxy (p. 446)
emission nebula (p. 437)
eruptive variable (star) (p. 436)
galactic cluster (p. 447)
Hertzsprung-Russell (H-R) diagram (p. 435)
Hubble's law (p. 448)
hydrogen burning (p. 440)
interstellar dust (p. 437)
irregular galaxy (p. 447)
light-year (p. 433)
Local Group (p. 447)
magnitude (p. 433)
main-sequence stars (p. 435)
nebula (p. 437)
neutron star (p. 443)
nova (p. 436)
planetary nebula (p.441)
protostar (p. 440)
pulsar (p. 444)
pulsating variable (star) (p. 435)
red giant (p. 435)
reflection nebula (p. 437)
spiral galaxy (p. 446)
stellar parallax (p. 432)
supergiant (p. 435)
supernova (p. 442)
white dwarf (p. 435)

Questions for Review

1. How far away in light-years is our nearest stellar neighbor, Proxima Centauri? Convert your answer to kilometers.
2. What is the most basic method of determining stellar distances?
3. Explain the difference between a star's apparent and absolute magnitudes. Which one is an intrinsic property of a star?
4. What is the ratio of brightness between a twelfth-magnitude star and a fifteenth-magnitude star?
5. Which information about a star can be determined from its color?
6. Which color are the hottest stars? Medium-temperature stars? Coolest stars?
7. Which property of a star can be determined from binary-star systems?
8. Make a generalization relating the mass and luminosity of main-sequence stars.
9. The disk of a star cannot be resolved telescopically. Explain the method that astronomers have used to estimate the size of stars.
10. Where on an H-R diagram does a star spend most of its lifetime?
11. How does the Sun compare in size and brightness to other main-sequence stars?
12. Why is interstellar matter important to stellar evolution?
13. Compare a bright nebula and a dark nebula.
14. Which element is the fuel for main-sequence stars? Red giants?
15. What causes a star to become a giant?
16. Why are less massive stars thought to age more slowly than more massive stars, even though they have much less "fuel"?
17. Enumerate the steps thought to be involved in the evolution of Sun-like stars.
18. What is the final state of a low-mass (red) main-sequence star?
19. What is the final state of a medium-mass (Sun-like) star?
20. How do the "lives" of the most massive stars end? What are the two possible products of this event?
21. Describe the general structure of the Milky Way.
22. Compare the three general types of galaxies.
23. Explain why astronomers consider elliptical galaxies more abundant than spiral galaxies, even though more spiral galaxies have been sighted.
24. How did Edwin Hubble determine that the Great Galaxy in Andromeda is located beyond our galaxy?
25. Describe the evidence that supports the Big Bang Theory?

Online Study Guide

The *Foundations of Earth Science* Web site uses the resources and flexibility of the Internet to aid in your study of the topics in this chapter. Written and developed by Earth science instructors, this site will help improve your understanding of Earth science. Visit **http://www.prenhall.com/lutgens** and click on the cover of *Foundations of Earth Science 5e* to find:

- Online review quizzes.
- Critical thinking exercises.
- Links to chapter-specific Web resources.
- Internet-wide key-term searches.

http://www.prenhall.com/lutgens

APPENDIX A METRIC AND ENGLISH UNITS COMPARED

Units

1 kilometer (km)	= 1000 meters (m)
1 meter (m)	= 100 centimeters (cm)
1 centimeter (cm)	= 0.39 inch (in.)
1 mile (mi)	= 5280 feet (ft)
1 foot (ft)	= 12 inches (in.)
1 inch (in.)	= 2.54 centimeters (cm)
1 square mile (mi^2)	= 640 acres (a)
1 kilogram (kg)	= 1000 grams (g)
1 pound (lb)	= 16 ounces (oz)
1 fathom	= 6 feet (ft)

Conversions

When you want to convert:	multiply by:	to find:
Length		
inches	2.54	centimeters
centimeters	0.39	inches
feet	0.30	meters
meters	3.28	feet
yards	0.91	meters
meters	1.09	yards
miles	1.61	kilometers
kilometers	0.62	miles
Area		
square inches	6.45	square centimeters
square centimeters	0.15	square inches
square feet	0.09	square meters
square meters	10.76	square feet
square miles	2.59	square kilometers
square kilometers	0.39	square miles
Volume		
cubic inches	16.38	cubic centimeters
cubic centimeters	0.06	cubic inches
cubic feet	0.028	cubic meters
cubic meters	35.3	cubic feet
cubic miles	4.17	cubic kilometers
cubic kilometers	0.24	cubic miles
liters	1.06	quarts
liters	0.26	gallons
gallons	3.78	liters

Masses and Weights

ounces	20.33	grams
grams	0.035	ounces
pounds	0.45	kilograms
kilograms	2.205	pounds

Temperature

When you want to convert degrees Fahrenheit (°F) to degrees Celsius (°C), subtract 32 degrees and divide by 1.8. When you want to convert degrees Celsius (°C) to degrees Fahrenheit (°F), multiply by 1.8 and add 32 degrees. When you want to convert degrees Celsius (°C) to kelvins (K), delete the degree symbol and add 273. When you want to convert kelvins (K) to degrees Celsius (°C), add the degree symbol and subtract 273.

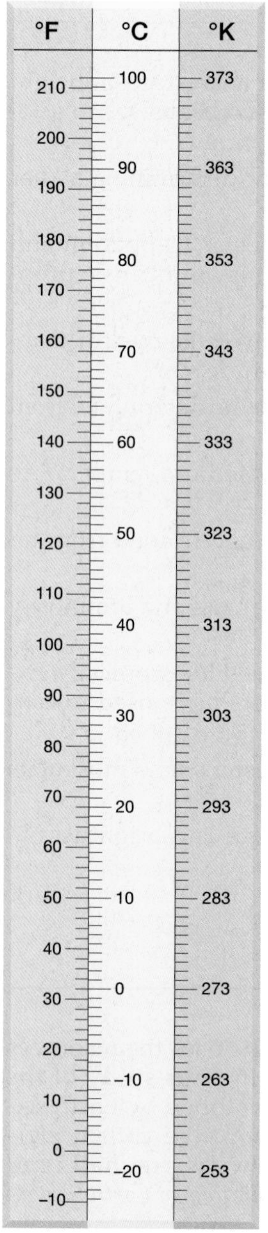

Figure A.1 A comparison of Fahrenheit, Celsius, and Kelvins temperature scales.

APPENDIX B MINERAL IDENTIFICATION KEY

Group I Metallic Luster			
Hardness	**Streak**	**Other Diagnostic Properties**	**Name (Chemical Composition)**
	Black streak	Black; magnetic; hardness = 6; specific gravity = 5.2; often granular	Magnetite (Fe_3O_4)
Harder than glass	Greenish-black streak	Brass yellow; hardness = 6; specific gravity = 5.2; generally an aggregate of cubic crystals	Pyrite (FeS_2) fool's gold
	Red-brown streak	Gray or reddish brown; hardness = 5–6; specific gravity = 5; platy appearance	Hematite (Fe_2O_3)
	Greenish-black streak	Golden yellow; hardness = 4; specific gravity = 4.2; massive	Chalcopyrite ($CuFeS_2$)
Softer than glass	Gray-black streak	Silvery gray; hardness = 2.5; specific gravity = 7.6 (very heavy); good cubic cleavage	Galena (PbS)
	Yellow-brown streak	Yellow brown to dark brown; hardness variable (1–6); specific gravity = 3.5–4; often found in rounded masses; earthy appearance	Limonite ($Fe_2O_3 \cdot H_2O$)
	Gray-black streak	Black to bronze; tarnishes to purples and greens; hardness = 3; specific gravity = 5; massive	Bornite (Cu_5FeS_4)
Softer than your fingernail	Dark gray streak	Silvery gray; hardness = 1 (very soft); specific gravity = 2.2; massive to platy; writes on paper (pencil lead); feels greasy	Graphite (C)

Group II Nonmetallic Luster (Dark-Colored)			
Hardness	**Cleavage**	**Other Diagnostic Properties**	**Name (Chemical Composition)**
	Cleavage present	Black to greenish black; hardness = 5–6; specific gravity = 3.4; fair cleavage, two directions at nearly 90 degrees	Augite (Ca, Mg, Fe, Al silicate)
		Black to greenish black; hardness = 5–6; specific gravity 3.2; fair cleavage, two directions at nearly 60 degrees and 120 degrees	Hornblende (Ca, Na, Mg, Fe, OH, Al silicate)
Harder than glass		Red to reddish brown; hardness = 6.5–7.5; conchoidal fracture; glassy luster	Garnet (Fe, Mg, Ca, Al silicate)
	Cleavage not prominent	Gray to brown; hardness = 9; specific gravity = 4; hexagonal crystals common	Corundum (Al_2O_3)
		Dark brown to black; hardness = 7; conchoidal fracture; glassy luster	Smoky quartz (SiO_2)
		Olive green; hardness = 6.5–7; small glassy grains	Olivine $(Mg, Fe)_2SiO_4$
	Cleavage present	Yellow brown to black; hardness = 4; good cleavage in six directions, light yellow streak that has the smell of sulfur	Sphalerite (ZnS)

(*continued*)

Group II Nonmetallic Luster (Dark-Colored) (Continued)			
Hardness	**Cleavage**	**Other Diagnostic Properties**	**Name (Chemical Composition)**
Softer than glass		Dark brown to black; hardness = 2.5–3, excellent cleavage in one direction; elastic in thin sheets; black mica	Biotite (K, Mg, Fe, OH, Al silicate)
	Cleavage absent	Generally tarnished to brown or green; hardness = 2.5; specific gravity = 9; massive	Native copper (Cu)
	Cleavage not prominent	Reddish brown; hardness = 1–5; specific gravity = 4–5; red streak; earthy appearance	Hematite (Fe_2O_3)
Softer than your fingernail		Yellow brown; hardness = 1–3; specific gravity = 3.5; earthy appearance; powders easily	Limonite ($Fe_2O_3 \cdot H_2O$)

Group III Nonmetallic Luster (Light Colored)			
Hardness	**Cleavage**	**Other Diagnostic Properties**	**Name (Chemical Composition)**
	Cleavage present	Salmon colored or white to gray; hardness = 6; specific gravity = 2.6; two directions of cleavage at nearly right angles	Potassium feldspar ($KAlSi_3O_8$)
Harder than glass			Plagioclase feldspar ($NaAlSi_3O_8$ to $CaAl_2Si_2O_8$)
	Cleavage absent	Any color; hardness = 7; specific gravity = 2.65; conchoidal fracture; glassy appearance; varieties; milky, rose, smoky, amethyst (violet)	Quartz (SiO_2)
		White, yellowish to colorless; hardness = 3; three directions of cleavage at 75 degrees (rhombohedral); effervesces in HCl; often transparent	Calcite ($CaCO_3$)
Softer than glass	Cleavage present	White to colorless; hardness = 2.5; three directions of cleavage at 90 degrees (cubic); salty taste	Halite (NaCl)
		Yellow, purple, green, colorless; hardness = 4; white streak; translucent to transparent; four directions of cleavage	Fluorite (CaF_2)
	Cleavage present	Colorless; hardness = 2–2.5; transparent and elastic in thin sheets; excellent cleavage in one direction; light mica	Muscovite (K, OH, Al silicate)
		White to transparent, hardness = 2; when in sheets, is flexible but not elastic; varieties: selenite (transparent, three directions of cleavage); satin spar (fibrous, silky luster); alabaster (aggregate of small crystals)	Gypsum ($CaSO_4 \cdot 2\,H_2O$)
Softer than your fingernail	Cleavage not prominent	White, pink, green; hardness = 1; forms in thin plates; soapy feel; pearly luster	Talc (Mg silicate)
		Yellow; hardness = 1–2.5	Sulfur (S)
		White; hardness = 2; smooth feel; earthy odor when moistened, has typical clay texture	Kaolinite (Hydrous Al silicate)
		Green; hardness = 2.5; fibrous; variety of serpentine	Asbestos (Mg, Al silicate)
		Pale to dark reddish brown; hardness = 1–3; dull luster; earthy; often contains spheroidal-shaped particles; not a true mineral	Bauxite (Hydrous Al oxide)

APPENDIX C

RELATIVE HUMIDITY AND DEW-POINT TABLES

Table C.1 Relative humidity (percent)

Depression of wet-bulb temperature (Dry-bulb temperature minus wet-bulb temperature = depression of the wet bulb)

Dry bulb (°C) — Dry-bulb (air) temperature	1	2	3	4	5	6	7	8	9	10	11	12	13	14	15	16	17	18	19	20	21	22
−20	28																					
−18	40																					
−16	48	0																				
−14	55	11																				
−12	61	23																				
−10	66	33	0																			
−8	71	41	13																			
−6	73	48	20	0																		
−4	77	54	32	11																		
−2	79	58	37	20	1																	
0	81	63	45	28	11																	
2	83	67	51	36	20	6																
4	85	70	56	42	27	14																
6	86	72	59	46	35	22	10	0														
8	87	74	62	51	39	28	17	6														
10	88	76	65	54	43	33	24	13	4													
12	88	78	67	57	48	38	28	19	10	2												
14	89	79	69	60	50	41	33	25	16	8	1											
16	90	80	71	62	54	45	37	29	21	14	7	1										
18	91	81	72	64	56	48	40	33	26	19	12	6	0									
20	91	82	74	66	58	51	44	36	30	23	17	11	5									
22	92	83	75	68	60	53	46	40	33	27	21	15	10	4	0							
24	92	84	76	69	62	55	49	42	36	30	25	20	14	9	4	0						
26	92	85	77	70	64	57	51	45	39	34	28	23	18	13	9	5						
28	93	86	78	71	65	59	53	45	42	36	31	26	21	17	12	8	4					
30	93	86	79	72	66	61	55	49	44	39	34	29	25	20	16	12	8	4				
32	93	86	80	73	68	62	56	51	46	41	36	32	27	22	19	14	11	8	4			
34	93	86	81	74	69	63	58	52	48	43	38	34	30	26	22	18	14	11	8	5		
36	94	87	81	75	69	64	59	54	50	44	40	36	32	28	24	21	17	13	10	7	4	
38	94	87	82	76	70	66	60	55	51	46	42	38	34	30	26	23	20	16	13	10	7	5
40	94	89	82	76	71	67	61	57	52	48	44	40	36	33	29	25	22	19	16	13	10	7

Relative humidity values

*To determine the relative humidity, find the air (dry-bulb) temperature on the vertical axis (far left) and the depression of the wet bulb on the horizontal axis (top). Where the two meet, the relative humidity is found. For example, when the dry-bulb temperature is 20°C and a wet-bulb temperature is 14°C, then the depression of the wet-bulb is 6°C (20°C − 14°C). From Table C.1, the relative humidity is 51 percent and from Table C.2, the dew point is 10°C.

Table C.2 Dew-point temperature (°C)*

Dew-point temperatures

Dry bulb (°C) — Dry-bulb (air) temperature	Dry-bulb temperature minus wet-bulb temperature = depression of the wet bulb																					
	1	**2**	**3**	**4**	**5**	**6**	**7**	**8**	**9**	**10**	**11**	**12**	**13**	**14**	**15**	**16**	**17**	**18**	**19**	**20**	**21**	**22**
−20	−33																					
−18	−28																					
−16	−24																					
−14	−21	−36																				
−12	−18	−28																				
−10	−14	−22																				
−8	−12	−18	−29																			
−6	−10	−14	−22																			
−4	−7	−12	−17	−29																		
−2	−5	−8	−13	−20																		
0	−3	−6	−9	−15	−24																	
2	−1	−3	−6	−11	−17																	
4	1	−1	−4	−7	−11	−19																
6	4	1	−1	−4	−7	−13	−21															
8	6	3	1	−2	−5	−9	−14															
10	8	6	4	1	−2	−5	−9	−14	−18													
12	10	8	6	4	1	−2	−5	−9	−16													
14	12	11	9	6	4	1	−2	−5	−10	−17												
16	14	13	11	9	7	4	1	−1	−6	−10	−17											
18	16	15	13	11	9	7	4	2	−2	−5	−10	−19										
20	19	17	15	14	12	10	7	4	2	−2	−5	−10	−19									
22	21	19	17	16	14	12	10	8	5	3	−1	−5	−10	−19								
24	23	21	20	18	16	14	12	10	8	6	2	−1	−5	−10	−18							
26	25	23	22	20	18	17	15	13	11	9	6	3	0	−4	−9	−18						
28	27	25	24	22	21	19	17	16	14	11	9	7	4	1	−3	−9	−16					
30	29	27	26	24	23	21	19	18	16	14	12	10	8	5	1	−2	−8	−15				
32	31	29	28	27	25	24	22	21	19	17	15	13	11	8	5	2	−2	−7	−14			
34	33	31	30	29	27	26	24	23	21	20	18	16	14	12	9	6	3	−1	−5	−12	−29	
36	35	33	32	31	29	28	27	25	24	22	20	19	17	15	13	10	7	4	0	−4	−10	
38	37	35	34	33	32	30	29	28	26	25	23	21	19	17	15	13	11	8	5	1	−3	−9
40	39	37	36	35	34	32	31	30	28	27	25	24	22	20	18	16	14	12	9	6	2	−2

APPENDIX D | EARTH'S GRID SYSTEM

A glance at any globe reveals a series of north-south and east-west lines that make up Earth's grid system, a universally used scheme for locating points on Earth's surface. The north-south lines of the grid are called **meridians** and extend from pole to pole (Figure D.1). All are halves of great circles. A **great circle** is the largest possible circle that can be drawn on a globe; if a globe were sliced along one of these circles, it would be divided into two equal parts called **hemispheres.** By viewing a globe or Figure D.1, it can be seen that meridians are spaced farthest apart at the equator and converge toward the poles. The east-west lines (circles) of the grid are known as **parallels.** As their names implies, these circles are parallel to one another (Figure D.1). While all meridians are parts of great circles, all parallels are not. In fact, only one parallel, the equator, is a great circle.

Latitude and Longitude

Latitude may be defined as distance, measured in degrees, *north* and *south* of the equator. Parallels are used to show latitude. Because all points that lie along the same parallel are an identical distance from the equator, they all have the same latitude designation. The latitude of the equator is 0 degrees, while the North and South poles lie 90 degrees N and 90 degrees S, respectively.

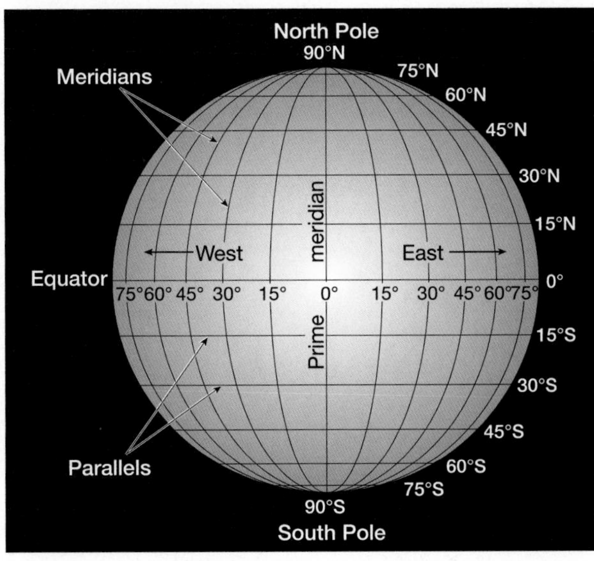

Figure D.1 Earth's grid system.

Longitude is defined as distance, measured in degrees, *east* and *west* of the zero or prime meridian. Because all meridians are identical, the choice of a zero line is obviously arbitrary. However, the meridian that passes through the Royal Observatory at Greenwich, England, is universally accepted as the reference meridian. Thus, the longitude for any place on the globe is measured east or west from this line. Longitude can vary from 0 degrees along the prime meridian to 180 degrees, halfway around the globe.

It is important to remember that when a location is specified, directions must be given—that is, north or south latitude and east or west longitude (Figure D.2). If this is not done, more than one point on the globe is being designated. The only exceptions, of course, are places that lie along the equator, the prime meridian, or the 180-degree meridian. It should also be noted that while it is not incorrect to use fractions, a degree of latitude or longitude is usually divided into minutes and seconds. A minute (′) is $\frac{1}{60}$ of a degree, and a second (″) is $\frac{1}{60}$ of a minute. When locating a place on a map, the degree of exactness will depend on the scale of the map. When using a small-scale world map or globe, it may be difficult to estimate latitude and longitude to the nearest whole degree or two. On the other hand, when a large-scale map of an area is used, it is often possible to estimate latitude and longitude to the nearest minute or second.

Distance Measurement

The length of a degree of longitude depends on where the measurement is taken. At the equator, which is a great circle, a degree of east-west distance is equal to approximately 111 kilometers (69 miles). This figure is found by dividing Earth's circumference—40,075 kilometers (24,900 miles)—by 360. However, with an increase in latitude, the parallels become smaller, and the length of a degree of longitude diminishes (Table D.1). Thus, at about latitude 60 degrees N and S, a degree of longitude has a value equal to about half of what it was at the equator.

As all meridians are halves of great circles, a degree of latitude is equal to about 111 kilometers (69 miles), just as is a degree of longitude along the equator. However, Earth is not a perfect sphere but is slightly flattened at the poles and

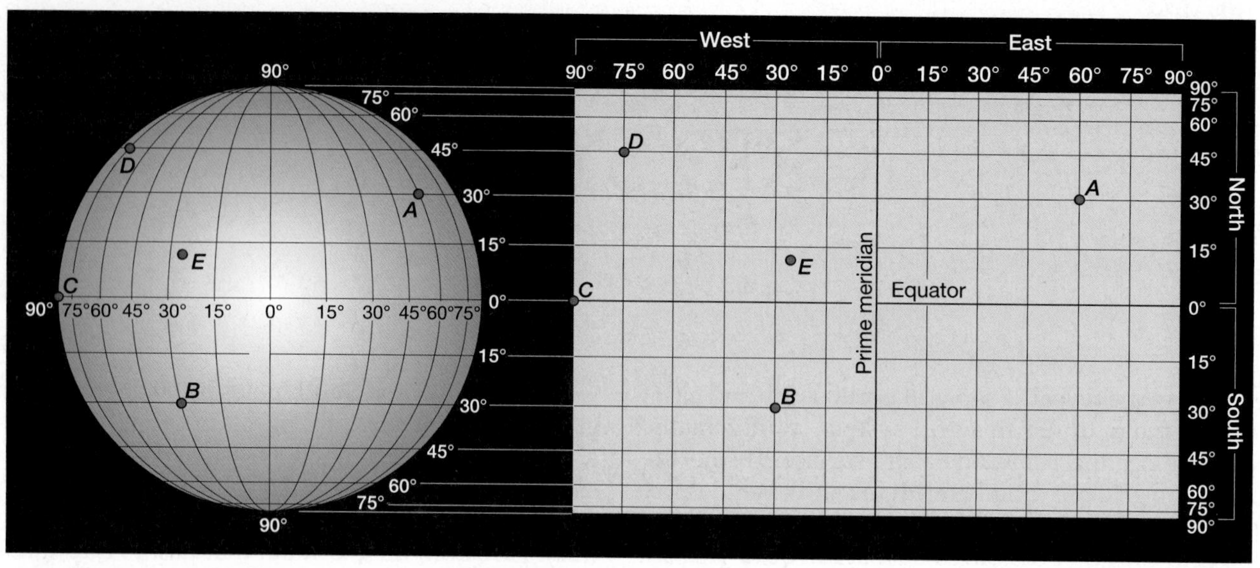

Figure D.2 Locating places using the grid system. For both diagrams: Point A is latitude 30 degrees N, longitude 60 degrees E; Point B is latitude 30 degrees S, longitude 30 degrees W; Point C is latitude 0 degrees, longitude 90 degrees W; Point D is latitude 45 degrees N, longitude 75 degrees W; Point E is approximately latitude 10 degrees N, longitude 25 degrees W.

bulges slightly at the equator. This causes small differences in the length of a degree of latitude.

Determining the shortest distance between two points on a globe can be done easily and fairly accurately using the "globe and string" method. It should be noted here that the arc of a great circle is the shortest distance between two points on a sphere. In order to determine the great circle distance (as well as observe the great circle route) between two places, stretch the string between the locations in question. Then, measure the length of the string along the equator (since it is a great circle with degrees marked on it) to determine the number of degrees between the two points. To calculate the distance in kilometers or miles, simply multiply the number of degrees by 111 or 69, respectively.

Table D.1

	Length of 1° Long.			Length of 1° Long.			Length of 1° Long.	
°Lat.	km	miles	°Lat.	km	miles	°Lat.	km	miles
0	111.367	69.172	30	96.528	59.955	60	55.825	34.674
1	111.349	69.161	31	95.545	59.345	61	54.131	33.622
2	111.298	69.129	32	94.533	58.716	62	52.422	62.560
3	111.214	69.077	33	93.493	58.070	63	50.696	31.488
4	111.096	69.004	34	92.425	57.407	64	48.954	30.406
5	110.945	68.910	35	91.327	56.725	65	47.196	29.314
6	110.760	68.795	36	90.203	56.027	66	45.426	28.215
7	110.543	68.660	37	89.051	55.311	67	43.639	27.105
8	110.290	68.503	38	87.871	54.578	68	41.841	25.988
9	110.003	68.325	39	86.665	53.829	69	40.028	24.862
10	109.686	68.128	40	85.431	53.063	70	38.204	23.729
11	109.333	67.909	41	84.171	52.280	71	36.368	22.589
12	108.949	67.670	42	82.886	51.482	72	34.520	21.441
13	108.530	67.410	43	81.575	50.668	73	32.662	20.287
14	108.079	67.130	44	80.241	49.839	74	30.793	19.126
15	107.596	66.830	45	78.880	48.994	75	28.914	17.959
16	107.079	66.509	46	77.497	48.135	76	27.029	16.788
17	106.530	66.168	47	76.089	47.260	77	25.134	15.611
18	105.949	65.807	48	74.659	46.372	78	23.229	14.428
19	105.337	65.427	49	73.203	45.468	79	21.320	13.242
20	104.692	65.026	50	71.727	44.551	80	19.402	12.051
21	104.014	64.605	51	70.228	43.620	81	17.480	10.857
22	103.306	64.165	52	68.708	42.676	82	15.551	9.659
23	102.565	63.705	53	67.168	41.719	83	13.617	8.458
24	101.795	63.227	54	65.604	40.748	84	11.681	7.255
25	100.994	62.729	55	64.022	39.765	85	9.739	6.049
26	100.160	62.211	56	62.420	38.770	86	7.796	4.842
27	99.297	61.675	57	60.798	37.763	87	5.849	3.633
28	98.405	61.121	58	59.159	36.745	88	3.899	2.422
29	97.481	60.547	59	57.501	35.715	89	1.950	1.211
30	96.528	59.955	60	55.825	34.674	90	0	0

GLOSSARY

Aa flow A type of lava flow that has a jagged blocky surface.

Abrasion The grinding and scraping of a rock surface by the friction and impact of rock particles carried by water, wind, or ice.

Absolute instability The condition of air that has an environmental lapse rate that is greater than the dry adiabatic rate (1°C per 100 meters).

Absolute magnitude The apparent brightness of a star if it were viewed from a distance of 32.6 light-years. Used to compare the true brightness of stars.

Absolute stability The condition of air that has an environmental lapse rate that is less than the wet adiabatic rate.

Abyssal plain Very level area of the deep-ocean floor, usually lying at the foot of the continental rise.

Accretionary wedge A large wedge-shaped mass of sediment that accumulates in subduction zones. Sediment is scraped from the subducting oceanic plate and accreted to the overriding crustal block.

Active continental margin Usually narrow and consisting of highly deformed sediments. They occur where oceanic lithosphere is being subducted beneath the margin of a continent.

Adiabatic temperature change Cooling or warming of air caused when air is allowed to expand or is compressed, not because heat is added or subtracted.

Advection fog A fog formed when warm, moist air is blown over a cool surface.

Aerosols Tiny solid and liquid particles suspended in the atmosphere

Aftershock A smaller earthquake that follows the main earthquake.

Air The mixture of gases and particles that make up Earth's atmosphere. Nitrogen and oxygen are most abundant.

Air mass A large body of air that is characterized by a sameness of temperature and humidity.

Air-mass weather The conditions experienced in an area as an air mass passes over it. Because air masses are large and fairly homogeneous, air-mass weather will be fairly constant and may last for several days.

Air pressure Force exerted by the weight of the air above.

Albedo The reflectivity of a substance, usually expressed as a percentage of the incident radiation reflected.

Alluvial fan A fan-shaped deposit of sediment formed when a stream's slope is abruptly reduced.

Alluvium Unconsolidated sediment deposited by a stream.

Alpine glacier A glacier confined to a mountain valley, which in most instances had previously been a stream valley.

Andesetic (intermediate) composition A compositional group of igneous rocks, in which the rock contains at least 25 percent dark silicate minerals. The other dominant mineral is plagioclase feldspar.

Aneroid barometer An instrument for measuring air pressure that consists of evacuated metal chambers that are very sensitive to variations in air pressure.

Angle of Repose The steepest angle at which loose material remains stationary without sliding downslope.

Angular unconformity An unconformity in which the strata below dip at an angle different from that of the beds above.

Anticline A fold in sedimentary strata that resembles an arch.

Anticyclone A high-pressure center characterized by a clockwise flow of air in the Northern Hemisphere.

Apparent magnitude The brightness of a star when viewed from Earth.

Aquifer Rock or soil through which groundwater moves easily.

Aquitard An impermeable bed that hinders or prevents groundwater movement.

Arctic (A) air mass A bitterly cold air mass that forms over the frozen Arctic Ocean.

Arête A narrow knifelike ridge separating two adjacent glaciated valleys.

Artesian well A well in which the water rises above the level where it was initially encountered.

Asteroids Thousands of small planetlike bodies, ranging in size from a few hundred kilometers to less than a kilometer, whose orbits lie mainly between those of Mars and Jupiter.

Asteroid belt The region in the solar system between the orbits of Mars and Jupiter where most asteroids are located.

Asthenosphere A subdivision of the mantle situated below the lithosphere. This zone of weak material exists below a depth of about 100 kilometers (62 miles) and in some regions extends as deep as 700 kilometers (435 miles). The rock within this zone is easily deformed.

Astronomical unit (AU) Average distance from Earth to the Sun; 1.5×10^8 km, or 93×10^6 miles.

Astronomy The scientific study of the universe; it includes the observation and interpretation of celestial bodies and phenomena.

Atmosphere The gaseous portion of a planet; the planet's envelope of air. One of the traditional subdivisions of Earth's physical environment.

Atom The smallest particle that exists as an element.

Atomic number The number of protons in the nucleus of an atom.

Backswamp A poorly drained area on a floodplain that results when natural levees are present.

Bar Common term for sand and gravel deposits in a stream channel or in near shore coastal area.

Barograph A recording barometer.

Barometer An instrument that measures atmospheric pressure.

Barometric tendency See *Pressure tendency.*

Barred spiral galaxy A galaxy having straight arms extending from its nucleus.

Barrier island A low, elongated ridge of sand that parallels the coast.

Basaltic composition A compositional group of igneous rocks in which the rock contains substantial dark silicate minerals and calcium-rich plagioclase feldspar.

Base level The level below which a stream cannot erode.

Basin A circular downfolded structure.

Batholith A large mass of igneous rock that formed when magma was emplaced at depth, crystallized, and subsequently exposed by erosion.

Bathymetry The measurement of ocean depths and the charting of the topography of the ocean floor.

Baymouth bar A sandbar that completely crosses a bay, sealing it off from the open ocean.

Beach drift The transport of sediment in a zigzag pattern along a beach caused by the uprush of water from obliquely breaking waves.

Beach nourishment Large quantities of sand added to the beach system to offset losses caused by wave erosion.

Bed load Sediment that is carried by a stream along the bottom of its channel.

Big Bang The theory that proposes that the universe originated as a single mass, which subsequently exploded.

Biogenous sediment Seafloor sediments consisting of material of marine-organic origin.

Biosphere The totality of life on Earth; the parts of the lithosphere, hydrosphere, and atmosphere in which living organisms can be found.

Black hole A massive star that has collapsed to such a small volume that its gravity prevents the escape of all radiation.

Blowout (deflation hollow) A depression excavated by the wind in easily eroded deposits.

Body wave A seismic wave that travels through Earth's interior.

Bowen's reaction series A concept proposed by N. L. Bowen that illustrates the relationships between magma and the minerals crystallizing from it during the formation of igneous rocks.

Braided stream A stream consisting of numerous intertwining channels.

Breakwater A structure protecting a nearshore area from breaking waves.

Bright nebula A cloud of glowing gas excited by ultraviolet radiation from hot stars.

Brittle failure (brittle deformation) The loss of strength by a material usually in the form of sudden fracturing.

Caldera A large depression typically caused by collapse or ejection of the summit area of a volcano.

Calorie The amount of heat required to raise the temperature of 1 gram of water 1°C.

Capacity The total amount of sediment a stream is able to transport.

Catastrophism The concept that Earth was shaped by catastrophic events of a short-term nature.

Cavern A naturally formed underground chamber or series of chambers most commonly produced by solution activity in limestone.

Celestial sphere An imaginary hollow sphere upon which the ancients believed the stars were hung and carried around Earth.

Cenozoic era A time span on the geologic time scale beginning about 65.5 million years ago following the Mesozoic era.

Chemical compound A substance formed by the chemical combination of two or more elements in definite proportions and usually having properties different from those of its constituent elements.

Chemical sedimentary rock Sedimentary rock consisting of material that was precipitated from water by either inorganic or organic means.

Chemical weathering The processes by which the internal structure of a mineral is altered by the removal and/or addition of elements.

Chinook A wind blowing down the leeward side of a mountain and warming by compression.

Cinder cone A rather small volcano built primarily of pyroclastics ejected from a single vent.

Circle of illumination The great circle that separates daylight from darkness.

Cirque An amphitheater-shaped basin at the head of a glaciated valley produced by frost wedging and plucking.

Cirrus One of three basic cloud forms; also one of the three high cloud types. They are thin, delicate ice-crystal clouds often appearing as veil-like patches or thin, wispy fibers.

Cleavage The tendency of a mineral to break along planes of weak bonding.

Climate A description of aggregate weather conditions; the sum of all statistical weather information that helps describe a place or region.

Cloud A form of condensation best described as a dense concentration of suspended water droplets or tiny ice crystals.

Clouds of vertical development A cloud that has its base in the low height range but extends upward into the middle or high altitudes.

Coarse-grained texture An igneous rock texture in which the crystals are roughly equal in size and large enough so that individual minerals can be identified with the unaided eye.

Cold front A front along which a cold air mass thrusts beneath a warmer air mass.

Color An obvious mineral characteristic that is often unreliable as a diagnostic property.

Columnar joints A pattern of cracks that forms during cooling of molten rock to generate columns that are generally six-sided.

Coma The fuzzy, gaseous component of a comet's head.

Comet A small body that generally revolves around the Sun in an elongated orbit.

Competence A measure of the largest particle a stream can transport; a factor dependent on velocity.

Composite cone A volcano composed of both lava flows and pyroclastic material.

Compound See Chemical compound.

Condensation The change of state from a gas to a liquid.

Condensation nuclei Tiny bits of particulate matter that serve as surfaces on which water vapor condenses.

Conditional instability The condition of moist air with an environmental lapse rate between the dry and wet adiabatic rates.

Conduction The transfer of heat through matter by molecular activity. Energy is transferred through collisions from one molecule to another.

Conduit A pipelike opening through which magma moves toward Earth's surface. It terminates at a surface opening called a vent.

Cone of depression A cone-shaped depression in the water table immediately surrounding a well.

Conformable Layers of rock deposited without interruption.

Contact (thermal) metamorphism Changes in rock caused by the heat from a nearby magma body.

Continental (c) air mass An air mass that forms over land; it is normally relatively dry.

Continental drift A theory that originally proposed that the continents are rafted about. It has essentially been replaced by the plate tectonics theory.

Continental margin That portion of the seafloor adjacent to the continents. It may include the continental shelf, continental slope, and continental rise.

Continental rise The gently sloping surface at the base of the continental slope.

Continental shelf The gently sloping submerged portion of the continental margin extending from the shoreline to the continental slope.

Continental slope The steep gradient that leads to the deep-ocean floor and marks the seaward edge of the continental shelf.

Continental volcanic arc Mountains formed in part by igneous activity associated with the subduction of oceanic lithosphere beneath a continent. Examples include the Andes and the Cascades.

Contour interval The difference I elevation between succeeding contour lines on a topographic map.

Contour line A line on a topographic map representing the same elevation in relation to sea level along its entire length.

Convection The transfer of heat by the movement of a mass or substance. It can take place only in fluids.

Convergence The condition that exists when the distribution of winds within a given area results in a net horizontal inflow of air into the area. Since convergence at lower levels is associated with an upward movement of air, areas of convergent winds are regions favorable to cloud formation and precipitation.

Convergent boundary A boundary in which two plates move together, causing one of the slabs of lithosphere to be consumed into the mantle as it descends beneath an overriding plate.

Coral reef Structure formed in a warm, shallow, sunlit ocean environment that consists primarily of the calcite-rich remains of corals as well as the limy secretions of algae and the hard parts of many other small organisms.

Core The innermost layer of Earth based on composition. It is thought to be largely an iron-nickel alloy with minor amounts of oxygen, silicon, and sulfur.

Coriolis effect The deflective force of Earth's rotation on all free-moving objects, including the atmosphere and oceans. Deflection is to the right in the Northern Hemisphere and to the left in the Southern Hemisphere.

Correlation Establishing the equivalence of rocks of similar age in different areas.

Covalent bond A chemical bond produced by the sharing of electrons.

Crater The depression at the summit of a volcano, or that which is produced by a meteorite impact.

Crevasse A deep crack in the brittle surface of a glacier.

Cross-beds Structures in which relatively thin layers are inclined at an angle to the main bedding. Formed by currents of wind or water.

Cross-cutting relationships A principle of relative dating. A rock or fault is younger than any rock (or fault) through which it cuts.

Crust The very thin outermost layer of Earth.

Cryovolcanism Volcanism that involves the eruption of magmas derived from the partial melting of ice rather than silicate rocks.

Crystal form The external appearance of a mineral as determined by its internal arrangement of atoms.

Crystallization The formation and growth of a crystalline solid from a liquid or gas.

Crystal settling During the crystallization of magma, the earlier formed minerals are denser than the liquid portion and settle to the bottom of the magma chamber.

Cumulus One of three basic cloud forms; also the name given one of the clouds of vertical development. Cumulus are billowy individual cloud masses that often have flat bases.

Cup anemometer An instrument used to determine wind speed.

Curie point The temperature above which a material loses its magnetization.

Cutoff A short channel segment created when a river erodes through the narrow neck of land between meanders.

Cyclone A low-pressure center characterized by a counterclockwise flow of air in the Northern Hemisphere.

Dark nebula A cloud of interstellar dust that obscures the light of more distant stars and appears as an opaque curtain.

Deep-ocean trench A narrow, elongated depression on the floor of the ocean.

Deep-sea fan A cone-shaped deposit at the base of the continental slope. The sediment is transported to the fan by turbidity currents that follow submarine canyons.

Deflation The lifting and removal of loose material by wind.

Deformation General term for the processes of folding, faulting, shearing, compression, or extension of rocks.

Degenerate matter Incomprehensibly dense material formed when stars collapse and form a white dwarf.

Delta An accumulation of sediment formed where a stream enters a lake or ocean.

Dendritic pattern A stream system that resembles the pattern of a branching tree.

Density The weight-per-unit volume of a particular material.

Deposition The process by which water vapor is changed directly to a solid without passing through the liquid state.

Desalination The removal of salts and other chemicals from seawater.

Desert One of the two types of dry climate; the driest of the dry climates.

Desert pavement A layer of coarse pebbles and gravel created when wind removed the finer material.

Detrital sedimentary rock Rock formed from the accumulation of material that originated and was transported in the form of solid particles derived from both mechanical and chemical weathering.

Dew point temperature The temperature to which air has to be cooled in order to reach saturation.

Diffused light Solar energy is scattered and reflected in the atmosphere and reaches Earth's surface in the form of diffuse blue light from the sky.

Dike A tabular-shaped intrusive igneous feature that cuts through the surrounding rock.

Dip-slip fault A fault in which the movement is parallel to the dip of the fault.

Discharge The quantity of water in a stream that passes a given point in a period of time.

Disconformity A type of unconformity in which the beds above and below are parallel.

Dissolved load That portion of a stream's load carried in solution.

Distributary A section of a stream that leaves the main flow.

Diurnal tidal pattern A tidal pattern exhibiting one high tide and one low tide during a tidal day; a daily tide.

Divergence The condition that exists when the distribution of winds within a given area results in a net horizontal outflow of air from the region. In divergence at lower levels, the resulting deficit is compensated for by a downward movement of air from aloft; hence, areas of divergent winds are unfavorable to cloud formation and precipitation.

Divergent boundary A region where the rigid plates are moving apart, typified by the mid-oceanic ridges.

Divide An imaginary line that separates the drainage of two streams; often found along a ridge.

Dome A roughly circular upfolded structure similar to an anticline.

Doppler effect The apparent change in wavelength of radiation caused by the relative motions of the source and the observer.

Doppler radar In addition to the tasks performed by conventional radar, this new generation of weather radar can detect motion directly and, hence, greatly improve tornado and severe storm warnings.

Drainage basin The land area that contributes water to a stream.

Drawdown The difference in height between the bottom of a cone of depression and the original height of the water table.

Drift See *Glacial drift.*

Drumlin A streamlined asymmetrical hill composed of glacial till. The steep side of the hill faces the direction from which the ice advanced.

Dry adiabatic rate The rate of adiabatic cooling or warming in unsaturated air. The rate of temperature change is 1°C per 100 meters.

Ductile deformation A type of solid-state flow that produces a change in the size and shape of a rock body without fracturing. Occurs at depths where temperatures and confining pressures are high.

Dune A hill or ridge of wind-deposited sand.

Dwarf planet Celestial bodies that orbit the Sun, are essentially spherical due to their gravity, and share their area in space with other similar objects. Examples include Pluto, Eris, a Kuiper belt object, and Ceres, the largest known asteroid.

Earthquake The vibration of Earth produced by the rapid release of energy.

Earth system A reference to the fact that our planet is a dynamic body with many separate but interacting parts. The atmosphere, hydrosphere, geosphere, and biosphere produce a complex and continuously interacting whole.

Echo sounder An instrument used to determine the depth of water by measuring the time interval between emission of a sound signal and the return of its echo from the bottom.

Elastic rebound The sudden release of stored strain in rocks that results in movement along a fault.

Electron A negatively charged subatomic particle that has a negligible mass and is found outside an atom's nucleus.

Element A substance that cannot be decomposed into simpler substances by ordinary chemical or physical means.

Elements of weather and climate Those quantities or properties of the atmosphere that are measured regularly and that are used to express the nature of weather and climate.

Elliptical galaxy A galaxy that is round or elliptical in outline. It contains little gas and dust; no disk or spiral arms; and few hot, bright stars.

Emergent coast A coast where land that was formerly below sea level has been exposed either because of crustal uplift or a drop in sea level or both.

Emission nebula A gaseous nebula that derives its visible light from the fluorescence of ultraviolet light from a star in or near the nebula.

End moraine A ridge of till marking a former position on the front of a glacier.

Energy levels Spherically-shaped negatively charged zones that surround the nucleus of an atom.

Environmental lapse rate The rate of temperature decrease with increasing height in the troposphere.

Eon The largest time unit on the geologic time scale, next in order of magnitude above era.

Ephemeral stream A stream that is usually dry because it carries water only in response to specific episodes of rainfall. Most desert streams are of this type.

Epicenter The location on Earth's surface that lies directly above the forces of an earthquake.

Epoch A unit of the geologic calendar that is a subdivision of a period.

Equatorial low A belt of low pressure lying near the equator and between the subtropical highs.

Equinox (spring or autumnal) The time when the vertical rays of the Sun are striking the equator. The length of daylight and darkness is equal at all latitudes at equinox.

Era A major division on the geologic calendar; eras are divided into shorter units called periods.

Erosion The incorporation and transportation of material by a mobile agent, such as water, wind, or ice.

Eruptive variable A star that varies in brightness.

Escape velocity The initial velocity an object needs to escape from the surface of a celestial body.

Esker Sinuous ridge composed largely of sand and gravel deposited by a stream flowing in a tunnel beneath a glacier near its terminus.

Estuary A funnel-shaped inlet of the sea that formed when a rise in sea level or subsidence of land caused the mouth of a river to be flooded.

Evaporation The process of converting a liquid to a gas.

Evaporite deposit A sedimentary rock formed of material deposited from solution by evaporation of the water.

External process Process such as weathering, mass wasting, or erosion that is powered by the Sun and transforms solid rock into sediment.

Extrusive (volcanic) Igneous activity that occurs outside the crust.

Eye A zone of scattered clouds and calm averaging about 20 kilometers (12 miles) in diameter at the center of a hurricane.

Eye wall The doughnut-shaped area of intense cumulonimbus development and very strong winds that surrounds the eye of a hurricane.

Fault A break in a rock mass along which movement has occurred.

Fault-block mountain A mountain formed by the displacement of rock along a fault.

Fault scarp A cliff created by movement along a fault. It represents the exposed surface of the fault prior to modification by weathering and erosion.

Felsic See *Granitic composition.*

Fetch The distance that the wind has traveled across the open water.

Fine-grained texture A texture of igneous rocks in which the crystals are too small for individual minerals to be distinguished with the unaided eye.

Fiord A steep-sided inlet of the sea formed when a glacial trough was partially submerged.

Fissure A crack in rock along which there is a distinct separation.

Fissure eruption An eruption in which lava is extruded from narrow fractures or cracks in the crust.

Flood The overflow of a stream channel that occurs when discharge exceeds the channel's capacity; the most common and destructive geologic hazard.

Flood basalts Flows of basaltic lava that issue from numerous cracks or fissures and commonly cover extensive areas to thicknesses of hundreds of meters.

Floodplain The flat, low-lying portion of a stream valley subject to periodic inundation.

Focus (earthquake) The zone within Earth where rock displacement produces an earthquake.

Fog A cloud with its base at or very near Earth's surface.

Fold A bent rock layer or series of layers that were originally horizontal and subsequently deformed.

Foliated texture A texture of metamorphic rocks that gives the rock a layered appearance.

Foreshocks Small earthquakes that often precede a major earthquake.

Fossil magnetism See *Paleomagnetism.*

Fossils The remains or traces of organisms preserved from the geologic past.

Fossil succession Fossil organisms succeed one another in a definite and determinable order, and any time period can be recognized by its fossil content.

Fracture Any break or rupture in rock along which no appreciable movement has taken place.

Fracture zone Linear zone of irregular topography on the deep-ocean floor that follows transform faults and their inactive extensions.

Freezing rain *See* Sleet.

Freezing The change of state from a liquid to a solid.

Front The boundary between two adjoining air masses having contrasting characteristics.

Frontal fog Fog formed when rain evaporates as it falls through a layer of cool air.

Frontal wedging Lifting of air resulting when cool air acts as a barrier over which warmer, lighter air will rise.

Fumarole A vent in a volcanic area from which fumes or gases escape.

Galactic cluster A system of galaxies containing from several to thousands of member galaxies.

Geocentric The concept of an Earth-centered universe.

Geologic time scale The division of Earth history into blocks of time, such as eons, eras, periods, and epochs. The time scale was created using relative dating principles.

Geology The science that examines Earth, its form and composition, and the changes it has undergone and is undergoing.

Geosphere The solid Earth, the largest of Earth's four major spheres.

Geostrophic wind A wind, usually above a height of 600 meters (2000 feet), that blows parallel to the isobars.

Geyser A fountain of hot water ejected periodically.

Glacial drift An all-embracing term for sediments of glacial origin, no matter how, where, or in what shape they were deposited.

Glacial erratic An ice-transported boulder that was not derived from bedrock near its present site.

Glacial striations Scratches and grooves on bedrock caused by glacial abrasion.

Glacial trough A mountain valley that has been widened, deepened, and straightened by a glacier.

Glacier A thick mass of ice originating on land from the compaction and recrystallization of snow that shows evidence of past or present flow.

Glassy texture A term used to describe the texture of certain igneous rocks, such as obsidian, that contain no crystals.

Glaze A coating of ice on objects formed when supercooled rain freezes on contact.

Graben A valley formed by the downward displacement of a fault-bounded block.

Gradient The slope of a stream; generally measured in feet per mile.

Granitic composition A compositional group of igneous rocks in which the rock is composed almost entirely of light-colored silicates.

Great circle The largest possible circle that can be drawn on a sphere. A great circle divides a sphere into two equal parts.

Greenhouse effect The transmission of short-wave solar radiation by the atmosphere coupled with the selective absorption of longer-wavelength terrestrial radiation, especially by water vapor and carbon dioxide.

Groin A short wall built at a right angle to the shore to trap moving sand.

Ground moraine An undulating layer of till deposited as the ice front retreats.

Groundwater Water in the zone of saturation.

Guyot A submerged flat-topped seamount.

Gyre The large, circular surface-current pattern found in each ocean.

Habit The characteristic shape (geometric form) of a crystal or aggregate of crystals.

Hail Nearly spherical ice pellets having concentric layers and formed by the successive freezing of layers of water.

Half-life The time required for one-half of the atoms of a radioactive substance to decay.

Halocline A layer of water in which there is a high rate of change in salinity in the vertical dimension.

Hanging valley A tributary valley that enters a glacial trough at a considerable height above its floor.

Hardness The resistance a mineral offers to scratching.

Hard stabilization Any form of artificial structure built to protect a coast or to prevent the movement of sand along a beach. Examples include groins, jetties, breakwaters, and seawalls.

Heat The kinetic energy of random molecular motion.

Heliocentric The view that the Sun is at the center of the solar system.

Hertzsprung–Russell (H–R) diagram A plot of stars according to their absolute magnitudes and spectral types.

High clouds Clouds that normally have their base above 6000 meters (20,000 feet); the base may be lower in winter and at high-latitude locations.

Horn A pyramid-like peak formed by glacial action in three or more cirques surrounding a mountain summit.

Horst An elongated, uplifted block of crust bounded by faults.

Hot spot A concentration of heat in the mantle capable of producing magma, which, in turn, extrudes onto Earth's surface. The intraplate volcanism that produced the Hawaiian Islands is one example.

Hot spring A spring in which the water is 6° to 9°C (10° to 15°F) warmer than the mean annual air temperature of its locality.

Hubble's law Relates the distance to a galaxy and its velocity.

Humidity A general term referring to water vapor in the air but not to liquid droplets of fog, cloud, or rain.

Hurricane A tropical cyclonic storm having winds in excess of 119 kilometers (74 miles) per hour.

Hydrogen burning The conversion of hydrogen through fusion to form helium.

Hydrogenous sediment Seafloor sediments consisting of minerals that crystallize from seawater. An important example is manganese nodules.

Hydrosphere The water portion of our planet; one of the traditional subdivisions of Earth's physical environment.

Hygrometer An instrument designed to measure relative humidity.

Hygroscopic nuclei Condensation nuclei having a high affinity for water, such as salt particles.

Hypothesis A tentative explanation that is tested to determine if it is valid.

Ice cap A mass of glacial ice covering a high upland or plateau and spreading out radially.

Ice sheet A very large, thick mass of glacial ice flowing outward in all directions from one or more accumulation centers.

Ice shelf A large, relatively flat mass of floating glacial ice that extends seaward from the coast but remains attached to the land along one or more sides.

Igneous rock A rock formed by the crystallization of molten magma.

Immature soil A soil lacking horizons.

Impact crater *See* Crater.

Incised meander Meandering stream channel that flows in a steep, narrow valley. These features form either when an area is uplifted or when base level drops.

Inclination of the axis The tilt of Earth's axis from the perpendicular to the plane of Earth's orbit.

Inclusion A piece of one rock unit contained within another. Inclusions are used in relative dating. The rock mass adjacent to the one containing the inclusion must have been there first in order to provide the fragment.

Index fossil A fossil that is associated with a particular span of geologic time.

Infiltration The movement of surface water into rock, sediment, or soil through cracks and pore spaces.

Infrared Radiation with a wavelength from 0.7 to 200 micrometers.

Inner core The solid innermost layer of Earth, about 1300 kilometers (800 miles) in radius.

Inner planets The innermost planets of our solar system, which include Mercury, Venus, Earth, and Mars. Also known as the terrestrial planets because of their Earth-like internal structure and composition.

Intensity (earthquake) A measure of the degree of earthquake shaking at a given locale based on the amount of damage.

Interior drainage A discontinuous pattern of intermittent streams that do not flow to the ocean.

Internal process Processes such as volcanic activity, earthquakes, and mountain building that derive their energy from Earth's interior.

Interstellar dust Dust and gases found between stars.

Intraplate volcanism Igneous activity that occurs within a tectonic plate away from plate boundaries.

Intrusive (plutonic) Igneous rock that formed below Earth's surface.

Ion An atom or molecule that possesses an electrical charge.

Ionic bond A chemical bond between two oppositely charged ions by the transfer of valence electrons from one atom to another.

Iron meteorite One of the three main categories of meteorites. This group is composed largely of iron with varying amounts of nickel (5–20 percent). Most meteorite finds are irons.

Irregular galaxy A galaxy that lacks symmetry.

Island arc See *Volcanic island arc.*

Isobar A line drawn on a map connecting points of equal atmospheric pressure, usually corrected to sea level.

Isotherms Lines connecting points of equal temperature.

Isotopes Varieties of the same element that have different mass numbers; their nuclei contain the same number of protons but different numbers of neutrons.

Jet stream Swift high-altitude winds (120 to 240 kilometers per hour [75 to 150 miles per hour]).

Jovian planet The Jupiter-like planets Jupiter, Saturn, Uranus, and Neptune. These planets have relatively low densities.

Kame A steep-sided hill composed of sand and gravel originating when sediment collected in openings in stagnant glacial ice.

Karst topography A topography consisting of numerous depressions called sinkholes.

Kettle Depressions created when blocks of ice became lodged in glacial deposits and subsequently melted.

Kuiper belt A region outside the orbit of Neptune where most short-period comets are thought to originate.

Laccolith A massive igneous body intruded between preexisting strata.

Lahar Mudflows on the slopes of volcanoes that result when unstable layers of ash and debris become saturated and flow downslope, usually following stream channels.

Lake-effect snow Snow showers associated with a cP air mass to which moisture and heat are added from below as it traverses a large and relatively warm lake (such as on of the Great Lakes), rendering the air mass humid and unstable.

Laminar flow The movement of water in straight-line paths that are parallel to the channel. Water moves downstream without mixing.

Land breeze A local wind blowing from land toward the water during the night in coastal areas.

Latent heat The energy absorbed or released during a change in state.

Lateral moraine A ridge of till along the sides of an alpine glacier composed primarily of debris that fell to the glacier from the valley walls.

Latitude Distance, measured in degrees, north and south of the equator.

Lava Magma that reaches Earth's surface.

Lightning A sudden flash of light generated by the flow of electrons between oppositely charged parts of a cumulonimbus cloud or between the cloud and the ground.

Light-year The distance light travels in a year; about 9.7 trillion kilometers (6 trillion miles).

Liquefaction A phenomenon, sometimes associated with earthquakes, in which soils and other unconsolidated materials containing abundant water are turned into a fluidlike mass that is not capable of supporting buildings.

Lithification The process, generally cementation and/or compaction, of converting sediments to solid rock.

Lithosphere The rigid outer layer of Earth, including the crust and upper mantle.

Local Group The cluster of 20 or so galaxies to which our galaxy belongs.

Localized convective lifting Unequal surface heating that causes localized pockets of air (thermals) to rise because of their buoyancy.

Loess Deposits of windblown silt, lacking visible layers, generally buff-colored, and capable of maintaining a nearly vertical cliff.

Longitude Distance, measured in degrees, east and west of the zero or *prime meridian.*

Longshore current A nearshore current that flows parallel to the shore.

Low clouds Clouds that form below a height of about 2000 meters (6500 feet).

Lower mantle *See* Mesosphere.

Lunar highlands Relatively light-colored regions on the surface of the Moon that are elevated several kilometers above the maria. Also called terrae.

Lunar regolith A thin, gray layer on the surface of the Moon, consisting of loosely compacted, fragmented material believed to have been formed by repeated meteoric impacts.

Luster The appearance or quality of light reflected from the surface of a mineral.

Mafic composition See *Basaltic composition.*

Magma A body of molten rock found at depth, including any dissolved gases and crystals.

Magmatic differentiation The process of generating more than one rock type from a single magma.

Magnetic reversal A change in the polarity of Earth's magnetic field that occurs over time intervals of roughly 200,000 years.

Magnetic time scale The history of magnetic reversals through geologic time.

Magnetometer A sensitive instrument used to measure the intensity of Earth's magnetic field.

Magnitude (earthquake) The total amount of energy released during an earthquake.

Magnitude (stellar) A number given to a celestial object to express its relative brightness.

Main-sequence stars A sequence of stars on the Hertzsprung-Russell diagram, containing the majority of stars, that runs diagonally from the upper left to the lower right.

Mantle The 2900-kilometer- (1800-mile-) thick layer of Earth located below the crust.

Mantle plume A source of some intraplate basaltic magma, these structures originate at great depth and, upon reaching the crust, spread laterally, creating a localized volcanic zone called a hot spot.

Maria (mare) The Latin name for the smooth areas of the Moon, formerly thought to be seas.

Marine terrace A wave-cut platform that has been exposed above sea level.

Maritime (m) air mass An air mass that originates over the ocean. These air masses are relatively humid.

Mass number The number of neutrons and protons in the nucleus of an atom.

Mass wasting The downslope movement of rock, regolith, and soil under the direct influence of gravity.

Meander A looplike bend in the course of a stream.

Mean solar day The average time between two passages of the Sun across the local celestial meridian.

Mechanical weathering The physical disintegration of rock, resulting in smaller fragments.

Medial moraine A ridge of till formed when lateral moraines from two coalescing alpine glaciers join.

Melting The change of state from a solid to a liquid.

Mercury barometer A mercury-filled glass tube in which the height of the mercury column is a measure of air pressure.

Meridian North-south lines (half circles) of Earth's geographic grid system. They extend from pole to pole.

Mesosphere (atmosphere) The layer of the atmosphere immediately above the stratosphere and characterized by decreasing temperatures with height.

Mesosphere (geology) The part of the mantle that extends from the core-mantle boundary to a depth of 660 kilometers (410 miles). Also known as the lower mantle

Mesozoic era A time span on the geologic time scale between the Paleozoic and Cenozoic eras—from about 251 to 65.5 million years ago.

Metamorphic rock Rock formed by the alteration of preexisting rock deep within Earth (but still in the solid state) by heat, pressure, and/or chemically active fluids.

Metamorphism The changes in mineral composition and texture of a rock subjected to high temperatures and pressures within Earth.

Meteor The luminous phenomenon observed when a meteoroid enters Earth's atmosphere and burns up; popularly called a "shooting star."

Meteorite Any portion of a meteoroid that survives its traverse through Earth's atmosphere and strikes Earth's surface.

Meteoroid Small, solid particles that have orbits in the solar system.

Meteorology The scientific study of the atmosphere and atmospheric phenomena; the study of weather and climate.

Meteor shower Many meteors appearing in the sky caused when Earth intercepts a swarm of meteoritic particles.

Middle clouds Clouds that occupy the height range from 2000 to 6000 meters (6500 to 20,000 feet).

Middle-latitude (midlatitude) cyclone Large low-pressure center with a diameter often exceeding 1000 kilometers (620 miles) that moves from west to east and may last from a few days to more than a week and usually has a cold front and a warm front extending from the central area of low pressure.

Mid-ocean ridge A continuous elevated zone on the floor of all the major ocean basins and varying in width from 500 to 5000 kilometers (300 to 3000 miles). The rifts at the crests of these ridges represent divergent plate boundaries.

Mineral A naturally occurring, inorganic crystalline material with a unique chemical composition.

Mineralogy The study of minerals.

Mineral resource All discovered and undiscovered deposits of a useful mineral that can be extracted now or at some time in the future.

Mixed tidal pattern A tidal pattern exhibiting two high tides and two low tides per tidal day with a large inequality in high-water heights, low-water heights, or both. Coastal locations that experience such a tidal pattern may also show alternating periods of diurnal and semidiurnal tidal patterns. Also called *mixed semidiurnal.*

Mixing ratio The mass of water vapor in a unit mass of dry air; commonly expressed as grams of water vapor per kilogram of dry air.

Model A term often used synonymously with hypothesis, but it is less precise because it is sometimes used to describe a theory as well.

Modified Mercalli Intensity Scale A 12-point scale developed to evaluate earthquake intensity based on the amount of damage to various structures.

Mohs scale A series of 10 minerals used as a standard in determining hardness.

Moment magnitude A more precise measure of earthquake magnitude than the Richter scale, it is derived from the amount of displacement that occurs along a fault zone.

Monsoon Seasonal reversal of wind direction associated with large continents, especially Asia. In winter, the wind blows from land to sea, in summer, from sea to land.

Mountain breeze The nightly downslope winds commonly encountered in mountain valleys.

Natural levees The elevated landforms that parallel some streams and act to confine their waters, except during floodstage.

Neap tide The lowest tidal range, occurring near the times of the first and third quarters of the Moon.

Nebula A cloud of interstellar gas and/or dust.

Nebular hypothesis The basic idea that the Sun and planets formed from the same cloud of gas and dust in interstellar space.

Neutron A subatomic particle found in the nucleus of an atom. The neutron is electronically neutral and has a mass approximately that of a proton.

Neutron star A star of extremely high density composed entirely of neutrons.

Nonconformity An unconformity in which older metamorphic or intrusive igneous rocks are overlain by younger sedimentary strata.

Nonfoliated texture Metamorphic rocks that do not exhibit foliation.

Nonrenewable resource A resource that forms or accumulates over such long time spans that it must be considered as fixed in total quantity.

Nonsilicate minerals Minerals that do not have the silicon-oxygen tetrahedron as their basic structure. These minerals are usually divided into classes based on the negatively charged ion or complex ion that the members have in common. Nonsilicates make up only about 8 percent of Earth's crust.

Normal fault A fault in which the rock above the fault plane has moved down relative to the rock below.

Normal polarity A magnetic field that is the same as the one that exists at present.

Nova A star that explosively increases in brightness.

Nucleus The small, heavy core of an atom that contains all of its positive charge and most of its mass.

Nuée ardente Incandescent volcanic debris buoyed up by hot gases that moves downslope in an avalanche fashion.

Numerical date Date that specifies the actual number of years that have passed since an event occurred.

Occluded front A front formed when a cold front overtakes a warm front. It marks the beginning of the end of a middle-latitude cyclone.

Oceanic plateau An extensive region on the ocean floor composed of thick accumulations of pillow basalts and other mafic rocks that in some cases exceed 30 kilometers in thickness.

Oceanic ridge See *Mid-ocean ridge.*

Oceanography The scientific study of the oceans and oceanic phenomena.

Oort cloud A spherical shell composed of comets that orbit the Sun at distances generally greater than 10,000 times the Earth-Sun distance.

Ore Usually, a useful metallic mineral that can be mined at a profit. The term is also applied to certain nonmetallic minerals, such as fluorite and sulfur.

Original horizontality Layers of sediments generally deposited in a horizontal or nearly horizontal position.

Orogenesis The processes that collectively result in the formation of mountains.

Orographic lifting Mountains acting as barriers to the flow of air force the air to ascend. The air cools adiabatically, and clouds and precipitation may result.

Outer core A layer beneath the mantle about 2200 kilometers (1364 miles) thick that has the properties of a liquid.

Outer planets The outermost planets of our solar system, which include Jupiter, Saturn, Uranus, and Neptune. These bodies are also known as the Jovian (Jupiter-like) planets.

Outgassing The escape of gases that had been dissolved in magma.

Outwash plain A relatively flat, gentle sloping plain consisting of materials deposited by meltwater streams in front of the margin of an ice sheet.

Overrunning Warm air gliding up a retreating cold air mass.

Oxbow lake A curved lake produced when a stream cuts off a meander.

Ozone A molecule of oxygen containing three oxygen atoms.

Pahoehoe flow A lava flow with a smooth-to-ropey surface.

Paleomagnetism The natural remnant magnetism in rock bodies. The permanent magnetization acquired by rock that can be used to determine the location of the magnetic poles and the latitude of the rock at the time it became magnetized.

Paleontology The systematic study of fossils and the history of life on Earth.

Paleozoic era A time span on the geologic time scale between Precambrian time and the Mesozoic era—from about 542 million to 251 million years ago.

Pangaea The proposed supercontinent, which 200 million years ago began to break apart and form the present landmasses.

Paradigm A theory that is held with a very high degree of confidence and is comprehensive in scope.

Parallels The east-west lines (circles) of the geographic grid system.

Parasitic cone A volcanic cone that forms on the flank of a larger volcano.

Parcel An imaginary volume of air enclosed in a thin elastic cover. It is considered to be a few hundred cubic meters in volume and is assumed to act independently of the surrounding air.

Partial melting The process by which most igneous rocks melt. Since individual minerals have different melting points, most igneous rocks melt over a temperature range of a few hundred degrees. If the liquid is squeezed out after some melting has occurred, a melt with a higher silica content results.

Passive continental margin Margin that consists of a continental shelf, continental slope, and continental rise. They are not associated with plate boundaries and, therefore, experience little volcanism and few earthquakes.

Period A basic unit of the geologic calendar that is a subdivision of an era. Periods may be divided into smaller units called epochs.

Periodic table An arrangement of the elements in which atomic number increases from left to right and elements with similar properties appear in columns called *families* or *groups*.

Permeability A measure of a material's ability to transmit water.

Perturbation The gravitational disturbance of the orbit of one celestial body by another.

Phanerozoic eon That part of geologic time represented by rocks containing abundant fossil evidence. The eon extending from the end of the Proterozoic eon (570 million years ago) to the present.

Physical environment The part of the environment that encompasses water, air, soil, and rock, as well as conditions such as temperature, humidity, and sunlight.

Piedmont glacier A glacier that forms when one or more alpine glaciers emerge from the confining walls of mountain valleys and spread out to create a broad sheet in the lowlands at the base of the mountains.

Pipe A vertical conduit through which magmatic materials have passed.

Planetary nebula A shell of incandescent gas expanding from a star.

Planetesimal A solid celestial body that accumulated during the first stages of planetary formation. Planetesmals aggregated into increasingly larger bodies, ultimately forming the planets.

Plate One of numerous rigid sections of the lithosphere that moves as a unit over the material of the asthenosphere.

Plate tectonics The theory that Earth's outer shell consists of individual plates that interact in various ways and thereby produce earthquakes, volcanoes, mountains, and the crust itself.

Playa lake A temporary lake in a playa.

Pleistocene epoch An epoch of the Quaternary period beginning about 1.6 million years ago and ending about 10,000 years ago. Best known as a time of extensive continental glaciation.

Plucking (quarrying) The process by which pieces of bedrock are lifted out of place by a glacier.

Pluton A structure that results from the emplacement and crystallization of magma beneath the surface of Earth.

Pluvial lake A lake formed during a period of increased rainfall. During the Pleistocene epoch, this occurred in some nonglaciated regions during periods of ice advance elsewhere.

Polar (P) air mass A cold air mass that forms in a high-latitude source region.

Polar easterlies In the global pattern of prevailing winds, winds that blow from the polar high toward the subpolar low. These winds, however, should not be thought of as persistent winds, such as the trade winds.

Polar front The stormy frontal zone separating air masses of polar origin from air masses of tropical origin.

Polar high Anticyclones that are assumed to occupy the inner polar regions and are believed to be thermally induced, at least in part.

Polar wandering As the result of paleomagnetic studies in the 1950s, researchers proposed that either the magnetic poles migrated greatly through time or the continents had gradually shifted their positions.

Porosity The volume of open spaces in rock or soil.

Porphyritic texture An igneous texture consisting of large crystals embedded in a matrix of much smaller crystals.

Positive-feedback mechanism As used in climatic change, any effect that acts to reinforce the initial change.

Precambrian All geologic time prior to the Paleozoic era.

Precession A slow motion of Earth's axis that traces out a cone over a period of 26,000 years.

Precipitation fog Fog formed when rain evaporates as it falls through a layer of cool air.

Pressure gradient The amount of pressure change occurring over a given distance.

Pressure tendency The nature of the change in atmospheric pressure over the past several hours. It can be a useful aid in short-range weather prediction.

Prevailing wind A wind that consistently blows from one direction more than from any other.

Primary pollutants Those pollutants emitted directly from identifiable sources.

Primary (P) wave A type of seismic wave that involves alternating compression and expansion of the material through which it passes.

Principal shells *See* Energy levels.

Proton A positively charged subatomic particle found in the nucleus of an atom.

Proton–proton chain A chain of thermonuclear reactions by which nuclei of hydrogen are built up into nuclei of helium.

Protoplanet A developing planetary body that grows by the accumulation of planetesimals.

Protostar A collapsing cloud of gas and dust destined to become a star.

Psychrometer A device consisting of two thermometers (wet-bulb and dry-bulb) that is rapidly whirled and, with the use of tables, yields the relative humidity and dew point.

Ptolemaic system An Earth-centered system of the universe.

Pulsar A variable radio source of small size that emits radio pulses in very regular periods.

Pulsating variable A variable star that pulsates in size and luminosity.

Pyroclastic flow A highly heated mixture, largely of ash and pumice fragments, traveling down the flanks of a volcano or along the surface of the ground.

Pyroclastic material The volcanic rock ejected during an eruption, including ash, bombs, and blocks.

Pyroclastic texture An igneous rock texture resulting from the consolidation of individual rock fragments that are ejected during a violent eruption.

Radial pattern A system of streams running in all directions away from a central elevated structure, such as a volcano.

Radiation or electromagnetic radiation The transfer of energy (heat) through space by electromagnetic waves.

Radiation fog Fog resulting from radiation heat loss by Earth.

Radiation pressure The force exerted by electromagnetic radiation from an object such as the Sun.

Radioactive decay (radioactivity) The spontaneous decay of certain unstable atomic nuclei.

Radiocarbon (carbon-14) dating The radioactive isotope of carbon is produced continuously in the atmosphere and used in dating events from the very recent geologic past (the last few tens of thousands of years).

Radiometric dating The procedure of calculating the absolute ages of rocks and minerals that contain radioactive isotopes.

Rain Drops of water that fall from a cloud and have a diameter of at least 0.5 millimeter.

Rain shadow desert A dry area on the lee side of a mountain range. Many middle-latitude deserts are of this type.

Rectangular pattern A drainage pattern characterized by numerous right-angle bends that develops on jointed or fractured bedrock.

Red giant A large, cool star of high luminosity; a star occupying the upper right portion of the Hertzsprung-Russell diagram.

Reflecting telescope A telescope that concentrates light from distant objects by using a concave mirror.

Reflection The process whereby light bounces back from an object at the same angle at which it encounters a surface and with the same intensity.

Reflection nebula A relatively dense dust cloud in interstellar space that is illuminated by starlight.

Refracting telescope A telescope that employs a lens to bend and concentrate the light from distant objects.

Regional metamorphism Metamorphism associated with the large-scale mountain-building processes.

Regolith The layer of rock and mineral fragments that nearly everywhere covers Earth's land surface.

Rejuvenation A change, often caused by regional uplift, that causes the force of erosion to intensify.

Relative dating Rocks placed in their proper sequence or order. Only the chronologic order of events is determined.

Relative humidity The ratio of the air's water-vapor content to its water-vapor capacity.

Relief The difference in elevation between two points.

Renewable resource A resource that is virtually inexhaustible or that can be replenished over relatively short time spans.

Reserve Already identified deposits from which minerals can be extracted profitably.

Retrograde motion The apparent westward motion of the planets with respect to the stars.

Reverse fault A fault in which the material above the fault plane moves up in relation to the material below.

Reverse polarity A magnetic field opposite to the one that exists at present.

Revolution The motion of one body about another, as Earth about the Sun.

Richter scale A scale of earthquake magnitude based on the motion of a seismograph.

Ridge push A mechanism that may contribute to plate motion. It involves the oceanic lithosphere sliding down the oceanic ridge under the pull of gravity.

Rift valley A region of Earth's crust along which divergence is taking place.

Rime A thin coating of ice on objects produced when supercooled fog droplets freeze on contact.

Rock A consolidated mixture of minerals.

Rock cycle A model that illustrates the origin of the three basic rock types and the interrelatedness of Earth's materials and processes.

Rock flour Ground-up rock produced by the grinding effect of a glacier.

Rotation The spinning of a body, such as Earth, about its axis.

Runoff Water that flows over the land rather than infiltrating into the ground.

Saffir-Simpson scale A scale, from 1 to 5, used to rank the relative intensity of a hurricane.

Salinity The proportion of dissolved salts to pure water, usually expressed in parts per thousand (‰).

Santa Ana The local name given a chinook wind in southern California.

Saturation The maximum possible quantity of water vapor that the air can hold at any given temperature and pressure.

Scale An expression of the relationship between distance or area on a map to the true distance or area on Earth's surface.

Scattering The redirecting (in all directions) of light by small particles and gas molecules in the atmosphere. The result is diffused light.

Scoria cone *See cinder cone.*

Sea arch An arch formed by wave erosion when caves on opposite sides of a headland unite.

Sea breeze A local wind blowing from the sea during the afternoon in coastal areas.

Seafloor spreading The process of producing new seafloor between two diverging plates.

Seamount An isolated volcanic peak that rises at least 1000 meters (3300 feet) above the deep-ocean floor.

Sea stack An isolated mass of rock standing just offshore, produced by wave erosion of a headland.

Seawall A barrier constructed to prevent waves from reaching the area behind the wall. Its purpose is to defend property from the force of breaking waves.

Secondary (S) wave A seismic wave that involves oscillation perpendicular to the direction of propagation.

Sediment Unconsolidated particles created by the weathering and erosion of rock, by chemical precipitation from solution in water, or from the secretions of organisms and transported by water, wind, or glaciers.

Sedimentary rock Rock formed from the weathered products of preexisting rocks that have been transported, deposited, and lithified.

Seismic reflection profile A method of viewing the rock structure beneath a blanket of sediment by using strong, low-frequency sound waves that penetrate the sediments and reflect off the contacts between rock layers and fault zones.

Seismic sea wave (tsunami) A rapidly moving ocean wave generated by earthquake activity that is capable of inflicting heavy damage in coastal regions.

Seismogram The record made by a seismograph.

Seismograph An instrument that records earthquake waves.

Seismology The study of earthquakes and seismic waves.

Semidiurnal tidal pattern A tidal pattern exhibiting two high tides and two low tides per tidal day with small inequalities between successive highs and successive lows; a semidaily tide.

Shield volcano A broad, gently sloping volcano built from fluid basaltic lavas.

Silicate Any one of numerous minerals that have the oxygen and silicon tetrahedron as their basic structure.

Silicon-oxygen tetrahedron A structure composed of four oxygen atoms surrounding a silicon atom that constitutes the basic building block of silicate minerals.

Sill A tabular igneous body that was intruded parallel to the layering of preexisting rock.

Sink hole (sink) A depression produced in a region where soluble rock has been removed by groundwater.

Slab pull A mechanism that contributes to plate motion in which cool, dense oceanic crust sinks into the mantle and "pulls" the trailing lithosphere along.

Slab suction One of the driving forces of plate motion, it arises from the drag of the subducting plate on the adjacent mantle. It is an induced mantle circulation that pulls both the subducting and overriding plates toward the trench.

Sleet Frozen or semifrozen rain formed when raindrops freeze as they pass through a layer of cold air.

Slip face The steep, leeward slope of a sand dune; it maintains an angle of about 34 degrees.

Snow A solid form of precipitation produced by sublimation of water vapor.

Solar nebula The cloud of interstellar gas and/or dust from which the bodies of our solar system formed.

Solstice (summer or winter) The time when the vertical rays of the Sun are striking either the Tropic of Cancer or the Tropic of Capricorn. Solstice represents the longest or shortest day (length of daylight) of the year.

Sonar Instrument that uses acoustic signals (sound energy) to measure water depths. Sonar is an acronym for *so*und *n*avigation and *r*anging.

Sorting The process by which solid particles of various sizes are separated by moving water or wind. Also the degree of similarity in particle size in sediment or sedimentary rock.

Source region The area where an air mass acquires its characteristic properties of temperature and moisture.

Specific gravity The ratio of a substance's weight to the weight of an equal volume of water.

Specific humidity The weight of water vapor compared with the total weight of the air, including the water vapor.

Spiral galaxy A flattened, rotating galaxy with pinwheel-like arms of interstellar material and young stars winding out from its nucleus.

Spit An elongated ridge of sand that projects from the land into the mouth of an adjacent bay.

Spring A flow of groundwater that emerges naturally at the ground surface.

Spring tide The highest tidal range. Occurs near the times of the new and full moons.

Stable air Air that resists vertical displacement. If it is lifted, adiabatic cooling will cause its temperature to be lower than that of the surrounding environment; if it is allowed, it will sink to its original position.

Stalactite The icicle-like structure that hangs from the ceiling of a cavern.

Stalagmite The columnlike form that grows upward from the floor of a cavern.

Stationary front A situation in which the surface position of a front does not move; the flow on either side of such a boundary is nearly parallel to the position of the front.

Steam fog Fog having the appearance of steam; produced by evaporation from a warm-water surface into the cool air above.

Stellar parallax A measure of stellar distance.

Steppe One of the two types of dry climate. A marginal and more humid variant of the desert that separates it from bordering humid climates.

Stony-iron meteorite One of the three main categories of meteorites. This group, as the name implies, is a mixture of iron and silicate minerals.

Stony meteorite One of the three main categories of meteorites. Such meteorites are composed largely of silicate minerals with inclusions of other minerals.

Storm surge The abnormal rise of the sea along a shore as a result of strong winds.

Strata (beds) Parallel layers of sedimentary rock.

Stratified drift Sediments deposited by glacial meltwater.

Stratosphere The layer of the atmosphere immediately above the troposphere, characterized by increasing temperatures with height due to the concentration of ozone.

Stratovolcano See *Composite cone.*

Stratus One of three basic cloud forms, also the name given one of the low clouds. They are sheets or layers that cover much or all of the sky.

Streak The color of a mineral in powdered form.

Stream valley The channel, valley floor, and the sloping valley walls of a stream.

Strike-slip fault A fault along which the movement is horizontal.

Subduction zone A long, narrow zone where one lithospheric plate descends beneath another.

Sublimation The conversion of a solid directly to a gas without passing through the liquid state.

Submarine canyon A seaward extension of a valley that was cut on the continental shelf during a time when sea level was lower, or a canyon carved into the outer continental shelf, slope, and rise by turbidity currents.

Submergent coast A coast with a form that is largely the result of the partial drowning of a former land surface either because of a rise of sea level or subsidence of the crust or both.

Subpolar low Low pressure located at about the latitudes of the Arctic and Antarctic circles. In the Northern Hemisphere, the low takes the form of individual oceanic cells; in the Southern Hemisphere, there is a deep and continuous trough of low pressure.

Subtropical high Not a continuous belt of high pressure but rather several semi-permanent, anticyclonic centers characterized by subsidence and divergence located roughly between latitudes 25 and 35 degrees both north and south of the equator.

Supergiant A very large star of high luminosity.

Supernova An exploding star that increases in brightness many thousands of times.

Superposition, Law of In any underformed sequence of sedimentary rocks or surface-deposited igneous materials, each bed is older than the layers above and younger than the layers below.

Surf A collective term for breakers; also the wave activity in the area between the shoreline and the outer limit of breakers.

Surface waves Seismic waves that travel along the outer layer of the Earth.

Suspended load The fine sediment carried within the body of flowing water.

Syncline A linear downfold in sedimentary strata; the opposite of anticline.

System Any size group of interacting parts that form a complex whole.

Tablemount See *Guyot*.

Temperature A measure of the degree of hotness or coldness of a substance.

Tenacity A mineral's toughness or resistance to breaking or deforming.

Terrae The extensively cratered highland areas of the Moon

Terrane A crustal block bounded by faults, whose geologic history is distinct from the histories of adjoining crustal blocks.

Terrestrial planet Any of the Earth-like planets, including Mercury, Venus, Mars, and Earth.

Terrigenous sediment Seafloor sediment derived from weathering and erosion on land.

Texture The size, shape, and distribution of the particles that collectively constitute a rock.

Theory A well-tested and widely accepted view that explains certain observable facts.

Thermal metamorphism See *Contact metamorphism*.

Thermocline A layer of water in which there is a rapid change in temperature in the vertical dimension.

Thermohaline circulation Movements of ocean water caused by density differences brought about by variations in temperature and salinity.

Thermosphere The region of the atmosphere immediately above the mesosphere and characterized by increasing temperatures due to absorption of very shortwave solar energy by oxygen.

Thrust fault A low-angle reverse fault.

Thunderstorm A storm produced by a cumulonimbus cloud and always accompanied by lightning and thunder. It is of relatively short duration and usually accompanied by strong wind gusts, heavy rain, and sometimes hail.

Tidal current The alternating horizontal movement of water associated with the rise and fall of the tide.

Tidal delta A deltalike feature created when a rapidly moving tidal current emerges from a narrow inlet and slows, depositing its load of sediment.

Tidal flat A marshy or muddy area that is covered and uncovered by the rise and fall of the tide.

Tide Periodic change in the elevation of the ocean surface.

Till Unsorted sediment deposited directly by a glacier.

Tombolo A ridge of sand that connects an island to the mainland or to another island.

Topographic map A map that shows the shape (changes in elevation) of the land surface.

Tornado A small, very intense cyclonic storm with exceedingly high winds, most often produced along cold fronts in conjunction with severe thunderstorms.

Tornado warning A warning issued when a tornado has actually been sighted in an area or is indicated by radar.

Tornado watch A forecast issued for areas of about 65,000 square kilometers (25,000 square miles), indicating that conditions are such that tornadoes may develop; it is intended to alert people to the possibility of tornadoes.

Trade winds Two belts of winds that blow almost constantly from easterly directions and are located on the equatorward sides of the subtropical highs.

Transform fault boundary A boundary in which two plates slide past one another without creating or destroying lithosphere.

Transform fault A major strike-slip fault that cuts through the lithosphere and accommodates motion between two plates.

Transpiration The release of water vapor to the atmosphere by plants.

Trellis pattern A system of streams in which nearly parallel tributaries occupy valleys cut in folded strata.

Trench An elongate depression in the seafloor produced by bending of oceanic crust during subduction.

Tropical (T) air mass A warm-to-hot air mass that forms in the subtropics.

Tropical depression By international agreement, a tropical cyclone with maximum winds that do not exceed 61 kilometers (38 miles) per hour.

Tropical storm By international agreement, a tropical cyclone with maximum winds between 61 and 119 kilometers (38 and 74 miles) per hour.

Tropic of Cancer The parallel of latitude, $23\frac{1}{2}$ degrees north latitude, marking the northern limit of the Sun's vertical rays.

Tropic of Capricorn The parallel of latitude, $23\frac{1}{2}$ degrees south latitude, marking the southern limit of the Sun's vertical rays.

Troposphere The lowermost layer of the atmosphere. It is generally characterized by a decrease in temperature with height.

Turbidity current A downslope movement of dense, sediment-laden water created when sand and mud on the continental shelf and slope are dislodged and thrown into suspension.

Turbulent flow The movement of water in an erratic fashion often characterized by swirling, whirlpool-like eddies. Most streamflow is of this type.

Ultramafic composition A compositional group of igneous rocks in which the rock contains mostly olivine and pyroxene.

Ultraviolet Radiation with a wavelength from 0.2 to 0.4 micrometer.

Unconformity A surface that represents a break in the rock record, caused by erosion or nondeposition.

Uniformitarianism The concept that the processes that have shaped Earth in the geologic past are essentially the same as those operating today.

Unsaturated zone Area above the water table where openings in soil, sediment, and rock are not saturated but filled mainly with air.

Unstable air Air that does not resist vertical displacement. If it is lifted, its temperature will not cool as rapidly as the surrounding environment, and so it will continue to rise on its own.

Upslope fog Fog created when air moves up a slope and cools adiabatically.

Upwelling The rising of cold water from deeper layers to replace warmer surface water that has been moved away.

Valence electrons The electrons involved in the bonding process; the electrons occupying the highest principal energy level of an atom.

Valley breeze The daily upslope winds commonly encountered in a mountain valley.

Valley glacier See *Alpine glacier*.

Valley train A relatively narrow body of stratified drift deposited on a valley floor by meltwater streams that issue from a valley glacier.

Vapor pressure That part of the total atmospheric pressure attributable to water-vapor content.

Vent A conduit that connects a magma chamber to a volcanic crater.

Viscosity A measure of a fluid's resistance to flow.

Visible light Radiation with a wavelength from 0.4 to 0.7 micrometer.

Volatiles Gaseous components of magma dissolved in the melt. Volatiles will readily vaporize (form a gas) at surface pressures.

Volcanic island arc A chain of volcanic islands generally located a few hundred kilometers from a trench where there is active subduction of one oceanic plate beneath another.

Volcanic bomb A streamlined pyroclastic fragment ejected from a volcano while molten.

Volcanic neck An isolated, steep-sided, erosional remnant consisting of lava that once occupied the vent of a volcano.

Volcano A mountain formed of lava and/or pyroclastics.

Warm front A front along which a warm air mass overrides a retreating mass of cooler air.

Water cycle The continuous movement of water from the oceans to the atmosphere (by evaporation), from the atmosphere to the land (by condensation and precipitation), and from the land back to the sea (via the flow of running water and groundwater). It is also known as the *hydrologic cycle*.

Water table The upper level of the saturated zone of groundwater.

Wave-cut cliff A seawater-facing cliff along a steep shoreline formed by wave erosion at its base and mass wasting.

Wave-cut platform A bench or shelf in the bedrock at sea level, cut by wave erosion.

Wave height The vertical distance between the trough and crest of a wave.

Wavelength The horizontal distance separating successive crests or troughs.

Wave period The time interval between the passage of successive crests at a stationary point.

Wave refraction The process by which the portion of a wave in shallow water slows, causing the wave to bend and tend to align itself with the underwater contours.

Weather The state of the atmosphere at any given time.

Weathering The disintegration and decomposition of rock at or near the surface of Earth.

Well An opening bored into the zone of saturation.

Westerlies The dominant west-to-east motion of the atmosphere that characterizes the regions on the poleward side of the subtropical highs.

Wet adiabatic rate The rate of adiabatic temperature change in saturated air. The rate of temperature change is variable, but it is always less than the dry adiabatic rate.

White dwarf A star that has exhausted most or all of its nuclear fuel and has collapsed to a very small size; believed to be near its final stage of evolution.

Wind Air flowing horizontally with respect to Earth's surface.

Wind vane An instrument used to determine wind direction.

Yazoo tributary A tributary that flows parallel to the main stream because a natural levee is present.

Zone of accumulation The part of a glacier characterized by snow accumulation and ice formation. Its outer limit is the snowline.

Zone of saturation Zone where all open spaces in sediment and rock are completely filled with water.

Zone of wastage The part of a glacier beyond the zone of accumulation, where all of the snow from the previous winter melts, as does some of the glacial ice.

INDEX

Aa flows, 198
Abrasion
 glacial, 107
 wave, 275–76
 wind, 123
Absolute magnitude, 433–34
Absolute stability/instability, 332, 333
Absorption, 308
Abyssal plains, 258
Accretionary wedge, 185, 256
Accumulation, zone of, 104, 105
Acid rain, 48
Active continental margins, 184, 252, 253–56
Adiabatic temperature changes, 328–29, 332, 333
Advection fog, 336, 339
Aerosols, 296, 297
Aftershocks, 163
Agate, 52, 55
Aggregate, 19, 20
Agriculture, irrigation for, 91, 92, 93
Air, 295–96. *See also* Atmosphere
 processes that lift, 329–31
 stability of, 331–34
 water vapor (humidity) in, 324–28
Air masses, 370–73
Air-mass thunderstorms, 379
Air-mass weather, 371, 372–73
Air pollution, 10, 11
Air pressure, 350–59
 measuring, 351–52
 wind and, 352–59
Air temperature
 atmosphere layers based on changes in, 298–300
 controls, 313–16
 data, 312–13
 dew point and, 326–27
 environmental lapse rate, 331–34
 global warming and, 309–12
 relative humidity and, 325
 world distribution, 316–17
Albedo, 307–8, 315–16
Alluvial channels, 75–77
Alluvial fan, 119, 120
Alluvium, 75
Almagest, 397
Alpha particles, 236
Alpine glaciers, 102–3, 105, 108
Altitude, 298, 301, 302, 314
Altocumulus clouds, 335, 336, 337
Altostratus clouds, 335, 336, 338
Amber, fossils preserved in, 233, 234
Amorphous solids, 18
Amphiboles, 20, 30, 142
Andean-type convergent zone, 184–85
Andean-type plate margins, 183
Andesite, 45
Andesitic magma, 196
Andesitic rocks, 44–45
Aneroid barometer, 351–52
Angle of repose, 68
Angular unconformity, 228–29, 230, 231
Anhydrite, 261
Antarctic Ice Sheet, 103, 114
Anthracite, 54
Anticline, 178, 179
Anticyclones, 356–59, 360, 378
Apparent magnitude, 433, 434
Aquifers, 88, 90, 92–93
Aquitards, 88, 89, 90
Archaeology, 230
Arctic (A) air masses, 371
Arctic Ocean, 246
Arêtes, 108–9
Argon, 295
Arid climate. *See* Deserts
Arroyo, 118
Artesian wells, 90–91
Artifacts, 230
Artificial cutoffs, 84
Artificial levees, 83–84, 85
Ash, volcanic, 41, 194, 199
Asteroids, asteroid belt, 410, 421–22, 425
Asthenosphere, 5, 6, 135, 177
Astronomical unit (AU), 401
Astronomy, 3. *See also* Solar system
 ancient, 396–98
 birth of modern, 398–404
Atlantic Ocean, 246
Atmosphere, 2–3, 5, 293–319
 composition, 295–98
 general circulation, 359–60
 global warming and, 309–12
 heating, 305–9. *See also* Air temperature
 height and structure, 298–300
 ozone layer, 296
 of planets, 408–9, 414, 417, 421
Atmospheric pressure, 298, 324
Atmospheric stability, 331–34
Atomic mass unit, 23
Atomic number, 19, 20–21, 236–37
Atoms, 20–23, 236
Autumnal equinox, 301, 302, 303
Avalanche, glowing, 205–6
Axis, inclination of, 301, 302
Axis of rotation, Earth's, 300, 396

Backscattering, 307
Backswamps, 80
Backwash, 274, 276, 277
Bacon, Francis, 364
Barograph, 351
Barometer, 351–52
Barometric tendency, 359
Barred spiral galaxies, 446, 448
Barrier islands, 280–81, 285
Bars, 80
Basalt, 44, 45, 209
Basaltic composition, 44, 45, 143
Basaltic lava flows, 197, 200–202
Basaltic magma, 143, 194, 196, 216
Base level, 77, 78, 80
Basin and Range, 118–19, 120, 180, 181
Basins, 178–79, 257–58
Batholiths, 184, 185, 211, 212–13
Bathymetry, 250–51
Baymouth bar, 279, 280, 282
Beach drift, 277, 278
Beaches, 274–78
Beach nourishment, 284
Bedding planes, 54
Bed load, 74, 119, 121
Bedrock channels, 75
Belt of soil moisture, 86
Bentonite, 29
Beta particle, 236–37
Betelgeuse, 434, 435
Big Bang Theory, 405, 448–50
Binary stars, 434–35, 444
Biochemical sediment, 51
Biogenous sediment, 260, 262
Biological activity, mechanical weathering by, 47, 49
Biosphere, 6
Biotite, 30
Bird-foot delta, 80
Bituminous coal, 54
Black dwarf, 443
Black holes, 443, 444, 445, 449
Blocks, volcanic, 199
Blowouts, 121
Body waves, 164–66
Bombs, volcanic, 199
Bonding, chemical, 21–23
Bowen's reaction series, 45, 46
Brahe, Tycho, 399–400, 432, 442
Braided channels, 76–77
Breakwater, 283
Breccia, 50–51, 53
Breezes, 362, 363. *See also* Wind
Bright nebula, 437
Brightness of star, 433–34
Brittle failure (brittle deformation), 178
Brittle rocks, 59
Bronze, 18
Bronze Age, 18
Budget of glacier, 104–6
Burrows, fossilized, 234

Calcareous ooze, 260, 262
Calcite, 31
Calcium bicarbonate, 94
Calcium carbonates, 260
Calderas, 200, 201, 207–8

Callisto (moon), 417, 418
Calorie, 323
Calving of glaciers, 106
Cambrian period, 240, 241
Capacity of stream, 75
Carbon, isotopes of, 23
Carbon-14, dating with, 238–39
Carbonaceous chondrite, 425
Carbonates, 27, 31
Carbon dioxide, 295, 309–11
Carbonic acid, 48, 93–94
Carbonization, 231–34
Cast, mold and, 231, 233
Catastrophism, 225
Caverns, 94–96
Celestial sphere, 396
Cementation, 50, 54
Cenozoic era, 240
Center of mass, 434, 435
Ceres, 421
Chalk, 52, 54
Changes of state of water, 322–24
Channelization, 84
Charon, 425
Chemical bonding, 21–23
Chemical compounds, 20, 21–23
Chemical fossils, 234
Chemical sedimentary rocks, 50, 51–54
Chemical weathering, 47–50, 247
Chert, 52, 55
Chikyu, 147
Chinooks, 362–63
Chlorofluorocarbons (CFCs), 297
Chrons, 150
Cinder cones, 202–4
Cinders, 199
Circle of illumination, 300
Circular orbital motion, 273–74
Circum-Pacific belt, 167
Cirques, 108–9
Cirrocumulus clouds, 335, 336, 337
Cirrostratus clouds, 335, 336, 337, 374
Cirrus clouds, 335, 336, 337, 374, 377
Cleavage, 25–26, 27, 29
Climate, 5, 294–95
  continental drift and ancient, 133–34
  dry, 117. *See also* Deserts
  elements, 295
  glaciers and, 102, 115
  ocean currents and 269–70, 271
  seafloor sediments and change in, 262
Cloudbursts, 341
Clouds, 335
  formation, 328–31, 334–36
  funnel, 383
  moisture in, 326–27
  storm, 374, 376–77, 379, 381
  temperature and cloud cover, 315–16
  types, 335–36, 337
  of vertical development, 336
Coal, 10, 50, 53–54, 56
Coarse-grained texture, 42, 44
Coastal upwelling, 270–71
Coastal zone, 271–86
  beaches and shoreline processes, 274–78
  classification, 285–86
  depositional features, 280–81
  emergent coasts, 285–86
  erosional features, 278–80
  evolving shore, 281–82
  shoreline as dynamic interface, 4, 271–72
  stabilizing, 282–85
  storm surge and damage to, 390
  submergent coasts, 285, 286
  temperature on windward vs. leeward, 314–15
  tides and, 286–89
Cold fronts, 374–75, 377
Collapse pits, 207
Collisional mountain ranges, 185–88
Collision-coalescence process of precipitation, 341
Color, 24
  of mineral, 24
  of stars, 434
Columbus, Christopher, 4
Columnar joints, 212
Coma, 422, 423
Comets, 396, 397, 421, 422–24
Compaction, 50, 54
Competence of stream, 74
Composite cones, 200, 202, 204–7
Compounds, chemical, 20, 21–23
Compressional forces, 178, 182
Conchoidal fracture, 26, 28
Concordant plutons, 210, 211–12
Condensation, 323, 325, 328–29, 334–36
Condensation nuclei, 335, 341
Conditional instability, 333–34
Conduction, 305, 307
Conduit, volcanic, 200, 209–10
Cone of depression, 90
Confining pressure, 58
Conformable rock layers, 228
Conglomerate, 50, 53, 54
Contact metamorphism, 56, 57
Continental (c) air masses, 372
Continental collisions, mountain building and, 186–88
Continental-continental convergence, 142, 143, 144
Continental crust, 177, 253, 256
Continental divide, 71
Continental drift, 130–34, 149–50
Continental lithosphere, 135
Continental margins, 184, 185, 252–57
Continental rifting, 139–40, 215
Continental rise, 253
Continental shelf, 252–53
Continental slope, 253
Continental tropical air masses, 372
Continental volcanic arc, 143, 183, 184–85, 214, 216, 258
Continents, 5, 131, 247, 360
Convection, 305, 307
  mantle-plate, 153–55
Convective lifting, localized, 329, 331
Convergence, 329, 330–31, 357, 377–78
Convergent boundaries, 137, 138, 141–43
  mountain building at, 143, 183–85
  volcanism at, 214, 216
Cool red stars, 441
Copernicus, Nicolaus, 398–99, 413
Coprolites, 234
Coquina, 52, 53
Core of Earth, 5, 6, 176, 177
Coriolis effect, 268, 269, 271, 353, 354, 356
Correlation of rock layers, 229–30, 232, 234–35
Corundum, 28
Covalent bond, 21, 23
Crab Nebula, 442, 444
Crater Lake-type calderas, 207, 208
Craters, 200, 201, 409–10, 411
Crests of ocean waves, 273
Crevasses, 104
Critical density, 449
Critical settling velocity, 75
Cross beds, 125
Cross-cutting relationships, 227–28, 231, 241
Crust of Earth, 5, 6, 28, 29, 176, 177
Cryovolcanism, 421
Crystallization, 29–31, 38–42, 45, 46, 59
Crystal settling, 45–46, 47
Crystal shape, 18, 19, 24, 25
Cumulonimbus clouds, 335, 336, 338, 341, 342, 374, 379, 381
Cumulus clouds, 335, 336, 338
Cumulus stage of thunderstorm development, 381
Cup anemometer, 363, 364
Curie point, 149
Currents, ocean, 268–71, 288–89, 317
Cut bank, 76
Cutoff, 76, 84
Cyclones, 356–59, 360, 361, 378–79
  middle-latitude, 330, 331, 375–78
Cygnus X–1, 444

Dams, 77, 78, 84
Dark matter, 450
Dark nebula, 437, 439
Darwin, Charles, 239
Dating
  numerical, 226, 241
  with radioactivity, 235–39, 241
  relative, 226–29
Daylight, length of, 302–4
Debris flow, 68
Decompression melting, 214, 215, 216
Deep-ocean basin, 257–58
Deep-ocean circulation, 271, 272
Deep-ocean trench, 141, 144, 257–58
Deep Sea Drilling Project, 146
Deep-sea fan, 253
Deferents in Ptolemaic system, 398
Deflation, 121
Deforestation, global warming and, 310
Deformation, rock, 162, 163, 177–78
Degenerate matter, 443, 449
Deltas, 80, 81, 279, 289
Dendritic drainage pattern, 82, 83
Density, 26
  critical, 449
  seawater, deep-ocean currents and, 271, 272
  of terrestrial vs. Jovian planets, 407–8
Deposition
  coastal zone, 280–81
  glacial, 109–13
  running water/stream, 71–72, 75
  unconformities and, 228, 231
  water's change of state and, 323, 324, 327
  wind deposits, 123–25
Depositional landforms, 80–82, 94–96, 109–13, 280–81

Depression, cone of, 90
Desert pavement, 121–22
Deserts, 116–25, 270
  distribution and causes, 116–17
  evolution, 118–19, 120
  rainshadow, 329, 330
  role of water, 117–18, 119
  wind erosion/deposits in, 119–25
Detachment fault, 180
Detrital sedimentary rocks, 50–51, 52, 53
Dew, 328, 335
Dew-point temperature, 326–27
Diamonds, 209–10
Differential stress, 58–59
Diffused light, 308
Dikes, 210–11, 227–28, 231
Diorite, 45
Dip-slip faults, 179–82
Discharge of stream, 72–73, 74
Disconformity, 229
Discordant plutons, 210–11
Dissolved load, 74
Distributaries, 80
Diurnal tidal pattern, 288
Divergence, 357, 358, 378
Divergent boundaries, 136, 138–40, 215, 216–17
Divide, 71
Domes, 178–79
Donga, 118
Doppler effect, 448, 449
Doppler radar, 385–86
Double refraction, 27, 28
Drainage basins, 71
Drainage patterns, 82, 83, 118–19
Drawdown, 90
Drift, glacial, 110–11
Drinking water, 85
Dripstone, 94–96
Drizzle, 341
Drumlins, drumlin fields, 112, 113
Dry adiabatic rate, 328, 329, 332, 333
Dry-bulb thermometer, 327
Dry climate, 117. *See also* Deserts
Ductile deformation, 178
Ductile rocks, 59
Dunes, sand, 124–25, 280
Dust Bowl, 121
Dwarf galaxies, 447
Dwarf planets, 407, 421, 425–27

Earth
  age of, determining, 225, 226, 235. *See also* Dating
  distribution of land and water, 246, 247
  interior, 5–6, 175–77
  magnetic field, 148–51, 152
  as planet, 398–99, 407
  spheres, 3–6
  as system, 6–8
Earthflow, 68
Earthquakes, 69, 160–75, 219
  destruction, 160–61, 169–74
  discovering cause of, 162–63
  distribution, 167
  Earth's interior and, 175–77
  elastic rebound, 162, 163
  epicenter, 161, 166–67
  expected world incidence, 169
  faults and, 161–63
  focus, 161
  foreshocks and aftershocks, 163
  intensity, 167–68
  locating, 166–67
  magnitude, 167, 168–69
  notable, 170
  seismic waves, 164–66, 175–77
Earth science, 2–3, 8, 13
Earth-Sun relationships, 300–304
Earth system, 6–8, 38–40
Earth system science, 7
Earthy luster, 24
Easterlies, 359, 360
Ebb currents, 289
Echo sounders, 250, 251
Elastic deformation, 178
Elastic rebound, 162, 163
Electric hygrometer, 328
Electromagnetic radiation, 305–7
Electromagnetic spectrum, 306
Electron capture, 236, 237
Electrons, 20–23, 236
Elements, 19–21
  atoms, 20–23
  economic value, 32–33
Elements of weather and climate, 295
Elliptical galaxies, 446–47
Elliptical orbits of planets, 400, 404, 405
Emergent coasts, 285–86
Emission nebulae, 437, 438
End moraines, 111–12, 113
Energy, 7, 304. *See also* Solar energy
  driving rock cycle, 40
  driving water cycle, 69
  fuels, 9, 10
  kinetic, 304, 305
  latent heat, 296, 322–24, 328, 388–89
  from ocean waves, 273, 274, 277, 282
  thermal, 304–5
Energy levels, 20
Environment, 8–11
  deductions about past, 235
Environmental lapse rate, 298–99, 331–34
Environmental problems, 10, 11
  global warming, 309–12
  groundwater, 91–93, 94
  shoreline erosion, 284–85
Eons, 239, 240
Ephemeral streams, 73, 117–18
Epicenter of earthquake, 161, 166–67
Epicycles in Ptolemaic system, 398, 399
Epochs, 240
Equal areas, Kepler's law of, 400, 401
Equatorial low, 359, 360, 361
Equinoxes, 301, 302–4
Eras, 240
Erastosthenes, 4
Erosion, 66
  glacial, 106–9
  groundwater, 86, 93–96
  on Mars, 416
  on Moon, 411
  running water/stream, 71, 74, 76, 77, 118, 119
  shoreline, 278–80, 284–85. *See also* Coastal zone
  soil, 10, 11
  unconformities and, 228, 231
  wave, 275–76
  wind, 119–23
Erosional agents, 40, 46, 50
Eruptive variables, 436
Escape velocity, 408
Eskers, 112, 113
Estuaries, 286
Evaporation, 52–53, 248, 249, 323, 325
Evaporation fogs, 339–40
Evaporites, 53, 261
Exfoliation domes, 47, 49
External processes, 66. *See also* Erosion; Mass wasting; Weathering
Extrusive igneous rocks, 40, 42, 45
Eye of hurricane, 387–88, 389
Eye wall, 387

Fathoms, 251
Fault-block mountains, 180, 181
Faults, 161–63, 179–82, 227–28
  transform fault boundaries, 137, 138, 143–45, 161
Fault scarps, 179, 180
Feldspars, 20, 29, 30, 43
Felsic rocks, 44, 45
Fetch, 273
Fine-grained texture, 42, 44
Fiords, 109
Fire, earthquake-caused, 173
Fire break, 173
Fissures and fissure eruptions, 209
Flash floods, 83, 118
Flood basalts, 209
Flood control, 83–85
Flood currents, 288
Flood deltas, 289
Floodplain, 79, 84
Floods, 75, 82–85, 118, 391
Fluorescence, 437
Focus of earthquake, 161
Fog, 328, 336–40
Folds, 178–79
Foliated texture, 59, 60
Footwall, 180, 181
Foreshocks, 163
Fossil assemblages, 234–35
Fossil fuels, 309–10
Fossil magnetism, 148–51, 152
Fossil record, 234
Fossils, 55, 56, 230–35
  continental drift and, 131–33, 134
  correlation and, 234–35
  index, 234–35
  preservation, 234
  types, 231–34
Fossil succession, 234
Fracture, 26, 28
Fracture zones, 104, 144
Franklin, Benjamin, 269
Freezing, 323
Freezing rain (glaze), 341, 342
Freshwater distribution in hydrosphere, 85
Friction, wind direction and, 355–56, 357
Frontal fog, 339–40
Frontal wedging, 329, 330
Fronts, 330, 359, 360, 373–75, 377
Frost, 324, 327
Frost wedging, 47, 48
Fujita intensity scale (F-scale), 384, 385
Fumaroles, 200
Funnel cloud, 383
Fusion, latent heat of, 323

Gabbro, 44, 45
Galactic clusters, 447, 448
Galactic disk, 444, 445
Galactic nucleus, 445
Galaxies, 405, 444–48, 449, 450
Galilei, Galileo, 351, 401–3, 404, 409, 417, 442, 444
Ganymede (moon), 417, 418
Gases
  in air, 295, 296
  greenhouse, 309
  liquid-gas change of state, 322, 323
  in magma, 194, 196–97, 198
  noble, 21
  in planets, 408
  as selective absorbers and radiators, 307, 308, 309
Gas molecules, behavior of, 351
Gastroliths, 234
Gems, 23, 28
Geocentric view, 396–98
Geographic poles, 148
Geographic position, temperature and, 314–15
Geologic structures, 177–82
Geologic time, 8, 223–43
  correlation of rock layers and, 229–30, 232
  dating with radioactivity, 235–39, 241
  fossils and, 230–35
  numerical dating, 226, 241
  relative dating, 226–29, 231
Geologic time scale, 8, 9, 224, 239–41
Geology, 2, 224–25
Geosphere, 5
Geostrophic wind, 354, 355
Geysers, 89
Glacial drift, 110–11
Glacial erratics, 110–11
Glacial striations, 107
Glacial trough, 107, 108, 109
Glaciers, 69–70, 102–15
  budget, 104–6
  climate and, 102, 115
  deposition, 109–13
  effects of Ice Age, 113–15
  erosion, 106–9
  maximum extent, 114
  movement, 104–6
  types, 102–4, 105, 108
Glass, 41
Glassy texture, 42, 44
Glaze, 341, 342
Global warming, 309–12
Glomar Challenger, 146
Glowing avalanches, 205–6
Gneiss, 60, 61
Gold karats, 21
Graben, 180, 181
Graded bedding, 253, 257
Gradient
  pressure, 352, 354, 362–63, 382, 387, 388
  stream, 72, 74, 78, 79
Grand Canyon, 66, 228, 229
Granite, 19, 20, 44, 45, 48–49
Granitic composition, 44, 45
Granitic magma, 194, 196
Granodiorite, 177
Graupel, 341
Gravity
  on Earth's Moon, 409
  formation of planets and, 405–7
  mass wasting and, 66–69
  Newton's law of universal gravitation, 404
  specific, 26–27
  stellar evolution and, 438, 439, 440
  tides and, 286–87
Great Dark Spot (Neptune), 420, 421
Great Galaxy in Andromeda, 445, 446, 447
Great Red Spot (Jupiter), 416, 417
Greenhouse effect, 309
Groins, 282–83
Ground moraine, 111, 113
Ground subsidence, earthquakes and, 173–75
Groundwater, 85–96
  artesian wells, 90–91
  distribution, 86–87
  environmental problems, 91–93, 94
  geologic roles, 86, 93–96
  importance, 85–86
  karst topography, 95, 96
  land subsidence and, 91–92
  springs, 88–89
  storage and movement, 87–88
  wells, 89–91
Gulf Stream, 269
Gullies, 74
Guyots (tablemounts), 258
Gypsum, 31, 52, 261
Gyres, 268–69

Habit, 25
Hail, 341, 342–43
Hale-Bopp, Comet, 396, 423, 424
Half-life, 237, 238
Halides, 31
Halite, 19, 31, 261
Halley's comet, 396, 397, 424
Hanging valleys, 107–8
Hanging wall, 180, 181
Hardness, 25
Hard stabilization, 282–83, 284
Hawaiian Islands, volcanoes of, 200–202, 217
Hawaiian-type calderas, 207
Headlands, wave refraction and, 277
Heat, 304–5
  latent, 296, 322–24, 388–89
  as metamorphic agent, 57–58
  specific, 314
  temperature and, 305
Heat balance, 270
Heat transfer, 305–7, 359–60
Helix Nebula, 441, 442
Herschel, William, 434
Hertzsprung-Russell (H-R) diagram, 435, 436, 440, 443
High clouds, 335, 336
Highlands, lunar, 409, 410–11
High-resolution multibeam sonar, 251
Highs, 117, 118, 356–59, 360, 378
Himalayas, 143, 144, 186–87, 188
Hipparchus, 433
Historical geology, 2
Hoar frost, 324, 327
Horizontal convergence, 357
Horizontality, principle of original, 227
Hornblende, 20, 30
Horns, 108–9
Horsehead Nebula, 439
Horsts, 180, 181
Hot spots, 147–48, 214, 215, 217, 218, 258
Hot springs, 89
Hoyle, Fred, 449
H-R diagram, 435, 436, 440, 443
Hubble, Edwin, 445, 448
Hubble's law, 448
Hubble Space Telescope, 425, 433, 448
Humidity, 324–28
Hurricanes, 370, 378, 379, 386–91
  destruction, 386, 389–91
  eye, 387–88, 389
  formation and decay, 388–89
  naming, 389
  profile, 386–88
Hutton, James, 224, 225, 228–29, 239
Hydrogen burning, 439, 440
Hydrogenous sediment, 260–61, 262
Hydrosphere, 4–5, 69, 85. *See also* Groundwater; Running water
Hygrometer, 327–28
Hygroscopic nuclei, 335, 341
Hypothesis, 12

Ice, 322, 323, 408
Ice Age, 113–15, 253
Icebergs, 106, 248, 249
Ice caps, 103
Ice crystal process of precipitation, 340–41
Ice nuclei, 340
Ice sheets, 102, 103
Ice shelves, 103
Igneous activity, 193–221
  intrusive, 210–13
  landforms, 207–10
  living with, 218–19
  materials extruded, 197–99
  nature of volcanic eruptions, 194–97
  plate tectonics and, 213–18
  volcanic structures and styles, 199–207
Igneous rock, 39, 40–46, 47, 241
  relative dating of, 227–28, 231
Illumination, circle of, 300
Impact craters, 409–10, 411
Impact hypothesis for Moon's origin, 411–12
Impression (fossil), 234
Incised meanders, 80
Inclination of axis, 301, 302
Inclusions, 228, 231
Index fossils, 234–35
Indian Ocean, 246, 269
Industrial rocks and minerals, 32
Inertia, 164, 404
Infiltration, 69
Infrared radiation, 306
Inner core, 176, 177
Inner planets, 407–8, 412–15
Inselbergs, 119, 120
Instability, 332, 333–34
Integrated Ocean Drilling Program (IODP), 147
Intensity scales, 167–68
Interface, 4, 272
Intergovernmental Panel on Climate Change (IPCC), 310–11
Interior drainage, 118–19
Intermediate rocks, 44–45
Intermittent streams, 73
Internal processes, 66. *See also* Mountain building; Volcanic eruptions
International Astronomical Union, 426
Interstellar dust, 437

Interstellar matter, 437
Intertropical convergence zone, 361
Intraplate volcanism, 214, 215, 217–18
Intrusive igneous activity, 210–13
Intrusive igneous rocks, 40, 42, 45, 227–28, 231
Io (moon), 417–18, 419
Ionic bonds, 21, 22–23
Ions, 22
Irons (meteorites), 425
Irregular galaxies, 447
Irrigation, groundwater depletion and, 91, 92, 93
Island arcs, 143, 183–84, 186, 187, 214, 216, 258
Islands, barrier, 280–81, 285
Isobars, 352, 353
Isotherms, 312, 313, 316, 317
Isotopes, 23, 236, 237, 238

Jasper, 52, 55
Jet streams, 355–56
JOIDES Resolution, 147
Jovian (outer) planets, 407–9, 416–21
Jupiter, 402, 407, 416–18, 419

Kames, 112–13
Karst topography, 95, 96
Katrina, Hurricane (2005), 370, 386, 387
Kepler, Johannes, 399, 400–401, 403
Kettles, 112
Kinetic energy, 304, 305
Kuiper belt, 423, 424, 425, 426

Laccoliths, 212
Lahars, 206–7, 218
Lake-effect snows, 372, 373
Lakes, 76, 77, 115, 119, 120, 248
Laminar flow, 72
Land, temperature control and, 313–14, 317
Land breeze, 362
Landslides, 173–75
Land subsidence, groundwater withdrawal and, 91–92
Lapilli, 199
Latent heat, 296, 322–24, 388–89
Lateral moraines, 111
Latitude, 301, 302–4
  air temperature and, 313, 314, 316–17
Lava, 40
Lava flow, 197–98, 200–202, 204
Lava plateaus, 209
Leeward coast, 314–15
Levees, 80–84, 85
Lightning, 379, 380
Light-year, 433
Lignite, 54
Limestone, 19, 31, 52
  caverns, 94–96
Liquefaction, 171, 172
Liquid water, 322, 323
Lithification, 39, 40, 54
Lithosphere, 5, 6, 135, 177. *See also* Plate boundaries
Lithospheric plates, 135–38
Local Group, 447
Localized convective lifting, 329, 331
Local winds, 362–63
Loess, 123–24
Long-period comets, 423–24
Longshore currents, 277, 278
Long-wave radiation, 307, 309
Low clouds, 335, 336
Lower mantle, 176, 177
Low-mass stars, death of, 441
Lows, 356–59, 360, 361. *See also* Cyclones
Lunar dust, 411
Lunar highlands, 409, 410–11
Lunar regolith, 411
Lunar surface, 409–11
Luster, 24

Mafic rocks, 44, 45
Magma, 38, 39, 40
  basaltic, 143, 194, 196, 216
  crystallization, 40, 41–42, 45
  dissolved gases in, 194, 196–97, 198
  viscosity, 194, 197
Magmatic differentiation, 45–46, 47
Magnetic field, Earth's, 148–51, 152
Magnetic poles, 148–50
Magnetic reversals, 150–51, 152
Magnetic time scale, 150
Magnetite, 27, 148
Magnetometers, 150, 152
Magnitude, 433
  earthquake, 167, 168–69
  stellar, 433–34
Main-sequence stars, 435, 439–40, 441
Manganese nodules, 260, 261
Mantle, 5, 6, 142–43, 176, 177
  plate-mantle convection, models of, 153–55
Mantle plume, 147–48, 214, 215, 217–18, 258
Marble, 31, 59, 60, 61
Maria (mare), 409, 410–11
Marine terrace, 278, 279
Maritime (m) air masses, 372–73, 377
Mars, 148, 398, 407, 414–16
Marsupials, 133
Mass
  center of, 434, 435
  stellar, 434–35, 441–43
Mass number, 23, 236–37
Mass wasting, 66–69
Meanders, 76, 77, 80
Mechanical weathering, 46–47, 48, 49
Medial moraines, 111
Melting, 322, 323
  decompression, 214, 215, 216
  partial, 142–43
Mercalli, Guiseppe, 167
Mercury barometer, 351
Mercury (planet), 407, 412–13
Mesocyclone, 382, 383, 386
Mesopause, 299–300
Mesosaurus, 132–33
Mesosphere, 299–300
Mesozoic era, 240
Metallic luster, 24
Metal sulfides, 260–61
Metamorphic rocks, 39, 40, 55–61, 241
Metamorphism, 55–59
Meteor, 424
Meteorites, 172, 424, 425
Meteoroids, 410, 421, 424–25
Meteorology, 2–3
Meteor showers, 424, 425
Methane, global warming and, 310
Micas, 30
Mica schists, 60
Microcontinents, 186
Micrometeorites, 424
Mid-Atlantic Ridge, 145, 148, 151
Middle clouds, 335, 336
Middle-latitude cyclone, 330, 331, 375–78. *See also* Severe weather
Middle-latitude deserts and steppes, 117
Mid-ocean ridge, 135, 138–39, 215, 216–17, 258–59
Milky Way Galaxy, 405, 444–45, 446
Millibar, 351
Mineralogy, 18
Minerals, 17–35
  characteristics, 18–19
  elements as building blocks of, 19–21
  magnetic, 148–49
  mineral resources, 32–33
  nonsilicate, 29, 31–32
  properties of, 23–28
  silicate, 28, 29–31, 43
Mist, 341
Mixed tidal pattern, 288
Mixing ratio, 324–25
Model, 12
Modified Mercalli Intensity Scale, 167, 168
Mohs scale of hardness, 25
Mold and cast, 231, 233
Moment magnitude, 169
Monsoons, 269, 360
Montreal Protocol, 297
Moon, Earth's, 409–12
  tides and, 286–87, 288
Moons, planetary, 402, 417–20, 421
Moraines, 111–12, 113
Motion
  circular orbital, 273–74
  Newton's laws of, 404
  retrograde, 398, 418
Mountain belts, 133, 182–85
Mountain breeze, 362, 363
Mountain building, 58, 143, 182–88
Mountains
  fault-block, 180, 181
  orographic lifting of air and, 329, 330
  rain shadow formed by, 117, 118
  as temperature control, 315
Mudflows, 206–7
Muscovite, 30

National Centers for Environmental Prediction (NCEP), 384
National Weather Service (NWS), 384, 385
Native elements, 31, 32
Natural hazards, 11, 386. *See also* Severe weather; Volcanic eruptions
Natural levees, 80–82
Neap tides, 287
Nebula, nebulae, 437, 438, 439, 441, 442
Nebular hypothesis, 405, 406
Neck, volcanic, 210
Neptune, 407, 420, 421
Neutrons, 20, 21, 23, 236
Neutron stars, 443–44
Newton, Isaac, 286, 403–4
Nimbostratus clouds, 335, 336, 338, 341, 374
Nitrogen, 295, 296, 308
Nitrous oxide, global warming and, 310
Noble gases, 21
Nonconformity, 228, 229
Nonfoliated texture, 59, 60–61

Nonmetallic luster, 24
Nonrenewable resources, 9, 91, 92
Nonsilicate minerals, 29, 31–32
Nor'easter, 373
Normal faults, 180, 181, 182
Normal lapse rate, 299
Normal polarity, 150
Northern Hemisphere, 246, 268–69, 302, 303, 304, 353
Nova, 436
Nuclear fusion, 405
Nucleus
  atomic, 20, 236–37
  galactic, 445
Nuée ardente, 205–6
Nullah, 118
Numerical dating, 226, 241

Obsidian, 19, 41, 42, 43
Occluded fronts, 375
Occulation, 421
Ocean currents, 268–71, 288–89, 317
Ocean drilling, seafloor spreading evidence from, 146–47
Ocean floor, 250–62
  continental margins, 252–57
  deep-ocean basin, 257–58
  mapping, 250–51
  provinces of, 252, 254–55
  seafloor sediments, 253, 256, 258, 259–62
  seafloor spreading, 139, 146–47, 150–51, 152, 216–17
  viewing from space, 251–52
Oceanic-continental convergence, 141–43
Oceanic crust, 177, 253, 256
Oceanic lithosphere, 135, 141
Oceanic-oceanic convergence, 142, 143
Oceanic plateaus, 186, 258
Oceanic ridge, 135, 138–39, 215, 216–17, 258–59
Oceanic volcanism, 214
Oceanography, 2, 246
Oceans, 4, 5, 245–65, 388
  circulation patterns, 268–69, 271, 272
  comparing continents to, 247
  composition of seawater, 247–49
  currents, 268–71, 288–89, 317
  geography of, 246–47
  layered structure, 249–50
  main ocean basins, 246
  shorelines, 4, 271–72. *See also* Coastal zone
  waves, 272–78, 282
Ogallala Formation, 88, 92
Olivine group, 30
Oort cloud, 423–24
Ooze, 260, 262
Opaque mineral, 24
Optical properties of minerals, 24–25, 27
Orbits
  of asteroids, 422
  elliptical, of planets, 400, 404, 405
  of Kuiper belt objects, 423
Ores, 32
Original horizontality, 227
Orogenesis, 182. *See also* Mountain building
Orographic lifting, 329, 330
Outer core, 176, 177
Outer planets, 407–9, 416–21
Outgassing, 248
Outwash plains, 112, 113

Overland flow, 71
Overrunning, 373
Oxbow lake, 76, 77
Oxidation, 47–48
Oxides, 31
Oxygen, 295, 296, 308
Ozone, 296–98

Pacific Ocean, 246
Pahoehoe flows, 198
Paleoclimatic evidence for continental drift, 133–34
Paleomagnetism, 148–51, 152
Paleontology, paleontologists, 2, 230
Paleozoic era, 240
Pangaea, 130–31, 134, 151–52, 153
Panthalassa, 149
Paradigms, 12
Parallax, stellar, 399–400, 432–33
Parasitic cone, 200
Parcel (volume of air), 328
Parent rock, 55
Partial melting, 142–43
Passive continental margin, 184, 185, 252–53, 256
Pearly luster, 24
Pele's hair, 202
Perched water table, 89
Peridotite, 177
Periodic table, 19, 21
Periods (geologic time scale), 240
Permeability, 87–88
Petrified fossils, 231, 233
Petroleum, 10
Phanerozoic eon, 239–40
Photochemical smog, 297
Physical environment, 8–9
Physical geology, 2
Piedmont glaciers, 103–4
Pipes, volcanic, 200, 209–10
Plagioclase feldspar, 30
Planetary albedo, 308
Planetary motion
  forces involved in, 403–4
  Kepler's laws of, 400–401
  Newton's explanation of, 404
  in Ptolemaic system, 397–98
Planetary nebulae, 441, 442
Planetesimals, 405–6
  asteroids, 410, 421–22, 425
  comets, 396, 397, 421, 422–24
Planets, 404–9
  atmospheres, 408–9, 414, 417, 421
  compositions, 408
  dwarf, 407, 421, 425–27
  Earth as, 398–99, 407
  extrasolar, 407
  formation, 405–7
  Greek geocentric view, 396–98
  Jovian (outer), 407–9, 416–21
  orbits, 400, 404, 405
  terrestrial (inner), 407–8, 412–15
Plate boundaries, 138–45
  convergent, 137, 138, 141–43, 183–85, 214, 216
  divergent, 136, 138–40, 215, 216–17
  earthquakes along faults associated with, 162–63
  transform fault, 137, 138, 143–45, 161
Plate-mantle convection models, 153–55
Plates, 135–38

Plate tectonics, 129–57, 413
  breakup of Pangaea, 151–52, 153
  continental drift hypothesis, 130–34
  forces driving plate motion, 152–53
  igneous activity and, 213–18
  mountain building and, 182–88
  plate boundaries, 138–45
  testing model, 145–51
Playa lake, 119, 120
Pleiades star cluster, 438
Pleistocene epoch, 114, 253
Plucking, 107
Pluto, 407, 421, 425–26, 427
Plutonic igneous rocks, 40, 42, 45, 227–28, 231
Plutons, 210–13
Pluvial lakes, 115
Point bars, 76
Polar air masses, 371, 372–73
Polar front, 359, 360
Polar high, 359, 360
Polarity, 150
Polar wandering, apparent, 149–50
Pollution
  air, 10, 11
  groundwater, 92–93, 94
Population, 11, 271
Pore space, 87, 88
Porosity, 87
Porphyritic texture, 42, 44
Potassium feldspar (orthoclase), 30
Potholes, 74
Precambrian time, 240, 241
Precipitates, chemical sedimentary rocks from, 51–52
Precipitation, 340–44. *See also* Weather
  formation, 340–41
  forms, 341–43. *See also* Snow
  measuring, 343–44
  water cycle and, 69–70
Precipitation fog, 339–40
Pressure, as metamorphic agent, 58–59
Pressure gradient, 352, 354, 362–63, 382, 387, 388
Pressure tendency, 359
Pressure zones, 359–60, 361
Prevailing wind, 363
Primary (P) waves, 164, 165–66
Primitive crust, 176
Principal shells, 20
Proglacial lakes, 115
Protons, 20–21, 23, 236
Protoplanets, 406–7
Protostar, 439
Protosun, 405
Ptolemaic system, 396–98, 403
Pulsar, 444
Pulsating variables, 435–36
Pumice, 19, 42, 43
P waves, 164, 165–66
Pyrite (fool's gold), 28
Pyroclastic flows, 205–6, 218
Pyroclastic materials, 197, 198–99, 202–4
Pyroxene group (augite), 30

Quartz, 19, 20, 29, 30, 52, 55
Quartzite, 61

Radar
  Doppler, 385–86
  weather, 344

Radar altimeters, 252
Radial drainage pattern, 82, 83
Radiation, 305–8
  absorption, 308
  albedo, 307–8, 315–16
  laws governing, 307
  scattering, 307, 308
  terrestrial, 307, 309
  ultraviolet (UV), 296, 297, 306, 308, 437
Radiation fog, 336–39, 340
Radiation pressure, 423
Radioactive decay, 23, 236–37
Radioactivity, 236
  dating with, 235–39
Radiocarbon dating, 238–39
Radiometric dating, 147, 148, 237–39, 241
Radiosonde, 299, 300
Rain, 341
Rainbows, 306
Rain gauge, standard, 343, 344
Rain shadow, 117, 118
Rainshadow desert, 329, 330
Rapids, 79
Recessional end moraines, 111
Recrystallization, 59
Rectangular drainage pattern, 82, 83
Red giants, 435, 436, 440
Red shifts, Doppler, 448
Reflection, 307–8
Reflection nebulae, 437, 438
Refraction
  double, 27, 28
  wave, 276–77
Regional metamorphism, 56
Relative dating, 226–29, 231
Relative humidity, 325–26
Relocation, 284
Renewable resources, 9
Repose, angle of, 68
Reserves, 32
Resources, 9–10, 32–33, 91, 92
Retrograde motion, 398, 418
Reverse faults, 180, 181, 182
Reverse polarity, 150
Revolution, 300
Rhyolite, 44, 45
Richter scale, 168–69
Ridge push, 152–53, 154
Rift valley, 138, 139, 259
Rills, 74
Rime, 341, 343
Ring of Fire, 204, 213, 216, 258
Ring systems, planetary, 418, 419–20, 421
Rip currents, 274
River systems, 71–72. *See also* Streams
Rock(s), 18–19, 37–63
  continental drift and similar types of, 133
  correlation of layers, 229–30, 232, 234–35
  dating, 226–29, 231, 241
  igneous, 39, 40–46, 47, 227–28, 231, 241
  metamorphic, 39, 40, 55–61, 241
  Moon, 410, 411
  in planets, 408
  sedimentary, 39, 40, 50–55, 57, 241
  weathering, 40, 46–50
Rock cycle, 38–40, 102
Rock deformation, 162, 163, 177–78
Rock flour, 107
Rock gypsum, 52
Rock salt, 52
Rockslide, 67, 68
Rotation of Earth, 300, 353, 354, 396
Ruby, 28
Running water, 70–85
  drainage basins and, 71
  erosion, 71, 74, 76, 77, 118, 119
  river systems, 71–72
  streamflow, 72–74
  work of, 74–75. *See also* Deposition; Streams
Runoff, 69, 248, 249

Saffir-Simpson scale, 389–90
Salinity, seawater, 247–49, 250
Saltation, 121
Salt flats, 53, 56
Salts, 247–49, 261
San Andreas Fault, 145, 146, 161, 182
Sand, 123, 276–78
Sand dunes, 124–25, 280
Sandstone, 51, 53
Santa Ana winds, 363
Sapphire, 28
Sargasso Sea, 269
Satellites, weather, 389, 390
Saturation, 324
  zone of, 87, 89
Saturn, 407, 418–20
Scattering, 307, 308
Schists, 60, 61
Scientific inquiry, 11–13
Scientific law, 12
Scientific methods, 12
Scientific revolution, 130
Scoria, 43
Scoria cones, 202–4
Sea arches, 279–80, 281
Sea breeze, 362
Seafloor spreading, 139
  evidence of, 146–47, 150–51
  igneous activity at divergent plate boundaries and, 216–17
  magnetic reversals and, 150–51, 152
Sea ice, 248, 249
Sea level
  air pressure at, 350
  effect of Ice Age on, 115
  global warming and, 311
  rising, shoreline erosion and, 284, 285
  as ultimate base level, 77
  world mean sea-level temperatures, 316, 317
Seamounts, 147–48, 258
Seasons, 300–304, 316, 317
Sea stacks, 279–80
Seawall, 275, 283
Seawater, composition of, 247–49, 250
Secondary waves, 164–66
Sedimentary rock, 39, 40, 50–55, 57, 241
  folds in, 178–79
Sediments, 40, 50, 51
  formation, 40, 46–50
  lithification, 39, 40, 54
  seafloor, 253, 256, 258, 259–62
  transport, 119–23, 253
Seismic reflection profiles, 251, 252
Seismic sea waves, 172–73, 174
Seismic waves, 164–66, 175–77
Seismogram, 164, 165, 166
Seismographs, 164, 168
Seismology, 164–66
Semidiurnal tidal pattern, 288
Severe thunderstorm outlooks, 385
Severe weather, 370, 378–91
  hurricanes, 370, 378, 379, 386–91
  thunderstorms, 378–81, 382, 385
  tornadoes, 370, 378, 381–86, 390–91
Shale, 51, 53
Sheet flow, 74
Sheeting, 47, 49
Shelf valleys, 253
Shield volcanoes, 200–202
Shoreline. *See* Coastal zone
Short-period comets, 422, 424
Short-wave solar radiation, 307, 309
Sidescan sonar, 251
Silicates, 28, 29–31, 43, 49–50, 194
Siliceous ooze, 260, 262
Silicon-oxygen tetrahedron, 29
Silky luster, 24
Sills, 211–12
Siltstone, 51
Sinkholes (sinks), 86, 96
Slab pull, 152, 154
Slab suction, 153, 154
Slate, 59, 60, 61
Sleet, 341, 342
Sling psychrometer, 327–28
Slip face, 125
Slump, 68
Smog, photochemical, 297
Snow, 104, 105, 316, 341, 344, 372, 373
Soda straw, 94, 95
Sodium chloride, 22, 247
Soil, 10, 11, 86, 171, 172. *See also* Sediments
Solar energy, 7, 300–304
  radiation of, 305–8
  wind and, 352, 353
Solar nebula, 405, 406
Solar system. *See also* Planets
  age, 439–40
  in ancient astronomy, 396–98
  asteroids, 410, 421–22, 425
  comets, 396, 397, 421, 422–24
  development of modern views of, 398–404
  Earth's Moon, 286–87, 288, 409–12
  meteoroids, 410, 421, 424–25
  origin, 405–7
Solar wind, 423
Solid/liquid/gas change of state, 322–24
Solstices, 301, 302–4
Sonar, 250–51
Sorting, 75
Source regions of air masses, 371–72
Southern Hemisphere, 246, 268, 269, 353
Specific gravity, 26–27
Specific heat, 314
Spiral galaxies, 445, 446, 447, 448
Spit, 279, 280, 281
Spring equinox, 301, 302, 303
Springs, 88–89
Spring tides, 287
Stability, atmospheric, 331–34
Stabilization of shore, 282–85
Stable air, 331, 334
Stalactites, 94, 95
Stalagmites, 95, 96
Stars, 432–44
  binary, 434–35, 444

evolution, 437–43
H-R diagrams, 435, 436, 440, 443
interstellar matter, 437
main-sequence, 435, 439–40, 441
neutron, 443–44
nuclear fusion, 405
properties, 432–35
variable, 435–36, 440
Stationary fronts, 375
Steam fog, 339, 340
Stellar brightness, 433–34
Stellar mass, 434–35, 441–43
Stellar parallax, 399–400, 432–33
Stellar remnants, 443–44
Steppe, 117
Stocks, 212
Stony-irons (meteorites), 425
Stony meteorites, 425
Storm Prediction Center (SPC), 384–85
Storms. *See* Severe weather
Storm surge, 390, 391
Strata of sedimentary rock, 51, 54
Stratified drift, 110
Stratocumulus clouds, 335, 336
Stratopause, 299
Stratosphere, 296–98, 299
Stratovolcanoes, 200, 202, 204–7
Stratus clouds, 335, 336, 374
Streak, 24–25
Stream channels, 72, 74, 75–77, 84
Streamflow, 72–74, 86
Streams, 71
base level, 77, 80
deposition, 71–72, 75, 80–82
discharge, 72–73, 74
drainage patterns, 82, 83
effects of Ice Age glaciers on, 114
ephemeral, 73, 117–18
erosion, 74, 77
floodplain, 79, 84
floods and flood control, 82–85
gradient, 72, 74, 78, 79
intermittent, 73
profile, 73–74
transportation of load, 74–75
trunk, 72
Stream valleys, 77–80
Strength, mineral, 25–26
Stress, differential, 58–59
Strike-slip faults, 182
Subduction zones, 141–43
igneous activity at, 214, 216
mountain building at, 183–85
Sublimation, 323
Submarine canyons, 253, 257
Submergent coasts, 285, 286
Submetallic luster, 24
Subpolar low, 359, 360
Subsidence, 91–92, 173–75
Subsystems, 7
Subtropical gyres, 268
Subtropical highs, 117, 118, 359, 360
Suction vortices, 382
Sulfates, 31
Sulfides, 31–32
Summer solstice, 301, 302, 303, 304
Sun. *See also* Solar system
Earth-Sun relationships, 300–304
energy, 7, 300–308, 352, 353
H-R diagram, 435, 436
magnitude, 433, 434
short-wave radiation, 307, 309
tides and, 287, 288
Sunspots, 402–3
Supercell thunderstorm, 382
Superclusters, 447
Supercooled water, 340–41
Supergiants, 435, 436
Supernova, 405, 441, 442–43, 444
Superposition, law of, 226, 231, 241
Surface ocean currents, 268–71
Surface waves, 164, 165–66
Surf and surf zone, 274, 275, 276
Suspended load, 74
Swash, 274, 276, 277
S waves, 164–66
Swells, 273
Synclines, 178, 179
System, 7
Earth as, 6–8, 38–40

Tablemounts (guyots), 258
Tabular plutons, 210–12
Tail of comet, 423
Talus or talus slopes, 47, 48
Tectonics, 130*n*
Telescope, invention of, 402, 403
Temperature, 305. *See also* Air temperature
adiabatic temperature changes, 328–29, 332, 333
controls, 313–16
dew-point, 326–27
global average surface, 5
heat and, 305
hurricane formation and ocean-water, 388
ocean currents and, 270
ocean layers based on, 249–50
stellar color and, 434
viscosity and, 194
Temperature gradient, 312
Tenacity, 25
Terminal end moraine, 111
Terrae, 409
Terranes, 185–86, 187
Terrestrial planets, 407–8, 412–15
escape velocities, 408–9
Terrestrial radiation, 307, 309
Terrigenous sediment, 260, 261–62
Texture, 42
igneous, 42, 43, 44
metamorphic, 59, 60
Theory, scientific, 12
Thermal energy, 304–5
Thermal metamorphism, 56, 57
Thermals, 331
Thermistor, 312
Thermocline, 249
Thermohaline circulation, 271, 272
Thermosphere, 300
Thrust faults, 180–82
Thunderstorms, 378–81, 382, 385
Tidal currents, 288–89
Tidal delta, 279, 289
Tidal flats, 287, 289
Tides, 286–89
Till, 110
Tombolo, 280, 281
Topaz, 23
Tornadoes, 370, 378, 381–86, 390–91
Tornado warning, 385
Tornado watches, 385
Trade winds, 359, 360
Transform fault, 182
Transform fault boundaries, 137, 138, 143–45, 161
Transition zone, 177
Transparent mineral, 24
Transpiration, 69
Travertine, 52
Trellis drainage pattern, 82, 83
Tributaries, 82, 118
Triton (moon), 421
Tropical air masses, 371, 372, 377
Tropical depression, 389
Tropical disturbances, 388, 389
Tropical storm, 389
Tropic of Cancer, 302, 303
Tropic of Capricorn, 302, 303
Tropopause, 299
Troposphere, 298–99
Troughs of ocean waves, 273
Trunk glaciers, 107
Trunk stream, 72
Tsunami, 172–73, 174
Turbidites, turbidity currents, 253, 257
Turbulent flow, 72
Typhoons, 387

Ultramafic rocks, 45
Ultraviolet radiation, 296, 297, 306, 308, 437
Unconformities, 228–29, 230, 231
Uniformitarianism, 225
Universal gravitation, Newton's law of, 404
Universe, 432, 448, 449
Big Bang Theory, 405, 448–50
Copernican (Sun-centered) view, 398–99, 401–3
Ptolemaic system, 396–98, 403
Unloading, 47
Unsaturated zone, 87
Unstable air, 331, 334
Upper mantle, 176, 177
Upslope fog, 339
Upwelling, 270–71
Uranus, 407, 419, 420–21
Ussher, James, 225

Valence electrons, 20, 21–23
Valley breeze, 362, 363
Valley glaciers, 102–3, 105, 108
Valleys, 77–80, 107–8, 109, 138, 139, 253, 259
Valley trains, 112
Vaporization, latent heat of 323
Vapor pressure, 324
Variable stars, 435–36, 440
Velocity
escape, 408
stream, 72–75
Vent, volcanic, 200
Venus, 402, 403, 407, 413–14
Vesicles, 42, 43
Viscosity of magma, 194, 197
Visible light, 306
Volatiles, 196–97, 198
Volcanic arc, continental, 143, 183, 184–85, 214, 216, 258
Volcanic ash, 41, 194, 199

Volcanic eruptions, 7, 194–205
  fissure eruptions, 209
  hazards, 218–19
  intrusive igneous activity, 210–13
  landforms, 207–10
  materials extruded, 197–99
  nature of, 194–97
  plate tectonics and, 213–18
  precursor, 219
  sea salts from, 247–48
  volcanic structures and eruptive styles, 199–205
Volcanic igneous rocks, 40, 42, 45, 227–28, 231
Volcanic island arcs, 143, 183–84, 186, 187, 214, 216, 258
Volcanic necks, 210
Volcanic pipes, 200, 209–10
Volcanism
  continental rifting and, 140
  cryovolcanism, 421
  global distribution, 213
  hot spots, 147–48, 214, 215, 217, 218, 258
  monitoring, 219
  on planets and moons, 410–11, 413, 414–15, 417, 418, 421
  zones of, 214–15, 216–18
Volcano(es)
  anatomy, 200
  highest, 196
  living with, 218–19
  locations of Earth's major, 213
  seamounts, 147–48, 258
  types, 200–208
Vortex, tornado, 381, 382
V-shaped valley, 78–79, 107–8, 109

Wadi, 118
Warm fronts, 373–74, 377
Wash, 118
Wastage, zone of, 104–6
Water. *See also* Groundwater; Running water
  in arid climates, role of, 117–18
  changes of state, 322–24
  chemical weathering agent, 47–48
  distribution of Earth's, 69
  drinking, 85
  on Mars, 414, 415, 416
  mass wasting and, 68
  pure vs. seawater, 247
  as temperature control, 313–14, 317
Water balance, 70
Water cycle, 69–70, 102
Waterfalls, 79
Water table, 87, 89, 90, 91
Water vapor, 296, 308, 309, 322, 323, 324–28, 334–36
Wave base, 274
Wave-cut cliffs, 278, 281
Wave-cut platform, 278–79, 281
Wave height, 273
Wavelength, 273, 306, 307
Wave period, 273
Wave refraction, 276–77
Waves
  ocean, 272–78, 282
  seismic, 164–72, 175–77
Weather, 5, 294–95, 370–78
  air-mass, 371, 372–73
  atmospheric stability and daily, 334
  elements, 295
  fronts, 330, 359, 360, 373–75, 377
  global warming and, 311, 312
  highs and lows, 356–59, 360, 361
  middle-latitude cyclone, 330, 331, 375–78
  severe, 370, 378–91
Weathering, 40, 46–50, 66, 247, 411
Weather radar, 344
Weather satellites, 389, 390
Wegener, Alfred, 130–34, 139, 149, 151
Welded tuff, 199
Wells, 89–91
Westerlies, 359, 360–62
Wet adiabatic rate, 328, 329, 332, 333
Wet-bulb thermometer, 327
Whitecaps, 273
White dwarfs, 435, 436, 441, 443
White frost, 324, 327
White light, 306
Whole-mantle convection, 154, 155
Wind, 352–64
  air pressure and, 352–59
  deposits, 123–25
  erosion, 119–23
  general circulation of atmosphere, 359–62
  geostrophic, 354, 355
  highs and lows, 356–59
  hurricane, 387, 390–91
  jet streams, 355–56
  labeling, 363
  local, 362–63
  measuring, 363–64
  in midlatitude cyclones, 377–78
  ocean waves and, 272–74
  on planets, 414, 416, 421
  prevailing, 363
  solar, 423
  surface ocean currents and, 268–71
  tornado, 382, 383–84, 385
  trade, 359, 360
  upwelling and, 270–71
Wind shadows, 124
Wind vane, 363, 364
Windward coast, 314–15
Windward mountain slopes, precipitation on, 329
Winter solstice, 301, 302, 303, 304
World Meteorological Organization (WMO), 295
World Weather Watch, 295

Yazoo tributaries, 82
Yellowstone-type calderas, 207–8

Zone of accumulation, 104, 105
Zone of fracture, 104
Zone of saturation, 87, 89
Zone of wastage, 104–6